Chittaranjan Kole
Editor

# Genomics and Breeding for Climate-Resilient Crops

## Vol. 2 Target Traits

 Springer

*Editor*
Prof. Chittaranjan Kole
Vice-Chancellor
Bidhan Chandra Krishi Viswavidyalaya
(Bidhan Chandra Agricultural University)
Mohanpur, Nadia, West Bengal, India

ISBN 978-3-642-37047-2      ISBN 978-3-642-37048-9 (eBook)
DOI 10.1007/978-3-642-37048-9
Springer Heidelberg New York Dordrecht London

Library of Congress Control Number: 2013939737

Printed on acid-free paper

Springer is part of Springer Science+Business Media (www.springer.com)

# Genomics and Breeding for Climate-Resilient Crops

*Dedicated to*
*Prof. Rajendra B. Lal*
*Vice-Chancellor*
*Sam Higginbottom Institute of Agriculture,*
*Technology and Sciences*
*(SHIATS),*
*Allahabad, India*

# Foreword by M. S. Swaminathan

I am very happy that Prof. Chittaranjan Kole and other eminent authors have prepared two books on Genomics and Breeding for Climate-Resilient Crops. These are timely publications since climate-smart agriculture is the need of the hour. Many of the crops formerly known as coarse cereals are both very nutrient rich and climate smart. It would therefore be more appropriate to refer to them as climate-smart nutricereals. In this connection I give below the views I expressed in an editorial which I wrote for *Science* under the title "Gene Banks for a Warming Planet" (Swaminathan 2009).

"At the International Congress of Genetics in New Delhi in 1983, I stressed the need for a conservation continuum, beginning with the revitalization of conservation of domesticated plants by farm families in all countries, and extending to the establishment of an international genetic resource repository maintained under permafrost conditions. Since then, thanks to the spread of participatory breeding and knowledge-management systems involving scientists and local communities, on-farm conservation and gene banks have become integral parts of national biodiversity conservation strategies. For example, there are now over 125,000 genetic strains of rice, of which over 100,000 are in a cryogenic gene bank maintained by the International Rice Research Institute (IRRI) in the Philippines. This gene pool is invaluable for adapting one of the world's most important cereal grains to the consequences of global climate change.

We now largely depend on a few crops such as rice, wheat, corn, soybeans, and potatoes to sustain global food systems. However, their genetic homogeneity increases their vulnerability to abiotic and biotic stresses. If their production is affected by a natural calamity, their prices will increase and food-deficient countries are likely to face riots and worse. Important publications such as *Lost Crops of the Incas* and *Lost Crops of Africa* document the historic role of agrobiodiversity in ensuring food and health security. It has therefore become an urgent task to save vanishing 'orphan crops.' We also know that millets, tubers, and grain legumes are rich in micronutrients but require less irrigation than the major crops. These plants and others are also sources of genes that confer tolerance to drought, floods, and the increased salinity of soils.

Although plant conservation on farms and in the wild is the ideal approach to preserving genetic diversity in crop plants, these methods are constantly jeopardized by invasive species, human destruction of habitat, and market factors. Therefore, other preservation strategies become essential. There are many cryogenic gene banks around the world resembling that at IRRI, but each is very expensive to maintain. Now, thanks to an initiative of the Government of Norway and the Global Biodiversity Trust that began in 2007, the Svalbard Gene Vault located near the North Pole will conserve over four million accessions without the need for expensive cryogenics. The remote isolation and capacity of this facility should be sufficient to preserve a sample of the existing genetic variability of all economically important plants, a vast resource generated over the past 10,000 years of agricultural evolution."

Mahatma Gandhi used to say that "nature provides for everybody's need, but not for everyone's greed." Thus, we find in nature halophytes which are salinity tolerant, xerophytes, which are drought resistant, and many other crops adapted to different agroecological conditions. We should conserve this genetic wealth of inestimable value. We should also promote anticipatory research in order to learn how to scientifically checkmate the adverse impact of unfavorable weather. This book provides guidelines for such work.

I thank Prof. Chittaranjan Kole for this labor of love in the cause of sustainable food security. I also congratulate all his authors. I hope these books will help to make life better for people everywhere.

Prof. M. S. Swaminathan<br>
Member of Parliament (Rajya Sabha),<br>
India and Emeritus Chairman,<br>
M S Swaminathan Research Foundation

# Reference

Swaminathan MS (2009) Gene banks for a warming planet. Science 325:31

# Foreword by Loren H. Rieseberg

In response to the remarkable rise of food prices in 2008, *The Economist* published an article titled, "Malthus, the false prophet: The pessimistic parson and early political economist remains as wrong as ever." The authors argue that neo-Malthusian worries about our ability to feed 9.2 billion people in 2050 are mistaken, and that advances in agricultural productivity will be sufficient to feed the world.

I am less optimistic. Growth in crop yields has been decelerating for some time, a trend that is likely to be exacerbated by climate change and regional water scarcity. Nonetheless, the Food and Agriculture Organization of the United Nations contends that 90 % of the necessary increase in crop production globally must come from higher yields, since there is little opportunity for expanding the agricultural land base.

Should we be worried about this? I think so. While I may not make it to 2050, I have two small children who will (I hope). My wife tells me that we should not worry about things we cannot change. However, as genomicists, agronomists, and plant breeders, we have the knowledge and tools to develop more productive, sustainable, and resilient crops (and thus perversely prove *The Economist* to be correct in their dismissal of Malthus and predictions of a food crisis later this century).

The *Genomics and Breeding for Climate-Resilient Crops* provides a blueprint for meeting this most important challenge. The first volume discusses how new genomic tools and resources can be used to accelerate breeding, with the overall goal of maximizing crop productivity while minimizing resource use and environmental damage. An especially promising approach, in my view, is the use of genomic tools to identify and introduce valuable alleles from the wild relatives of crops into elite cultivars. It is this untapped variation in wild species—housed in seed banks around the world—that has the greatest chance of providing quantum jumps in yield.

The second volume is a natural extension of the first, focusing on the key traits (drought tolerance, heat tolerance, water use efficiency, disease resistance, nitrogen use efficiency, nitrogen fixation, and carbon sequestration) necessary for climate-resilient agriculture. Any hope we have of ameliorating the impact of climate change on crop productivity rests on our ability to manipulate these traits.

This twin book project is timely, as the world is slowly waking up to the fact that a global food crisis of enormous proportions is brewing (indeed, I suspect that it will arrive long before 2050). As a consequence, these volumes are likely to form the basis of new courses on climate change and agriculture at academic institutions, to influence policy-makers worldwide, and to provide motivation and guidance to funding agencies. With sufficient investment in agricultural research and public breeding programs, I hope that my worries about filling 9.2 billion bellies are unfounded and that human ingenuity will once again trump Malthusian pessimism.

Loren H. Rieseberg
Canada Research Chair in Plant Evolutionary Genomics
University of British Columbia, Vancouver, Canada

# Foreword by Calvin O. Qualset

We are in the midst of a new era in crop improvement by genetic means with unprecedented ability to manipulate genes beyond Harlan's genepool compatibility circles. Still, wide hybridizations are difficult to produce, but the ability to introduce genes by parasexual means widens the genepool to include "any gene from any species" in the breeders' repertoire. Genomics provides the bases for such events. On the other side of the technological advances is the realization that our globe is under change due to anthropological causes, and perhaps by other forces. Climate change is a reality, but also is the reality that climate change is not predictable on a day-to-day or even annual basis. Agriculturalists will tell you that climate change occurs every year—sometimes for a succession of years. Our natural resources include atmospheric properties that effect climate change and our other natural resources—soil, biological, and water—are subjected to perturbations that can be detrimental and ultimately affect the sustainability of humankind.

It is a pleasure to see this comprehensive two-volume treatment dedicated to the subject of modifying the resiliency of crops to mitigate the impacts of long-term climate changes. But how about short-term climatic effects? These are the reality in agriculture, and we cannot say that crop breeders have been unaware of the short-term effects. Research programs over many decades have been dedicated to fitting crops to their environments, much as natural selection has adapted organisms to their environments. For example, latitudinal adaptation of crops or forest trees is genetically determined. The genetic bases are generally understood and those genetic effects may be exploited to produce climate-resilient crops. As early as 1921 Mooers' classic paper "Agronomic Placement of Varieties" (Mooers 1921) showed how crop management decisions could stabilize and maximize the performance of crop varieties that differed in their genetic potentials, such as maize. He pioneered what is now known as the regression approach to visualizing genotype × environment interaction that has become a mainstay for characterizing genotype performance in variable environments. He was able to show that certain varieties had greater resilience than others and that "placement" of varieties should match their potential and the environmental potential that he called soil quality.

Coping with the vagaries of environment has long been an issue in food, feed, and fiber production and these two volumes are relevant to the current concerns about climate change and to the stabilization of sustainable yields. Now, more than ever, integration of scientific approaches is necessary to mitigate the impacts of climate on crop production, including soil science, pest control, water management, human nutrition requirements, recognition of climatic variances, and, yes, socio-economic factors. Most of these topics are found in these two remarkable volumes, written by global experts.

I encourage the reading of the chapters on history, principles, and methods in Volume 1, as well as the innovative chapters on critical crop traits for resilience in Volume 2.

Calvin O. Qualset
Professor Emeritus
University of California
Davis, California, USA

# Reference

Mooers CA (1921) J Am Soc Agron 13:337–353

# Foreword by Ronald L. Phillips

As crop scientists, we are inherently charged with preserving, protecting, and defending the world's food supply, much like the oath that a US President may take when assuming office. Our profession takes seriously the need to *preserve* the germplasm that exists around the world—evidenced by the major seed banks for a large variety of crops. We also do our best to *protect* the food supply by responding to various biotic and abiotic stresses via conventional breeding and genetic engineering approaches. *Defending* the world's food supply takes various forms such as policies that provide open access to valuable materials or via intellectual property rights that encourage the development of new types of products, including those for feed, fiber, biofuels, nutritionally improved types, and even medical applications.

The environment causes many difficult situations—and always has—such as the dust bowl in the US during the 1930s. Drought still plagues many areas of the globe and is one of the myriad of conditions that cannot be well-predicted in advance. Then we have the complication that the genotype being grown interacts with the environment in a very complex fashion making difficult the development of improved types.

Genomic approaches add to our understanding of many of the environmental issues. Data from genomic sequencing now provide the basic framework on which to gain information for enhancing the understanding of such phenomena as genotype × environment interactions, gene expression related to environmental stresses, and provide clues as to how to apply the information for the betterment of agriculture. Marker-assisted selection is being employed in most modern plant-breeding programs with considerable success. For example, the *Sub1a* gene that provides impressive flooding tolerance in rice can be transferred from one variety to another by marker-assisted conventional breeding in 2.5 years or less.

In addition to various challenges traditionally faced by crop scientists, we now have the specter of even more extreme and variable conditions under which crops are grown. Climate change is a reality; however one of the debatable questions deals with whether it is man-made. Do you believe it or not? My question is "Does it matter?" Underpinning any belief about climate change is the fact that climate determines the very future of the world's food supply. Research must be focused on mitigating the often devastating effects of climate, which appear to be increasingly serious.

In this unique book project, the need is emphasized for meaningful and efficient phenotyping, including innovative techniques, as well as the use of model organisms such as *Arabidopsis*, *Medicago truncatula*, and *Lotus japonicas*. The effectiveness of plant breeding is well established; improved methods of phenotyping and testing for various traits will only enhance the contributions of plant breeding. Attention to the thousands of accessions of various crops in germplasm banks in terms of phenotyping would speed the movement of such materials into prebreeding programs. As with the search for durable disease resistance, breeding for "durable" resilience to important climate traits will require detailed understanding of the underlying genetics reviewed in these volumes. The shift of flowering time is a trait of interest with considerable underlying genetic information. Selection for flowering in the cooler part of the day, for example, may avoid some of the dramatic effects of high temperature on grain production.

These volumes on Genomics and Breeding for Climate-Resilient Crops are carefully designed to provide up-to-date insights on research accomplished as well as what is needed to preserve, protect, and defend our food supply. There is a logical and systematic progression of thought throughout the two volumes and numerous ideas are presented on climate change topics. As the Earth's temperature is rising, huge ice caps are melting, highs and lows of rainfall are hitting extremes, and carbon dioxide is changing the pH of oceans, we must take action on developing a comprehensive program to reduce the effects of climate threats on our food supply. International cooperation is needed, and these volumes reflect the international interest in the goal of developing climate-resilient crops. The environmental events that bring dramatic headlines in news programs and magazines demand that crop scientists employ the most effective technologies to circumvent the reduction in food supply due to climate. This is especially important given that another billion people exist on this planet every 14 years, and they must have an adequate food supply. This two-book project is a welcome contribution to the future of genomics and breeding for climate-resilient crops.

Ronald L. Phillips
Regents Professor Emeritus
University of Minnesota, USA

# Foreword by J. Perry Gustafson

The UN projects that by 2050 World food production will need to increase by a minimum of 70 % to feed a projected World population of more than nine billion. This 70 % increase will not be enough to improve the diets of the one billion hungry people in the world; it will only be enough to keep the same diet the world has today. It is clear that extraordinary improvements in agricultural productivity will be necessary. World food production has been steadily increasing from approx. 2.94 billon metric tons (BT) in 1961 to approx. 8.27 BT in 2007. Most importantly, this dramatic increase in food production was produced on approximately the same amount of land currently under production as was under production in 1950. Thus, the increase in production was the result of improved crop cultivars, crop technology advances, and better management practices. Projections indicate that world food production between now and 2050 can be increased to meet population demands for improved dietary standards provided that existing and newly developed technology is utilized to genetically improve cultivars, and that the world works very hard to improve crop management. This all needs to be accomplished without causing any additional adverse effects on the environment, and while clearly avoiding the cultivation of new land. Plant breeders will have to pay close attention to the effects of global warming on food production. However, the chapters in this book clearly show that current advances in technology are capable of doing so and of dramatically decreasing the time to deliver genetic improvements into the field. These include techniques, which bypass some traditional approaches to seed production. Introducing gene complexes by genetic manipulation from related species has a long and successful history and with the addition of new genetic engineering techniques will continue into the future. Arguably, the carefully coordinated application of all existing and new technologies discussed in the two volumes will be critically important to feed an ever-increasing population, sustaining the productivity of arable lands, and maintaining our fragile environment.

J. Perry Gustafson
Adjunct Professor of Genetics
University of Missouri, Columbia, Missouri, USA

# Preface

Climate change is expected to enormously affect life on the Earth. It will cause drastic changes in the environment and ecology and thus will severely impact agriculture. Therefore, it poses a serious challenge for global food security. It is expected to cause drastic changes in agroclimatic conditions including temperature, rainfall, soil nutrients and health, and incidence of pathogens and pests leading to striking reduction in crop yields due to global warming, water scarcity, changes of rainfall patterns resulting in increasingly frequent drought and flood, and other extreme weather events. Plant pathogens and pests may also evolve quickly with more virulent pathotypes and biotypes and so may extend their geographical spread leading to epidemics and severity due to climate change. Furthermore, elevated $CO_2$ levels will also reduce the nutritional quality of most crops and some crops may even become more toxic due to changes in the chemical composition of their tissues. Climate change will also cause elevation of greenhouse gas emission. The most grave and still unknown concern, however, involves the critical effects that interactions among various biotic and abiotic excesses or paucities will have on crops and cropping systems making the task of feeding a world population of nine billion by 2050 extremely challenging.

Several eminent scientists from different parts of the world are planning to put significant effort into combating or mitigating the threat to food security due to climate change. We organized an international workshop on Climate Change during the 20th International Conference on the Status of Plant and Animal Genome Research (PAG conference) held during January 2012 in San Diego, California and established the International Climate Resilient Crop Genomics Consortium (ICRCGC) with a membership of over 30 active scientists from over ten countries (http://www.climatechangegenomics.org). Recently, many more scientists have become interested in this critical topic and we organized a special international workshop on Genomics and Breeding of Climate-Resilient Crops for Future Food Security during the 6th International Crop Science Congress held during August 6–10, 2012 in Bento Goncalves, Brazil followed by a brain-storming discussion to formulate the future strategies and work plans for combating climate change. We recently organized another two workshops on this subject in January 2013 in

San Diego, California during the 21st PAG conference. ICRCGC is now preparing a white paper for more serious and broad dialogs to initiate international and multi-disciplinary efforts to combat climate change using genetic resources and advanced genomics and breeding tools.

The central strategy of combating climate change will obviously involve the development of climate-resilient crop cultivars with broader genome plasticity allowing wider adaptability, broader genome elasticity with potential for high response to phenotypic, chemotypic, and molecular selection and above all durable and robust resistance to biotic and abiotic stresses. However, genomics and breeding for climate-resilient crops are relatively new fields of study and research and given the seriousness of the threats of climate change will obviously be included in course curricula in academic institutes and in frontier programs of agricultural research organizations at national and international levels. This topic is expected to be of increasing interest to policy-makers, social activists, and both public and private sector agencies supporting agricultural research. There are a few critical and comprehensive reviews on this subject and a number of publications are also available on the assessment of the impact of climate change on agriculture and suggested strategies to circumvent the severe effects to ensure food security. These deliberations are, however, scattered over the pages of newspapers, newsletters, journals, and web sites. Hence, a compilation of narratives on the concepts, strategies, tools, and related issues was felt to be lacking and this was the guiding force behind the inception of this two-volume work on Genomics and Breeding for Climate-Resilient Crops.

## Volume 1 "Genomics and Breeding for Climate-Resilient Crops: Concepts and Strategies"

Volume 1 of this book deliberates on the basic concepts and strategies of genomics and breeding for developing climate-resilient crop varieties. In recent years considerable gains have been made in our understanding of plant genome organization and gene expression. In large part this has been achieved through the study of "model species," i.e., species in which genetics and genomics are more tractable than in many crop plants. The best known and most developed of these models is *Arabidopsis thaliana*, the DNA sequence of which was published in 2000. Subsequently, a number of different model species have been developed and the number of crop species that feed the world with sequenced genomes has also increased.

Progress in developing the resources, tools, and approaches to allow more rapid development of improved crops has been significant in the last decade. These include genomics, transcriptomics and metabolomics as well as nondestructive, dynamic high-throughput phenotyping (phenomics), and novel approaches to germplasm characterization and population improvement. It is timely therefore to provide a detailed analysis of where we are now and what progress can be expected in the near future. This description of the "state of the art" is presented in Volume 1.

The first chapter of Volume 1 by Abberton provides an introduction to the potential damaging effects of climate change on agriculture leading to future food insecurity and narrates the required integrated approaches to address this serious threat. It elucidates on the role of plant improvement and the requirement of judicious deployment of genomic tools for utilization of genetic diversity and precision high-throughput phenotyping as supplementary strategies for developing climate-resilient crop varieties.

At the heart of crop improvement are genetic resources, their collection, characterization, and utilization. The contribution of Pignone and Hammer (Chap. 2) shows ways in which the use of genetic resources can be targeted at the challenges of a changing climate. Genetic resources offer a vast reservoir of important novel traits and allelic variation for traits. Increasingly, genomics tools are being brought to bear on the variation collected in genebanks.

De Pace and coauthors in Chap. 3 detail modern methods of identifying traits for incorporation into selection criteria for new cultivars and to uncover the underlying genetic control of variation in these traits. They present reasons as to why paleoclimate and vegetation-type reconstruction from fossil records and species vicariance help in understanding the long-term dynamics of plant features and trait evolution associated with dispersal and climate changes. Comparative genomics demonstrated as alleles for those plant features (i.e., plant morphology and phenophase alteration), and for biotic (response to bacterial and fungal pathogens) and abiotic (i.e., drought, flooding) stress resistance are still part of the standing genetic endowment of the living gene pools of the crop and forest plant species and allied wild relatives. Therefore, recapitulation of evolutionary and domestication processes using various genomic tools will provide innovative molecular breeding methods to explore and select the genetic variants needed for forest and crop adaptation to climate change pressures.

Subsequent chapters deal more explicitly with how modern genetics and genomics can be used within crop improvement programs. Dwiyanti and Yamada (Chap. 4) describe genetic mapping and identification of quantitative trait loci (QTL) for traits involved in responses to changing climate. Generally, QTLs related to climate change response are complex and largely affected by environmental conditions. Nevertheless, researchers have succeeded in identifying several major QTLs and applying the knowledge in crop breeding. Various genomics tools are now available in crop species that have been the subject of most research, particularly maize but now increasingly rice and other cereals. Progress in wheat has been slow due to the limited extent of genetic variation in the crop and its hexaploid nature, but next-generation sequencing (NGS) approaches are changing this rapidly. Genome sequencing with ever-increasing speed and reducing costs brings with it the potential for "genotyping by sequencing," which when allied with sophisticated statistical approaches is likely to allow the potential of genome-wide or genomic selection (GWS) to be realized more effectively. Knowledge of QTLs in model species and advances in genomic tools can be applied to crop plants with limited genomic information.

Adaptation to climate change would require convergence of appropriate technologies, policies, and institutional innovations and Chap. 5 by Prasanna and coauthors focuses on some of the promising genomic tools and strategies that can enhance time- and cost-effectiveness of breeding for climate-resilient major cereal crops, particularly maize. They deliberate on employment of modern breeding strategies such as high-density genotyping, whole-genome re-sequencing, high-throughput and precise phenotyping, and genomics-assisted breeding including genome-wide association studies, breeder-ready marker development, rapid-cycle genomic selection, marker-assisted recurrent selection, and crop modeling.

The strategy for climate-resilient agriculture should be to maximize crop production with minimal or no damage to the environment. This would demand changes in the approaches to crop improvement, and in the deployment of recent techniques involving genomics in crop research. Designer crops have to be developed with enhanced efficiency in the use of radiation energy, nutrients, and water; they also have to fit the system of conservation agriculture including zero tillage. Talukdar and Talukdar (Chap. 6) describe recent changes in strategies and approaches to crop breeding highlighting the progress of genomic tools in meeting the challenges of climate change in agriculture.

Of course a major modern tool for the improvement of specific traits is genetic engineering. To date, this technology has been widely employed but only for a few traits in a small number of crops. Developments in genetic engineering have been instrumental for commercial application in the production of transgenic plants for biotic and abiotic stresses. It has tremendous potential in agriculture and especially for important traits related to climate change. Yadav and coauthors (Chap. 7) address the challenge of deploying this technique for important complex traits that are central to climate adaptation.

New approaches to selection are required to take full advantage of the pace at which new genomic knowledge is being acquired. Ceccarelli and coauthors (Chap. 8) and Murphy and coauthors (Chap. 9) explain how participatory breeding, where farmers are involved in the selection process, can be focused on adaptation.

In Chap. 8, Ceccarelli and coauthors show how climate changes have affected humanity for a long time, and how crops and people have reacted and adjusted to changes. As biodiversity has sometimes been negatively affected by modern crop production, they discuss how participatory plant breeding, by exploiting specific adaptation and farmers' knowledge, can positively contribute not only to crops' adaptation to climate changes but to increase biodiversity and to increase production directly in the hands of the farmers, thus also improving the accessibility and availability of food.

Murphy and coauthors (Chap. 9) explain how participatory breeding can be focused on adaptation. Evolutionary breeding has been shown to efficiently increase fitness, disease resistance, and related yield components in self-pollinating cereals while maintaining a broad-based genetic diversity in the field. Through buffering of biotic and abiotic stresses, increased on-farm genetic diversity can be important to maintaining yield stability across time and space. Inclusion of on-farm selection of diverse evolutionary breeding populations within and across fluctuating environments has the potential to greatly increase the scope of genotypically plastic cultivars capable of adaptation to unpredictable climate-induced environmental change.

Modern crop improvement in both its genomics and phenomics components is increasingly data rich. Edwards (Chap. 10) summarizes the role and importance of bioinformatics in integrating these data and converting them to usable knowledge has been emphasized greatly over the past two decades and is now being increasingly reflected in the public sector and commercial programs throughout the world.

Chapter 11 by Lybbert and coauthors deals with the critical issue of international collaboration and the importance of international funding to facilitate the exchange of knowledge and germplasm required to achieve success in global crop development. The benefits of climate-resilient crops are often complex and the adoption of these crops will require international collaboration and coordination of public and private sectors. Support for networking and funding of collaborative research and extension activities will determine global success in the development of climate-resilient crops and their impact on food security.

However, the importance of regulation and intellectual property is pervasive in the area of modern crop improvement and not restricted to genetic engineering. These issues are considered by Blakeney (Chap. 12). Similarly, sociopolitical consideration extends across the whole of the application of genomics, its translation into the development of cultivars and farmers' access to them and the ability to incorporate them into their operations as enumerated by Hughes and coauthors in Chap. 13.

## Volume 2 "Genomics and Breeding for Climate-Resilient Crops: Target Traits"

In many cases breeders seek improvements in yield and yield per unit of input, however, factors limiting yield are many and various and often these factors must be addressed directly in targeted approaches. Volume 2 elucidates the genomic and breeding approaches for genetic improvement in the major target traits covered in its 13 chapters.

In the majority of crop species, the timing of flowering represents a key adaptive trait, with a major impact on yield. In Chap. 1, Bentley and coauthors review the genetic determinants and environmental cues that influence flowering time in a range of crop species. They deliberate on the consequences of climate change on crop adaptation mediated by flowering and discuss the breeding targets and methodologies to mitigate detrimental yield.

Plant growth and development are largely dependent on the root system due to its crucial role in water and mineral uptake affecting overall plant growth and architecture. Most agricultural crops have a remarkable level of genetic variation in root morphology that can be harnessed for improving crop adaptation to several abiotic stresses. The importance of roots and roots traits (size, architecture, interactions with soil, exudation, etc.) has long been recognized, but progress has been slow due to difficulty in phenotyping and screening techniques. However, genomics approaches allied to the development of noninvasive dynamic imaging techniques capable of phenotyping root traits and largely effecting QTL identification to facilitate marker-assisted selection will bring significant new opportunities for crop improvement as enunciated in Chap. 2 by Silvas and coauthors.

Two of the major abiotic stresses involve responses to low or high temperatures. Because of the need to feed an ever-increasing global population, attempts at agricultural production are being extended into marginal locations, including those at higher altitudes where growth conditions are suboptimal due to the cold stresses commonly encountered. Amongst the traits associated with survival under such conditions, the acquisition of genes for freezing tolerance is considered of primary importance with gene improvement targeted specifically at the timeliness of engagement, the maintenance, and the subsequent release of cold acclimation mechanisms necessary to ensure a winter survival appropriate to the region of crop growth, followed by rapid recovery to ensure good crop yields once the threats of encountering further freezing temperatures are diminished. Humphreys and Gasior (Chap. 3) deliberate the detailed effects of cold on crop growth and development and ultimately on crop yield, and the strategies and tools for developing crop varieties with improved tolerance to winter and freezing.

Porch and Hall (Chap. 4) describe crop improvement with respect to heat tolerance. Many current environments experience high temperatures that reduce crop yield, and projected increases in temperature could reduce grain or fruit yield by about 10 % per °C increase in temperature. Yet, relatively little effort has been devoted to breeding for heat tolerance. However, for a few crop species, heat-resistant cultivars have been bred by conventional hybridization and selection for heat tolerance during reproductive development. The successes that have been achieved are described and provide blueprints, whereby heat-resistant cultivars could be bred for many annual crop species. Molecular approaches to enhance breeding for heat tolerance are also discussed.

A major focus of this volume is on water: drought stress and water use efficiency important for interactions with soil including effects on flooding propensity. Crop improvement has had limited success in developing new cultivars with enhanced adaptation to drought-prone environments, although it has been pursued for various decades. Research on the mechanisms underlying the efficient use of water by crops and water productivity remains essential for succeeding in this endeavor. They may be improved through genetic enhancement. Advances in genetics, "omics," precise phenotyping and physiology coupled with new developments in bioinformatics and phenomics can provide new insights into traits that enhance adaptation to water scarcity. Chapter 5 by Ortiz provides an update on research advances and breeding main grain crops for drought-prone environments.

For several decades now, plant breeders have been selecting for high water use efficiency as a way to increase agricultural productivity in water-limited environments. Water use efficiency is the ratio between carbon gain (photosynthesis) and water loss (transpiration), which inherently occurs during stomatal opening. High water use efficiency is generally associated with greater drought tolerance, but this does not always equate to greater productivity. There are few studies that have demonstrated improvements in water use efficiency that lead to improved yields. Chapter 6 by Bramley and coauthors describe the multifarious hydraulic and biochemical processes controlling plant water loss and photosynthesis. They show that water use efficiency is predominantly driven by plant hydraulic properties, and genes that are mostly involved in gas exchange. Genetic variation for these properties exists in agricultural crops, but research needs to be directed towards examining the influence of high water use efficiency on yield production in targeted environments.

The enormous diversity in rice and its adaptation to contrasting hydrological and edaphic conditions made it one of the crops most acquiescent to genetic manipulation to keep up with the increasing adversities of climate change, including the increasing flood incidences predicted by several climate models. A classical example is presented in Chap. 7 by Ismail, summarizing the progress in breeding for flood tolerance in rice and prospects for future improvements to cope with further deteriorations projected in rainfed lowlands of the tropics. A case was presented where deployment of the *SUB1A* gene that confers tolerance to submergence during the vegetative stage resulted in considerable impacts in farmers' fields in flood-prone areas, with yield advantages of 1 to over 3 t $ha^{-1}$ following 4–18 days of complete submergence. This is a classic example of the use of genomic tools to resolve current issues and cope with further adversities of climate change, while keeping up with the rising demands for more food.

Clearly responses to biotic stresses, and postharvest losses, are crucial aspects in maintaining yield in many environments and often diseases and insects are a major cause of biomass loss.

Chapter 8 by Bariana and coauthors summarizes the role of genomics in the breeding of disease-resistant crop cultivars. Various selection technologies used in cereal and pulse improvement programs are discussed. Current information on disease response linked markers is reviewed in light of their implementation. The potential use of whole genome molecular scanning in breeding disease-resistant crops is also explored. The role of information management in programs aimed at the application of genomics to crop improvement is emphasized.

Agricultural research has for decades focused on gathering crucial information on the biochemical, genetic, and molecular realms that deal with plant–insect interactions in changing ecosystems. Environmental conditions, which include the overall conditions of climate change, are a reality that needs to be considered as one of the crucial phenomena of changing ecosystems when planning future crop improvement, security, and/or pest management strategies. In the context of climate change in Chap. 9 Emani and Hunter attempt to integrate past and present research in classical and molecular breeding, transgenic technology, and pest management. The integrated approach will direct present research efforts that aim at creating plant–insect pest interaction climate change models that reliably advise future strategies to develop improved insect-resistant, climate-resilient plant varieties.

One of the foundations of the increases in crop productivity in the past has been improved nutrient availability, especially nitrogen and phosphorus. The increases in crop production that are needed to meet the demands of a growing world population will require greater supply and uptake of essential nutrients by plants. Even in the absence of climate change, there is a need to improve the nutrient use efficiency of our present cropping systems to make better use of nonrenewable resources and to minimize the adverse environmental effects of the over-use of fertilizers. Moreover, in many parts of the world, the nutrient concentration of staple food crops is low, and significant gains in the health of communities can be achieved by biofortification of grain. Breeding crop varieties that are better able to use poorly available sources of nutrients in the soil or which can respond better to inputs of nutrients is an important aspect of increasing nutrient use efficiency. The uptake and use of nitrogen and phosphorus by crop plants are complex processes and to date, there has been limited success in improving nutrient use efficiency using conventional approaches. Advanced genetic techniques allied to traditional plant breeding may play an important role in improving nutrient use efficiency. McDonald and coauthors present a review on the importance of improving nitrogen use efficiency and on research accomplished so far for that purpose in Chap. 10.

Nitrogen assimilation and fixation underlie the advances of the green revolution and limit the impact of both agriculture and environmental plant productivity on carbon sequestration. For a number of years the potential importance of legumes in crop rotations across many agroecosystems has been recognized. However, the limited extent to which this potential has been realized has been disappointing. Legumes do not just contribute in terms of food, feed, and fertility but are also important as fuel wood and could help more with respect to carbon sequestration. However, their key attribute and a major reason why they are so important for the future of world agriculture is the nitrogen-fixing symbioses they form with nodulating bacteria. Genetic and genomic tools have been applied powerfully in recent years to understand the control of the legume–rhizobia interaction utilizing model legumes, particularly *Medicago truncatula* and *Lotus japonicus* and the great challenge is to deploy this information in the improvement of the major grain and forage legumes (Lightfoot, Chap. 11). However, the world's major nitrogen sink crops are the grasses and cereals. Equally, the world's major carbon sink crops are trees and wetland plants. These crops and plants could be made more efficient in their use of nitrogen and consequently, water. Scientists have tried many transgenes in many crops to achieve these improvements with good rates of success. Stacks of genes, and new sets of transgenes hold promise to deliver significant improvements across the world's major crops. On the basis of these discoveries efforts to develop genetically, or even transgenically altered environmental plants might also help to slow global warming.

Carbon sequestration in plants has been proposed as a possible moderator or solution to the rising levels of atmospheric carbon dioxide ($CO_2$) threatening to alter global temperature and climate. Chapter 12 by Cseke and coauthors examines the different mechanisms of carbon sequestration within the Earth's natural carbon cycle with a special focus on events associated with plant development. This chapter outlines the specific chemical and biological processes that allow plants to capture, allocate, and provide for long-term storage of $CO_2$ in the form of both above-ground and below-ground biomass. They specifically examine the contribution of mycorrhizal and other soil community-level interactions as an important reminder that healthy soils are required for the uptake of nutrients needed for efficient carbon sequestration. This chapter provides a perspective on molecular approaches to enhancing carbon sequestration in biological systems.

Gases that trap heat in the atmosphere are referred to as greenhouse gases. Four major greenhouse gases that are abundant in the atmosphere today are carbon dioxide, methane, nitrous oxide, and fluorocarbons. These are naturally occurring greenhouse gases because they are potentially essential to keeping the Earth's temperature warm. However, man-made activities have increased the number of these gases, resulting in more heat getting trapped in the atmosphere. Among these gases, carbon dioxide has the highest concentration in the atmosphere. Chapter 13 by Abberton deliberates on greenhouse gas emission and carbon sequestration.

The chapters of Volume I "Genomics and Breeding for Climate-Resilient Crops: Concepts and Strategies" were contributed by 38 scientists from 14 countries including Australia, China, France, Germany, India, Italy, Japan, Kenya, Nigeria, Netherlands, Philippines, UK, USA, and Zimbabwe. The chapters of Volume 2 "Genomics and Breeding for Climate-Resilient Crops: Target Traits" were contributed by 48 scientists from nine countries including Australia, Brazil, Germany, India, Japan, Philippines, Sweden, USA, and UK. Altogether 84 scientists from 16 countries authored the 26 chapters of the two volumes. I wish to extend my thanks and gratitude to all these eminent scientists for their excellent contributions and constant cooperation.

I have been working with Springer since 2006 and edited several book series, and I have developed a cordial relationship with all the staff involved. I wish to thank Dr. Christina Eckey and Dr. Jutta Lindenborn for their constant guidance and cooperation right from the planning up to the completion of this book project. It was highly enriching and comfortable to work with them all along.

I must thank my wife and colleague Phullara, our son Sourav, and our daughter Devleena for their patience over several months as I put in extra time on this project and also for assisting me with the editing of these two books as always.

I feel myself fortunate that several pioneering scientists of the field of plant and agricultural sciences from around the world have been so kind to me over the last three decades of my professional career. Five such legendary scientists, Profs. M.S. Swaminathan, Loren H. Rieseberg, Calvin O. Qualset, Ronald L. Phillips, and J. Perry Gustafson, have kindly penned the forewords for this work and given readers a rare opportunity to learn from their precise assessment of the problem of climate change and vision for the future road map of genomic and breeding research required to attain sustainable food security in the future. I express my deep regards and gratitude to them for sharing their wisdom and philosophy and also for all the generosity, affection, encouragement, and inspiration they showered upon me!

Chittaranjan Kole
Mohanpur, Nadia, West Bengal, India

# Abbreviations

| | |
|---|---|
| AAT | Alanine aminotransferase |
| ABA | Abscisic acid |
| ABP9 | ABA-responsive cis-elements binding protein 9 |
| ABRE | ABA-responsive elements |
| ADH | Alcohol dehydrogenase |
| ADP | Adenosine diphosphate |
| AFP | Antifreeze protein |
| *AG* | *AGAMOUS* gene |
| AGI | Arabidopsis Genome Initiative |
| *AGL24* | *AGAMOUS-LIKE 24* gene |
| AHB1 | *Arabidopsis* hemoglobin |
| ALAT | Alanine aminotransferase |
| AM | Arbuscular mycorrhizae |
| AM | Association mapping |
| ANN | Artificial neural network |
| ANOVA | Analysis of variance |
| AOC | Allene oxide cyclase |
| AON | Autoregulation of nodulation |
| AOS | Allene oxide synthase |
| *AP1* | *APETALA1* gene |
| *AP2* | *APETALA2* gene |
| *AP3* | *APETALA3* gene |
| APase | Acid phosphatase |
| APX | Ascorbate peroxidase |
| AQP | Aquaporin |
| ARF | Auxin-responsive factor |
| ASE | Allele-specific expression |
| ASI | Anthesis-silking interval |
| ATP | Adenosine triphosphate |
| *At*SAP5 | *Arabidopsis thaliana* stress-associated protein 5 |
| AVRDC | Asian Vegetable Research and Development Center |

| | |
|---|---|
| BAC | Bacterial artificial chromosome |
| BC | Backcross |
| BCA | Belowground carbon allocation |
| BGMV | Bean golden mosaic virus |
| *bHLH* | Basic helix-loop-helix |
| BIL | Backcross inbred line |
| BLB | Bacterial leaf blight |
| BNI | Biological nitrification inhibition |
| BR | Brassinosteroid |
| *Bt* | *Bacillus thuringiensis* |
| *BTC1* | *BOLTING TIME CONTROL 1* gene |
| BVOC | Biogenic volatile compounds |
| bZIP | Basic leucine zipper protein |
| CA | Cold acclimation |
| *CAB2* | *CHLOROPHYLL A/B BINDING PROTEIN 2* gene |
| *CAL* | *CAULIFLOWER* gene |
| CAM | Carboxylic acid metabolism |
| CAM | Crassulacean acid metabolism |
| CaMV | Cauliflower mosaic virus |
| CAT | Catalase |
| CAX1 | Calcium exchanger 1 |
| CBF | C-repeat binding factor |
| *CCA1* | *CIRCADIAN CLOCK-ASSOCIATED 1* gene |
| *CCR2* | *COLD AND CIRCADIAN-REGULATED 2* gene |
| CDF | Cation diffusion facilitator |
| *CDF1* | *CYCLING DOF FACTOR 1* gene |
| cDNA | Complementary DNA |
| CDPK | Calcium-dependent protein kinase |
| CE | Cation efflux |
| *CEN* | *CENTRORADIALIS* gene |
| CGIAR | Consultative Group on International Agricultural Research |
| CID | Carbon isotope discrimination |
| CIMMYT | International Maize and Wheat Improvement Center |
| CIPK | CBL-interacting protein kinase |
| CIRAD | Centre International de Reseaux Agriculture and Development |
| CLE | Clavata3/ESR-related peptide |
| CLV3 | Clavata3 |
| *CO* | *CONSTANS* gene |
| Col | Columbia |
| COR | Cold-regulated |
| CP | Crude protein |
| CRT | C-repeat |
| *CRY1* | *CRYPTOCHROME 1* gene |
| *CRY2* | *CRYPTOCHROME 2* gene |
| CSP | Cold-shock protein |

| CSSLs | Chromosome segment substitution lines |
|---|---|
| CT | Condensed tannins |
| CTD | Canopy temperature depression |
| CV | Coefficient of variation |
| DAM | Dormancy-associated MADS-box |
| *DAM* | Dormancy-associated MADS-box genes |
| DArT | Diversity array technology |
| DEG | Differentially expressed gene |
| *DF1* | *DELAYED FLOWERING 1* gene |
| DH | Doubled haploid |
| DHL | Doubled haploid line |
| DIMBOA | 2,4-Dihydroxy-7-methoxy-2H-1,4-benzoxazin-3-one |
| DREB | Dehydration-responsive element-binding protein |
| DREB1 | Dehydration-responsive element-binding protein 1 |
| ds-cDNA | Double-stranded cDNA |
| *DTH8* | *DAYS TO HEADING 8* gene |
| *EBS* | *EARLY BOLTING IN SHORT DAYS* gene |
| ECM | Ectomycorrhizae |
| *Ef7* | *Early flowering 7* gene |
| *Ehd1* | *Early heading date 1* gene |
| *Ehd2* | *Early heading date 2* gene |
| *Ehd3* | *Early heading date 3* gene |
| EIN | Expression interaction network |
| *ELF3* | *EARLY FLOWERING 3* gene |
| EPA | Environmental Protection Agency |
| ERF | Ethylene-responsive factor |
| ESK1 | *ESKIMO1* mutation |
| EST | Expressed sequence tag |
| ET | Evapotranspiration |
| EU | European Union |
| FACE | Free-air carbon dioxide enrichment |
| FAO | Food and Agricultural Organization |
| *FAR1* | *FAR-RED IMPARED RESPONSE 1* gene |
| FGAS | Functional genomics of abiotic stress |
| *FHY3* | *FAR-RED ELONGATED HYPOCOTYL 3* gene |
| *FKF1* | *FLAVIN-BINDING, KELCH REPEAT, F-BOX 1* gene |
| *FLC* | *FLOWERING LOCUS C* gene |
| *FpCBF* | *Festuca pratensis CBF* gene |
| *FPF1* | *FLOWERING PROMOTING FACTOR 1* |
| *Fr-A1* | Frost resistance QTL (on chromosome 5A in *Triticum* spp.) |
| *Fr-A2* | Frost resistance QTL (on chromosome 5A in *Triticum* spp.) |
| *Fr-Am2* | Frost resistance QTL (on chromosome 5A in *Triticum monococcum*) |
| *Fr-H2* | Frost resistance QTL (on chromosome 5A in *Barley*) |
| *FRI* | *FRIGIDA* gene |

| | |
|---|---|
| FRV-N | Function-restricted value neighborhood |
| *FT* | *FLOWERING LOCUS T* gene |
| FT | Freezing tolerance |
| *FT1* | *FT-like 1* gene |
| *Ful* | *FRUITFULL* gene |
| G × E | Genotype × environment interaction |
| GA | Gibberellic acid |
| *GAI* | *GIBBERELLIN INSENSITIVE* gene |
| GBS | Genotyping-by-sequences/ing |
| GCP | Generation Challenge Program |
| GDH | Glutamate dehydrogenase |
| GDP | Guanosine diphosphate |
| GEBV | Genomic estimated breeding value |
| GFP | Green fluorescent protein |
| *Ghd7* | *Grain number, plant height and heading date 7* gene |
| *Ghd8* | *Grain number, plant height and heading date 8* gene |
| GHG | Greenhouse gas |
| *GI* | *GIGANTEA* gene |
| GM | Genetically modified |
| GO | Gene ontology |
| GOGAT | Glutamine:2-oxoglutarate aminotransferase |
| GPX | Glutathione peroxidase |
| GRAS | GAI, RGA, and SCR proteins |
| GRN | Genetic regulatory network |
| GS | Genomic selection |
| GS | Glutamine synthetase |
| GSH | Glutathione |
| GTL1 | GT-2-LIKE1 |
| GTP | Guanosine triphosphate |
| GWAS | Genome-wide association scan |
| GWAS | Genome-wide association selection |
| GWP | Global-warming potential |
| H3Ac | Histone H3 acetylation |
| H3K27 | Histone 3 lysine 27 |
| H3K4me3 | H3 lysine K4 trimethylation |
| *Hd1* | *Heading date 1* gene |
| *Hd2* | *Heading date 2* gene |
| *Hd3a* | *Heading date 3a* gene |
| *Hd4* | *Heading date 4* gene |
| *Hd5* | *Heading date 5* gene |
| *Hd6* | *Heading date 6* gene |
| *Hd9* | *Heading date 9* gene |
| *Hd16* | *Heading date 16* gene |
| *Hd17* | *Heading date 17* gene |
| *HDA6* | *HISTONE DEACETYPLASE 6* gene |

| | |
|---|---|
| HDAC | Human histone deacetylase complex |
| HK | Histidine kinase |
| HOS | High expression of osmotically responsive gene |
| HPL | Hydroperoxide lyase |
| HPLC | High-performance liquid chromatography |
| HRE | Hypoxia-responsive element |
| HSP | Heat shock protein |
| *Hsp101* | *Heat shock protein 101* gene |
| HUP | Hypoxia-responsive unknown protein |
| *HvCBF3* | *Hordeum* CBF3 gene |
| *HvCBF4* | *Hordeum* CBF 4 gene |
| *HvCBF8* | *Hordeum* CBF8 gene |
| IAA | Indole acetic acid |
| IARI | Indian Agricultural Research Institute |
| ICARDA | International Center for Agricultural Research in the Dry Areas |
| ICE1 | Interleukin-1β-converting enzyme 1 |
| ICIS | International Crop Information System |
| ICRISAT | International Crops Research Institute for the Semi-Arid Tropics |
| *ID1* | *INDETERMINATE 1* gene |
| IITA | International Institute of Tropical Agriculture |
| ILs | Isogenic lines |
| InDel | Insertion/deletion |
| IPCC | Intergovernmental Panel on Climate Change |
| IPM | Integrated pest management |
| IRGSP | International Rice Genome Sequencing Project |
| IRI | Ice-recrystallization inhibition |
| IRRI | International Rice Research Institute |
| JA | Jasmonic acid |
| JA-Ile | Jasmonic acid isoleucine |
| kin1 | Protein kinase 1 |
| KNOX1 | KNOTTEDI-like homeobox |
| LC | Liquid chromatography |
| LCM | Laser capture micro |
| LD | Long day |
| $LD_{50}$ | Lethal dose for 50% of plants |
| LDH | Lactate dehydrogenase |
| LEA | Late-embryogenesis abundant |
| L*er* | Landsberg *erecta* |
| *LFY* | *LEAFY* gene |
| LG | Linkage group |
| *LHY* | *LATE ELONGATED HYPOCOTYL* gene |
| LM | Laser microdissection |
| LOES | Low oxygen escape syndrome |
| LOX | Lipoxygenase |

| | |
|---|---|
| *LpCBF* | *Lolium perenne CBF DREB*-related gene |
| LR | Lateral roots |
| LRL | Lateral root length |
| LRR | Leucine-rich repeat |
| *Ma* | *Maturity* gene |
| *Ma1* | *Maturity 1* gene |
| MAB | Marker-assisted breeding |
| MADS-box | *MCM1/AGAMOUS/DEFICIENS/SRF*-box |
| MAGIC | Multiparent advanced genetic intercross |
| MAPK | Mitogen-activated protein kinase |
| MARS | Marker-assisted recurrent selection |
| MAS | Marker-aided/assisted selection |
| MDA | Malondialdehyde |
| MDU | Methylene diurea |
| ME | Malic enzyme |
| MEF | Managed environment facilities |
| miR156 | MicroRNA 156 |
| miR172 | MicroRNA 172 |
| miRNA | MicroRNA |
| MLP | Major latex protein |
| MRL | Main/maximum root length |
| mRNA | Messenger RNA |
| MS | Mass spectrometry |
| MTME | Multitrait multienvironment |
| MTP | Metal tolerance protein |
| MYC | Myelocytomatosis |
| NA | Nonacclimated |
| NADP | Nicotinamide adenine dinucleotide phosphate |
| NAM | No apical meristem |
| NAM | Nested association mapping |
| NBM | Net biome productivity |
| NCED | 9-Cis-epoxycarotenoid dioxygenase 1 |
| NEE | Net ecosystem exchange |
| NIL | Near-isogenic lines |
| NIP | Nodulin-like intrinsic protein |
| npcRNA | Nonprotein coding RNA |
| NPN | Nonforage protein |
| NPP | Net plant productivity |
| NPQ | Nonphotochemical quenching |
| NR | Nitrate reductase |
| NRN | Nodal root number |
| NSF | National Science Foundation |
| NTUE | Nutrient-use efficiency |
| NUE | Nitrogen-use efficiency |
| NUpE | N uptake efficiency |

| | |
|---|---|
| NUtE | N utilization efficiency |
| OA | Optimal allocation |
| OAA | Oxaloacetate |
| OGRO | Overview of functionally characterized genes in rice online database |
| OPDA | Oxophytodienoic acid |
| OPR | Xo-phytodienoatereductase |
| *OsFD* | *Oryza sativa FD* gene |
| *OsGI* | *Oryza sativa GIGANTEA* gene |
| *OsId* | *Oryza sativa Indeterminate* gene |
| *OsLFL1* | *Oryza sativa LEC1* and *FUSCA-LIKE1* genes |
| *OsPRR37* | *Oryza sativa PSEUDO-RESPONSE REGULATOR 37* gene |
| PAGE | Polyacrylamide gel electrophoresis |
| PAP | Phosphatidic acid phosphatase |
| PAR | Photosynthetically active radiation |
| PC | Phytochelatin |
| PCD | Programmed cell death |
| PDC | Pyruvate decarboxylase |
| PEG | Polyethylene glycol |
| *PEP* | *PERPETUAL FLOWERING* gene |
| PEP | Phosphoenolpyruvate |
| *PEP1* | *PERPETUAL FLOWERING 1* gene |
| PEP-CK | Phosphoenolpyruvate carboxykinase |
| Pfam | Protein family |
| PGA | 3-Phosphoglycerate |
| PH | Plant height |
| PHD | Plant homeodomain |
| PHGPX | Phospholipid hydroperoxide glutathione peroxidase |
| *PhyA* | *PHYTOCHROME A* gene |
| *PhyB* | *PHYTOCHROME B* gene |
| *PI* | *PISTILLATA* gene |
| PIP | Plasma-membrane intrinsic protein |
| PLRV | Potato leaf roll virus |
| PPD | Photoperiod |
| *PPD-1* | *PHOTOPERIOD-1* gene |
| *Ppd-H1* | *Photoperiod H1* gene |
| PPDK | Pyruvate orthophosphate dikinase |
| PPI | Protein-protein interaction |
| PPO | Polyphenol oxidase |
| PR | Pathogenesis-related |
| PRC2 | POLYCOMB REPRESSIVE COMPLEX 2 |
| PRL | Primary root length |
| *PRR* | *PSEUDO-RESPONSE REGULATOR* gene |
| PSI | Photosystem I |
| PSII | Photosystem II |

| PSII | Photosystem II |
| PSR | Phytosulfokine receptor kinase |
| *PTLF* | *Populus trichocarpa LEAFY* homolog |
| PUE | Phosphorus-use efficiency |
| PVC | Polyvinyl chloride |
| PVY | Potato virus Y |
| QC | Quiescent center |
| *QFt5F-2/* *QWs5F-1* | QTL for freezing tolerance on chromosome 5 |
| Q-TARO | QTL Annotation Rice Online Database |
| QTL | Quantitative trait locus/loci |
| QTL× E | QTL× environment interaction |
| R/FR | Red/far red |
| RABID | Rapid bulk inbreeding |
| RAM | Root apical meristem |
| RAPD | Random(ly) amplified DNA polymorphism |
| *RCN1* | *Rice TERMINAL FLOWER 1* gene |
| *RCN2* | *Rice CENTRORADIALIS 1 and 2* gene |
| RFLP | Restriction fragment length polymorphism |
| *RFT1* | *RICE FLOWERING LOCUS T 1* gene |
| *RGA* | *REPRESSOR OF GAI-3* gene |
| *RGL1* | *RGA-LIKE1* gene |
| RH | Relative humidity |
| *RID1* | *Rice Indeterminate 1* gene |
| RIL | Recombinant inbred line |
| RLD | Root length density |
| RLK | Receptor-like kinase |
| RLP | Receptor-like protein |
| RN | Root number |
| RNAi | RNA interference |
| RNAseq | RNA sequencing |
| ROS | Reactive oxygen species |
| RPF | Root pulling force |
| RSA | Root system architecture |
| RT-PCR | Real-time PCR |
| Rubisco | Ribulose 1,5-bisphosphate carboxylase/oxygenase |
| RuBP | Ribulose 1,5-bisphosphate |
| RV | Root volume |
| RW | Root weight |
| SAM | Shoot apical meristem |
| SAM | S-adenosyl-L-methionine-synthase |
| SAP | Stress-associated protein |
| SD | Short day |
| SDS | Sodium dodecyl sulfate |
| *Se5* | *Photoperiodic sensitivity 5* gene |

| | |
|---|---|
| SF | Stagnant flooding |
| SIP | Small basic intrinsic protein |
| SIPK | Salicylic acid-induced protein kinase |
| siRNA | Small RNA |
| SIWI | Stockholm International Water Institute |
| SLA | Specific leaf area |
| SNP | Single nucleotide polymorphism |
| SOC | Soil organic carbon |
| SOD | Superoxide dismutase |
| SOM | Soil organic matter |
| SRN | Seminal root number |
| SSD | Single-seed descent |
| SSH | Suppression subtractive hybridization |
| SSR | Simple sequence repeat |
| *STK* | *SEESTICK* gene |
| *SUB1* | *Submergence tolerance 1* gene |
| SUS | Sucrose synthase |
| *SVP* | *SHORT VEGETATIVE PHASE* gene |
| SW | Shoot weight |
| TE | Transpiration efficiency |
| *TEM* | *TEMPPRANILLO* gene |
| TF | Transcription factor |
| *TFL1* | *TERMINAL FLOWER 1* gene |
| *TFS* | *TWIN SISTER OF FT* gene |
| TH | Thermal hysteresis |
| TILLING | Targeting induced local lesions in genomes |
| TIP | Tonoplast intrinsic protein |
| TIR | Toll/interleukin receptor |
| TM | Trial mean |
| *TOC1* | *TIMING OF CAB EXPRESSION 1* gene |
| TPP | Trehalose-6-phosphate phosphatase |
| TPS | Trehalose-6-phosphate synthase |
| UNFCC | United Nations Framework Convention on Climate Change |
| UV | Ultraviolet |
| *VGT1* | *VEGETATIVE TO GENERATIVE TRANSITION 1* gene |
| *VIN3* | *VERNALIZATION INSENSITIVE 3* gene |
| VPD | Vapor pressure deficit |
| VRN | Vernalization |
| *VRN* | *VERNALIZATION* gene |
| *VRN1* | *VERNALIZATION 1* gene |
| *Vrn-1* | *Vernalization gene* (found originally on chromosome 5 in *Triticum* spp.) |
| *VRN2* | *VERNALIZATION 2* gene |
| *VRN3* | *VERNALIZATION 3* gene |
| WANA | West Asia and North Africa |

| WIPK | Wound-induced protein kinase |
| WMVI | Watermelon mosaic virus II |
| WP | Water productivity |
| WSC | Water soluble carbohydrate |
| WUE | Water-use efficiency |
| XET | Xyloglucan endotransglycosylase |
| XHT | Xyloglucan hydrolase |
| XIP | X-intrinsic protein |
| *ZTL* | *ZEITLUPE* gene |
| ZYMV | Zucchini yellow mosaic virus |

# Contents

# Chapter 1
# Flowering Time

**A.R. Bentley, E.F. Jensen, I.J. Mackay, H. Hönicka, M. Fladung, K. Hori,
M. Yano, J.E. Mullet, I.P. Armstead, C. Hayes, D. Thorogood, A. Lovatt,
R. Morris, N. Pullen, E. Mutasa-Göttgens, and J. Cockram**

**Abstract** Adaptation genes have a major role to play in the response of plants to
environmental changes. Flowering time is a key adaptive trait, responding to
environmental and endogenous signals that ensure reproductive growth and development
occurs under favorable environmental conditions. Under a climate change
scenario, temperature and water conditions are forecast to change and/or fluctuate,
while photoperiods will remain constant at any given latitude. By assessing the
current knowledge of the flowering-time pathways in both model (*Arabidopsis
thaliana*) and key cereal (rice, barley, wheat, maize), temperate forage and biofuel
grasses (perennial ryegrass, *Miscanthus*, sugarcane), root (sugar beet), and tree
(poplar) crop species, it is possible to define key breeding targets for promoting
adaptation and yield stability under future climatic conditions. In *Arabidopsis*, there
are four pathways controlling flowering time, and the genetic and/or epigenetic
control of many of the steps in these pathways has been well characterized. Despite

A.R. Bentley • I.J. Mackay • E. Mutasa-Göttgens • J. Cockram (✉)
The John Bingham Laboratory, NIAB, Huntingdon Road, Cambridge CB3 0LE, UK
e-mail: james.cockram@niab.com

E.F. Jensen • I.P. Armstead • C. Hayes • D. Thorogood • A. Lovatt
Institute of Grassland and Environmental Research, Plas Gogerddan, Aberystwyth SY23 3EB, UK

H. Hönicka • M. Fladung
Johann Heinrich von Thünen Institute, Institute of Forest Genetics, Sieker Landstr. 2, 22927
Grosshansdorf, Germany

K. Hori • M. Yano
National Institute of Agrobiological Sciences, 2-1-2 Kannondai, Tsukuba, Ibaraki 305-8602, Japan

J.E. Mullet
Department of Biochemistry and Biophysics, Texas A&M University, College Station, TX, USA

R. Morris • N. Pullen
Computational and Systems Biology Department, John Innes Centre, Norwich Research Park,
Colney, Norwich NR4 7UH, UK

C. Kole (ed.), *Genomics and Breeding for Climate-Resilient Crops*, Vol. 2,
DOI 10.1007/978-3-642-37048-9_1, © Springer-Verlag Berlin Heidelberg 2013

this, even in this model species, there is little published information on the molecular basis of adaptation to the environment. In contrast, in crop and tree species, flowering time has been continually selected, either directly or indirectly as breeders and growers have selected the material that best suits a particular location. Understanding the genetic basis of this adaptive selection is now being facilitated via cloning of major genes, the mapping of QTL, and the use of marker-assisted breeding for specific flowering targets. In crop species where the genetic basis of flowering is not well understood (i.e., in the emerging biofuel grass, *Miscanthus*), such work is in its infancy. In cases where the genetic basis is well established, however, there are still grounds for important discovery, via new and emerging methods for mapping and selecting for flowering-time traits (i.e., QTL mapping in MAGIC populations, RABID selection), as well as methods for creating new genetic combinations with potentially novel flowering-time phenotypes (i.e., via targeted mutagenesis). In the future it is likely that computational modeling approaches which incorporate gene networks and the range of phenological response to measurable environmental conditions will play a central role in predicting the resilience of crop and tree species under climate change scenarios.

## 1.1   Introduction

The efficiency with which flowering plants sexually reproduce and pass on their genetic information to future generations is critical to their adaptive success. Floral transition, the switch between the vegetative and reproductive phases of plant development, is a crucial step in the life history of all crop species. The timing of flowering determines seed set and dry matter production (Cleland et al. 2007; Jung and Muller 2009) and must coincide with optimal local climatic conditions (Buckler et al. 2009) so as to avoid abiotic and biotic stress and maximize yield (Jung and Muller 2009; Craufurd and Wheeler 2009). It is, therefore, a key determinant of regional adaptation and a crucial component in the delivery of sustainable increases in crop production worldwide.

Climate is a key driver of environmental change with impact for crop production capacity (Rosenzweig et al. 2008). Global climate change, as a result of accumulation of $CO_2$ and other greenhouse gases, is forecast to cause a multitude of environmental effects. Given that regional climatic changes, especially increasing temperatures, are responsible for observable changes in natural plant systems above natural internal variability on all continents (Rosenzweig et al. 2008), it is likely that they will also have impacts on global crop production.

As population growth and climate change have the potential to cause wide-scale food shortages, there is a need to provide useful information to farmers and foresters as to what varieties will be best adapted to maintain and increase yields in new climatic conditions (Cleland et al. 2007; Craufurd and Wheeler 2009). This means there is a need to focus on key adaptive traits, including flowering time, in order to maintain and increase crop productivity in increasingly unpredictable environments (Reynolds and Ortiz 2010). Adaptive strategies can be either agronomic (i.e., modifying the time of planting to avoid stress in the field) or trait based (i.e., manipulating phenology so critical growth stages do not coincide with periods of stress) (Ainsworth and Ort 2010; Reynolds et al. 2010). A major plant breeding challenge is catering for the range of environmental factors that interact with the expression of key adaptive traits such as flowering time. This is necessary in order to recognize gains at multiple locations across years. Within this context, a major target is to take advantage of the positive aspects of climate change while offsetting the negative impacts, with genetic combinations of traits that are robust to inter- and intra-seasonal variation in water and temperature and that perform not only in suboptimal conditions but remain responsive in favorable years (Reynolds et al. 2010). Conventional breeding programs have a strong track record in both marginal and favorable environments, with approximately 25 % of the global wheat production increase since the 1990s due to improved production in marginal environments (Lantican et al. 2003).

The reproductive stage is a major determinant of yield in crop species. Temperature influences the timing of developmental stages, alone and in concert with day length (photoperiod) (Cleland et al. 2007; Ainsworth and Ort 2010), which controls floral transition through specific genetic pathways. Under many climate change scenarios, ambient temperatures will rise, while photoperiods will remain unchanged at given latitudes. Heat accelerates growth and development, with annual species having shorter life cycles, less opportunity to intercept radiation (Reynolds et al. 2010) and perform photosynthesis coupled with a shorter reproductive phase resulting in lower yields (Craufurd and Wheeler 2009; Ainsworth and Ort 2010). High temperatures also detrimentally affect a number of key reproductive stages including pollen viability, female gametogenesis, pollen-pistil interaction, fertilization, and grain formation, resulting in severe decreases in yield (Ainsworth and Ort 2010; Lukac et al. 2012). However, there is very little information available on the relationship between photoperiod and temperature in supraoptimal conditions (Craufurd and Wheeler 2009; Ainsworth and Ort 2010).

Global warming should accelerate spring by warming the coldest days of late winter, resulting in longer growing seasons (Cleland et al. 2007). Rosenzweig et al. (2008) reported that increased temperature resulted in earlier flowering and spring arrival in natural plant communities, although late-flowering species were less affected by warming compared to early species. Concentrations of atmospheric $CO_2$ oscillate annually, dipping during summer in the Northern Hemisphere, which is the largest area of terrestrial vegetation that is actively fixing carbon through photosynthesis (Cleland et al. 2007). Recent observations have shown that the amplitude of this cycle has increased and shifted earlier, supporting earlier growing seasons (Cleland et al. 2007). In wild plant species, Wolkovich et al. (2012) reported that spring leafing and flowering are advanced by 5–6 days per °C, although it is not known whether species that flower earlier are more or less sensitive to temperature than late-flowering species. The phenology of agricultural species is, however, thought to change at lower rates and is more able to respond to elevated $CO_2$ (Cleland et al. 2007).

The Intergovernmental Panel on Climate Change (IPCC) predicts that cereal yields in mid- to high-latitude regions will increase with global warming, promoting crop production despite reduced yields at lower latitudes (Ainsworth and Ort 2010). However, this prediction is based on maintenance of existing levels of yield improvement, which are currently around 1 % per year for cereal crops (Mackay et al. 2011), although rates of yield increase for all major cereal crops are declining (Fischer and Edmeades 2010). Rice, maize, and sorghum are grown extensively in tropical regions, being well adapted to high temperatures due to breeding to develop temperature resilient varieties for these environments. In tropical ecosystems, flowering phenology is less sensitive to temperature and photoperiod than it is to changes in precipitation, with yield instability in dry conditions (Cleland et al. 2007; Reynolds et al. 2010). As many developing countries are located in subtropical and tropical regions, it is important to more fully understand the sensitivity of the production systems and crop species relied upon to climate change (Rosenzweig et al. 2008). Africa, which relies heavily on low-input rainfed crop production systems with an inherently low adaptive capacity, is most likely to be markedly affected by climate change (Mertz et al. 2009). Crop models also predict that rice and wheat in Southeast Asia and maize in southern Africa are likely to be negatively affected by changing climates in the absence of adaptation strategies (Ainsworth and Ort 2010).

It is also essential to consider the impacts of climate change in temperate and cold regions of the world where flowering is a function of both winter cold and spring heat (Yu et al. 2010). Cereals grow across a wide range of semiarid environments in temperate regions, but show marked reductions in productivity (Reynolds et al. 2010) and yield (Lobell and Field 2007) at high temperatures. Increased temperatures in the winter may delay the fulfillment of vernalization requirement (a prolonged period of cold, nonfreezing temperatures required for subsequent competence to flower) resulting in later flowering, although the increased spring temperatures could mask or offset this (Yu et al. 2010). In areas of high latitude/altitude, the effect could be exacerbated, as plants in these regions are particularly sensitive to temperature cues. For example, on the Tibetan Plateau in western China, Yu et al. (2010) reported delayed fulfillment of chilling due to increased winter temperatures conferring later onset of the growing season.

The genetic control of flowering time has been extensively studied in model species, as well as in a number of important field and tree crop species. Knowledge of both genetic and phenotypic variation for flowering time is potentially useful in providing information relevant under climate change scenarios. In this chapter, we describe the control of flowering time in key model, cereal, biofuel, root, and tree crop species and discuss key targets relevant to climate change.

## 1.2   Control of Flowering Time and Key Flowering-Time Targets for Climate Change

### *1.2.1*   **Arabidopsis thaliana**

Due to the sessile nature of plants, they have evolved a complex system of genetic pathways that utilize environmental cues such as light quality, light duration, and temperature to modulate the transition from vegetative to reproductive growth. In temperate species, coordination of flowering with prevalent seasonal conditions minimizes risk of frost damage to cold-sensitive reproductive tissues and synchronizes developmental growth stages with changes in available light, temperature, and water resources. Furthermore, the ability to modulate flowering time helps a single species to display considerable ecogeographical range. This is exemplified by the model dicotyledonous plant *Arabidopsis thaliana*, whose natural distribution spans the semiarid regions in the South Mediterranean to cold and wet environments within the Arctic Circle. Adaptation to such diverse conditions means that *Arabidopsis* flowering time shows almost continuous variation, a large proportion of which is genetically inherited. The effects of climate change are predicted to impact on plant adaptation, of which flowering time is a widely presumed to be a major contributor. Indeed, phenotypic variation and genetic adaptive values associated with plant flowering time have already been shown to be affected by climate change (e.g., Bradshaw and Holzapfel 2001; Franks et al. 2007).

Due to the wealth of natural variation in flowering time, combined with its many experimental advantages, knowledge of the converging genetic pathways and networks that control flowering time is best understood in *Arabidopsis* (reviewed by Amasino 2010), making it a suitable model for investigating the genetic components that control adaptation to current and future environments. Ultimately, *Arabidopsis* floral pathways converge to regulate the development of the shoot apical meristem (SAM), triggering the transition from vegetative to reproductive phase mediated by floral meristem identity genes such as *LEAFY* (*LFY*), *APETALA 1* (*AP1*), and *CAULIFLOWER* (*CAL*) (reviewed by Irish 2010). To date over 180 *Arabidopsis* flowering-time genes have been characterized (http://arabidopsis-reactome.org/), predominantly allocated to one of four major pathways. The vernalization and photoperiod pathways mediate signals from the environment, while the autonomous pathway monitors endogenous cues from the developmental state of the plants. In addition, the gibberellic acid (GA) pathway forms a fourth distinctive promotive pathway (Putterill et al. 1995). These pathways interact at specific points, enabling a combined model describing the genetic basis of the floral transition to be made.

### 1.2.1.1   The Vernalization Pathway and the Memory of Winter

The process by which exposure to extended periods of cold nonfreezing temperatures renders plants competent to flower is termed vernalization (Chouard 1960). Physiological experiments in *Arabidopsis* have shown that vernalization is perceived in the SAM and that the "memory" of winter is mitotically stable, rendering the SAM competent to undergo subsequent vegetative–reproductive transition (Burn et al. 1993). Thus, vernalization does not directly cause plants for flower; rather it produces an epigenetic change to a state in which plants are able to respond to subsequent flowering signals such as inductive photoperiod.

Late-flowering *Arabidopsis*–ecotypes: the majority of *Arabidopsis* accessions possess a vernalization requirement, behaving as winter annuals whereby flowering is prevented in the autumn, allowing plants to overwinter vegetatively before flowering under inductive photoperiods the following spring. However, some accessions [and most lab strains such as Landsberg *erecta* (L*er*) and Columbia (Col)] lack a vernalization response and readily flower in the absence of vernalization. Genetic mapping found early and late ecotypes to commonly differ at two loci, *FRIGIDA* (*FRI*) and *FLOWERING LOCUS C* (*FLC*) (Burn et al. 1993; Lee et al. 1993; Clark and Dean 1994). *FRI* encodes a protein with two coiled-coil domains and acts to increase mRNA abundance of the floral repressor, *FLC*, that encodes a MADS-box transcription factor (Michaels and Amasino 1999; Sheldon et al. 1999). Almost all early-flowering *Arabidopsis* ecotypes have mutations that disrupt the *FRI* open reading frame, indicating inactivation of *FRI* provides the basis for the evolution of many early-flowering ecotypes from ancestral late-flowering accessions (Johanson et al. 2000; Gazzani et al. 2003; Shindo et al. 2005). Analysis of natural variants also found *FLC* to play a central role in the vernalization pathway, with *FLC* mRNA and protein expressed at much higher levels in vernalization requiring ecotypes compared to early-flowering annual varieties (Michaels and Amasino 1999; Sheldon et al. 1999, 2000). Complementation studies in which *FLC* is constitutively expressed in the early-flowering line L*er* show *FLC* expression is sufficient to produce a later flowering phenotype that is responsive to vernalization. Thus, *FLC* acts as a strong floral repressor that negatively regulates genes that promote floral transition. During vernalization, *FLC* expression is reduced (Fig. 1.1), and the extent of this reduction is proportional to both the duration of the cold treatment and to the extent flowering is promoted, providing a molecular explanation for the quantitative nature of vernalization treatment (Sheldon et al. 2000). Although a number of early-flowering accessions have been identified which possess mutations within *FLC*, unlike those found in *FRI*, they do not represent null alleles or possess mutations within the predicted protein. Rather, they possess "weak" *FLC* alleles, which display low levels of expression (Gazzani et al. 2003; Michaels et al. 2003).

Late-flowering mutants—epigenetic changes to *FLC* chromatin: although it is not known how the cold stimulus is perceived, analysis of mutants generated in vernalization-responsive lines has identified genes that mediate the vernalization

**Fig. 1.1** Expression of major vernalization pathway genes in vernalization-responsive lines during and after vernalization treatment in (**a**) *Arabidopsis* and (**b**) temperate cereals

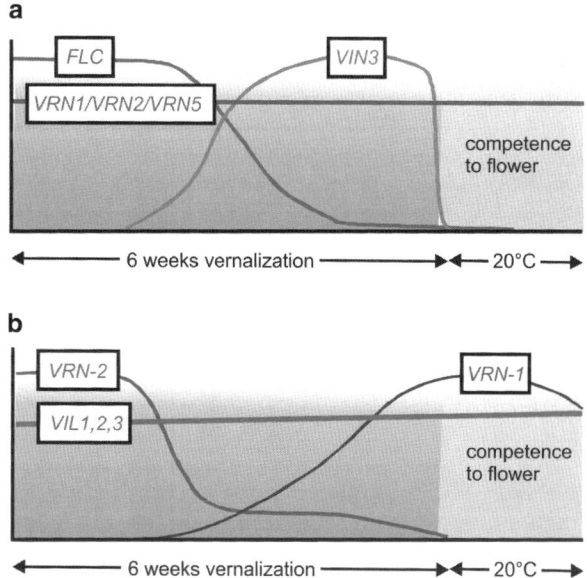

response, termed *VERNALIZATION* (*VRN*) genes (Chandler et al. 1996). All *VRN* genes cloned to date have been implicated in the epigenetic control of *FLC* expression. Prior to cold treatment, *FLC* chromatin contains epigenetic marks characteristic of active chromatin: histone H3 acetylation (H3Ac) and H3 lysine K4 trimethylation (H3K4me3) (Bastow et al. 2004; Finnegan et al. 2005). After vernalization, these epigenetic marks are reduced, while histone modifications characteristic of inactive chromatin are increased: H3K9me2 and H3K27me2. Thus, vernalization-mediated inactivation of *FLC* expression results in the competence to flower under inductive photoperiods. The first *VRN* genes identified were *VRN1* and *VRN2*. When exposed to low temperatures, expression levels of *FLC* in *vrn1* and *vrn2* mutants are reduced to levels comparable to those seen in wild-type plants. However, once moved to warmer growth conditions, these mutants are not able to maintain low *FLC* transcript levels (Gendall et al. 2001; Levy et al. 2002). Thus, *VRN1* and *VRN2* are required for the maintenance but not the establishment of *FLC* repression. More recently, mutations in the plant homeodomain (PHD) gene, *VERNALIZATION INSENSITIVE 3* (*VIN3*), have been shown to result in a late-flowering phenotype, due to expression of *FLC* remaining elevated during vernalization of *vin3* mutants (Sung and Amasino 2004). Unlike *VRN1* and *VRN2* whose expression is constitutive, *VIN3* expression is initiated during vernalization treatment, reaching maximal expression after several weeks of exposure, and returns to pre-vernalized levels on transfer to normal ambient growth temperatures (Fig. 1.1). Thus, vernalization-driven upregulation of *VIN3* is thought to contribute to the mechanism by which the duration of vernalization is measured, ensuring flowering is not promoted by short periods of cold. *VIN3* belongs to a small family of related genes, of which *VRN5* (also known as *VIN3-LIKE 1, VIL1*) has also been shown to be involved in the repression of *FLC* (Sung et al. 2006; Greb et al. 2007). Like *VIN3*, *VNR5* is expressed in the shoot and root meristem. However, *VRN5* expression remains unaffected by exposure to vernalization (Fig. 1.1). Collectively, VIN3, VRN2, VRN5, and VEL1 are thought to form components of the plant homeodomain-POLYCOMB REPRESSIVE COMPLEX 2 (PHD-PRC2), mediating vernalization-induced epigenetic silencing of *FLC* chromatin by trimethylation of histone 3 lysine 27 (H3K27) (Sung et al. 2006; Wood et al. 2006; Greb et al. 2007; De Lucia et al. 2008; Angel et al. 2011). While it is undetermined whether *VRN1* forms part of the PHD-PRC2 complex, *vrn1* mutants do not accumulate H3K9 methylation during vernalization (Bastow et al. 2004; Sung and Amasino 2004).

### 1.2.1.2   The Photoperiod Pathway

*Arabidopsis* is a facultative long-day (LD) species, with floral induction accelerated under LD photoperiods (typically ~16 h light), although flowering will eventually occur under short-day (SD) conditions (~8 h light). Photoperiod, and other components of light quality, is perceived in the leaves and transmitted to the SAM via a mobile flowering signal "florigen," now thought to be FLOWERING LOCUS T (FT) (reviewed by Wigge 2011). Central to the plants response to the duration of daily light periods is an endogenous timer, which generates a 24 h timekeeping mechanism that is entrained by light and temperature signals to daily cycles of light. Thus, circadian-regulated outputs can be modified in response to changes in the environment. Outputs from the central oscillator go on to control varied circadian-regulated responses such as stomatal opening, leaf movement, and flowering (Dunlap 1999). Anticipation of day/night cycles enhances growth and survival, thus conferring a fitness advantage. Early reports supporting this hypothesis showed that when grown under SDs, arrhythmic plants were less viable (Green et al. 2002). Subsequently, Dodd et al. (2005) found arrhythmic mutants grown under LDs fixed less carbon, contained less chlorophyll, grew slower, and had lower survival compared to plants in which the clock period matched the environment. Indeed, analysis of a range of circadian rhythm periods has recently found that lines possessing a clock that matches the environment produce more offspring and are thus adaptive (Yerushalmi et al. 2011). In order to attain such adaptive benefits for the plant, the circadian clock needs to be updated with the daily changes in light qualities that allow the plant to monitor and respond to seasonal changes. Initially, the light signal is perceived by three main groups of photoreceptor: the phytochromes (red/far-red), cryptochromes (blue), and phototropins (blue) (reviewed by Heijde and Ulm 2012; Li et al. 2012). Ultimately, light signals from the photoreceptors are passed on via intermediaries such as FAR-RED ELONGATED HYPOCOTYL 3 (FHY3) and FAR-RED IMPAIRED RESPONSE 1 (FAR1) (Lin et al. 2007). The ~24 h circadian rhythms of the plant are generated by an interlocked transcriptional-translational feedback loop known as the "circadian clock." More than 25 genes have been implicated in clock function (reviewed by Nakamichi 2011). However, the central components of the clock are the morning-phased MYB transcription factors *CIRCADIAN CLOCK-ASSOCIATED 1* (*CCA1*) and *LATE ELONGATED HYPOCOTYL* (*LHY*), whose proteins repress evening-phased genes such as *TIMING OF CAB EXPRESSION 1* (*TOC1*) (Alabadi et al. 2001). In turn, TOC1 prevents activation of morning-expressed genes during the night phase (Huang et al. 2012a), thus completing the loop. Outputs of the circadian clock allow plant responses to light quality to be transmitted.

Floral promotion by photoperiod was first found to be controlled by *CONSTANS* (*CO*), mutations of which are insensitive to inductive LD photoperiods (Putterill et al. 1995). *CO* acts as a floral promoter and its expression is regulated by the circadian clock, with low expression early in the day, followed by rapid increase in expression ~10 h after dawn, peaking at around 15 h (Suárez-López et al. 2001). Stabilization of CO protein occurs in the light, while CO degradation occurs in the dark (Valverde et al. 2004). Due to the diurnal oscillation of *CO* expression peaking late in the day in LDs, but during the dark in SDs, CO protein stabilization (and hence floral promotion) occurs in a LD-specific manner. Circadian phasing of *CO* expression is modulated by the activity of the products of additional genes, whose transcription are also circadianly regulated: *CYCLING DOF FACTOR 1* (*CDF1*), *FLAVIN-BINDING, KELCH REPEAT, F-BOX 1* (*FKF1*), and *GIGANTEA* (*GI*). *CDF1* shows peak expression soon after dawn when *CO* expression is low, with the CDF1 protein found to bind to the *CO* promoter, negatively regulating *CO* expression (Imaizumi et al. 2005). CDF1-mediated *CO* repression is overcome by the proteins encoded by *GI* and *FKF1*, and whose transcription is upregulated late in the day. GI and FKF1 proteins interact and upregulate *CO* expression via interaction of FKF1-GI complexes with CDF1 and other members of the CDF family (Imaizumi et al. 2005; Fornara et al. 2009). In addition to transcriptional control of *CO*, its protein stability and accumulation also shows circadian oscillation. While CO accumulates in white, blue, and far-red light, it is degraded in red light or in the dark (Valverde et al. 2004). Numerous proteins post-translationally regulate CO. For example, the photoreceptor PHYB promotes CO degradation early in the day, while CO stabilization late in the day is mediated by PHYA, CRYPTOCHROME 1 (CRY1), and CRY2 (Valverde et al. 2004), with CO degradation thought to be mediated by a protein complex formed between COP1 and members of the SUPPRESSOR OF PHYA-105 (SPA) family (Hoecker and Quail 2001). Ultimately, regulation of CO enables LDs (in which CO protein accumulates towards the end of the light period) to be distinguished from SD photoperiods (under which CO is not stably produced) (Srikanth and Schmid 2011).

### 1.2.1.3 The Autonomous Pathway

Photoperiod pathway genes were identified as mutants with delayed flowering under LDs but not SDs, demonstrating they are "blind" to photoperiod. A second group of mutations are late flowering under both LDs and SDs, which is overcome by vernalization or growth in far-red-enriched light (Martinez-Zapater and Somerville 1990). This class of genes therefore promote flowering in a photoperiod-independent or "autonomous" manner and belong to the autonomous pathway. The early-flowering phenotype of these genes can largely be ascribed to their repression of *FLC* transcription (Michaels and Amasino 2001), explaining why their mutation results in a robust vernalization requirement. The autonomous pathway includes *FCA, FLOWERING LOCUS Y (FY), FP, FVE, FLOWERING LOCUS D (FLD), FLOWERING LOCUS K HOMOLOGY DOMAIN (FLK)*, and *LUMINIDEPENDENS (LD)*. Of these, *FCA, FLK, FPA,* and *FY* encode (or are predicted to encode) proteins involved in RNA breakdown and processing (Macknight et al. 1997; Simpson et al. 2003; Lim et al. 2004; Manzano et al. 2009). The remaining genes are thought to be involved in chromatin structure and control of transcription. *FVE* and *FLD* encode proteins homologous to components of the human histone deacetylase complex (HDAC) and are thought to be required for deacetylation of *FLC* chromatin and repress its expression (He et al. 2003; Ausín et al. 2004). Indeed, FLD has recently been shown to directly interact with the chromatin modulation protein HISTONE DEACETYLASE 6 (HDA6) (Yu et al. 2011). *FLD* encodes a homeodomain protein with homology to transcriptional regulators and is involved in RNA metabolism (Lee et al. 1994). Collectively, autonomous pathway genes are thought to control flowering time relative to the endogenous age of the plant, mediated via the regulation of chromatin modification and RNA processing. As such, many autonomous pathway mutants and mutant combinations display additional pleiotropic phenotypes.

### 1.2.1.4    The Role of Ambient Temperature on Flowering

As observed in crop species, ambient temperature affects *Arabidopsis* flowering, with low temperatures delaying flowering, especially under SDs (Blázquez et al. 2003; Balasubramanian et al. 2006a). Light perception has been implicated in the ambient temperature response, with the early-flowering phenotypes of *phyB* and *cry2* mutants found to be temperature dependent (Blázquez et al. 2003; Halliday et al. 2003). Interestingly, Blázquez et al. (2003) also found the autonomous pathway genes *FCA* and *FVE* to be involved nonredundantly in the ambient temperature-dependent control of flowering, in an *FLC*-independent manner. Elevation of *FT* expression in wild-type strains, and lack of floral promotion in *ft* mutants at higher ambient temperatures, suggests that this pathway acts via the floral integrator, *FT* (Balasubramanian et al. 2006a). However, as *ft* loss-of-function mutants are still temperature responsive, *FT* is not the only pathway output. More recently, *SHORT VEGETATIVE PHASE (SVP)*, a MADS-box gene that acts downstream of *FCA* and *FVE*, was found to delay flowering under lower ambient temperatures via repression of both *FT* and *SOC1* (Lee et al. 2007). Protein–protein interaction between SVP and FLC is required for the repression of *FT* and *SOC1*, indicating that the ambient and vernalization response pathways converge at these pathway integrators (Li et al. 2008).

### 1.2.1.5 Phytohormone Control of Flowering

While all known phytohormones have been reported to influence flowering time (reviewed by Davis 2009), the GA pathway is the most studied. Initially, exogenous application of GA was found to accelerate flowering in wild-type *Arabidopsis* strains (Langridge 1957), as well as more recently in late-flowering mutants *gi*, *fca*, *fld*, *fe*, *co*, *fpa*, *ft*, *fve*, and *fwa* (Chandler and Dean 1994). In contrast, mutants that block GA biosynthesis (*ga1*, *ga5*) and signaling pathways [*GIBBERELLIN INSENSITIVE (GAI)*, *REPRESSOR OF GAI-3 (RGA)*, and *RGA-LIKE1 (RGL1)*] have delayed flowering, especially under SD photoperiods when other floral promotion pathways are not active. These results are mirrored by a delay in flowering after application of a GA-biosynthesis inhibitor to wild-type lines, while supplementing the inhibitor with exogenous GA restored normal flowering (Hisamatsu and King 2008). Numerous studies have implicated GA signaling in the upregulation of the central flowering-time pathway genes *FT* and *SOC1*, as well as the meristem identity gene *LFY* (Blázquez et al. 1998; Moon et al. 2003; Liu et al. 2008). Additionally, the brassinosteroid (BR) phytohormone signaling pathway is thought to represent an additional floral-regulating pathway (reviewed by Li et al. 2010). Numerous mutations causing BR insensitivity and BR deficiency result in late flowering (Azpiroz et al. 1998; Chory et al. 1991; Domagalska et al. 2007). Exogenous application of BR has been shown to shorten the circadian rhythms of *CHLOROPHYLL A/B-BINDING PROTEIN 2 (CAB2)*, *COLD AND CIRCADIAN-REGULATED 2 (CCR2)*, and *CCA1* (Hanano et al. 2006), indicating that BR affects flowcring via modulation of the circadian clock. Furthermore, mutants of *BRASSINOSTEROID INSENSITIVE 1* (a component of cellular BR receptor complexes) have delayed flowering in autonomous pathway mutants through enhancement of *FLC* expression (Domagalska et al. 2007). While the interaction of BRs with flowering is clear, further investigations are needed to clarify the underlying molecular mechanisms.

### 1.2.1.6 Pathway Integration

The flowering pathways in *Arabidopsis* transmit environmental and endogenous signals that converge to regulate the expression of a common set of floral meristem organ identity genes that specify the differentiation of the floral organs (reviewed by Causier et al. 2010), with signal strengths depending on the environmental and endogenous conditions perceived. The floral pathway integrators include *FLC*, *FT*, *SOC1*, and *LF* and, with the exception of *FLC*, represent an intermediary level of control between the flowering-time genes and the meristem identity genes. As discussed above, the vernalization and autonomous pathways converge at the floral repressor *FLC*, while stabilized CO protein under inductive LD photoperiods results in upregulation of the floral promoter, *FT*, and its downstream targets. *FT* is expressed in the leaf phloem (An et al. 2004; Wigge et al. 2005) and encodes a component of the mobile flowering signal "florigen" which is transported from the leaf phloem to the SAM, where it induces floral transition. Outputs of the vernalization/autonomous pathways merge with those of the photoperiod pathway via FLC-mediated inhibition of *FT* via direct binding to the *FT* promoter (Heliwell et al. 2006). *FT* expression is negatively regulated by additional pathways, further safeguarding against premature flowering (reviewed by Yant et al. 2009). Of particular prominence is SVP, with interaction between SVP and FLC thought to integrate floral signals from the autonomous, thermosensory, and gibberellin pathways by regulation of *FT* and *SOC1* expression (Li et al. 2008). Recently, *TEMPRANILLO* (*TEM*) genes have been found to link the photoperiod and GA-dependent floral pathways, controlling floral transition via repression of GA biosynthesis and *FT* (Osnato et al. 2012). Once transported to the SAM, FT activates the MADS-box transcription factor *SOC1*, which as part of a protein complex with AGAMOUS-LIKE 24 (AGL24) is thought to directly promote *LFY* expression (Lee et al. 2008). Additional FT protein complexes also occur in the SAM, where in conjunction with the bZIP transcription factor, it is thought to activate expression of the floral meristem identity genes *SOC1* and *AP1* (Abe et al. 2005). Establishment and maintenance of SAM floral identity by SOC1 and AP1 lead to floral organ production. Thus, the interplay between floral promotion and repression mediated by components of the flowering-time pathways converges to regulate the floral pathway integrators, leading to floral transition in response to environmental and endogenous signals.

### 1.2.1.7 Seasonal Flowering-Time Pathway Predominance and Current Ecogeographic Adaptation

The predominance of the inputs from the various flowering pathways alters according to seasonal environmental changes (reviewed by Simpson and Dean 2002). Considering the biennial winter form, prevailing temperature and photoperiod conditions on germination in the autumn are much like those in the spring. However, floral transition is prevented by vernalization requirement, which maintains expression of the repressor, *FLC*. Although prolonged exposure to vernalization during the winter removes *FLC*-mediated floral repression, the absence of inductive LD photoperiods means that floral transition is blocked. However, during spring, the "memory" of exposure to vernalizing temperatures, along with the lengthening day lengths and increasing ambient temperatures, leads to upregulation of the floral pathway integrators and subsequent flowering. Mutations leading to loss or reduction of vernalization requirement allow adaptation to different environments and lead to a change in the prevalence of floral pathways. Within this model, it is immediately evident how altered environmental variables due to climate change may affect flowering: environmental cues triggering floral induction may cease to be reliable if they occur at altered points within the life cycle, thereby eliciting a maladaptive or novel flowering responses (Nicotra et al. 2010). Of course, such environmental changes will have different effects on flowering time depending on plant genotype. Understanding the adaptive advantages conferred by flowering-time gene allelic variants in current environments is an important step towards predicting the effects of climate change on flowering.

While flowering time is presumed to be involved in adaptation to environment, the molecular bases of such adaptation is only just beginning to be investigated. Understanding how genetic variation in flowering pathway genes adapts plants to current environments allows experimental and modeling approaches to investigate the environmental effects of climate change on specific plant ideotypes. The depth of knowledge of *Arabidopsis* flowering-time gene networks, combined with extensive natural variation in flowering time in local accessions, makes it a good model with which to investigate such goals. However, a major barrier to understanding the adaptive roles of flowering time is the difficulty in demonstrating how specific allelic variants affect fitness within a given environment (Gaut 2012). Investigations have been predominantly based on common garden phenotypic experiments of geo-referenced accessions combined with genome-wide or candidate gene association analyses. Analysis of the vernalization pathway genes *FLC* and *FRI* has been by far the most common target for candidate gene analyses, with allelic variants at both genes associated with flowering and latitude (and its environmental components, altitude, minimum temperature, and precipitation) (Stinchcombe et al. 2004; Shindo et al. 2005; Méndez-Vigo et al. 2011; Sánchez-Bermejo et al. 2012). These results indicate that *FLC* and *FRI* may confer climatic adaptation by modulating vernalization sensitivity in accordance with local ecogeographic conditions. Genome-wide association scans (GWAS) in geo-referenced *Arabidopsis* lines have also identified *FRI* as a significant genetic component of flowering time (Aranzana et al. 2005). Such results suggest that combinations of allelic series at *FLC* and *FRI* may approach a near continuum of vernalization response that adapts plants to local conditions. It should be noted that natural variation at *FLC* is associated with variation in temperature-dependent seed germination, identifying a broader adaptive significance of the locus (Chiang et al. 2009). Latitudinal clines in light responses have been identified (Stenøien et al. 2002), with allelic variation at the light perception loci *PHYB* and *PHYC* subsequently found to be significantly associated with flowering and environmental variables such as latitude or precipitation (Balasubramanian et al. 2006b; Fournier-Level et al. 2011; Méndez-Vigo et al. 2011). In order to help translate knowledge of *Arabidopsis* floral adaptation to crop species, the next challenge will be to identify and characterize additional loci in other floral pathways that play a role in local adaptation and to gain further insight into the underlying reasons behind increased fitness in different environments. To date, GWAS experiments investigating adaptive variation have been relatively underpowered, predominantly due to lack of accessions investigated (typically 100–200). However, as larger sample sizes of known geographic provenance begin to be utilized in conjunction with higher numbers of genetic markers (e.g., Horton et al. 2012, >1,000 accessions, 250,000 SNPs) or whole-genome sequence (http://1001genomes.org/), genomic regions and, ultimately, allelic variants involved in local adaptation should soon be identified. The central role flowering time plays in synchronizing plant development with local conditions means that many of these are likely to be resolved as variants within flowering-time genes or their *cis*-regulatory regions. Incorporating such data into increasingly sophisticated flowering-time models will allow better predictions on how the environmental change predicted by climate change will affect plant fitness. However, it is widely acknowledged that results from controlled environment conditions, on which *Arabidopsis* models are largely based, do not necessarily directly equate to "real world" results, with outdoor experiments in which individual environmental components are altered to simulate the effect of climate change on flowering and phenology having only been applied in crop species to date.

## *1.2.2  Domesticated Cereals*

### 1.2.2.1  Rice

While the model species *Arabidopsis* has proven invaluable in the genetic dissection of floral genetic pathways in dicotyledonous plants (including crops such as members of the Brassicaceae), the ancient divergence of the dicot and monocot clades and their subsequent independent evolutionary trajectories means that the monocot model species rice (*Oryza sativa*) has proven invaluable for investigation of the genetics of flowering in grass crops. Unlike in *Arabidopsis*, to date this has largely been achieved by analysis of natural variation in flowering-time genes. Like all crops, the ability of rice to restrict flowering to the seasons best suited for their reproduction depends mainly on the accurate measurement of seasonal changes in day length and temperature. Rice is a SD plant, flowering when the photoperiod is shorter than a critical day length (Izawa 2007). Recent molecular genetic studies have identified a number of quantitative trait loci (QTLs) and genes responsible for the photoperiodic regulation of flowering in rice. Studies of gene functions and interactions among such genes have revealed a complex system of photoperiodic flowering regulation (Tsuji et al. 2011).

*Oryza* species, including cultivated rice and its wild relatives, show a wide range of variation in flowering time and photoperiod sensitivity. Many QTLs associated with flowering time have been mapped in detail by using segregating populations derived from crosses among rice cultivars and wild relatives (Gu and Foley 2007; Uga et al. 2007; Nonoue et al. 2008; Matsubara et al. 2008a; Maas et al. 2010; Ebana et al. 2011; Shibaya et al. 2011; see also the Q-TARO [Yonemaru et al. 2010] and Gramene databases [Youens-Clark et al. 2011]). Some of these QTLs have been cloned using map-based strategies (Yano et al. 2000; Takahashi et al. 2001; Kojima et al. 2002; Doi et al. 2004; Xue et al. 2008; Wei et al. 2010; Yan et al. 2011a; Matsubara et al. 2012). Rice mutants have also been used to isolate flowering-time genes and investigate their functions (Izawa et al. 2000; Kuromori et al. 2009; Saito et al. 2012). Since the successful map-based cloning of

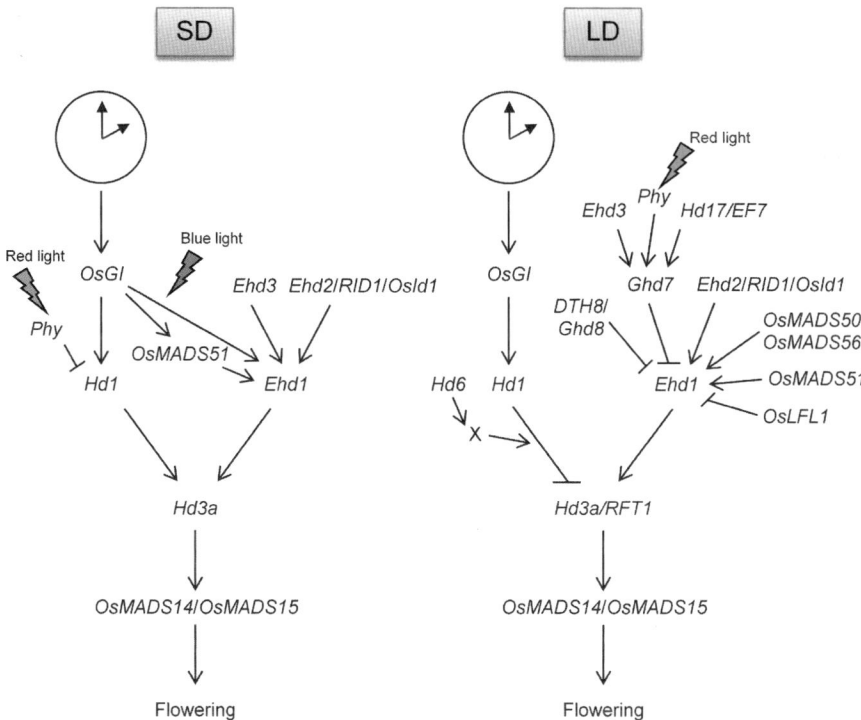

**Fig. 1.2** Regulatory network of flowering-time genes in rice. The clocks at the *top* indicate the circadian clock. *SD* short day, *LD* long day. See text for details

*Photoperiodic sensitivity 5* (*Se5*) by Izawa et al. (2000) and of *Heading date 1* (*Hd1*) by Yano et al. (2000), more than 40 genes associated with flowering time have been isolated by using mutants and natural variations (QTLs) in rice (see the OGRO database, Yamamoto et al. 2012).

The molecular genetic pathway for control of flowering time in rice has been well characterized (Fig. 1.2; Hayama and Coupland 2004; Izawa 2007; Tsuji et al. 2011). Under SD photoperiods, flowering is promoted by the transcription of *Heading date 3a* (*Hd3a*), which is activated independently by the products of *Hd1* and *Early heading date 1* (*Ehd1*) (Yano et al. 2000; Kojima et al. 2002; Doi et al. 2004; Tamaki et al. 2007). *Hd3a* is a homolog of the *Arabidopsis* floral pathway integrator *FT*, and Hd3a protein acts as a mobile flowering signal (florigen) (Tamaki et al. 2007). *Hd1* is a homolog of the *Arabidopsis* photoperiod pathway gene, *CO* (Yano et al. 2000), while *OsGI* is homologous to an output of the *Arabidopsis* circadian clock, *GI* (Hayama et al. 2002). *OsGI* expression is regulated by the rice circadian clock and activates *Hd1* expression (Hayama et al. 2003). While the OsGI-Hd1-Hd3a signaling pathway is evolutionarily related to the GI-CO-FT pathway of *Arabidopsis*, the rice photoperiod pathway genes *Ehd1*, *Ehd2/RID1/OsId1* (*Early heading date 2/Rice Indeterminate 1*), and *Ehd3* (*Early heading date 3*) have no homologs in *Arabidopsis*. *Ehd1* expression is upregulated by the products of *Ehd2/RID1/OsId1*, *Ehd3*, and *OsMADS51*. *Ehd2/RID1/OsId1* is an ortholog of maize *indeterminate1* (*id1*) and shows strong flowering promotion activity (Matsubara et al. 2008b). OsGI protein also strongly activates *Ehd1* expression directly or through the intermediary *OsMADS51* (Itoh et al. 2010).

Under noninductive LD photoperiods, transcriptional activation of *Hd3a* is lower than under SD conditions; consequently, rice flowering is suppressed. Regulation of *Hd3a* expression comes from functional conversion of Hd1. Although Hd1 activates *Hd3a* expression under SD conditions, it attenuates *Hd3a* expression under LDs. This day-length-dependent conversion of Hd1 activity is caused by phytochrome-mediated signaling (Hayama and Coupland 2004; Izawa 2007). Hd1 repressor function is enhanced by the kinase activity encoded by *Hd6* (*Heading date 6*) and mediated by an unknown gene (Takahashi et al. 2001; Ogiso et al. 2010). *RFT1* (*RICE FLOWERING LOCUS T 1*) is a paralog of *Hd3a*, located within 11.5 kb of *Hd3a* on rice chromosome 6. Komiya et al. (2008) found *RFT1* expression to increase under LDs, and RFT1 protein was shown to move from the leaves to the shoot SAM. These results strongly indicate that RFT1 is the LD-specific florigen and that flowering-time control in rice involves two florigen genes, *Hd3a* under SD conditions and *RFT1* under LD conditions (Komiya et al. 2008, 2009). *RFT1* expression is promoted by the product of *Ehd1* (Doi et al. 2004). *Ehd1* expression is induced by the products of *OsMADS50*, *OsMADS51*, and *Ehd2/RID1/OsId1* but is repressed by the products of *Ghd7* (*Grain number, plant height, and heading date 7*), *DTH8* (*DAYS TO HEADING 8*)/*Ghd8* (*GRAIN NUMBER, PLANT HEIGHT, AND HEADING DATE 8*), and *OsLFL1* (*Oryza sativa LEC1* and *FUSCA-LIKE1*). The products of *OsMADS50* and *OsMADS56* might form a protein complex that regulates *Ehd1* (Ryu et al. 2009). DTH8/Ghd8 might form a protein complex with Hd1 to control flowering (Wei et al. 2010; Yan et al. 2011a). OsLFL1 protein has been proposed to downregulate *Ehd1* expression (Peng et al. 2008). Ghd7 protein functions as a flowering repressor together with Hd1 under LD conditions (Xue et al. 2008; Itoh et al. 2010), and it has no counterpart in *Arabidopsis*. Expression of *Ghd7* is induced by the products of *Ehd3* and *Hd17/Ef7* (*Heading date 17/Early flowering 7*). *Hd17/Ef7* is a homolog of *Arabidopsis* *EARLY FLOWERING 3* (*ELF3*), which plays important roles in maintaining circadian rhythms in *Arabidopsis* (Matsubara et al. 2012; Saito et al. 2012).

Flowering-time regulation of genes downstream of *Hd3a* and *RFT1* is the same under both SD and LD conditions. Hd3a interacts with OsFD (homologous to the product of the *Arabidopsis* autonomous pathway gene, *FD*) and GF14c, which encodes a rice 14-3-3 protein isoform, to initiate floral transition in the SAM (Purwestri et al. 2009). OsFD and GF14c induce the expression of *OsMADS14* and *OsMADS15*, which are two rice homologs of *Arabidopsis AP1* and *FUL* (Lim et al. 2000). The products of *RCN1* and *RCN2* (Rice *TERMINAL FLOWER 1/CENTRORADIALIS 1* and *2*) also control the phase transition of the SAM from vegetative to reproductive (Nakagawa et al. 2002).

The expression of many rice flowering-time genes is strictly controlled by a critical day length (Izawa 2007; Tsuji et al. 2011). For example, *Hd3a* and *Ehd1* are expressed in the morning under SDs, whereas *Ghd7* is expressed in the morning under LDs. *OsGI* expression shows a daily circadian oscillation with a peak at the end of the light period (Hayama et al. 2002). Recent studies indicate that expression of some flowering-time genes is regulated by the circadian gating of light responses through phytochromes (red light) and cryptochromes (blue light) (Itoh et al. 2010). *Ghd7* expression is induced through phytochrome signaling, and the sensitivity to red light is gated at the beginning of the light period during LDs. *Ehd1* expression resulted in clear gating responses to blue light pulses (Itoh et al. 2010). In *Arabidopsis*, *ELF3* is required to gate light input into the circadian oscillator, altering the light sensitivity of the central oscillator at a particular point in the circadian cycle (Hicks et al. 2001). However, a rice mutant of *Hd17/Ef7* (a homolog of *ELF3*) was more sensitive to red light than wild-type plants, but the gating of *Ghd7* expression was unchanged (Saito et al. 2012). Thus, the gating mechanisms of flowering-time pathways may differ between *Arabidopsis* and rice. Additional molecular genetic and physiological analyses are needed to further understand these mechanisms. Recent progress has demonstrated that many flowering-time genes are evolutionarily conserved among cereals, including maize, wheat, barley, sorghum, and *Brachypodium* (Greenup et al. 2009; Distelfeld et al. 2009a; Higgins et al. 2010). As regulation of flowering time enables crop cultivars to adapt to particular environments (Izawa 2007; Higgins et al. 2010), appropriate combinations of flowering genes could increase the adaptability of crop species to future climatic conditions. Therefore, elucidation of rice flowering-time pathways should reveal interesting targets for increasing crop production in rice and other cereal crops.

Recent changes in the global climate have had negative effects on rice production around the world. In particular, higher temperatures have caused reductions in both yield potential and grain quality (Yamakawa et al. 2007; Jagadish et al. 2008; Ishimaru et al. 2012). In addition, water shortages have become more severe in tropical and subtropical areas, resulting in unstable rice production (O'Toole 1982; Fukai and Cooper 1995). To cope with these problems, several strategies have been considered. One of these is manipulation of flowering time to avoid severe stresses during flowering and ripening, which are the most sensitive developmental stages in rice. The timing of flowering is one of the most important factors affecting rice growth and yield, so large changes can have detrimental effects on rice production, such as short growth period (leading to low yield) and failure of seed set under unfavorable environmental conditions. Thus, any flowering-time changes must be relatively small, allowing adaptation to regional cultivation systems and environmental conditions. Therefore, fine-tuning of flowering time is a major objective of rice breeding programs. In general, alterations in the function of floral pathway genes often cause not only changes in flowering time but also other morphological changes having negative effects on rice growth and yield potential. Therefore, it is impractical to use such genes for the alteration of flowering time in commercial cultivars. On the other hand, a few genes affecting flowering time in rice have been identified and used in rice breeding programs for many years. To facilitate appropriate modification of flowering time for use in commercial cultivars, natural allelic variations in these genes needs to be characterized. QTL analyses were performed in 12 populations derived from crosses of the *japonica* cultivar "Koshihikari," a common parental line, with diverse cultivars originating from various regions in Asia (Ebana et al. 2011). These analyses revealed a comprehensive series of QTLs involved in natural variation in flowering time and clearly demonstrated that a limited number of loci [*Hd1*, *Hd2*, *Hd6*, *RFT1*, *Ghd7*, *DTH8* (*Ghd8:Hd5*), and *Hd16*] explain a large part of the varietal differences in flowering time (Ebana et al. 2011). Sequence analyses have revealed loss-of-function alleles for all of these loci except *Hd2* and *Hd16*, which have not yet been cloned. These results indicate that a large portion of the phenotypic variation in heading date and day-length response can be generated by combinations of different alleles at these loci (Ebana et al. 2011; Shibaya et al. 2011). By combining the appropriate alleles, it should be possible to develop rice lines with flowering times ranging from extremely early to extremely late.

Although several major flowering-time QTLs were successfully detected within $F_2$ populations, it is very likely that additional QTLs with minor effects are also involved in the phenotypic variation in these populations. To detect these "hidden" QTLs, approaches such as genetic analysis using advanced backcross progeny ($BC_4F_2$) are being deployed (Fukuoka et al. 2010). Preliminary results clearly suggest that a limited number of additional QTLs are involved in natural variation in flowering time (K. Hori and M. Yano, unpublished data). Previously, QTLs with minor effects at the locations of *Hd3a*, *Hd4* (likely to be allelic to *Ghd7*), and *Hd9* have been identified (Kojima et al. 2002; Lin et al. 2002, 2003). Recently, an additional minor QTL, *Hd17*, has been cloned by using a map-based strategy (Matsubara et al. 2012). These types of minor-effect QTLs will be necessary to fine-tune flowering time in rice.

On the basis of early studies of flowering time in rice, a set of isogenic lines (ILs) of "Koshihikari" were developed with different alleles of flowering-time genes by using marker-assisted selection (MAS) (Takeuchi et al. 2006). ILs with a loss-of-function allele at *Hd1* (designated Kanto IL1) and gain-of-function alleles at *Hd4* (Wakei 367), *Hd5* (Wakei 371), and *Hd6* (Wakei 370) have been successfully developed (Fig. 1.3). In each IL (except for that containing *Hd4*), less than 650 kb of donor genome was introgressed. Compared to "Koshihikari," which flowered at 107 days, the days to heading of Kanto IL1 was 95 days, and those of Wakei 367, Wakei 370, and Wakei 371 were about 110 days, 117 days, and 118 days, respectively. The phenotypes of the gain-of-function ILs developed in this study were similar to those of "Koshihikari." However, the culm length and cool-temperature tolerance of Kanto IL1 were different from those of "Koshihikari" (shorter culm, less yield potential, and lower level of cool-temperature tolerance) (Takeuchi et al. 2006). It is unknown whether this trait association was caused by the pleiotropic effects of a single gene or by linked genes. However, a recent large-scale QTL analysis of agronomic trait selection in rice breeding programs (Hori et al. 2012) clearly demonstrated that flowering-time genes can have pleiotropic effects on morphological traits such as culm length, panicle length, and eating quality. It is important to watch for these types of effects when introgressing genes to modify flowering time.

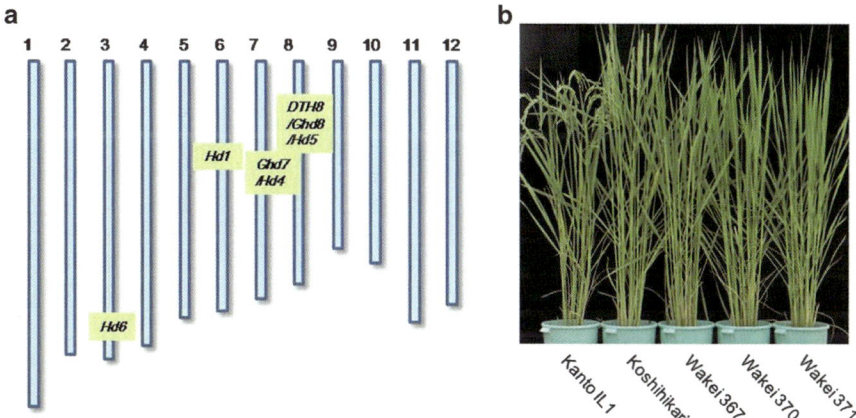

**Fig. 1.3** Development of a series of isogenic lines (ILs) of Japanese elite cultivar "Koshihikari" with different flowering times. (**a**) Chromosomal locations of target loci for the manipulation of flowering time in rice. (**b**) Plant features of ILs at normal flowering time for "Koshihikari." "Koshihikari," recurrent parent; Kanto IL1, IL with loss of function at *Hd1*; Wakei 367, IL with gain of function at *Hd4* (possibly allelic to *Ghd7*); Wakei 370, IL with gain of function at *Hd6*, Wakei 371, IL with gain of function at *DTH8/Ghd8/Hd5*

These results demonstrate that MAS for flowering time can be effective in rice. Thus, MAS could be performed on cultivars grown in subtropical areas, where plants are at risk of both high-temperature stress and drought stress. Most of the genes involved in flowering time have been cloned, making it possible to predict which genes should be introgressed to change the flowering time of particular elite cultivars and facilitating the development of markers to follow the genes through the breeding program.

In addition to the effects of single genes, gene interactions also play an important role in flowering-time determination (Yano et al. 2001; Izawa 2007). For example, loss of function of *Hd1* sometimes diminishes or weakens the effect of other genes involved in day-length response (Lin et al. 2003; Ogiso et al. 2010). Therefore, the genetic background of a cultivar must be considered when planning which genes to combine (pyramid) to alter flowering time. Additionally, the effects of flowering-time genes are often dependent on day length. In general, strong day-length response is conferred by functional alleles of photoperiod-sensitivity loci, which can cause dramatic changes in flowering time depending on the day length in the region of cultivation. For example, cultivars with strong day-length response flower late in temperate areas, which have relatively long days during the growing season, but flower early in tropical areas, which have short-day conditions. These factors need to be considered when designing experiments to manipulate flowering time in particular cultivars.

### 1.2.2.2 Wheat and Barley

The closely related Triticeae cereal crops wheat (*Triticum aestivum*) and barley (*Hordeum vulgare* ssp. *vulgare*) are grown in a wide range of agricultural environments. Barley is predominantly grown for brewing and animal feed, while wheat provides ~40 % of the world's food, with worldwide production of bread wheat steadily increasing from 3.1 billion tons in 1970 to 6.8 billion tons in 2008 (http://faostat.fao.org/). Unlike the model grass crop rice, wheat and barley are LD responsive and a possess vernalization requirement (in winter cultivars). Historic intensification of wheat and barley cultivation worldwide and empirical selection over multiple environments and years has allowed optimization of flowering time to ensure maximum yields within a given climatic zone. While barley is a diploid crop, the process of domestication for bread (hexaploid, AABBDD genomes) and durum wheat (tetraploid, AABB) involved a series of hybridization events between diploid progenitors. Historically, the spread of wheat and barley is characterized by adaptation to an increasingly wider range of climatic conditions, as early farmers cultivated areas progressively distant from the centers of origin. Subsequent farmer selections were influenced by the response of germplasm to environmental factors, with wheat and barley landraces (locally adapted types that predate industrial breeding approaches) showing maturation times that are strongly correlated to the mean temperature and latitude of the areas in which they were adapted. In temperate regions, early flowering avoids abiotic and biotic stresses such as drought conditions and pathogen attack late in the season, whereas late flowering promotes biomass accumulation where there is ample water for crop growth (Jung and Muller 2009).

To achieve optimum environmental and geographical adaptation, farmers and breeders have selected a complex of genes controlling the transition to flowering, through selection for differing vernalization requirement (*VRN* genes), day-length sensitivity (*PPD* genes), as well as additional smaller effect earliness *per se* (*Eps*) loci, which together account for much of the genetic variation for flowering time. Analogous to biennial *Arabidopsis* accessions discussed previously, the majority of autumn-sown wheat and barley varieties require a period of vernalization in order for flowering the following spring to take place under inductive LD photoperiods. Spring-sown cultivars (and some autumn-sown cultivars sown in warmer regions such as the southern Mediterranean) do not require vernalization and vary in the magnitude of response to LD photoperiods (Cockram et al. 2007a). *Eps* effects, which are relatively small and are thought to act independently of environmental cues, can be observed once the effects of photoperiod and vernalization response have been accounted for.

The genetic control of heading is under strong but complex selection in temperate cereals (Borras-Gelonch et al. 2012), and although much is known of the effects of individual major loci on floral initiation, the interaction between known loci and the influence of modifier genes of smaller effect is less well understood. Similarly, there is a paucity of information on the physiological and developmental effects of flowering-time genes on the complete life cycle and its individual components (Gonzalez et al. 2005).

Cloning of *VRN* and *PPD* genes and identification of the influence of variation at these loci have led to greater precision in the optimization of flowering time. Matching the appropriate photoperiod response to agro-environment is of importance, as illustrated by an estimated 35 % yield advantage in wheat associated with the photoperiod-insensitive *Ppd-D1a* mutation in southern European environments (Worland 1996).

Vernalization requirement in wheat and barley is quantitative and controlled predominantly by orthologous *VERNALIZATION 1* (*VRN-1*), *VRN-2*, and *VRN-3* loci located at colinear chromosomal locations (Cockram et al. 2007a), with early-flowering vernalization nonresponsive alleles at all three loci epistatic to late-flowering vernalization-responsive alleles. The first to be characterized was *VRN-$A^m1$* in the diploid wheat *T. monococcum*, found to encode a MADS-box transcription factor with sequence homology to the *Arabidopsis* meristem identity gene, *AP1* (Yan et al. 2003). Orthologous *AP1*-like genes have subsequently been identified at the colinear loci in barley (*VRN-H1*), bread wheat (*VRN-A1*, *VRN-B1*, *VRN-D1*), and durum wheat (*VRN-A1*, *VRN-B1*), with deletions within putative regulatory regions within the promoter or intron 1, predominantly created via illegitimate recombination (Cockram et al. 2007b), thought to confer vernalization insensitivity due to removal of putative *cis*-regulatory sites (reviewed by Trevaskis 2010). Cereal *VRN-1* expression gradually increases with vernalization, with longer cold treatment leading to increased expression (Danyluk et al. 2003; Trevaskis et al. 2003; Yan et al. 2003) and earlier flowering. Thus, *VRN-1* genes appear to be activators of flowering, which are upregulated by vernalization, allowing the plant to respond to inductive photoperiods. As in *Arabidopsis*, epigenetic modification has also been implicated in the "memory" of vernalization in cereal crops, with changes in histone methylation of *VRN-1* intron 1 and promoter chromatin associated with its stable expression and early flowering after vernalization (Oliver et al. 2009). While cereal homologs of *Arabidopsis VIN3*-like genes have been identified and shown to be weakly upregulated on vernalization, none show the characteristic expression pattern displayed by *VIN3* (Fu et al. 2007).

*VRN-1* is thought to positively regulate expression of cereal *FT-like 1* (*FT1*) genes, which show high homology to the *Arabidopsis* floral pathway integrator, *FT*. Indeed, colinear cereal *VRN-3* vernalization response loci have recently been found to be encoded by mutations within *FT1* genes, which are upregulated under inductive LD photoperiods once the vernalization requirement has been met (Turner et al. 2005; Yan et al. 2006; Hemming et al. 2008). The natural *FT1* mutations underlying dominant *Vrn-3* alleles lead to very early flowering irrespective of exposure to vernalization and are largely deployed in agricultural environments with short growing seasons. The gene underlying the colinear cereal *VRN-2* loci was first identified in *T. monococcum* (Yan et al. 2004) and found to encode a predicted protein with a CCT domain (also found in *Arabidopsis* flowering pathway genes, such as *CO* and *TOC1*). In barley and durum wheat, the *VRN-2* loci are encoded by tandemly duplicated *ZCCT* genes, with recessive mutations (point mutations or whole gene deletions) within *ZCCT* genes associated with vernalization insensitivity (Yan et al. 2004; Karsai et al. 2005; Distelfeld et al. 2009a). In vernalization-responsive cereal lines grown in the absence of vernalization, *VRN-2* is highly expressed under LDs but not under SDs, indicating that *VRN-2* represses flowering under inductive photoperiods. *VRN-2* expression during vernalization mirrors that of the *Arabidopsis* vernalization pathway gene *FLC*, showing progressive downregulation, proportional to the duration of cold treatment. Current cereal flowering-time models suggest that *VRN-2* encodes a repressor of *FT1*, with VRN-1 thought to repress *VRN-2* expression (Trevaskis 2010). In this model, increasing expression of *VRN1* during vernalization removes VRN2-mediated repression of *FT1* in the leaves. VRN3 is exported to the shoot apex, presumably via the phloem, where under inductive photoperiods it further promotes *VRN-1* transcription above thresholds required for reproductive apex transitioning (Distelfeld et al. 2009b).

Response to photoperiod in bread/durum wheat is largely determined by orthologous *PHOTOPERIOD-1 (PPD-1)* loci encoding a member of the *PSEUDO-RESPONSE REGULATOR (PRR)* gene family, which in *Arabidopsis* act close to the central oscillator. Flowering in LDs is thought to be mediated by activation of cereal *CO1* expression through the circadian clock. In barley, the photoperiod-insensitive allele is due to a nonsynonymous point mutation within the CCT domain (Turner et al. 2005), while in durum (*Ppd-A1a*) (Wilhelm et al. 2009) and bread wheat (*Ppd-A1a, -D1a*), insensitive alleles are thought to be due to deletions within the promoter and are associated with its misexpression (Beales et al. 2007; Wilhelm et al. 2009; Shaw et al. 2012). In both wheat and barley, photoperiod-insensitive alleles lead to the upregulation of *FT1* during the light phase under inductive LDs (Turner et al. 2005; Shaw et al. 2012). An increased number of copies of *Ppd-B1* have been shown to alter gene expression in a similar way, producing a moderate photoperiod-insensitive early-flowering effect (Bentley et al. 2011; Diaz et al. 2012). Putative null alleles have also been identified in bread wheat, but without confirmed loss of function (Beales et al. 2007). Recent transgenic studies in barley support the assumption that cereal *CO1* genes promote flowering by upregulation of *FT1*, with *PPD-H1* regulating *FT1* independently of *CO1* (Campoli et al. 2012).

The hexaploid nature of wheat results in complex interaction between homoeologous flowering-time loci, and the cloning of *PPD-1* loci has led to recent characterization of their relative effects on flowering time. Flowering-time analysis of known photoperiod-insensitive mutant alleles in near-isogenic bread wheat lines has shown that the *Ppd-D1a* allele markedly reduces flowering time in SDs, with the *Ppd-A1a* durum allele also having a strong effect, followed by a moderate earliness conferred by the *Ppd-B1a* copy number variant (Bentley et al. 2011, 2013). The *Ppd-A1a* allele also gives a strong reduction in flowering in short days (Wilhelm et al. 2009), as well as in field conditions (Clarke et al. 1998). Further work has shown that combining the *Ppd-D1a* and *Ppd-A1a* alleles produces earlier flowering than either *Ppd-D1a* or *Ppd-A1a* in isolation in near-isogenic lines assessed in a short photoperiod (Shaw et al. 2012). Barley possesses an additional major photoperiod response locus, termed *PPD-H2*. *PPD-H2* modulates flowering under SDs and is thought to be encoded by the *FT*-like gene *HvFT3* (Faure et al. 2007). Deletion of *HvFT3* is associated with recessive *ppd-H2* alleles that delay flowering under SD photoperiods. Thus, *PPD-H2* augments the delay of flowering conferred by the vernalization pathway during the winter. While orthologous loci are not known in bread wheat, this may be due to the hexaploid nature of wheat masking phenotypic effect or recessive alleles. If *HvFT3* orthologs are present in wheat, the identification and consolidation of recessive alleles at all three homoeologous genes into one genetic background would provide a novel source of useful flowering-time variation of particular relevance to a warming climate, allowing floral repression in the winter without the need for a strong vernalization response. Indeed, "alternative" barley varieties (essentially cold hardy spring types that can be planted in the autumn for harvest the following summer) appear to possess flowering-time gene haplotypes that invariably include recessive *ppd-H2* alleles, thus preventing premature flowering (Cockram unpublished).

Additionally, *Eps* loci that promote flowering independently of environmental cues have been identified (Griffiths et al. 2009; Bennett et al. 2012), although most remain ill defined. However, the effects of the *Eps-A$^m$1* locus in diploid wheat have been relatively well investigated and found to display distinct allelic effects on vegetative and spike development phases in a thermosensitive manner (Lewis et al. 2008), suggesting that detailed investigation of *Eps* loci is likely to show interaction with environment.

Despite current understanding of the flowering-time genes, genetics, and pathways in wheat and barley, the complex interactions between major genes and QTLs, and their interplay with climatic conditions, still need integration. In addition, knowledge of the effect of individual alleles and combinations of alleles on the various stages of vegetative and reproductive development beyond what is currently known (Gonzalez et al. 2005) is required to determine whether the alleles can be used in fine-tuning key stages of plant development. The availability of wheat near-isogenic lines for flowering pathway genes (e.g., Bentley et al. 2011, 2013) is a useful resource for further refining understanding of the role of flowering time on development and as a component of yield in a range of environments and growing conditions.

There are many challenges facing worldwide wheat and barley production under a climate change scenario. Although photoperiods will remain unchanged, conditions, particularly at the end of the growing season, are likely to become drier, with terminal drought stress and high heat both being detrimental to reproductive development (Lukac et al. 2012) and grain fill, both of which contribute to yield. The ability to manipulate gross flowering time through the use of major flowering-time alleles in breeding programs provides the basis for a genetics and breeding approach to adapt elite varieties to the prospect of shorter growing seasons. However, work is still required to understand the effect of early flowering time on yield potential across environments. In addition, the potential utility of loci of minor effect and/or which affect various stages of reproductive development could offer the ability to shorten or lengthen various phases of the process, thereby fine-tuning flowering to suit particular regional climatic conditions, and to adapt to any changes in these conditions. For example, the Indo-Gangetic Plains of India today represent a high-potential wheat environment supplying food to over 200 million people. Under predictions of climate change, by 2050 the region will be heat stressed with a short productive season (Ainsworth and Ort 2010) resulting in drastic food shortages in this area of the developing world. It is therefore crucial that breeding seeks to adapt elite local germplasm (i.e., by shortening the life cycle) to match the predicted growing conditions while attempting to maintain or, ideally, increase the yield potential.

### 1.2.2.3 Maize

Maize is a $C_4$ grass species that is grown widely in both food-secure and food-insecure regions of the world, making it a priority crop for adaptation to changing climate (Ainsworth and Ort 2010). The natural variation in maturity time of maize landraces is 2–11 months, and maize was originally a SD photoperiod-responsive crop of tropical origin. Selection of specific flowering-time alleles means that maize is now adapted to a wide range of environments and elevations in tropical and temperate regions of the world (Buckler et al. 2009). In tropical areas, maize is grown for seed and forage, but in northern Europe and northern America, where the period of favorable temperatures is short, only early-flowering varieties are grown. In such agricultural environments, maize generally experiences drought stress shortly before and during flowering, with plants particularly sensitive to drought during the anthesis to silking period (Jung and Muller 2009). Forage maize is cultivated at more extreme latitudes within the EU where seasons are too short for yielding seed production. Maize is also recognized as a cost competitive biofuel crop, with plant biomass fermented to methane. In order to optimize fuel production, maximum biomass is achieved by modifying the phenology to later flowering, hence extending the vegetative phase (Jung and Muller 2009) and avoiding seed development.

Climate change models suggest that areas previously unsuitable for grain maize cultivation could become viable in the near future. However, modified environmental conditions could alter crop competitiveness in the areas currently cultivated. With these pressures in mind, the development of new maize hybrid cultivars better adapted to more northern latitudes, on one side, and to drier environments, on the other, is a major priority. Northward expansion will present the challenge of suboptimal adaptation of the existing elite germplasm to new photoperiodic conditions, e.g., extreme LDs. Such conditions could negatively influence both grain yield and/or biomass. Additionally, more northern cultivation is likely to be exposed to unpredictable extreme weather events such as cold spells during the early and late phases of cultivation. Adaptation of maize to drier environments, due to reduced rainfall or reduced availability of irrigation, can be mitigated by including changes in maize flowering time as breeding objectives. Maize flowering time has been a major trait under tight selection in modern maize breeding.

Genetic variability for flowering time in maize is essentially quantitative (Salvi et al. 2007; Buckler et al. 2009), i.e., it is attributable to the cumulative effects of numerous QTLs, each of which individually has a small effect. It has been proposed that selection in this inbreeding species favored additive QTLs of small effect to ensure partially synchronous flowering, thereby maintaining fitness (Buckler et al. 2009). Despite the recent availability of a genome sequence assembly, relatively few maize genes controlling flowering have been cloned. Maize flowering-time mutants include *INDETERMINATE 1 (ID1)* (Colasanti et al. 1998) (orthologous to rice flowering-time gene *EHD2*) and *DELAYED FLOWERING 1 (DF1)* (Muszynski et al. 2006). Both mutants produce more leaves than the wild type and act within the autonomous pathway. In addition, the flowering-time QTL on maize chromosome 8, *VEGETATIVE TO GENERATIVE TRANSITION 1 (VGT1)*, has been positionally cloned using a combined biparental and association mapping approach (Salvi et al. 2007). *VGT1* is encoded by naturally occurring variation within a 2 kb intergenic region upstream of *ZmRAP2.7* (an *AP2*-like transcription factor homologous to the *Arabidopsis* photoperiod gene *TOE1*). This noncoding region approximately 70 kb upstream of *ZmRAP2.7* shows sequence conservation between multiple grass species and acts as a cis-regulatory element controlling expression. A second maize flowering-time QTL, which gives a 6-day delay in flowering in response to photoperiod, has also been fine mapped to chromosome 10 (Ducrocq et al. 2009). The 170 kb region contains a single gene encoding a CCT domain protein orthologous to the rice gene *Gdh7*, both of which are related by evolutionary descent to the *ZCCT* genes which underlie the bread/durum *VRN-2* vernalization response loci, duc to segmental duplication in the ancestral cereal genome (Cockram et al. 2010a, 2012). Additional candidate genes for maize flowering time have also been identified through QTL meta-analysis (e.g., Xu et al. 2012), and association analyses have identified previously validated and novel loci (Buckler et al. 2009). Until recently, details of the maize circadian clock components have been poorly defined. However, recent work utilizing the 13k feature Affymetrix GeneChip found 10 % of maize genes to be circadianly regulated, including maize homologs of *Arabidopsis* flowering-time pathway genes, such as *CO*, *GI*, *FCA*, *PRR7*, *LUX*, and *LDL1* (Khan et al. 2010).

Further understanding of the genetic control of maize flowering will allow breeding programs to introduce favorable alleles from tropical photoperiod-sensitive germplasm into breeding programs for temperate zones and vice versa to meet future climate and production needs. It will also allow elucidation of the relationship between flowering time and biomass production, relevant to its continued utility in forage and biofuel production. The coincidence of flowering with midsummer drought is frequent in maize production areas, and this will presumably increase with the effects of climate change. From this perspective, breeding for early-flowering hybrids could mitigate the impact of drought events, which, in the absence of strong water constraint, could be counterbalanced by selecting for later flowering.

## 1.2.3    Temperate Forage and Biofuel Grasses

### 1.2.3.1    Perennial Ryegrass

Perennial ryegrass (*Lolium perenne*) is the most widely grown pasture grass species in temperate areas of the world and is also a major constituent of many amenity grass mixtures used in lawns and sports turfs. Additionally, because of the success of ryegrass breeders in producing high-sugar varieties, ryegrass has now become a potentially economically viable source of bioethanol, a product of post-harvest microbial fermentations of the available plant carbohydrates. The seasonal timing of flowering of a ryegrass population has a profound effect on resource utilization and allocation within that population, impacting directly on field performance in terms of overall biomass, quality traits associated with animal and biofuel production, and on seed yield, which is necessary for the propagation and marketing of ryegrasses. Thus, an understanding of the phenology of flowering in ryegrass is not only fundamental to developing an understanding of ryegrass biology but is of vital importance in maintaining and adapting the agronomic viability of ryegrasses in the context of an uncertain climatic future.

Once flowering is initiated, internode elongation raises the developing inflorescences above the grass canopy. The rapid increase in growth rates associated with inflorescence development leads to an increase in the overall biomass of the grass sward, and this increase is even observed under repeated grazing or cutting. Seasonal patterns of production are similar for all temperate forage species and cultivars, although shifts to a later peak of production by several days occur with later flowering cultivars. Seasonal and environmental variables such as temperature and moisture availability will influence yield and its seasonal distribution (Anslow and Green 1967), largely affected by the timing and intensity of flowering of individual cultivars. Crop digestibility decreases during this period of rapid reproductive growth (Green et al. 1971), and, in practice, farmers aim to cut silage for conservation before maximum yield is achieved but where digestibility has not fallen below around 70 % of total dry matter. After inflorescence development and emergence, new grass shoots are largely vegetative, although some genotypes do flower a second time albeit less profusely. Herbage yield declines steadily towards zero during the winter months.

Flowering in perennial forage grasses is analogous to that in winter varieties of wheat and barley that require vernalization, followed by LDs to initiate flowering. Homologs of wheat, barley, and rice flowering genes have been identified in perennial ryegrass and allelic variation for vernalization response and flowering time identified. A homolog of the temperate cereal *VRN-1* gene was shown to be located on *Lolium perenne* linkage group 4 (Jensen et al. 2005) in a mapping family developed between two parent plants with contrasting vernalization requirements, coinciding with a QTL accounting for over a quarter of the total observed variation in flowering time. Other QTLs on linkage groups 2, 6, and 7 accounted for between 5 and 19 % of the variation for flowering time.

Genes with homology to the rice *Hd3a*/cereal *VRN-3* genes (*FT*) (Armstead et al. 2004) and rice *Hd1* (*CO*) (Armstead et al. 2005; Skøt et al. 2007) have been found to underlie a major QTL for ear emergence (the time at which the terminal spikelet of the inflorescence emerges above the flag leaf sheath/blade junction) in *L. perenne*, located on linkage group 7. Therefore, the genetic control of flowering in *L. perenne* and, by association, other temperate perennial grasses appears to be governed by the same genes as temperate cereals with a vernalization requirement. As the vernalization and day-length flowering requirements of temperate perennial forage grasses are similar to winter cereals, it is likely that homologs of further flowering control genes will likely be identified. If the flowering model described by Distelfeld et al. (2009b) is to be applied to perennial forage grasses, it is possible that as new tillers are produced during the LD summer period, *VRN-2* orthologs are once more expressed, blocking the expression of *VRN-3* orthologs, the production of which can only be made following activation of *VRN-1* under vernalizing conditions. Thus, floral initiation of new tillers only occurs once the LD threshold has been reached the following spring and not the previous autumn. Under climate change scenarios, vernalization conditions may be reduced, but, in grasses, the vernalization requirement can be supplemented by SD photoperiods (see Heide 1994).

Floral initiation is controlled by day length, with further development being a growth process varying annually under the influence of temperature, moisture, and nutrient availability. Plasticity and adaptation are therefore inbuilt, ensuring flowering is synchronized with optimum growing conditions, suggesting robust adaptation capacity under a changing climate.

For energy grasses and forage crops, the breeding priority is biomass yield (leaves and stem), with the additional target of digestibility where sugar availability is important (e.g., for forage and biofuel production). *L. perenne* is cut prior to flowering in the spring when digestibility is highest, allowing for a second cut for silage and for the sward to be left for grazing. Farmers prefer a mixed sward of varieties to mitigate disease risks and these mixtures will comprise varieties with varied heading dates.

Unlike cereals, ryegrass has not been through the major genetic bottlenecks associated with agricultural domestication, and it consequently retains most of the characteristics and variability (potential phenotypic plasticity—i.e., the ability to change phenotype in response to environmental stimuli) of a nonagricultural grass species—a variability which is maintained by their generally obligate outbreeding nature. This is reflected in the range of flowering phenotypes that can be observed across geographical clines. The variation is mediated by the length and intensity of seasonal temperature fluctuations in combination with a range of other abiotic parameters which vary across and between geographic regions. The combination of allelic variation, outbreeding nature, and a wide, often continuous geographic distribution means that populations and individuals retain extensive adaptive potential.

In ryegrasses, a combination of QTL analyses based upon biparental crosses, the development of association genetics panels, and comparative genome analyses with model and crop species has provided some insights into the genetic architecture of flowering time. Measurements of flowering time in relation to genetic population structure have been carried out in a number of studies on *L. perenne* and related grasses such as *L. multiflorum* (Italian ryegrass) and *Festuca pratensis* (meadow fescue) (Armstead et al. 2004, 2005, 2008; Inoue et al. 2004; Yamada et al. 2004; Jensen et al. 2005; Shinozuka et al. 2005; Skøt et al. 2005, 2007, 2011; Ergon et al. 2006). While these indicate that there are many QTLs, which can influence flowering time distributed across the genome, two particular regions on chromosomes 4 and 7 are frequently associated with significant QTLs for flowering time.

The region on chromosome 7 is of particular interest as comparative genomics has suggested the presence of a strong candidate gene in that region of *L. perenne*, an ortholog of *FT*-like genes that control flowering in *Arabidopsis*, barley, wheat, and rice (Armstead et al. 2004). The protein products of *FT* and *OsFTL2* have been experimentally determined to be directly involved in inducing the transition from vegetative to reproductive growth at the meristem (Corbesier et al. 2007; Tamaki et al. 2007). The presence in this region of the *L. perenne* ortholog of this gene (designated *LpFT3*) has been confirmed by both genetic mapping and sequencing of the candidate region (Skøt et al. 2011).

*LpFT3* is considered to be a potentially major factor in determining the plasticity of the flowering response in different ryegrass populations. To test this, the diversity of the allelic variation at the *LpFT3* locus has been investigated across a set of nine European ecotypes and agricultural collections of ryegrasses, which represent a full range of flowering times (Skøt et al. 2011). The resulting marker assays have distinguished six major alleles for *LpFT3* within the ecotype collection and highly significant associations between *LpFT3* allele type and flowering time. Interestingly, the most significant allele/phenotype relationships were not determined by variation at the protein-coding level (*LpFT3* is highly conserved) but with the variation present in the 5′ noncoding region. This indicates that the phenotypic plasticity observed in ryegrass populations that is associated with *LpFT3* is not a consequence of direct structural variation of the protein product, but rather of the regulation of the expression of this key signal protein in flowering induction. Over 90 % of the *LpFt3* alleles present in the population consisted of just three types (the *a*, *b*, and *c* variants). Generally speaking, the *a* allele was associated with late flowering times, the *b* allele with intermediate flowering times, and the *c* allele with early flowering times, with the majority of significant effects being associated with the *a* and *c* alleles.

Many successful ryegrass breeding programs are based on a recurrent population selection model. The breeder's aim is to shift the population mean values for target traits by a gradual accumulation of favorable alleles. These results in populations/ varieties that can deliver agronomic performance in the heterogeneous pasture environment over a number of years—which itself can only be achieved by retaining sufficient phenotypic plasticity within the population to cope with diverse environmental challenges. Currently, ryegrass breeding is reliant on phenotypic selection. However, it is likely that in the future molecular breeding for ryegrasses may exploit developing genomic selection protocols. The objective of genomic selection is to be able to estimate the breeding value by simultaneous estimation of the effect of thousands of genome-wide markers in one step (Meuwissen et al. 2001; Jannink et al. 2010; Heslot et al. 2012). The principle is that by flanking "all possible QTLs" within a population (i.e., having dense and even molecular marker coverage of the genome), the detailed relationship between positive changes in target phenotypes and shifts in allele frequencies on a training population is observed. Prediction models based on this information are then used to make selections on the breeding population based on the molecular marker data alone. The major advantage of this approach is that molecular marker assays are becoming increasingly available and economic and can be used at any growth stage, whereas phenotyping remains expensive and time-consuming and is, by definition, tied to plant development. Thus, genomic selection is likely to represent the best way forward for capturing and delivering the phenotypic plasticity which is and will be required by ryegrass varieties that are to be successful in a changing environment.

### 1.2.3.2 *Miscanthus*

*Miscanthus* is a $C_4$ perennial that has been grown for bioenergy since the 1990s. *Miscanthus* is mostly combusted, either directly in biomass boilers or cofired with coal, or it is used for animal bedding. On a commercial scale, only the naturally occurring sterile hybrid *M.* × *giganteus* is used for bioenergy. It is fast growing with excellent cold tolerance, and its moisture content at harvest is more amenable for combustion than wood chip. *M.* × *giganteus* requires low to zero inputs and remains productive under much lower temperatures than its grass relatives maize, sorghum, and sugarcane. However, the sterility of *M.* × *giganteus* means propagation must be via rhizome splitting. This elevates establishment costs, limits development of new varieties, and presents a disease risk when grown on a large scale. Expansion and development of the *Miscanthus* industry requires novel hybrid production through the creation of intra- and interspecific hybrids, requiring suitable germplasm and knowledge of flowering requirements. Fortunately the *Miscanthus* genus has an extensive natural distribution over a broad range of Asian and South Pacific latitudes (Hodkinson et al. 1997). Germplasm collections have included some highly productive genotypes where growth exceeds 7 m in native environments (Chen and Renvoize 2006).

Flowering also impacts biomass quantity and quality: flowering signals the end of vegetative growth and thus biomass accumulation, but is also expected to be, at least partially, coupled with senescence and nutrient remobilization to the underground rhizome, promoting sustainability for the following year's growth. Senescence also lowers moisture content for harvest. As a new and undomesticated crop, knowledge of the flowering requirements for *Miscanthus* is limited. *M.* × *giganteus* flowers late or not at all in the UK (Lewandowski et al. 2000; Jensen et al. 2011a), while its progenitor species, *M. sinensis* and *M. sacchariflorus* (Hodkinson and Renvoize 2001), show contrasting flowering requirements, where

*M. sinensis* genotypes appear to flower much more readily compared to *M. sacchariflorus* genotypes in European (Clifton-Brown et al. 2001; Jensen et al. 2011a) and Chinese (Yan et al. 2011b) field conditions.

Flowering time for *M. sinensis* genotypes has been partially explained by growing season rainfall, degree days, and mean temperature (Jensen et al. 2011a), and observed delays in heading dates of *M. sinensis* between 2008 and 2009 appeared to be associated with moisture availability (Jensen et al. 2011b). Recent reanalysis of European *Miscanthus* trials showed that flowering occurred with fewer degree days in northern European than in southern locations, sometimes by as much as 400 °Cd (above a base of 10 °C) (Jensen et al. 2011a). A role of vernalization and/or LD photoperiods on flowering times in some *M. sinensis* genotypes may therefore yet be determined.

In six *M. sacchariflorus* genotypes from diverse latitudes, a quantitative SD flowering response has recently been demonstrated (Jensen unpublished). Here, flowering was delayed under LDs by a minimum of 51 and maximum of 83 days. An effect of warmer temperatures promoting flowering time was also noted in some genotypes. This study showed that flowering time followed a latitudinal cline where genotypes from the northernmost locations (39°N) headed at day lengths of approximately 14 h and 570 °Cd (above a base of 10 °C) and those from further south (28°N) headed at day lengths of about 12 h and 1,000 °Cd. A single genotype showed anomalous flowering time to this pattern, but this was later explained by localized geographical climate from the collection site. There appeared to be a lower degree day requirement for flowering in types from more northerly latitudes. In the above study, flowering phenology in *M. sacchariflorus* appeared similar to that reported for sorghum and sugarcane. This is encouraging as the first high-resolution genetic map for *Miscanthus* has shown extensive synteny with sorghum (Ma et al. 2012), suggesting that the identification of genes controlling flowering time in *Miscanthus* will be aided by recent progress in sorghum.

*M. sacchariflorus* genotypes exhibit different levels of sensitivity to SDs, with some effect of temperature, while in *M. sinensis* genotypes flowering response is better explained by degree days and moisture availability (the impact of moisture stress on *M. sacchariflorus* genotypes has not yet been reported). The increased temperatures associated with climate change are likely to promote flowering, when moisture deficits allow, so that later or nonflowering genotypes will be required to maximize the growing season. The development of markers for the determination of flowering times will greatly facilitate breeding, although modeling studies have indicated the existence of flowering-time trends that correspond to the collection location of the germplasm. For a biomass crop such as *Miscanthus*, late or non-flowering is essential for the reasons described above. Removing seed production may also prevent invasive propagation, especially important with a species nonnative to Europe and the USA (Rounsaville et al. 2011).

Varieties requiring SD photoperiods to flower may be grown in warmer southern climates where flowering can be induced for seed production. The resultant progeny can be used as a biomass crop in cooler northern climates where flowering will be induced late or not at all. However, synchronization of material for crosses can be difficult in a genus where flowering times can be so varied. Additionally, plants which thrive in warmer climates may not have the adaptations required to flourish under colder conditions. An alternative is therefore to make selections of good material in the climate where the product is to be aimed and induce flowering under artificial conditions. Preferably, progeny from a *Miscanthus* breeding program will be triploid, and thus sterile, to ameliorate concerns over invasiveness. This necessitates hybridizations between tetraploid and diploid parents, but requires the appropriate conditions to permit synchronous flowering. Since *Miscanthus* genotypes and accessions flower best under climatic conditions similar to those of their natural habitat, modified glasshouse compartments are used in breeding programs to allow the simulation of average temperature regimes and photoperiods found in Asian collection locations. Synchronization of flowering for tetraploid *M. sacchariflorus*, diploid *M. sacchariflorus*, and diploid *M. sinensis* genotypes has now been achieved using climatic conditions mirroring those in central Japan between 21 July and 29 October (Fig. 1.4). These genotypes had been selected from European field trials as elites in terms of biomass potential, and crosses have now been made possible both between and within species and ploidy groups providing a previously unexplored germplasm resource for exploitation in bioenergy production.

### 1.2.3.3  Sorghum

Energy sorghums are annual crops designed to accumulate sugar (sweet sorghum) and/or high biomass (Rooney et al. 2007). Delayed flowering in energy sorghum is an important trait because long vegetative growth duration increases biomass yield. Delayed flowering also increases yield of energy sorghum by improving the efficiency of radiation capture, radiation use efficiency, and biomass partitioning (Olson et al. 2012). Delayed flowering in sweet sorghum increases biomass and sugar yield; however, flowering is important as a mechanism to trigger accumulation of sucrose in sweet sorghum stems prior to the end of the season (Lingle 1987). In contrast, grain sorghum has been selected to flower early (60–80 days) to enhance production of grain over biomass and to avoid drought, insect pressure, and cold temperatures that often occur later in the season with adverse impact on grain production. Flowering time in sorghum is regulated by photoperiod, gibberellins, temperature, length of the juvenile phase, and other factors (Quinby 1967; Major et al. 1990; Foster and Morgan 1995; Craufurd et al. 1999).

Sorghum is a SD plant; therefore, photoperiod-sensitive genotypes show delayed flowering when plants are exposed to day lengths that exceed their critical photoperiod. In SD (10 h day lengths), most sorghum genotypes will flower within 60–80 days (Major et al. 1990). However, when grown in 14 h day lengths, highly photoperiod-sensitive sorghum genotypes flower in >175 days or not at all (Craufurd et al. 1999). Quinby and colleagues identified four major loci that modified the flowering time of sorghum and termed them maturity or *Ma* loci ($Ma_1$–$Ma_4$) (Quinby 1967). Rooney and Aydin (1999) subsequently identified $Ma_5$

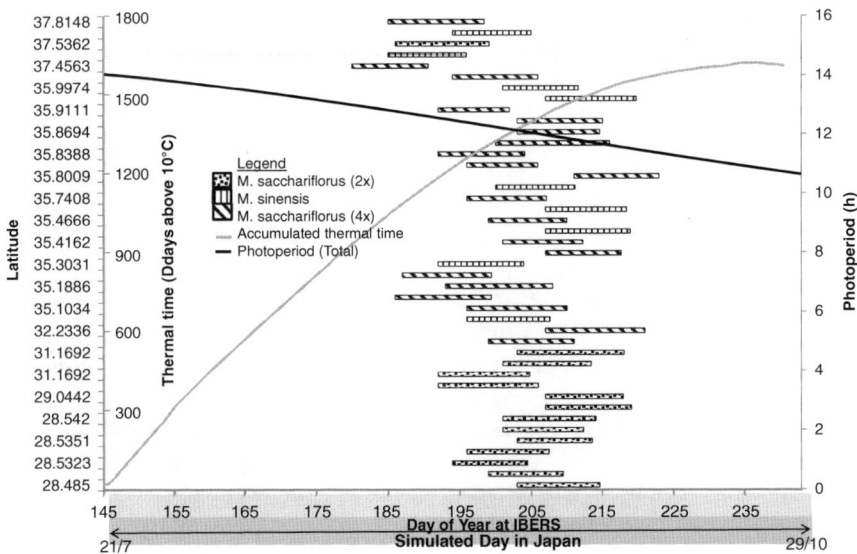

**Fig. 1.4** Start and duration of flowering in $n = 43$ genotypes grown in controlled glasshouse conditions set to photoperiods equivalent to 35° latitude and a corresponding temperature regime

and $Ma_6$, loci that are especially important modifiers of photoperiod sensitivity. Sorghum genotypes dominant for $Ma_1$–$Ma_6$ are very photoperiod sensitive and will not undergo floral initiation until day lengths are less than 12.2 h. In most northern temperate regions where energy sorghum is planted in the spring when day lengths exceed 12.2 h and are increasing, plants will not flower until October after 5–6 months of growth.

The molecular genetic basis of photoperiod sensitivity in sorghum has recently been clarified. $Ma_3$, the first sorghum maturity locus cloned, encodes *PHYB* (Childs et al. 1997). Sorghum plants with null versions of *PHYB* ($ma_3^R$) flower in less than 60 days regardless of day length demonstrating that light input via PHYB is required to delay flowering in response to increasing day length. $Ma_1$ was identified as *PRR37* (Murphy et al. 2011), an ortholog of the cereal *PPD-1* genes of barley and wheat (Turner et al. 2005; Beales et al. 2007). Sorghum *PRR37* expression is regulated by the circadian clock and light through a mechanism consistent with the external coincidence model (Murphy et al. 2011).

Breeding of energy sorghum hybrids utilizes knowledge of $Ma_1$–$Ma_6$ alleles that modulate the activity of the photoperiod-sensitive flowering regulatory pathway. Male and female inbred lines that flower early are used to generate hybrid seed in a manner similar to grain sorghum. However, the parents of grain sorghum hybrids are normally both recessive for $Ma_1$ and for one or more other maturity loci, so that grain sorghum hybrids flower early in the season to maximize grain yield. In contrast, energy sorghum inbreds flower early but are recessive for different *Ma* loci. As a result, the recessive alleles are complemented in $F_1$ plants; therefore, energy sorghum hybrids are photoperiod sensitive and flower late, increasing biomass yield.

## 1.2.4   Root Crop Species: Sugar Beet

Sugar beet is the most important of the *Beta vulgaris* subsp. *vulgaris* crops, which include spinach, Swiss chard, fodder, and garden beet. Domesticated from wild sea beet, *Beta vulgaris* subsp. *maritima*, it became regularly cultivated in European fields in the seventeenth century mainly as cattle fodder. The successful crystallization of beet sugar by the German chemist Andreas Sigismund Marggraaf in the mid-1700s was a key turning point for the crop and industrial-scale beet sugar production in Europe (Frances 2006). Today, sugar beet is cultivated throughout the world, accounting for about 30 % of world sugar production and is becoming a strong contender for biofuel production (Panella 2010). It is mainly grown in temperate regions (latitudes of 30–60°N), where sugarcane is not a viable option and is cultivated as either a spring or winter crop in the cooler (e.g., northern European) and warmer (e.g., Californian Imperial Valley) climates, respectively. Only a few subtropical countries including Morocco, Egypt, and India are known to grow and process beet alongside cane (Biancardi and Lewellen 2010).

Although the distribution of sugar beet cultivation worldwide has historically been driven by political agendas, it also represents the ability of breeders to produce varieties that can be grown economically across a range of climatic conditions. Much of the breeding success has depended on manipulation of morphological and anatomical characters to improve the root size and shape as well as physiological characters to optimize yield by extending the growing season (Bosemark 2006). Flowering-time control is a major breeding target for an extended growing reason and requires a detailed understanding of the physiological response of sugar beet to external (light, temperature) and internal (hormonal) environment cues and of the genetic basis of the floral transition.

*Beta vulgaris* flowers in LD and includes both biennial and annual types. The annual growth habit is characterized by bolting without vernalization and is controlled by dominant alleles of the bolting gene *B* (Munerati 1931). This is now known to be a pseudo-response regulator of the *PRR3/PRR7* type and renamed *BOLTING TIME CONTROL 1* (*BvBTC1*) (Pin et al. 2012). Biennial types which include all sugar beet cultivars have an obligate requirement for vernalization and include all sugar beet cultivars. A number of nonsynonymous SNPs and a large insertion (~28 kb) in the *BvBTC1* promoter have been associated with the biennial growth habit and, consistent with accepted nomenclature, are referred to as the *Bvbtc1* recessive form (Pin et al. 2012). Although *BvBTC1* is now widely accepted as encoding the originally described bolting gene *B*, other dominant bolting loci have been identified and mapped in sugar beet (Buttner et al. 2010).

It has long been recognized that, in biennial sugar beets, a complex relationship exists between vernalization and photoperiod (Owen and Stout 1940). In the sugar beet wild crop relative, *B. vulgaris* subsp. *maritima*, it has now been demonstrated that a compensatory relationship exists between vernalization and inductive photoperiods (Van Dijk 2009). The critical threshold vernalization requirement can therefore be reduced by vernalizing plants in LD. A possible mechanism for this relationship is starting to emerge based on the observation that the annual *BvBTC1* integrates LD cues to upregulate the floral promoter *BvFT2*, a homolog of the *Arabidopsis FT* gene (Pin et al. 2010), whereas the biennial *Bvbtc1* must first be upregulated by vernalization before it can effectively upregulate *BvFT2* (Pin et al. 2012). In respect of its own photoperiodic response and downstream regulation of *BvFT2*, a coincidence model for *BvBTC1/Bvbtc1* similar to that observed for *CO* has been proposed (Pin et al. 2012). This may be possible although the existence of a *BvCO* cannot be ruled out because a functional homolog of *CO*, the so-called *BvCOL-1*, has been identified together with 13 other *CONSTANS-like* gene family members in beet (Chia et al. 2008). Regulation of the sugar beet *FLOWERING LOCUS T* (*BvFT2*) by *BvBTC1* is mediated by the paralogous *BvFT1*, a direct repressor of *BvFT2* (Pin et al. 2010), which is itself constitutively downregulated in the presence of *BvBTC1* and by *Bvbtc1* only after vernalization.

Transgenic overexpression of *BvFT1* suppresses bolting, and it has been proposed as a potentially better target for vernalization control than *Bvbtc1*. This is supported further by the observation that in sugar beet, attainment of critical threshold vernalization requirement coincides with upregulation of *BvFT2* (Chiurugwi et al. 2012). The repression of *BvFT2* is therefore necessary for bolting resistance and requires increased expression of *BvFT1* and low levels of *BvBTC1/ Bvbtc1*.

The picture that is emerging from the molecular characterization of the many sugar beet flowering genes analyzed so far and compared with Arabidopsis homologs suggests functional conservation at the protein levels and not necessarily always at the physiological level; the roles of some beet proteins are divergent (Reeves et al. 2007; Abou-Elwafa et al. 2010). The same may be true for GA whose role in the orchestration of some aspects of reproductive growth developmental processes in sugar beet may be distinct from *Arabidopsis*. For example, unlike *Arabidopsis*, bolting responses in sugar beet are dependent on vernalization, such that the plants are not able to respond to applied GA in noninductive photoperiod without prior vernalization (Mutasa-Göttgens et al. 2010). Similar behavior exists in vernalization-dependent grasses and cereals (reviewed by Mutasa-Göttgens and Hedden 2009). In sugar beet, GA accumulates in shoot apices during vernalization and is depleted immediately prior to bolting (Debenham 1999) and has been shown to be necessary for bolting and pollen development but not flower induction (Mutasa-Göttgens et al. 2008). Further, it has been demonstrated that GA activates bolting independently of the bolting gene *B* locus and most probably also *BvBTC1* (Mutasa-Göttgens et al. 2010). In *Arabidopsis*, GA has recently been demonstrated to activate the transcription of *FT* and *TWIN SISTER OF FT* (*TFS*) in the vasculature to promote flowering in LDs (Porri et al. 2012). A similar role in the transcriptional regulation of both the sugar beet *BvFT1* and *BvFT2* control flowering time is conceivable, based on the abundance of GA-responsive elements in regulatory ($5'$ UTR and intronic) sequences of both genes (Mutasa-Göttgens unpublished data).

The availability of genome data has contributed significantly to the speed and accuracy of gene identification, cloning, and characterization in sugar beet. A public collection sugar beet ESTs (http://compbio.dfci.harvard.edu/cgi-bin/tgi/gimain.pl?gudb=beet) and a number of reference genomes have been assembled although not yet published, with the closest to publication being the GABI BeetSeq project (http://www.gabi.de/projekte-alle-projekte-neue-seite-144.php). Currently, the only published annotated reference is from the transcriptome of vernalized sugar beet apices with and without GA treatment (Mutasa-Göttgens et al. 2012). This has enabled the first comprehensive genome-scale analysis of candidate regulatory genes in the sugar beet vernalization pathway. For example, bioinformatic analyses of these data revealed a homolog of *AP1* to be significantly upregulated in apices of vernalized, nonbolted apices. Upregulation of *VRN-1* in cereals is well characterized, and it has been recently demonstrated that in wheat, *VRN-1* directly upregulates *FT1* (*VRN-3*) immediately before flowering (Shimada et al. 2009). Based on these data, it is not unreasonable to expect that the *BvAP1-like* MADS gene could also be required to upregulate the floral promoter *BvFT2*. A novel and previously unsuspected role for the RAV1-like AP2/AP3 domain protein in vernalization was also revealed by its threefold increase in expression in vernalized plant apices, which was enhanced to ~6-fold by GA. Most powerfully, these data sets enabled a global assessment of genes that were positively negatively coregulated with the *BvAP1-like* and *BvRAV1-like* (or any other selected gene) to provide the most comprehensive set of putative candidate genes in the same regulatory networks (Mutasa-Göttgens et al. 2012). In future, functional analyses of these and many of the other potential vernalization-responsive genes will assist with the task of deciphering floral regulatory networks in sugar beet, yielding further breeding targets for fine-tuning flowering time and deciding targets for changing climatic conditions. Besides prevention of premature flowering in crops, breeders also need to synchronize flowering time in order to produce high-quality hybrid seeds. Based on existing published data and assumptions from knowledge of defined gene functions in model plants, we can already begin to piece together some parts of the jigsaw of the sugar beet flowering-time gene regulatory network as indicated in Fig. 1.5.

A fully bolting resistant sugar beet variety insensitive to vernalization and developed for autumn sowing and winter cultivation is anticipated by 2022, with a winter beet biotech trait being ranked as a top priority by the major sugar beet breeding companies (personal communication, Elisabeth Wremerth-Weich, Syngenta). Until then, early spring-sown sugar beet in cool temperate climates, in which weather conditions can be unpredictable and often unseasonably cold, remains vulnerable to vernalization, premature bolting and eventually flowering as days get longer. This is problematic not only due to yield loss but also weed beet problems in subsequent crops, arising from the shed seeds

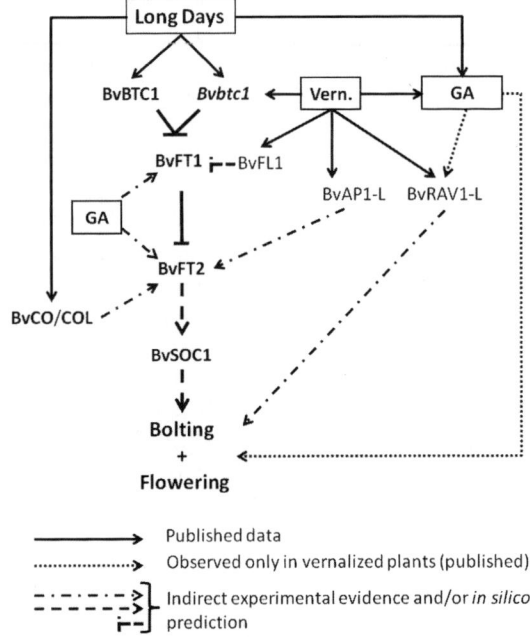

**Fig. 1.5** Conceptual model of the floral regulatory network in *Beta vulgaris* ssp. *vulgaris*

## 1.2.5 Tree Crop Species: Poplar

### 1.2.5.1 Poplar Flowering-Time Pathway Genes

The genus *Populus* (family Salicaceae) includes between 22 and 85 different species of deciduous plants native to the Northern Hemisphere (Eckenwalder 1996). Poplars, like other perennial plants, display successive years of vegetative growth before reaching sexual maturity. After this time, shoot meristems begin cyclical transitions between vegetative and reproductive growth (Hsu et al. 2011). Floral initiation in poplars occurs between mid-May and mid-June, and the effects of photoperiod on flowering time, growth cessation, and bud set in the fall have been clearly shown (Böhlenius et al. 2006). The role of different critical day lengths has also been shown to be necessary for growth cessation in European aspen (*Populus tremula*) trees originating from different latitudes (Böhlenius et al. 2006). Variations in temperature and light intensity during the suitable period for flower bud initiation could explain the large year-to-year variations in flowering of aspen and many other tree species (Owens 1995). Flowering time seems to be determined by the accumulated amount of heat or heat plus light that the plant receives during a period preceding the floral initiation (Poethig 2003). A study of adult willow cuttings in controlled environments indicated that both photoperiod and temperature affect floral initiation (Junttila 1980). Willow (*Salix*) and poplars are both members of the family Salicaceae and similar influence has been proposed for poplars (Brunner and Nilsson 2004).

Flowering genes discovered in *Arabidopsis* have been the basis for genetic studies on flowering regulation in poplar with genetic transformation with *FT* genes from poplar, *PtFT1* (Böhlenius et al. 2006) and *PtFT2* (Hsu et al. 2006), inducing early flowering in poplar. The overexpression of *LFY* (Weigel and Nilsson 1995) and its poplar homolog *PTLF* (Rottmann et al. 2000) also resulted in precocious flowering in hybrid aspen. Most *Arabidopsis* flowering genes have been shown to have two or more counterparts in poplar (Leseberg et al. 2006). Comparative analysis of multiple angiosperm genomes has implicated gene duplication in the expansion and diversification of many gene families (Rodgers-Melnick et al. 2012). However, empirical data and theory suggest that whole-genome and small-scale duplication events differ with respect to the types of genes preserved as duplicate pairs (Rodgers-Melnick et al. 2012). When duplicated genes accumulate mutations independently, a new or different function can develop in one of the duplicated genes (Force et al. 1999; Leseberg et al. 2006). In some cases, a gene may lose its function and accumulate mutations as a pseudogene (Leseberg et al. 2006). The prevalent duplication of poplar ABC MADS-box genes suggests rich modifications of the genetic network that underline the development of the unisexual, two-whorled flowers (Leseberg et al. 2006). There are two poplar homologs each for *Arabidopsis AP1*, *PISTILLATA* (*PI*), *APETALA3* (*AP3*), *AGAMOUS* (*AG*), and *SEESTICK* (*STK*) (Brunner et al. 2000; Leseberg et al. 2006). Some groups of MADS-box genes were further expanded, e.g., four homolog genes of *SOC1* and eight for *SVP/AGL24* (Leseberg et al. 2006). While no convincing *FLC* homologs are found in poplar (Leseberg et al. 2006), poplar homologs of *Arabidopsis FT*, *PtFT1* (Böhlenius et al. 2006) and *PtFT2* (Hsu et al. 2006), seem to coordinate the repeated cycles of vegetative and reproductive growth in poplar (Hsu et al. 2011). Reproductive onset is determined by *PtFT1* in response to winter temperatures, whereas vegetative growth and inhibition of bud set are promoted by *PtFT2* in response to warm temperatures and LDs (Hsu et al. 2011).

The *Populus trichocarpa CO* gene *PtCO2* shows a similar expression pattern as *CO* in *Arabidopsis* (Böhlenius et al. 2006). A model explaining the different critical day lengths for growth cessation and bud set for different provenances would be that *PtCO2* expression starts increasing earlier after dawn in the more southern provenances (Böhlenius et al. 2006). Phytochrome genes have been implicated in SD-induced bud set and growth cessation in poplars (Howe et al. 1998; Ingvarsson et al. 2006). *Populus* has three phytochrome genes, *PHYA*, *PHYB*, and *PHYB2* (Howe et al. 1998). *PHYB2* has been mapped to a linkage group containing QTL for bud set and bud flush in several experiments (Frewen et al. 2000; Chen et al. 2002). Heterologous overexpression of the *PHYA* gene from oat (*Avena sativa*) in hybrid aspen resulted in insensitivity to photoperiod (Olsen et al. 1997).

Genetic studies have revealed very complex interactions between regulators of different developmental processes in plants. For instance, the overexpression of

*BpMADS4* from birch (Elo et al. 2001) has been found to induce early flowering in birch (*Betula pendula*) (Elo et al. 2007) and apple (*Malus × domestica*) (Flachowsky et al. 2007). However, overexpression of *BpMADS4* as well as the *Arabidopsis* homolog gene *FUL* does not induce early flowering in poplar but avoids normal senescence and winter dormancy (Hoenicka et al. 2008). Overexpression of *FLOWERING PROMOTING FACTOR 1* (*FPF1*) induced early flowering in *Arabidopsis* (Kania et al. 1997), with a similar approach showing a strong effect on wood formation but no effect on flowering time in hybrid aspen (*Populus tremula × P. tremuloides*) (Hoenicka et al. 2012a).

Experiments with transgenic poplars have shown that downregulation of *PopCEN1* and its close paralog, *PopCEN2*, both members of the *CENTRORADIALIS (CEN)/TERMINAL FLOWER 1* (*TFL1*) subfamily, induce accelerated onset of mature tree characteristics, including age of first flowering, number of inflorescences, and proportion of short shoot (Mohamed et al. 2010). Terminal vegetative meristems remained indeterminate in *PopCEN1*-RNAi trees, suggesting the possibility that florigen signals are transported to auxiliary meristems rather than the shoot apex (Mohamed et al. 2010). However, the auxiliary inflorescences (catkins) of *PopCEN1*-RNAi trees contained fewer flowers than did wild-type catkins, suggesting a possible role in maintaining the indeterminacy of the inflorescence apex. Expression of *PopCEN1* was significantly correlated with delayed spring bud flush in multiple years, and in controlled environment experiments, 35S::*PopCEN1* and RNAi transgenics required different chilling times to release dormancy (Mohamed et al. 2010).

Coping with their long-living sessile lifestyle, trees have become masters of adaptation, improving their chances of survival and reproduction in a potentially changing environment (Rohde and Junttila 2008). Perennials species arrest their growth, form buds, enter a dormant state, and induce cold hardiness (Bielenberg 2011). Phenological data collected in a large geographical area has confirmed the advancement of phenological timing of poplar in Finland and Canada (Beaubien and Freeland 2000; Beaubien and Hamann 2011). *Populus tremuloides* showed bloom dates advanced by two weeks between 1936 and 2006 in Alberta, Canada (Beaubien and Hamann 2011). During past episodes of climate change, many plant species experienced large-scale expansions (Keller et al. 2011), which likely encountered strong selection as they colonized new environments. Keller et al. (2011) examined the extent to which populations of balsam poplar (*Populus balsamifera* L.) have become locally adapted as the species expanded into its current range since the last glaciation. Comparison of molecular and quantitative trait divergence across a range-wide sample of 20 balsam poplar populations revealed selection as the primary source process structuring variability in ecophysiological and phenological traits at both regional and local spatial scales, suggesting strong adaptive evolutionary responses to historical changes in climate (Keller et al. 2011). The largest divergence estimates were found for the phenology traits bud flush and bud set. These traits determine the start and end of the period of active stem.

Understanding the genetic basis of phenology and ecological adaptation is critical for development of breeding strategies for poplar and other forest tree species (Keller et al. 2012). Several candidate genes for adaptive phenology have been studied in poplars (Keller et al. 2011, 2012). In balsam poplar, homologs of the *Arabidopsis* vernalization pathway gene *FRI* seem to play a role in the temperature-sensitive timing of seasonal development (Keller et al. 2011). The central circadian clock gene *GI5* showed in other study the strongest and most consistent evidence for ecological adaptation (Keller et al. 2012). *GI* is directly involved with light signaling to the circadian clock and is also an upstream regulator of the major floral genes *CO* and *PtFT1* (Keller et al. 2012). The strongest evidence for local selection in *FRI* is perhaps largely due to polymorphism at synonymous and intron sites (Keller et al. 2011). Members of the phytochrome family of light-signaling genes also appear to be repeat target of local selection in some taxa, although they do not appear to be under local selection in *P. balsamifera* (Keller et al. 2011). In *P. tremula* L. *PHYB2* shows excess polymorphism as well as strong latitudinal clines (Ingvarsson et al. 2006. In *Arabidopsis*, *ZEITLUPE* (*ZTL*) genes are hypothesized to be light photoreceptors, known to interact functionally with the circadian clock genes *GI* and *TOC1* (Kim et al. 2007). In balsam poplar, *ZTL* paralogs (*PbZTL1* and *PbZTL2*) showed evidence of selection across multiple analyses (Keller et al. 2011). *PbZTL1* showed a combination of low diversity and evidence of a balanced polymorphism. This was due to a single synonymous SNP at the end of exon 2 segregating at high frequency, which also showed differentiation between northern and southern accessions. Poplar homologs of *LFY* and *EARLY BOLTING IN SHORT DAYS* (*EBS*) also showed evidence of local selection (Keller et al. 2011).

Given the size and range of *Populus* distribution, it is difficult to make predictions on the response of poplars to climate change. Species-specific surveys are required to provide information on inappropriate timing of growth cessation, increased susceptibility to injury (due to abiotic or biotic stress) and potential effects on the subsequent dormancy status of overwintering buds (Tanino et al. 2010; Bielenberg 2011). Climate change is likely to have a profound impact on tree–pathogen interactions, representing a worldwide interdisciplinary challenge for long-term sustainability (Jeger and Pautasso 2008) and adaptability of forest tree ecosystems. Predicting the response of biodiversity in forest ecosystems to climate change has become an important field of research (Bellard et al. 2012). Because of climate change, tree species may no longer be optimally adapted to the environmental conditions in a given region. Trees are long-living organisms, which have experienced decades or even centuries of climate change. Little is known about the effects of the predicted environmental changes on the physiology and growth of old trees (Phillips et al. 2008). The response of plants that developed under preindustrial conditions to current environmental conditions can provide insights into long-term vegetation responses to predicted global environmental change scenarios (Ward and Strain 1999; Phillips et al. 2008).

Several studies have indicated an advancement of the timing of the spring events during the past decades in poplars (Linkosalo et al. 2009; Wu et al. 2009). Warming estimates derived from the phenological trends of several tree species suggested that, for example, in Finland, the mean spring temperature increase has been 1.8 °C per century, which is very close to the value of 1.5 °C per century indicated by long-term temperature records (Linkosalo et al. 2009). The extreme weather events proposed under global warming may affect growth, reproduction, and survival of poplars (Augspurger 2009). One very damaging event for many deciduous tree species is freezing temperatures in spring causing frost damage to freshly developed leaves and flowers. Climate predictions include not only an increase in mean temperature but also greater variability of temperatures (Rigby and Porporato 2008; Augspurger 2009). In poplars, active growing organs formed in early spring, e.g., expanding shoots, leaves, and flowers, are susceptible to frost (Augspurger 2009). Spring bud flush occurs in response to warming temperature, whereas bud set is primarily controlled by photoperiod (Böhlenius et al. 2006; Luquez et al. 2008). Global warming may have also an impact on gametogenesis and flower bud break in temperate trees as chilling winter temperatures are required for the onset of both processes (Julian et al. 2011; Leida et al. 2012; Hoenicka and Fladung unpublished data).

The floral pathway integrator genes *PtFT1/PtFT2*, the central circadian clock gene *GIGANTEA 5*, and the photoreceptor gene *PHYB2* have clearly been related to genetic regulation of phenology in *Populus* (Ingvarsson et al. 2006; Hall et al. 2011; Keller et al. 2011, 2012). Therefore, they represent important targets for breeding on climate change. Dormancy-associated MADS-box genes (DAM genes) (Yamane et al. 2011; Leida et al. 2012) may also be important breeding targets in poplars. However, breeding for flowering-time adaptation to climate change is very difficult in poplars due to the long phase of reproductive incompetence common in forest tree species. Genetic regulation of floral induction in perennials is much more complex than in other plant species. Perennial plants have also to pass from the juvenile to the adult stage, but unlike annual herbaceous plants, juvenility in tree species can last even decades until the reproductive stage is reached. Poplars usually require more than 7 years to develop first flowers, a severe impediment for breeding. Development of efficient early-flowering transgenic poplar lines may help to overcome this in the future.

Genetic transformation with different flowering-time genes has produced early-flowering transgenic lines in different tree species, e.g., poplar (Weigel and Nilsson 1995; Böhlenius et al. 2006; Zhang et al. 2010) and apple (Flachowsky et al. 2009). In poplar, the genetic transformation with a heat-inducible *FT* gene from *Arabidopsis* promotes the more efficient flowering induction (Zhang et al. 2010). However, hitherto available early-flowering poplar systems often show a disturbed microsporogenesis (Hoenicka et al. 2012b). Development of fertile early-flowering trees was possible in apple and citrus plants. Downregulation of miR156 and gene stacking approaches based on *FT* expressing transgenic poplar lines may produce early-flowering plants with a better performance. On the other hand, release of transgenic trees is still difficult in most countries due to biosafety concerns. Tree breeding using early-flowering transgenic lines followed by selection of genetic modification (GM)-free trees before release has been proposed to overcome those biosafety concerns (Flachowsky et al. 2009). However, plant epigenetics has recently gained unprecedented interest, not only as a subject of basic research but also as a possible new source of beneficial traits for plant breeding (Mirouze and Paszkowski 2011). Epigenetic breeding may overcome limitations caused by the prolonged vegetative phase on forest tree breeding. It has been shown that height, growth, and bud phenology are influenced by the temperature during zygotic embryogenesis in *Picea abies* (Kvaalen and Johnsen 2007). Although no such studies have been reported for other plant species, similar results in poplar and other forest tree species cannot be excluded. Furthermore, DNA methylation was found to play an important role during flower development in andromonoecious poplar (Song et al. 2012). As an obligate outbreeder, populations of *Populus* contain large amounts of genetic variation. Human-aided movement of tree species in reforestation programs has been proposed as a practical and cost-effective climate change adaptation strategy (Gray et al. 2011).

## 1.3 Breeding for Flowering-Time Adaptation to Climate Change: Modeling in *Arabidopsis* and Crop Species

High-throughput technologies such as genomics, microarrays, proteomics, transcriptomics, and metabolomics have revolutionized plant biology. Progress in these areas has been rapid and has transformed the way biologists tackle new problems, providing a wealth of easily accessible and searchable information such as annotated genomes, phylogenetic relationships, function and structure predictions, expression and co-expression patterns, and metabolic profiles. Sequence-based bioinformatics has become a key part of plant biology and one that is likely to gain importance with the ever-increasing ease and speed of genome sequencing. Modeling of mechanisms has played second fiddle to the wave of genetic and bioinformatics discoveries that have been prevalent in the field of plant biology in the past two decades, and our understanding is lagging behind the data accumulation rate. More recently, however, there has been recognition that systems approaches that include computational modeling will have a key role to play in elucidating many aspects of plant development and the interactions between a plant and its environment (Hammer et al. 2004; Yin and Struik 2010). Some areas are already well advanced such as circadian clock modeling (Doyle et al. 2002; Edwards et al. 2006). Flowering-time control is another area that has benefited from modeling. Early approaches were based on theoretical considerations about the governing behavior of the floral transition and involved biochemical switches of hypothesized components. These mathematical models had a number of assumptions to make solving the problems analytically tractable (e.g., Charles-Edwards et al. 1979. With the development and availability of computers and software packages, these simple models have been generally superseded by much more complicated systems of many variables, which are solved numerically (Thornley and Johnson 1990).

While modeling of the floral transition in *Arabidopsis* was just developing a decade ago, the modeling of flowering time in crop species was in full bloom. However, the types of modeling used in crop and model species are quite different, and it is interesting to compare these approaches. Crop modeling has been goal oriented in terms of making useful predictions for agriculture, whereas model species approaches have targeted a more gene-based understanding of the system. Crop modeling has very much been based on empirical studies, using data such as observed time to flowering or fruit production, to restrain predictions that were built using regression models. With regression statistics and ANOVA, it is possible to account for factors like $CO_2$ emissions, location, and light intensity that can vary hugely and are of importance to plant breeders and growers alike. Correct timing of crop production is essential for many producers, and the efficacy of these models is testament to their strength. These data-driven approaches are very successful and highlight the need to reduce the inherent complexity of the system in order to use the power of data to guide predictions.

In *Arabidopsis* research there has been a focus on ordinary differential equations, which allow the dynamic system of interacting components to be tracked. Converting a genetic regulatory network (GRN) into mathematical formulae follows standard conventions (Alon 2007), which allows for the analysis of a mechanistic model with kinetic parameters having a biological meaning. Unfortunately these parameters are often unknown experimentally and so have to be estimated. The goal is often to gain an understanding of genetic control elements and infer molecular mechanisms. In order to make the approach tractable, many factors are excluded from such GRN-based models that are relevant to those with a more agricultural interest.

Many crop models use QTL analysis for traits of interest. This operates at a level above genes by linking phenotypic and genotypic data. These complex traits are often non-transferable between species, yet genes are frequently highly similar/homologous (especially enzymes) and likely to carry out the same functions, motivating gene models. Interestingly it is general genetic motifs that are often most conserved between species. Thus, the knowledge of the workings of one motif in a species is likely applicable to another species. With our increasing understanding of gene networks coupled with QTL analysis drilling down to individual genes or even SNPs, transferable gene-level models that cross scales and integrate up to the environment level are within grasp.

There are many climate change models for $CO_2$, water availability, and temperature for the years ahead. These are key factors for plant development, and the challenge is to incorporate these predictions into plant breeding tools (Soussana et al. 2010). Here we review a few examples of modeling in crops and *Arabidopsis* and reflect on where a multiscale approach could lead to the combining of the phenotype-based work in crops with molecular level research. We suggest that this may be a crucial step in ensuring future plant breeding to successfully incorporate knowledge of climate change at the same time supporting a growing global population.

White et al. (2008) include two major flowering regulators of bread wheat in their approach that used genetic information from 29 spring and winter wheat lines. Data from multiple locations worldwide were split into either a calibration or evaluation set for the gene-based model parameters. A linear regression approach was used to estimate the genetic effects of the *VRN-1* loci on vernalization requirements and the *PPD-D1* locus on photoperiod sensitivity. The use of a specific simulation environment is common to these types of models, and White et al. (2008) use CSM-Cropsim-CERES-Wheat (Jones et al. 2003), which can simulate the development of many stages of wheat growth and also incorporate strain-specific factors. The conventionally estimated parameters in the simulations predicted almost all of the variation in time to flowering for the calibration data, with a modest reduction for the evaluation set, as expected. Results from using gene-based coefficients reduced the accuracy only slightly indicating the possibility that using genetic information in wheat modeling together with the more conventional phenotypic data has potential. The quantity and quality of data is a constraining factor at present, especially in terms of understanding the loci effects, but also from environmental data, for example, the accuracy of the reported weather conditions, which have a large effect.

Messina et al. (2006) use a similar approach in soybean. Taking a simulation model, CROPGRO-Soybean (Boote et al. 1998), they then used linear functions to predict cultivar-specific parameters, which were then used to estimate flowering time, as well as post-flowering development stages and yield. The model was evaluated using field trial data from other locations and was shown to predict maturity date particularly well for most varieties. Interestingly the results are stated to be comparable to those from common bean, which is encouraging for the development of gene-based modeling across species.

Yin et al. (2005a) developed a model for spring barley using reciprocal photoperiod transfer experiments to estimate the genotype-specific parameters, which were again evaluated in independent field trials. Additionally they performed a sensitivity analysis on their four parameters, showing they were all important for predicting inter-genotype differences in flowering time. They also found that the importance of their four parameters could be ranked, with the minimum number of days to flowering at optimal temperature and photoperiodic conditions being the most important, followed by photoperiod sensitivity. This regression-based model gave a reasonably good prediction of variation in time to flowering across both genotypes and environments. Yin et al. (2005b) then progressed this by developing a QTL base to the original model (Yin et al. 2005a). Changing the parameters to QTL effects from the genotype-specific parameters used previously has reduced the accuracy of the model by 9 % (to 72 %).

A recent model developed by Uptmoor et al. (2012), based on an earlier model (Uptmoor et al. 2008), used genotype-specific parameters and QTL effects as the inputs to a model for predicting flowering time in *Brassica oleracea*. In this model the predictability of flowering time using genotype-specific parameters was reduced by unfavorably high temperatures. This suggests that noisy environmental conditions, which can be filtered by using an integrated network approach, are not fully taken into account with this modeling framework. Using QTL effects as the parameters instead further reduced the ability of the model to capture inter-genotype variability under both low and high temperatures.

Incorporating QTL effects into models at present seems to produce unsatisfactory results, but the exact reasons are unknown. This could be because of undetected minor QTLs (Yin et al. 2005b) or poor estimation of their effects (Uptmoor et al. 2009). Sampling more plants, and at a finer resolution, should result in data that can give a more precise idea of the effects of QTLs. Nevertheless, the results using genotype-specific model parameters can give good predictions of flowering time, but the use of more complex models should, for an extra computational cost, give consistently better predictions. Wurr et al. (2004) considered the effects of climate change on winter cauliflower production using simulations of four different scenarios for future global greenhouse gas emissions. All forecasts predicted a rise in temperature. In the model this increase in temperature led to shorter juvenile and curd growth phases, but longer curd induction in most cases. Importantly location effect was found to dominate the time to maturity, raising questions for both breeders and growers.

Welch et al. (2003) employed a neural-network approach to quantifying flowering time for a number of *Arabidopsis* genotypes. Neural networks are composed of interlinked nodes, each with a number of inputs, and an output to a subsequent node, with the network structure decided by the modeler. The links between nodes have an associated weight, which adjusts the value between the output and input nodes. The weights are established through a training procedure using experimental data, typically using a least squares residual. Welch et al. (2003) looked at the inflorescence transition in *Arabidopsis* and how they were specifically controlled by the autonomous and photoperiod pathways. Their network can reproduce the floral transition of many mutant genotypes both at 16 °C and at 24 °C. At the lower temperature, the rate of *Arabidopsis* development is much reduced. Intriguingly they found the order of inflorescence transition between two loss-of-function genotypes switches between the two temperatures. Many crop simulation models would not be able to show this result, which demonstrates how using network-based methods could potentially do more than just predict flowering time. This sophisticated machine-learning approach can be viewed as regression using a network-based model that in principle could take advantage of known gene networks. Mapping gene networks onto neural-network structures is, however, not straightforward. Gene network diagrams are qualitative in nature and do not give any idea of time or space, thus missing potentially interesting dynamics. Simulating these networks therefore can lead to new theories and suggest new experiments that can lead to increased understanding in all physiological networks.

The work by Jaegar et al. (unpublished) is based upon the GRN of the floral transition in *Arabidopsis*. This is based on the simplifying step of grouping genes with common function into regulatory hubs. With this slight abstraction, direct molecular relevance is lost. However, gains are made in terms of qualitative predictions that can be tested experimentally. To parameterize this model the model is fit to flowering-time data from a range of genotypes. In *Arabidopsis* the start of the floral transition is clearly seen by its bolting and the production of cauline leaves instead of rosette leaves. On completion of the floral transition, flowers are made instead of cauline leaves, and the transition from a vegetative to a reproductive state is complete. The number of rosette and cauline leaves can be counted for each genotype, and using parameter space exploration algorithms estimates the parameters in the model. The input to this model is just *FT* levels; if it was known how environmental factors affected *FT* expression, leaf numbers could be predicted, and therefore developmental time, for many genotypes. If this type of approach was developed in important crop species, it is possible it will prove very helpful for predictive breeding and crop scheduling.

All models rely on a number of parameters. Parameter determination or estimation is thus a key step towards predictions. Model validation is crucial, irrespective of the species studied, and must include separate experiments for independent evaluation of parameters.

Current crop models for predicting flowering time are highly valuable; however, to fully exploit the wealth of genomic information that is becoming available, these models need to bridge scales. Using gene network-based approaches should be able to calculate flowering time accurately and predict interesting dynamics as a function of different inputs, as summarized in Fig. 1.6. If it was known how environmental factors affected inputs to these models, then the dynamics modeled in silico could give important information to plant breeders. The challenge is thus to characterize perturbations of the control variables such as temperature, $CO_2$, and water availability on the key inputs to the genetic networks (e.g., *FLC*, *FT*, and others in *Arabidopsis*) and then to drive the change of these variables by climate models.

## 1.4 Novel Methods for Selecting for Flowering Time

The high heritability of flowering time makes it an ideal target for marker-assisted breeding approaches aimed at modulating flowering time for current and future agricultural environments. Traditionally, selection for flowering time in crops has been based on phenotyping studies, relying on natural variation in crop and wild relative gene pools (Jung and Muller 2009). Identification of genes and functional markers now allows for the use of MAS for tracking and selecting flowering-time alleles in crossing programs. Where there is insufficient variation for this trait, novel genetic variation can be achieved via mutagenesis, e.g., TILLING (Jung and Muller 2009). Targeted genetic modification, which is currently restricted to a few key regulatory genes, can be achieved through transformation, including overexpression/suppression of gene activity (Jung and Muller 2009). Detection of flowering-time loci is also aided by the availability and cost reductions of genotyping platforms, as well as increasingly sophisticated mapping populations

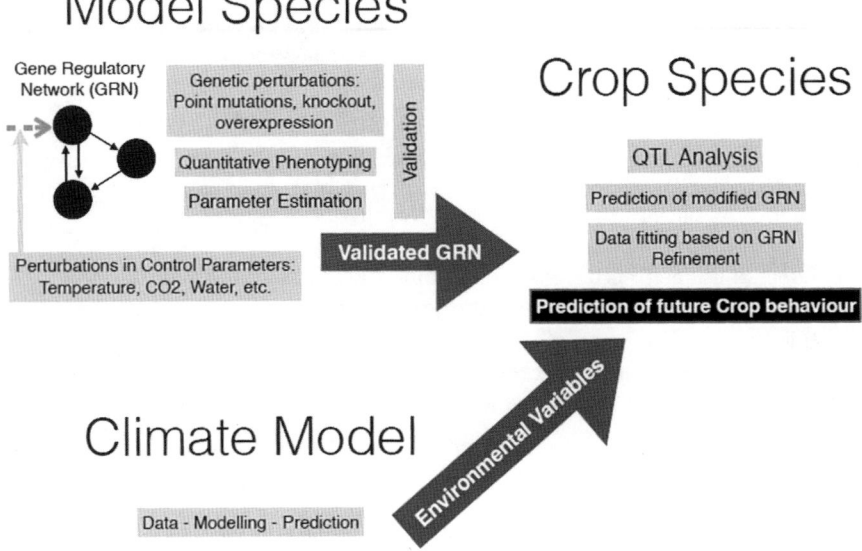

**Fig. 1.6** Overview of the potential use of gene regulatory networks to transfer information from model species to crop species and integrate with climate modeling

(Paux et al. 2011). For example, multiparent advanced genetic intercross (MAGIC) is a direct development of the advanced intercross (AIC) (Darvasi and Soller 1995). It was shown that greater precision in mapping would be achieved if the $F_2$ of a biparental cross was first intermated for several generations before use in mapping. The additional cycles of crossing reduce the extent of linkage disequilibrium within the population such that significant marker-trait associations are indicative of close linkage. An extension to the AIC has also been developed in which populations were established from multiple founder lines, intermated, and then used in mapping (Mott et al. 2000). In a mouse population established from eight inbred lines, they were subsequently able to map 843 QTLs for 97 traits to intervals with an average 95 % confidence interval of 2.8 Mb (Valdar et al. 2006). More recently, 15 papers have been published together, reporting the first results from a similar approach, termed the "collaborative cross" in which a large set of mouse inbred lines have been derived from an intercross among eight founders (The Collaborative Cross Consortium 2012). In plants, Mackay and Powell (2007) retermed this approach as MAGIC and advocated its use in crop genetics. MAGIC populations benefit from high allelic inputs, increasing the allelic combinations that can be studied within a unified germplasm platform. Such populations have a number of benefits over traditional biparental populations (1) high levels of recombination resulting in greater mapping precision and (2) greater genetic diversity such that more QTLs could be detected for multiple traits, thus increasing the number of phenotypic traits for analysis within a single population. The latter permits detailed analysis of genotype × genotype interaction between related traits. There are a number of

MAGIC populations that are complete or under development in bread wheat (NIAB, UK; CSIRO, Australia), durum (UNBO, Italy), sorghum (ICRISAT, India), rice (IRRI, Philippines), cowpea (IITA, Nigeria), and oats (IBERS, UK). Because MAGIC populations take several generations of crossing and selfing to establish, their application in mapping projects is only just beginning, with the first set of results emerging from an Australian wheat population (Huang et al. 2012b). A population in *Arabidopsis* has also been created and used successfully to map QTLs for germination and bolting time (Kover et al. 2009). MAGIC populations should allow the identification of novel flowering-time QTLs across the range of crop species for which they are being developed.

Other diverse population resources are also suitable for fine mapping, such as nested association mapping (NAM) (Yu et al. 2008). A NAM population is a set of biparental crosses with a common parent. Linkage analysis can be carried out within each cross in standard fashion, but an increase in precision comes by exploiting linkage disequilibrium over the interlinked crosses. Buckler et al. (2009) used a NAM population to detect QTLs for flowering time in maize. NAM and MAGIC are distinguished from conventional biparental mapping populations not only by their increased diversity but also by their size, e.g., the maize NAM population consists of 200 recombinant inbred lines (RILs) from 25 crosses, and a MAGIC population should consist at least 1,000 RILs (typically derived from an 8- or 16-way intercross). Like MAGIC populations, association mapping (AM) populations benefit from multiple parental inputs and recombination events. The advantages of AM are well documented (e.g., Mackay and Powell 2007). Although relatively recently applied to plants, the use of association mapping panels in model species (e.g., Atwell et al. 2011) and cereal crops (e.g., Cockram et al. 2010b) is more established than the use of MAGIC and NAM. Specifically, AM of flowering time has been demonstrated in various species (Aranzana et al. 2005; Skøt et al. 2005), either on its own or in support of additional mapping approaches.

Genomic selection (GS) (Goddard and Hayes 2009; Jannink et al. 2010) uses newly developed statistical methods to estimate trait effects simultaneously for all markers in a high-density set. Selection then occurs for several generations on predicted trait value, termed genomic estimated breeding value (GEBV), before marker effects are reestimated to account for allele frequency changes. Thus, selections for the target trait can be made over successive years without the need for phenotypic evaluation. GS is now used commercially by animal breeders, and although experimental support for some of these methods is emerging in crops (e.g., Crossa et al. 2010), the opportunities to increase rates of response to selection using GS have yet to be widely exploited in plants. Rapid advances in the availability of genomics tools are occurring for a number of key crops (e.g., maize sequence assembly v5a.59, bread wheat physical map construction, high-throughput SNP detection and genotyping platforms, whole-genome sequence surveys, and the prospect of affordable genotyping-by-sequencing approaches), while their cost of implementation is reducing. Therefore, the validation of GS methodologies in crops would be timely, helping to ensure the maximal exploitation of emerging high-throughput genotyping platforms, thus leveraging increased efficiency for selection for flowering time within crop breeding programs.

The availability of cheap marker systems is also opening up novel applications of marker-assisted breeding, which may result in a reevaluation of some established breeding practices. For example, to generate RILs rapidly, breeders of autogamous crops commonly use single seed descent (SSD) or doubled haploid (DH) production. The latter is quickest and creates fully inbred progenies but is also more expensive and in many species is technically challenging. For many breeding programs therefore, the production of inbred lines by SSD is the major method for generating new inbred lines as candidates for selection. However, bulk inbreeding, in which at each generation of selfing all seed are harvested as a bulk, is as fast as SSD but cheaper: there is no requirement to sow single seeds harvested from individual plants, and harvesting and sowing can be automated. However, bulk inbreeding is not commonly used since the lines that it creates vary in coancestry, with many lines originating from common ancestral plants in the $F_2$ or subsequent generations. This sampling process inevitably results in loss of genetic variation. However, results (Mackay unpublished data) have shown by computer simulation that by screening the final bulk with between 50 and 100 markers, it is possible to select a set of lines, which are at least as unrelated to each other as those produced by SSD. This process is termed rapid bulk inbreeding (RABID). It is possible to use RABID to devise breeding programs in which more crosses are progressed speculatively, with selection among crosses occurring later in the inbreeding process than is currently practical. RABID can be easily extended to include selection for traits of high heritability, such as flowering time, during the bulk inbreeding process. The speeding up of the breeding cycle which RABID allows, and use of markers in selecting among lines within a cross, may also result in simple integration into schemes of genomic selection in inbreeding species. For crop research purposes, RABID also allows cheaper production of mapping populations.

Novel methods can also be applied to phenotyping studies. For example, artificial warming experiments are heavily relied on for estimating plant responses to global climate change, although this method is usually on a small scale compared to observation experiments over time and/or space (Wolkovich et al. 2012). The underlying assumption is that plant response under artificial warming is a proxy for a long-term response to global warming (Wolkovich et al. 2012). The effect of changing $CO_2$ concentrations on development has also been studied in a number of plant species (reviewed in Craufurd and Wheeler 2009) using controlled environment chambers and field-based free air carbon enrichment (FACE) facilities. However, the majority of studies have revealed little direct effect of increased $CO_2$ concentration alone on flowering time in any of the $C_3$ or $C_4$ species tested.

## 1.5   Future Directions and Challenges

Integration of genetic, crop modeling and physiological approaches should define the selective landscape through which breeders must navigate (Cooper et al. 2009). A key area in the manipulation of crop adaptation lies in the regulation of developmental phases to produce germplasm tailored to different agri-environments and end uses. The importance of phenology has meant that systems for the prediction of flowering time are of great interest to agronomists, breeders, crop physiologists, and crop modelers. If flowering time is unreliable, then yield will be poorly predicted, as simulated growth will occur under different conditions from those observed. This is illustrated by maize, where a single day of drought during flowering can decrease yield by 8 %. Physiological models are being created to define new ideotypes and to direct breeders' attention to appropriate targets for selection. However, to date, these models have not adequately taken into account genetic relationships among traits and QTLs, while crop models commonly require such detailed information that they are inadequate tools for breeders, who screen thousands of new genotypes each year. Equally, genetic modeling has not taken into account the ecophysiological consequences of changes in trait values, including flowering.

As global temperatures increase and precipitation levels fluctuate, annual variation in photoperiod will remain the same. In order to sustain crop yields within these parameters, flowering time must be tailored across the range of economically significant crop species grown worldwide with a great degree of precision. Using existing and future knowledge of the genetic control of flowering time will aid breeders in delivering this precision.

**Acknowledgements** Authors were supported by the following grants: AB (BB/I002561/1), EJ (BB/E014933/1), CH (BBSRC/DEFRA LK0863 GIANT), IA and DT (BB/J004405/1), MY and KH (Grants from the Ministry of Agriculture, Forestry and Fisheries of Japan - Integrated Research Project for Plant, Insect, and Animal using Genome Technology, IP1001, and Genomics for Agricultural Innovation, GPN0001), MF and HH (Federal Office for Agriculture and Food (BLE) grant #: 511-06.01-28-1-45.005-10 and German Research Foundation (DFG) grant #:FL 263/21-1), JC (BB/J002542/1).

## References

Abe M, Kobayashi Y, Yamamoto S, Daimon Y, Yamaquchi A, Ikeda Y, Ichinoki H, Notaquchi M, Goto K, Araki T (2005) FD, a bZIP protein mediating signals from the floral pathway integrator. J Exp Bot 61:2247–2254

Abou-Elwafa SF, Büttner B, Chia T, Schulze-Buxloh G, Hohmann U, Mutasa-Göttgens E, Jung C, Müller AE (2010) Conservation and divergence of autonomous pathway genes in the flowering regulatory network of *Beta vulgaris*. J Exp Bot 62:3359–3374

Ainsworth EA, Ort DR (2010) How do we improve crop production in a warming world? Plant Physiol 154:526–530

Alabadi D, Oyama T, Yanovsky MJ, Harmon FG, Más P, Kay SA (2001) Reciprocal regulation between *TOC1* and *LHY/CCA1* within the *Arabidopsis* circadian clock. Science 293:880–883

Alon U (2007) An introduction to systems biology: design principles of biological circuits, vol 10. Chapman & Hall/CRC, London

Amasino R (2010) Seasonal and developmental timing of flowering. Plant J 61:1001–1013

An H, Roussot C, Suárez-López P, Corbesier L, Vincent C, Piñeiro M, Hepworth S, Mouradov A, Justin S, Turnbull C, Coupland G (2004) CONSTANS acts in the phloem to regulate a systemic signal that induces photoperiodic flowering of *Arabidopsis*. Development 131:3615–3626

Angel A, Song J, Dean C, Howard M (2011) A polycomb-based switch underlying quantitative epigenetic memory. Nature 476:105–109

Anslow RC, Green JO (1967) The seasonal growth of pasture grasses. J Agric Sci Camb 68:109–122

Aranzana MJ, Kim S, Zhao K, Bakker E, Horton M, Jakob K, Lister C, Molitor J et al (2005) Genome-wide association mapping in *Arabidopsis* identifies previously known flowering time and pathogen resistance genes. PLoS Genet 1:e60

Armstead IP, Turner LB, Farrell M, Skot L, Gomez P, Montoya T, Donnison IS, King IP, Humphreys MO (2004) Synteny between a major heading-date QTL in perennial ryegrass (*Lolium perenne* L.) and the *Hd3* heading-date locus in rice. Theor Appl Genet 108:822–828

Armstead IP, Skot L, Turner LB, Skot K, Donnison IS, Humphreys MO, King IP (2005) Identification of perennial ryegrass (*Lolium perenne* (L.)) and meadow fescue (*Festuca pratensis* (Huds.)) candidate orthologous sequences to the rice *Hd1* (*Se1*) and barley *HvCO1 CONSTANS-like* genes through comparative mapping and microsynteny. New Phytol 167:239–247

Armstead IP, Turner LB, Marshall AH, Humphreys MO, King IP, Thorogood D (2008) Identifying genetic components controlling fertility in the outcrossing grass species perennial ryegrass (*Lolium perenne*) by quantitative trait loci analysis and comparative genetics. New Phytol 178:559–571

Atwell S, Huang YS, Vilhjálmsson BJ, Willems G, Horton M, Li Y, Meng D, Platt A, Tarone AM, Hu TT, Jiang R, Muliyat NW, Zhang X, Amer MA, Baxter I, Brachi B, Chory J, Dean C, Debieu M, de Meaux J, Ecker JR, Faure N, Kniskern JM, Jones JDG, Michael T, Nemri A, Roux F, Salt DE, Tang C, Todesco M, Traw MB, Weigel D, Marjoram P, Borevitz JO, Ausín I, Nordborg M (2011) Genome-wide association study of 107 phenotypes in a common set of *Arabidopsis thaliana* inbred lines. Nature, 465:627–631

Augspurger CK (2009) Spring 2007 warmth and frost: phenology, damage and refoliation in a temperate deciduous forest. Funct Ecol 23:1031–1039

Ausín I, Alonso-Blanco C, Jarillo JA, Ruiz-García L, Martinez-Zapater JM (2004) Regulation of flowering time by FVE, a retinoblastoma-associated protein. Nat Genet 36:162–166

Azpiroz R, Wu Y, Locascio JC, Feldmann KS (1998) An *Arabidopsis* brassinosteroid-dependent mutant is blocked in cell elongation. Plant Cell 10:219–230

Balasubramanian S, Sureshkumar S, Lempe J, Weigel D (2006a) Potent induction of Arabidopsis thaliana flowering by elevated growth temperature. PLoS Genet 2:e106

Balasubramanian S, Sureshkumar S, Agrawal M, Michael TP, Wessinger C, Maloof JN, Clark R, Warthmann N, Chory J, Weigel D (2006b) The *PHYTOCHROME C* photoreceptor gene mediates natural variation in flowering and growth responses of *Arabidopsis thaliana*. Nat Genet 38:711–715

Bastow R, Mylne JS, Lister C, Lippman Z, Martienssen RA, Dean C (2004) Vernalization requires epigenetic silencing of *FLC* by histone methylation. Nature 427:164–167

Beales J, Turner A, Griffiths S, Snape JW, Laurie DA (2007) A *Pseudo-Response Regulator* is misexpressed in the photoperiod insensitive *Ppd-D1a* mutant of wheat (*Triticum aestivum* L.). Theor Appl Genet 115:721–733

Beaubien EG, Freeland HJ (2000) Spring phenology trends in Alberta, Canada: links to ocean temperature. Int J Biometeorol 44:53–59

Beaubien EG, Hamann A (2011) Plant phenology networks of citizen scientists: recommendations from two decades of experience in Canada. Int J Biometeorol 55:833–841

Bellard C, Bertelsmeier C, Leadley P, Thuiller W, Courchamp F (2012) Impacts of climate change on the future of biodiversity. Ecol Lett 15:365–377

Bennett D, Izanloo A, Edwards J, Kuchel H, Chalmers K, Tester M, Reynolds M, Schnurbusch T, Langridge P (2012) Identification of novel quantitative trait loci for days to ear emergence and flag leaf glaucousness in bread wheat (*Triticum aestivum* L.) population adapted to southern Australian conditions. Theor Appl Genet 124:697–711

Bentley AR, Turner AS, Gosman N, Leigh FJ, Maccaferri M, Dreisigacker S, Greenland A, Laurie DA (2011) Frequency of the photoperiod-insensitive *Ppd-A1a* alleles in tetraploid, hexaploid and synthetic hexaploid wheat germplasm. Plant Breed 130:10–15

Bentley AR, Horsnell R, Werner CP, Turner AS, Rose GA, Bedard C, Howell P, Wilhelm EP, Mackay IJ, Howells RM, Greenland A, Laurie DA, Gosman N (2013) Short, natural, and extended photoperiod response in BC2F4 lines of bread wheat with different Photoperiod-1 (Ppd-1) alleles. J Exp Bot. doi:10.1093/jxb/ert038

Biancardi E, Lewellen RT (2010) Foreword. Sugar Technol 12:179–180

Bielenberg DG (2011) Knowing when not to grow. New Phytol 189:3–5

Blázquez MA, Green R, Nilsson O, Sussman MR, Weigel D (1998) Gibberellins promote flowering of Arabidopsis by activating the *LEAFY* promoter. Plant Cell 10:791–800

Blázquez MA, Ahn JH, Weigel D (2003) A thermosensory pathway controlling flowering time in Arabidopsis thaliana. Nat Genet 33:168–171

Böhlenius H, Huang T, Charbonnel-Campaa L, Brunner AM, Jansson S, Strauss SH, Nilsson O (2006) *CO/FT* regulatory module controls timing of flowering and seasonal growth cessation in trees. Science 312:1040–1043

Boote K, Jones J, Hoogenboom G, Pickering N (1998) The CROPGRO model for grain legumes. In: Tsuji GY, Hoogenboom G, Thornton PK (eds) Understanding options for agricultural production. Kluwer Academic, Dordrecht, pp 99–128

Borras-Gelonch G, Rebetzke GJ, Richards RA, Romagosa I (2012) Genetic control of duration of pre-anthesis phases in wheat (*Triticum aestivum* L.) and relationships to leaf appearance, tillering, and dry matter accumulation. J Exp Bot 63:69–89

Bosemark NO (2006) Genetics and breeding. In: Draycott AP (ed) Sugar beet. Wiley-Blackwell, Chicester, pp 50–88

Bradshaw WE, Holzapfel CM (2001) Genetic shift in photoperiodic response correlated with global warming. Proc Natl Acad Sci USA 98:14509–14511

Brunner A, Nilsson O (2004) Revisiting tree maturation and floral initiation in the poplar functional genomics era. New Phytol 164:43–51

Brunner AM, Rottmann WH, Sheppard LA, Krutovskii K, DiFazio SP, Leonardi S, Strauss SH (2000) Structure and expression of duplicate *AGAMOUS* orthologues in poplar. Plant Mol Biol 44:619–634

Buckler ES, Holland JB, Bradbury PJ, Acharya CB, Brown PJ, Browne C, Ersoz E, Flint-Garcia S, Garcia A, Glaubitz JC, Goodman MM, Harjes C, Guill K, Kroon DE, Larsson S, Lepak NK, Li H, Mitchell SE, Pressoir G, Peiffer JA, Rosas MO, Rocheford TR, Romay MC, Romero S, Salvo S, Villeda HS, Silva HS, Sun Q, Tian F, Upadyayula N, Ware D, Yates H, Yu J, Zhang Z, Kresovich S, McMullen MD (2009) The genetic architecture of maize flowering time. Science 325:714–718

Burn JE, Bangall DJ, Metzeger JD, Dennis ES, Peacock WJ (1993) DNA methylation, vernalization, and the initiation of flowering. Proc Natl Acad Sci USA 90:287–291

Buttner B, Abou-Elwafa SF, Zhang W, Jung C, Müller AE (2010) A survey of EMS-induced biennial *Beta vulgaris* mutants reveals a novel bolting locus which is unlinked to the bolting gene *B*. Theor Appl Genet 121:1117–1131

Campoli C, Drosse B, Searle I, Coupland G, von Korff M (2012) Functional characterization of *HvCO1*, the barley (*Hordeum vulgare*) flowering time ortholog of *CONSTANS*. Plant J 69:868–880

Causier B, Schwarz-Sommer Z, Davies B (2010) Floral organ identity: 20 years of ABCs. Semin Cell Dev Biol 21:73–79

Chandler J, Dean C (1994) Factors affecting the vernalization response and flowering time of late flowering mutants of *Arabidopsis thaliana* L. Heynh. J Exp Bot 278:1279–1288

Chandler J, Wilson A, Dean C (1996) *Arabidopsis* mutants showing altered response to vernalization. Plant J 10:637–644

Charles-Edwards DA, Cockshull KE, Horridge JS, Thornley JHM (1979) A model of flowering in chrysanthemum. Ann Bot 44:557–566

Chen SL, Renvoize SA (2006) Miscanthus. In: Wu ZY, Raven PH, Hong DY (eds) Flora of China, vol 22. Science/Missouri Botanical Garden Press, Beijing/St Louis, pp 581–583

Chen TH, Howe GT, Bradshaw HD (2002) Molecular genetic analysis of dormancy–related traits in poplars. Weed Sci 50:232–240

Chia TYP, Müller AE, Jung C, Mutasa-Göttgens ES (2008) Sugar beet contains a large *CONSTANS-LIKE* gene family including a *CO* homologue that is independent of the early bolting (*B*) gene locus. J Exp Bot 59:2735–2748

Chiang GCK, Barua D, Kramer EM, Amasino RM, Donohue K (2009) Major flowering time gene, *FLOWERING LOCUS C*, regulates seed germination in *Arabidopsis thaliana*. Proc Natl Acad Sci USA 106:11661–11666

Childs KL, Miller FR, Cordonnier-Pratt MM, Pratt LH, Morgan PW, Mullet JE (1997) The sorghum photoperiod sensitivity gene, *Ma3*, encodes a phytochrome B. Plant Physiol 113:611–619

Chiurugwi T, Holmes HF, Qi A, Chia TYP, Hedden P, Mutasa-Göttgens ES (2012) Development of new quantitative physiological and molecular breeding parameters based on the sugar-beet vernalization intensity model. J Agric Sci. doi:10.1017/S0021859612000573

Chory J, Nagpal P, Peto CA (1991) Phenotypic and genetic analysis of *det2*, a new mutant that affects light-regulated seedling development in *Arabidopsis*. Plant Cell 3:445–459

Chouard P (1960) Vernalization and its relationship to dormancy. Annu Rev Plant Physiol 11:191–238

Clark JH, Dean C (1994) Mapping *FRI*, a locus controlling flowering time and vernalization response in *Arabidopsis thaliana*. Mol Gen Genet 242:81–89

Clarke JM, DePauw RM, Thiessen LL (1998) Registration of wheat genetic stocks near-isogenic for photoperiod sensitivity. Crop Sci 38:882

Cleland EE, Chuine I, Menzel A, Mooney HA, Schwartz MD (2007) Shifting plant phenology in response to global change. Trends Ecol Evol 22:357–365

Clifton-Brown JC, Lewandowski I, Andersson B, Basch G, Christian DG, Kjeldsene JB, Jørgensene U, Mortensene JV, Riched AB, Schwarze K-U, Tayebic K, Teixeirac F (2001) Performance of 15 *Miscanthus* genotypes at five sites in Europe. Agron J 93:1013–1019

Cockram J, Jones H, Leigh FJ, O'Sullivan D, Powell W, Laurie DA, Greenland AJ (2007a) Control of flowering time in temperate cereals: genes, domestication, and sustainable productivity. J Exp Bot 58:1231–1244

Cockram J, Mackay IJ, O'Sullivan DM (2007b) The role of double-stranded break repair in the creation of phenotypic diversity at cereal *VRN1* loci. Genetics 177:1–5

Cockram J, Howells R, O'Sullivan DM (2010a) Segmental duplication harbouring group IV *CONSTANS*-like genes in cereals. Genome 53:231–240

Cockram J, White J, Zuluaga DL, Smith D, Comadran J, Macaulay M, Luo Z, Kearsey MJ, Werner P, Harrap D, Tapsell C, Liu H, Hedley PE, Steine N, Schulte D, Steuernagel B, Marshall DF, Thomas WTB, Ramsay L, Mackay I, Balding DJ, The AGOUEB Consortium, Waugh R, O'Sullivan DM (2010b) Genome-wide association mapping to candidate polymorphism resolution in the unsequenced barley genome. Proc Natl Acad Sci USA 107:21611–21616

Cockram J, Thiel T, Steuernagel B, Stein N, Taudien S, Bailey P, O'Sullivan P (2012) Genome dynamics explain the evolution of flowering time CCT domain gene families in the Poaceae. PLoS One 7:e45307

Colasanti J, Yuan Z, Sundaresan V (1998) The *indeterminate* gene encodes a zinc finger protein and regulates a leaf-generated signal required for the transition to flowering in maize. Cell 93:593–603

Collaborative Cross Consortium (2012) The genome architecture of the Collaborative Cross mouse genetic reference population. Genetics 190:389–401

Cooper M, van Eeuwijk FA, Hammer G, Podlich DW, Messina C (2009) Modelling QTL for complex traits: detection and context for plant breeding. Curr Opin Plant Biol 12:231–240

Corbesier L, Vincent C, Jang SH, Fornara F, Fan QZ, Searle I, Giakountis A, Farrona S, Gissot L, Turnbull C, Coupland G (2007) FT protein movement contributes to long-distance signaling in floral induction of *Arabidopsis*. Science 316:1030–1033

Craufurd PQ, Wheeler TR (2009) Climate change and the flowering time of annual crops. J Exp Bot 60:2529–2539

Craufurd PQ, Mahalakshmi V, Bidinger FR, Mukuru SZ, Chantereau J, Omanga PA, Qi A, Roberts EH, Ellis RH, Summerfield RJ, Hammer GL (1999) Adaptation of sorghum: characterisation of genotypic flowering responses to temperature and photoperiod. Theor Appl Genet 99:900–911

Crossa J, de los Campos G, Pérez P, Gianola D, Burgueño J, Araus JL, Makumbi D, Singh RP, Dreisigacker S, Yan J, Arief V, Banziger M, Braun J-J (2010) Prediction of genetic values of quantitative traits in plant breeding using pedigree and molecular markers. Genetics 186:713–724

Danyluk J, Kane NA, Breton G, Limin AE, Fowler DB, Sarhan F (2003) *TaVRT-1*, a putative transcription factor associated with vegetative to reproductive transition in cereals. Plant Physiol 132:1849–1860

Darvasi A, Soller M (1995) Advanced intercross lines, an experimental population for fine genetic mapping. Genetics 141:1199–1207

Davis SJ (2009) Integrating hormones into the floral-transition pathway of *Arabidopsis thaliana*. Plant Cell Environ 32:1201–1210

De Lucia F, Crevillen P, Jones AME, Greb T, Dean C (2008) A PHD-polycomb repressive complex 2 triggers the epigenetic silencing of *FLC* during vernalization. Proc Natl Acad Sci USA 105:16831–16836

Debenham GB (1999) Bolting and flowering mechanisms in sugar beet, *Beta vulgaris*, spp *vulgaris* (L). PhD Thesis, University of Nottingham, Nottingham

Diaz A, Zikhali M, Turner AS, Isaac P, Laurie DA (2012) Copy number variation affecting the *Photoperiod-B1* and *Vernalisation-A1* genes is associated with altered flowering time in wheat (*Triticum aestivum*). PLoS One 7:e33234

Distelfeld A, Tranquilli G, Li C, Yan L, Dubcovsky J (2009a) Genetic and molecular characterization of the *VRN2* loci in tetraploid wheat. Plant Physiol 149:245–257

Distelfeld A, Li C, Dubcovsky J (2009b) Regulation of flowering in temperate cereals. Curr Opin Plant Biol 12:178–184

Dodd AN, Salathia N, Hall A, Kevei E, Toth R, Nagy F, Hibberd JM (2005) Plant circadian clocks increase photosynthesis, growth and survival, and competitive advantage. Science 309:630–633

Doi K, Izawa T, Fuse T, Yamanouchi U, Kubo T, Shimatani Z, Yano M, Yoshimura A (2004) *Ehd1*, a B-type response regulator in rice, confers short-day promotion of flowering and controls *FT-like* gene expression independently of *Hd1*. Genes Dev 18:926–936

Domagalska MA, Schomburg FM, Amasino RM, Vierstra RD, Nagy F, Davis SJ (2007) Attenuation of brassinosteroid signalling enhances *FLC* expression and delays flowering. Development 134:2841–2850

Doyle MR, Davis SJ, Bastow RM, McWatters HG, Kozma-Bognar L, Nagy F, Millar AJ, Amasino RM (2002) The *ELF4* gene controls circadian rhythms and flowering time in *Arabidopsis thaliana*. Nature 419:74–77

Ducrocq S, Giauffret C, Madur D, Combes V, Dumas F, Jouanne S, Coubriche D, Jamin P, Moreau L, Charcosset A (2009) Fine mapping of haplotype structure analysis of a major flowering time quantitative trait locus on maize chromosome 10. Genetics 183:1555–1563

Dunlap JC (1999) Molecular bases for circadian clocks. Cell 96:271–290

Ebana K, Shibaya T, Wu J, Matsubara K, Kanamori H, Yamane H, Yamanouchi U, Mizubayashi T, Kono I, Shomura A, Ito S, Ando T, Hori K, Matsumoto T, Yano M (2011) Uncovering of major genetic factors generating naturally occurring variation in heading date among Asian rice cultivars. Theor Appl Genet 122:1199–1210

Eckenwalder JE (1996) Systematics and evolution of *Populus*. In: Stettler RF, Bradshaw HD Jr, Hellman PE, Hinckley TM (eds) Biology of *Populus* and its implications for management and conservation. NRC Research, Ottawa, ON, pp 7–32

Edwards KD, Anderson PE, Hall A, Salathia NS, Locke JCW, Lynn JR, Straume M, Smith JQ, Millar AJ (2006) *FLOWERING LOCUS C* mediates natural variation in the high-temperature response of the *Arabidopsis* circadian clock. Plant Cell 18:639–650

Elo A, Lemmetyinen J, Turunen ML, Tikka L, Sopanen T (2001) Three MADS-box genes similar to *APETALA1* and *FRUITFULL* from silver birch (*Betula pendula*). Physiol Plant 11:95–103

Elo A, Lemmetyinen J, Novak A, Keinonen K, Porali I, Hassinen M, Sopanen T (2007) *BpMADS4* has a central role in the inflorescence initiation in silver Birch (*Betula pendula* Roth). Physiol Plant 131:149–158

Ergon Å, Fang C, Jørgensen Ø, Aamlid TS, Rognli OA (2006) Quantitative trait loci controlling vernalisation requirement, heading time and number of panicles in meadow fescue (*Festuca pratensis* Huds). Theor Appl Genet 112:232–242

Faure S, Higgins J, Turner A, Laurie DA (2007) The FLOWERING LOCUS T-like gene family in barley (Hordeum vulgare). Genetics 176:599–609

Finnegan EJ, Kovac KA, Jaligot E, Sheldon CC, Peacock WJ, Dennis ES (2005) The downregulation of *FLOWERING LOCUS C* (*FLC*) expression in plants with low levels of DNA methylation and by vernalization occurs by distinct mechanisms. Plant J 44:420–432

Fischer RA, Edmeades GO (2010) Breeding and cereal yield progress. Crop Sci 50:s85–s98

Flachowsky H, Peil A, Sopanen T, Elo A, Hanke V (2007) Overexpression of *BpMADS4* from silver birch (*Betula pendula*) in apple (*Malus* x *domestica*) induces early flowering. Plant Breed 126:137–145

Flachowsky H, Hanke M-V, Peil A, Strauss SH, Fladung M (2009) A review on transgenic approaches to accelerate breeding of woody plants. Plant Breed 128:217–226

Force A, Lynch M, Pickett FB, Amores A, Yan YL, Postlethwait J (1999) Preservation of duplicate genes by complementary, degenerative mutations. Genetics 151:1531–1545

Fornara F, Panigrahi KCS, Gissot L, Sauerbrunn N, Rühl M, Jarillo J, Coupland G (2009) *Arabidopsis* DOF transcription factors act redundantly to reduce *CONSTANS* expression and are essential for a photoperiodic flowering response. Dev Cell 17:75–86

Foster KR, Morgan PW (1995) Genetic regulation of development in *Sorghum bicolor*. IX. The *ma3R* allele disrupts diurnal control diurnal control of gibberellin biosynthesis. Plant Physiol 108:337–343

Fournier-Level A, Korte A, Cooper MD, Nordborg M, Schmitt J, Wilczek AM (2011) A map of local adaptation in *Arabidopsis thaliana*. Science 334:86–89

Frances SA (2006) Development of sugar beet. In: Draycott AP (ed) Sugar beet. Wiley-Blackwell, Chicester, pp 9–29

Franks SJ, Sim S, Wels AE (2007) Rapid evolution of flowering time by an annual plant in response to a climate fluctuation. Proc Natl Acad Sci USA 104:1278–1282

Frewen BE, Chen TH, Howe GT, Davis J, Rohde A, Boerjan W, Bradshaw HD Jr (2000) Quantitative trait loci and candidate gene mapping of bud set and bud flush in *Populus*. Genetics 154:837–845

Fu D, Dunbar M, Dubcovsky J (2007) Wheat *VIN3*-like PHD finger genes are up-regulated by vernalization. Mol Genet Genomics 277:301–313

Fukai S, Cooper M (1995) Development of drought-resistant cultivars using physio-morphological traits in rice. Field Crops Res 40:67–86

Fukuoka S, Nonoue Y, Yano M (2010) Germplasm enhancement by developing advanced plant materials from diverse rice accessions. Breed Sci 60:509–517

Gaut B (2012) *Arabidopsis thaliana* as a model for the genetics of local adaptation. Nat Genet 44:115–116

Gazzani S, Gendall AR, Lister C, Dean C (2003) Analysis of the molecular basis of flowering time variation in *Arabidopsis* accessions. Plant Physiol 132:1107–1114

Gendall AR, Levy YY, Wilson A, Dean C (2001) The *VERNALIZATION 2* gene mediates the epigenetic regulation of vernalization in *Arabidopsis*. Cell 107:525–535

Goddard ME, Hayes BJ (2009) Mapping genes for complex traits in domestic animals and their use in breeding programmes. Nat Rev Genet 10:381–391

Gonzalez FG, Slafer GA, Miralles DJ (2005) Pre-anthesis development and number of fertile florets in wheat as affected by photoperiod sensitivity genes *Ppd-D1* and *Ppd-B1*. Euphytica 146:253–269

Gray LK, Gylander T, Mbogga MS, Chen PY, Hamann A (2011) Assisted migration to address climate change: recommendations for aspen reforestation in western Canada. Ecol Appl 21:1591–1603

Greb T, Mylne JS, Crevillen P, Geraldo N, An H, Gendall AR, Dean C (2007) The PHD finger protein VRN5 functions in the epigenetic silencing of *Arabidopsis FLC*. Curr Biol 17:73–78

Green JO, Corall AJ, Terry RA (1971) Grass species and varieties. Relationships between stage of growth, yield and forage quality. GRI Technical Report No 8. Grassland Research Institute, Hurley, Maidenhead, Berkshire

Green RM, Tingay S, Wang ZY, Tobin EM (2002) Circadian rhythms confer a higher level of fitness to *Arabidopsis* plants. Plant Physiol 129:576–584

Greenup A, Peacock WJ, Dennis ES, Trevaskis B (2009) The molecular biology of seasonal flowering-responses in *Arabidopsis* and the cereals. Ann Bot 103:1165–1172

Griffiths S, Simmonds J, Leverington M, Wang Y, Fish L, Sayers L, Alibert L, Orford S, Wingen L, Herry L, Faure S, Laurie DA, Bilham L, Snape J (2009) Meta-QTL analysis of the genetic control of ear emergence in elite European winter wheat germplasm. Theor Appl Genet 119:383–395

Gu XY, Foley M (2007) Epistatic interactions of three loci regulate flowering time under short and long daylengths in a backcross population of rice. Theor Appl Genet 114:745–754

Hall D, Ma XF, Ingvarsson PK (2011) Adaptive evolution of the *Populus tremula* photoperiod pathway. Mol Ecol 20:1463–1474

Halliday KJ, Salter MG, Thinglass E, Whitelam GC (2003) Phytochrome control of flowering is temperature sensitive and correlates with expression of the floral integrator *FT*. Plant J 33:875–885

Hammer GL, Sinclair TR, Chapman SC, van Oosterom E (2004) On systems thinking, systems biology, and the *in silico* plant. Plant Physiol 134:909–911

Hanano S, Domagalska MA, Nagy F, Davis SJ (2006) Multiple phytohormones influence distinct parameters of the plant circadian clock. Genes Cells 11:1381–1392

Hayama R, Coupland G (2004) The molecular basis of diversity in the photoperiodic flowering responses of *Arabidopsis* and rice. Plant Physiol 135:677–684

Hayama R, Izawa T, Shimamoto K (2002) Isolation of rice genes possibly involved in the photoperiodic control of flowering by a fluorescent differential display method. Plant Cell Physiol 43:494–504

Hayama R, Yokoi S, Tamaki S, Yano M, Shimamoto K (2003) Adaptation of photoperiodic control pathways produces short-day flowering in rice. Nature 422:719–722

He Y, Michaels SD, Amasino RM (2003) Regulation of flowering time by histone acetylation in *Arabidopsis*. Science 302:1751–1754

Heide OM (1994) Control of flowering and reproduction in temperate grasses. New Phytol 128:347–362

Heijde M, Ulm R (2012) UV-B photoreceptor-mediated signalling in plants. Trends Plant Sci 17:230–237

Heliwell CA, Wood CC, Robertson M, James Peacock W, Dennis ES (2006) The *Arabidopsis* FLC protein interacts directly in vivo with *SOC1* and *FT* chromatin and is part of a high-molecular-weight protein complex. Plant J 46:183–192

Hemming MN, Peacock WJ, Dennis ES, Trevaskis B (2008) Low-temperature and daylength cues are integrated to regulate *FLOWERING LOCUS T*, in barley. Plant Physiol 46:183–192

Heslot N, Yang H-P, Sorrells ME, Jannink J-L (2012) Genomic selection in plant breeding: a comparison of models. Crop Sci 52:146–160

Hicks KA, Albertson TM, Wagner DR (2001) *EARLY FLOWERING3* encodes a novel protein that regulates circadian clock function and flowering in *Arabidopsis*. Plant Cell 13:1281–1292

Higgins JA, Bailey PC, Laurie DA (2010) Comparative genomics of flowering time pathways using *Brachypodium distachyon* as a model for the temperate grasses. PLoS One 5:e10065

Hisamatsu T, King RW (2008) The nature of floral signals in *Arabidopsis*. II. Roles for *FLOWERING LOCUS T (FT)* and gibberellin. J Exp Bot 59:3821–3829

Hodkinson TR, Renvoize S (2001) Nomenclature of *Miscanthus* x *giganteus* (Poaceae). Kew Bull 56:759–760

Hodkinson TR, Renvoize SS, Chase MW (1997) Systematics of *Miscanthus*. Asp Appl Biol 49:189–198

Hoecker U, Quail P (2001) The phytochrome A-specific signalling intermediate SPA1 interacts directly with COP1, and constitutive repressor of light signalling in *Arabidopsis*. J Biol Chem 276:38173–38178

Hoenicka H, Nowitzki O, Hanelt D, Fladung M (2008) Heterologous overexpression of the birch *FRUITFULL-like* MADS-box gene *BpMADS4* prevents normal senescence and winter dormancy in *Populus tremula* L. Planta 227:1001–1011

Hoenicka H, Lautner S, Klingberg A, Koch G, El-Sherif F, Lehnhardt D, Zhang B, Burgert I, Odermatt J, Melzer S, Fromm J, Fladung M (2012a) Influence of over-expression of the *Flowering Promoting Factor 1* gene *(FPF1)* from *Arabidopsis* on wood formation in hybrid poplar (*Populus tremula* L. × *P. tremuloides* Michx.). Planta 235:359–373

Hoenicka H, Lehnhardt D, Polak O, Fladung M (2012b) Early flowering and genetic containment studies in transgenic poplar. iForest 5:138–146

Hori K, Kataoka T, Miura K, Yamaguchi M, Saka N, Nakahara T, Sunohara Y, Ebana K, Yano M (2012) Variation in heading date conceals quantitative trait loci for other traits of importance in breeding selection of rice. Breed Sci 62(3):223–234

Horton MW, Hancock AM, Huang YS, Toomajian C, Atwell S, Auton A, Muliyati NW, Platt A, Seperone FG, Vilhjálmsson BJ, Nordburg M, Borevitz JO, Bergelson J (2012) Genome-wide patterns of genetic variation in worldwide *Arabidopsis thaliana* accessions from the RegMap panel. Nat Genet 44:212–216

Howe GT, Bucciaglia PA, Hackett WP, Furnier GR, Cordonnier-Pratt MM, Gardner G (1998) Evidence that the phytochrome gene family in black cottonwood has one *PHYA* locus and two *PHYB* loci but lacks members of the *PHYC/F* and *PHYE* subfamilies. Mol Biol Evol 15:160–175

Hsu CY, Liu Y, Luthe DS, Yuceer C (2006) Poplar *FT2* shortens the juvenile phase and promotes seasonal flowering. Plant Cell 18:1846–1861

Hsu CY, Adams JP, Kim H, No K, Ma C, Strauss SH, Drnevich J, Vandervelde L, Ellis JD, Rice BM, Wickett N, Gunter LE, Tuskan GA, Brunner AM, Page GP, Barakat A, Carlson JE, DePamphilis CW, Luthe DS, Yuceer C (2011) *FLOWERING LOCUS T* duplication coordinates reproductive and vegetative growth in perennial poplar. Proc Natl Acad Sci USA 108:10756–10761

Huang W, Pérez-García P, Pokhilko A, Millar AJ, Antoshechkin I, Riechmann JL, Mas P (2012a) Mapping the core of the *Arabidopsis* circadian clock defines the network structure of the oscillator. Science 336:75–79

Huang BE, George AW, Forrest KL, Kilian A, Hayden MJ, Morell MK, Cavanagh CR (2012b) A multiparent advanced generation inter-cross population for genetic analysis in wheat. Plant Biotechnol J 10:826–839

Imaizumi T, Schultz TF, Harmon FG, Ho LA, Kay SA (2005) FKF1 F-box protein mediates cyclic degradation of a repressor of *CONSTANS* in *Arabidopsis*. Science 309:293–297

Ingvarsson PK, García MV, Hall D, Luquez V, Jansson S (2006) Clinal variation in *phyB2*, a candidate gene for day-length-induced growth cessation and bud set, across a latitudinal gradient in European aspen (*Populus tremula*). Genetics 172:1845–1853

Inoue M, Gao ZS, Hirata M, Fujimori M, Cai HW (2004) Construction of a high-density linkage map of Italian ryegrass (*Lolium multiflorum* Lam.) using restriction fragment length polymorphism, amplified fragment length polymorphism, and telomeric repeat associated sequence markers. Genome 47:57–65

Irish VF (2010) The flowering of *Arabidopsis* flower development. Plant J 61:1014–1028

Ishimaru T, Hirabayashi H, Kuwagata T, Ogawa T, Kondo M (2012) The early-morning flowering trait of rice reduces spikelet sterility under windy and elevated temperature conditions at anthesis. Plant Prod Sci 15:19–22

Itoh H, Nonoue Y, Yano M, Izawa T (2010) A pair of floral regulators sets critical day length for *Hd3a* florigen expression in rice. Nat Genet 42:635–638

Izawa T (2007) Adaptation of flowering-time by natural and artificial selection in Arabidopsis and rice. J Exp Bot 58:3091–3097

Izawa T, Oikawa T, Tokutomi S, Okuno K, Shimamoto K (2000) Phytochromes confer the photoperiodic control of flowering in rice (a short-day plant). Plant J 22:391–399

Jagadish SVK, Craufurd PQ, Wheeler TR (2008) Phenotyping parents of mapping populations of rice for heat tolerance during flowering. Crop Sci 48:1140–1146

Jannink JL, Lorenz AJ, Iwata H (2010) Genomic selection in plant breeding: from theory to practice. Brief Funct Genomics 9:166–177

Jeger MJ, Pautasso M (2008) Plant disease and global change – the importance of long-term data sets. New Phytol 177:8–11

Jensen LB, Andersen JR, Frei U, Xing YZ, Taylor C, Holm PB, Lubberstedt TL (2005) QTL mapping of vernalization response in perennial ryegrass (*Lolium perenne* L.) reveals co-location with an orthologue of wheat *VRN1*. Theor Appl Genet 110:527–536

Jensen E, Farrar K, Thomas-Jones S, Hastings A, Donnison I, Clifton-Brown J (2011a) Characterization of flowering time diversity in *Miscanthus* species. Glob Change Biol Bioenergy 3:387–500

Jensen E, Squance M, Hastings A, Jones S, Farrar K, Huang L, King R, Clifton-Brown J, Donnison I (2011b) Understanding the value of hydrothermal time on flowering in *Miscanthus* species. Asp Appl Biol 112:181–189

Johanson U, West J, Lister C, Michaels S, Amasino R, Dean C (2000) Molecular analysis of *FRIGIDA*, a major determinant of natural variation in *Arabidopsis* flowering time. Science 290:344–347

Jones JW, Hoogenboom G, Porter CH, Boote KJ, Batchelor WD, Hunt LA, Wilkens PW, Singh U, Gijsman AJ, Ritchie JT (2003) The DSSAT cropping system model. Eur J Agron 18:235–265

Julian C, Rodrigo J, Herrero M (2011) Stamen development and winter dormancy in apricot (*Prunus armeniaca*). Ann Bot 108:617–625

Jung C, Muller AE (2009) Flowering time control and applications in plant breeding. Trends Plant Sci 14:563–573

Junttila O (1980) Flower bud differentiation in *Salix pentandra* as affected by photoperiod, temperature and growth regulators. Physiol Plant 49:127–134

Kania T, Russenberger D, Peng S, Apel K, Melzer S (1997) *FPF1* promotes flowering in Arabidopsis. Plant Cell 9:1327–1338

Karsai I, Szűcs P, Mészáros K, Filichkina T, Hayes PM, Skinner JS, Láng L, Bedö Z (2005) The *Vrn-H2* locus is a major determinant of flowering time in a facultative x winter growth habit barley (*Hordeum vulgare* L.) mapping population. Theor Appl Genet 110:1458–1466

Keller SR, Levsen N, Ingvarsson PK, Olson MS, Tiffin P (2011) Local selection across a latitudinal gradient shapes nucleotide diversity in balsam poplar, *Populus balsamifera* L. Genetics 188:941–952

Keller SR, Levsen N, Olson MS, Tiffin P (2012) Local adaptation in the flowering time gene network of balsam poplar, *Populus balsamifera* L. Mol Biol Evol 29:3143–3152

Khan S, Rowe SC, Harmon FG (2010) Coordination of the maize transcriptome by a conserved circadian clock. BMC Plant Biol 10:126

Kim WY, Fujiwara S, Suh SS, Kim J, Kim Y, Han L, David K, Putterill J, Nam HG, Somers DE (2007) ZEITLUPE is a circadian photoreceptor stabilized by GIGANTEA in blue light. Nature 449:356–360

Kojima S, Takahashi Y, Kobayashi Y, Monna L, Sasaki T, Araki T, Yano M (2002) *Hd3a*, a rice ortholog of the Arabidopsis *FT* gene, promotes transition to flowering downstream of *Hd1* under short-day conditions. Plant Cell Physiol 43:1096–1105

Komiya R, Ikegami A, Tamaki S, Yokoi S, Shimamoto K (2008) *Hd3a* and *RFT1* are essential for flowering in rice. Development 135:767–774

Komiya R, Yokoi S, Shimamoto K (2009) A gene network for long-day flowering activates *RFT1* encoding a mobile flowering signal in rice. Development 136:3443–3450

Kover PX, Valdar W, Trakalo J, Scarcelli N, Ehrenreich IM, Purugganan MD, Durrant C, Mott R (2009) A multiparent advanced generation inter-cross to fine-map quantitative traits in *Arabidopsis thaliana*. PLoS Genet 5:e1000551

Kuromori T, Takahashi S, Kondou Y, Shinozaki K, Matsui M (2009) Phenome analysis in plant species using loss-of-function and gain-of-function mutants. Plant Cell Physiol 50:1215–1231

Kvaalen H, Johnsen O (2007) Timing of bud set in *Picea abies* is regulated by a memory of temperature during zygotic and somatic embryogenesis. New Phytol 177:49–59

Langridge J (1957) The aseptic culture of *Arabidopsis thaliana* (L.) Heynh. Aust J Biol Sci 10:243–252

Lantican MA, Pingali PL, Rajaram S (2003) Is research on marginal lands catching up? The case of unfavourable wheat growing environments. Agric Econ 29:353–361

Lee I, Bleecker A, Amasino R (1993) Analysis of naturally occurring late flowering in *Arabidopsis thaliana*. Mol Gen Genet 237:171–176

Lee I, Aukerman M, Gore S, Lohman K, Michaels S, Weaver L, John M, Feldmann K, Amasino R (1994) Isolation of *LUMINIDEPENDENS*: a gene involved in the control of flowering time in *Arabidopsis*. Plant Cell 6:75–83

Lee J, Yoo S, Park S, Hwang I, Lee J, Ahn J (2007) Role of *SVP* in the control of flowering time by ambient temperature in *Arabidopsis*. Genes Dev 21:397–402

Lee J, Oh M, park H, Lee I (2008) SOC1 translocated to the nucleus by interaction with AGL24 directly regulates *LEAFY*. Plant J 55:832–843

Leida C, Romeu JF, García B, Ríos G, Badenes ML (2012) Gene expression analysis of chilling requirements for flower bud break in peach. Plant Breed 131:329–334

Leseberg CH, Li A, Kang H, Duvall M, Mao L (2006) Genome-wide analysis of the MADS-box gene family in *Populus trichocarpa*. Gene 378:84–94

Levy YY, Mesnage S, Mylne JS, Gendall AR, Dean C (2002) Multiple roles of *Arabidopsis VRN1* in vernalization and flowering time control. Science 297:243–246

Lewandowski I, Clifton-Brown JC, Scurlock JMO, Huisman W (2000) *Miscanthus*: European experience with a novel energy crop. Biomass Bioenergy 19:209–227

Lewis S, Faricelli ME, Appendino ML, Valárik M, Dubcovsky J (2008) The chromosome region including the earliness per se locus $Eps-A^m1$ affects the duration of early development phases and spikelet number in diploid wheat. J Exp Bot 59:3595–3607

Li D, Liu C, Shen L, Robertson M, Helliwell CA, Ito T, Meyerowitz E, Yu H (2008) A repressor complex governs the integration of flowering signals in *Arabidopsis*. Dev Cell 15:110–120

Li J, Li Y, Chen S, An L (2010) Involvement of brassinosteroid signals in the floral-induction network of *Arabidopsis*. J Exp Bot 61:4221–4230

Li J, Terzaghi W, Deng XW (2012) Genomic basis for light control of plant development. Protein Cell 3:106–116

Lim J, Moon YH, An G, Jang SK (2000) Two rice MADS domain proteins interact with *OsMADS1*. Plant Mol Biol 44:513–527

Lim MH, Kim J, Kim YS, Chung KS, Sea YH, Lee I, Hong CB, Kim HJ, Park CM (2004) A new *Arabidopsis* gene, *FLK*, encodes an RNA binding protein with K homology motifs and regulates flowering time via *FLOWERING LOCUS C*. Plant Cell 16:731–740

Lin HX, Ashikari M, Yamanouchi U, Sasaki T, Yano M (2002) Identification and characterization of a quantitative trait locus, *Hd9*, controlling heading date in rice. Breed Sci 52:35–41

Lin HX, Liang ZW, Sasaki T, Yano M (2003) Fine mapping and characterization of quantitative trait loci *Hd4* and *Hd5* controlling heading date in rice. Breed Sci 53:51–59

Lin R, Ding L, Casola C, Ripoll DR, Feschotte C, Wang H (2007) Transposase-derived transcription factors regulate light signalling in *Arabidopsis*. Science 318:1302–1305

Lingle SE (1987) Sucrose metabolism in the primary culm of sweet sorghum during development. Crop Sci 27:1214–1219

Linkosalo T, Häkkinen R, Terhivuo J, Tuomenvirta H, Hari P (2009) The time series of flowering and leaf bud burst of boreal trees (1846–2005) support the direct temperature observations of climatic warming. Agric Meteorol 149:453–461

Liu C, Chen H, Er H, Soo H, Kumar P, Han J, Liou Y, Yu H (2008) Direct interaction of AGL24 and SOC1 integrates flowering signals in *Arabidopsis*. Development 135:1481–1491

Lobell DB, Field CB (2007) Global scale climate-crop yield relationships and the impacts of recent warming. Environ Res Lett 2:014002

Lukac M, Gooding MJ, Griffiths S, Jones HE (2012) Asynchronous flowering and within-plant flowering diversity in wheat and the implications for crop resilience to heat. Ann Bot 109:843–850

Luquez V, Hall D, Albrectsen BR, Karlsson J, Ingvarsson P, Jansson S (2008) Natural phenological variation in aspen (*Populus tremula*): the SwAsp collection. Tree Genet Genomes 4:279–292

Ma X-F, Jensen E, Alexandrov N, Troukhan M, Zhang L, Thomas-Jones S, Farrar K, Clifton-Brown J, Donnison I, Swaller T, Flavell R (2012) High resolution genetic mapping by genome sequencing reveals genome duplication and tetraploid genetic structure of the diploid *Miscanthus sinensis*. PLoS One 7:1–11

Maas L, McClung A, McCouch S (2010) Dissection of a QTL reveals an adaptive, interacting gene complex associated with transgressive variation for flowering time in rice. Theor Appl Genet 120:895–908

Mackay I, Powell W (2007) Methods for linkage disequilibrium mapping in crops. Trends Plant Sci 12:57–63

Mackay I, Horwell A, Garner J, White J, McKee J, Philpott H (2011) Reanalyses of the historical series of UK variety trials to quantify the contributions of genetic and environmental factors to trends and variability in yield over time. Theor Appl Genet 122:225–238

Macknight R, Bancroft I, Page T, Lister C, Schmidt R, Love K, Westphal L, Murphy G, Sherson S, Cobbett C, Dean C (1997) *FCA*, a gene controlling flowering time in *Arabidopsis*, encodes a protein containing RNA-binding domains. Cell 89:737–745

Major DJ, Rood SB, Miller FR (1990) Temperature and photoperiod effects mediated by the sorghum maturity genes. Crop Sci 30:305–310

Manzano D, Marquardt S, Jones AM, Barule I, Li F, Dean C (2009) Altered interactions within FY/AtCPSF complexes required for *Arabidopsis* DAC-mediated chromatin silencing. Proc Natl Acad Sci USA 106:8772–8777

Martinez-Zapater J, Somerville CR (1990) Effect of light quality and vernalization on late-flowering mutants of *Arabidopsis thaliana*. Plant Physiol 92:770–776

Matsubara K, Kono I, Hori K, Nonoue Y, Ono N, Shomura A, Mizubayashi T, Yamamoto S, Yamanouchi U, Shirasawa K, Nishio T, Yano M (2008a) Novel QTLs for photoperiodic flowering revealed by using reciprocal backcross inbred lines from crosses between *japonica* rice cultivars. Theor Appl Genet 117:935–945

Matsubara K, Yamanouchi U, Wang ZX, Minobe Y, Izawa T, Yano M (2008b) *Ehd2*, a rice ortholog of the maize *INDETERMINATE1* gene, promotes flowering by up-regulating *Ehd1*. Plant Physiol 148:1425–1435

Matsubara K, Ogiso-Tanaka E, Hori K, Ebana K, Ando T, Yano M (2012) Natural variation in *Hd17*, a homolog of Arabidopsis *ELF3* that is involved in rice photoperiodic flowering. Plant Cell Physiol 53:709–716

Méndez-Vigo B, Picó FX, Ramiro M, Martínez-Zapater JM, Alonso-Blanco C (2011) Altitudinal and climatic adaptation is mediated by flowering traits and *FLC*, *FRI* and *PHYC* genes in *Arabidopsis*. Plant Physiol 157:1942–1955

Mertz O, Halsnaes K, Olesen JE, Rasmussen K (2009) Adaptation to climate change in developing countries. Environ Manag 43:743–752

Messina CD, Jones JW, Boote KJ, Vallejos CE (2006) A gene-based model to simulate soybean development and yield responses to environment. Crop Sci 46:456–466

Meuwissen THE, Hayes BJ, Goddard ME (2001) Prediction of total genetic value using genome-wide dense marker maps. Genetics 157:1819–1829

Michaels SD, Amasino RM (1999) *FLOWERING LOCUS C* encodes a novel MADS domain protein that acts as a repressor of flowering. Plant Cell 13:935–941

Michaels SD, Amasino RM (2001) Loss of *FLOWERING LOCUS C* activity eliminates the late-flowering phenotype of *FRIGIDA* and autonomous pathway mutants but not responsiveness to vernalization. Plant Cell 13:935–941

Michaels SD, He Y, Scortecci KS, Amasino RM (2003) Attenuation of *FLOWERING LOCUS C* activity as a mechanism for the evolution of a summer-annual flowering behaviour in *Arabidopsis*. Proc Natl Acad Sci USA 100:10102–10107

Mirouze M, Paszkowski J (2011) Epigenetic contribution to stress adaptation in plants. Curr Opin Plant Biol 14:267–274

Mohamed R, Wang CT, Ma C, Shevchenko O, Dye SJ, Puzey JR, Etherington E, Sheng X, Meilan R, Strauss SH, Brunner AM (2010) *Populus CEN/TFL1* regulates first onset of flowering, axillary meristem identity and dormancy release in *Populus*. Plant J 62:674–688

Moon J, Suh S, Lee H, Choi K, Hong C, Paek N, Kim S, Lee I (2003) The *SOC1* MADS-box gene integrates vernalization and gibberellins signals for flowering in *Arabidopsis*. Plant J 35:613–623

Mott R, Talbot CJ, Turri MG, Collins AC, Flint J (2000) A method for fine mapping quantitative trait loci in outbred animal stocks. Proc Natl Acad Sci USA 97:12649–12654

Munerati O (1931) L'eredita della tendenza alla annualita nella commune barbabietola coltivata. Z Zuchtung Reihe A, Pflanzenzucht 17:84–89

Murphy RL, Klein RR, Morishige DT, Brady JA, Rooney WL, Miller FR, Dugas DV, Klein PE, Mullet JE (2011) Coincident light and clock regulation of *pseudoresponse regulator protein 37* (*PRR37*) controls photoperiodic flowering in sorghum. Proc Natl Acad Sci USA 108:16469–16474

Muszynski MG, Dam T, Li B, Shirbroun DM, Hou Z, Breggermann E, Archibald R, Ananiev EV, Danilevskaya OA (2006) *Delayed flowering 1* encodes a basic leucine zipper protein that mediates floral inductive signals at the shoot apex in maize. Plant Physiol 142:1523–1536

Mutasa-Göttgens ES, Hedden P (2009) Gibberellin as a factor in floral regulatory networks. J Exp Bot 60:1979–1989

Mutasa-Göttgens E, Qi A, Mathews A, Thomas S, Phillips A, Hedden P (2008) Modification of gibberellin signalling (metabolism and signal transduction) in sugar beet: analysis of potential targets for crop improvement. Transgenic Res 18:301–308

Mutasa-Göttgens ES, Qi A, Zhang W, Schulze-Buxloh G, Jennings A, Hohmann W, Müller AE, Hedden P (2010) Bolting and flowering control in sugar beet: relationships and effects of gibberellin, the bolting gene B and vernalization. AoB Plants 2010:plq012

Mutasa-Göttgens ES, Joshi A, Holmes HF, Hedden P, Göttgens B (2012) A new RNASeq-based reference transcriptome for sugar beet and its application in transcriptome-scale analysis of vernalization and gibberellin responses. BMC Genomics 13:99

Nakagawa M, Shimamoto K, Kyozuka J (2002) Overexpression of *RCN1* and *RCN2*, rice *TERMINAL FLOWER 1/CENTRORADIALIS* homologs, confers delay of phase transition and altered panicle morphology in rice. Plant J 29:743–750

Nakamichi N (2011) Molecular mechanisms underlying the *Arabidopsis* circadian clock. Plant Cell Physiol 52:1709–1718

Nicotra AB, Atkin OK, Bonser SP, Davidson AM, Finnegan EJ, Mathesius U, Poot P, Purugganan MD, Richards CL, Valladares F, van Kleunen M (2010) Plant phenotypic plasticity in a changing climate. Trends Plant Sci 15:684–692

Nonoue Y, Fujino K, Hirayama Y, Yamanouchi U, Lin S, Yano M (2008) Detection of quantitative trait loci controlling extremely early heading in rice. Theor Appl Genet 116:715–722

O'Toole JC (1982) Adaptation of rice to drought-prone environments. Drought resistance in crops with emphasis on rice. International Rice Research Institute, Los Baños, pp 195–213

Ogiso E, Takahashi Y, Sasaki T, Yano M, Izawa T (2010) The role of casein kinase II in flowering time regulation has diversified during evolution. Plant Physiol 152:808–820

Oliver SN, Finnegan EJ, Dennis ES, Peacock WJ, Trevaskis B (2009) Vernalization-induced flowering in cereals is associated with changes in histone methylation at the *VERNALIZATION1* gene. Proc Natl Acad Sci USA 106:8386–8391

Olsen JE, Juntilla O, Nilsen J, Eriksson ME, Martinussen I, Olsson O, Sandberg G, Moritz T (1997) Ectopic expression of oat *phytochrome A* in hybrid aspen changes critical daylength for growth and prevents cold acclimatization. Plant J 12:1339–1350

Olson SN, Ritter K, Rooney W, Kemanian A, McCarl BA, Zhang Y, Hall S, Packer D, Mullet J (2012) High biomass yield energy sorghum: developing a genetic model for C4 grass bioenergy plants. Biofuel Bioprod Biorefin. doi: 10.1002/bbb.1357

Osnato M, Castillejo C, Matías-Hernández L, Pelaz S (2012) *TEMPRANILLO* genes link photoperiod and gibberellin pathways to control flowering time in *Arabidopsis*. Nat Commun 3:808

Owen FV, Stout M (1940) Photothermal induction of flowering in sugar beet. J Agric Res 61:101–124

Owens JN (1995) Constraints to seed production: temperate and tropical forest trees. Tree Physiol 15:477–484

Panella L (2010) Sugar beet as an energy crop. Sugar Technol 12:288–293

Paux E, Sourdille P, Mackay I, Feuillet C (2011) Sequence-based marker development in wheat: advances and applications to breeding. Biotechnol Adv 30(5):1071–1088

Peng LT, Shi ZY, Li L, Shen GZ, Zhang JL (2008) Overexpression of transcription factor *OsLFL1* delays flowering time in *Oryza sativa*. J Plant Physiol 165:876–885

Phillips NG, Buckley TN, Tissue DT (2008) Capacity of old trees to respond to environmental change. J Integr Plant Biol 50:1355–1364

Pin PA, Benlloch R, Bonnet D, Wremerth-Weich E, Kraft T, Gielen JJL, Nilsson O (2010) An antagonistic pair of *FT* homologs mediates the control of flowering time in sugar beet. Science 330:1397–1400

Pin PA, Zhang W, Vogt SH, Dally N, Büttner B, Schulze-Buxloh G, Jelly NS, Chia TYP, Mutasa-Göttgens ES, Dohm JC, Himmelbauer H, Weisshaar B, Kraus J, Gielen JJL, Lommel M, Weyens G, Wahl B, Schechert A, Nilsson O, Jung C, Kraft T, Müller AE (2012) A pseudo-response regulator gene controls life cycle adaptation in beet. Curr Biol 22:1–7

Poethig RS (2003) Phase change and the regulation of developmental timing in plants. Science 301:334–336

Porri A, Torti S, Romera-Branchat M, Coupland G (2012) Spatially distinct regulatory roles for gibberellins in the promotion of flowering of *Arabidopsis* under long photoperiods. Development 139:2198–2209

Purwestri YA, Ogaki Y, Tamaki S, Tsuji H, Shimamoto K (2009) The 14-3-3 protein *GF14c* acts as a negative regulator of flowering in rice by interacting with the florigen *Hd3a*. Plant Cell Physiol 50:429–438

Putterill J, Robson F, Lee K, Simon R, Coupland G (1995) The *CONSTANS* gene of *Arabidopsis* promotes flowering and encodes a protein showing similarities to zinc finger transcription-factors. Cell 80:847–857

Quinby JR (1967) The genetic control of flowering and growth in sorghum. Adv Agron 25:125–162

Reeves PA, He Y, Schmitz RJ, Amasino RM, Panella LW, Richards CM (2007) Evolutionary conservation of the *FLOWERING LOCUS C*-mediated vernalization response: evidence from the sugar beet (*Beta vulgaris*). Genetics 176:295–307

Reynolds MP, Ortiz R (2010) Adapting crops to climate change: a summary. In: Reynolds MP (ed) Climate change and crop production. CABI, Wallingford, Oxfordshire, pp 1–8

Reynolds MP, Hays D, Chapman S (2010) Breeding for adaptation to heat and drought stress. In: Reynolds MP (ed) Climate change and crop production. CABI, Wallingford, Oxfordshire, pp 71–91

Rigby JR, Porporato A (2008) Spring frost risk in a changing climate. Geophys Res Lett 35: L12703

Rodgers-Melnick E, Mane SP, Dharmawardhana P, Slavov GT, Crasta OR, Strauss SH, Brunner AM, Difazio SP (2012) Contrasting patterns of evolution following whole genome versus tandem duplication events in *Populus*. Genome Res 22:95–105

Rohde A, Junttila O (2008) Remembrances of an embryo: long-term effects on phenology traits in spruce. New Phytol 177:2–5

Rooney WL, Aydin S (1999) Genetic control of a photoperiod-sensitive response in *Sorghum bicolor* (L.) Moench. Crop Sci 39:397–400

Rooney WL, Blumenthal J, Bean B, Mullet JW (2007) Designing sorghum as a dedicated bioenergy feedstock. Biofuel Bioprod Biorefin 1:147–157

Rosenzweig C, Karoly D, Vicarelli M, Neofotis P, Wu Q, Casassa G, Menzel A, Root TL, Estrella N, Seguin B, Tryjanowski P, Liu C, Rawlins S, Imeson A (2008) Attributing physical and biological impacts to anthropogenic climate change. Nature 453:353–358

Rottmann WH, Meilan R, Sheppard LA, Brunner AM, Skinner JS, Ma C, Cheng S, Jouanin L, Pilate G, Strauss SH (2000) Diverse effects of overexpression of *LEAFY* and *PTLF*, a poplar (*Populus*) homolog of *LEAFY/FLORICAULA*, in transgenic poplar and *Arabidopsis*. Plant J 22:235–245

Rounsaville TJ, Touchell DH, Ranney TC (2011) Fertility and reproductive pathways in diploid and triploid *Miscanthus sinensis*. HortScience 46:1353–1357

Ryu CH, Lee S, Cho LH, Kim SL, Lee YS, Choi SC, Jeong HJ, Yi J, Park SJ, Han CD, An G (2009) *OsMADS50* and *OsMADS56* function antagonistically in regulating long day (LD)-dependent flowering in rice. Plant Cell Environ 32:1412–1427

Saito H, Ogiso-Tanaka E, Okumoto Y, Yoshitake Y, Izumi H, Yokoo T, Matsubara K, Hori K, Yano M, Inoue H, Tanisaka T (2012) *Ef7* encodes an ELF3-like protein and promotes rice flowering by negatively regulating the floral repressor gene *Ghd7* under both short- and long-day conditions. Plant Cell Physiol 53:717–728

Salvi S, Sponza G, Morgante M, Tomes D, Niu X, Fengler KA, Meeley R, Ananiev EV, Svitashev S, Bruggemann Em Li B, Hainey CF, Radocic S, Zaina G, Rafalski J-A, Tingey SV, Miao G-H, Phillips RL, Tuberosa R (2007) Conserved noncoding genomic sequences associated with a flowering-time quantitative trait locus in maize. Proc Natl Acad Sci USA 104:11376–11381

Sánchez-Bermejo E, Méndez-Vigo B, Picó FZ, Martínez-Zapater JM, Alonso-Blanco C (2012) Novel natural alleles at *FLC* and *LVR* loci account for enhanced vernalization responses in *Arabidopsis thaliana*. Plant Cell Environ 35(9):1672–1684

Shaw LM, Turner AS, Laurie DA (2012) The impact of photoperiod insensitive *Ppd-1a* mutations on the photoperiod pathway across the three genomes of hexaploid wheat (*Triticum aestivum*). Plant J 71:71–84

Sheldon CC, Burn JE, Perez PP, Metzger J, Edwards JA, Peacock WJ, Dennis ES (1999) The *FLF* MADS box gene: a repressor of flowering in *Arabidopsis* regulated by vernalization and methylation. Plant Cell 11:445–458

Sheldon CC, Rouse DT, Finnegan EJ, Peacock WJ, Dennis ES (2000) The molecular basis of vernalization: the central role of *FLOWERING LOCUS C* (*FLC*). Proc Natl Acad Sci USA 97:3753–3758

Shibaya T, Nonoue Y, Ono N, Yamanouchi U, Hori K, Yano M (2011) Genetic interactions involved in the inhibition of heading by heading date QTL, *Hd2* in rice under long-day conditions. Theor Appl Genet 123:1133–1143

Shimada S, Ogawa T, Kitagawa S, Suzuki T, Ikari C, Shitsukawa N, Abe T, Kawahigashi H, Kikuchi R, Handa H, Murai K (2009) A genetic network of flowering-time genes in wheat leaves, in which an APETALA1/FRUITFULL-like gene, VRN1, is upstream of FLOWERING LOCUS T. Plant J 58:668–681

Shindo C, Aranzana MJ, Lister C, Baxter C, Nicholls C, Nordburg M, Dean C (2005) Role of *FRIGIDA* and *FLOWERING LOCUS C* in determining variation in flowering time of *Arabidopsis*. Plant Physiol 138:1163–1173

Shinozuka H, Hisano H, Ponting RC, Cogan NOI, Jones ES, Forster JW, Yamada T (2005) Molecular cloning and genetic mapping of perennial ryegrass casein protein kinase 2 alpha-subunit genes. Theor Appl Genet 112:167–177

Simpson GG, Dean C (2002) Arabidopsis, the Rosetta Stone of flowering time? Science 296:285–289

Simpson GG, Dijkwel PP, Quesada V, Henderson I, Dean C (2003) FY is and RNA 3' end-processing factor that interacts with FCA to control the *Arabidopsis* floral transition. Cell 113:3–25

Skøt L, Humphreys MO, Armstead I, Heywood S, Skøt KP, Sanderson R, Thomas ID, Chorlton KH, Hamilton NRS (2005) An association mapping approach to identify flowering time genes in natural populations of *Lolium perenne* (L.). Mol Breed 15:233–245

Skøt L, Humphreys J, Humphreys MO, Thorogood D, Gallagher J, Sanderson R, Armstead IP, Thomas ID (2007) Association of candidate genes with flowering time and water-soluble carbohydrate content in *Lolium perenne* (L.). Genetics 177:535–547

Skøt L, Thomas A, Skøt K, Thorogood D, Latypova G, Asp T, Armstead I (2011) Allelic variation in the *Lolium perenne* L. (perennial ryegrass) *FLOWERING LOCUS T* (*LpFT3*) gene is associated with changes in flowering time across a range of germplasm populations. Plant Physiol 155:1013–1022

Song Y, Ma K, Bo W, Zhang Z, Zhang D (2012) Sex-specific DNA methylation and gene expression in andromonoecious poplar. Plant Cell Rep 31:1393–1405

Soussana J-F, Graux A-I, Tubiello FN (2010) Improving the use of modelling for projections of climate change impacts on crops and pastures. J Exp Bot 61:2217–2228

Srikanth A, Schmid M (2011) Regulation of flowering time: all roads lead to Rome. Cell Mol Life Sci 68:2013–2037

Stenøien HK, Fenster CB, Kuittinen H, Savolainen O (2002) Quantifying latitudinal clines to light responses in natural populations of *Arabidopsis thaliana* (Brassicaceae). Am J Bot 89:1064–1068

Stinchcombe JR, Weinig C, Ungerer M, Olsen KM, Mays C, Halldorsdottir SS, Purugganan MD, Schmitt J (2004) A latitudinal cline in flowering time in *Arabidopsis thaliana* modulated by the flowering time gene *FRIGIDA*. Proc Natl Acad Sci USA 101:4712–4717

Suárez-López P, Wheatley K, Robson F, Onouchi H, Valverde F, Coupland G (2001) *CONSTANS* mediates between the circadian clock and the control of flowering in *Arabidopsis*. Nature 410:1116–1120

Sung S, Amasino RM (2004) Vernalization in *Arabidopsis thaliana* is mediated by the PHD finger protein VIN3. Nature 427:159–164

Sung S, Schmitz RJ, Amasino RM (2006) A PHD finger protein involved in both the vernalization and photoperiod pathways in *Arabidopsis*. Genes Dev 20:3244–3248

Takahashi Y, Shomura A, Sasaki T, Yano M (2001) *Hd6*, a rice quantitative trait locus involved in photoperiod sensitivity, encodes the α subunit of protein kinase CK2. Proc Natl Acad Sci USA 98:7922–7927

Takeuchi Y, Ebitani T, Yamamoto T, Sato H, Ohta H, Hirabayashi H, Kato H, Ando I, Nemoto H, Imbe T, Yano M (2006) Development of isogenic lines of rice cultivar Koshihikari with early and late heading by marker-assisted selection. Breed Sci 56:405–413

64 A.R. Bentley et al.

Tamaki S, Matsuo S, Wong HL, Yokoi S, Shimamoto K (2007) Hd3a protein is a mobile flowering signal in rice. Science 316:1033–1036
Tanino KK, Kalcsits L, Silim S, Kendall E, Gray GR (2010) Temperature-driven plasticity in growth cessation and dormancy development in deciduous woody plants: a working hypothesis suggesting how molecular and cellular function is affected by temperature during dormancy induction. Plant Mol Biol 73:49–65
Thornley JHM, Johnson IR (1990) Plant and crop modelling: a mathematical approach to plant and crop physiology. Clarendon/Oxford University Press, Oxford/New York
Trevaskis B (2010) The central role of the *VERNALIZATION1* gene in the vernalization response of cereals. Funct Plant Biol 37:479–487
Trevaskis B, Bagnall DJ, Ellis MH, Peacock WJ, Dennis ES (2003) MADS box genes control vernalization-induced flowering in cereals. Proc Natl Acad Sci USA 100:13099–13104
Tsuji H, Taoka KI, Shimamoto K (2011) Regulation of flowering in rice: two florigen genes, a complex gene network, and natural variation. Curr Opin Plant Biol 14:45–52
Turner A, Beales J, Faure S, Dunford RP, Laurie DA (2005) The pseudo-response regulator *Ppd-H1* provides adaptation to photoperiod in barley. Science 310:1031–1034
Uga Y, Nonoue Y, Liang Z, Lin H, Yamamoto S, Yamanouchi U, Yano M (2007) Accumulation of additive effects generates a strong photoperiod sensitivity in the extremely late-heading rice cultivar 'Nona Bokra'. Theor Appl Genet 114:1457–1466
Uptmoor R, Schrag T, Stützel H, Esch E (2008) Crop model based QTL analysis across environments and QTL based estimation of time to floral induction and flowering in *Brassica oleracea*. Mol Breed 21:205–216
Uptmoor R, Osei-Kwarteng M, Gürtler S, Stützel H (2009) Modelling the effects of drought stress on leaf development in a *Brassica oleracea* doubled haploid population using two-phase linear functions. J Am Soc Hortic Sci 134:543–552
Uptmoor R, Li J, Schrag T, Stützel H (2012) Prediction of flowering time in *Brassica oleracea* using a quantitative trait loci-based phenology model. Plant Biol 14:179–189
Valdar W, Solberg LC, Gauguier D, Burnett S, Klenerman P, Cookson WO, Taylor MS, Rawlins JNP, Mott R, Flint J (2006) Genome-wide genetic association of complex traits in heterogeneous stock mice. Nat Genet 38:879–887
Valverde F, Mouradov A, Soppe W, Ravenscroft D, Samach A, Coupland G (2004) Photoreceptor regulation of CONSTANS protein in photoperiodic flowering. Science 303:1003–1006
Van Dijk H (2009) Evolutionary change in flowering phenology in the iteroparous herb *Beta vulgaris* ssp. *maritima*: a search for the underlying mechanisms. J Exp Bot 60:3143–3155
Ward JK, Strain BR (1999) Elevated $CO_2$ studies: past, present and future. Tree Phys 19:211–220
Wei X, Xu J, Guo H, Jiang L, Chen S, Yu C, Zhou Z, Hu P, Zhai H, Wan J (2010) *DTH8* suppresses flowering in rice, influencing plant height and yield potential simultaneously. Plant Physiol 153:1747–1758
Weigel D, Nilsson O (1995) A developmental switch sufficient for flower initiation in diverse plants. Nature 377:495–500
Welch SM, Roe JL, Dong Z (2003) A genetic neural network model of flowering time control in *Arabidopsis thaliana*. Agron J 95:71–81
White JW, Herndl M, Hunt LA, Payne TS, Hoogenboom G (2008) Simulation-based analysis of effects of *Vrn* and *Ppd* loci on flowering in wheat. Crop Sci 48:678–687
Wigge PA (2011) FT, a mobile developmental signal in plants. Curr Biol 21:R374–R378
Wigge PA, Kim MC, Jaeger KE, Busch W, Schmid M, Lohmann JU, Weigel D (2005) Integration of spatial and temporal information during floral induction in *Arabidopsis*. Science 309:1056–1059
Wilhelm EP, Turner AS, Laurie DA (2009) Photoperiod insensitive *Ppd-A1a* mutations in tetraploid wheat (*Triticum durum* Desf.). Theor Appl Genet 118:285–294
Wolkovich EM, Cook BI, Allen JM, Crimmins TM, Betancourt JL, Travers SE, Pau S, Regetz J, Davies TJ, Kraft NJB, Ault TR, Bolmgren K, Mazer SJ, McCabe GJ, McGill BJ, Parmesan C,

Salamin N, Schwartz MD, Cleland EE (2012) Warming experiments underpredict plant phenological responses to climate change. Nature 485:494–497

Wood CC, Robertson M, Tanner G, Peacock WJ, Dennis EJ, Hellwell CA (2006) The *Arabidopsis thaliana* vernalization response requires a polycomb-like protein complex that also includes VERNALIZATION INSENSITIVE 3. Proc Natl Acad Sci USA 103:14631–14636

Worland AJ (1996) The influence of flowering time genes on environmental adaptability in European wheat. Euphytica 89:49–57

Wu RF, Shen JG, Yan WX, Zhang H (2009) Impact of climate warming on phenophase of *Populus tomentosa* in Inner Mongolia. Chin J Appl Ecol 20:785–790

Wurr DCE, Fellows JR, Fuller MP (2004) Simulated effects of climate change on the production pattern of winter cauliflower in the UK. Sci Hortic 101:359–372

Xu J, Liu Y, Liu J, Cao M, Wang J, Lan H, Xu L, Lu Y, Pan G, Rong T (2012) The genetic architecture of flowering time and photoperiod sensitivity in maize as revealed by QTL review and meta-analysis. J Integr Plant Biol 54:358–373

Xue W, Xing Y, Weng X, Zhao Y, Tang W, Wang L, Zhou H, Yu S, Xu C, Li X, Zhang Q (2008) Natural variation in *Ghd7* is an important regulator of heading date and yield potential in rice. Nat Genet 40:761–767

Yamada T, Jones ES, Cogan NOI, Vecchies AC, Nomura T, Hisano H, Shimamoto Y, Smith KF, Hayward MD, Forster JW (2004) QTL analysis of morphological, developmental, and winter hardiness-associated traits in perennial ryegrass. Crop Sci 44:925–935

Yamakawa H, Hirose T, Kuroda M, Yamaguchi T (2007) Comprehensive expression profiling of rice grain filling-related genes under high temperature using DNA microarray. Plant Physiol 144:258–277

Yamamoto E, Yonemaru JI, Yamamoto T, Yano M (2012) OGRO: overview of functionally characterized genes in rice online database. Rice 5:26. doi:10.1186/1939-8433-5-26

Yamane H, Ooka T, Jotatsu H, Hosaka Y, Sasaki R, Tao R (2011) Expressional regulation of *PpDAM5* and *PpDAM6*, peach (*Prunus persica*) dormancy-associated MADS-box genes, by low temperature and dormancy-breaking reagent treatment. J Exp Bot 62:3481–3488

Yan L, Loukoianov A, Tranquilli G, Helguera M, Fahima T, Dubcovsky J (2003) Positional cloning of the wheat vernalization gene *VRN1*. Proc Natl Acad Sci USA 100:6263–6268

Yan L, Loukoianov A, Blechl A, Tranquilli G, Ramakrishna W, SanMiguel P, Bennetzen JL, Echenique V, Dubcovsky J (2004) The wheat *VRN2* gene is a flowering repressor down-regulated by vernalization. Science 303:1640–1644

Yan L, Fu D, Li C, Blechl A, Tranquilli G, Bonafede M, Sanchez A, Valarik M, Yasuda S, Dubcovsky J (2006) The wheat and barley vernalization gene *VRN3* is an orthologue of *FT*. Proc Natl Acad Sci USA 103:19581–19586

Yan WH, Wang P, Chen HX, Zhou HJ, Li QP, Wang CR, Ding ZH, Zhang YS, Yu SB, Xing YZ, Zhang QF (2011a) A major QTL, *Ghd8*, plays pleiotropic roles in regulating grain productivity, plant height, and heading date in rice. Mol Plant 4:319–330

Yan J, Chen W, Luo FAN, Ma H, Meng A, Li X, Zhu M, Li S, Zhou H, Zhu W, Han BIN, Ge S, Li J, Sang TAO (2011b) Variability and adaptability of *Miscanthus* species evaluated for energy crop domestication. GCB Bioenergy 4:49–60

Yano M, Katayose Y, Ashikari M, Yamanouchi U, Monna L, Fuse T, Baba T, Yamamoto K, Umehara Y, Nagamura Y, Sasaki T (2000) *Hd1*, a major photoperiod sensitivity quantitative trait locus in rice, is closely related to the *Arabidopsis* flowering time gene *CONSTANS*. Plant Cell 12:2473–2484

Yano M, Kojima S, Takahashi Y, Lin HX, Sasaki T (2001) Genetic control of flowering time in rice, a short-day plant. Plant Physiol 127:1425–1429

Yant L, Mathieu J, Schmid M (2009) Just say no: floral repressors help *Arabidopsis* bide time. Curr Opin Plant Biol 12:580–586

Yerushalmi S, Yakir E, Green RM (2011) Circadian clocks and adaptation in *Arabidopsis*. Mol Ecol 20:1155–1165

Yin X, Struik PC (2010) Modelling the crop: from system dynamics to systems biology. J Exp Bot 61:2171–2183

Yin X, Struik PC, Tang J, Qi C, Liu T (2005a) Model analysis of flowering phenology in recombinant inbred lines of barley. J Exp Bot 56:959–965

Yin X, Struik PC, van Eeuwijk FA, Stam P, Tang J (2005b) QTL analysis and QTL-based prediction of flowering phenology in recombinant inbred lines of barley. J Exp Bot 56:967–976

Yonemaru JI, Yamamoto T, Fukuoka S, Uga Y, Hori K, Yano M (2010) Q-TARO: QTL annotation rice online database. Rice 3:194–203

Youens-Clark K, Buckler E, Casstevens T, Chen C, DeClerck G, Derwent P, Dharmawardhana P, Jaiswal P, Kersey P, Karthikeyan AS, Lu J, McCouch SR, Ren L, Yu C-W, Liu X, Luo M, Chen C, Lin X, Tian G, Lu Q, Cui Y, Wu K (2011) HISTONE DEACETYLASE6 interacts with FLOWERING LOCUS D and regulates flowering time in Arabidopsis. Plant Physiol 156:173–184

Yu J, Holland JB, McMullen MD, Buckler ES (2008) Genetic design and statistical power of nested association mapping in maize. Genetics 178:539–551

Yu H, Luedeling E, Xu J (2010) Winter and spring warming result in delayed spring phenology on the Tibetan Plateau. Proc Natl Acad Sci USA 107:22151–22156

Yu C-W, Liu X, Luo M, Chen C, Lin X, Tian G, Lu Q, Cui Y, Wu K (2011) HISTONE DEACETYLASE6 interacts with FLOWERING LOCUS D and regulates flowering in Arabidopsis. Plant Physiol 156:173–184

Zhang H, Harry DE, Yuceer C, Hsu CY, Vikram V, Shevchenko O, Etherington E, Strauss SH (2010) Precocious flowering in trees: the *FLOWERING LOCUS T* gene as a research and breeding tool in *Populus*. J Exp Bot 61:2549–2560

# Chapter 2
# Root Characters

Silvas J. Prince, Raymond N. Mutava, Camila Pegoraro, Antonio Costa de Oliveira, and Henry T. Nguyen

**Abstract** Environmental stresses as major threats to global food security in the twenty-first century and improving biomass production and seed yield per area of crop plants become very critical. Plants are sessile organisms that cannot escape from environmental constraints, and as a result, they have evolved numerous adaptive responses to cope with environmental stresses. However, many biotic and abiotic responses start on roots through the sensing and response to environmental cues. Root-related abiotic stress factors such as drought, flooding, and soil salinity are already causing significant agricultural yield losses and will become even more prevalent in the coming decades due to the unpredictable climatic changes. Many scientists are starting to see roots as central to their efforts to produce crops with better yield efforts under stress environments. Alteration/manipulation of root system architecture through any approaches holds the potential to increase plant yield and optimize the agricultural land use. Most environmental stresses share common effects and responses such as reduction of growth and photosynthesis, oxidative damage, hormonal changes, and the accumulation of numerous stress-related proteins. However, no root-related gene or QTL has been commercially deployed in crop plants for drought-tolerance improvement. As breeding programs rely on high-throughput strategies to select genotypes with desirable trait variation that are easy to apply, reliable, and affordable, genetic improvement of root traits is slow in progress. The results from several QTL studies also highlight the feasibility of marker-aided selection as an alternative to conventional labor-intensive, phenotypic screening of drought-avoidance root traits in crop plants. In addition, various applications of DNA sequencing technologies were also

S.J. Prince • R.N. Mutava • H.T. Nguyen (✉)
National Center for Soybean Biotechnology, University of Missouri, 25 Agriculture Building, Columbia, MO 65211-7140, USA
e-mail: nguyenhenry@missouri.edu

C. Pegoraro • A.C. de Oliveira
Plant Genomics and Breeding Center, School of Agronomy Eliseu Maciel, Federal University of Pelotas, 96010-610 Pelotas, RS, Brazil

C. Kole (ed.), *Genomics and Breeding for Climate-Resilient Crops*, Vol. 2,
DOI 10.1007/978-3-642-37048-9_2, © Springer-Verlag Berlin Heidelberg 2013

applied to understand the mechanism of root traits in relation to abiotic and biotic stress tolerance in crop plants. Thus recent studies in model and crop plants aid the researchers to explore the genetic mechanisms/genes involved in root development and its expression pattern under various biotic and abiotic stresses. Various research groups across the globe are actively involved in advancing the understanding of the root biology using multidisciplinary approaches. Recent advances in candidate gene approaches and genetic engineering of crops have shown promising improvements in drought tolerance of crops such as rice, maize, canola, and soybean. However, genomic approaches are allied to understand the genetic regulation of root development by identification and introgression of major gene/QTL, and advancement in novel technologies brings significant new opportunities to improve crop tolerance to biotic and abiotic stresses.

## 2.1 Introduction

Global climate change is predicted to lead to extreme temperatures and severe drought in some parts of the world, while other parts will suffer from heavy storms and periodic flooding (Marshall et al. 2012). In addition, increase in greenhouse gas emissions has resulted in altered precipitation, increase in arid land, desertification, and finally reduction in crop productivity. Global climate change is likely to increase the problems of food insecurity, hunger, and malnutrition for millions of people, particularly in South Asia, sub-Saharan Africa, and small islands (IPCC 2007). Many recent studies have identified environmental stresses as major threats to global food security in the twenty-first century (Ashmore et al. 2006; Battisti and Naylor 2009). Therefore, improving biomass production and seed yield per area under suboptimal water availability due to drought and other abiotic stresses by improving the plants themselves is now even more pressing. FAO estimates that the current soybean worldwide production (about 220 million tons) must increase by a staggering 140 % to meet food demands of a growing human population by 2050 (Bruinsma 2009). With decreasing availability of well-watered agricultural areas, attempts to reach such future production levels will require the use of existing or new cropping areas having limited water supply (Sinclair et al. 2010). Serraj et al. (2009) reviewed for rice (*Oryza sativa* L.) the challenges of increasing the land area and yields in view of water limitations. Root-related abiotic stress factors such as drought, flooding, and soil salinity are already causing significant agricultural yield losses and will become even more prevalent in the coming decades due to the effects of global climate change (Ashmore et al. 2006; Ortiz et al. 2008; Battisti and Naylor 2009; Feng and Kobayashi 2009; Fuhrer 2009; Wassmann et al. 2009a).

## 2.2   Importance of Root Traits in Plant Development

Roots are very important for plant development, since they act as anchors, providing mechanical support, and also as chemical extractors for the plant. Roots can also act as storage organs, serve as food sources, and sense soil changes such as soil water and nutrient content. Root sensing of water deficit is linked to one of the most severe stresses occurring in plants—drought. However, many biotic and abiotic responses start on roots through the sensing and response to environmental cues. Many scientists are starting to see roots as central to their efforts to produce crops with better yield efforts that go beyond the Green Revolution. Root system architecture (RSA) (Smith and De Smet 2012), for example, holds the potential for the exploitation and manipulation of root characteristics to both increase plant yield and optimize agricultural land use.

Roots started to receive higher attention when the model species *Arabidopsis* provided the first root affecting mutations (Dolan et al. 1993; Scheres et al. 1996). The first understanding on root development came from the fact that division, enlargement, and differentiation of cells occur in a spatially divided way. The root apical meristem (RAM) is a region of continuous cell division at the root tip. New cells go through a process of elongation, enlarging by a factor of 100, and by the time they reach a maturation size, they undergo differentiation steps into various cell types (reviewed in Costa de Oliveira and Varshney 2011). When roots grow, a series of lateral roots are formed, resulting in a branching pattern that covers higher volumes of soil as it branches. Shallow versus deep roots are some of the differential strategies of plants to adapt to their environments, but some residual variability can be seen in different species, which are amenable to selection and could help unravel root genetics. The root traits commonly characterized in quantitative trait loci (QTL) mapping studies to identify their functional roles are listed in Table 2.1.

Water and nutrient uptake and nitrogen fixation are among the most important roles of roots in plant growth and development. The efficient use of water is regarded as the key for crop production in areas subjected to drought occurrences. Water shortages prevent large-scale irrigation, and under scarce water regimes, further fertilizer application has a dual impact on yield and environment, besides being economically unviable in Third World countries (Denning et al. 2009). Drought causes more losses than any other abiotic stress on crops, and 70 % of the world's freshwater demand is for agricultural use (SIWI-IWMI 2004; Xiong et al. 2006). To determine if a genotype is drought resistant, non-root traits such as stomata conductance, water leakage through the cuticle, and nocturnal stomata opening have been described, but whole plant traits such as osmotic adjustments have also played a role in drought response (Blum 2011). Another important process with direct participation of roots is nitrogen fixation. The availability of nitrogen is critical to sustained plant growth and reproduction, and yet atmospheric nitrogen is unavailable to most organisms. In legumes, a symbiotic relationship with soil bacteria (rhizobia) provides the formation of nodules in organs capable of fixing atmospheric nitrogen by the action of a nitrogenase enzyme complex of the

**Table 2.1** Root traits and their functional characteristics

| Root trait | Functional characteristics |
| --- | --- |
| Maximum root depth | Potential for absorption of soil moisture and nutrients in deeper soil layer |
| Root to shoot ratio | Assimilate allocation |
| Root volume | The ability to permeate a large volume of soil |
| Root number | Physical strength, potential for root system architecture |
| Root diameter | Potential for penetration ability, branching, hydraulic conductivity |
| Deep root to shoot ratio | Vertical root growth, potential for absorption of soil moisture and nutrients in deeper soil layers |
| Root length/weight density | Rate of water and nutrient uptake |
| Root branching | Power of soil exploration (the major contribution to total root length) |
| Total root length/ surface area | The total system size: the size of contact with soil (major determinant of water and nutrient uptake as an entire root system) |
| Specific root length | Degree of branching, density of root materials, porosity due to aerenchyma development |
| Hardpan penetration ability | Ability to penetrate subsurface hardpans |

*Source*: Gowda et al. (2011)

endocytotic bacteria. Recent advances in the understanding of root nodulation regulation in legumes indicate that systemic signaling ensures a balance between the formation of nodules and energy requirements. Autoregulation of nodulation (*AON*), which involves *CLE* (*CLV3/ESR*-related peptide family), *CLV2* (*CLAVATA2*), and *KLV* (*KLAVIER*), has shown interesting similarities between nodule regulation and other *CLE* peptide ligand–receptor systems, particularly in *CLAVATA* signaling pathway (Reid et al. 2011). The study of root development has progressed in model species, taking advantage of their simpler root architecture and available genome sequences (AGI 2000; IRGSP 2005). The *Arabidopsis thaliana* root has 15 distinct cell types surrounding a radial axis (Iyer-pascuzzi et al. 2009). Several miRNAs that are involved in plant root development are shared between rice and *Arabidopsis* (Meng et al. 2010). These similarities seem to break when one looks at root system patterns between monocots and dicots. In *Arabidopsis*, the gene *Gnom1* is required for the formation of the lateral primordium, through an asymmetrical division of pericycle cells (Coudert et al. 2010). Two mutations in orthologous gene in rice, *crown rootless4* (*crl4*) and *OsGnom1*, produce crown rootless phenotypes. *GNOM1* is a membrane-associated guanine nucleotide exchange factor of the ADP-ribosylation factor G protein (*ARF_GEF*) that acts on the regulation of *PIN1* (*PINFORMED1*) auxin efflux carrier proteins that regulate auxin transport (Kitomi et al. 2008; Liu et al. 2009).

The dynamics of root behavior changes in response to low soil phosphorus status and the identification of neighboring crop plants. When intercropped with its kin, maize or soybean roots grew close to each other. However, when maize *GZ1* was grown with soybean *HX3*, the roots on each plant tended to avoid each other, becoming shallower (Fang et al. 2011). These preliminary findings indicate a mechanism of exudate sensing between different taxa could give a whole new meaning to intercropping systems and weed control.

## 2.3 Identification of Major Root Traits: Morphological and Physiological

### 2.3.1 Root System Architecture

The root system architecture (RSA) describes the organization of the primary and lateral roots. The presence of other types of roots and their distribution are key for the understanding of nutrient- and water-use efficiency in plants (Smith and De Smet 2012). RSA and root hydraulics have a considerable impact on water capture in drought-prone environments (Tuberosa et al. 2002a; Zhao et al. 2005; Dorlodot et al. 2007). Recent studies have reported that even small improvements in water uptake (7 %) can lead to near 30 % increase of deep root length and translate into higher grain yield under drought (Bernier et al. 2005). The flow of uptake is thought to be controlled by the plant under most but in very dry soil conditions (Hopmans and Bristow 2002). However, root length density (RLD) sets the proportion of water uptake under wet conditions (Gardner 1965). Therefore, there seems to be a difference on the strategy used in drought-prone and wet environments (Draye et al. 2010).

### 2.3.2 Root Hair and Root Epidermis

There are at least 39 genes required for the initiation and growth of root hairs in *Arabidopsis*. Among the most known genes are *TRANSPARENT TESTA GLABRA1* (*TTG1*), *GLABRA3* (*GL3*), *ENHANCER OF GLABRA3* (*EGL3*), and *GLABRA2* (*GL2*), which have been well described (Galway et al. 1994; Walker et al. 1999; Bernhardt et al. 2003). The mutants *TTG1* and *GL2* have root hairs at nearly all root epidermal cells, while the mutants *GL3* and *EGL3* have lower numbers of atrichoblasts. The domains *Tg.1* encodes a protein with WD40 repeats localized in the nuclei of trichomes at all the developmental stages. It appears that *GL2* is a direct target of *GL3* and *EGL3*, and *TTG1* is directly regulated by *GL1* (Zhao et al. 2008).

### 2.3.3 Tissue Patterning Genes

This category comprehends the cortex and endodermis layer, which originate from a common initial cell near the quiescent center (QC) (Scheres et al. 2002). The number of layers outside the endodermis is variable according to species, ranging from one thick walled in *Arabidopsis* to 4–6 in barley and 8–15 in rice and maize (Scheres et al. 2002; Hochholdinger et al. 2004b). Mutations affecting the patterning of the ground tissue, such as *SCARECROW* (*SCR*) and *SHORTROOT* (*SHR*), have a single layer instead of cortex and endodermis. These are putative transcription factors (TF) of the *GRAS* family responsible for specifying QC and for controlling the periclinal cell division of the first daughter cell which originates from the two adjacent layers (Ueda et al. 2005).

### 2.3.4 Stele Development

Monocotyledonous roots consist of thickened cell walls in stele and sclerenchyma that can be observed in the outer cortex (Briggs 1978). However, the root radial pattern in *Arabidopsis* is formed by one xylem axis and two phloem poles, surrounded by one pericycle layer (Scheres et al. 2002). The mutations involved in stele patterning in *Arabidopsis*, which have been described, are *WOODEN-LEG* (*WOL*) and *ALTERED PHLOEM DEVELOPMENT* (*APL*), causing defects in the vascular system (Bonke et al. 2003; Sieberer et al. 2003).

### 2.3.5 Root Meristems

The organization of root meristem tissues is such that longitudinal cell files are formed. From tip upwards, three regions, the division, elongation, and differentiation zone, can be observed. During embryogenesis, first the primary or embryonic radicle and few seminal roots are formed in both mono- and dicots, respectively (reviewed in Orman et al. 2011). On the other hand, lateral roots (LRs) are formed from existing roots postembryonically. In *Arabidopsis*, LRs come from pericycle cells (Scheres et al. 2002), but in monocots, these come from pericycle and endodermis cells (Hochholdinger et al. 2004b).

Nearly Two hundred genes have been described affecting longitudinal pattern in *Arabidopsis*. One of the most studied mutants is the *Gnom* (*GN*), a lack of root phenotype, which encodes an *ARF GDP/GTP* exchange factor involved in the formation of embryonic axis and polar localization of *PIN1* (Geldner et al. 2004). Other important mutants are *BODENLOS* (*BDL*) and *MONOPTEROS* (*MP*), which encode *IAA12* (*INDOLE ACETIC ACID-INDUCED PROTEIN 12*) and *ARF5* (*AUXIN RESPONSE FACTOR 5*), genes involved in auxin that cause a lack of

**Root system development and growth**
(Genes involved in parenthesis)

**ROOT SYSTEM**

**I. Root meristematic activity**
*A. QC and stem cell activation*
[AP2 TFs; Iyer-  Pascuzzi and Benfey 2009]
*B. Cell division and proliferation*
(Cytochrome b5 like heme/steroid binding domain; Ikeyama et al. 2010)
*C. Cell division maintenance*
(Homeobox protein; Nakajima et al. 2001, ARR and GRAS transcription factors; Hirch and Oldroyd 2009)

**LATERAL ROOTS**

*A. Pericylcle Differentiation*
*Auxin transport*
(GNOM1; Liu et al.2009, Multidrug-resistance/P-glycoproteins; Terasaka et al. 2005)
*Auxin signaling protein*
(Aux/IAA gene family; Kang et al. 2013)
*Polyamines*
(Arginine and ornithine decarboxylase; Watson et al. 1998)
*B. Initiation and Development*
*LBD proteins*
(Auxin response factor; Casimiro et al. 2001, NAC genes; Hao et al. 2011)
*Cell cycle genes*
(CDC2; Doerner et al. 1996 and Cyclin genes-G2/M phase; Himanen et al. 2002)

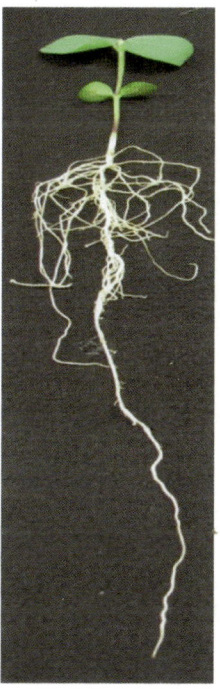

**Root traits important under abiotic stress**
(Genes involved in parenthesis)

**ROOT LENGTH**

*I. Increase in cell number and size*
(BR; Goncalez-Garcia et al. 2011, MADS-BOX genes; Zhang and Forde 1998)
*II. Cell maturation*
(Callose synthase; Chen et al. 2009)

**MAINTENANCE OF ROOT TIP GROWTH**

*A. Cell wall loosening*
(Expansin; Wu et al. 1996, Endo-1, 4-b-D-glucanases; Inukai et al. 2012)
*B. Apoplastic ROS*
(Oxalase oxidase, POX; Carol and Dolan 2006, Chalcone synthase; McKhann and Hirch 1994)
*C. Phloem flux maximization*
(Sucrose hydrolysis pathway; Giaquinta et al. 1983)
*D. Gravity induced auxin gradient*
(Auxin efflux protein PIN3; Blilou et al. 2005, Adenosine kinase- S-adenosyl-L-Met cycle; Young et al. 2006, ABA accumulation; Sharp and Lenoble 2002)

**ROOT HAIR DEVELOPMENT**
(Jasmonate and Ethylene pathways; Zhu et al. 2006)

**Fig. 2.1** Cellular mechanisms/hormones involved in root system development and its importance under abiotic stress

primary root (Shevell et al. 2000). The cellular mechanisms/hormones involved in root system development and various root traits important under abiotic stress are illustrated in Fig. 2.1.

Even though a few mutants are known in *Arabidopsis* affecting root development, not much is known considering the vast universe of interactions faced by roots with abiotic and biotic factors present in the soils.

## 2.4  Screening Techniques for Root-Related Abiotic Stresses

The characterization of genetic diversity for root traits related to abiotic stresses such as the resistance to heavy metals, drought, salinity, and cold and the identification of novel resistant germplasm sources is the first breeding step. This will be used regardless of conventional, marker-assisted selection, or functional analyses of gene associated to abiotic stress response pathways are used (Lu et al. 2012).

The detection of new resistance sources depends on the development of screening methods that are fast, simple, cost-effective, and reproducible under the targeted environmental conditions. Many field and laboratory techniques have been used with success for the screening of plants resistant to different abiotic stresses, which will be discussed in this chapter.

In this section, the described screening techniques will focus on the rice crop (*Oryza sativa* L.), since it is the second cereal in cultivated area, one of the main human sources of food, and a model species among the Poaceae, presenting a significant degree of synteny and colinearity with other cereals.

For the selection of rice plants resistant to Fe, Asch et al. (2005) developed a fast and efficient screening method. In this method, rice seeds are soaked in water overnight and kept in Petri dishes for 60 h inside a growth chamber. Three-day-old seedlings are transferred to translucent pots containing 150 mL of 0.5× Yoshida's nutrient solution added with 1 % agar. The root zone is covered with aluminum foil and the vases are placed in a controlled chamber for 4 weeks. Twenty eight days after sowing, Fe is added as ferrous sulfate ($FeSO_4$) for a final concentration of 2,000 mg $L^{-1}$ for 3 days. The response to excess Fe (bronzing) is evaluated visually for the whole plant and expressed as the percentage of the affected leaf area. Besides, the tissue iron content must also be evaluated.

Another technique of screening for the selection of tolerant genotypes to Fe excess consists in the selection under hydroponic culture. In this technique, seeds are germinated in wet Whatman paper, maintained in chamber at 26 °C, 16/8 h light/dark photoperiod and 100 % relative humidity for 72 h. After this period, the uniform seedlings are transferred to a nylon mesh adapted to the lid of a plastic pot containing nutrient solution (Camargo and Oliveira 1981). The recipients must be arranged in a chamber with temperature of 25 ± 1 °C and 16 h light/8 h dark. The seedlings develop under these conditions for 14 days and, subsequently, are transferred to the recipients containing the Fe stress treatment (500 mg $L^{-1}$ of $FeSO_4 \cdot 7H_2O$ at pH 4.0 ± 0.1) remaining under these conditions for up to 12 days, without aeration and with daily pH measures. These variables discriminate better genotypes with desired root and shoot length, shoot dry matter, and Fe content (Bresolin 2010).

For the selection of rice plants tolerant to Ni, Samantaray et al. (1997) proposed a technique where the seeds from different genotypes are sown in nets submerged in nutrient solution and maintained at 25 °C and 16 h of light/8 h dark regimen. The stress by Ni is obtained through the addition of 18 μg $dm^3$ nickel sulfate for 13 days. The characters evaluated were germination rate, primary root and shoot length, and total biomass. For the selection of drought-tolerant rice genotypes, a field screening technique has been proposed by Sakai et al. (2010). For this technique, 23-day-old seedlings are transferred to the field, with a distance of 25 cm between plants and 40 cm between rows. The drought stress conditions were simulated through the suspension of irrigation at the 26th day after transplanting, draining the soil and keeping off rainfall using the rain-out shelter. Furthermore, in order to prevent water movement from outside the experimental block, a transparent vinyl sheet was placed 60 cm deep into the soil. Later, the plants are irrigated 2–3 times per week (approximately 420 L water irrigation for 57.8 $m^2$). These conditions were maintained until 1 week before the samples are harvested.

On another study, a field technique was also studied, where the soil was flooded, drained, and sprinkler irrigated for up to 50 days after sowing, when the irrigation was stopped to induce drought tolerance (Pantuwan et al. 2004). For selecting drought-tolerant rice plants during the reproductive stage, Garrity and O'Toole (1994) proposed a field technique in which the sowing is performed at different dates and the irrigation stopped during reproductive stage. This can be achieved in dry seasons where no interference by rain is detected.

For drought tolerance, in vitro screening techniques have also been proposed (Biswas et al. 2002; Joshi et al. 2011). In this technique, rice calli are induced in MS media with different concentrations of high-molecular-weight polyethylene glycol, in order to simulate the drought stress. Characters evaluated were germination, calli formation, and plant regeneration.

The commonly evaluated characters when screening for drought are flowering date, chlorophyll fluorescence, panicle water potential, spikelet fertility, grain yield root system, or a visual estimate of plant water status such as leaf rolling or death (Fukai et al. 1999; Sayed 2003).

For selection of rice plants tolerant to salinity stress, a simple, efficient, and reproducible technique was described (Aslam et al. 1993). In this technique, 14-day-old rice seedlings are transferred to foam-plugged holes in polystyrene (thermopal) sheets floated over 100 $dm^3$ of nutrient solution in painted galvanized-iron growth tanks lined with plastic ($120 \times 90 \times 30$ cm). Three days after the transfer, NaCl is added to the nutrient solution in 25 mol $m^{-3}$ dose at each 24 h, until reaching the desired concentration. Six plants of each genotype or line are transplanted and maintained under high salt (desired concentration) for 15 days. Another simple method has been described that measures the germination rate under salt stress (Sumith and Abeysiriwardena 2004). In this technique, seeds are sown under high-salt conditions. In vitro embryo culture has also been used as a screening method for selecting salt-tolerant rice plants (Shanthi et al. 2010). In this method, callus induction from a range of genotypes is performed in MS plus different NaCl concentrations (50 mM, 100 mM, and 150 mM). The percentage of calli formed is calculated, and the genotypes presenting higher rates are transferred to a regeneration medium containing the same NaCl concentrations. The percentage of regeneration indices are used for selection.

Among the characters used on the screening techniques, one can highlight leaf $Na^+$ concentration (Yeo et al. 1988), leaf area/total dry matter weight ratio (Akita and Cabuslay 1990), shoot $Na^+$ concentration, leaf to leaf partitioning, tissue tolerance and plant vigor (Yeo et al. 1990), fresh and dry weight, root and shoot length, number of tillers and mortality rates (Aslam et al. 1993), $K^+/Na^+$ ratio, grain yield (Asch et al. 2000), relative water content in the leaves (Suriya-arunroj et al. 2004), and floret fertility (Rao et al. 2008). Recently, the use of bypass flow has been suggested for the selection of salinity-resistant genotypes (Faiyue et al. 2012).

## 2.5   Effect of Abiotic Stresses on Root Development and Performance

Plants are sessile organisms that cannot escape from environmental constraints, and as a result, they evolved numerous adaptive responses to cope with environmental stresses. Agricultural yield losses due to abiotic environmental stresses are obvious and have been well documented. The awareness of the growing impact of environmental stress has led to worldwide efforts in adapting agricultural production to such adverse environmental conditions (Lobell et al. 2008; Ortiz et al. 2008; Wassmann et al. 2009b). Attempts to breed for or genetically engineer abiotic stress tolerance into important crop plants have been largely unsuccessful (Sinclair 2011). The exposure to stress leads to numerous physiological changes in crop plants, such as changes in the photosynthetic gas exchange and assimilate translocation (Martin and Ruiztorres 1992; Morgan et al. 2004), altered water uptake and evapotranspiration (Rivelli et al. 2002; Katerji et al. 2010), effects on nutrient uptake and translocation (Hu and Schmidhalter 2005; Sanchez-Rodriguez et al. 2010), antioxidant reactions (Blokhina et al. 2003; Apel and Hirt 2004), programmed cell death (Kangasjärvi et al. 2005), and altered gene expression and enzyme activity (Yamakawa et al. 2007; Guo et al. 2009; Frei et al. 2010). These changes are usually the result of tissue dehydration (Dobra et al. 2010), which occurs when there is an imbalance between root water uptake and leaf transpiration (Jackson et al. 2003). Water deficit occurs in tissues under drought, low-temperature, heat, or salt stress conditions (Aroca et al. 2001, 2007; Wahid and Close 2007). Under some environmental conditions, dehydration is the first signal that induces the plant to respond (Christmann et al. 2007), and the importance of the hydration state of tissues in the response of the plant to different stresses is well supported by experimental evidence (Aroca et al. 2001; Bouchabke-Coussa et al. 2008; Matsuo et al. 2009). When leaves begin to dehydrate, plants generally start closing their stomata; however, under some environmental situations or in specific plant genotypes, modification of root water uptake capacity plays a more important role compared with stomatal closure in avoiding stress-induced growth reduction (Matsuo et al. 2009). Proposed strategies to face these challenges include improved agronomic management and breeding of novel crop genotypes adapted to stress environments. Molecular breeding projects have been initiated to adapt novel varieties of major cereal crops to various abiotic stresses including heat (Pinto et al. 2010), drought (Araus et al. 2008; Fleury et al. 2010), salinity (Ren et al. 2005), and tropospheric ozone (Frei et al. 2008). The focus of most agronomic and breeding efforts has been on mitigating quantitative yield losses caused by environmental stress factors (Wang and Frei 2011).

## 2.6  Effect of Abiotic Stresses on Root System

The root system has been recently viewed as an important trait to improve crop productivity because of its pivotal roles during plant growth, e.g., acquisition of water and nutrients, soil anchorage, response to biotic and abiotic stresses, and interaction with symbiotic organisms (Gewin 2010; Herder et al. 2010). During the last decade, significant progress has been achieved in understanding the genes and gene regulatory networks involved in the root system of the model plant *Arabidopsis* (Benková and Hejátko 2009; Fukaki and Tasaka 2009; Iyer-Pascuzzi et al. 2009). In addition, a high-resolution spatiotemporal map of *Arabidopsis* root based on cell type and developmental stage-specific transcriptome data (Birnbaum et al. 2003; Brady et al. 2007) provides useful clues on the molecular mechanisms of root development and functions (Brown et al. 2005; Petersson et al. 2009; Parizot et al. 2010; Brady et al. 2011). However, not much is known about the root system of other plant species, including rice, which differ from *Arabidopsis* in anatomy and morphology, particularly in radial tissue patterning, root hair patterning, and cell types for lateral root formation (Hochholdinger et al. 2004b; Rebouillat et al. 2009). The root system exposed to various abiotic stresses including excess salt, heavy metals, improper nutrient balance, and temperature extremes shows that similar morphological responses and mechanisms are involved. Plants roots will form aerenchyma during water logging, drought, and nutrient deficiency (Jackson and Armstrong 1999; Zhu et al. 2010; Postma and Lynch 2011). It has been proposed that modifying programmed cell death and cytokinin levels to prevent early senescence may be helpful in increasing plant abiotic stress tolerance (Xu et al. 2004; Huynh et al. 2005). Another root trait, adventitious root formation, is a common response of plants to excess water, but it is also induced during other stresses such as phosphorus starvation (Miller et al. 2003). Recently, the root system has been hailed as the key to a new "Green Revolution," and therefore its improvement is an important strategy for crop breeding to overcome various stress conditions and achieve increasing yield worldwide (Gewin 2010; Herder et al. 2010).

## 2.7  Quantitative Trait Loci Mapping of Root Traits

Root system architecture (RSA) is the spatial configuration of a root system in the soil and plays a major role in regulating the acquisition of soil resources such as nutrients and water (Wang et al. 2006a). Proper establishment of RSA is, therefore, of great importance with respect to fulfilling a plant's functional requirements, particularly in agronomically important crops such as cereals, which account for 70 % of food production worldwide (Chandler and Brendel 2002). Manipulation of RSA is a key strategy aimed at achieving a new Green Revolution and further boosting up crop yields (de Dorlodot et al. 2007; Lynch 2007; Den Herder et al. 2010).

The importance of root characteristics as a selection criterion in a breeding program has long been recognized, but the complex nature of root architecture and interactions of roots with the surrounding rhizosphere has made it difficult to select for root-associated traits in the field during the breeding cycle. A deep root system is essential for crops to utilize nitrate and water in deeper soil layers, especially under abiotic stress conditions (Jordan et al. 1983; Wiesler and Horst 1994). However, because of the higher phosphorus availability in surface soil strata, a shallow root system with enhanced adventitious rooting is important in the absorption of phosphorus (Lynch 2011).

As breeding programs rely on high-throughput strategies to select genotypes with desirable trait variation that are easy to apply, reliable, and affordable, genetic improvement has to date been driven largely by selection for yield associated with traits that control the growth and development of aboveground plant parts. Several studies have described QTLs that provide access to valuable genetic diversity for the morphophysiological features that characterize root functionality for different model and crop plants.

## 2.7.1  Arabidopsis

*Arabidopsis thaliana* provides a scientifically attractive and simple model for studying root growth and architecture because of the simplicity of its root system and the relative ease with which it can be visualized in vitro on vertical agar plates. Several groups have taken advantage of the natural variation that exists in *Arabidopsis* accessions to map quantitative trait loci (QTLs) controlling its root system development (Kobayashi and Koyama 2002; Rauh et al. 2002; Hoekenga et al. 2003; Loudet et al. 2005). Loudet et al. (2005) used this natural variation and mapped QTLs for primary root length, lateral root number and density, and total length of the lateral root system in a Bay-0 × Shahdara population. QTL analyses in *Arabidopsis* using a variety of mapping populations (Loudet et al. 2002; El-Lithy et al. 2004; Balasubramanian et al. 2009; Kover et al. 2009) have detected QTLs involved in either constitutive traits related to root growth (Loudet et al. 2005) or focused on QTLs associated with root growth responses to the environment, such as responses to low phosphate (Reymond et al. 2006), low nitrogen (Rauh et al. 2002), and osmotic stress (Xiong et al. 2006; Gerald et al. 2006).

Some QTLs for growth-related traits in response to environmental changes have already been cloned in *Arabidopsis*. For example, the differential response of root growth of some *Arabidopsis* accessions to phosphate starvation led to the identification of allelic differences responsible for this phenotype (Reymond et al. 2006; Svistoonoff et al. 2007). Mouchel et al. (2004) cloned a major gene (BREVIS RADIX) explaining most of the phenotypic variation for primary root elongation in a cross between accessions, Uk-1 and Sav-0. The QTL for primary root growth arrest has also been cloned and identified as the gene *LPR1* underlying a multi-copper oxidase that is involved in the perception of low phosphate status in the root cap (Svistoonoff et al. 2007). Even though cloning of QTLs has been done in *Arabidopsis*, very little has been achieved in transferring this to the crop plants.

## 2.7.2  Soybean

Deeper rooting might be the underlying mechanism influencing drought tolerance in soybean (*Glycine max* (L) Merr.) genotypes because it is positively related to yield under drought stress (Brown et al. 1985; Cortes and Sinclair 1986; Hudak and Patterson 1995). Root plasticity is also associated with drought resistance. Other studies have shown that a fibrous root system is more efficient for water absorption in soybean because of the increase in surface area and the number of root tips (Hudak and Patterson 1995; Pantalone et al. 1996). Genetic diversity in soybean root traits has been reported for root elongation, root nodule formation, fibrous root, and root mass in response to various abiotic/biotic stresses, such as aluminum toxicity (Bianchi-Hall et al. 2000; Villagarcia et al. 2001), flooding (Bacanamwo and Purcell 1999), soybean cyst nematode infection (*Heterodera glycines*) (Concibido et al. 1994; Miltner et al. 1991), iron deficiency (Charlson et al. 2005; Lin et al. 1997; Wang et al. 2008), manganese deficiency (Kassem et al. 2004), and phosphorus deficiency (Liang et al. 2010; Zhang et al. 2009a). Although many studies on the soybean root system have focused on the morphogenesis (Hudak and Patterson 1995, 1996; Pantalone et al. 1996; Sloane et al. 1990), physiological characteristics (King ct al. 2009; Sinclair et al. 2008; Manavalan et al. 2009), and nutrient uptake (Zhang et al. 2010), there have been efforts focusing on genetic improvement, particularly mapping QTLs.

Abdel-Haleem et al. (2011) identified five QTLs on chromosomes Gm01 (Satt383), Gm03 (Satt339), Gm04 (Sct_191), Gm08 (Satt429), and Gm20 (Sat_299) that explained up to 51 % of the phenotypic variation for fibrous root scores. These QTLs were colocalized with QTLs related to root morphology, suggesting that the fibrous root QTLs might be associated with other morphophysiological traits and seed yield in soybean. Even though none of these QTLs are considered a major fibrous root QTL, genetic dissection of the fibrous root trait at the individual marker loci can enable marker-assisted selection and development of soybean genotypes with *enhanced* levels of fibrous roots. Lu et al. (2010) reported a QTL for root dry weight and root weight (RW)/shoot weight (SW) ratio on chromosome 9, as well as a QTL for main root length (MRL) and RW/SW ratio on chromosome 17. Cui et al. (2007) identified three QTLs for RW/SW ratio on chromosomes 3, 10, and 18, which explained 4.8–9.6 % of the phenotypic variation. Wang et al. (2006b) detected three QTLs for root weight at the maturity stage on chromosomes 6 and 11 explaining 6.8–26.3 % of the phenotypic contributions, while Liu et al. (2005) identified six QTLs for drought tolerance on chromosome 6 at V4 stage, including three major QTLs that explained 22.0–24.7 % of phenotypic variation (relative dry weight of root, relative total length of root, and relative root volume). Zhou et al. (2009) detected six QTLs for root traits including

(RW, root volume (RV), RW/SW, MRL/plant weight (PW), and RW/PW) on chromosomes 11 and 14 and identified some novel QTLs on the hot chromosomes, such as one QTL for MRL and one QTL for RW/PW on chromosome 6 and one QTL for MRL/RW on chromosome 8. Another study by Brensha et al. (2012) aiming at evaluating QTLs that underlie several root and shoot traits in the "Essex" by "Forrest" (E × F, $n = 94$) recombinant inbred lines (RILs) found a total of 12 QTLs: three for maximum root length, one for lateral root number, one for basal root thickness, two for root fresh weight, and two for root dry weight. Using an RIL population derived from crosses between the cultivated soybean cultivar Jackson and wild soybean JWS156-1, Tuyen et al. (2010) detected a major QTL for alkaline salt tolerance on the molecular linkage group D2 (chromosome 17), which the wild accession allele contributed 50.2 % of the total variation for salt tolerance rating. With the prospects of changing environmental conditions, drought will become a major factor limiting plant growth and crop productivity in many parts of the world (Boyer 1982; Specht et al. 2001). Therefore, mapping QTLs and finding genes that control root and shoot traits is of extreme importance (Specht et al. 2001; Li et al. 2005a). The root and shoot trait QTLs found are therefore important when introduced in soybean breeding programs to produce improved drought-tolerant cultivars or germplasm lines with competitive yields.

### 2.7.3   Chickpea

The prolific root system in chickpeas (*Cicer arietinum*) contributes to grain yield under terminal drought conditions (Kashiwagi et al. 2006). Reports on the relationship of other morphophysiological traits to grain yield under drought conditions are variable. Thus, breeding efforts using any of these traits as criteria for drought tolerance are few. Although the importance of a prolific root system in terminal drought tolerance is well recognized, only limited efforts have been made to breed for improved root traits. This is because screening for root traits is a destructive and labor-intensive process that is difficult to use in large segregating populations. Combining different drought-resistance mechanisms is a potential strategy for enhancing the levels of drought resistance. Efforts have been made to combine the large root trait with few leaflets, and breeding lines have been developed combining these traits (Saxena 2003). It is well recognized that molecular markers linked to the major genes controlling root traits can facilitate marker-assisted breeding (MAB) for those traits. A major QTL contributing one third of the phenotypic variation for root length and root biomass has been identified (Chandra et al. 2004), and efforts are being made to identify additional QTLs for root traits. MAB for root traits in chickpeas is in progress.

The RILs of ICC 4958 × Annigeri have been extensively studied for root traits. A simple sequence repeat (SSR) marker (TAA 170) was found to the linked to a major QTL that accounted for 33 % of the phenotypic variation for root weight and 33 % of the phenotypic variation for root length (Chandra et al. 2004). Some preliminary screening of the chickpea mini-core germplasm collection for root proliferation and depth in cylinder culture has indicated that contrasting parents are available with wider variation for these traits than what was found between ICC 4958 and Annigeri (Krishnamurthy et al. 2003). RILs are now being developed from two new crosses (ICC 283 × ICC 8261 and ICC 4958 × ICC 1882) after selecting contrasting parents of similar maturity.

## 2.7.4   Common Beans

Common bean (*Phaseolus vulgaris* L.) is the most popular food legume of the East and Southern Africa and the Americas. It is a source of proteins, vitamins, and minerals in the diet of the poor (Broughton et al. 2003). About 60 % of common beans produced globally are grown in the marginal regions subjected to either intermittent or terminal drought risk (Thung and Rao 1999). There are genetic differences within common bean genotypes for adaptation to drought stress (Acosta-Gallegos and Adams 1991; Acosta-Gallegos and Kohashi-Shibata 1989; Ramirez-Vallejo and Kelly 1998; Munoz-Perea et al. 2006), and this variation has been exploited in breeding for drought tolerance (Teran and Singh 2002; Beebe et al. 2008; Singh et al. 2008). Plant mechanisms that improve drought adaptation in common beans include a deep and balanced rooting system that increases extraction of soil moisture from lower soil depth (Sponchiado et al. 1989; White and Castillo 1992).

Even though the importance of root traits for adaptation to drought stress is well recognized in common bean breeding (Sponchiado et al. 1989; Rao 2001; Beebe et al. 2008), interactions with different soils make root observation difficult to study. Selection for drought tolerance using QTL analysis would pave the way for the application of marker-assisted selection for root traits related to drought avoidance that are relatively more difficult and also time-consuming to evaluate phenotypically.

Asfaw and Blair (2012) used an RIL population of DOR 364 × BAT 477 to evaluate rooting pattern traits. One of the parents, BAT 477, is a drought-tolerant genotype, while the other parent, DOR 364, is a commercial cultivar developed for irrigated conditions and not adapted to drought. This study identified a total of 15 putative QTLs for seven rooting pattern traits (rooting depth, total root length, fine root length, thicker root length, specific root length, root volume, and root length distribution) and four shoot traits (leaf chlorophyll content, leaf area, leaf total nitrogen content, and stem total nitrogen content) that were scattered over 5 of the 11 linkage groups. The QTLs detected for all these root traits, except total root length and fine root length, were main-effect QTLs and did not interact with the level of water supply. Other QTLs for total root length, fine roots, thick roots, root volume, and root biomass were colocalized and explained relatively more pheno-typic variance. This suggests that the QTLs affecting root traits in common beans are based on the constitutive expression of genes and that drought avoidance based on deep rooting, longer root length, thicker roots, increasing root length distribution with depth, root volume, and root biomass can be used in molecular breeding. The results from such studies highlight the feasibility of marker-aided selection as an alternative to conventional labor-intensive, phenotypic screening of drought-avoidance root traits.

## 2.7.5 Maize

The importance of root system on grain yield in maize (*Zea mays*) was recognized many years ago (Wilson 1930). Tuberosa et al. (2002b) reported significant, although low, positive correlation between root traits and grain yield under water stress conditions. Similarly, in a recent study by Cai et al. (2012), a significant association was observed between grain yield and RSA. This study also revealed that the relationship between grain yield and RSA was greater when roots were established at an early plant developmental stage. Many studies have also revealed that the putative QTLs of root traits on chromosome region bins 1.06/1.07, 6.02, and 10.04 were associated with adaptation to abiotic stress, such as low nitrogen stress (Liu et al. 2008a, 2011), low phosphorus stress (Zhu et al. 2006b; Chen et al. 2008), and low water stress (Ribaut et al. 1996; Tuberosa et al. 2002b; Landi et al. 2010). Although it is still not possible to distinguish between pleiotropy and close linkage of RSA and abiotic stress tolerance, this could point to a possible genetic association between root growth and adaptation of plants to abiotic stress.

The design of ideotype for root architecture to attain optimal maize productivity relies on a thorough understanding of the genetic basis of RSA. Significant genetic variation in RSA exists among maize genotypes, and this could be tapped to enhance nutrient efficiency, drought tolerance, and lodging resistance (Jenison et al. 1981; Hebert et al. 1992; Landi et al. 1998; Tuberosa et al. 2003; Chun et al. 2005). Root traits are genetically controlled by many small-effect loci that strongly interact with the environment (de Dorlodot et al. 2007). Many QTLs that regulate RSA have been identified in several maize mapping populations, particularly in response to different environmental factors (e.g., nitrogen and phosphorus deficiency and drought stress) (Lebreton et al. 1995; Guingo et al. 1998; Kaeppler et al. 2000; Landi et al. 2002; Tuberosa et al. 2002b; Hund et al. 2004; Zhu et al. 2005, 2006a, b; Liu et al. 2008a, b; Hochholdinger and Tuberosa 2009; Messmer et al. 2009; Trachsel et al. 2009; Ruta et al. 2010a, b). Moreover, Hund et al. (2011) performed a meta-analysis of maize root QTLs using data from 15 QTL studies of nine mapping populations, and several putative consensus root QTL clusters were located in bins 1.07, 2.04, 2.08, 3.06, 6.05, and 7.04.

Although maize root QTL analysis has attracted much attention, several constraints that limit the progress of QTL discovery remain. These include the method of RSA phenotyping and the strategy of QTL mapping. Given that evaluation of maize RSA directly in the soil is difficult, RSA characterization is usually performed with seedlings grown in paper rolls, in hydroponic systems, or on gel-based plates (Kaeppler et al. 2000; Tuberosa et al. 2002b; Liu et al. 2008a, b; Hund et al. 2009a, b). Root traits expressed in young seedlings, such as root growth angle and seminal and lateral root length and number, can be evaluated easily, but more complex traits expressed at later stages of development, such as shoot-borne roots formation, are rarely evaluated (Lynch 2011; Trachsel et al. 2011). Furthermore, RSA phenotypes of seedlings grown in controlled environments might not accurately reflect root growth under field conditions (Zhu et al. 2011). To overcome these limitations, maize root QTL analyses are required on the basis of RSA characterization in plants grown in the field at different developmental stages.

Lebreton et al. (1995) identified QTLs for root traits influencing seminal root numbers (SRN), nodal root number (NRN) at the base of the stem, and root pulling force (RPF) measured at the end of the season. Four of these QTLs were for RPF and four were for both SRN and NRN. Some studies conducted using a maize population obtained from the single cross Lo9643 × Lo1016 highlighted the role of a major QTL on bin 1.06, which influenced root traits of seedlings grown in hydroponics (Tuberosa et al. 2002b) and of adult plants grown in the field (Landi et al. 2002). These two studies also showed that the same QTL region on bin 1.06 influenced grain yield under both well-watered and water-stressed conditions, thus leading Landi et al. (2002) to hypothesize that this QTL might concurrently be able to affect root features and grain yield.

## 2.7.6 Wheat

Wheat (*Triticum aestivum*) has a dense fibrous root system that is difficult for breeders to quantify and to select directly. Therefore, mapping QTLs for root traits and developing MAS will help breeders select root traits desirable for efficient acquisition of water and nutrients from soils.

In wheat, numerous QTLs for root traits have been detected, with QTL numbers varying from two to six, depending on the traits (Ren et al. 2012). In this study colocation of QTLs for primary and lateral root parameters was identified on chromosome 1A (between SSR markers Xcfa21581 and Xwmc329.2) and on chromosome 2B (between SSR markers Xgwm210 and Xbarc1138.2) for maximum root length (MRL) and lateral root length (LRL). These two loci might confer genes regulating both primary and lateral root development. It has been reported in *Arabidopsis* that mutations of some genes regulating RSA (such as *SHORTROOT* and *SCARECROW*) cause defects in both primary and lateral roots (Wysocka-Diller et al. 2000; Helariutta et al. 2000). Another QTL between Xgwm570 and Xgwm169.2 on chromosome 6A explained 30.5 % and 24.5 % of phenotypic variations in LRL and RN, respectively. These major loci showed linkage with previously reported QTLs for yield component and nutrient uptake. The existence of major QTLs for root trait and their linkage with agronomic traits and nutrient uptake will facilitate the design of root morphology for better yield performance and efficient nutrient use. The locus between Xbarc1005 and Xbarc257.2 on chromosome 7B was detected to control MRL and primary root elongation. This locus has been reported to control grain yield of durum wheat (*Triticum durum* Desf.) under a wide range of water availability (Maccaferri et al. 2008).

Deeper roots, especially deeper seminal roots, are considered important for wheat growth under drought conditions (Sanguineti et al. 2007; Araus et al. 2008). Mapped QTLs for primary root length (PRL) and maximum root length (MRL) might therefore provide candidate loci in breeding wheat with deeper root features and better performance under drought stress.

### 2.7.7 Sorghum

Nodal root angle in sorghum (*Sorghum bicolor* (L) Moench) influences vertical and horizontal root distribution in the soil profile and is therefore relevant to drought adaptation. Mace et al. (2012) mapped four QTLs for nodal root angle (qRA) in sorghum and three QTLs for root dry weight. Nodal root angle QTLs explained 58.2 % of the phenotypic variance and were validated across a range of diverse inbred lines. Three of the four nodal root angle QTLs showed homology to previously identified root angle QTLs in rice and corn, whereas all four QTLs colocated with previously identified QTL for stay-green in sorghum. A putative association between nodal root angle QTLs and grain yield was identified through single marker analysis on field testing data from a subset of the mapping population grown in hybrid combination with three different tester lines. Furthermore, a putative association between nodal root angle QTLs and stay-green was identified using datasets from selected sorghum nested association mapping populations segregating for root angle. Heritability for nodal root angle has been shown to be considerably greater than for plant size traits, such as root dry weight and shoot dry weight (Mace et al. 2012). This high heritability was also seen in durum wheat accessions (Sanguineti et al. 2007). The consistency of the nodal root angle phenotype of parents across experiments and the high broad-sense heritability (repeatability) indicate that nodal root angle is predominately influenced by genotype. The identification of nodal root angle QTLs presents new opportunities for improving drought adaptation mechanisms via molecular breeding to manipulate a trait for which selection has previously been difficult.

### 2.7.8 Rice

There has been considerable progress in mapping QTLs for root traits, and several of them are consistent across genetic backgrounds and environments in rice (*Oryza sativa*) (Kamoshita et al. 2008; Courtois et al. 2009; Khowaja et al. 2009). Introgression of these root QTLs into elite varieties through MAB might have an impact on the sustainability of rice production in drought-prone, rainfed environments

(Venuprasad et al. 2011). These populations have been screened for different drought-resistance traits grown under different stress protocols. To start with, QTLs for leaf rolling, root-related traits, and osmotic adjustment have been mapped in recombinant inbred (RI) lines of Co39 × Moroberekan (Champoux et al. 1995; Lilley et al. 1996; Ray et al. 1996). QTLs for root and certain physio-morphological traits were mapped using RI lines of Bala × Azucena (Price et al. 2000). QTLs for root traits and plant production traits have been mapped using doubled haploid (DH) lines of IR64 × Azucena (Yadav et al. 1997; Courtois et al. 2000; Hemamalini et al. 2000; Venuprasad et al. 2002). A DH line population derived from CT9993-5-10-1 and IR62266-42-6-2 has been extensively used by many researchers in mapping QTLs for root traits, osmotic adjustment (Kamoshita et al. 2002a; Nguyen et al. 2004), cell membrane stability (Tripathy et al. 2000), and epicuticular wax (Srinivasan et al. 2008). Kamoshita et al. (2002a) reported QTLs for root depth, penetrated root thickness, deep root-to-shoot ratio, deep root dry weight, deep root per tiller, and deep root mass to be associated with RM212 on chromosome 1 in CT9993/IR62266 DH lines. Also, a QTL for osmotic adjustment was reported to be close to this region in IR62266/IR60080 backcross progenies (Robin et al. 2003). This region was found to be associated with root volume (Qu et al. 2008) and basal root thickness in IRAT109/Yuefi RI lines (Li et al. 2005b) and leaf drying in Zhenshen/IRAT109 RI lines in rice (Yue et al. 2006).

Attempts have been made to improve yield under drought with the introgression of root QTLs into elite lines through MAB (Shen et al. 2001), in order to develop a few introgression lines with an improved root phenotype (Steele et al. 2006) and yield under rainfed environments (Steele et al. 2007). However, it is not clear whether the higher yields of the introgression lines were primarily a result of an improved root phenotype because certain other lines without the root QTLs also gave higher yields under rainfed conditions (Steele et al. 2007). The small phenotypic effect (some <10 %) and large intervals (3.7–7.3 Mb) of the root QTLs used in MAB are more likely to confer linkage drag of unwanted effects (Shen et al. 2001; Steele et al. 2006, 2007). Further, low heritability of certain root traits is also a common breeding concern (Gowda et al. 2011). Identifying large-effect QTLs with narrow intervals is critical for improving the efficacy of MAB (Bernier et al. 2008). In most rainfed lowland rice-growing areas of Asia, the soil density increases under drought, leading to soil compaction (Ingram et al. 1994), and root elongation is limited by both water deficit and mechanical impedance (Bengough et al. 2011). Rice plants with thicker roots are able to better penetrate hardpans and access soil moisture at greater depth (Yu et al. 1995). A large-effect QTL (37.6 %) for basal root thickness was detected on chromosome 4 in CT9993/IR62266 DH lines (Zhang et al. 2001). The same QTL was also linked to root pulling force in these DH lines (Nguyen et al. 2004). Root pulling force had positive correlations with plant water status and yield stability under drought in rice (Kumar et al. 2004). High root pulling force is associated with ability of the rice plant to develop deep and large-diameter roots with greater penetration ability (Lafitte et al. 2001; Clark et al. 2008). The QTLs for basal root thickness and root pulling force on chromosome 4 are colocated with QTLs for grain yield under drought stress in these DH lines (Babu et al. 2003). QTLs for putative traits overlapping with QTLs for yield under drought stress are good candidates for MAB for drought resistance (O'Toole 2004). The colocation of QTLs for putative root traits with QTLs for yield provides clues on their causal relationships (Hochholdinger and Tuberosa 2009) and can assist breeders in identifying the best QTL alleles for a successful MAB program (Landi et al. 2010). A QTL for penetrated root thickness was mapped on chromosome 9 in CT9993/IR62266 DH lines (Zhang et al. 2001). These QTLs on chromosomes 4 and 9 were found to be meta-QTLs (Courtois et al. 2009) and are thought to improve rice yields under drought through increased water uptake by roots (Kamoshita et al. 2008). Developing and testing NILs for these QTLs will help to verify their agronomic value and also to understand the mechanism of drought resistance in rice. Recently, Suji et al. (2012) observed that considerable variation in drought response and grain yield under rainfed condition in TPE among the IR20 NILs. Five out of 41 NILs tested gave higher yields under rainfed and irrigated conditions, as compared to IR20. Two NILs, viz., 212 and 297, with three and two root QTLs, respectively, had thicker and longer nodal roots and higher total and deep nodal root weights than IR20. In addition, NIL 297 had more nodal root volume and surface area, while NIL 212 had a higher number of nodal roots compared to IR20. The list of root trait QTLs reported in different rice mapping populations under different experimental conditions is summarized in Table 2.2, while Table 2.3 gives a list of root trait QTLs identified in different crops.

QTL studies on Al tolerance have been reported in rice using six different inter- and intraspecific mapping populations (Ma et al. 2002; Nguyen et al. 2003; Wu et al. 2000; Xue et al. 2006, 2007). Together, these studies report a total of 33 QTLs, located on all the 12 chromosomes, with three intervals (on chromosomes 1, 3, and 9) being detected in multiple studies. Nguyen et al. (2002) detected 20 QTLs controlling root growth under AL stress in rice. These QTLS were distributed over 10 of the 12 rice chromosomes suggesting multigenic control. Two QTLs with the largest effect on root length ratio (qALRR-1-1 and qALRR-8) were localized on chromosomes 1 and 8, respectively.

## 2.8   Transcriptome Analysis

The application of genomic techniques to plant research has yielded a multitude of discoveries concerning plant cellular biology, development, and evolution. Now, the sudden rise of relatively low cost and rapid "next-generation" DNA sequencing technologies is dramatically advancing our ability to comprehensively describe the nucleic acid-based information in a cell at unparalleled resolution and depth. Already this technology has been employed to study genome sequence variation, ancient DNA, cytosine DNA methylation, protein–DNA interactions, transcriptomes, alternative splicing, small RNA populations, and mRNA regulation, with a number of these applications being effectively applied to plant systems. Current deep sequencing technologies produce many giga bases of single-base resolution information and

**Table 2.2** List of root traits evaluated in different rice mapping population

| Genetic background | Parent type | Root traits evaluated | Number of QTLs (range of phenotypic variation (R2)) | Screening methods and treatments | Reference |
|---|---|---|---|---|---|
| CO39 × Moroberekan (RILs) | I × TJ | Root thickness, root/shoot ratio, root dry weight | 56 (6.0–33.0 %) | PVC cylinders; well watered and drought stress | Champoux et al. (1995) |
| | | Root penetration ability | 29 (6.0–19.0 %) | Wax-petrolatum layer system; well watered | Ray et al. (1996) |
| IR64 × Azucena (DH) | I × TJ | Root thickness, maximum root length, total root weight, deep root weight, deep root weight per tiller, and deep root to shoot ratio | 39 (4.0–22.3 %) | PVC cylinders; aerobic | Yadav et al. (1997) |
| | | Root penetration ability and root thickness | 12 (4.0–35.0 %) | Wax-petrolatum layer system; well watered | Zheng et al. (2000) |
| | | Root volume, root dry weight, root thickness, root length | 3 (12.9–30.7 %) | PVC cylinders; well watered and drought stress | Venuprasad et al. (2002) |
| | | Total root length, root volume, root thickness, root number per plant, root dry weight, and root/shoot ratio | 29 (11.9–26.7 %) | PVC cylinders; drought stress | Hemamalini et al. (2000) |
| CT9993 × IR62266 (DH) | TJ × I | Osmotic adjustment, root penetration index, root thickness, root pulling force, root dry weight, and root length | 36 (8.0–37.0 %) | Wax-petrolatum layer system and field; aerobic | Zhang et al. (2001) |
| | | Deep root morphology (deep root mass, deep root ratio), rooting depth, root thickness | 38 (3.6–51.8 %) | PVC cylinders; drought stress | Kamoshita et al. (2002a) |
| Bala × Azucena (RILs) | I × TJ | Root penetration ability | 17 (5.0–18.0 %) | Wax layer system; well watered | Price et al. (2000) |
| | | Root morphology and distribution | 25 (5.4–28.0) | Glass-sided chambers; drought stress | Price et al. (2002) |
| | I × TJ | Root length, root thickness | 28 (5.4–13.5 %) | Soil box method; well watered and drought stress | MacMillan et al. (2006) |
| | | | 23 (11.4–20.0 %) | | |

(continued)

**Table 2.2** (continued)

| Genetic background | Parent type | Root traits evaluated | Number of QTLs (range of phenotypic variation (R2)) | Screening methods and treatments | Reference |
|---|---|---|---|---|---|
| IR1552 × Azucena (RILs) | | Seminal root length, adventitious root number, lateral root length, lateral root number | | Pot culture, well watered and drought stress | Zheng et al. (2003) |
| IR58821 × IR52561 (RILs) | I × I | Total root number, root penetration ability, root penetration index, penetrated root thickness, and penetrated root length | 28 (6.0–27.0 %) | Wax petroleum layer system; well watered | Ali et al. (2000) |
| | | Deep root morphology and root thickness | 20 (5.7–29.9 %) | PVC cylinders; drought stress | Kamoshita et al. (2002b) |
| IAC165 × Co39 (RILs) | TJ × I | Maximum root length, root thickness, root dry weight | 29 (6.3–24.4 %) | PVC cylinders; aerobic | Courtois et al. (20C3) |
| IRAT 109 × Yuefu (DH) | TJ × TJ | Basal root thickness, total root number, maximum root length, root fresh weight, root dry weight | 51 (1.1–25.6 %) | PVC cylinders, lowland field; well watered and drought | Li et al. (2005a) |

*Source:* Gowda et al. (2011) (modified)

**Table 2.3** QTLs for various root traits reported in response to drought stress

| Trait | Species | Population | Type | No. of QTLs | Reference |
|---|---|---|---|---|---|
| Root size | Barley | Derkado × B83-12/21/5 | DHL | Many | Chloupek et al. (2006) |
| Roots traits and yield | Corn | Lo964 × Lo1016 | $F_3$ | <11 | Tuberosa et al. (2002b) |
| Roots traits and yield | Corn | Lo964 × Lo1016 | NIL | 1 | Landi et al. (2010) |
| Root growth traits | Rice | Azucena × Bala | RIL | <24 | Price et al. (2002) |
| Root length | Rice | Kalinga III × Azucena | NILS | 1 | Steele et al. (2006) |
| Root depth and length | Rice | IR64 × Kinandang Patong | RIL | 1 major | Uga et al. (2011) |
| Root length, number, thickness, penetration index | Rice | IR58821 × IR52561 | RIL | 28 | Ali et al. (2000) |
| Root morphology and root distribution | Rice | IR64 × Azucena | DHL | 39 | Yadav et al. (1997) |
| Root morphology, root cell length | Rice | Azucena × Bala | $F_2$ | 24 | Price and Tomos (1997) |
| Root thickness, root penetration index | Rice | CT9993 × IR62266 | DHL | 5 | Zhang et al. (2001) |
| Root thickness, root penetration index | Rice | IR64 × Azucena | DHL | 12 | Zheng et al. (2000) |
| Root traits | Rice | CT9993 × IR62266 | DHL | 8 | Kamoshita et al. (2002a) |
| Root traits | Rice | IR1552 × Azucena | RIL | | Zheng et al. (2003) |
| Deep root morphology, root thickness | Rice | IR58821/IR52561 | RIL | <12 | Kamoshita et al. (2002b) |
| Nodal root angle | Sorghum | | RIL | 4 | Mace et al. (2012) |
| Root aerenchyma formation | Corn | B64 × teosinte | F2 | 4 | Mano et al. (2007) |
| Boron efficiency and tolerance in yield | Arabidopsis | | RIL | 5 | Zeng et al. (2008) |
| Phosphorus deficiency and root elongation | Arabidopsis | Bay-O × Shahdara | RIL | 1 | Reymond et al. (2006) |
| Phosphorus deficiency and root morphology | Soybean | BD2 × BX10 | RIL | 3 cluster of QTLs | Liang et al. (2010) |
| Phosphorus deficiency resistance | Soybean | Kefeng × Nanong 1138–2 | RIL | 7 | Li et al. (2005a) |
| Aluminum tolerance | Soybean | Essex × Forrest | RIL | 11 | Sharma et al. (2011) |

(continued)

**Table 2.3** (continued)

| Trait | Species | Population | Type | No. of QTLs | Reference |
|---|---|---|---|---|---|
| Manganese tolerance (root necrosis) | Soybean | Essex × Forrest | RIL | 1 | Kassem et al. (2004) |

*Source*: Plant stress.com (modified)

*DHL* doubled haploid, *RIL* recombinant inbred lines, *BC* backcross, *BIL* backcross recombinant inbred lines, *NIL* near-isogenic lines

can perform multiple genome-scale experiments in a single experimental run. They are, therefore, effective in the analysis of many plant genome equivalents. With this technology, the instrument massively parallelizes individual reactions, sequencing hundreds of thousands to hundreds of millions of distinct relatively short (50–400 bases) DNA sequences in a single run. However, it should be noted that some significant challenges remain in the employment of this new technology. The most evident are informatics and data processing issues that arise from the generation of such large volumes of data (terabytes per run is not unusual). The reviews by Lister et al. (2009), Jackson et al. (2011), and Paterson et al. (2010) will be useful to understand the advancements in sequencing technologies, applications, and its insights in comparing plant genomes. This section focuses on various applications of DNA sequencing technologies to understanding the mechanism of root traits in relation to abiotic and biotic stress tolerance in crop plants.

## 2.8.1 Arabidopsis

There have been a number of transcript-level studies of plant roots under low-$O_2$ conditions. Exogenously applied sucrose greatly enhances the anoxia tolerance of *A. thaliana*, so Loreti et al. (2005) tested root transcript response to low $O_2$ with and without sucrose. In particular, they found that alcohol dehydrogenase (ADH), pyruvate decarboxylase 1 and 2 (PDC1 and 2), and *Arabidopsis* nonsymbiotic hemoglobin 1 (AHB1) are induced within 40 min of anoxia, and there is a prolonged ability to accumulate ADH and PDC1 when exogenous sucrose is available. Sucrose synthase (SUS) is usually induced under low-$O_2$ conditions, but here it was not induced under anoxia when exogenous sucrose was supplied. Some general trends under anoxia noted by the authors were that lipid metabolism decreased, glycolysis enzymes increased, and several heat-shock protein (HSP) transcripts increased. Under low $O_2$, plants shift metabolism to generate energy via glycolysis and fermentation of pyruvate (reviewed in Bailey-Serres and Voesenek 2008), and the transcript changes here are consistent with those cellular programs. In a combined transcriptome and metabolome study of *A. thaliana* roots under hypoxia, Van Dongen et al. (2009) found similar trends to Loreti et al. (2005).

Transcripts of *ADH*, *PDC*, *SUS*, and *AHB1* were induced. Generally, there is an increase in transcripts that serve to generate ATP under low $O_2$ and a decrease in transcripts of proteins associated with ATP-consuming processes. In metabolite analyses, they noted a strong increase in carbon flux to glycolysis brought on by the switch to fermentative metabolism. Some specific metabolites that were induced were proline and GABA, similar to what was seen in root drought studies.

To bring the functional genomic data on plants under low $O_2$ together, Narsai et al. (2011) performed a synthesis study of transcripts and metabolites under low $O_2$. Common changes in carbon metabolism are noted. This was also seen in another synthesis study by Mustroph et al. (2010) that looked at transcript responses to low $O_2$ over four kingdoms of life. Narsai et al. (2011) saw that a small number of genes orthologous between rice and *A. thaliana* were regulated in opposite manners, including zinc finger domain transcription factor and kinases. Also, nonsymbiotic hemoglobin transcript levels increased in *A. thaliana,* though they decreased in rice and poplar (*Populus trichocarpa*). This might have to do with nitric oxide (NO) scavenging and signaling. Mustroph et al. (2010) noted that NO reductases are induced in plants and bacteria during hypoxia. Therefore, NO signaling is an area that deserves further consideration in future low-$O_2$ studies. There are some additional insights on transcriptional regulation under low $O_2$ from Mustroph et al. (2010). The authors saw that most organisms induce heat-shock protein and ROS network genes under low-$O_2$ stress, including some heat-shock transcription factors. Specifically in plants, several other transcription factor families had members that were DEGs, including ERF, MYB, CCCH-type zinc finger, WRKY, bZIP, NAC, and MADS. The DEGs in plants also included ethylene biosynthesis and detection transcripts. This is not surprising considering the roles this hormone is known to play in signaling for promotion or repression of under-water elongation, as well as the development of aerenchyma. Because part of a plant's response to low $O_2$ can be elongation, cell wall expansins and XETs were also DEGs, as they were for roots under low water stress. Another interesting aspect of the DEGs analyzed by Mustroph et al. (2010) is the number of transcripts for proteins of unknown function. Four hundred fifty of these hypoxia-responsive unknown proteins (HUPs) are conserved across species, and 89 out of the 200 most highly induced genes are higher plant-specific HUPs. These HUPs make interesting targets for future functional studies of low-$O_2$-responsive genes. Indeed, overexpression of a subset of the HUPs in submerged *A. thaliana* increased survival, although the authors note more precise regulation might give better results (Lee et al. 2011). Certain transcription factors and signaling components important for species-specific responses to low $O_2$ have been verified in functional studies. As already discussed, expression of ERF transcription factors SUB1A or SNORKELs has roles in determining rice response to submergence (Xu et al. 2006; Hattori et al. 2009). Another ERF transcription factor, *hypoxia-responsive element 1* (*HRE1*), confers tolerance to low oxygen in *A. thaliana* when it is overexpressed (Licausi et al. 2010). A heat-shock transcription factor transcript (*HSFA2*) is also induced by anoxia in *A. thaliana*. Subsequently, its protein levels rise, conferring protection to plant tissues under stress (Banti et al. 2010). This is significant because there is often a disconnect between transcript induction and protein levels under anoxia. This might be because translational efficiency during low $O_2$ can be significantly impaired, possibly because of low ATP and changes in posttranscriptional regulation (Branco-Price et al. 2005, 2008).

Michal et al. (2009) profiled gene expression in *Arabidopsis* roots 2.5 and 5 h after shoot exposure to low air-relative humidity. The expressions of some aquaporins were found to be induced under low RH, in addition to multiple genes with diversified biological roles. Transcription of two aquaporins was localized to leaf trichomes and the roots, especially to cells around the vascular system and cells of the differentiation zone. These results suggest shoot–root communication with plant roots perceiving the low RH stimulus from shoots through a sensing mechanism(s). This leads to distinct plant transcriptional responses, potentially reflecting the activation of various biological processes. Transcriptome analysis of gene expression during the hydrotropic response in *Arabidopsis* seedlings (Moriwaki et al. 2010) revealed the transcript levels of 793 genes were significantly changed 1 or 2 h after hydrotropic stimulation. A large number of hydrostimulation-responsive genes were found to be similar with abscisic acid (ABA) or water stress-responsive genes. In contrast, a little overlap of transcript abundance between hydrostimulation-responsive and gravistimulation-responsive genes was also observed. These results suggest that ABA and water stress responses are important signal transduction mechanisms involved in the root hydrotropic response, and the signaling pathways involved in hydrotropism differ from those of gravitropism.

Kovalchuk et al. (2005) tested the effect of two heavy metals, cadmium (Cd) and lead (Pb), on *Arabidopsis* and reported induced root growth in plants exposed to Pb. They noticed that plants grown on mediums with a concentration of 200 μM lead to roots twice as long as roots of plants grown under control conditions. Based on global genome expression analysis, they also observed a significant percent of transport-related genes to be upregulated by Cd and Pb, 13 % and 11 %, respectively. The following genes were found to be differentially regulated on exposure to heavy metal stress (Cd and Pb): germin-like, cytochrome P450, major latex protein (MLP). Germin-like protein genes were shown to be induced by various stresses, including heavy metal stress (Patnaik and Khurana 2001; Nakata et al. 2002). A similar study to understand the transcriptional regulation in roots and leaves of *Arabidopsis* to cadmium treatment was reported by Herbette et al. (2006) using the whole-genome CATMA microarray. This study demonstrates the existence of a regulatory network that differentially modulates gene expression in a tissue- and kinetic-specific manner in response to cadmium. The response in roots was observed with induction of genes involved in sulfur assimilation–reduction and glutathione (GSH) metabolism. In addition, HPLC analysis of GSH and phytochelatin (PC) content shows a transient decrease of GSH in roots after 2 and 6 h of metal treatment correlates with an increase of PC contents. Altogether, their results suggest that to cope with cadmium, plants activate the sulfur assimilation pathway by increasing transcription of related genes to provide an enhanced supply of GSH for PC biosynthesis. This transcription regulation could be the first step of an adaptive response required to ensure a sufficient supply of sulfur compounds during Cd-induced

PC synthesis. In addition, rapid responses in the transcriptome of *Arabidopsis* roots to a decreased iron (Fe) supply were studied using DNA microarrays, revealing nodulin-like candidate gene families with putative roles in Fe homeostasis (Gollhofer et al. 2011).

Functional genomic analysis of plant responses to nutrient deficiency has been mainly based on transcriptional profiling coupled with the use of T-DNA mutants, overexpression, and RNAi strategies. Pi starvation led to the differential expression of several hundred genes, of which 40–50 % was repressed when Pi was resupplied in *Arabidopsis* (Misson et al. 2005). *PLDZ1* and *PLDZ2* are strongly induced in Pi-deprived plants, and mutational analysis showed that the encoded phospholipases release Pi for other cellular activities and regulation of primary root growth by degrading phospholipids. This might be because of reduced Pi recycling or because the biosynthesis of phospholipid-derived signaling molecules is compromised (Li et al. 2006). In addition to genes involved in metabolic processes, four TFs induced by Pi deprivation were shown to play roles in root architecture and/or root hair formation. Overexpression of *AtZAT6* altered root architecture (Devaiah et al. 2007a), while mutant analysis of *AtbHLH32* reduced root hair number and resulted in higher anthocyanin and Pi content (Chen et al. 2007). RNAi suppression of AtWRKY75 resulted in an increase in LR and RH number and length (Devaiah et al. 2007b). Genomic analysis also led to the discovery of microRNAs involved in Pi homeostasis. Pi starvation rapidly induces the expression of microRNA399 (miR399), which mediates the degradation of the mRNA of PHO2. *Arabidopsis* pho2 mutants accumulate excessive amounts of Pi in the shoot. Suppression of PHO2 by the transcriptional activation of miR399 leads to increased Pi acquisition and translocation to the shoot. miR399 acts as a systemic signal because it undergoes long-distance movement from shoot to roots (Lin et al. 2008; Pant et al. 2008). Low-potassium (K) availability also regulates root architecture and leads to a reduction in the number and length of LRs. Microarray analysis showed that genes related to reactive oxygen species (ROS) are activated, while the TF AtMYB77 is repressed in K-limiting conditions. Further analysis revealed that hydrogen peroxide ($H_2O_2$) is involved in K signaling and leads to increased K uptake (Shin and Schachtman 2004) and MBY77 modulates auxin sensitivity and LR formation by interacting with auxin-responsive factors (*ARFs*). This suggests that reduction of LR formation under low-K availability might be due to a reduction in auxin response (Shin et al. 2007).

In addition to the nutrients discussed above, genome-wide nitrogen (N) responses have been examined in *Arabidopsis*, maize, and *Medicago truncatula*. Over 1,000 genes were found to be responsive in the three plant species, and responses to N deprivation were more rapid than for other nutrients (Scheible et al. 2004; Liu et al. 2008b). Systemic signals were also shown to trigger specific transcriptome responses depending upon the nitrogen source available to the plants (Ruffel et al. 2008). Although transcription profiling identified many N-responsive genes, including several dozen TFs, the functional characterization of genes responsive to local or systemic N signals is still lacking. Two recent studies give us the first look at environmental effects on the transcription network in specific cell types.

Dinneny et al. (2008) examined the response to high salt and iron deficiency in specific cell types and developmental stages of the root. Most responses to salt and iron were cell type specific and were dependent on specific environmental conditions. The 244 genes unaffected by either stress were enriched for genes functioning in cell-type specification, suggesting that genes necessary for cell identity are environment independent. Unexpectedly, genes coregulated by salt and iron are not ubiquitously expressed. This indicates that cell-type-specific genes might be targets for stress regulation. The level of conservation among biological functions enriched in cell types under high-salt and low-iron conditions varied according to cell type and stress, highlighting the need for additional studies in this area. A similar study illustrates the power of these datasets to reveal novel connections in the root. Gifford et al. (2008) examined the effect of adding back nitrogen (N) on the transcriptional profile of different cell types in N-depleted roots. Their analysis revealed that miR167a,b and its target, ARF8, regulate lateral root development in response to nitrogen. The authors identified 126 putative targets of ARF8 and showed that N status coordinately regulates these genes. Several recent reports analyze the effects of stress, nutrients, and genotype on the *Arabidopsis* metabolome (Cook et al. 2004; Hirai et al. 2004; Nikiforova et al. 2005; Keurentjes et al. 2006; Rowe et al. 2008), but these have focused primarily on aerial organs or whole plants. Cell-type-specific proteomic and metabolic data are currently in development (Benfey lab unpublished data). The combination of transcriptional, proteomic, and metabolomic cell-type-specific data will facilitate not only the functional identification of networks controlling root growth and development but also an understanding of how these networks interact with each other and the emergent properties of the root.

Cell-type-specific profiling reveals more transcriptional complexity than whole organ profiling because of dilution of cell-type-specific genes in whole organ experiments. Transcriptional profiling of nearly all cell types in the *Arabidopsis* root, coupled with 13 longitudinal sections, created a complete spatiotemporal map of the root (Brady et al. 2007). This analysis revealed dominant expression patterns between ontologically unrelated cell types and their fluctuation in developmental time. This study also predicted new transcription factor targets from coexpressed genes for auxin biosynthesis functions and MYB-binding sites and *Altered Tryptophan Regulation 1* (*ATR1*). Understanding the functional importance of these dominant expression patterns will contribute to our understanding of the networks that guide root development in space and time. Another set of studies has examined the transcriptome of different organs and developmental stages of *Arabidopsis*, including the root, in response to more than 40 conditions (Schmid et al. 2005; Kilian et al. 2007; Goda et al. 2008). Specifically, the cell-type-specific resolution AtGenExpress data provide a powerful tool for coexpression analyses, reverse genetics, and functional genomics approaches. Yang et al. (2011) made a global analysis of gene-level microRNA expression using deep sequencing data to study the miRNA biogenesis at different *Arabidopsis* tissues. They determined that less than 40 % of annotated MIR genes are expressed in shoot, root, or inflorescence. These results provide conservative but accurate gene-level expression information on the miRNA transcriptome in *Arabidopsis*. These authors also demonstrated that miRDeep can be adapted to plants with specifically modified parameters. In the near future, these expanded expression profile studies with higher cellular resolution should offer desired information that can be used to trace evolutionary changes in the cisregulatory elements and subfunctionalization of the MIR genes. In addition to gene expression, proteins and metabolites contribute to an organism's molecular phenotype. A recent global analysis of the *Arabidopsis* whole root proteome identified approximately 5,159 proteins in 10-day-old roots and 4,466 in 23-day-old roots (Baerenfaller et al. 2008). Proteins in GO categories for intracellular protein transport, response to oxidative stress, and toxin catabolic processes were overrepresented in their analyses.

## 2.8.2   Soybean

For transcript analyses, there have been several recent studies looking at how the expression levels of specific gene families are affected by drought stress. This includes genes involved in the synthesis of isoflavonoids (Gutierrez-Gonzalez et al. 2010), protein phosphorylation kinases (Le et al. 2011a), and NAC domain transcription factors (Hao et al. 2011; Le et al. 2011b). Gutierrez-Gonzalez et al. (2010) showed that progressive water deficit during seed-filling stages caused a decrease in seed isoflavone accumulation. The only isoflavone biosynthetic transcript significantly changed by drought in this study was the first in the pathway, *PAL1*, which is upregulated. The authors hypothesize that during severe water deficit, phenylpropanoid metabolism is shifted towards compounds other than isoflavones. In particular, it might be directed towards creating a precursor for several antioxidant compounds. Signal transduction pathways are also commonly altered by phosphorylation during abiotic stresses, including drought (for review, see Yang et al. 2010). Le et al. (2011a) studied transcript levels of genes involved in His-to-Asp phosphorelay during dehydration stress. Many of the 83 genes in this study had high levels of root expression or were even root specific. Among them, 51 of the transcripts were found to be dehydration responsive, and 25 of those had changes in the root tissue. This study again emphasizes tissue specificity of transcript changes and that, in some cases, opposite effects are seen in root versus shoot tissue. Le et al. (2011b) found 152 NAC transcription factors in soybean and predicted 58 of them to be stress-induced based on homology with *A. thaliana* NACs. Further investigation with a subset of the NACs found that 29 of them were reduced and six of them were repressed by dehydration stress. Twenty-two of the transcript changes varied by tissue type, and two of them were root specific. Hao et al. (2011) focused on two of the soybean NACs, *GmNAC20* and *GmNAC11*. They did not test for drought stress, but they found that overexpression of either gene in *A. thaliana* improved root growth and survival under salt stress. *GmNAC20* overexpression also led to enhanced cold tolerance, and even with no stress applied, upregulation of this NAC gave increased lateral root formation, probably through auxin signaling. A similar lateral root promotion was seen in *A. thaliana* with *AtNAC2* overexpression (He et al. 2005). The *GmNAC20* and *AtNAC2* protein sequences share 70 % identity. It would be of interest to see if overexpression of *GmNAC20* also improves low water stress tolerance and discover how root-specific overexpression affects soybeans under drought.

There has also been a proteomic analysis of soybean roots under low water stress (Yamaguchi et al. 2010). In this study, protein levels in three defined regions of the root tip were observed under water stress conditions in comparison with developmental and temporal controls. In total, they detected 35 proteins with abundance changes in response to water stress, and the majority of the changes were conserved between the first two root-tip segments. Three major categories of proteins were discovered in this analysis, including isoflavonoid biosynthesis proteins, proteins involved in lignin accumulation, and proteins regulating ferritin/iron distribution. Isoflavones and the sequestering of free iron in cells could be working to lower ROS accumulation during low water stress. Lignification might be triggered by ROS buildup, but it has possible roles in aiding roots under water stress conditions by targeting water transport to growing tissues while preventing water loss. Isoflavones might have an additional role in determining the distribution of auxin in the root tip, thereby modulating root growth patterns. However, the precise functions for auxin in roots during low water stress need further evaluation.

In a series of proteomic experiments on waterlogged soybean roots, changes in protein involved in fermentation, glycolysis, isoflavonoid production, and increased cell wall extensibility have also been observed (Komatsu et al. 2009; Alam et al. 2010). Included in these proteins is an *ADH* gene, *ADH2*, which responds specifically to flooding stress in the root of soybean (Komatsu et al. 2011). In Alam et al. (2010), they noted an increase in coproporphyrinogen oxidase, which again points to the importance of hemes and oxygen signaling during waterlogging stress. They also saw increases in a number of proteins linked to PCD and therefore possibly to aerenchyma formation. There was a decrease in SAM in this study, as seen in tomato (Ahsan et al. 2007). In a cell wall-specific proteome study of flooded wheat roots, there was a notable decrease in proteins involved in cell wall elongation and an increase in chitinases, which could be involved in abiotic stress signaling (Kong et al. 2010). The variable results in these proteomic studies with respect to cell wall composition and extensibility point to the need for further, region-specific protein analysis to find the broad picture of cell wall activity during low oxygen stress. For the metabolomic portion of the Narsai et al. (2011) analysis, they found that plant species commonly shift to fermentative pathways under low $O_2$, leading to lactate and ethanol production. Other metabolites that commonly increased were alanine, GABA, succinate, lysine, and tyrosine. Conversely, decreases in aspartate were common. Some proline accumulation was noticeable, especially in later time points of the studies. This study included the metabolite data from Van Dongen et al. (2009), and therefore the observations in both are similar. In addition, Van Dongen et al. (2009) noted that changes in many metabolite levels upon low oxygen stress can be transient. This might be due in part to an initial increase followed by a decrease, when the stress induces the breakdown of sucrose and glycolysis.

To identify genes responsible for multiple stress tolerance, the transcriptional profiles of genes in leaves and roots of seedlings (two-leaf stage) of the soybean inbred line HJ-1 were examined after 48 h under various stress conditions: salt (120 mM NaCl), saline/alkali (70 mM NaCl and 50 mM NaHCO$_3$), and drought (2 % PEG 8000). Gene expression at the transcriptional level was investigated by Fan et al. (2012) and also identified 874, 1,897, and 535 genes were upregulated in leaves under salt, saline/alkali, and drought conditions, respectively. They also found that 1,822, 1,731, and 1,690 genes were upregulated in roots under salt, saline/alkali, and drought stress, respectively. Other comparisons among salt, saline/alkali, and drought stress yielded similar results in terms of the percentage of genes classified into each GO category. Moreover, 69 genes differentially expressed in both organs with similar expression patterns are clustered together across all conditions. Furthermore, comparison of gene expression among salt-, saline-/alkali-, and drought-treated plants revealed that genes associated with calcium signaling and nucleic acid pathways were upregulated in the responses to all three stresses, indicating a degree of cross talk among these pathways.

Qiao-Ying et al. (2012) recently made an attempt to construct two small RNA libraries and two degradome libraries from the roots of Al-treated and Al-free *G. soja* seedlings. In this study, they identified 30 miRNA and 86 target genes of the known miRNAs in response to aluminum stress. In addition, five genes were found to be the targets of novel miRNA. Most of the target genes cleaved by conserved miRNA families (52 genes) might play roles in the regulation of transcription. Additionally, some genes that are known to be responsive to stress, such as those for the *Auxin Response Factor* (*ARF*), domain-containing disease resistance protein (*NB-ARC*), *leucine-rich repeat and toll/interleukin-1 receptor-like* protein (*LRR–TIR*) domain protein, cation transporting ATPase, Myb transcription factors, and the no apical meristem (NAM) protein, were found to be cleaved under Al stress conditions.

### 2.8.3 Chickpea

With chickpea, as with many other under-researched crops, microarray-based expression analyses have not been performed. Instead, open-architecture transcription profiling technologies, such as deep SuperSAGE (Matsumura et al. 2012), were applied. Using SuperSAGE, Molina (2008) described 80,238 regulated unique 26 bp tags (UniTags) representing 17,493 transcripts from roots of the drought-tolerant chickpea variety ILC588 early after onset of desiccation. Millan et al. (2006) comprehensively characterized at the molecular level drought stress response in chickpea and served as a prerequisite for the identification of expression markers useful for high-throughput germplasm and expression-QTL (e-QTL) analysis at the onset of knowledge-based breeding for stress tolerance. In addition, a follow-up study on responses of chickpea roots to salinity stress by Molina et al. (2011) also provided evidence for an important role of oxylipins in desiccation stress responses and eventually also for dehydration tolerance. Varshney et al. (2009) studied the implication of some key enzymes of the lipoxygenase pathway in response to drought stress in the roots from the drought-tolerant chickpea variety ICC 4958 and the sensitive variety ICC 1882. Recently, Domenico et al. (2012) evaluated changes in gene expression of the target genes, specifically for two lipoxygenases (*LOX 1* and *LOX 2*), two hydroperoxide lyases (*HPL 1* and *HPL 2*), an allene oxide synthase (*AOS*), an allene oxide cyclase (*AOC*), and an oxophytodienoate reductase (*OPR*) at different time (0, 2, 24, 48, and 72 h) after the onset of drought stress. They further showed that gene overexpression positively correlates with the levels of major oxylipin metabolites from the *AOS* branch of the pathway, which finally leads to the synthesis of jasmonates. Higher levels of jasmonic acid (JA), its precursor 12-oxophytodienoic acid (OPDA), and the active form JA-isoleucine (JA-Ile) were especially detected in the root tissues of the tolerant variety, prompting the authors to assume jasmonates have a role in the early signaling of drought stress in chickpea and its involvement in the tolerance mechanism of the drought-tolerant variety.

Comparative analyses of exudates from 6-day-old roots of the three legume species white lupin (*Lupinus albus*), soybean (*Glycine max*), and cowpea (*Vigna sinensis*) were performed (Liao et al. 2012) under axenic conditions, and their constitutively secreted proteomes were analyzed. Between 42 and 93 unique root extracellular proteins with two or more different peptide fragments per protein were identified by LC–MS/MS. Functional annotation of these proteins classified them into 14–16 different functional categories. Among those, 14 homologous proteins were common among two legume species. Among the unique proteins, 58 in white lupin, 85 in soybean, and 31 in cowpea were specific for each plant species, and many of them were classified in the same functional categories. Interestingly, in contrast to soybean and cowpea, two protein bands of approximately 16 and 30 kDa were present on the SDS-PAGE gel of white lupin. The identification of these bands revealed pathogenesis-related proteins, class III chitinase, and a thaumatin-like protein. These results imply that root extracellular proteins play important roles in the cross talk between plant roots and the rhizosphere.

## 2.8.4   Groundnut

Using suppression subtractive hybridization (SSH), cDNA libraries that were enriched with differentially expressed ESTs from root knot nematode-challenged root tissues from near-isogenic tolerant (NemaTEM) and susceptible (Florunner) peanut cultivars were collected and analyzed. Seventy ESTs (36 from NemaTEM- and 34 from Florunner-specific libraries) were identified and annotated into seven functional categories (stress responses, metabolism, transcriptional regulation, protein synthesis and/or modification, transport functions, cellular architecture, and proteins with unknown functions). However, significant transcriptional reprogramming and upregulation of genes in roots implicated in modification of cellular architecture, adhesion, and proliferation marked an early onset of compatible host–pathogen interaction (Tirumalaraju et al. 2011).

## 2.8.5 Rice

Using an Affymetrix rice genome array chip, Wang et al. (2011a, b) profiled control and stressed roots at each of two developmental stages and found 1,154 and 1,114 differentially expressed genes (DEGs) at tillering and panicle elongation stages, respectively. Root-specific upregulated genes included many transcription factors, as well as genes involved in metabolism, cell wall expansion, and response to phytohormones. This study also looked at DEGs in leaves and the panicle under drought. They found a group of transcripts commonly induced in all three tissues studied, including late-embryogenesis abundant (LEA) proteins, dehydrin family, protein phosphatase 2C family, and abiotic stress response proteins. Yang et al. (2004) found 121 transcripts affected by water deficit in various sections of the rice root system, and Rabello et al. (2008) found 84 transcripts expressed exclusively in their drought-tolerant genotype. For both studies, there is much overlap with the functional categories found with genome-wide profiling. The highest number of transcripts in Yang et al. (2004) is in the category of cell organization and cell wall biogenesis. This group includes expansins, several of which were also DEGs in Wang et al. (2011a, b) and have been considered as candidate genes for rice root QTLs (Vinod et al. 2006). The candidate genes associated with maximum root length (MRL) between specific markers SBH14 (Sharma et al. 2003) and RM201 (Shashidhar et al. 2000), reported earlier, were confirmed with CT9993/IR62266 NILs (Prabuddha et al. 2008). A critical finding of the Wang et al. (2011a, b) study was that there were several genes regulated in opposing manners depending on tissue type and/or stage. This finding highlights the importance of temporal tissue and even cell-specific studies instead of making generalizations about abiotic stress response based on a single tissue type at one point in time. Moumeni et al. (2011) profiled the root-specific gene expression pattern and surveyed for differentially expressed genes (cell growth systems, hormone biosynthesis, amino acid metabolism, transport system, transcription factors, etc.) in IR64 NILs under different levels of drought stress. The activated genes under drought stress were mostly involved in secondary metabolism, response to stimulus, defense response, transcription, and signal transduction, and downregulated genes were involved in photosynthesis and cell wall growth. Some of the specific TF genes, such as NAC (*LOC_Os02g57650*), SNFs (*LOC_Os02g32570*, *LOC_Os04g47830*), and *bZIP* (*LOC_Os09g13570*), were specifically activated in root tissue of tolerant NILs. They also identified gibberellic acid and auxin cross talk as a factor to modulate lateral root formation in tolerant NILs.

Currently, there are only a few examples when altering transcript levels of specific root-related rice genes which have led to increased drought tolerance. The overexpression of a metallothionein expressed highly in roots, *OsMT1a*, allowed plants to grow better under osmotic stress and dehydration (Yang et al. 2009). The authors believe the increased *OsMT1a* might activate antioxidant enzymes and zinc finger transcription factors leading to enhanced ROS scavenging and, therefore, drought tolerance. Both Rabello et al. (2008) and Wang et al. (2011a, b) showed changes in transcript levels of metallothionein-family proteins under drought stress. Other overexpression experiments involved NAC domain transcription factors, which are plant-specific and known to act during biotic and abiotic stress. One *NAC* gene, *OsNAC10*, was overexpressed specifically in roots. It increased root diameter over nontransgenic plants and significantly increased grain yield under drought conditions in the field (Jeong et al. 2010). Overexpression of another NAC gene that has high root expression levels, *OsNAC45*, increased seedling recovery from desiccation stress (Zheng et al. 2009). Globally, the expression levels of several NAC domain transcription factors were shown to be differentially regulated upon drought stress (Wang et al. 2011a, b).

A genome-wide transcriptome study in rice also characterized the arsenic-responsive genes and found a large set of differentially expressed genes belonging to the defense and stress response in plants, in addition to being involved in general metabolism (Chakrabarty et al. 2009). However, arsenate (AsV) stress led to the induction of a larger set of responsive genes in comparison to arsenite (AsIII), indicating the occurrence of an AsV-specific response in rice. Among the expressed genes under arsenic stress, 72 and 27 genes are unique to differential regulation by two different inorganic forms of arsenic (AsV) and (AsIII), respectively. However, expression of one of the *Cytochrome P450* genes (*Os01g43740*) in rice root was induced by arsenate (AsV), but not by other heavy metals. Thus, this gene can be used as an important biomarker for arsenate (AsV) stress in rice. A number of differentially expressed genes were found in Cd-exposed rice roots (Zhang et al. 2012), including 28 upregulated genes and 19 downregulated genes. They were found to be involved in diverse biological processes, such as metabolism, stress response, ion transport and binding, protein structure and synthesis, as well as signal transduction. Notably, a number of known functional genes were identified encoding membrane proteins and stress-related proteins, such as heat-shock proteins, monosaccharide transporters, CBL-interacting serine/threonine-protein kinases, and metal tolerance proteins. In this study, they found that several cDNA clones derived from Cd-exposed rice seedlings encode membrane proteins, Nramps (natural resistance-associated macrophage proteins) involved in metal ion transport. The altered expression of Nramp and the carrier membrane protein gene might be an adaptive reaction of the rice root cells to Cd stress. In addition, the cation diffusion facilitator (CDF) family, also called the cation efflux (CE) family, belongs to the MTP (metal tolerance protein) group. The upregulation of putative copper amine oxidase by Cd stress triggers the accumulation of ROS in rice root cells, thus altering the ROS status and causing a change in the expression of ROS-signaling molecules. Thus, it would be helpful in eliminating ROS in rice root cells and might aid in the protection of plant cells from Cd stress. Laser capture microdissection (LCM) is frequently the method of choice for cell-type-specific analyses in species where GFP-marked lines are not readily available. An LCM transcriptome atlas for rice contains 40 cell types, including 13 from the root (Jiao et al. 2009). As with *Arabidopsis*, GO category enrichment in specific cell types correlated with known biological functions, and new functions could be predicted. Comparisons between *Arabidopsis* and rice roots showed interesting parallels and dissimilarities. Finally, overexpression of a rice bHLH TF, *OsPTF1*, led to an increase in root biomass and a concomitant increase in tiller number and Pi content in rice (Yi et al. 2005).

### 2.8.6  Maize

Maize root system architecture is significantly shaped by its interaction with the soil environment and its adaptability to changing environmental conditions. Recently, several studies analyzed transcriptome profiles of maize roots after exogenous stimulation. RNA-level changes at low water potential have been looked at in defined sections of the maize root tip (Spollen et al. 2008). The root-tip sections were decided based on observations in Sharp et al. (2004) that 3 mm from the root tip, the expansion rate of root cells under water stress begins to decelerate until it completely ceases around 5 mm from the tip. These dynamics of longitudinal expansion are different from those seen in control root cells and show a reduced zone of growth in maize root tips under water stress. By comparing transcript changes in the different root-tip regions under control and stress conditions, Spollen et al. (2008) identify several categories of transcripts thought to be involved in primary response to low water stress. The largest group was involved in ROS metabolism, followed by carbon metabolism, signaling, membrane transport, transcription factors, and cell wall structure. In the ROS group, transcripts involved in both consumption and production of these molecules were upregulated, showing that a fine balance is likely required to prevent damage from accumulation yet still allow for localized growth. Several metallothionein-like transcripts were upregulated in maize, as reported in rice (Rabello et al. 2008; Yang et al. 2009; Wang et al. 2011a, b). A number of additional DEGs identified in this study might be involved in ABA response, including transcripts for a CIPK3-like protein, PP2C proteins, bZIP family transcription factors, ABA-response element-binding protein 3 (homolog of *OsDREB1a*), and dehydrins. In other categories, expansins and inositol phosphate transcripts were increased.

A cell wall proteomic analysis of root tips under water stress was carried out using the same defined sections as in Spollen et al. (2008) (Zhu et al. 2007). In Zhu et al. (2007), low water-responsive proteins were categorized into ROS metabolism, defense/detoxification, hydrolases, carbohydrate metabolism, and other/unknown. In corroboration with the Spollen et al. (2008) study, certain oxalate oxidase and peroxidase levels were increased. Again, a balance of ROS levels is hypothesized to be critical for stress response. There were also changes in several xyloglucan endotransglycosylase/hydrolase protein levels (XHT/XET). These are proteins involved in cell wall expansion that have been analyzed in other water stress studies, but here, as with ROS metabolism proteins, localized increases and decreases in their levels might be important in stress response. Root-tip invertase and other carbon metabolism protein levels were altered under stress, as were transcript levels in this category in Spollen et al. (2008). Carbon metabolic enzymes can direct the flow of resources for growth to the root system as they elongate in search of water. It should be noted that some cell wall proteins, such as expansins, are too tightly bound to have been present in extracts of this study. So, although expansin transcript levels have been shown to be upregulated under water stress in several studies, different solubilization methods will be needed to look at corresponding protein levels other than those used in Zhu et al. (2007).

There are also a few transcriptome studies of roots under waterlogging. Water-logged roots of maize had many DEGs involved in signaling, including kinases, ethylene biosynthesis, hemoglobin synthesis, and the transcript of oxygen sensing *prolyl 4-hydrolase* (*P4H*) (Zou et al. 2010). The roles of *P4H* genes in low-$O_2$ sensing in *A. thaliana* have been more extensively studied, and they might be required for regulation of certain transcription factors as they are in mammals (Vlad et al. 2007). Zou et al. (2010) also saw increases in transcripts that lead to accumulation of compounds like alanine and GABA. They aligned their transcriptomic results with QTL maps for traits of submerged maize seedlings, including root length and dry mass (Qiu et al. 2007), but the candidate genes they came up with require further evaluation. Vantoai et al. (2009) looked specifically at expression of certain transcription factors in soybean roots under waterlogging. They looked at an intolerant line and variety that was rated tolerant to flooding in field trials. The tolerant variety showed an enhancement of ethylene biosynthesis genes over the intolerant one. Also differentially expressed between the two varieties throughout all time points tested were a MYB domain transcription factor, a leucine zipper transcription factor, and a hemoglobin gene. Proteome analyses of roots under waterlogging stress display similar cellular trends to transcriptome studies of roots under low oxygen or excess water stress. In a study of waterlogged tomato (*Solanum lycopersicum*) roots, Ahsan et al. (2007) found increases in enzymes such as ADH and enolase, pointing to the well-characterized shift to fermentative metabolism and increased glycolytic flux upon low oxygen stress. There were also increases in proteins involved in the synthesis of GA and brassinosteroids, both plant hormones that affect development. Other increases included cell wall degradation proteins and phenylpropanoid/flavonoid synthesis proteins. There was a decrease in S-adenosyl-L-methionine-synthase (SAM), which is likely to reduce ethylene biosynthesis but might also lead to increased accumulation of polyamines via the linkage of these processes in the AdoMet cycle (reviewed in Roje 2006). The survival of maize plants depends on the adaptation to anaerobic conditions when the root system is submerged or subjected to flooding of the soil. In a comparative microarray study of roots grown under low-oxygen conditions, the expression of 39 miRNAs was significantly altered (Zhang et al. 2008). Several targets of these miRNAs were transcription factors that were also induced upon submergence of the maize roots. Other target genes were related to carbohydrate and energy metabolism, as well as ROS removal. These results suggest that submergence-responsive miRNAs are involved in the regulation of metabolic, physiological, and morphological adaptations of maize roots at the posttranscriptional level. In recent years, proteomic studies have provided new insights into the regulation of maize root development (Hochholdinger et al. 2006). These experiments either generated reference maps of the most abundant proteins of a particular stage of root development or compared differential protein accumulation levels between different genotypes or treatments of a particular root type.

Most of these studies analyzed complete roots (reviewed in Hochholdinger and Tuberosa 2009) while others focused on defined longitudinal zones (Zhu et al. 2006b), tissues, cell types (Dembinsky et al. 2007), or subcellular fractions (Zhu et al. 2006a, 2007) of maize roots. Reference maps provided initial cues on the most abundant proteins in the maize primary root and highlighted considerable changes in primary root proteome composition during development (Hochholdinger et al. 2006). Similarly, a reference map of the maize pericycle revealed the most abundant proteins in this cell type that is involved in lateral root initiation (Dembinsky et al. 2007). Comparative proteome studies aim to identify differentially expressed proteins between different genotypes or treatments. Such analyses can reveal functional categories or biochemical pathways related to form, development, or function of particular root types. The results of selected studies related to this category are summarized here. First, the preferential accumulation of proteins in the primary root of the mutant *lrt1* (*lateral rootless 1*), which does not initiate lateral roots, demonstrated the influence of lateral roots on the proteome composition of the maize primary root (Hochholdinger et al. 2004a). Moreover, comparative analyses of wild-type and rum1 primary roots, which do not initiate lateral roots, suggested posttranscriptional manifestation of protein accumulation differences by detecting similar levels of the corresponding transcripts in the two genotypes (Liu et al. 2006). Furthermore, analysis of shoot-borne root initiation via the mutant rtcs revealed subtle developmental regulation of auxin-related genes during shoot-borne root formation (Sauer et al. 2006). Additionally, a comparative proteome analysis of maize hybrids and their parental-inbred lines suggested that nonadditive protein accumulation in maize hybrids could be related to the manifestation of heterosis in maize seedling roots (Hoecker et al. 2008). Finally, proteomic comparisons of roots before and after a specific treatment highlighted proteins that might be associated with phosphorus depletion (Li et al. 2007, 2008) and water deficit (Zhu et al. 2007), revealing complex and quantitative changes in protein synthesis during these processes.

Feng et al. (2009) demonstrated that maize hybrid seedlings showed higher levels of heterosis in root traits. But the previous studies showed that heterosis is associated with differential gene expression between hybrids and their parents, but the responsible molecular mechanisms have not been determined. Long nonprotein coding RNAs (npcRNAs) represent an emerging class of riboregulators, which either act directly in this long form or are reprocessed into shorter miRNAs and siRNAs. Xing et al. (2010) reported a novel npcRNA upregulated in maize seedling roots of hybrid Zong3/87-1. It was identified and named ZmUHR, and it comprised multiple members arranged in clusters of 1–4 copies throughout the maize draft genome. These genes showed consistent and preferential expression in young and vigorous growth tissues, with the promoter sequences related to meristem-specific activation in maize. A transcriptome study was performed in reciprocal maize (*Zea mays* L.) hybrids of parental-inbred lines B73 and Mo17 to study the developmental manifestation of heterotic root traits (Paschold et al. 2012). Nearly 70 % of all expressed genes were differentially expressed between the two parents and 42–57 % of expressed genes were differentially expressed between one of the parents and one of the hybrids. In both hybrids, ~12 % of expressed genes exhibited nonadditive gene expression. Consistent with the dominance model (i.e., complementation) for heterosis, 865 genes that were expressed in the hybrids were expressed in only one of the two parents. They also identified that alleles from the inactive inbred were activated in the hybrid, presumably via interactions with regulatory factors from the active inbred. Finally, in hybrids, ~14 % of expressed genes exhibited allele-specific expression (ASE) levels that differed significantly from the parental-inbred expression ratios, providing further evidence for interactions of regulatory factors from one parental genome with target genes from the other parental genome.

Root hairs are unicellular extensions of specialized epidermis cells. Under limiting conditions, they significantly increase the water and nutrient uptake capacity of plants by enlarging their root surface. Thus far, little is known about the initiation and growth of root hairs in the monocot model species maize. To gain a first insight into the protein composition of root hair specialized cells, Nestler et al. (2011) identified and analyzed the most abundant proteins of maize root hairs attached to 4-day-old primary roots of the inbred line B73 by combining 1DE with nanoLC-MS/MS in a shotgun proteomic experiment. Among the identified proteins, homologs of 252 proteins have been previously associated with root hair formation and development in other species. Comparison of the root hair reference proteome of the monocot species maize with the previously published root hair proteome of the dicot species soybean revealed conserved, but also unique, protein functions in root hairs of these two major groups of flowering plants.

Based on microarray experiments, Liu et al. (2008b) made an effort to demonstrate and study local nitrate-induced lateral root formation in maize, interactions among hormonal pathways, and local nitrate signaling pathways. Moreover, genes related to nitrate uptake and assimilation, sugar transport and utilization, and cell division and expansion were induced by local nitrate application, implying a role of these transcripts in the root system response to nitrate. Pi starvation led to the differential expression of several hundred genes, of which 40–50 % were repressed when Pi was resupplied in maize (Calderon-Vazquez et al. 2008). Additionally, comparative proteome analysis of Pi responses in maize indicates that organic acid secretion, sugar metabolism, and root cell proliferation are important components of high tolerance to low-Pi conditions (Li et al. 2007). Together, these studies demonstrate differences in gene and protein expression between dissimilar root structures and cell types and highlight the need for additional work in this area.

## 2.8.7  Wheat

Wheat is an excellent species to study freezing tolerance and other abiotic stresses. However, the sequence of the wheat genome has not been completely characterized because of its complexity and large size. To circumvent this obstacle, Houde et al. (2006) selected various stage-specific tissues to identify genes involved in cold acclimation and associated stresses, based on a large-scale EST sequencing approach that was undertaken by the Functional Genomics of Abiotic Stress (FGAS) project. Based on Gene Ontology (GO) derived from comparative analysis of ESTs from FGAS and NSF–DuPont, no prominent differences were found for GO classes for biological processes, transcription, and protein metabolism. However, GOs for enzyme regulator activity and nutrient reservoir activity had a lower representation, while GOs for transcription factor activity, nuclease activity, plasma membrane, secondary metabolism, response to external stimulus, carbohydrate binding, response to abiotic stimulus, cell–cell signaling, development, and behavior were more abundant in the FGAS dataset. Digital expression analyses of these datasets provide a genomic resource with an overview of metabolic changes and specific pathways that are regulated under stress conditions in wheat and other cereals.

Wild wheat relatives have also been used to look at root transcript differences under drought. Ergen and Budak (2009) surveyed a panel of two wild emmer wheat lines (*Triticum turgidum* spp. *dicoccoides*) with contrasting responses under water-limited conditions. These drought-tolerant and sensitive lines of emmer wheat were profiled on Affymetrix arrays under desiccation stress (Ergen et al. 2009). In root tissue, 926 probes had differential expression between the two genotypes during the stress. Among the 30 DEGs that varied the most between the genotypes were transcription factors, heat-shock proteins, cold-regulated proteins, dehydrins, lipases, and a lysine decarboxylase-like protein. For the transcription factors, several NAC domain proteins as well as other major families like WRKY and MYB had altered expressions levels.

## 2.8.8  Barley

Global transcriptome analysis on barely mutant lines by Kwasniewski et al. (2010) reveals new candidate genes involved in root hair morphogenesis in barley. Expression profiling of the root-hairless mutant rhl1.a and its wild-type parent variety "Karat" revealed 10 genes potentially involved in the early step of root hair formation in barley. The identified genes encode proteins associated with the cell wall and membranes, including one gene for xyloglucan endotransglycosylase, three for peroxidase enzymes, and five for arabinogalactan protein; extensin; leucine-rich-repeat protein; phosphatidylinositol phosphatidylcholine transfer protein; and a Rho GTPase GDP dissociation inhibitor, respectively. The expression levels of these genes were strongly reduced in roots of the *root-hairless mutant rhl1.a* compared to the parent variety. The expression of all these genes was similar in another mutant, rhp1.b, that has lost its ability to develop full root hairs but still forms hairs blocked at the primordium stage, along with its wild-type relative.

## 2.8.9 Cotton

Global gene expression responses to waterlogging in roots and leaves of cotton revealed the expression of 1,012 genes (4 % of genes assayed) in root tissue as early as 4 h after flooding (Christianson et al. 2010). Many of these genes were associated with cell wall modification and growth pathways, glycolysis, fermentation, mito- chondrial electron transport, and nitrogen metabolism. It also altered global gene expression in leaf tissues, significantly changing the expression of 1,305 genes (5 % of genes assayed) after 24 h of flooding. Genes affected were associated with cell wall growth and modification, tetrapyrrole synthesis, hormone response, starch metabolism, and nitrogen metabolism. Transcriptome analysis in salt stress reveals its regulated biological processes and key pathways in roots (Yao et al. 2011). In this study, microarray analysis showed that 1,503 probe sets were upregulated and 1,490 probe sets were downregulated in plants exposed for 3 h to 100 mM NaCl, and RT-PCR analysis validated 42 relevant/related genes. The distribution of enriched gene ontology terms showed that important processes, such as the response to water stress and pathways of hormone metabolism and signal transduc- tion, were induced by the NaCl treatment. Some key regulatory gene families involved in abiotic and biotic factors of stress, such as WRKY, ERF, and JAZ, were differentially expressed.

## 2.8.10 Foxtail Millet

A study with 21-day-old seedlings of two foxtail millet (*Setaria italica*) cultivars differing in salt tolerance (Puranik et al. 2011) was found to also differ in lipid peroxidation, ion balance, and activity of antioxidative enzymes (glutathione reductase and catalase) under short-term salinity stress. Comparative transcriptome analysis of these contrasting foxtail millet (Prasad and Lepakshi) cultivars showed that 159 (63.9 %) were differentially expressed ($\geq$2-fold) in response to salinity stress, with 115 (72.3 %) up- and 44 (27.7 %) downregulated. A data search of transcriptional profiling under salinity stress in other species revealed that 81 (51 %) of the 159 differentially expressed transcripts found in foxtail millet have not been reported in previous studies. Hence, these new transcripts might represent untapped gene sources allowing specific responses to short-term salt stress in an orphan crop known to possess a natural adaptation capacity to abiotic stress. The marked induction of genes, specifically in cultivar Prasad, involved in signal transduction, transcription, transport, protein degradation, metabolism, stress/ defense, and of unknown function suggests that they are part of this cultivar's response to salinity stress. The presence of these inducible genes in a salinity-tolerant cultivar might improve its capacity to maintain cellular homeostasis in the face of salinity stress. In a similar tolerant line, Sreenivasulu et al. (2004) reported earlier that transcripts of hydrogen peroxide-scavenging enzymes such as phospholipid hydroperoxide glutathione peroxidase (PHGPX), ascorbate peroxidase (APX), and catalase 1 (CAT1), in addition to some genes of cellular metabolism, were found to be especially upregulated at high salinity.

## 2.9    Major Research Groups Working on Roots

Although roots have been neglected since the onset of genetics in the beginning of the twentieth century, a few research groups have paved the way for the advance of root biology and genomics. One of the pioneers working on root-related traits is Jonathan Lynch, at the Pennsylvania State University (USA). His group focuses on the understanding of the genetic, physiological, and ecological basis of plant adaptation to infertile soils. Model species used are maize, common bean and Lupinus (Burton et al. 2012; Lynch and Brown 2012; Postma and Lynch 2012). Much of the understanding of root architecture and dynamics comes from the group of Philip Benfey of the Duke University (USA). This research group addresses the question of how cells acquire their identities. Using a combination of genetics, molecular biology, and genomics, genes that regulate formation of the root in the plant model system *Arabidopsis thaliana* are functionally characterized. The characterization of mutants SHORTROOT and SCARECROW and their pathways has contributed to a great understanding of radial patterning of the root (Benfey and Mitchell-Olds 2008; Ingram et al. 2012). On the more applied side, one of the outstanding groups working on roots is the group of Roberto Tuberosa of the University of Bologna (Italy). His group focuses on QTL traits affecting roots in wheat and maize, aiming to improve drought tolerance. His group has demonstrated the success of QTL cloning of *Vgt1*, a major QTL for flowering time (Salvi et al. 2007; Tuberosa et al. 2011). On the same line of research, but focusing on legume crops, Rajeev Varshney's group of ICRISAT (India) has greatly contributed to the understanding of root-related QTLs and MAS of root-related traits (Mir et al. 2012). Many other groups have also contributed for the solution of root-related biological challenges on the current black boxes that is the root biological system (Gruber et al. 2011). Other root research groups include the University of Missouri (USA) (Henry Nguyen and Robert Sharp) with a focus on genetics and physiology of soybean and maize root growth and adaptation to drought and Cornell University (USA) (Leon Kochian) with a focus on aluminum tolerance genes and mechanisms in cereals. Also, a group at the Federal University of Pelotas (Brazil) (Antonio Costa de Oliveira) has a focus on root development in relation to iron stress in rice.

## 2.10 Future Prospects

The identification of markers or genes associated with root growth and architecture would be particularly useful for breeding programs to improve root traits by molecular marker-assisted selection (MAS). Recent advances in candidate gene approaches and genetic engineering of crops have shown promising improvements in drought tolerance of crops such as rice, maize, canola, and soybean. Few papers have described work on the identification of QTLs for root traits in wheat. Uga et al. (2011) identified a major QTL Dro1 to increase rooting depth and proved under upland field conditions in rice. Ma et al. (2005) found a QTL in wheat for root growth rate under Al treatment. QTLs of root traits (primary/lateral root length and number, root dry matter) under control conditions and during nitrogen deficiency were identified in wheat (Laperche et al. 2006). Relative root growth was also used by Jefferies et al. (1999) to map QTL for tolerance to toxic levels of soil boron. However, QTLs corresponding to root architecture in dry environments are yet to be discovered in wheat and barley (Fleury et al. 2010).

In sorghum, Al-activated citrate exudation from root apices of an Al-tolerant sorghum line is controlled at the AltSB locus, which explains more than 80 % of the phenotypic variation in Al tolerance in the mapping populations studied (Magalhaes et al. 2004, 2007). *SbMATE*, the gene that underlies the *AltSB* locus, encodes a plasma membrane-localized citrate transporter that belongs to the multidrug and toxic compound extrusion (MATE) family. *SbMATE* is expressed primarily in the root apices of Al-tolerant lines and is induced by Al, ultimately being responsible for the observed Al-activated root citrate exudation. Sutton et al. (2007) identified a gene (*Bot1*) underlying the tolerance QTL on chromosome 4H of barley, that is, a putative integral transmembrane boron transporter with similarity to bicarbonate transporters in animals (Sutton et al. 2007).

The tools of genomics offer the means to produce comprehensive datasets on changes in gene expression, protein profiles, and metabolites that result from exposure to various abiotic and biotic stresses. Comparison of gene expression in *Arabidopsis* and rice showed that the two species share many common stress-inducible genes (Shinozaki and Yamaguchi-Shinozaki 2007). Abiotic stress tolerance involves similar transcription factors in both dicotyledonous and monocotyledonous plants, and some molecular mechanisms of drought tolerance have been extensively described (reviewed by Yamaguchi-Shinozaki and Shinozaki 2006). It includes signal transduction cascade and activation/regulation of transcription, functional protection of proteins by late-embryogenesis abundant proteins (e.g., dehydrins) and chaperone proteins (e.g., heat-shock proteins), accumulation of osmolytes (proline, glycine betaine, trehalose, mannitol, myo-inositol), induction of chemical antioxidants (ascorbic acid and glutathione), and enzymes reducing the toxicity of reactive oxygen species (superoxide dismutase, glutathione S-transferase). Despite the existence of common regulatory mechanisms across species, the conservation of the molecular response to dehydration across experiments (Mohammadi et al. 2007; Aprile et al. 2009) is low due to variation in stress dynamics, stage of development, and tissue analyzed. Recently, Jeong et al. (2013) reported that the increased expression of *NAC* genes in rice roots caused the enlargement of the root tissues, thereby enhancing the tolerance to drought stress at the reproductive stages and increased yield as well.

However, no root-related gene or QTL has been commercially deployed in crop plants for drought-tolerance improvement. As another possible strategy, given the wealth of genomic information and examples in model crop species, stacking of genes involved in a particular pathway related to dehydration tolerance/root depth should be considered. Further progress in breeding for drought resistance or other root-related stresses will depend on our ability to identify subtle cultivar level differences in expression of the gene networks involved in stress adaptation. The success of any selection strategy would ultimately be determined by the reproductive success, thus, by the final grain yield under field conditions. This must be part of an integrated, multidisciplinary approach spanning plant molecular biology, physiology, and breeding (Manavalan et al. 2009) to achieve the needed improvement in the genetic potential for grain yield. Biotechnology has provided new opportunities such as the possibility of manipulating transcription factors associated with stress response, which regulate a large number of genes. Such transcription factors have been used in genetic engineering for tolerance to abiotic and biotic stresses (Agarwal et al. 2006). Although, conventional breeding has mainly focused in selecting lines for dehydration avoidance strategies over dehydration tolerance (Blum 2009), several reports on transgenic lines have evaluated the dehydration tolerance approach—high survival after the stress or even dehydration of seedling grown hydroponically (Nakashima et al. 2007). Surprisingly only few genetic engineering studies identified genes that showed a role in root growth. Considering the importance of roots in drought avoidance and water uptake, more emphasis should be given to develop transgenic crop plants with improved root system architecture.

# References

Abdel-Haleem H, Lee GJ, Boerma RH (2011) Identification of QTL for increased fibrous roots in soybean. Theor Appl Genet 122:935–946

Acosta-Gallegos JA, Adams MW (1991) Plant traits and yield stability of dry bean (*Phaseolus vulgaris*) cultivars under drought stress. J Agric Sci 117:213–219

Acosta-Gallegos JA, Kohashi-Shibata V (1989) Effect of water stress on growth and yield of indeterminate dry bean (*Phaseolus vulgaris*) cultivars. Field Crops Res 20:81–90

Agarwal PK, Agarwal P, Reddy MK, Sopory SK (2006) Role of DREB transcription factors in abiotic and biotic stress tolerance in plants. Plant Cell Rep 25:1263–1274

AGI (2000) The Arabidopsis genome initiative. Nature 408:796–815

Ahsan N, Lee DG, Lee SH, Lee KW, Bahk JD, Lee BH (2007) A proteomic screen and identification of waterlogging-regulated proteins in tomato roots. Plant Soil 295:37–51

Akita S, Cabuslay GS (1990) Physiological basis of differential response to salinity in rice cultivars. Plant Soil 123:277–294

Alam I, Lee DG, Kim KH, Park CH, Sharmin SA, Lee H, Oh KW, Yun BW, Lee BH (2010) Proteome analysis of soybean roots under waterlogging stress at an early vegetative stage. J Biosci 35:49–62

Ali ML, Pathan M, Zhang J, Bai G, Sarkarung S, Nguyen HT (2000) Mapping QTL for root traits in a recombinant inbred population from two *indica* ecotypes in rice. Theor Appl Genet 101:756–766

Apel K, Hirt H (2004) Reactive oxygen species: metabolism, oxidative stress, and signal transduction. Annu Rev Plant Biol 55:373–399

Aprile A, Mastrangelo AM, De Leonardis AM, Galiba G, Roncaglia E, Ferrari F, De Bellis L, Turchi L, Giuliano G, Cattivelli L (2009) Transcriptional profiling in response to terminal drought stress reveals differential responses along the wheat genome. BMC Genomics 10:279

Araus JL, Slafer GA, Royo C, Dolores SM (2008) Breeding for yield potential and stress adaptation in cereals. Crit Rev Plant Sci 27:377–412

Aroca R, Tognoni F, Irigoyen JJ, Sanchez-Diaz M, Pardossi A (2001) Different root low temperature response of two maize genotypes differing in chilling sensitivity. Plant Physiol Biochem 39:1067–1073

Aroca R, Porcel R, Ruiz-Lozano JM (2007) How does arbuscular mycorrhizal symbiosis regulate root hydraulic properties and plasma membrane aquaporins in *Phaseolus vulgaris* under drought, cold or salinity conditions? New Phytol 173:808–816

Asch F, Dingkuhn M, Dorffling K, Miezan K (2000) Leaf K/Na ratio predicts salinity induced yield loss in irrigated rice. Euphytica 113:109–118

Asch F, Becker M, Kpongor DS (2005) A quick and efficient screen for resistance to iron toxicity in lowland rice. J Plant Nutr Soil Sci 168:764–773

Asfaw A, Blair MW (2012) Quantitative trait loci for rooting pattern traits of common beans grown under drought stress versus non-stress conditions. Mol Breed 30:681–695

Ashmore M, Toet S, Emberson L (2006) Ozone – a significant threat to future world food production? New Phytol 170:201–204

Aslam M, Qureshi RH, Ahmed N (1993) A rapid screening technique for salt tolerance in rice (*Oryza sativa* L.). Plant Soil 150:97–107

Babu RC, Nguyen BD, Chamarerk V, Shanmugasundaram P, Chezhian P, Jeyaprakash P, Ganesh SK, Palchamy A, Sadasivam S, Sarkarung S, Wade LJ, Nguyen HT (2003) Genetic analysis of drought resistance in rice by molecular markers: association between secondary traits and field performance. Crop Sci 43:1457–1469

Bacanamwo M, Purcell LC (1999) Soybean root morphological and anatomical traits associated with acclimation to flooding. Crop Sci 39:143–149

Baerenfaller K, Grossmann J, Grobei MA, Hull R, Hirsch-Hoffmann M, Yalovsky S, Zimmermann P, Grossniklaus U, Gruissem W, Baginsky S (2008) Genome-scale proteomics reveals *Arabidopsis thaliana* gene models and proteome dynamics. Science 320:938–941

Bailey-Serres J, Voesenek LACJ (2008) Flooding stress: acclimations and genetic diversity. Annu Rev Plant Biol 59:313–339

Balasubramanian S, Schwartz C, Singh A, Warthmann N, Kim MC, Maloof JN, Loudet O, Trainer GT, Dabi T, Borevitz JO, Chory J, Weigel D (2009) QTL mapping in new *Arabidopsis thaliana* advanced intercross recombinant inbred lines. PLoS One 4:e4318

Banti V, Mafessoni F, Loreti E, Alpi A, Perata P (2010) The heat-inducible transcription factor HsfA2 enhances anoxia tolerance in Arabidopsis. Plant Physiol 152:1471–1483

Battisti DS, Naylor RL (2009) Historical warnings of future food insecurity with unprecedented seasonal heat. Science 323:240–244

Beebe S, Rao IM, Cajiao C, Grajales M (2008) Selection for drought resistance in common bean also improves yield in phosphorus limited and favorable environments. Crop Sci 48:582–592

Benfey PN, Mitchell-Olds T (2008) From genotype to phenotype: systems biology meets natural variation. Science 320:495–497

Bengough AG, McKenzie BM, Hallett PD, Valentine TA (2011) Root elongation, water stress and mechanical impedance: a review of limiting stresses and beneficial root tip traits. J Exp Bot 62:59–68

Benková E, Hejátko J (2009) Hormone interactions at the root apical meristem. Plant Mol Biol 69:383–396

Bernhardt C, Lee MM, Gonzales A, Zhang V, Lloyd A, Schiefelbein J (2003) The gHLH genes *GLABRA3* (*GL3*) and *ENHANCER OD GLABRA 3* (*EGL3*) specify epidermal cell fate in the Arabidopsis root. Development 130:6431–6643

Bernier PY, Robitaille G, Rioux D (2005) Estimating the mass density of fine roots of trees for minirhizotron-based estimates of productivity. Can J Forest Res 35:1708–1713

Bernier J, Atlin G, Serraj R, Kumar A, Spaner D (2008) Breeding upland rice for drought resistance. J Sci Food Agric 88:927–930

Bianchi-Hall CM, Carter TE Jr, Bailey MA, Mian MAR, Rufty TW, Ashley DA, Boerma HR, Arellano C, Hussey RS, Parrott WA (2000) Aluminum tolerance associated with quantitative trait loci derived from soybean PI 416937 in hydroponics. Crop Sci 40:538–545

Birnbaum K, Shasha DE, Wang JY, Jung JW, Lambert GM, Galbraith DW, Benfey PN (2003) A gene expression map of the Arabidopsis root. Science 302:1956–1960

Biswas J, Chowdhury B, Bhattacharya A, Mandal AB (2002) *In vitro* screening for increased drought tolerance in rice. In Vitro Cell Dev Biol-Plant 38:525–530

Blilou I, Xu J, Wildwater M, Willemsen V, Paponov I, Friml J, Heidstra R, Aida M, Palme K, Scheres B (2005) The PIN auxin efflux facilitator network controls growth and patterning in Arabidopsis roots. Nature 433:39–44

Blokhina O, Virolainen E, Fagerstedt KV (2003) Antioxidants, oxidative damage and oxygen deprivation stress: a review. Ann Bot 91:179–194

Blum A (2009) Effective use of water (EUW) and not water-use efficiency (WUE) is the target of crop yield improvement under drought stress. Field Crops Res 112:119–123

Blum A (2011) Plant breeding for water-limited environments. Springer, Berlin

Bonke M, Thitamadee S, Mahonen AP, Hauser M-T, Helariutta Y (2003) APL regulates vascular tissue identity in Arabidopsis. Nature 426:181–186

Bouchabke-Coussa O, Quashie ML, Seoane-Redongo J, Fortabat M, Gery G, Yu A, Linderme D, Trouverie J, Granier F, Teoule E, Durand-Tardif M (2008) ESKIMO1 is a key gene involved in water economy as well as cold acclimation and salt tolerance. BMC Plant Biol 8:125

Boyer JS (1982) Plant productivity and environment. Science 218:443–448

Brady SM, Orlando DA, Lee JY, Wang JY, Koch J, Dinneny JR, Mace D, Ohler U, Benfey PN (2007) A high-resolution root spatiotemporal map reveals dominant expression patterns. Science 318:801–806

Brady SM, Zhang L, Megraw M, Martinez NJ, Jiang E, Yi CS, Liu W, Zeng A, Taylor-Teeples M, Kim D, Ahnert S, Ohler U, Ware D, Walhout AJ, Benfey PN (2011) A stele-enriched gene regulatory network in the Arabidopsis root. Mol Syst Biol 7:459–468

Branco-price C, Kawaguchi CR, Ferreira RB, Bailey-serres J (2005) Genome-wide analysis of transcript abundance and translation in Arabidopsis seedlings subjected to oxygen deprivation. Ann Bot 96:647–660

Branco-price C, Kaiser KA, Jang CJ, Larive CK, Bailey-Serres J (2008) Selective mRNA translation coordinates energetic and metabolic adjustments to cellular oxygen deprivation and reoxygenation in *Arabidopsis thaliana*. Plant J 56:743–755

Brensha W, Kantartzi SK, Meksem K, Grier RL, Barakat A, Lightfoot DA, Kassem A (2012) Genetic analysis of root and shoot traits in the 'Essex' by 'Forrest' recombinant inbred line (RIL) population of soybean [*Glycine max* (L.) Merr.]. J Plant Genome Sci 1:1–9

Bresolin APS (2010) Morphological characterization and gene expression analysis in rice (*Oryza sativa* L.) under iron stress. PhD Thesis, University of Federal de Pelotas, Pelotas, 144 p

Briggs DE (1978) Barley. Wiley, New York, pp 429–496

Broughton WJ, Hernandez G, Blair MW, Beebe SE, Gepts P, Vanderleyden J (2003) Beans (*Phaseolus* spp.)—model food legumes. Plant Soil 252:55–128

Brown EA, Cavines CE, Brown DA (1985) Response of selected soybean cultivars to soil moisture deficit. Agron J 77:274–278

Brown DM, Zeef LAH, Ellis J, Goodacre R, Turner SR (2005) Identification of novel genes in Arabidopsis involved in secondary cell wall formation using expression profiling and reverse genetics. Plant Cell 17:2281–2295

Bruinsma J (2009) The resource outlook to 2050: by how much do land, water and crop yield need to increase by 2050? In: FAO expert meeting on how to feed the world in 2050, Rome, Italy

Burton AL, Williams M, Lynch JP, Brown KM (2012) RootScan: software for high-throughput analysis of root anatomical traits. Plant Soil 357:189–203

Cai H, Chen F, Mi G, Zhang F, Maurer HP, Liu W, Reif JC, Yuan L (2012) Mapping QTLs for root system architecture of maize in the field at different developmental stages. Theor Appl Genet 125:1313–1324

Carol RJ, Dolan L (2006) The role of reactive oxygen species in cell growth: lessons from root hairs. J Exp Bot 57:1829–1834

Casimiro I, Marchant A, Bhalerao RP, Beeckman T, Dhooge S, Swarup R, Graham N, Inzé D, Sandberg G, Casero PJ, Bennett M (2001) Auxin transport promotes Arabidopsis lateral root initiation. Plant Cell 13:843–852

Calderon-Vazquez C, Ibarra-Laclette E, Caballero-Perez J, Herrera-Estrella L (2008) Transcript profiling of Zea mays roots reveals gene responses to phosphate deficiency at the plant and species-specific levels. J Exp Bot 59:2479–2497

Camargo CEO, Oliveira OF (1981) Tolerance of wheat cultivars to different aluminum levels in hydroponic culture and soil. Bragantia 40:21–23

Chakrabarty D, Trivedi PK, Misra P, Tiwari M, Shri M, Shukla D, Kumar S, Rai A, Pandey A, Nigam D, Tripathi RD, Tuli R (2009) Comparative transcriptome analysis of arsenate and arsenite stresses in rice seedlings. Chemosphere 74:688–702

Champoux MC, Wang G, Sarkarang S, Mackill DJ, O'Toole JC, Huang N, McCouch SR (1995) Locating genes associated with root morphology and drought avoidance in rice via linkage to molecular markers. Theor Appl Genet 90:961–981

Chandler VL, Brendel V (2002) The corn genome sequencing project. Plant Physiol 130:1594–1597

Chandra S, Buhariwalla HK, Kashiwagi J, Hari Krishna S, Rupa Sridevi K, Krishnamurthy L, Serraj R, Crouch JH (2004) Identifying QTL-linked markers in marker-deficient crops. In: Proceedings of the 4th international crop science congress, Brisbane, Australia, 26 Sept–1 Oct 2004. http://www.regional.org.au/au/cs/2004/poster/3/4/1/795_chandras.html

Charlson DV, Grant D, Bailey TB, Cianzio SR, Shoemaker RC (2005) Molecular marker Satt481 is associated with iron deficiency chlorosis resistance in a soybean breeding population. Crop Sci 45:2394–2399

Chen ZH, Nimmo GA, Jenkins GI, Nimmo HG (2007) BHLH32 modulates several biochemical and morphological processes that respond to P-i starvation in Arabidopsis. Biochem J 405:191–198

Chen J, Xu L, Cai Y, Xu J (2008) QTL mapping of phosphorus efficiency and relative biologic characteristics in corn (Zea mays L.) at two sites. Plant Soil 313:251–266

Chen X, Liu L, Lee E, Han X, Rim Y, Chu H, Kim S, Sack F, Kim J (2009) The Arabidopsis callose synthase gene GSL8 is required for cytokinesis and cell patterning. Plant Physiol 150:105–113

Chloupek O, Forster BP, Thomas WT (2006) The effect of semi-dwarf genes on root system size in field-grown barley. Theor Appl Genet 112:779–786

Christmann A, Weiler EW, Steudle E, Grill E (2007) A hydraulic signal in root-to-shoot signalling of water shortage. Plant J 52:167–174

Christianson JA, Llewellyn DJ, Dennis ES, Wilson LW (2010) Global gene expression responses to waterlogging in roots and leaves of cotton. Plant Cell Physiol 51:21–37

Chun L, Mi G, Li J, Chen F, Zhang F (2005) Genetic analysis of corn root characteristics in response to low nitrogen stress. Plant Soil 276:369–382

Clark LJ, Price AH, Steele KA, Whalley WR (2008) Evidence from near-isogenic lines that root penetration increases with root diameter and bending stiffness in rice. Funct Plant Biol 35:1163–1171

Concibido VC, Denny RL, Boutin SR, Hautea R, Orf JH, Young ND (1994) DNA marker analysis of loci underlying resistance to soybean cyst-nematode (Heterodera glycines Ichinohe). Crop Sci 34:240–246

Cook D, Fowler S, Fiehn O, Thomashow MF (2004) A prominent role for the CBF cold response pathway in configuring the low-temperature metabolome of Arabidopsis. Proc Natl Acad Sci USA 101:15243–15248

Cortes PM, Sinclair TR (1986) Gas-exchange of field-grown soybean under drought. Agron J 78:454–458

Costa de Oliveira A, Varshney R (2011) Introduction to root genomics. In: Costa de Oliveira A, Varshney R (eds) Root genomics. Springer, Berlin, pp 1–10

Coudert Y, Perin C, Courtois B, Khong NG, Gantet P (2010) Genetic control of root development in rice, the model cereal. Trends Plant Sci 15:219–226

Courtois B, McLaren G, Sinha PK, Prasad K, Yadav R, Shen L (2000) Mapping QTLs associated with drought avoidance in upland rice. Mol Breed 6:55–66

Courtois B, Shen L, Petalcorin W, Carandang S, Mauleon R, Li Z (2003) Locating QTLs controlling constitutive root traits in the rice population IAC 165×Co39. Euphytica 134:335–345

Courtois B, Ahmadi N, Khowaja F, Price AH, Rami JF, Frouin J, Hamelin C, Ruiz M (2009) Rice root genetic architecture: meta-analysis from a drought QTL database. Rice 2:115–128

Cui SY, Geng LY, Meng QC, Yu DY (2007) QTL mapping of phosphorus deficiency tolerance in soybean (*Glycine max* L.) during seedling stage. Acta Agron Sin 33:378–383

Cui K, Huang J, Xing Y, Yu S, Xu C, Peng S (2008) Mapping QTLs for seedling characteristics under different water supply conditions in rice (*Oryza sativa*). Physiol Plant 132:53–68

de Dorlodot S, Forster B, Pages L, Price A, Tuberosa R, Draye X (2007) Root system architecture: opportunities and constraints for genetic improvement of crops. Trend Plant Sci 12:474–481

Dembinsky D, Woll K, Saleem M, Liu Y, Fu Y, Borsuk LA, Lamkemeyer T, Fladerer C, Madlung J, Barbazuk B, Nordheim A, Nettleton D, Schnable PS, Hochholdinger F (2007) Transcriptomic and proteomic analyses of pericycle cells of the maize (*Zea mays* L.) primary root. Plant Physiol 145:575–588

Den Herder G, Van Isterdael G, Beeckman T, De Smet I (2010) The roots of a new green revolution. Trends Plant Sci 15:600–607

Denning G, Kabambe P, Sanchez P, Malik A, Flor R, Harawa R, Nkhoma P, Zamba C, Banda C, Magombo C, Keating M, Wangila J, Sachs J (2009) Input subsidies to improve smallholder maize productivity in Malawi: toward an African green revolution. PLoS Biol 7(1):e1000023

Devaiah BN, Nagarajan VK, Raghothama KG (2007a) Phosphate homeostasis and root development in Arabidopsis are synchronized by the zinc finger transcription factor ZAT6. Plant Physiol 145:147–159

Devaiah BN, Karthikeyan AS, Raghothama KG (2007b) WRKY75 transcription factor is a modulator of phosphate acquisition and root development in Arabidopsis. Plant Physiol 143:1789–1801

Dinneny JR, Long TA, Wang JY, Jung JW, Mace D, Pointer S, Barron C, Brady SM, Schiefelbein J, Benfey PN (2008) Cell identity mediates the response of Arabidopsis roots to abiotic stress. Science 320:942–945

Dobra J, Motyka V, Dobrev P, Malbeck J, Prasil IT, Haisel D, Gaudinova A, Havlova M, Gubis J, Vankova R (2010) Comparison of hormonal responses to heat, drought and combined stress in tobacco plants with elevated proline contents. J Plant Physiol 167:1360–1370

Doerner P, Jørgensen JE, You R, Steppuhn J, Lamb C (1996) Control of root growth and development by cyclin expression. Nature 380:520–523

Dolan L, Janmaat K, Willemsen V, Linstead P, Poethig S, Roberts K, Scheres B (1993) Cellular organization of the *Arabidopsis thaliana* root. Development 119:71–84

Domenico SD, Bonsegna S, Horres R, Pastor V, Taurino M, Poltronieri P, Imtiaz M, Kahl G, Flors V, Winter P, Santino A (2012) Transcriptomic analysis of oxylipin biosynthesis genes and chemical profiling reveal an early induction of jasmonates in chickpea roots under drought stress. Plant Physiol Biochem 61:115–122

Dorlodot S, Foster B, Pages L, Price A, Tuberosa R, Draye X (2007) Root system architecture: opportunities and constraints for genetic improvement of crops. Trends Plant Sci 12:474–481

Draye X, Kim Y, Lobert G, Javaux M (2010) Model-assisted integration of physiological and environmental constrains affecting the dynamic and spatial patterns of root water uptake from soils. J Exp Bot 61:2145–2155

El-Lithy ME, Clerkx EJ, Ruys GJ, Koornneef M, Vreugdenhil D (2004) Quantitative trait locus analysis of growth-related traits in a new Arabidopsis recombinant inbred population. Plant Physiol 135:444–458

Ergen NZ, Budak H (2009) Sequencing over 13000 expressed sequence tags from six subtractive cDNA libraries of wild and modern wheat following slow drought stress. Plant Cell Environ 32:220–236

Ergen NZ, Thimmapuram J, Bohnert HJ, Budak H (2009) Transcriptome pathways unique to dehydration tolerant relatives of modern wheat. Funct Integr Genomics 9:377–396

Faiyue B, Aal-azzawi MJ, Flowers TJ (2012) A new screening technique for salinity resistance in rice (Oryza sativa L.) seedlings using bypass flow. Plant Cell Environ 35:1099–1108

Fan X, Wang J, Yang N, Dong Y, Liu L, Wang F, Wang N, Chen N, Liu W, Sun Y, Wu J, Li H (2012) Gene expression profiling of soybean leaves and roots under salt, saline–alkali and drought stress by high-throughput illumina sequencing. Gene 512(2):392–402

Fang S, Gao X, Deng Y, Chen X, Liao H (2011) Crop root behavior coordinates phosphorus status and neighbors: from field studies to three-dimensional reconstruction of root system architecture. Plant Physiol 1555(3):1277–1285

Feng ZZ, Kobayashi K (2009) Assessing the impacts of current and future concentrations of surface ozone on crop yield with meta-analysis. Atmos Environ 43:1510–1519

Feng WJ, Zhang YR, Yao YY, Guo GG, Zhang GP, Ni ZF, Sun QX (2009) Comparative proteomic expression profile in seedling roots of maize hybrid and parents. Progr Nat Sci 19:619–627

Fleury D, Jefferies S, Kuchel H, Langridge P (2010) Genetic and genomic tools to improve drought tolerance in wheat. J Exp Bot 61:3211–3222

Frei M, Tanaka JP, Wissuwa M (2008) Genotypic variation in tolerance to elevated ozone in rice: dissection of distinct genetic factors linked to tolerance mechanisms. J Exp Bot 59:3741–3752

Frei M, Tanaka JP, Chen CP, Wissuwa M (2010) Mechanisms of ozone tolerance in rice: characterization of two QTLs affecting leaf bronzing by gene expression profiling and bio-chemical analyses. J Exp Bot 61:1405–1417

Fuhrer J (2009) Ozone risk for crops and pastures in present and future climates. Naturwis-senschaften 96:173–194

Fukai S, Pantuwan G, Jongdee B, Cooper M (1999) Screening for drought resistance in rainfed lowland rice. Field Crops Res 64:61–74

Fukaki H, Tasaka M (2009) Hormone interactions during lateral root formation. Plant Mol Biol 69:437–449

Galway ME, Masucci JD, Lloyd AM, Walbot V, Davis RW, Shiefelbein JW (1994) The TTG gene is required to specify epidermal cell fate and cell patterning in the Arabidopsis root. Dev Biol 166:740–754

Gardner WR (1965) Dynamic aspects of soil water availability to plants. Science 143:1460–1462

Garrity DP, O'Toole JC (1994) Screening rice for drought resistance at the reproductive phase. Field Crops Res 39:99–110

Geldner N, Richter S, Vieten A, Marquardt S, Torres-Ruiz RA, Mayer U, Jurgens G (2004) Partial loss-of-function alleles reveal a role for Gnom in auxin transport-related, post-embryonic development of Arabidopsis. Development 131:389–400

Gerald FJN, Lehti-Shiu MD, Ingram PA, Deak KI, Biesiada T, Malamy JE (2006) Identification of quantitative trait loci that regulate Arabidopsis root system size and plasticity. Genetics 172:485–498

Gewin V (2010) Food: an underground revolution. Nature 466:552–553

Giaquinta RT, Lin W, Sadler NL, Franceschi VR (1983) Pathway of phloem unloading of sucrose in corn roots. Plant Physiol 72:362–367

Gifford ML, Dean A, Gutierrez RA, Coruzzi GM, Birnbaum KD (2008) Cell specific nitrogen responses mediate developmental plasticity. Proc Natl Acad Sci USA 105:803–808

Goda H, Sasaki E, Akiyama K, Maruyama-Nakashita A, Nakabayashi K, Li W, Ogawa M, Yamauchi Y, Preston J, Aoki K, Kiba T, Takatsuto S, Fujioka S, Asami T, Nakano T, Kato H, Mizuno T, Sakakibara H, Yamaguchi S, Nambara E, Kamiya Y, Takahashi H, Hirai MY, Sakurai T, Shinozaki K, Saito K, Yoshida S, Shimada Y (2008) The AtGenExpress hormone and chemical treatment data set: experimental design, data evaluation, model data analysis and data access. Plant J 55:526–542

Gollhofer J, Schläwicke C, Jungnick N, Schmidt W, Buckhout TJ (2011) Members of a small family of nodulin-like genes are regulated under iron deficiency in roots of Arabidopsis thaliana. Plant Physiol Biochem 49:557–564

González-García M, Vilarrasa-Blasi J, Zhiponova M, Divol F, Mora-García S, Russinova E, Caño-Delgado AI (2011) Brassinosteroids control meristem size by promoting cell cycle progression in *Arabidopsis* roots. Development 138:849–859

Gowda VRP, Henry A, Yamauchi A, Shashidhar HE, Serraj R (2011) Root biology and genetic improvement for drought avoidance in rice. Field Crops Res 122:1–13

Gruber V, Zahaf O, Diet A, de Zelicourt A, de Lorenzo L, Crespi M (2011) Impact of the environment on root architecture in dicotyledonous Plants. In: Costa de Oliveira A, Varshney R (eds) Root genomics. Springer, Berlin, pp 113–132

Guingo E, Hebert Y, Charcosset A (1998) Genetic analysis of root traits in corn. Agronomie 18:225–235

Guo PG, Baum M, Grando S, Ceccarelli S, Bai GH, Li RH, von Korff M, Varshney RK, Graner A, Valkoun J (2009) Differentially expressed genes between drought-tolerant and drought-sensitive barley genotypes in response to drought stress during the reproductive stage. J Exp Bot 60:3531–3544

Gutierrez-Gonzalez JJ, Guttikonda SK, Tran IS, Aldrich DL, Zhong R, Yu O, Nguyen HT, Sleper DA (2010) Differential expression of isoflavone biosynthetic genes in soybean during water deficits. Plant Cell Physiol 51:936–948

Hao YJ, Wei W, Song QX, Chen HW, Zhang YQ, Wang F, Zou HF, Lei G, Tian AG, Zhang WK, Ma B, Zhang JS, Chen SY (2011) Soybean NAC transcription factors promote abiotic stress tolerance and lateral root formation in transgenic plants. Plant J 68:302–313

Hattori Y, Nagai K, Furukawa S, Song XJ, Kawano R, Sakakibara H, Wu J, Matsumoto T, Yoshimura A, Kitano H, Matsuoka M, Mori H, Ashikari M (2009) The ethylene response factors SNORKEL1 and SNORKEL2 allow rice to adapt to deep water. Nature 460:1026–1030

He XJ, Mu LR, Cao WH, Zhang ZG, Zhang JS, Chen SY (2005) AtNAC2, a transcription factor downstream of ethylene and auxin signaling pathways, is involved in salt stress response and lateral root development. Plant J 44:903–916

Hebert Y, Barriere Y, Bertholeau JC (1992) Root lodging resistance in forage corn: genetic variability of root system and aerial part. Maydica 37:173–183

Helariutta Y, Fukaki H, Wysocka-Diller J, Nakajima K, Jung J, Sena G, Hauser MT, Benfey PN (2000) The SHORTROOT gene controls radial patterning of the Arabidopsis root through radial signaling. Cell 101:555–567

Hemamalini GS, Shashidhar HE, Hittalmani S (2000) Molecular marker assisted tagging of morphological and physiological traits under two contrasting moisture regimes at peak vegetative stage in rice (*Oryza sativa* L.). Euphytica 112:69–78

Herbette S, Taconnat L, Hugouvieux V, Piette L, Magniette MLM, Cuine S, Auroy P, Richaud P, Forestier C, Bourguignon J, Renou JP, Vavasseur A, Leonhardt N (2006) Genome-wide transcriptome profiling of the early cadmium response of Arabidopsis roots and shoots. Biochimie 88:1751–1765

Herder GD, Van Isterdael G, Beeckman T, De Smet I (2010) The roots of a new green revolution. Trends Plant Sci 15:600–607

Himanen K, Boucheron E, Vanneste S, de Almeida EJ, Inzé D, Beeckman T (2002) Auxin-mediated cell cycle activation during early lateral root initiation. Plant Cell 14:2339–2351

Hirai MY, Yano M, Goodenowe DB, Kanaya S, Kimura T, Awazuhara M, Arita M, Fujiwara T, Saito K (2004) Integration of transcriptomics and metabolomics for understanding of global responses to nutritional stresses in *Arabidopsis thaliana*. Proc Natl Acad Sci USA 101:10205–10210

Hirsch S, Oldroyd GED (2009) GRAS-domain transcription factors that regulate plant development. Plant Signal Behav 4:698–700

Hochholdinger F, Tuberosa R (2009) Genetic and genomic dissection of maize root development and architecture. Curr Opin Plant Biol 12:172–177

Hochholdinger F, Guo L, Schnable PS (2004a) Lateral roots affect the proteome of the primary root of maize (*Zea mays* L.). Plant Mol Biol 56:397–412

Hochholdinger F, Park WJ, Sauer M, Woll K (2004b) From weeds to crops: genetic analysis of root development in cereals. Trends Plant Sci 9:42–48

Hochholdinger F, Sauer M, Dembinsky D, Hoecker N, Muthreich N, Saleem M, Liu Y (2006) Proteomic dissection of plant development. Proteomics 6:4076–4083

Hoecker N, Lamkemeyer T, Sarholz B, Paschold A, Fladerer C, Madlung J, Wurster K, Stahl M, Piepho H-P, Nordheim A, Hochholdinger F (2008) Analysis of non-additive protein accumulation in young primary roots of a maize (*Zea mays* L.) $F_1$-hybrid compared to its parental inbred lines. Proteomics 8:3882–3894

Hoekenga OA, Vision TJ, Shaff JE, Monforte AJ, Lee GP et al (2003) Identification and characterization of aluminum tolerance loci in Arabidopsis (Landsberg erecta 3 Columbia) by quantitative trait locus mapping. A physiologically simple but genetically complex trait. Plant Physiol 132:936–948

Hopmans JW, Bristow KL (2002) Current capabilities and future needs of root water and nutrient uptake modeling. Adv Agron 77:103–183

Houde M, Belcaid M, Ouellet F, Danyluk J, Monroy AF, Dryanova A, Gulick P, Bergeron A, Laroche A, Links MG, MacCarthy L, Crosby WL, Sarhan F (2006) Wheat EST resources for functional genomics of abiotic stress. BMC Genomics 7:149

Hu YC, Schmidhalter U (2005) Drought and salinity: a comparison of their effects on mineral nutrition of plants. J Plant Nutr Soil Sci 168:541–549

Hudak CM, Patterson RP (1995) Vegetative growth analysis of a drought-resistant soybean plant introduction. Crop Sci 35:464–471

Hudak CM, Patterson RP (1996) Root distribution and soil moisture depletion pattern of a drought-resistant soybean plant introduction. Agron J 88:478–485

Hund A, Fracheboud Y, Soldati A, Frascaroli E, Salvi S, Stamp P (2004) QTL controlling root and shoot traits of corn seedlings under cold stress. Theor Appl Genet 109:618–629

Hund A, Ruta N, Liedgens M (2009a) Rooting depth and water use efficiency of tropical corn inbred lines, differing in drought tolerance. Plant Soil 318:311–325

Hund A, Trachsel S, Stamp P (2009b) Growth of axile and lateral roots of corn: one development of a phenotying platform. Plant Soil 325:335–349

Hund A, Reimer R, Messmer R (2011) A consensus map of QTLs controlling the root length of corn. Plant Soil 344:143–158

Huynh LN, Vantoai T, Streeter J, Banowetz G (2005) Regulation of flooding tolerance of SAG12: ipt Arabidopsis plants by cytokinin. J Exp Bot 56:1397–1407

Ikeyama Y, Tasaka M, Fukaki H (2010) RLF, a cytochrome b5-like heme/steroid binding domain protein, controls lateral root formation independently of ARF7/19-mediated auxin signaling in Arabidopsis thaliana. Plant J 62:865–875

Ingram KT, Bueno FD, Namuco OS, Yambao EB, Beyrouty CA (1994) Rice roots for drought tolerance and their genetic variation. In: Kirk GRD (ed) Rice roots: nutrient and water use. International Rice Research Institute, Manila, pp 67–77

Ingram PA, Zhu J, Shariff A, Davis IW, Benfey PN, Elich T (2012) High-throughput imaging and analysis of root system architecture in Brachypodium distachyon under differential nutrient availability. Philos Trans R Soc B Biol Sci 367:1559–1569

Inukai Y, Sakamoto T, Morinaka Y, Miwa M, Kojima M, Tanimoto E, Yamamoto H, Sato K, Katayama Y, Matsuoka M, Kitano H (2012) ROOT GROWTH INHIBITING, a Rice Endo-1,4-b-D-Glucanase, regulates cell wall loosening and is essential for root elongation. J Plant Growth Regul 31:373–381

IPCC (2007) Climate change 2007: synthesis report. In: Pachauri RK, Reisinger A (eds) Contribution of working groups I, II and III to the fourth assessment report of the intergovernmental panel on climate change. IPCC, Geneva

IRGSP (International Rice Genome Sequencing Project) (2005) The map-based sequence of the rice genome. Nature 436:793–800

Iyer-Pascuzzi A, Simpson J, Herrera-Estrella L, Benfey PN (2009) Functional genomics of root growth and development in Arabidopsis. Curr Opin Plant Biol 12:165–171

Iyer-Pascuzzi AS, Benfey PN (2009) Transcriptional networks in root cell fate specification. Biochim Biophys Acta 1789:315–325

Jackson MB, Armstrong W (1999) Formation of aerenchyma and the processes of plant ventilation in relation to soil flooding and submergence. Plant Biol 1:274–287

Jackson MB, Saker LR, Crisp CM, Else MA, Janowiak F (2003) Ionic and pH signaling from roots to shoots of flooded tomato plants in relation to stomatal closure. Plant Soil 253:103–113

Jackson SA, Iwata A, Lee S, Schmutz J, Shoemaker R (2011) Sequencing crop genomes: approaches and applications. New Phytol 191:915–925

Jefferies SP, Barr AR, Karakousis A, Kretschmer JM, Manning S, Chalmers KJ, Nelson JC, Islam AKMR, Langridge P (1999) Mapping of chromosome regions conferring boron toxicity tolerance in barley. Theor Appl Genet 98:1293–1303

Jenison JR, Shank DB, Penny LH (1981) Root characteristics of 44 corn inbreds evaluated in four environments. Crop Sci 21:233–237

Jeong JS, Kim YS, Baek KH, Jung H, Ha SH, Do CY, Kim M, Reuzeau C, Kim JK (2010) Root-specific expression of OsNAC10 improves drought tolerance and grain yield in rice under field drought conditions. Plant Physiol 153:185–197

Jeong JS, Kim YS, Redillas MC, Jang G, Jung H, Bang SW, Choi YD, Ha S, Reuzeau C, Kim JK (2013) OsNAC5 overexpression enlarges root diameter in rice plants leading to enhanced drought tolerance and increased grain yield in the field. Plant Biotechnol J 11:101–114

Jiao Y, Tausta LS, Gandotra N, Sun N, Liu T, Clay NK, Ceserani T, Chen M, Ma L, Holford M, Zhang H, Zhao H, Deng X, Nelson T (2009) A transcriptome atlas of rice cell types reveals cellular, functional, and developmental hierarchies. Nat Genet 41:258–263

Jordan WR, Dugas WA Jr, Shouse PJ (1983) Strategies for crop improvement for drought-prone regions. Agric Water Manag 7:281–299

Joshi R, Shukla A, Sairam RK (2011) In vitro screening of rice genotypes for drought tolerance using polyethylene glycol. Acta Physiol Plant 33:2209–2217

Kaeppler SM, Parke JL, Mueller SM, Senior L, Stuber C, Tracy WF (2000) Variation among corn inbred lines and detection of quantitative trait loci for growth at low phosphorus and responsiveness to arbuscular mycorrhizal fungi. Crop Sci 40:358–364

Kamoshita A, Wade IJ, Yamauchi A (2000) Genotypic variation in response of rainfed lowland rice to drought and rewatering: III. Water extraction during the drought period. Plant Prod Sci 3:189–196

Kamoshita A, Jingxian Z, Siopongco J, Sarkarung S, Nguyen HT, Wade LJ (2002a) Effects of phenotyping environment on identification of quantitative trait loci for rice root morphology under anaerobic conditions. Crop Sci 42:255–265

Kamoshita A, Wade L, Ali M, Pathan M, Zhang J, Sarkarung S, Nguyen HT (2002b) Mapping QTLs for root morphology of a rice population adapted to rainfed lowland conditions. Theor Appl Genet 104:880–893

Kamoshita A, Babu RC, Boopathi NM, Fukai S (2008) Phenotypic and genotypic analysis of drought-resistance traits for development of rice cultivars adapted to rainfed environments. Field Crops Res 109(103):1–23

Kangasjärvi J, Jaspers P, Kollist H (2005) Signaling and cell death in ozone-exposed plants. Plant Cell Environ 28:1021–1036

Kang B, Zhang Z, Wang L, Zheng L, Mao W, Li M, Wu Y, Wu P, Mo X (2013) OsCYP2, a chaperone involved in degradation of auxin-responsive proteins, plays crucial roles in rice lateral root initiation. Plant J 74:86–97. doi:10.1111/tpj.12106

Kashiwagi J, Krishnamurthy L, Crouch JH, Serraj R (2006) Variability of root length density and its contributions to seed yield in chickpea (Cicer arietinum L) under terminal drought stress. Field Crops Res 95:171–181

Kassem MA, Meksem K, Kang CH, Njiti VN, Kilo V, Wood AJ, Lightfoot DA (2004) Loci underlying resistance to manganese toxicity mapped in a soybean recombinant inbred line population of 'Essex' × 'Forrest'. Plant Soil 260:197–204

Katerji N, Rana G, Mastrorilli M (2010) Modeling of actual evapotranspiration in open top chambers (OTC) at daily and seasonal scale: multi-annual validation on soybean in contrasted conditions of water stress and air ozone concentration. Eur J Agron 33:218–230

Keurentjes JJB, Fu JY, deVos CHR, Lommen A, Hall RD, Bino RJ, van der Plas LHW, Jansen RC, Vreugdenhil D, Koornneef M (2006) The genetics of plant metabolism. Nat Genet 38:842–849

Khowaja FS, Norton GJ, Courtois B, Price AH (2009) Improved resolution in the position of drought-related QTLs in a single mapping population of rice by meta-analysis. BMC Genomics 10:276

Kilian J, Whitehead D, Horak J, Wanke D, Weinl S, Batistic O, D'Angelo C, Bornberg-Bauer E, Kudla J, Harter K (2007) The AtGenExpress global stress expression data set: protocols, evaluation and model data analysis of UV-B light, drought and cold stress responses. Plant J 50:347–363

King CA, Purcell LC, Brye KR (2009) Differential wilting among soybean genotypes in response to water deficit. Crop Sci 49:290–291

Kitomi Y, Ogawa A, Kitano H, Inukai Y (2008) CRL4 regulates crown root formation through auxin transport in rice. Plant Root 2:19–28

Kobayashi Y, Koyama H (2002) QTL analysis of Al tolerance in recombinant inbred lines of Arabidopsis thaliana. Plant Cell Physiol 43:1526–1533

Komatsu S, Yamamoto R, Nanjo Y, Mikami Y, Yunokawa H, Sakata K (2009) A comprehensive analysis of the soybean genes and proteins expressed under flooding stress using transcriptome and proteome techniques. J Proteome Res 8:4766–4778

Komatsu S, Thibaut D, Hiraga S, Kato M, Chiba M, Hashiguchi A, Tougou M, Shimamura S, Yasue H (2011) Characterization of a novel flooding stress-responsive alcohol dehydrogenase expressed in soybean roots. Plant Mol Biol 77:309–322

Kong FJ, Oyanagi A, Komatsu S (2010) Cell wall proteome of wheat roots under flooding stress using gel-based and LC MS/MS-based proteomics approaches. Biochim Biophys Acta 1804:124–136

Kovalchuk I, Titov V, Hohn B, Kovalchuk O (2005) Transcriptome profiling reveals similarities and differences in plant responses to cadmium and lead. Mutat Res 570:149–161

Kover PX, Valdar W, Trakalo J, Scarcelli N, Ehrenreich IM, Purugganan MD, Durrant C, Mott R (2009) A multiparent advanced generation inter-cross to fine-map quantitative traits in Arabidopsis thaliana. PLoS Genet 5:e1000551

Krishnamurthy L, Kashiwagi J, Upadhyaya HD, Serraj R (2003) Genetic diversity of drought avoidance root traits in the mini-core germplasm collection of chickpea. Int Chickpea Pigeonpea Newsl 10:21–24

Kumar R, Sanjeev M, Srivastava MN (2004) Evaluation of morphophysiological traits associated with drought tolerance in rice. Indian J Plant Physiol 9:305–307

Kwasniewski M, Janiaka A, Mueller-Roeber B, Szarejko I (2010) Global analysis of the root hair morphogenesis transcriptome reveals new candidate genes involved in root hair formation in barley. J Plant Physiol 167:1076–1083

Lafitte HR, Champoux MC, McLaren G, O'Toole JC (2001) Rice root morphological traits are related to isozyme group and adaptation. Field Crops Res 71:57–70

Landi P, Albrecht B, Giuliani MM, Sanguineti MC (1998) Seedling characteristics in hydroponic culture and field performance of corn genotypes with different resistance to root lodging. Maydica 43:111–116

Landi P, Sanguineti M, Darrah L, Giuliani M, Salvi S, Conti S, Tuberosa R (2002) Detection of QTLs for vertical root pulling resistance in corn and overlap with QTLs for root traits in hydroponics and for grain yield under different water regimes. Maydica 47:233–243

Landi P, Giuliani S, Salvi S, Ferri M, Tuberosa R, Sanguineti MC (2010) Characterization of root-yield-1.06, a major constitutive QTL for root and agronomic traits in corn across water regimes. J Exp Bot 61:3553–3562

Laperche A, Devienne-Barret F, Maury O, Le Gouis J, Ney B (2006) A simplified conceptual model of carbon/nitrogen functioning for QTL analysis of winter wheat adaptation to nitrogen deficiency. Theor Appl Genet 113:1131–1146

Le DT, Nishiyama R, Watanabe Y, Mochida K, Yamaguchi-shinozaki K, Shinozaki K, Tran LS (2011a) Genome-wide expression profiling of soybean two-component system genes in soybean root and shoot tissues under dehydration stress. DNA Res 18:17–29

Le DT, Nishiyama R, Watanabe Y, Mochida K, Yamaguchi-shinozaki K, Shinozaki K, Tran LS (2011b) Genome-wide survey and expression analysis of the plant-specific NAC transcription factor family in soybean during development and dehydration stress. DNA Res 18:263–276

Lebreton C, Lazic-Jancic V, Steed A et al (1995) Identification of QTL for drought responses in corn and their use in testing causal relationships between traits. J Exp Bot 46(288):853–865

Lee SC, Mustroph A, Sasidharan R, Vashisht D, Pedersen O, Oosumi T, Voesenek LACJ, Bailey-Serres J (2011) Molecular characterization of the submergence response of the Arabidopsis thaliana ecotype Columbia. New Phytol 190:457–471

Li YD, Wang YJ, Tong YP, Gao JG, Zhang JS, Chen SY (2005a) QTL mapping of phosphorus deficiency tolerance in soybean. Euphytica 142:137–142

Li Z, Mu P, Li C, Zhang H, Li Z, Gao Y, Wang X (2005b) QTL mapping of root traits in a doubled haploid population from a cross between upland and lowland japonica rice in three environments. Theor Appl Genet 110:1244–1252

Li MY, Qin CB, Welti R, Wang XM (2006) Double knockouts of phospholipases D zeta 1 and D zeta 2 in Arabidopsis affect root elongation during phosphate-limited growth but do not affect root hair patterning. Plant Physiol 140:761–770

Li K, Xu C, Zhang K, Yang A, Zhang J (2007) Proteomic analysis of roots growth and metabolic changes under phosphorus deficit in maize (Zea mays L.) plants. Proteomics 7:1501–1512

Li K, Xu C, Li Z, Zhang K, Yang A, Zhang J (2008) Comparative proteome analyses of phosphorus responses in maize (Zea mays L.) roots of wild-type and a low-P-tolerant mutant reveal root characteristics associated with phosphorus efficiency. Plant J 55:927–939

Li J, Xie Y, Dai A, Liu L, Li Z (2009) Root and shoot traits responses to phosphorus deficiency and QTL analysis at seedling stage using introgression lines of rice. J Genet Genomics 36:173–183

Liang Q, Cheng X, Mei M, Yan X, Liao H (2010) QTL analysis of root traits as related to phosphorus efficiency in soybean. Ann Bot 106:223–234

Liao C, Hochholdinger F, Li C (2012) Comparative analyses of three legume species reveals conserved and unique root extracellular proteins. Proteomics 12:3219–3228

Licausi F, Van Dongen JT, Giuntoli B, Novi G, Santaniello A, Geigenberger P, Perata P (2010) HRE1 and HRE2, two hypoxia-inducible ethylene response factors, affect anaerobic responses in Arabidopsis thaliana. Plant J 62:302–315

Lin S, Cianzio SR, Shoemaker RC (1997) Mapping genetic loci for iron deficiency chlorosis in soybean. Mol Breed 3:219–229

Lin SI, Chiang SF, Lin WY, Chen JW, Tseng CY, Wu PC, Chiou TJ (2008) Regulatory network of microRNA399 and PHO2 by systemic signaling. Plant Physiol 147:732–746

Lilley JM, Ludlow MM, McCouch SR, O'Toole JC (1996) Locating QTL for osmotic adjustment and dehydration tolerance in rice. J Exp Bot 47:1427–1436

Lister R, Gregory BD, Ecker JR (2009) Next is now: new technologies for sequencing of genomes, transcriptomes, and beyond. Curr Opin Plant Biol 12:107–118

Liu Y, Gai JY, Lü HN, Wang YJ, Chen SY (2005) Identification of drought tolerance germplasm and inheritance and QTL mapping of related root traits in soybean. Acta Genet Sin 32:855–863

Liu Y, Lamkemeyer T, Jakob A, Mi G, Zhang F, Nordheim A, Hochholdinger F (2006) Comparative proteome analyses of maize (*Zea mays* L.) primary roots prior to lateral root initiation reveal differential protein expression in the lateral root initiation mutant rum1. Proteomics 6:4300–4308

Liu J, Li J, Chen F, Zhang F, Ren T, Zhuang Z, Mi G (2008a) Mapping QTLs for root traits under different nitrate levels at the seedling stage in corn (*Zea mays* L.). Plant Soil 305:253–265

Liu JX, Han LL, Chen FJ, Bao J, Zhang FS, Mi GH (2008b) Microarray analysis reveals early responsive genes possibly involved in localized nitrate stimulation of lateral root development in maize. Plant Sci 175:272–282

Liu S, Wang J, Wang L, Xue Y, Wu P, Shou H (2009) Adventitious root formation in rice requires OsGNOM1 and is mediated by the OsPINs family. Cell Res 19:1110–1119

Liu JC, Cai HG, Chu Q, Chen XH, Chen FJ, Yuan LX, Mi GH, Zhang FS (2011) Genetic analysis of vertical root pulling resistance (VRPR) in corn using two genetic populations. Mol Breed 28:463–474

Lobell DB, Burke MB, Tebaldi C, Mastrandrea MD, Falcon WP, Naylor RL (2008) Prioritizing climate change adaptation needs for food security in 2030. Science 319:607–610

Loreti E, Poggi A, Novi G, Alpi A, Perata P (2005) A genome-wide analysis of the effects of sucrose on gene expression in *Arabidopsis* seedlings under anoxia. Plant Physiol 137:1130–1138

Loudet O, Chaillou S, Camilleri C, Bouchez D, Daniel-Vedele F (2002) Bay-0 x Shahdara recombinant inbred line population: a powerful tool for the genetic dissection of complex traits in Arabidopsis. Theor Appl Genet 104:1173–1184

Loudet O, Gaudon V, Trubuil A, Daniel-Vedele F (2005) Quantitative trait loci controlling root growth and architecture in *Arabidopsis thaliana* confirmed by heterogeneous inbred family. Theor Appl Genet 110:742–753

Lu CX, Guo JQ, Wang Y, Leng JT, Yang GM, Hou WS, Wu CX, Han TF (2010) Identification, inheritance analysis, and QTL mapping of root and shoot traits in soybean variety PI471938 with tolerance to wilting. Acta Agron Sin 36:1476–1483

Lu M, Ying S, Zhang D, Shi Y, Song Y, Wang T, Li Y (2012) A maize stress responsive NAC transcription factor, Zm SNAC1, confers tolerance to dehydration in transgenic *Arabidopsis*. Plant Cell Rep 31:1701–1711

Lynch JP (2007) Turner review no. 14. Roots of the second green revolution. Aust J Bot 55:493–512

Lynch JP (2011) Root phenes for enhanced soil exploration and phosphorus acquisition: tools for future crops. Plant Physiol 156:1041–1049

Lynch JP, Brown K (2012) New roots for agriculture – exploiting the root phenome. Philos Trans R Soc 367(1595):1598–1604

Ma J, Shen R, Zhao Z, Wissuwa M, Takeuchi Y, Ebitani T, Yano M (2002) Response of rice to Al stress and identification of quantitative trait loci for Al tolerance. Plant Cell Physiol 43:652

Ma HX, Bai GH, Carver BF, Zhou LL (2005) Molecular mapping of a quantitative trait locus for aluminum tolerance in wheat cultivar Atlas 66. Theor Appl Genet 112:51–57

Maccaferri M, Sanguineti MC, Corneti S, Ortega JL, Salem MB, Bort J, DeAmbrogio E, del Moral LF, Demontis A, El-Ahmed A, Maalouf F, Machlab H, Martos V, Moragues M,

Motawaj J, Nachit M, Nserallah N, Ouabbou H, Royo C, Slama A, Tuberosa R (2008) Quantitative trait loci for grain yield and adaptation of durum wheat (*Triticum durum* Desf.) across a wide range of water availability. Genetics 178:489–511

Mace ES, Singh V, Van Oosterom EJ, Hammer GL, Hunt CH, Jordan DR (2012) QTL for nodal root angle in sorghum (*Sorghum bicolor* L. Moench) co-locate with QTL for traits associated with drought adaptation. Theor Appl Genet 124:97–109

Magalhaes JV, Garvin DF, Wang Y, Sorrells ME, Klein PE, Schaffert RE, Li L, Kochian LV (2004) Comparative mapping of a major aluminum tolerance gene in sorghum and other species in the *Poaceae*. Genetics 167:1905–1914

Magalhaes JV, Liu J, Guimaraes CT, Lana UG, Alves VM, Wang YH, Schaffert RE, Hoekenga OA, Pineros MA, Shaff JE, Klein PE, Carneiro NP, Coelho CM, Trick HN, Kochian LV (2007) A gene in the multidrug and toxic compound extrusion (MATE) family confers aluminum tolerance in sorghum. Nat Genet 39:1156–1161

MacMillan K, Emrich K, Piepho HP, Mullins CE, Price AH (2006) Assessing the importance of genotype×environment interaction for root traits in rice using a mapping population II: conventional QTL analysis. Theor Appl Genet 113:953–964

Manavalan LP, Guttikonda SK, Tran LP, Nguyen HT (2009) Physiological and molecular approaches to improve drought resistance in soybean. Plant Cell Physiol 50:1260–1276

Mano Y, Omori F, Takamizo T, Kindiger B, Bird RM, Loaisiga CH, Takahashi H (2007) QTL mapping of root aerenchyma formation in seedlings of a maize×rare teosinte "*Zea nicaraguensis*" cross. Plant Soil 295:103–113

Marshall A, Aalen RB, Audenaert D, Beeckman T, Broadley MR, Butenko MA, Cano-Delgado AI, Vries SD, Dresselhaus T, Felix G, Graham NS, Foulkes J, Granier C, Greb KT, Grossniklaus U, Hammond JP, Heidstra R, Hodgman C, Hothorn M, Inze D, Ostergaard L, Russinova E, Simon R, Skirycz A, Stahl Y, Zipfel C, Smete ID (2012) Tackling drought stress: RECEPTOR-LIKE KINASES present new approaches. Plant Cell 24:2262–2278

Martin B, Ruiztorres NA (1992) Effects of water-deficit stress on photosynthesis, its components and limitations, and on water-use efficiency in wheat (*Triticum aestivum* L.). Plant Physiol 100:733–739

Matsumura H, Molina C, Kruger DH, Terauchi R, Kahl G (2012) DeepSuperSAGE: high-throughput transcriptome sequencing with now- and next-generation sequencing technologies. In: Harbers M, Kahl G (eds) Tag-based next generation sequencing. Wiley-VCH, KGaA, Weinheim, pp 3–22

Matsuo N, Ozawa K, Mochizuki T (2009) Genotypic differences in root hydraulic conductance of rice (*Oryza sativa* L.) in response to water regimes. Plant Soil 316:25–34

McKhann HI, Hirsch AM (1994) Isolation of chalcone synthase and chalcone isomerase cDNAs from alfalfa (Medicago sativa L.): highest transcript levels occur in young roots and root tips. Plant Mol Biol 24(5):767–777

Meng Y, Ma X, Chen D, Wu P, Chen M (2010) MicroRNA-mediated signaling involved in plant root development. Biochem Biophys Res Commun 393:345–349

Messmer R, Fracheboud Y, Banziger M, Vargas M, Stamp P, Ribaut J-M (2009) Drought stress and tropical corn: QTL-by-environment interactions and stability of QTLs across environments for yield components and secondary traits. Theor Appl Genet 119:913–930

Michal L, Nathalie R, Yogev R, Igor K, Smadar W, JorgeHugo L, Shabtai C, Gadi G, Hinanit K, Yoram K (2009) Transcriptional profiling of Arabidopsis thaliana plants' response to low relative humidity suggests a shoot-root communication. Plant Sci 177:450–459

Millan T, Clarke H, Siddique K, Buhariwalla H, Gaur P, Kumar J, Gil J, Kahl G, Winter P (2006) Chickpea molecular breeding: new tools and concepts. Euphytica 147:81–103

Miller CR, Ochoa I, Nielsen KL, Beck D, Lynch JP (2003) Genetic variation for adventitious rooting in response to low phosphorus availability: potential utility for phosphorus acquisition from stratified soils. Funct Plant Biol 30:973–985

Miltner ED, Karnok KJ, Hussey RS (1991) Root response of tolerant and intolerant soybean cultivars to soybean cyst nematode. Agron J 83:571–576

Mir RR, Zaman-Allah M, Sreenivasulu N, Trethowan R, Varshney RK (2012) Integrated geno-mics, physiology and breeding approaches for improving drought tolerance in crops. Theor Appl Genet 125:625–645

Misson J, Raghothama KG, Jain A, Jouhet J, Block MA, Bligny R, Ortet P, Creff A, Somerville S, Rolland N, Doumas P, Nacry P, Herrerra-Estrella L, Nussaume L, Thibaud M (2005) A genome-wide transcriptional analysis using *Arabidopsis thaliana* Affymetrix gene chips deter-mined plant responses to phosphate deprivation. Proc Natl Acad Sci USA 102:11934–11939

Mohammadi M, Kav NN, Deyholos MK (2007) Transcriptional profiling of hexaploid wheat roots identifies novel, dehydration-responsive genes. Plant Cell Environ 30:630–645

Molina C (2008) High-throughput genome-wide expression analysis of a non model organism: the chickpea root and nodule transcriptome under salt and drought stress. Dissertation, 230 p. http://www.ub.uni-frankfurt.de/dissertation

Molina C, Zaman-Allah M, Khan F, Fatnassi N, Horres R, Rotter B, Steinhauer D, Amenc L, Drevon JJ, Winter P, Kahl G (2011) The salt-responsive transcriptome of chickpea roots and nodules via deepSuperSAGE. BMC Plant Biol 14:11–31

Morgan PB, Bernacchi CJ, Ort DR, Long SP (2004) An *in vivo* analysis of the effect of season-long open-air elevation of ozone to anticipated 2050 levels on photosynthesis in soybean. Plant Physiol 135:2348–2357

Moriwaki T, Miyazawa Y, Takahashi H (2010) Transcriptome analysis of gene expression during the hydrotropic response in Arabidopsis seedlings. Environ Exp Bot 69:148–157

Mouchel CF, Briggs GC, Hardtke CS (2004) Natural genetic variation in Arabidopsis identifies BREVIS RADIX, a novel regulator of cell proliferation and elongation in the root. Genes Dev 18:700–714

Moumeni A, Satoh K, Kondoh H, Asano T, Hosaka A, Venuprasad R, Serraj R, Kumar A, Leung H, Kikuchi S (2011) Comparative analysis of root transcriptome profiles of two pairs of drought-tolerant and susceptible rice near-isogenic lines under different drought stress. BMC Plant Biol 11:174

Munoz-Perea CG, Teran H, Allen RG, Wright JL, Westermann DT, Singh SP (2006) Selection for drought resistance in dry bean landraces and cultivars. Crop Sci 46:2111–2120

Mustroph A, Lee SC, Oosumi T, Zanetti ME, Yang H, Ma K, Yaghoubi-Masihi A, Fukao T, Bailey-Serres J (2010) Cross-kingdom comparison of transcriptomic adjustments to low-oxygen stress highlights conserved and plant-specific responses. Plant Physiol 152:1484–1500

Nakajima K, Sena G, Nawy T, Benfey PN (2001) Intercellular movement of the putative transcription factor SHR in root patterning. Nature 413:307–311

Nakashima K, Tran LS, Van Nguyen D, Fujita M, Maruyama K, Todaka D, Ito Y, Hayashi N, Shinozaki K, Yamaguchi-Shinozaki K (2007) Functional analysis of a NAC-type transcription factor OsNAC6 involved in abiotic and biotic stress-responsive gene expression in rice. Plant J 51:617–630

Nakata M, Shiono T, Watanabe Y, Satoh T (2002) Salt stress-induced dissociation from cells of a germin-like protein with Mn-SOD activity and an increase in its mRNA in a moss, *Barbula unguiculata*. Plant Cell Physiol 43:1568–1574

Narsai R, Rocha M, Geigenberger P, Whelan J, Van Dongen JT (2011) Comparative analysis between plant species of transcriptional and metabolic responses to hypoxia. New Phytol 190:472–487

Nestler J, Schütz W, Hochholdinger F (2011) Conserved and unique features of the maize (*Zea mays* L.) root hair proteome. J Proteome Res 10:2525–2537

Nguyen VT, Nguyen BD, Sarkarung S, Martinez C, Paterson AH, Nguyen HT (2002) Mapping of genes controlling aluminum tolerance in rice: comparison of different genetic backgrounds. Mol Genet Genomics 267:772–780

Nguyen BD, Brar DS, Bui BC, Nguyen TB, Pham LN, Nguyen HT (2003) Identification and mapping of the QTL for aluminum tolerance introgressed from the new source, *Oryza rufipogon* Griff., into indica rice (*Oryza sativa* L.). Theor Appl Genet 106:583–593

Nguyen TT, Klueva N, Chamareck V, Aarti A, Magpantay G, Millena ACM, Pathan MS, Nguyen HT (2004) Saturation mapping of QTL regions and identification of putative candidate genes for drought tolerance in rice. Mol Genet Genomics 272:35–46

Nikiforova VJ, Kopka J, Tolstikov V, Fiehn O, Hopkins L, Hawkesford MJ, Hesse H, Hoefgen R (2005) Systems rebalancing of metabolism in response to sulfur deprivation, as revealed by metabolome analysis of arabidopsis plants. Plant Physiol 138:304–318

O'Toole JC (2004) Rice and water: the final frontier. In: First international conference on rice for the future. Rockefeller Foundation, Bangkok, p 26

Orman B, Ligeza A, Szarejko I, Maluszynski M (2011) Introduction to root genomics. In: Costa de Oliveira A, Varshney R (eds) Root genomics. Springer, Berlin, pp 11–72

Ortiz R, Sayre KD, Govaerts B, Gupta R, Subbarao GV, Ban T, Hodson D, Dixon JA, Ortiz-Monasterio JI, Reynolds M (2008) Climate change: can wheat beat the heat? Agric Ecosyst Environ 126:46–58

Pant BD, Buhtz A, Kehr J, Scheible WR (2008) MicroRNA399 is a long distance signal for the regulation of plant phosphate homeostasis. Plant J 53:731–738

Pantalone VR, Rebetzke GJ, Burton JW, Carter TE Jr (1996) Phenotypic evaluation of root traits in soybean and applicability to plant breeding. Crop Sci 36:456–459

Pantuwan G, Fukai S, Cooper M, Rajatasereekul S, O'Toole JC, Basnayake J (2004) Yield response of rice (*Oryza sativa* L.) genotypes to drought under rainfed lowlands. 4. Vegetative stage screening in the dry season. Field Crops Res 89:281–297

Parizot B, De RB, Beeckman T (2010) VisuaLRTC: a new view on lateral root initiation by combining specific transcriptome data sets. Plant Physiol 153:34–40

Paschold A, Jia Y, Marcon C, Lund S, Larson NB, Yeh CT, Ossowski S, Lanz C, Nettleton D, Schnable PS, Hochholdinger F (2012) Complementation contributes to transcriptome complexity in maize (*Zea mays* L.) hybrids relative to their inbred parents. Genome Res 22 (12):2445–2454

Paterson AH, Freeling M, Tang H, Wang X (2010) Insights from the comparison of plant genome sequences. Annu Rev Plant Biol 61:349–372

Patnaik D, Khurana P (2001) Germins and germin like proteins: an overview. Indian J Exp Biol 39:191–200

Petersson SV, Johansson AI, Kowalczyk M, Makoveychuk A, Wang JY, Moritz T, Grebe M, Benfey PN, Sandberg G, Ljung K (2009) An auxin gradient and maximum in the Arabidopsis root apex shown by high-resolution cell-specific analysis of IAA distribution and synthesis. Plant Cell 21:1659–1668

Pinto RS, Reynolds MP, Mathews KL, McIntyre CL, Olivares-Villegas JJ, Chapman SC (2010) Heat and drought adaptive QTL in a wheat population designed to minimize confounding agronomic effects. Theor Appl Genet 121:1001–1021

Postma JA, Lynch JP (2011) Root cortical aerenchyma enhances the growth of maize on soils with suboptimal availability of nitrogen, phosphorus, and potassium. Plant Physiol 156:1190–1201

Postma JA, Lynch JP (2012) Complementarity in root architecture for nutrient uptake in ancient maize/bean and maize/bean/squash polycultures. Ann Bot 110(2):521–534

Prabuddha HR, Manjunatha K, Venuprasad R, Vinod MS, Jureifa JH, Shashidhar HE (2008) Identification of near-isogenic lines: an innovative approach, validated for root and shoot morphological characters in a mapping population of rice. Euphytica 160:357–368

Price AH, Tomos AD (1997) Genetic dissection of root growth in rice II: mapping quantitative trait loci using molecular markers. Theor Appl Genet 95:143–152

Price AH, Steele KA, Moore BJ, Barraclough PB, Clark LJ (2000) A combined RFLP and AFLP map of upland rice (Oryza sativa) used to identify QTL for root-penetration ability. Theor Appl Genet 100:49–56

Price AH, Steele KA, Moore BJ, Jones RGW (2002) Upland rice grown in soil-filled chambers exposed to contrasting water-deficit regimes. II. Mapping quantitative trait loci for root morphology and distribution. Field Crops Res 76:25–43

Puranik S, Jha S, Srivastava PS, Sreenivasulu N, Prasad M (2011) Comparative transcriptome analysis of contrasting foxtail millet cultivars in response to short-term salinity stress. J Plant Physiol 168:280–287

Qiao-Ying Z, Cun-Yi Y, Qi-Bin M, Xiu-Ping L, Wen-Wen D, Hai N (2012) Identification of wild soybean miRNAs and their target genes responsive to aluminum stress. BMC Plant Biol 2012 (12):182

Qiu F, Zheng Y, Zhang Z, Xu S (2007) Mapping of QTL associated with water logging tolerance during the seedling stage in maize. Ann Bot 99:1067–1081

Qu Y, Mu P, Zhang H, Chen CY, Gao Y, Tian Y, Wen F, Li Z (2008) Mapping QTLs of root morphological traits at different growth stages in rice. Genetica 133:187–200

Rabello AR, Guimaraes CM, Rangel PH, Da SFR, Seixas D, Souza ED, Brasileiro AC, Spehar CR, Ferreira ME, Mehta A (2008) Identification of drought-responsive genes in roots of upland rice (*Oryza sativa* L). BMC Genomics 9:485

Ramirez-Vallejo P, Kelly JD (1998) Traits related to drought resistance in common bean. Euphytica 99:127–138

Rao IM (2001) Role of physiology in improving crop adaptation to abiotic stresses in the tropics: the case of common bean and tropical forages. In: Pessarakli M (ed) Handbook of plant and crop physiology. Marcel Dekker, New York, pp 583–613

Rao PS, Mishra B, Gupta SR, Rathore A (2008) Reproductive stage tolerance to salinity and alkalinity stresses in rice genotypes. Plant Breed 127:256–261

Ray JD, Yu LX, McCouch SR, Champoux MC, Wang G, Nguyen HT (1996) Mapping quantitative trait loci associated with root penetration ability in rice. Theor Appl Genet 92:627–636

Rauh B, Basten C, Buckler E (2002) Quantitative trait loci analysis of growth response to varying nitrogen sources in Arabidopsis thaliana. Theor Appl Genet 104:743–750

Rebouillat J, Dievart A, Verdeil JL, Escoute J, Giese G, Breitler JC, Gantet P, Espeout S, Guiderdoni E, Périn C (2009) Molecular genetics of rice root development. Rice 2:15–34

Reid DE, Ferguson BJ, Hayashi S, Lin Y-H, Gresshoff PM (2011) Molecular mechanisms controlling legume autoregulation of nodulation. Ann Bot 108:789–795

Ren ZH, Gao JP, Li LG, Cai XL, Huang W, Chao DY, Zhu MZ, Wang ZY, Luan S, Lin HX (2005) A rice quantitative trait locus for salt tolerance encodes a sodium transporter. Nat Genet 37:1141–1146

Ren Y, He X, Liu D, Li J, Zhao X, Li B, Tong Y, Zhang A, Li Z (2012) Major quantitative trait loci for seminal root morphology of wheat seedlings. Mol Breed 30:139–148

Reymond M, Svistoonoff S, Loudet O, Nussaume L, Desnos T (2006) Identification of QTL controlling root growth response to phosphate starvation in *Arabidopsis thaliana*. Plant Cell Environ 29:115–125

Ribaut JM, Hoisington DA, Deutsch JA, Jiang C, Gonzalez-De-Leon D (1996) Identification of quantitative trait loci under drought conditions in tropical corn. 1. Flowering parameters and the anthesis-silking interval. Theor Appl Genet 92:905–914

Rivelli AR, James RA, Munns R, Condon AG (2002) Effect of salinity on water relations and growth of wheat genotypes with contrasting sodium uptake. Funct Plant Biol 29:1065–1074

Robin S, Pathan MS, Courtois B, Lafitte R, Carandang S, Lanceras S, Amante M, Nguyen HT, Li Z (2003) Mapping osmotic adjustment in an advanced backcross inbred population of rice. Theor Appl Genet 107:1288–1296

Roje S (2006) S-Adenosyl-l-methionine: beyond the universal methyl group donor. Phytochemistry 67:1686–1698

Rowe HC, Hansen BG, Halkier BA, Kliebenstein DJ (2008) Biochemical networks and epistasis shape the *Arabidopsis thaliana* metabolome. Plant Cell 20:1199–1216

Ruffel S, Freixes S, Balzergue S, Tillard P, Jeudy C, Martin-Magniette ML, van der Merwe MJ, Kakar K, Gouzy J, Fernie AR, Udvardi M, Salon C, Gojon A, Lepetit M (2008) Systemic signaling of the plant nitrogen status triggers specific transcriptome responses depending on the nitrogen source in *Medicago truncatula*. Plant Physiol 146:2020–2035

Ruta N, Liedgens M, Fracheboud Y, Stamp P, Hund A (2010a) QTLs for the elongation of axile and lateral roots of corn in response to low water potential. Theor Appl Genet 120:621–631

Ruta N, Stamp P, Liedgens M, Fracheboud Y, Hund A (2010b) Collocations of QTLs for seedling traits and yield components of tropical corn under water stress conditions. Crop Sci 50:1385–1392

Sakai T, Duque MC, Cabrera FAV, Martínez CP, Ishitani M (2010) Establishment of drought screening protocols for rice under field conditions. Acta Agron 59:338–346

Salvi S, Sponza G, Morgante M, Tomes D, Niu X, Fengler KA, Meeley R, Ananiev EV, Svitashev S, Bruggemann E, Li B, Hainey CF, Radovic S, Zaina G, Rafalski JA, Tingey SV, Miao GH, Phillips RL, Tuberosa R (2007) Conserved noncoding genomic sequences associated with a flowering-time quantitative trait locus in maize. Proc Natl Acad Sci USA 104(27):11376–11381

Samantaray S, Rout GR, Das P (1997) Tolerance of rice to nickel in nutrient solution. Biol Plant 40:295–298

Sanchez-Rodriguez E, Rubio-Wilhelmi MD, Cervilla LM, Blasco B, Rios JJ, Leyva R, Romero L, Ruiz JM (2010) Study of the ionome and uptake fluxes in cherry tomato plants under moderate water stress conditions. Plant Soil 335:339–347

Sanguineti MC, Li S, Maccaferri M, Corneti S, Rotondo F, Chiari T, Tuberosa R (2007) Genetic dissection of seminal root architecture in elite durum wheat germplasm. Ann Appl Biol 151:291–305

Sauer M, Jakob A, Nordheim A, Hochholdinger F (2006) Proteomic analysis of shoot-borne root initiation in maize (*Zea mays* L.). Proteomics 6:2530–2541

Saxena NP (2003) Management of drought in chickpea: a holistic approach. In: Saxena NP (ed) Management of agricultural drought – agronomic and genetic options. Oxford & IBH Publishing, New Delhi, pp 103–122

Sayed OH (2003) Chlorophyll fluorescence as a tool in cereal crop research. Photosynthetica 41:321–330

Scheible WR, Morcuende R, Czechowski T, Fritz C, Osuna D, Palacios-Rojas N, Schindelasch D, Thimm O, Udvardi MK, Stitt M (2004) Genome-wide reprogramming of primary and secondary metabolism, protein synthesis, cellular growth processes, and the regulatory infrastructure of Arabidopsis in response to nitrogen. Plant Physiol 136:2483–2499

Scheres B, McKhann HI, van den Berg C (1996) Roots redefined: anatomical and genetic analysis of root development. Plant Physiol 111:959–964

Scheres B, Benfey PN, Dolan L (2002) Root development. http://www.aspb.org/downloads/Arabidopsis./Scheres.pdf

Schmid M, Davison TS, Henz SR, Pape UJ, Demar M, Vingron M, Scholkopf B, Weigel D, Lohmann JU (2005) A gene expression map of *Arabidopsis thaliana* development. Nat Genet 37:501–506

Serraj R, Kumar A, McNally KL, Slamet-Loedin I, Bruskiewich R, Mauleon R, Cairns J, Hijmans RJ (2009) Improvement of drought resistance in rice. Adv Agron 103:41–99

Shanthi P, Jebaraj S, Geetha S (2010) *In vitro* screening for salt tolerance in Rice (*Oryza sativa*). J Plant Breed 1:1208–1212

Sharma N, Shashidhar HE, Hittalmani S (2003) Root length specific SCAR marker in rice. Rice Genet Newsl 19:47–48

Sharma AD, Sharma H, Lightfoot DA (2011) The genetic control of tolerance to aluminum toxicity in the 'Essex' by 'Forrest' recombinant inbred line population. Theor Appl Genet 122:687–694

Sharp RE, Lenoble ME (2002) ABA, ethylene and the control of shoot and root growth under water stress. J Exp Bot 53:33–37

Sharp RE, Poroyko V, Hejlek IG, Spollen WG, Springer GK, Bohnert HJ, Nguyen HT (2004) Root growth maintenance during water deficits: physiology to functional genomics. J Exp Bot 55:2343–2351

Shashidhar HE, Sharma N, Venuprasad R, Toorchi M, Hittalmani S (2000) Identification of traits and molecular markers associated with components of drought resistance in rainfed lowland rice. In: International rice genetics symposium. IRRI, Los Banos

Shen L, Courtois B, McNally KL, Robin S, Li Z (2001) Evaluation of near isogenic lines of rice introgressed with QTLs for root depth through marker-aided selection. Theor Appl Genet 103:75–83

Shevell DE, Kunkel T, Chua N-H (2000) Cell wall alterations in the Arabidopsis emb30 mutant. Plant Cell 12:2047–2059

Shin R, Schachtman DP (2004) Hydrogen peroxide mediates plant root cell response to nutrient deprivation. Proc Natl Acad Sci USA 101:8827–8832

Shin R, Burch AY, Huppert KA, Tiwari SB, Murphy AS, Guilfoyle TJ, Schachtman DP (2007) The Arabidopsis transcription factor MYB77 modulates auxin signal transduction. Plant Cell 19:2440–2453

Shinozaki K, Yamaguchi-Shinozaki K (2007) Gene networks involved in drought stress response and tolerance. J Exp Bot 58:221–227

Sieberer T, Hauser M-T, Seifert GJ, Luschnig C (2003) PROPORZ1, a putative Arabidopsis transcriptional adaptor protein, mediates auxin and cytokinin signals in the control of cell proliferation. Curr Biol 13:837–842

Sinclair TR (2011) Challenges in breeding for yield increase for drought. Trends Plant Sci 16:289–293

Sinclair TR, Zwieniecki MA, Holbrook NM (2008) Low leaf hydraulic conductance associated with drought tolerance in soybean. Acta Physiol Plant 132:446–451

Sinclair TR, Messina CD, Beatty A, Samples M (2010) Assessment across the United States of the benefits of altered soybean drought traits. Agron J 102:475–482

Singh SP, Teran H, Lema M, Dennis MF, Hayes R, Robinson C (2008) Breeding for slow-darkening, high-yielding broadly adapted dry bean pinto 'Kimberly' and 'Shoshone'. J Plant Regul 2:180–186

SIWI-IWMI (2004) Water-more nutrition per drop, towards sustainable food production and consumption patterns in a rapidly changing world. Stockholm International Water Institute, Stockholm

Sloane RJ, Patterson RP, Carter TE Jr (1990) Field drought tolerance of a soybean plant introduction. Crop Sci 30:118–123

Smith S, De Smet I (2012) Root system architecture: insights from Arabidopsis and cereal crops. Philos Trans R Soc B Biol Sci 367:1441–1452

Specht JE, Chase K, Macrander M, Graef GL, Chung J, Markwell JP, Germann M, Orf JH, Lark KG (2001) Soybean response to water: a QTL analysis of drought tolerance. Crop Sci 41:493–509

Spollen WG, Tao W, Valliyodan B, Chen K, Hejlek IG, Kim J, LeNoble ME, Zhu J, Bohnert HJ, Henderson D, Schachtman DP, Davis GE, Springer GK, Sharp RE, Nguyen HT (2008) Spatial distribution of transcript changes in the maize primary root elongation zone at low water potential. BMC Plant Biol 8:32

Sponchiado B, White J, Castillo J, Jones P (1989) Root growth of four common bean cultivars in relation to drought tolerance in environments with contrasting soil types. Exp Agric 25:249–257

Srinivasan S, Gomez SM, Kumar SS, Ganesh SK, Biji KR, Senthil A, Babu RC (2008) QTLs linked to leaf epicuticular wax, physiomorphological and plant production traits under drought stress in rice. Plant Growth Regul 56:245–256

Sreenivasulu N, Miranda M, Prakash HS, Wobus U, Weschke W (2004) Transcriptome changes in foxtail millet genotypes at high salinity: identification and characterization of a PHGPX gene specifically upregulated by NaCl in a salt-tolerant line. J Plant Physiol 161:467–477

Steele KA, Price AH, Shashidhar HE, Witcombe JR (2006) Marker-assisted selection to introgress rice QTLs controlling root traits and aroma into an Indian upland rice variety. Theor Appl Genet 112:208–221

Steele KA, Virk DS, Kumar R, Prasad SC, Witcombe JR (2007) Field evaluation of upland rice lines selected for QTLs controlling root traits. Field Crops Res 101:180–186

Suji KK, Prince KSJ, Mankhar PS, Kanagaraj P, Poornima R, Amutha K, Kavitha S, Biji KR, Michael Gomez S, Chandra Babu R (2012) Evaluation of rice (*Oryza sativa* L.) near iso-genic lines with root QTLs for plant production and root traits in rainfed target populations of environment. Field Crops Res 137:89–96

Sumith D, Abeysiriwardena Z (2004) A simple screening technique for salinity tolerance in rice: germination rate under stress. Rice Research and Development Institute, Ibbagamuwa

Suriya-arunroj D, Supapoj N, Toojinda T, Vanavichit A (2004) Relative leaf water content as an efficient method for evaluating rice cultivars for tolerance to salt stress. Sci Asia 30:411–415

Sutton T, Baumann U, Hayes J, Collins NC, Shi B, Schnurbusch T, Hay A, Mayo G, Pallota M, Tester M, Langridge P (2007) Boron-toxicity tolerance in barley arising from efflux transporter amplification. Science 30:1446–1449

Svistoonoff S, Creff A, Reymond M, Sigoillot-Claude C, Ricaud L, Blanchet A, Nussaume L, Desnos T (2007) Root tip contact with low-phosphate media reprograms plant root architecture. Nat Genet 39:792–796

Teran H, Singh SP (2002) Comparison of sources and lines selected for drought resistance in common bean. Crop Sci 42:64–70

Terasaka K, Blakeslee JJ, Titapiwatanakun B, Peer WA, Bandyopadhyay A, Makam SN, Lee OR, Richards EL, Murphy AS, Sato F, Yazaki K (2005) PGP4, an ATP binding cassette P-Glycoprotein, catalyzes auxin transport in *Arabidopsis thaliana* roots. Plant Cell 17:2922–2939

Thung M, Rao IM (1999) Integrated management of abiotic stresses. In: Singh SP (ed) Common bean improvement in the twenty-first century. Kluwer, Dordrecht, pp 331–370

Tirumalaraju SV, Jain M, Gallo M (2011) Differential gene expression in roots of nematode-resistant and -susceptible peanut cultivars in response to early stages of peanut root-knot nematode parasitization. J Plant Physiol 168:481–492

Trachsel S, Messmer R, Stamp P, Hund A (2009) Mapping of QTLs for lateral and axile root growth of tropical corn. Theor Appl Genet 119:1413–1424

Trachsel S, Kaeppler S, Brown K, Lynch J (2011) Shovelomics: high throughput phenotyping of corn (*Zea mays* L.) root architecture in the field. Plant Soil 341:75–87

Tripathy JN, Zhang J, Robin S, Nguyen TT, Nguyen HT (2000) QTLs for cell-membrane stability mapped in rice under drought stress. Theor Appl Genet 100:1197–1202

Tuberosa R, Salvi S, Giuliani S, Sanguineti MC, Landi P, Maccaferri M, Conti S (2002a) Mapping QTLs regulating morpho-physiological traits and yield in drought-stressed maize: case studies, shortcomings and perspectives. Ann Bot 89:941–963

Tuberosa R, Sanguineti MC, Landi P, Giuliani S, Salvi S, Conti S (2002b) Identification of QTLs for root characteristics in corn grown in hydroponics and analysis of their overlap with QTLs for grain yield in the field at two water regimes. Plant Mol Biol 48:697–712

Tuberosa R, Salvi S, Sanguineti MC, Maccaferri M, Giuliani S, Landi P (2003) Searching for quantitative trait loci controlling root traits in corn: a critical appraisal. Plant Soil 255:35–54

Tuberosa R, Salvi S, Sanguineti MC, Frascaroli E, Conti S, Landi P (2011) Genomics of root architecture and functions in maize. In: Costa de Oliveira A, Varshney R (eds) Root genomics. Springer, Berlin, pp 179–204

Tuyen DD, Lal SK, Xu DH (2010) Identification of a major QTL allele from wild soybean (*Glycine soja* Sieb. & Zucc.) for increasing alkaline salt tolerance in soybean. Theor Appl Genet 121:229–236

Ueda M, Koshino-Kimura Y, Okada K (2005) Stepwise understanding of root development. Curr Opin Plant Biol 8:71–76

Uga Y, Okuno K, Yano K (2011) Dro1, a major QTL involved in deep rooting of rice under upland field conditions. J Exp Bot 62:2485–2494

Van Dongen JT, Fröhlich A, Ramírez-aguilar SJ, Schauer N, Fernie AR, Erban A, Kopka J, Clark J, Langer A, Geigenberger P (2009) Transcript and metabolite profiling of the adaptive response to mild decreases in oxygen concentration in the roots of arabidopsis plants. Ann Bot 103:269–280

Vantoai T, Alves J, Valliyodan B, Goulart P, Lee J, Fritschi F, Mohammed R, Grover S, Nguyen HT (2009) Expression of root-related transcription factors associated with flooding tolerance of soybean (*Glycine max*). CBFV, Brazil

Varshney RK, Hiremath PJ, Lekha PT, Kashiwagi J, Balaji J, Deokar AA, Vadez V, Xiao Y, Srinivasan R, Gaur PM, Siddique KH, Town CD, Hoisington DA (2009) A comprehensive resource of drought- and salinity-responsive ESTs for gene discovery and marker development in chickpea. BMC Genomics 10:523

Venuprasad R, Shashidhar HE, Hittalmani S, Hemamalini GS (2002) Tagging quantitative trait loci associated with grain yield and root morphological traits in rice under contrasting moisture regimes. Euphytica 128:293–300

Venuprasad R, Impa SM, Gowda RPV, Atlin GN, Serraj R (2011) Rice near isogenic lines contrasting for grain yield under lowland drought stress. Field Crops Res 123:38–46

Villagarcia MR, Carter TE, Rufty TW, Niewoehner AS, Jennette MW, Arrellano C (2001) Genotypic rankings for aluminum tolerance of soybean roots grown in hydroponics and sand culture. Crop Sci 41:1499–1507

Vinod MS, Sharma N, Manjunatha K, Kanbar A, Prakash NB, Shashidar HE (2006) Candidate genes for drought tolerance and improved productivity in rice. J Biosci 31:69–74

Vlad F, Spano T, Vlad D, Daher FB, Ouelhadj A, Fragkostefanakis S, Kalaitzis P (2007) Arabidopsis prolyl 4-hydroxylases are differentially expressed in response to hypoxia, anoxia and mechanical wounding. Acta Physiol Plant 130:471–483

Wahid A, Close TJ (2007) Expression of dehydrins under heat stress and their relationship with water relations of sugarcane leaves. Biol Plant 51:104–109

Walker AR, Davison PA, Bolognesi-Winfield AC, James CM, Srinivasan N, Blundell TL, Esch JJ, Marks MD, Gray JC (1999) The *TRANSPARENT TESTA GLABRA1* locus which regulates trichome differentiation and anthocyanin biosynthesis in Arabidopsis, encodes a WD40 repeat protein. Plant Cell 11:1337–1349

Wang Y, Frei M (2011) Stressed food – the impact of abiotic environmental stresses on crop quality. Agric Ecosyst Environ 141:271–286

Wang H, Inukai Y, Yamauchi A (2006a) Root development and nutrient uptake. CRC Crit Rev Plant Sci 25:279–301

Wang HL, Yu DY, Wang YJ, Chen SY, Gai JY (2006b) Mapping QTLs of soybean root weight with RIL population NJRIKY. Hereditas 26:333–336

Wang J, McLean PE, Lee R, Goos RJ, Helms T (2008) Association mapping of iron deficiency chlorosis loci in soybean (*Glycine max* L. Merr.) advanced breeding lines. Theor Appl Genet 116:777–787

Wang D, Pan Y, Zhao X, Zhu L, Fu B, Li Z (2011a) Genome-wide temporal-spatial gene expression profiling of drought responsiveness in rice. BMC Genomics 12:149

Wang WS, Pan Y, Zhao X, Dwivedi D, Zhu L, Ali J, Fu B, Li Z (2011b) Drought-induced site-specific DNA methylation and its association with drought tolerance in rice (*Oryza sativa* L.). J Exp Bot 62:1951–1960

Wassmann R, Jagadish SVK, Heuer S, Ismail A, Redona E, Serraj R, Singh RK, Howell G, Pathak H, Sumfleth K (2009a) Climate change affecting rice production: the physiological basis for possible adaptation strategies. Adv Agron 101:59–122

Wassmann R, Jagadish SVK, Sumfleth K, Pathak H, Howell G, Ismail A, Serraj R, Redona E, Singh RK, Heuer S (2009b) Regional vulnerability of climate change impacts on Asian rice production and scope for adaptation. Adv Agron 102:91–133

Watson MB, Emory KK, Piatak RM, Malmberg RL (1998) Arginine decarboxylase (polyamine synthesis) mutants of *Arabidopsis thaliana* exhibit altered root growth. Plant J 13:231–239

White JW, Castillo JA (1992) Evaluation of diverse shoot genotypes on selected root genotypes of common bean under soil water deficits. Crop Sci 32:762–765

Wiesler F, Horst WJ (1994) Root growth and nitrate utilization of corn cultivars under field conditions. Plant Soil 163:267–277

Wilson HK (1930) Plant characters as indices in relation to the ability of corn strains to withstand lodging. J Am Soc Agron 22:453–458

Wu Y, Sharp RE, Durachko DM, Cosgrove DJ (1996) Growth maintenance of the maize primary root at low water potentials involves increases in cell-wall extension properties, expansin activity, and wall susceptibility to expansins. Plant Physiol 111:765–772

Wu P, Liao CD, Hu B, Yi KK, Jin WZ, Ni JJ, He C (2000) QTLs and epistasis for aluminum tolerance in rice (*Oryza sativa* L.) at different seedling stages. Theor Appl Genet 100:1295–1303

Wysocka-Diller J, Helariutta Y, Fukaki H, Malamy J, Benfey P (2000) Molecular analysis of SCARECROW function reveals a radial patterning mechanism common to root and shoot. Development 127:595–603

Xing G, Guo G, Yao Y, Peng H, Sun Q, Ni Z (2010) Identification and characterization of a novel hybrid upregulated long non-protein coding RNA in maize seedling roots. Plant Sci 179:356–363

Xiong L, Wang RG, Mao G, Koczan JM (2006) Identification of drought tolerance determinants by genetic analysis of root response to drought stress and abscisic acid. Plant Physiol 142:1065

Xu P, Rogers SJ, Roossinck MJ (2004) Expression of antiapoptotic genes bcl-xL and ced-9 in tomato enhances tolerance to viral-induced necrosis and abiotic stress. Proc Natl Acad Sci USA 101:15805–15810

Xu K, Xu X, Fukao T, Canlas P, Maghirang-rodriguez R, Heuer S, Ismail AM, Bailey-Serres J, Ronald PC, Mackill DJ (2006) Sub1A is an ethylene-response-factor-like gene that confers submergence tolerance to rice. Nature 442:705–708

Xue Y, Wan J, Jiang L, Wang C, Liu L, Zhang YM, Zhai HQ (2006) Identification of quantitative trait loci associated with aluminum tolerance in rice (*Oryza sativa* L.). Euphytica 150:37–45

Xue Y, Jiang L, Su N, Wang J, Deng P, Ma JF, Zhai HQ, Wang JM (2007) The genetic basic and fine mapping of a stable quantitative-trait loci for aluminium tolerance in rice. Planta 227:255–262

Yadav R, Courtois B, Huang N, McLaren G (1997) Mapping genes controlling root morphology and root distribution on a double-haploid population of rice. Theor Appl Genet 94:619–632

Yamaguchi M, Valliyodan B, Zhang J, Lenoble ME, Yu O, Rogers EE, Nguyen HT, Sharp RE (2010) Regulation of growth response to water stress in the soybean primary root. I. Proteomic analysis reveals region-specific regulation of phenylpropanoid metabolism and control of free iron in the elongation zone. Plant Cell Environ 33:223–243

Yamaguchi-Shinozaki K, Shinozaki K (2006) Transcriptional regulatory networks in cellular responses and tolerance to dehydration and cold stresses. Annu Rev Plant Biol 57:781–803

Yamakawa H, Hirose T, Kuroda M, Yamaguchi T (2007) Comprehensive expression profiling of rice grain filling-related genes under high temperature using DNA microarray. Plant Physiol 144:258–277

Yang L, Zheng B, Mao C, Qi X, Liu F, Wu P (2004) Analysis of transcripts that are differentially expressed in three sectors of the rice root system under water deficit. Mol Genet Genomics 272:433–442

Yang Z, Wu Y, Li Y, Ling HQ, Chu C (2009) OsMT1a, a type 1 metallothionein, plays the pivotal role in zinc homeostasis and drought tolerance in rice. Plant Mol Biol 70:219–229

Yang S, Vanderbeld B, Wan J, Huang Y (2010) Narrowing down the targets: towards successful genetic engineering of drought-tolerant crops. Mol Plant 3:469–490

Yang X, Zhang H, Li L (2011) Global analysis of gene-level microRNA expression in Arabidopsis using deep sequencing data. Genomics 98:40–46

Yao D, Zhang X, Zhao X, Liu C, Wang C, Zhang Z, Zhang C, Wei Q, Wang Q, Yan H, Li F, Su Z (2011) Transcriptome analysis reveals salt-stress-regulated biological processes and key pathways in roots of cotton. Genomics 98:47–55

Yeo AR, Yeo ME, Flowers TJ (1988) Selection of lines with high and low sodium transport from within varieties of an inbreeding species; rice (*Oryza sativa* L.). New Phytol 110:13–19

Yeo AR, Yeo ME, Flowers SA, Flowers TJ (1990) Screening of rice (*Oryza sativa* L.) genotypes for physiological characters contributing to salinity resistance, and their relationship to overall performance. Theor Appl Genet 79:377–384

Young LS, Harrison BR, Narayana Murthy UM, Moffatt BA, Gilroy S, Masson PH (2006) Adenosine kinase modulates root gravitropism and cap morphogenesis in Arabidopsis. Plant Physiol 142:564–573

Yi KK, Wu ZC, Zhou J, Du LM, Guo LB, Wu YR, Wu P (2005) OsPTF1, a novel transcription factor involved in tolerance to phosphate starvation in rice. Plant Physiol 138:2087–2096

Yue B, Xue W, Xiong L, Yu X, Luo L, Cui K, Jin D, Xing Y, Zhang Q (2006) Genetic basis of drought resistance at reproductive stage in rice: separation of drought tolerance from drought avoidance. Genetics 172:1213–1228

Yu L, Ray JD, O'Toole JC, Nguyen HT (1995) Use of wax-petrolatum layers for screening rice root penetration. Crop Sci 35:684–687

Zeng C, Han Y, Shi L, Peng L, Wang Y, Xu F, Meng J (2008) Genetic analysis of the physiological responses to low boron stress in *Arabidopsis thaliana*. Plant Cell Environ 31:112–122

Zhang J, Zheng HG, Aarti A, Pantuwan G, Nguyen TT, Tripathy JN, Sarial AK, Robin S, Babu RC, Nguyen BD, Sarkarung S, Blum A, Nguyen HT (2001) Locating genomic regions associated with components of drought resistance in rice: comparative mapping within and across species. Theor Appl Genet 103:19–29

Zhang H, Forde BG (1998) An Arabidopsis MADS Box gene that controls nutrient-induced changes in root architecture. Science 279:407–409

Zhang Z, Wei L, Zou X, Tao Y, Liu Z, Zheng Y (2008) Submergence responsive microRNAs are potentially involved in the regulation of morphological and metabolic adaptations in maize root cells. Ann Bot 102:509–519

Zhang D, Cheng H, Geng LY, Kan GZ, Cui SY, Meng QC, Gai JY, Yu DY (2009a) Detection of quantitative trait loci for phosphorus deficiency tolerance at soybean seedling stage. Euphytica 167:313–322

Zhang YL, Jia JH, Zhao YQ, Gu SY, Xu JG (2010) Screening index for low phosphorus tolerance at seedling stage. Agric Sci Technol 11(87–89):97

Zhang M, Liu X, Yuan L, Wu K, Duan J, Wang X, Yang L (2012) Transcriptional profiling in cadmium-treated rice seedling roots using suppressive subtractive hybridization. Plant Physiol Biochem 50:79–86

Zhao CX, Deng XP, Shan L, Steudle E, Zhang SQ, Ye Q (2005) Changes in root hydraulic conductivity during wheat evolution. J Integr Plant Biol 47:302–310

Zhao M, Morohashi K, Hatlestad G, Grotewold E, Lloyd A (2008) The TTG1-bHLH-MYB complex controls trychome cell fate and patterning through direct targeting of regulatory loci. Development 135:1991–1999

Zheng HG, Babu RC, Pathan MS, Ali L, Huang N, Courtois B, Nguyen HT (2000) Quantitative trait loci for root-penetration ability and root thickness in rice: comparison of genetic backgrounds. Genome 43:53–61

Zheng BS, Yang L, Zhang WP, Mao CZ, Wu YR, Yi KK, Liu FY, Wu P (2003) Mapping QTLs and candidate genes for rice root traits under different water-supply conditions and comparative analysis across three populations. Theor Appl Genet 107:1505–1515

Zheng X, Chen B, Lu G, Han B (2009) Overexpression of a NAC transcription factor enhances rice drought and salt tolerance. Biochem Biophys Res Commun 379:985–989

Zhou R, Wang XZ, Chen HF, Zhang XJ, Shan ZH, Wu XJ, Cai SP, Qiu DZ, Zhou XA, Wu JS (2009) QTL analysis of lodging and related traits in soybean. Acta Agron Sin 35:57–65

Zhu JK (2002) Salt and drought stress signal transduction in plants. Annu Rev Plant Biol 53:247–273

Zhu J, Kaeppler SM, Lynch JP (2005) Mapping of QTLs for lateral root branching and length in corn (*Zea mays* L.) under differential phosphorus supply. Theor Appl Genet 111:688–695

Zhu C, Gan L, Shen Z, Xia K (2006) Interactions between jasmonates and ethylene in the regulation of root hair development in Arabidopsis. J Exp Bot 57:1299–1308

Zhu J, Chen S, Alvarez S, Asirvatham VS, Schachtman DP, Wu Y, Sharp RE (2006a) Cell wall proteome in the maize primary root elongation zone. I. Extraction and identification of water soluble and lightly ionically bound proteins. Plant Physiol 140:311–325

Zhu J, Mickelson SM, Kaeppler SM, Lynch JP (2006b) Detection of quantitative trait loci for seminal root traits in corn (*Zea mays* L.) seedlings grown under differential phosphorus levels. Theor Appl Genet 113:1–10

Zhu J, Alvarez S, Marsh EL, Lenoble ME, Cho I, Sivagura M, Chen S, Nguyen HT, Wu Y, Schachtman DP, Sharp RE (2007) Cell wall proteome in the maize primary root elongation zone. II. Region-specific changes in water soluble and lightly ionically bound proteins under water deficit. Plant Physiol 145:1533–1548

Zhu J, Brown KM, Lynch JP (2010) Root cortical aerenchyma improved the drought tolerance of maize. Plant Cell Environ 33:740–749

Zhu J, Ingram PA, Benfey PN, Elich T (2011) From lab to field, new approaches to phenotyping root system architecture. Curr Opin Plant Biol 14:310–317

Zou X, Jiang Y, Liu L, Zhang Z, Zheng Y (2010) Identification of transcriptome induced in roots of maize seedlings at the late stage of water logging. BMC Plant Biol 10:189

# Chapter 3
# Cold Tolerance

**Mike Humphreys and Dagmara Gasior**

**Abstract** Considerations of preparations for a changing climate will generate thoughts of mitigating a rise in temperature and greenhouse gas emission and a change in water availability. Accordingly, reduced prioritization on future research objectives aimed at crop adaptations sufficient to instigate and sustain cold tolerance or winter hardiness expression at any specific location might be deemed by some as the logical outcome. However, such a conclusion would be a grave mistake. With increasing frequency, crops of high agricultural value are being grown at locations beyond their natural ranges of adaptation, a consequence in part of farmers attempting to seize new opportunities to exploit some positive scenarios of climate change that might provide more profitable agricultural output. A second and even greater driver is man's response to the ever-increasing requirement to feed a growing global population, and with only limited and finite land available that is deemed suitable for agricultural use. For the latter, there is increased use of marginal locations for agricultural production, which will include those locations at high altitude where temperatures are frequently suboptimal for crop production and, in many cases, likely to challenge crop persistency over winter months. In certain temperate locations, where winter temperatures are considered generally moderate, crop growing seasons are becoming extended, encouraged frequently by national policy makers seeing economic advantages in management practices that can achieve an all-year-round cropping potential, but with a great risk. The maintenance of crop growth is the consequence of failure, at least in part, of the initiation and subsequent expression of the appropriate adaptive responses necessary to assure a high probability of winter survival which include growth cessation. Such scenarios place crops at risk of total collapse following any sudden temperature drop and especially onsets of frost conditions. In situations of fluctuating winter temperatures, assured crop survival requires the maintenance of the required adaptive response in place until such time as there is little likelihood of

M. Humphreys (✉) • D. Gasior
IBERS, Aberystwyth University, Gogerddan, Aberystwyth SY23 3EE, Ceredigion, Wales, UK
e-mail: mkh@aber.ac.uk

C. Kole (ed.), *Genomics and Breeding for Climate-Resilient Crops*, Vol. 2,
DOI 10.1007/978-3-642-37048-9_3, © Springer-Verlag Berlin Heidelberg 2013

any further incidence of frosts. For optimal crop production, the subsequent appropriate timing of the cessation of the adaptive responses is also essential to enable crop growth to proceed fully as soon as possible, once growth advantageous spring conditions arise.

Cold tolerance and winter survival are complex traits, each having distinct genetic controls and involving responses to the many interacting stresses, their relative importance dependent on crop location. Frost tolerance is considered the trait of main priority with the understanding and manipulation of the factors necessary to optimize initiation of the appropriate cold acclimation responses sufficient to retain cell membrane integrity and prevent desiccation, the most appropriate objectives in crop improvement. Gene expression sufficient to initiate frost tolerance has many equivalent requirements and responses to those required to combat other abiotic stresses that can induce cell desiccation such as prolonged exposures to conditions of drought or salinity. Some of the major aspects and their relative importance are reviewed herein.

## 3.1   Introduction

Unlike animals that can seek shelter during winter, land plants are stationary and exposed to all that the weather might throw at them; to survive, they must either adapt to the climate native to their specific location or alternatively avoid the worst conditions via their ontogeny either through entry into vegetative dormancy for perennial species or in the case for summer annuals, by completing their life cycles and reproductive phase prior to the full onset of winter conditions, overwintering as seed.

Greaves (1996) defines suboptimal temperature stress as "*any reduction in growth or induced metabolic, cellular or tissue injury that results in limitations to the genetically determined yield potential caused as a direct result of exposure to temperatures below the thermal thresholds for optimal biochemical and physiological activity or morphological development.*"

The human population is increasing at an alarming rate and is anticipated to rise globally to nine billion by 2050, while at the same time agricultural productivity in many regions is decreasing due to the effects of increasing environmental stresses. While the impact of droughts on crop yields receives with some justification particular attention, cold stress is also a serious threat to the sustainability of crop yields and can lead to major crop losses that can include detrimental effects on their quality and post-harvest life.

With population growth, a demand for increased food production will follow, and already an increasing use of marginal lands for agriculture is being observed together with the frequent use of crop species not adapted ideally to the growth conditions they are likely to encounter. It would be incorrect to assume that cold stress was an issue restricted only to agricultural systems in extreme northern and southern latitudes. In countries where land suitable for agriculture is limited or fully utilized, increasing use is being made of alternative marginal regions at altitudes higher than previously employed for agriculture where low temperatures are encountered that may and frequently do provide constraints to efficient crop persistency and production.

At higher latitudes, low temperature is the most important limiting factor and is especially important in northern regions such as Norway where forage species are the major components of agricultural land. In such locations, winters are long and growing seasons very short. In less extreme conditions than those found in northern Europe, the increased use of marginal regions for crop production has led to the expanded use of crops such as maize and rice. In part, this has been encouraged by farmers making use of rising temperatures having resulted from climate change to grow crops considered previously completely unsuitable to their farming locations. For example, in northern Britain, increased temperatures have led to the use of forage maize in areas close to and beyond their safe margins of acclimatization bringing potential risk as frequently temperature extremes and weather patterns fluctuate wildly in such locations over very short time periods. Any small deterioration in anticipated weather conditions can result in potentially total crop failures. In addition, with expanded use of certain crops and due to rising temperatures, plant pathogens previously uncommon in certain locations are becoming more prevalent and elsewhere appearing earlier in the growing season than in previous years, e.g., crown rust in northern Britain (Roderick et al. 2003). Alternative biotic and abiotic stresses interact frequently to exacerbate crop loss. Furthermore, the growth of "new" crops by farmers previously untrained and unfamiliar to their use and their requirements for optimal growth will only further exacerbate through poor farming practice, the dangers to crop production.

Chilling or cold stress (temperatures 0–15 °C) and frost stress (<0 °C) are distinct from one another and should be considered, at least in part, as different stresses as they harbor alternative mechanisms for plant resistance, and these controlled unsurprisingly by different genetic determinants. Rice may suffer chilling injury even at temperatures as high as 15 °C. Protection against cold stress can be achieved through the use of the correct agronomic methodologies and also by breeding and the use of the best adapted varieties. One example of the use of appropriate farming practice is the late sowing of warm-season crops to avoid encountering the lowest expected temperatures at the germination stage when crops are especially vulnerable and, in reverse, a strategy that uses early varieties to avoid an encounter with cold at their maturation stage (Caradus and Christie 1998).

Most temperate plants acquire chilling and freezing tolerance upon prior exposure to sublethal low temperature and reduced day-length, a process called cold acclimation (CA). Many physiological and biochemical changes occur during CA including slowed or arrested growth; reduced water content; protoplasm viscosity; alterations to photosynthetic pigments; reduced ATL levels (Levitt 1980); transient increases in abscisic acid (ABA) (Chen et al. 1983); changes in membrane lipid composition (Uemura and Steponkus 1994); accumulation of compatible solutes including proline, betaine, polyols, and soluble sugars; and the accumulation of antioxidants (Tao et al. 1998).

Considerable resources are necessary to sustain and protect plant metabolism under low temperature stress and for subsequent recovery following the onset of more benign growth conditions. Our best known temperate grasses, cereals, and many other crops like artichokes, asparagus, leeks, and onions all store fructans, a soluble polymer of fructose molecules capable of rapid polymerization and depolymerization. The partitioning of solutes is important because survival from freezing depends on survival of apices, particularly the lateral buds rather than mature leaf tissue (Eagles et al. 1993).

Many agronomically important crops are incapable of CA. Cold stress affects virtually all aspects of cellular function in plants. The cold stress signal is transduced through several components of complex signal transduction pathways. The main components are calcium, reactive oxygen species, protein kinase, protein phosphatase, and lipid signaling cascades. The plant hormone abscisic acid (ABA) also mediates the response of cold stress and functions in many plant developmental processes including bud dormancy. It also has a major role in drought tolerance where it is produced in plant roots in response to the onset of decreased soil water potential. The cold stress signal leads to regulation of transcription factors and effector genes known collectively as cold-regulated (*COR*) genes. The effector genes encoding proteins under this category include chaperones, late embryogenesis abundant (LEA) proteins, osmotin, antifreeze proteins (AFPs), mRNA-binding proteins, and key enzymes for osmolyte biosynthesis such as proline, water channel proteins, sugar and proline transporters, detoxification enzymes, enzymes for fatty acid metabolism, proteinase inhibitors, ferritin, and lipid-transfer proteins. The transcription factors involved during cold stress response are inducers of C-repeat binding factor expression-1, C-repeat binding factors, myeloblasts, and mitogen-activated protein kinase. Analyses of the expression of *COR* genes indicate the presence of multiple signal transduction pathways between the initial stress signals and the outcome of gene expression. Use of these genes and transcription factors in genetic modification of agricultural crops can improve cold tolerance and productivity.

Various phenotypic symptoms in response to cold stress include poor germination, stunted seedlings, yellowing of leaves (chlorosis), reduced leaf expansion, and wilting, and these may lead to death of tissue. Cold stress also severely hampers the reproductive development of plants.

## 3.2  Winter Hardiness

Winter hardiness in temperate plant species is a complex trait comprising the ability to survive a range of abiotic and biotic stresses, which include freezing, ice encasement, waterlogging, soil heave, desiccation by dry winds, starvation, and snow mold. Winter hardiness in a plant has three components: acclimation, midwinter hardiness, and de-acclimation. The acclimation process is triggered late in the growing season by decreasing photoperiods as day-lengths shorten and as temperatures decline. These environmental cues induce physiological and biochemical changes in the plant that then result in greater cold tolerance.

In short, plant winter hardiness is the outcome of a seasonal shift between growth, quiescence, and assimilate storage in response to a cool temperate climate. Its level of effectiveness will vary and will be dependent on the location and on genotype × environment (G × E) interactions. As with breeding for drought resistance, prior to any accurate assessment of crop performance and winter hardiness, it is necessary that new varieties should be assessed under all or as many as practical the environmental scenarios they are likely to encounter if they are to be used widely commercially. G × E interactions are complex. For example, a winter hardy plant growing in a maritime environment will not necessarily have an equivalent adaptation when transferred to a continental climate and vice versa.

It should be noted that combinations of fluctuating temperatures and changes to light intensities common to what are considered mild maritime climates such as those found in the UK can be as hazardous to a plant's survival as those that are encountered following persistent severe subzero temperatures such as those found commonly in extreme northern or southern latitudes and will require an alternative adaptive response. However, as a general requirement when plants are exposed to subzero temperature, there is a need for resistance to desiccation through the maintenance of the integrity of the plant cell membrane.

After CA, certain species can withstand extreme low temperatures. Non-acclimated (NA) birch and dogwood are injured at $-10\,°C$ but after CA can survive experimental freezing to $-196\,°C$ and survive $-40\,°C$ to $-50\,°C$ under natural conditions. NA wheat and rye are killed at $-5\,°C$ to $-10\,°C$ but after CA, wheat can survive to $-15\,°C$ and rye to $-30\,°C$ (Thomashow 1990). Among grasses, the most freezing-tolerant forage grass used in European agriculture is Timothy (*Phleum pratense*) whose minimum temperature survival has been measured as $-25\,°C$.

Possibly the most important component of winter hardiness in Northern Europe is the ability to withstand periods of ice encasement. Ice-encasement tolerance expressed as $LD_{50}$ (lethal dose for 50 % of plants) varied from 50 days at $-1\,°C$ in Berings hairgrass (*Deschampsia beringensis*) to 2 days in a cultivar of orchard grass (*Dactylis glomerata*) (Humphreys and Humphreys 2005).

In general, a tolerance to freezing temperatures is the most important component for winter survival, but also of considerable importance as described earlier is the capability to withstand combinations of stresses due to desiccation, which is influenced also by wind and occurrence of ice encasement. Other factors affecting winter survival is resistance to mechanical heaving and also low light, snow cover, winter pathogens, and fluctuating temperatures, with the relative importance of each depending on the location.

Cold and freezing tolerance are complex traits governed by many gene loci of greater or lesser importance and involving epigenetic and pleiotropic effects. It has been demonstrated that in certain crops such as the outbreeding forage grasses of the *Lolium–Festuca* complex, the potential of a plant genome for cold-tolerance expression may in certain cases not be achieved fully until obstacles to their full expression are removed or suppressed (Rapacz et al. 2005). In this case, through gene segregation at meiosis, cultivars and plant genotypes that were previously considered to have low winter hardiness and poor snow mold resistance were able to produce novel progeny by androgenesis that expressed both traits with extreme efficiency to extents well in excess of their parent genotypes. In another study, the authors proved that freezing tolerance (FT) was only loosely correlated with tolerance of ice encasement in ten forage grass species (Gudleifsson et al. 1986).

In Icelandic grasslands, the physical properties of the soils affect survival rates; snow mold was not considered as important a factor as was ice encasement, which was the main cause of plant damage (Gudleifsson 1971). Andrews and Gudleifsson (1983) found no correlation between FT and ice-encasement tolerance in cultivars of Timothy and concluded that the greater winter hardiness of Timothy relative to other species was due to its high resistance to ice encasement. Spring ground cover studies in Nordic countries (Nordic Gene Bank 1996) showed that Timothy cultivars were more damaged by ice (~50 % damage) than snow cover (~25 % damage), with frost alone causing intermediate damage (~30 %).

The rate and extent of de-hardening is a critical factor in winter survival. Overwintering plants are particularly susceptible to freezing damage in the spring if the de-acclimation process occurs either prematurely or too rapidly, or if unpredictable temperature fluctuations occur (Levitt 1980; Gay and Eagles 1991). Eagles (1989) suggested that the nature of an adaptive CA process will vary with the stability and predictability of winter conditions in a particular environment. In stable and predictably cold continental climates where the onset of freezing temperatures is rapid, a photoperiod-triggered and rapid acclimation process is desirable, while in the more variable and less severe conditions of a maritime climate, a temperature-dependent response might enable plants to exploit a mild autumn or spring by continuing to grow. However, in cultivars adapted to maritime climates, de-acclimation may occur in response to fluctuations in winter and spring temperature with a risk of damage by subsequent frosts (Eagles 1994).

A changing climate that generates a change in plant ontogeny will require a change in strategy by the plant breeder and will require a complex and holistic approach to counter interacting plant stresses. At locations where winter temperatures are increasing as a consequence of climate change, so that continued plant growth is encouraged, where previously it was prevented due to low temperatures that induced winter dormancy, and also where precipitation is decreasing, that priorities for crop designs for resistance to stresses other than those associated normally with winter such as drought-tolerance, previously considered important only at other times of the year and at alternative growth stages, will become major requirements necessary to safeguard crop yields. A decrease in winter rainfall that causes unseasonal winter droughts is becoming more commonplace and this will require a new breeding strategy that encompasses a common stress tolerance to what was considered previously solely as winter and summer stress factors.

Simulations predict the following changes in the Norwegian climate towards the year 2100: changes in mean yearly temperatures of 2.5–3.5 °C, most pronounced in continental and northern regions; milder winters with increased minimum temperatures of 2.5–4 °C, more frequent incidents of freezing-thawing cycles; and less snow cover in continental regions (Rognli pers. com–RegClim project). The increasing mean temperatures will lead to prolonged growth seasons and give increased biomass production, especially in annual crop plants but also in perennial crops provided that they are well adapted to the changing winter climates of which increased freezing stress possibly will be the most challenging.

Ice encasement, waterlogging, and soil compaction all include a component of hypoxic or anoxic stress. In waterlogged soils, respiration by roots and soil microflora depletes dissolved oxygen and toxic by-products of anaerobic respiration accumulate (Pulli 1989). In colder climates, soil and root respiration is slowed but freezing and ice encasement may completely seal soil and plant surfaces against penetration by oxygen. Injury to cells may result from ethanol self-poisoning, cytoplasmic acidosis, insufficient adenosine triphosphate (ATP) generation, and metabolic lesions caused by reentry of oxygen (Vartapetian and Jackson 1997). The shoot suffers an impeded supply of water, minerals, and root hormones.

Metabolic adaptations are important in short-term survival of anaerobic stress (Vartapetian and Jackson 1997). Increased activity in both the glycolytic and fermentation pathways is observed, along with increased expression of enzymes in this pathway.

Avoidance of cytoplasmic acidosis by early switching from production of acidic lactate to neutral ethanol may underlie root tolerance of reduced oxygen (Davies 1980). In this model, transient lactate fermentation acidifies the cytoplasm at the onset of anaerobiosis, suppresses lactate dehydrogenase (LDH) activity, and triggers pyruvate decarboxylase (PDC). Transient acidosis of the cytoplasm has been shown in excised maize root tips within 20 min of transfer to anaerobic conditions (Roberts et al. 1982), and direct manipulation of cell acidity substantiated the idea of a pH switch (Fox et al. 1995).

Anatomical adaptations in roots may promote survival by allowing avoidance of anoxia (Jackson 1990) and include root aerenchyma, which transports oxygen to, and removes volatiles from, the roots. Dormancy is exhibited by shoot tissue of many species exposed to prolonged anaerobic conditions as in the rhizomes of wetland species (Brändle 1991). Stomatal closure, epinastic leaf curvature, slowed leaf expansion, and enhanced leaf senescence are also triggered by soil waterlogging, as a means of reducing transpiration (Else et al. 1995).

## 3.3  Genetic Adaptations for Cold Acclimation and Improved Winter Hardiness

Altered gene expression during CA (Guy et al. 1985) has been demonstrated in a range of crop species (Hughes and Dunn 1996) and the model species *Arabidopsis thaliana* (Thomashow 1998). Cold-responsive genes are involved in biochemical and physiological changes required for growth and development at low temperature or are directly involved in freezing tolerance (FT) (Thomashow 1998).

The major negative effect of cold stress is that it induces severe membrane damage due largely to the acute dehydration associated with freezing. Cold stress is perceived by the receptor at the cell membrane. Then a signal is transduced to switch on the cold-responsive genes and transcription factors for mediating stress tolerance. Understanding the mechanism of cold stress tolerance and genes involved in the cold stress signaling network is a vital step for crop improvement.

Several studies have established that major genes, or gene clusters, involved in the control of frost and drought tolerance are located on a region of the long arm of Triticeae group 5 chromosomes. Traits like winter hardiness (Hayes et al. 1993; Pan et al. 1994), vernalization response and frost tolerance (Sutka and Snape 1989; Galiba et al. 1995; Laurie et al. 1995), cold- and drought-induced ABA (abscisic acid) production (Galiba et al. 1993; Quarrie et al. 1997), and osmotic stress tolerance (Galiba et al. 1992) have all been mapped to this region. Across the grasses and cereals, this chromosome region has been a major focus for genome study and for crop improvement.

Development of winter hardiness requires exposure of plants to low nonfreezing temperatures, typically 0–10 °C, and a shortened photoperiod. The majority of research studies for crop winter survival have focused on genetic changes that affect the key stage of CA either by natural breeding, often employing gene transfers from wild crop relatives (e.g., Humphreys et al. 2007). For many countries, due to concerns with the use of genetic modification (GM) technologies, plant breeding offers the only opportunity for an improved crop design, but transgenic technologies may be employed either as "proof of principle" of gene function or directly in crop improvement when restrictions on the use of GM technologies are loosened (Sanghera et al. 2011). In both approaches, efforts have concentrated on inclusion of functional genes necessary for the induction of appropriate physiological mechanisms required to withstand exposures to freezing temperatures. For all crops to survive the winter, plants must engage mechanisms whereby sensitive tissues can avoid freezing or undergo cold hardening compatible with the normal variations of the local climate, coordinate the induction of the tolerance at the appropriate time, maintain adequate tolerance during times of risk, and properly time the loss of tolerance and resumption of growth when the risk of freezing has passed (Guy 1990). For some locations and latitudes, should winter temperatures continue to rise following climate change, it may be necessary for breeding efforts to concentrate more on adaptation to short day-length rather than to a tolerance of low temperature in order to both achieve winter hardiness and also to avoid a vernalization requirement otherwise necessary for flower induction, of course an essential prerequisite for seed production and a crop yield.

Considerable resources are necessary to sustain and protect plant metabolism under low temperature stress and for recovery subsequent to the onset of more benign growth conditions. CA and freezing tolerance are the result of a complex interaction between low temperature, light, and photosystem II (PSII) excitation pressure. The redox state of PSII reflects fluctuations in the photosynthetic energy balance and so acts as a sensor of any environmental stresses that disturb that balance. Changes to the redox state of PSII, triggered by a low-temperature shift, were proposed to be a temperature-sensing mechanisms involved in cold acclimation (Rapacz et al. 2004). Humphreys et al. (2006) reported how non-photochemical quenching (NPQ) mechanisms for expulsion of excess light energy during CA are found in the forage grass species *Festuca pratensis*, which is adapted to northern latitudes, but are not expressed similarly in its close relatives, the major agricultural grasses *Lolium perenne* and *Lolium multiflorum,* which are adapted to lower latitudes. As will be described below, alien gene transfers between *F. pratensis* and *Lolium* spp. have enhanced PSII adaptation to freezing temperatures and have led to improved CA efficiency and to freezing tolerance (Humphreys et al. 2007).

### 3.3.1  A Novel Introgression-Mapping Approach for PSII Adaptations to Freezing

Modern breeding methods can make use of evolved adaptations to cold stress in wild-type relatives to improve their freezing tolerance (FT). The technique introgression mapping is employed currently in the *Lolium–Festuca* complex for trait "dissection" and transfers of key alleles from donor to recipient species by natural plant breeding. The grass complex provides the main grass species used for livestock agriculture and offers excellent opportunities for analyzing the genetic determinants of both simple (e.g., Moore et al. 2005) and complex traits (e.g., Humphreys et al. 2005). The highly heterogeneous genotypes and phenotypes typical of *Lolium* and *Festuca* species are a consequence of excessive genome reorganization due to the promiscuous chromosome recombination that occurs in these obligate outbreeding species. Their heterogeneous nature has provided *Lolium* and *Festuca* species with considerable allelic variation and adaptive capabilities for the successful colonization and establishment as distinct ecotypes evolved to growth in contrasting climates within temperate grasslands throughout the world. *Festuca* species are generally better adapted than *Lolium* to marginal areas where they may be exposed to the more extremes of abiotic stresses (Humphreys et al. 1998). *F. pratensis* (meadow fescue) is a particularly good source of genetic variants for cold tolerance (Alm et al. 2011). In terms of CA, a very important recent development was the discovery of a potential role for photoreceptors in *Lolium* and *Festuca* responses to low temperatures (Rapacz et al. 2004).

Cold acclimation and freezing tolerance (FT) are the results of a complex interaction between low temperature, light, and photosystem II (PSII) excitation pressure. At low temperatures, plants have two principal difficulties: the maintenance of cell membranes in a fluid state, and the thermo-dependency of photosynthetic electron transport and carbon fixation which are slowed at low temperature (Guy 1990; Huner et al. 1996; Thomashow 1999). The PSII reaction center is the key site for regulation of light energy and also the main site of photoinhibitory damage. The D1/D2 protein dimer at the core of PSII appears to be crucial in maintaining the integrity of the complex (Mattoo et al. 1989). Indeed, Canter et al. (2000) demonstrated in *F. pratensis* that repair measures to the chloroplast are CA induced. A representation difference analysis was used to amplify selectively upregulated cDNA fragments from cold-induced *F. pratensis* seedlings. The gene *psba*, which codes for the D1 protein of PSII, was prominent among the upregulated cDNA fragments and was recovered following 4 days of CA.

Photoinhibition (the light-induced reduction in the photosynthetic capacity) is related to the redox state of PSII expressed as excitation pressure. The redox state of PSII reflects fluctuations in the photosynthetic energy balance and so acts as a sensor of any environmental stresses. Changes that disturb that balance include those triggered by a low-temperature shift and have been proposed to be one of the temperature-sensing mechanisms involved in CA (Rapacz 2002).

During autumn and winter, plants are subjected to excess light compared to the energy demand of dark photosynthetic reactions. This can cause photoinhibition. When the rate of PSII damage exceeds the rate of repair, photoinhibition occurs reflected in decreased $F_v/F_m$ (maximum quantum yield of PSII). Rapacz et al. (2004) using plant populations derived from *L. multiflorum* × *F. pratensis* cultivars reported a relationship between cold tolerance and $F_v/F_m$. There was a strong negative correlation between maximum quantum yields of PSII ($F_v/F_m$) before winter and winter survival, with plants with higher $F_v/F_m$ having lower winter survival. It was found among *L. multiflorum* × *F. pratensis* hybrid plants that those that were winter hardy were also more resistant to cold-induced inactivation of PSII. This was due primarily to increases in non-photochemical quenching (NPQ) during CA where excess energy was dissipated by heat via the xanthophyll cycle. Humphreys et al. (2006) reported a significant increase in NPQ activity by *F. pratensis* genotypes in response to CA conditions, a response that was not repeated by *Lolium* species thereby providing one possible explanation why *F. pratensis* has more efficient CA capabilities.

A set of seven chromosome substitution lines where *F. pratensis* chromosomes replaced homoeologous *Lolium perenne* chromosomes (King et al. 2002a, b) in an otherwise undisturbed *Lolium* genome were used by the current authors to locate the source of the *F. pratensis* genes responsible for NPQ expression. The substitution lines were assessed for changes in PSII during a CA period of 2 weeks and for NPQ at two contrasting light intensities (PAR 175 $\mu$mol m$^{-2}$ s$^{-1}$, and PAR 250 $\mu$mol m$^{-2}$ s$^{-1}$). All CA substitution lines were subsequently freeze tested and the $LT_{50}$ (freezing temperature necessary to induce 50 % tiller mortality) determined.

The chlorophyll fluorescence studies on the different chromosome substitution lines demonstrated that *L. perenne* in combination with different *F. pratensis* chromosomes responded to CA and light of different intensities, in different ways. However, it was the PSII response by the *F. pratensis* chromosome 4 substitution line that was associated primarily with efficient induction of NPQ expression under high light intensity, and this corresponded with enhanced freezing tolerance. In support of this new research outcome, Humphreys et al. (2006) provided preliminary evidence that the presence of an *F. pratensis* translocation at the end of *L. multiflorum* chromosome 4 led to increased CA-induced NPQ, and this was frequently associated with an improvement in freezing tolerance. Simple sequence repeat (SSR) markers were associated with the alien gene transfers suitable for future use in marker-assisted breeding (Mike Humphreys, unpublished).

Using the extensive macrosynteny between related grass crops (rice, Triticeae) among the Pooideae, a comparative crop species study for mapping orthologous gene loci led to the publication of the first genetic linkage map in *F. pratensis* for quantitative trait loci (QTL) associated with abiotic stress resistance (Alm et al. 2011). The study showed that major structural differences exist between *Lolium/Festuca* and *Triticeae* chromosomes 4 and 5, which are especially relevant to the acquisition of freezing tolerance. It is considered that two major frost tolerance and winter survival QTLs on *Festuca* chromosome 5 correspond to the *Fr-A1* and *Fr-A2* loci on homoeologous Group 5A in wheat (Sutka and Snape 1989; Vagujfalvi et al. 2003). However, an important QTL for frost tolerance was located at a terminal region of *Lolium/Festuca* chromosome 4, which contains the vernalization locus *Vrn-1* (Jensen et al. 2005), which in the Triticeae is located on chromosome 5. In a study of near-isogenic lines (NILs) of wheat, it was shown that the *Vrn-1-Fr-1* interval accounted for 70–90 % of the difference in the FT of the NILs substantiating the importance of this locus in CA (Storlie et al. 1998). An interaction between the vernalization and CA regulatory gene networks are thought to operate by extending the CA period and increasing the frost tolerance by delaying the induction of *Vrn-1* and thereby the transition to the reproductive stage (Galiba et al. 2009). The structural difference between chromosomes 4 and 5 of *Festuca* and Triticeae makes it possible to separate the effects of vernalization and frost tolerance/winter survival.

The identification of *F. pratensis* genes on chromosome 4 thought to also contain the *Vrn-1* locus and now with proven involvement in a CA response by PSII for the first time provides us with both a mechanism for improved FT, and through introgression mapping, the means by which to access the *F. pratensis* genes responsible to improve winter hardiness and freezing tolerance in *Lolium*.

## *3.3.2   Cold-Regulated Genes*

The C-repeat binding factor (*CBF*) genes are key regulators of the expression of *COR* (cold-regulated genes), which are conserved among diverse plant lineages such as dicots and monocots. The CBF transcription factors recognize the cis-acting CRT/DRE (C-repeat/dehydration-responsive element) element in the regulatory regions of *COR* genes (Stockinger et al. 1997). Twenty *CBF* genes have been identified in barley (*Hordeum vulgare*), of which 11 are found in two tight tandem clusters on the long arm of chromosome 5H in the same region as the *Fr-H2* frost tolerance locus (Skinner et al. 2006; Francia et al. 2007). An orthologous genomic region in *Triticum monococcum* contains similar *CBF* gene clusters located at the *Fr-A$^m$2* frost tolerance QTL (Vagujfalvi et al. 2003; Miller et al. 2006). In *Lolium perenne*, Tamura and Yamada (2007) mapped four *LpCBF* genes in a short interval on *Lolium* LG5 most likely syntenic with regions on Triticeae group 5 chromosomes. Studies of the organization of the *CBF* cluster in barley and wheat have shown that the number of *CBF* genes at the *Fr-H2/Fr-A1* locus may vary among cultivars with winter forms having a higher copy number of some *CBF*s (Francia et al. 2007; Knox et al. 2010). The cosegregation of the *CBF* gene clusters with the barley *Fr-H2* and wheat *Fr-A$^m$2* frost tolerance loci, their role in cold acclimation (Stockinger et al. 1997), and the association of transcript levels of *CBF* genes with *FT* loci (Vagujfalvi et al. 2003) make them obvious candidates for one of the two major frost tolerance QTLs on Triticeae group 5 chromosomes. The locations of two frost tolerance/winter survival QTLs on the chromosome 5F of the forage grass *Festuca pratensis* correspond most likely to the *Fr-A1* and *Fr-A2* loci on wheat homoeologous group 5A chromosomes; one of these QTLs (*QFt5F-2/QWs5F-1*) has *FpCBF6* as a candidate gene and shown to be rapidly upregulated during CA (Alm et al. 2011).

Using a targeted approach of achieving understanding of key regulatory mechanisms through crops of common ancestry and using conserved genome regions and synteny, knowledge achieved through studies made in fully sequenced model crops and organisms such as rice, *Brachypodium,* or *Arabidopsis* may be used in crop genome studies where researchers lack access to equivalent resources.

A similar approach may be applied to the photoperiodic response where Triticeae Group 1, 2, and 7 chromosomes have known genes and QTLs for photoperiodic response (Welsh et al. 1973; Law et al. 1978; Scarth and Law 1984; Hayes et al. 1993; Pan et al. 1994; Laurie et al. 1995; Bezant et al. 1996; Sourdille et al. 2000) influencing adaptation and survival and which may be sought in other crop species.

As implied earlier, from a plant "perspective," many stresses that contribute to suboptimal growth, yield and persistency, and the relevant plant-initiated responses aimed at mitigating their most damaging effects should not be considered in isolation, even though many researchers frequently do just that by becoming specialists focused in aspects of a single abiotic stress without due consideration of other contributory stress factors. Aspects of stresses such as drought, cold, or salinity have a similar detrimental effect on the plant and they or indeed others may interact to enhance a stress effect and complicate all endeavors to initiate an adaptive response. Abiotic stresses, notably extremes in temperature, photon irradiance, and supplies of water and inorganic solutes, frequently limit growth and productivity of major crop species such as wheat. These often operate in conjunction: extreme temperature and high photon irradiance often accompany low water supply, which can in turn be exacerbated by subsoil mineral toxicities that constrain root growth. Furthermore, one abiotic stress can decrease a plant's ability to resist a second abiotic, or indeed a biotic, stress.

Most cereals are moderately sensitive to a wide range of abiotic stresses, and variability in the gene pool generally appears to be relatively small providing only few opportunities for major step changes in tolerance. Although of only moderate benefit, examples exist where cold tolerance has been enhanced following introgression from tolerant landraces into commercial lines using marker-assisted breeding (Dubcovsky 2004). The exploitation of fully sequenced model grass crops such as rice and *Brachypodium* provides candidate genes for transfers to enhance stress tolerance. Taken together with steadily increasing transformation frequencies for many grasses, the functional genomics approach to the study and manipulation of abiotic stresses in grasses is becoming a feasible objective. Need for use of the pioneer model plant *Arabidopsis thaliana* for such work is decreasing steadily as more relevant knowledge is becoming obtainable, firstly through model crops of greater synteny with the target crops, and finally of course from genomic studies made on those crops themselves. In other words, for forage grasses and cereals, model grasses *Brachypodium distachyon* and rice are more relevant for genetic studies of abiotic stress resistance than the dicot model species *Arabidopsis thaliana*. Of course, the reverse would apply for dicotyledonous crops such as those among the Brassiceae. Many of the mechanisms of tolerance to abiotic stresses can have fundamentally different characteristics between monocots and dicots, so transferring knowledge from *Arabidopsis* to the major crops are often of limited value. Having said that, a large part of the fundamental research concerning plant cold stress response and FT was carried out using the model species, *Arabidopsis thaliana,* as key mechanisms and transcription factors are found both in dicots and monocots and are involved in the regulation of expression of many cold (and drought) stress response genes (Gilmour et al. 1998; Ito et al. 2006). In the dicot model plant *Arabidopsis,* three *CBF* genes exist (Shinwari et al. 1998), while in the monocot cereal species, 10–20 *CBF* genes have been identified (Miller et al. 2006; Skinner et al. 2006; Francia et al. 2007). Even though dicot and monocot *CBF* gene function is conserved in the sense that *CBF* genes all appear to be involved in abiotic stress response pathways, there have been divergence in the *CBF* family size and hence probably also *CBF* gene function during the evolution of dicot and monocot lineages (Sandve and Fjellheim 2010). Such lineage-specific evolution can provide a direct inference of gene function based on homology between model plants and agriculturally important species. Moreover, neither *Arabidopsis* nor rice is adapted to a perennial life in extreme winter climates. This is important because adaptation to a perennial life history in harsh winter climates must have required changes at the genetic level, which cannot be studied using an annual model species. Hence, if we only use model plants such as *Arabidopsis* or rice to do research on cold and frost stress response, this might provide only limited insights into the genetic mechanisms underlying FT in important agricultural species.

The Pooideae is a large and economically important subfamily including cereals (Triticeae tribe) and forage grasses (Poaeae tribe). Divergence of temperate grasses from the most recent common ancestor shared with rice is thought to have happened ~46–42 million years ago (Gaut 2002; Sandve et al. 2008). Parallel to the origin and early evolution of the Pooideae group, the global climate became gradually cooler

(Zachos et al. 2001). As opposed to rice, which is adapted to warm and humid environments, Pooideae grasses radiated in cooler environments (Barker et al. 2001). This is reflected by the present distribution of Pooideae species which is extremely skewed towards cooler environments (Hartley 1973). Thus, evolution of cold and frost stress responses, either through fine tuning of ancient abiotic stress responses or evolution of novel adaptations to cold environments, must have been central for adaptation, colonization, and speciation in the Pooideae subfamily.

## 3.4 Effects of Freezing Temperatures on Plant Physiology

Each plant has an optimum set of temperatures for its growth and development, but these will differ from one species or species-ecotype to another. Plants native to warm habitats such as maize (*Zea mays*), soybean (*Glycine max*), cotton (*Gossypium hirsutum*), and tomato (*Lycopersicon esculentum*) exhibit symptoms of chilling injury upon exposure to even nonfreezing temperatures below 10–15 °C (Guy 1990; Lynch 1990).

Cold stress-induced injury in plants may appear after 48–72 h of stress exposure. Plants exposed to cold stress show various phenotypic symptoms that include reduced leaf expansion, wilting, and chlorosis (yellowing of leaves) and may lead to necrosis. Cold stress also severely affects the reproductive development of plants, and this has been seen in rice plants at the time of anthesis and leads to sterility in flowers. The effects of cold stress at the reproductive stage of plants delay heading and result in pollen sterility, which is thought to be one of the key factors responsible for the reduction in grain yield of crops (Suzuki et al. 2008). As mentioned earlier, the major adverse effect of cold stress in plants has been seen in terms of plasma membrane damage. This has been documented due to cold stress-induced dehydration (Steponkus 1984; Steponkus et al. 1993). The plasma membrane is made up of lipids and proteins. Lipids in the plasma membrane are made up of two kinds of fatty acids: unsaturated and saturated fatty acids. Unsaturated fatty acids have one or more double bonds between two carbon atoms, whereas saturated fatty acids are fully saturated with hydrogen atoms. Lipids containing more saturated fatty acids solidify faster and at temperatures higher than those containing unsaturated fatty acids. Therefore, the relative proportion of these two types of fatty acids in the lipids of the plasma membrane determines the fluidity of the membrane (Steponkus et al. 1993). At the transition temperature, a membrane changes from a semifluid state into a semicrystalline state. Cold-sensitive plants usually have a higher proportion of saturated fatty acids in their plasma membrane. Therefore, cold-sensitive plants have a higher transition temperature. On the other hand, cold-resistant plants have a higher proportion of unsaturated fatty acids and thereby a lower transition temperature.

Ice formation is the major cause of plant damage. Ice formation in plant tissues during cold stress leads to dehydration. Ice is formed in the apoplast, the free diffusional space outside the plasma membrane, because it has relatively lower solute concentration. It is known that the vapor pressure of ice is much lower than water at any given temperature. Therefore, ice formation in the apoplast establishes a vapor pressure gradient between the apoplast and surrounding cells. As a result of this gradient, the cytoplasmic water migrates down the gradient from the cell cytosol into the apoplast. This adds to existing ice crystals in the apoplastic space and causes a mechanical strain on the cell wall and plasma membrane, leading to cell rupture (McKersie and Bowley 1997; Olien and Smith 1997; Uemura and Steponkus 1997).

In addition to the well-established harmful effect of cold stress alterations in lipid composition of the biomembranes, affecting their fluidity (Senser and Beck 1982; Quinn 1985; Williams 1990; Welti et al. 2002), certain additional factors may also contribute to damage induced by cold stress. This includes synthesis and accumulation of compatible solutes, the synthesis of cold acclimation-induced proteins (Close 1997; Shinozaki and Yamaguchi-Shinozaki 2000), changes in the carbohydrate metabolism (Hansen and Beck 1994; Hansen et al. 1997; Liu et al. 1998; Frankow-Lindberg 2001), and increases in the radical scavenging potential of the cells (Tao et al. 1998; Hernández-Nistal et al. 2002; Baek and Skinner 2003). Taken together, cold stress results in loss of membrane integrity, leading to solute leakage. Further, cold stress disrupts the integrity of intracellular organelles, leading to the loss of compartmentalization. Exposure of plants to cold stress also causes reduction and prevention of photosynthesis, protein assembly, and general metabolic processes.

Recently, attempts have also been directed towards analyzing the effect of cold stress on the whole-plant metabolome. Metabolic profiling of *Arabidopsis* plants revealed that CA increases 75 % of the 434 analyzed metabolites (Cook et al. 2004; Kaplan et al. 2004). The role of such metabolites in plants has been known as osmoprotectants. In addition to their role as osmoprotectants and osmolytes, certain metabolites (individual metabolites or the redox state) induced during CA might act as signals for reconfiguring gene expression. For example, cold stress induces the accumulation of proline, a well-known osmoprotectant. Microarray and RNA gel blot analyses have shown that proline can induce the expression of many genes, which have the proline-responsive element sequence ACTCAT in their promoters (Satoh et al. 2002; Oono et al. 2003).

## 3.4.1   Impacts of Cold Stress on Crop Production

Each crop has an optimal thermal threshold. As mentioned earlier, tropical and subtropical crops may be damaged at above 0 °C temperatures, while temperate crops depending on species and ecotype can acclimatize and resist temperatures from −5 °C to −30 °C (Thomashow 1998). In temperate crops, suboptimal temperatures during spring lead to decreased productivity and yield stability (Stamp 1984).

## 3.4.2  Chilling Injury (Temperatures 0–15 °C)

Resistance to low but nonfreezing temperatures takes the form of physiological and morphological adaptations that allow plants to tolerate or avoid cold stress and may arise either at the cellular, tissue, or whole plant level. Cold avoidance mechanisms are found in both annual and perennial plants and include seed production or vegetative dormancy, the latter also an avoidance strategy in temperate frost tolerant crops such as many tree and forage species.

For crops grown outside their regions of evolved adaptation such as maize, exposure to low temperatures may cause reductions in growth or induce damage in some form. Maize being moderately sensitive to chilling temperatures may have cellular or tissue injury on exposure to temperatures below 5 °C and reduced growth at temperatures below 15 °C. Chilling stress can be assessed at different developmental stages such as at germination and during growth and also in terms of photosynthetic activity, development of reproductive organs, and fruit ripening (Blum 1988).

Cold tolerance as with drought tolerance can alter at different stages of plant ontogeny. For example, in rice, cold tolerance may reduce at the reproductive growth stages compared with the vegetative when demands on plant resources in terms of grain filling reduce resilience against low temperatures. In maize, a susceptibility to low temperatures early in the life cycle may be explained by injuries incurred during seed, embryo, and seedling development. For maize, temperature for germination should not fall below 10 °C (Herczegh 1970). Normal leaf development in maize requires temperatures above 15 °C (Nie et al. 1992). Other crops such as tomato may suffer reduced growth and fruit formation and over more prolonged periods of death following exposures to low temperatures (Nieuwhof et al. 1997).

In all cases plant ontogeny affects tolerance to cold (and especially freezing temperatures). Seed imbibition is very sensitive to chilling. In maize, chilling results in exudation of amino acids and carbohydrates from kernels as a consequence of cell membrane damage (Miedema 1982). Low temperatures not only affect germination rate but also subsequent seedling growth. The spring temperatures of more northerly latitudes are usually too low for the favorable germination of crops such as maize. Most maize plants germinate poorly at temperatures below 6 °C (Eagles 1982).

Soil temperatures are also critical to plant growth with any decrease of potential damage to root growth and water and nutrient uptake. A small reduction of 2–3 °C soil temperatures may in maize affect root growth. The effect of root temperature on growth in maize is complex and may also affect shoot growth, but after tassel formation and stem elongation, shoot growth is affected more by air rather than soil temperature (Ellis et al. 1992). As a breeding strategy, reliable germination and rapid growth in cold soils should raise crop performance across environments and extend crop range and should facilitate early sowing. In general leaf extension is controlled more by the temperature at the shoot apex rather than at the root (Watts

1972). After emergence, leaves appear successively, but young leaves are more susceptible to cold than mature leaves.

Cold sensitivity can be manipulated by changes in the levels of desaturation of fatty acids in membrane lipids. The importance of the role of membrane lipid desaturation in cold tolerance has been demonstrated in transgenic experiments (Nishida and Murata 1996). Protective proteins are induced by cold temperatures (Guy 1990). The effects of low temperatures on protein synthesis associated with cell membrane protection has been studied in maize and shown to be high in winter but to subsequently decline. Similar responses have been reported in many other crops such as spinach, rapeseed, and rice (Guy et al. 1985; Meza-Basso et al. 1986; Hahn and Walbot 1989). Proline has been shown to improve CT and aid cell structure protection in many crops, such as maize (Xin and Li 1993), potato, wheat and barley, and in *L. perenne* was shown to improve osmotic adjustment during CA (Thomas and James 1993).

Cold phases cause a reaction in the plant that prevents sugar entry into the pollen and as a consequence no subsequent starch accumulation to provide the energy necessary for pollen germination thereby limiting pollination and seed set.

## 3.5 Biotechnology

As an alternative strategy to plant breeding, biotechnological approaches, although not suitable for all countries due to current restrictions in their use, offer new strategies that can be used to develop transgenic crop plants with improved tolerance to cold stress and also if plant hybridization is possible, a means to test the efficacy of candidate genes for use in plant breeding programs. As cold tolerance is a quantitative trait, the transfer of individual gene variants will in many cases have only a limited benefit and will not reproduce generally the overall tolerance observed in the donor genotype. However, rapid advance in recombinant DNA technology and development of precise and efficient gene transfer protocols have resulted in efficient transformation and generation of transgenic lines in a number of crop species (Wani et al. 2008). A number of genes have been isolated and characterized that are responsive to freezing stress.

When a plant is subjected to chilling or freezing stress, a suite of genes are engaged, resulting in increased levels of several metabolites and proteins, some of which may be responsible for conferring a certain degree of protection. Since many aspects of CA are under transcriptional control, transcription regulatory factors are considered suitable for use in transgenic technologies where they may "trigger" a stress response and provide some tolerance. As mentioned previously, modifications in lipid composition that stabilize cell membranes and prevent cellular leakage lead to CA. Therefore, the overexpression of glycerol-3-phosphate acyltransferase led to the alteration to the unsaturation of fatty acids and conferred chilling tolerance in transgenic rice (Ariizumi et al. 2002) and tomato (Sui et al. 2007) plants.

Over the last 20 years, advancement in plant biotechnology has led to the identification and isolation of a number of transcription factor(s) related to cold stress tolerance and in many cases aided understanding of those genes with major relevance to the acquisition of cold and freezing tolerance. These include encoding enzymes that are required for the biosynthesis of various osmoprotectants for modifying membrane lipids, late embryogenesis abundant (LEA) proteins, and detoxification enzymes. In these studies, either a single gene for a protective protein or an enzyme was overexpressed under the control of the constitutive 35S cauliflower mosaic virus (CaMV), promoter in transgenic plants, although several genes have been shown to function in environmental stress tolerance and response (Shinozaki et al. 2003).

The C-repeat binding factor (CBF) genes represent one of the most significant discoveries in the field of low-temperature adaptation and signal transduction and are present in all major crops. The expression of many low temperature-inducible genes is regulated by *CBF/DREB1* (dehydration-responsive element binding) transcription factors. Three *CBF/DREB1* genes (*CBF3/DREB1a*, *CBF1/DREB1b*, and *CBF2/DREB1c*) belonging to the *AP2/DREBP* family of DNA-binding proteins have been identified in *Arabidopsis* (Jaglo-Ottosen et al. 2001). Transgenic *Arabidopsis* plants constitutively overexpressing a cold-inducible transcription factor (*CBF1*; *CRT/DRE*-binding protein) showed tolerance to freezing without any negative effect on the development and growth characteristics (Jaglo-Ottosen et al. 1998). Furthermore, overexpression of *Arabidopsis CBF1* has been shown to activate *COR* (cold-regulated) homologous genes at non-acclimating temperatures (Jaglo-Ottosen et al. 2001). Several stress-induced *COR* genes such as *rd29A*, *COR15A*, *kin1*, and *COR6.6* are triggered in response to cold treatment (Thomashow 1998).

Various *COR* genes isolated from *Arabidopsis* have protective roles against dehydration. Overexpression of *CBF1/DREB1b* and *CBF3/DREB1a* enhances cold tolerance by inducing *COR* genes (Jaglo-Ottosen et al. 1998; Liu et al. 1998). It also leads to the accumulation of sugar and proline (Gilmour et al. 2000). *CBF/DREB1* genes are thought to be activators that integrate several components of the CA response by which plants increase their tolerance to low temperatures after exposure to low but nonfreezing conditions. In recent times, extensive research efforts have been undertaken to identify and characterize *COR* genes, and a number of orthologs of the *Arabidopsis CBF* cold-response pathway have been found (Yamaguchi-Shinozaki and Shinozaki 2006). Many of these putative orthologs have been structured, analyzed, and functionally tested. The expression patterns of the *CBFs* and *CORs* in response to low temperature are similar in a variety of plants species, involving rapid cold-induced expression of the *CBFs* followed by expression of *CBF*-targeted genes that increase freezing tolerance.

Recently, a *CBF1* gene was introduced into tomato under the control of a CaMV35S promoter and that resulted in transgenic plants showing improved tolerance to chilling and higher activity of superoxide dismutase (SOD), higher non-photochemical quenching (NPQ), and lower malondialdehyde (MDA) content.

This would suggest that CBF1 protein plays an important role in protection of PSII and PSI during low temperature stress (Zhang et al. 2011). The relevance in forage grasses of CA adaptations for freezing tolerance that led to high NPQ expression was described earlier (Rapacz et al. 2005; Humphreys et al. 2007).

The importance of *CBF*-independent pathways is also supported by analysis of mutants that have increased freezing tolerance, for example, mutations in eskimo1 (*ESK1*), a protein of unknown function, and result in constitutive freezing tolerance. The *Eskimo1* mutation was first identified as a mutation conferring frost survival without recourse to a CA period (Xin and Browse 1998). Subsequently, Bouchabke-Coussa et al. (2008) demonstrated that *ESKIMO1* mutants are more tolerant to freezing, but only after acclimation. The genes that are affected by the *ESK1* mutation are distinct from those of the *CBF* regulon (Xin et al. 2007).

Various studies using transcriptome data have demonstrated that additional cold-regulatory pathways exist in addition to those cold-responsive genes regulated by *CBFs* (Flower and Thomashow 2002; Kreps et al. 2002). At least 28 % of the cold-responsive genes are not regulated by *CBF* transcription factors indicating that these genes are members of different low-temperature regulons, including 15 encoding known or putative transcription factors (Vogel et al. 2005).

The expression of related cold shock proteins (CSPs) from bacteria, *CspA* from *Escherichia coli,* and *CspB* from *Bacillus subtilis* promotes stress adaptation in multiple plant species. Transgenic rice plants expressing *CspA* and *CspB* show improved stress tolerance against various stresses, including cold, heat, and drought.

When overexpressed in *Arabidopsis* and tobacco, the soybean gene *SCOF-1*, which encodes for a zinc-finger protein, can activate *COR* gene expression and increase freezing tolerance in non-acclimated transgenic plants. The *SCOF-1* gene may regulate the activity of SGBF-1 as a transcription factor in inducing *COR* gene expression and interacts with a G-box binding *bZIP* protein, *SGBF-1*. The *SGBF-1* protein can activate ABRE-driven reporter gene expression in *Arabidopsis* leaf protoplasts (Kim et al. 2001).

Forward genetic analysis in *Arabidopsis* identified two transcription factors and high expression of osmosis-responsive genes, *HOS9* and *HOS10*, which are required for basal freezing tolerance (Zhu et al. 2004, 2005). Similarly, microarray analysis led to the identification of the cold stress-inducible *AP2* family transcription factor gene related to *ABI3/VP1* (*RAV1*) (Flower and Thomashow 2002; Vogel et al. 2005) that might regulate plant growth under cold stress.

The overexpression of genes encoding *LEA* proteins can improve the stress tolerance of transgenic plants. For example, the freezing tolerance of strawberry leaves was enhanced by expression of the wheat dehydrin gene *WCOR410* (Houde et al. 2004). Trehalose is a nonreducing disaccharide that is present in diverse organisms ranging from bacteria and fungi to invertebrates, in which it serves as an energy source, osmolyte, or protein/membrane protectant. Various studies have revealed regulatory roles of trehalose-6-phosphate, a precursor of trehalose, in sugar metabolism and growth and development in plants (Iordachescu and Imai 2008). Trehalose levels are generally quite low in plants but may alter in response to environmental stresses. Although the involvement of trehalose metabolism in stress tolerance is indubitable, our understanding of how it exactly interacts and acts upon stress pathways is far from complete. Studies of individual trehalose biosynthesis genes will help us to precisely assess their specific roles in the abiotic stress context and may enable us to develop new strategies to enhance abiotic stress tolerance of crop plants. Plants synthesize trehalose in a pathway that is common to most organisms, through the production of the intermediate trehalose-6-phosphate.

### 3.5.1   Microbial Trehalose Biosynthesis Genes

Major advances in the study of trehalose biosynthesis in plants have been made in the past decade. Transgenic plants overexpressing microbial trehalose biosynthesis genes have been shown to contain increased levels of trehalose and display drought, salt, and cold tolerance. In silico expression profiling of all *Arabidopsis* trehalose-6-phosphate synthases (TPSs) and trehalose-6-phosphate phosphatases (TPPs) revealed that certain classes of TPS and TPP genes are differentially regulated in response to a variety of abiotic stresses. These studies point to the importance of trehalose biosynthesis in stress responses. Since trehalose is an osmoprotectant (Müller et al. 1995) and membrane and protein integrity protectant (Crowe et al. 1984), different groups with varied success attempted to create stress tolerant plants by introducing microbial trehalose biosynthetic genes by use of transgenic technologies (Iordachescu and Imai 2008). Pramanik and Imai (2005) and Shima et al. (2007) isolated and characterized two rice TPPs (*OsTPP1* and *OsTPP2*). Both rice TPPs were induced transiently by cold, salt, and drought stress, as well as exogenous ABA applications; however, transient induction of *OsTPP1* occurred generally earlier than that of *OsTPP2* suggesting a tight regulation of trehalose biosynthesis in response to multiple abiotic stresses (Shima et al. 2007). Trehalose was also transiently induced following chilling stress, its increase being correlated with the increase of *OsTPP1* transcript and OsTPP1 activity (Pramanik and Imai 2005). Moreover, accumulation of trehalose in response to chilling stress coincided with the phase change of glucose and fructose levels (Pramanik and Imai 2005).

Many temperate crops such as the forage ryegrasses accumulate fructan as their main vegetative carbon and energy reserve, and it is a major nutrient for livestock on grazed pastures and receiving conserved feeds. Accumulation of fructan, rather than starch, has been suggested to have evolved to meet a need to adapt to cold winters and dry summers. Consequently, fructans are widely believed, and very frequently quoted, to be important in determining tolerance to environmental stresses such as cold and drought. Carbohydrates are well recognized as having multiple roles in integrating a wide range of plant growth and environmental responses. However, many correlations between fructan and stress tolerance are circumstantial, and evidence of direct relationships is poor. It has been widely reported that fructan content increases during exposure to many abiotic stresses, but accumulation purely as a by-product of a reduced growth rate when photosynthesis remains unaffected is very different from an active, functional role in stress tolerance. Prior to winter, temperate plant species acclimatize, during which their metabolism is redirected towards synthesis of cryoprotectant molecules such as soluble sugars (saccharose, raffinose, stachyose, trehalose), in addition to sugar alcohols (sorbitol, ribitol, inositol) and low molecular weight nitrogenous compounds (proline, glycine betaine). These, in conjunction with dehydrins, COR proteins and heat-shock proteins (HSPs) act to stabilize both membrane phospholipids and proteins, and cytoplasmic proteins, maintain hydrophobic interactions and ion homeostasis, and scavenge reactive oxygen species (ROS); other solutes released from the symplast serve to protect the plasma membrane from ice adhesion and subsequent cell disruption (Hare et al. 1998; Iba 2002; Wang et al. 2003; Gusta et al. 2004; Chen and Murata 2008). The process of solute release, especially of vacuolar fructans to the extracellular space, is vesicle-mediated and tonoplast-derived (Valluru and Van den Ende 2008). Fructans are transported to the apoplast by postsynthesis mechanisms, probably in response to cold stress (Valluru et al. 2008). The activity of fructan exohydrolase, which generates increased sugar (glucose, fructose, sucrose) content, is an important part of the hardening process. Symplastic and apoplastic soluble sugar—not only fructan precursors but also trehalose, raffinose, as well as fructo- and gluco-oligosaccharides—contributes directly to membrane stabilization (Livingston et al. 2006).

In barley, trehalose induces the expression and activity of fructan biosynthesis enzymes. However, for fructan accumulation, glucose or mannitol is also required (Wagner et al. 1986; Müller et al. 2000). From a microarray analysis following trehalose treatment, Bae et al. (2005) showed that the expression of a wide range of other genes was also influenced by trehalose. A role for trehalose and trehalose-6-phosphate in abiotic and biotic stress signaling has been confirmed by the observation that coordinated changes occur in transcript levels of the enzymes involved in their metabolism, especially after exposure to cold, osmotic, and salinity stresses (Iordachescu and Imai 2008).

## 3.6 The *CBF/DREB* Pathway

As previously mentioned, the *CBF/DREB*-responsive pathway provides a major route for the production of cold-responsive proteins. CBF1, 2, and 3 are all responsive to low temperature, and their encoding genes are present in tandem on *Arabidopsis* chromosome 4. *CBF2/DREB1C* is a negative regulator of both *CBF1/DREB1B* and *CBF3/DREB1A*. *CBF3* is thought to regulate the expression level of *CBF2* (Novillo et al. 2004; Chinnusamy et al. 2006). Thus, the function(s) of *CBF1* and *CBF3* differs from those of *CBF2* and act additively to induce the set of *CBF*-responsive genes required to complete the process of CA (Novillo et al. 2007). Upstream of *CBF* lie both *ICE1* (inducer of *CBF* expression), a positive regulator of *CBF3*, and *HOS1* (high expression of osmotically sensitive), a negative regulator of *ICE1*. Because of the rapid induction of *CBF* transcripts following plant exposure to low temperature, *ICE1* is unlikely to require fresh synthesis, but is already present in the absence of cold stress and is only activated when temperature is lowered (Chinnusamy et al. 2003). The *LOS1* (low expression of osmotically responsive genes) product is a translation elongation factor 2-like protein, which negatively regulates *CBF* expression.

The likely regulators of *CBF1* and *CBF2* are bHLH proteins other than ICE1. In addition to *ICE1*, a further positive regulator of *CBF* expression is LOS4, an RNA helicase-like protein (Gong et al. 2002). *CAX1* (cation exchanger), which plays a role in returning cytosolic $Ca_2+$ concentrations to basal levels following a transient increase in response to low-temperature stress, is a negative regulator of CBF1, 2, and 3 (Hirschi 1999; Catalá et al. 2003).

Knowledge gained from the study of *Arabidopsis* has proven to be largely, but not completely transferable to crop plants. A better model for cereal and grass crop species of the Pooideae will be rice and especially *Brachypodium* with which they share closer syntenic relations (Moore et al. 1995). In barley, the *CBF* genes *HvCBF3, HvCBF4,* and *HvCBF8* are all components of the frost resistance QTL located on chromosome 5H (Francia et al. 2004).

A contrasting approach has targeted comparisons between spring- and winter-sown cereal cultivars. For example, Monroy et al. (2007) observed that spring and winter wheat share the same initial rapid expression of cold-inducible genes, but that their transcriptional profiles diverge widely during CA. While in winter cultivars the expression of CA genes continues over time, in spring cultivars, their levels of expression decline and the CA process is overridden by the transition from the vegetative to the reproductive stage.

## 3.7 Antifreeze Proteins and Ice Recrystallization Inhibition Genes

Antifreeze proteins were first isolated from Antarctic fish and were shown to depress the freezing point of blood serum to −2 °C, one degree below that at which the serum melts (Duman et al. 1993). This noncolligative effect, termed thermal hysteresis (TH), is evident in other aqueous solutions and appears to result from the adsorption of the proteins to the surface of ice crystals and inhibition of further crystal growth.

In general, plant AFPs exhibit low levels of TH and are present at much lower concentrations than fish AFPs (Duman 1994; Hon et al. 1995). AFPs may modify the activity of ice nucleators. Indeed, AFPs infiltrated into the extracellular space of leaves of *Solanum tuberosum, Brassica napus,* and *Arabidopsis* depressed the freezing point by up to 1.8 °C (Cutler et al. 1989). No ice crystals were formed showing that the AFP acted as an anti-ice nucleator.

Functional molecular studies support a link between ice recrystallization inhibition (*IRI*) genes and adaptation to cold stress (John et al. 2009). Transcription of the

*IRI* genes is controlled by exposure to cold temperature, and they encode proteins that bind to ice crystals to change their ability to increase in size in vitro. In vivo protein function is however still debated. The current leading hypothesis is that IRI activity hinders ice expansion in the apoplast and thus protects plant cells from mechanical damage, elevating plant frost tolerance. But *IRI* genes also encode an LRR domain that is homologous to pathogenesis-related proteins, and this raises the question if IRI proteins have dual functions in cold stress response.

Frost tolerance adaptations are, in many organisms, associated with the evolution of AFPs (Zachariassen and Kristiansen 2000). AFPs can affect freezing and ice crystallization-related stress via different mechanisms. TH depresses the freezing point at which ice crystallization initiates, which renders it possible for organisms to remain unfrozen yet survive under freezing temperatures. IRI on the other hand does not hinder ice crystallization but manipulates the growth of the ice crystals such that small ice crystals grow at the expense of larger ice crystals. Even though the functional significance of IRI in vivo has been demonstrated, IRI proteins are believed to prevent or minimize the cellular damage in plants (Smallwood and Bowles 2002).

Animal AFPs generally possess high TH characteristics and lower ice crystallization initiation temperatures by 1–5 °C (Barrett 2001; Griffith and Yaish 2004). Plant AFPs on the other hand have low TH activity but exhibit strong ice recrystallization inhibition (IRI) activity (Griffith and Yaish 2004). AFPs have been isolated from different plants and there are at least five isoforms of AFPs within the cold tolerant grasses; full-length nucleotide sequences are available for genes encoding five AFPs (Griffith and Yaish 2004; Middleton et al. 2009). In addition, AFPs have been identified in carrot (Worrall et al. 1998). Pooideae subfamily-specific *IRI* gene homologs have so far been identified and isolated in perennial ryegrass (Sandve et al. 2008), wheat (Tremblay et al. 2005), and Antarctic hair grass (John et al. 2009). Most likely these genes have evolved from a common ancestor gene of the leucine-rich repeat phytosulfokine receptor kinase (*LRR-PSR*)-like genes present in rice and *Arabidopsis* (Sandve et al. 2008).

## 3.8  Conclusion

To summarize all the above, a simple illustration is presented (Fig. 3.1) of the impacts of subzero temperatures on a cell of a non-acclimated or cold-sensitive grass genotype and the response by an equivalent genotype once acclimated fully having achieved a frost tolerance sufficient to safeguard against cell damage by the winter stresses endured.

By 2050, it is estimated that the Earth's human population will be 9.07 billion. Of these, 62 % will live in Africa, Southern Asia, and Eastern Asia; numerically the same as if the world's current population lived solely in these regions. These overpopulated regions will suffer the greatest from the outcomes of climate change and in particular high temperatures and shortage of nonsalinized water suitable for

**Fig. 3.1** The consequences on a non-acclimated plant cell when exposed to freezing temperatures, and high irradiance, and the resilience found to safeguard cellular integrity in an equivalent adapted and cold-acclimated genotype

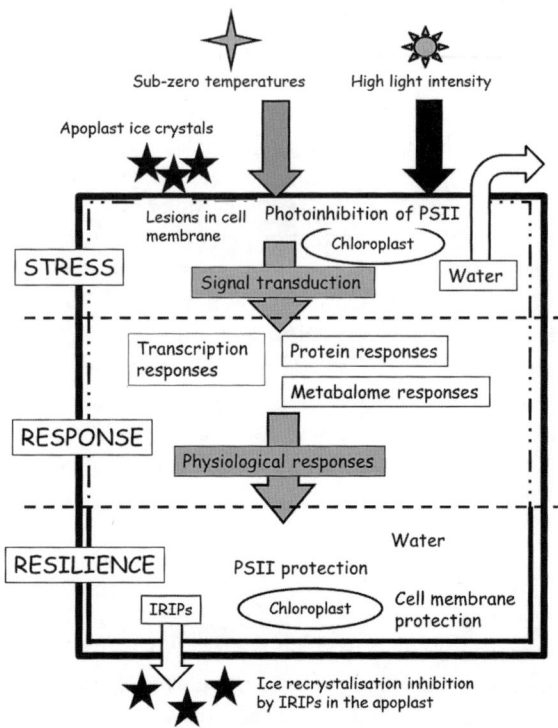

human consumption and food production. In more northern and southern regions, climate predictions indicate opportunities for some increases in crop yields will likely following climate change. However, it will still require plant breeders to provide the crops best adapted to these regions. Weather extremes fluctuate with increasing frequency and crop design must respond to include some plasticity to provide appropriate responses to the diverse stresses crops will likely encounter.

In Ireland, where livestock agriculture predominates, increasing temperatures have led to national policies aimed at an all-year growing season. Such policies in locations prone to fluctuating weather patterns bring obvious dangers to sustainable crop production, as does the cultivation of marginal regions such as uplands and mountain regions subject to low temperatures and usage of non-frost adapted high-yielding crop species.

The cold tolerance trait is complex but many attributes of stress tolerance require common gene complexes for resistance to desiccation, and as described herein, these will often span requirements to resist freezing, drought, salinity, and high winds. Specific to cold tolerance, it will be necessary to target those genes that control the main regulatory systems and in particular those concerned with cold acclimation and photoperiod response.

In summary, while growth of many crops must from necessity be restricted to their current climatic zones of adaptation, for others including many of the forage and energy crops and cereals, a holistic approach to plant breeding is now required that provides crops with the capability to withstand a wide range of abiotic and biotic stresses. New high throughput genotype and phenotype technologies such as the incorporation of genome-wide selection in crop improvement programs and use of hyperspectral imaging phenomics facilities provide opportunities to assemble correct crop genotypes for a predictable and desirable phenotype. Such technologies for the first time make possible complex redesigns of crops to provide where deemed necessary new adaptive variation derived either directly from wild-type relatives or via transgenic technologies. To increase crop production, the incorporation of nonnative and even tropical crops, including $C_4$ plants with more efficient photosynthesis, in cooler temperate regions will in some cases become a feasible option. Such strategies require careful preparation by the plant breeder to ensure sufficient and appropriate adaptive variation is available. A good example at IBERS Aberystwyth University is the development and use of the biomass elephant grass (*Miscanthus giganteus*). This fast growing $C_4$ biomass plant, a native to Eastern Asia, is grown commercially as a natural triploid hybrid generated from progenitor species *M. sinensis* and *M. sacchariflorus*. Ecotypes of the progenitor species' show vast diversity in location and in biomass and adaptive traits, and by careful plant collection and selection of the progenitor species, new synthetic *M. giganteus* hybrids are being developed to include winter adaptive traits that match the growth conditions they will encounter in Europe far from their natural habitat (Robson et al. 2012). While equivalent breeding strategies may not be available to all food crops, it will require to feed future generations ingenuity such as this to achieve the maximum possible crop growth potential in all those marginal locations available for agricultural use.

# References

Alm V, Busso CM, Ergon A, Rudi H, Larsen A, Humphreys MW, Rognli OA (2011) QTL analysis and comparative genetic mapping of frost tolerance, winter survival and drought tolerance in meadow fescue (*Festuca pratensis* Huds.). Theor Appl Genet 123(3):369–382

Andrews CJ, Gudleifsson BE (1983) A comparison of cold hardiness and ice encasement tolerance of timothy grass and winter wheat. Can J Plant Sci 63:429–435

Ariizumi T, Kishitani S, Inatsugi R, Nishida I, Murata N, Toriyama K (2002) An increase in unsaturation of fatty acids in phosphatidylglycerol from leaves improves the rates of photo-synthesis and growth at low temperatures in transgenic rice seedlings. Plant Cell Physiol 43:751–758

Bae HH, Herman E, Bailey B, Bae HJ, Sicher R (2005) Exogenous trehalose alters Arabidopsis transcripts involved in cell wall modification, abiotic stress, nitrogen metabolism, and plant defence. Physiol Plant 125:114–126

Baek K-H, Skinner DZ (2003) Alteration of antioxidant enzyme gene expression during cold acclimation of near-isogenic wheat lines. Plant Sci 165:1221–1227

Barker NP, Clark LG, Davis JI, Duvall MR, Guala GF, Hsiao C, Kellogg EA, Linder HP, Mason-Gamer RJ, Mathews SY, Simmons MP, Soreng RJ, Spangler RE (2001) Phylogeny and subfamilial classification of the grasses (*Poaceae*). Ann MO Bot Gard 88:373–457

Barrett J (2001) Thermal hysteresis proteins. Int J Biochem Cell Biol 33:105–117

Bezant J, Laurie D, Pratchett N, Chojecki J, Kearsey M (1996) Marker regression mapping of QTL controlling flowering time and plant height in a spring barley (*Hordeum vulgare* L) cross. Heredity 77:64–73

Blum A (1988) Plant breeding for stress environments. CRC, Boca Raton, FL

Bouchabke-Coussa O, Quashie M, Seoane-Redondo J, Fortabat M, Gery C, Yu A, Linderme D, Trouverie J, Granier F, Téoulé E, Durand-Tardif M (2008) *ESKIMO1* is a key gene involved in water economy as well as cold acclimation and salt tolerance. BMC Plant Biol 8:125–151

Brändle RA (1991) Flooding resistance of rhizomatous amphibious plants. In: Jackson MB, Davies DD, Lambers H (eds) Plant life under oxygen deprivation. SPB Academic, Hague, pp 35–46

Canter PH, Bettany AJE, Donnison I, Timms E, Humphreys MW, Jones RN (2000) Expressed sequence tags (ESTs) during cold-acclimation in *Festuca pratensis* include a homologue of the chloroplast gene *PsbA*. J Exp Bot 51:72

Caradus JR, Christie BR (1998) Winterhardiness and artificial frost tolerance of white clover ecotypes and selected breeding lines. Can J Plant Sci 78:251–255

Catalá R, Santos E, Alonso JM, Ecker JR, Martínez-Zapater JM, Salinas J (2003) Mutations in the Ca2+/H+ transporter CAX1 increase CBF/DREB1 expression and the cold-acclimation response in Arabidopsis. Plant Cell 15:2940–2951

Chen TH, Murata N (2008) Glycinebetaine: an effective protectant against abiotic stress in plants. Trends Plant Sci 13:499–505

Chen H-H, Li PH, Brenner ML (1983) Involvement of abscisic acid in potato cold acclimation. Plant Physiol 71:362–365

Chinnusamy V, Ohta M, Kanrar S, Lee B, Hong X, Agarwal M, Zhu JK (2003) ICE1: a regulator of cold-induced transcriptome and cold tolerance in Arabidopsis. Genes Dev 17:1043–1054

Chinnusamy V, Zhu J, Zhu JK (2006) Gene regulation during cold acclimation in plants. Physiol Plant 126:52–61

Close TJ (1997) Dehydrins: a commonalty in the response of plants to dehydration and low temperature. Physiol Plant 100:291–296

Cook D, Fowler S, Fiehn O, Thomashow MF (2004) A prominent role for the CBF cold response pathway in configuring the low-temperature metabolome of Arabidopsis. Proc Natl Acad Sci USA 101:15243–15248

Crowe JH, Crowe LM, Chapman D (1984) Preservation of membranes in anhydrobiotic organisms: the role of trehalose. Science 223:701–703

Cutler AJ, Saleem M, Kendall E, Gusta LV, Georges F, Fletcher GL (1989) Winter under antifreeze proteins improve the cold hardiness of plant tissues. J Plant Physiol 135:351–354

Davies DD (1980) Anaerobic metabolism and the production of organic acids. In: Davies DD (ed) The biochemistry of plants, vol 2. Academic, London, pp 581–611

Dubcovsky J (2004) Marker assisted selection in public breeding programs: the wheat experience. Crop Sci 44:1895–1898

Duman JG (1994) Purification and characterization of a thermal hysteresis protein from a plant, the bittersweet nightshade *Solanum dulcamara*. Biochim Biophys Acta 1206:129–135

Duman JG, Wu DW, Olsen TM, Urrutia M, Tursman D (1993) Thermal-hysteresis proteins. In: Steponkus PL (ed) Advances in low temperature biology, vol 2. JAI, London, pp 131–182

Eagles HA (1982) Inheritance of emergence time and seedling growth at low temperatures in four lines of maize. Theor Appl Genet 62:81–87

Eagles CF (1989) Temperature-induced changes in cold tolerance of *Lolium Perenne*. J Agric Sci Camb 113:339–347

Eagles CF (1994) Temperature, photoperiod and dehardening of forage grasses and legumes. In: Crop adaptation to cool climates. Proceedings of the COST 814 workshop, Hamburg, Germany, pp 75–82

Eagles CF, Williams J, Louis DV (1993) Recovery after freezing in *Avena sativa* L., *Lolium perenne* L. & *L. multiflorum* Lam. New Phytol 123:477–483

Ellis RH, Summerfield RJ, Edmeades GO, Roberts EH (1992) Photoperiod, temperature and the interval from sowing to tassle initiation in diverse cultivars of maize. Crop Sci 32:125–132

Else MA, Davies WJ, Malone M, Jackson MB (1995) A negative hydraulic message from oxygen-deficient roots of tomato plants? Influence of soil flooding on leaf water potential, leaf expansion and the synchrony of stomatal conductance and root hydraulic conductivity. Plant Physiol 109:1017–1024

Flower S, Thomashow MF (2002) Arabidopsis transcriptome profiling indicates that multiple regulatory pathways are activated during cold acclimation in addition to the CBF cold response pathway. Plant Cell 14:1675–1690

Fox GG, McCallan GG, Ratcliffe RG (1995) Manipulating cytoplasmic pH under anoxia: a critical test of the role of pH in the switch from aerobic to anaerobic metabolism. Planta 195:324–330

Francia E, Rizza F, Cattivelli L, Stanca AM, Galiba G, Tóth B, Hayes PM, Skinner JS, Pecchioni N (2004) Two loci on chromosome 5H determine low-temperature tolerance in a "Nure" (winter) "Tremois" (spring) barley map. Theor Appl Genet 108:670–680

Francia E, Barabaschi D, Tondelli A, Laido G, Rizza F, Stanca AM, Busconi M, Fogher C, Stockinger EJ, Pecchioni N (2007) Fine mapping of a *HvCBF* gene cluster at the frost resistance locus *Fr-H2* in barley. Theor Appl Genet 115(8):1083–1091

Frankow-Lindberg BE (2001) Adaptation to winter stress in nine white clover populations: changes in non-structural carbohydrates during exposure to simulated winter conditions and 'spring' regrowth potential. Ann Bot 88:745–751

Galiba G, Simonsarkadi L, Kocsy G, Salgo A, Sutka J (1992) Possible chromosomal location of genes determining the osmoregulation of wheat. Theor Appl Genet 85(4):415–418

Galiba G, Tuberosa R, Kocsy G, Sutka J (1993) Involvement of chromosome 5A and 5D in cold-induced abscisic acid accumulation in and frost tolerance of wheat calli. Plant Breed 110:237–242

Galiba G, Quarrie SA, Sutka J, Morgounov A, Snape JW (1995) RFLP mapping of the vernaliza-tion (Vrn1) and frost-resistance (Fr1) genes on chromosome 5a of wheat. Theor Appl Genet 90 (7–8):1174–1179

Galiba G, Vagujfalvi A, Li CX, Soltesz A, Dubcovsky J (2009) Regulatory genes involved in the determination of frost tolerance in temperate cereals. Plant Sci 176:12–19

Gaut BS (2002) Evolutionary dynamics of grass genomes. New Phytol 154:15–28

Gay AP, Eagles CF (1991) Quantitative analysis of cold hardening and dehardening in *Lolium*. Ann Bot 67:339–346

Gilmour SJ, Zarka DG, Stockinger EJ, Salazar MP, Houghton JM, Thomashow MF (1998) Low temperature regulation of the Arabidopsis CBF family of AP2 transcriptional activators as an early step in cold-induced COR gene expression. Plant J 16:433–442

Gilmour SJ, Sebolt AM, Salazar MP, Everard JD, Thomashow MF (2000) Overexpression of the Arabidopsis CBF3 transcriptional activator mimics multiple biochemical changes associated with cold acclimation. Plant Physiol 124:1854–1865

Gong Z, Lee H, Xiong L, Jagendorf A, Stevenson B, Zhu JK (2002) RNA helicase-like protein as an early regulator of transcription factors for plant chilling and freezing tolerance. Proc Natl Acad Sci USA 99:11507–11512

Greaves A (1996) Improving suboptimal temperature tolerance in maize-the search for variation. J Exp Bot 47:307–323

Griffith M, Yaish MW (2004) Antifreeze proteins in overwintering plants: a tale of two activities. Trends Plant Sci 9(8):399–405

Gudleifsson BE (1971) Um kal og kalskemmdir. I. Raektun og nytiar t6na og hrif pessara ptta kal [Winter damages I. The influence of cultivation and grassland management on winter damage]. Arsrit Raktunarfelags Norburlands 68:73–93

Gudleifsson BE, Andrews CJ, Bjornsson H (1986) Cold hardiness and ice tolerance of pasture grasses grown and tested in controlled environments. Can J Plant Sci 66:601–608

Gusta LV, Wisniewski M, Nesbitt NT, Gusta ML (2004) The effect of water, sugars, and proteins on the pattern of ice nucleation and propagation in acclimated and nonacclimated canola leaves. Plant Physiol 135:1642–1653

Guy CL (1990) Cold acclimation and freezing stress tolerance: role of protein metabolism. Annu Rev Plant Physiol Plant Mol Biol 41:187–223

Guy CL, Niemi KJ, Brambl R (1985) Altered gene expression during cold acclimation of spinach. Proc Natl Acad Sci USA 82:3673–3677

Hahn M, Walbot V (1989) Effects of cold-treatment on protein synthesis and mRNA levels in rice leaves. Plant Physiol 91:930–938

Hansen J, Beck E (1994) Seasonal changes in the utilization and turnover of assimilation products in a 8-year-old Scots pine (*Pinus sylvestris* L.) trees. Trees 8:172–182

Hansen J, Türk R, Vogg G, Heim R, Beck E (1997) Conifer carbohydrate physiology: updating classical views. In: Rennenberg H, Eschrich W, Ziegler H (eds) Trees – contributions to modern tree physiology. Backhuys, Leiden, pp 97–108

Hare PD, Cress WA, Van Staden J (1998) Dissecting the roles of osmolyte accumulation during stress. Plant Cell Environ 21:535–553

Hartley H (1973) Studies on the origin, evolution, and distribution of the gramineae. V. The subfamily *Festucoideae*. Aust J Bot 21:201–234

Hayes PM, Blake T, Chen THH, Tragoonrung S, Chen F, Pan A, Liu B (1993) Quantitative trait loci on barley (*Hordeum vulgare* L) chromosome-7 associated with components of winterhardiness. Genome 36(1):66–71

Herczegh M (1970) Some problems of cold tolerance. In: Kovacs I (ed) Some methodological achievements of the hungarian hybrid maize breeding. Akademiai Kiado, Budapest, pp 271–281

Hernández-Nistal J, Dopico B, Labrador E (2002) Cold and salt stress regulates the expression and activity of a chickpea cytosolic Cu/Zn superoxide dismutase. Plant Sci 163:507–514

Hirschi KD (1999) Expression of Arabidopsis CAX1 in tobacco: altered calcium homeostasis and increased stress sensitivity. Plant Cell 11:2113–2122

Hon WC, Griffith M, Mlynarz A, Kwok YA, Yang DCS (1995) Antifreeze proteins in winter rye are similar to pathogenesis-related proteins. Plant Physiol 109:879–889

Houde M, Dallaire S, N'dong D, Sarhan F (2004) Overexpression of the acidic dehydrin WCOR410 improves freezing tolerance in transgenic strawberry leaves. Plant Biotechnol 2:381–387

Hughes MA, Dunn MA (1996) The molecular biology of plant acclimation to low temperature. J Exp Bot 47:291–305

Humphreys MO, Humphreys MW (2005) Breeding for stress resistance: general principles. In: Ashraf M, Harris PJC (eds) Abiotic stresses: plant resistance through breeding and molecular approaches, C hap 2. Pub Haworth, New York, pp 19–46

Humphreys MW, Pasakinskiene I, James AR, Thomas H (1998) Physical mapping quantitative traits for stress resistance in the forage grasses. J Exp Bot 49:1611–1618

Humphreys J, Harper JA, Armstead IP, Humphreys MW (2005) Introgression-mapping of genes for drought resistance transferred from *Festuca arundinacea* var. *glaucescens* into *Lolium multiflorum*. Theor Appl Genet 11(3):579–587

Humphreys MW, Yadav RS, Cairns AJ, Turner LB, Humphreys J, Skot L (2006) A changing climate for grassland research. New Phytol 169(1):9–26

Humphreys MW, Gasior D, Lesniewska-Bocianowska A, Zwierzykowski Z, Rapacz M (2007) Androgenesis as a means of dissecting complex genetic and physiological controls: selecting useful gene combinations for breeding freezing-tolerant grasses. Euphytica 158(3):337–345

Huner NPA, Maxwell DP, Gray GR, Savitch LV, Krol M, Ivanov AG (1996) Sensing environmental change: PSII excitation pressure and redox signalling. Physiol Plant 8:358–364

Iba K (2002) Acclimative response to temperature stress in higher plants: approaches of gene engineering for temperature tolerance. Annu Rev Plant Biol 53:225–245

Iordachescu M, Imai R (2008) Trehalose biosynthesis in response to abiotic stresses. J Int Plant Biol 50(10):1223–1229

Ito Y, Katsura K, Maruyama K, Taji T, Kobayashi M, Seki M, Shinozaki K, Yamaguchi-Shinozaki K (2006) Functional analysis of rice DREB1/CBF-type transcription factors involved in cold-responsive gene expression in transgenic rice. Plant Cell Physiol 47:141–153

Jackson MB (1990) Hormones and developmental change in plants subjected to submergence or soil waterlogging. Aquat Bot 38:49–72

Jaglo-Ottosen KR, Gilmour SJ, Zarka DG, Schabenberger O, Thomashow MF (1998) Arabidopsis CBF1 overexpression induces COR genes and enhances freezing tolerance. Science 280:104–106

Jaglo-Ottosen KR, Kleff S, Amundsen KL, Zhang X, Haake V, Zhang JZ, Detis T, Thomashow MF (2001) Components of the Arabidopsis C-repeat/dehydration-responsive element binding factor cold-response pathway are conserved in Brassica napus and other plant species. Plant Physiol 127:910–917

Jensen LB, Andersen JR, Frei U, Xing YZ, Taylor C, Holm PB, Lubberstedt TL (2005) QTL mapping of vernalization response in perennialryegrass (Lolium perenne L.) reveals co-location with an orthologue of wheat VRN1. Theor Appl Genet 110:527–536

John UP, Polotnianka RM, Sivakumaran KA, Chew O, MacKin L, Kuiper MJ, Talbot JP, Nugent GD, Mautord J, Schrauf GE (2009) Ice recrystallization inhibition proteins (IRIPs) and freeze tolerance in the cryophilic Antarctic hair grass Deschampsia antarctica E. Desv. Plant Cell Environ 32 (4):336–348

Kaplan F, Kopka J, Haskell DW, Zhao W, Schiller KC, Gatzke N, Sung DY, Guy CL (2004) Exploring the temperature stress metabolome of Arabidopsis. Plant Physiol 136:4159–4168

Kim JC, Lee SH, Cheong YH, Yoo CM, Lee SI, Chun J, Yun DJ, Hong JC, Lee SY, Lim CO, Cho MJ (2001) A novel cold-inducible zinc finger protein from soybean, SCOF-1, enhances cold tolerance in transgenic plants. Plant J 25:247–259

King J, Armstead IP, Donnison IS, Thomas HT, Jones RJ, Kearsey MJ, Roberts LA, Thomas A, Morgan WG, King IP (2002a) Physical and genetic mapping in the grasses Lolium perenne and Festuca pratensis. Genetics 161:307–314

King J, Roberts LA, Kearsey MJ, Thomas HM, Jones RN, Huang L, Armstead IP, Morgan WG, King IP (2002b) A demonstration of a 1:1 correspondence between chiasma frequency and recombination using a Lolium perenne/Festuca pratensis substitution line. Genetics 161:315–324

Knox AK, Dhillon T, Cheng HM, Tondelli A, Pecchioni N, Stockinger EJ (2010) CBF gene copy number variation at Frost Resistance-2 is associated with levels of freezing tolerance in temperate-climate cereals. Theor Appl Genet 121(1):21–35

Kreps JA, Wu Y, Chang HS, Zhu T, Wang X, Harper JF (2002) Transcriptome changes for Arabidopsis in response to salt, osmotic, and cold stress. Plant Physiol 130:2129–2141

Laurie DA, Pratchett N, Bezant JH, Snape JW (1995) RFLP mapping of five major genes and eight quantitative trait loci controlling flowering time in a winter x spring barley (Hordeum vulgare L.) cross. Genome 38(3):575–585

Law CN, Sutka J, Worland AJ (1978) Genetic study of day-length response in wheat. Heredity 41:185–191

Levitt J (1980) Responses of plants to environmental stress, vol I, 2nd edn, Chilling, freezing, and high temperature stresses. Academic, New York

Liu Q, Ksauga M, Sakuma Y, Abe H, Miura S, Yamaguchi-Shinozaki K, Shinozaki K (1998) Two transcription factors, DREB1and DREB2, with an EREBP/AP2 DNA binding domain separate two cellular signal transduction pathways in drought and low temperature-responsive gene expression, respectively, in Arabidopsis. Plant Cell 10:1391–1406

Livingston DP, Premakumar R, Tallury SP (2006) Carbohydrate partitioning between upper and lower regions of the crown in oat and rye during cold acclimation and freezing. Cryobiology 52:200–208

Lynch DV (1990) Chilling injury in plants: the relevance of membrane lipids. In: Katterman F (ed) Environmental injury to plants. Academic, New York, pp 17–34

Mattoo AK, Marder J, Edelman M (1989) Dynamics of the photosystem II reaction center. Cell 56:241–246

McKersie BD, Bowley SR (1997) Active oxygen and freezing tolerance in transgenic plants. In: Li PH, Chen THH (eds) Plant cold hardiness: molecular biology, biochemistry, and physiology. Plenum, New York, pp 203–214

Meza-Basso L, Alberdi M, Raynal M, Ferrero-Cadinanos ML, Delseny M (1986) Changes in protein synthesis in Rapeseed (*Brassica napus*) seedlings during a low temperature treatment. Plant Physiol 82:733–738

Middleton AJ, Brown AM, Davies PL, Walker VK (2009) Identification of the ice-binding face of a plant antifreeze protein. FEBS Lett 583:815–819

Miedema P (1982) The effects of low temperatures on *Zea mays*. Adv Agron 35:93–128

Miller AK, Galiba G, Dubcovsky J (2006) A cluster of 11 CBF transcription factors is located at the frost tolerance locus Fr-A(m)2 in *Triticum monococcum*. Mol Genet Genomics 275 (2):193–203

Monroy AF, Dryanova A, Malette B, Oren DH, Farajalla MR, Liu W, Danyluk J, Ubayasena LWC, Kane K, Scoles GJ, Sarhan F, Gulick PJ (2007) Regulatory gene candidates and gene expression analysis of cold acclimation in winter and spring wheat. Plant Mol Biol 64:409–423

Moore G, Devos KM, Wang Z, Gale MD (1995) Grasses, line up and form a circle. Curr Biol 5:737–739

Moore BJ, Donnison IS, Harper JA, Armstead IP, King J, Thomas H, Jones RN, Jones TH, Thomas HM, Morgan WG, Thomas A, Ougham HJ, Huang L, Fentem T, Roberts LA, King IP (2005) Molecular tagging of asenescence gene by introgression mapping of a mutant stay-green locus from *Festuca pratensis*. New Phytol 165:801–806

Müller J, Boller T, Wiemken A (1995) Trehalose and trehalase inplants: recent developments. Plant Sci 112:1–9

Müller J, Aeschbacher RA, Sprenger N, Boller T, Wiemken A (2000) Disaccharide-mediated regulation of sucrose: fructan-6-fructosyltransferase, a key enzyme of fructan synthesis in barley leaves. Plant Physiol 123:265–274

Nie G-Y, Long SP, Baker NR (1992) The effects of development at sub-optimal growth temperatures on photosynthetic capacity and susceptibility to chilling-dependent photoinhibition in *Zea mays*. Physiol Plant 85:554–560

Nieuwhof M, Keizer LCP, Oeveren JC (1997) Effects of temperature on growth and development of adult plants of genotypes of tomato (*Lycopersicon esculentum* Mill.). J Genet Breed 51 (3):185–193

Nishida I, Murata N (1996) Chilling sensitivity in plants and cyanobacteria: the crucial contribution of membrane lipids. Annu Rev Plant Physiol Plant Mol Biol 47:541–568

Novillo F, Alonso JM, Ecker JR, Salinas J (2004) *CBF2/DREB1C* is a negative regulator of *CBF1/ DREB1B* and *CBF3/DREB1A* expression and plays a central role in stress tolerance in *Arabidopsis*. Proc Natl Acad Sci USA 101:3985–3990

Novillo F, Medina J, Salinas J (2007) *Arabidopsis CBF1* and *CBF3* have a different function than *CBF2* in cold acclimation and define different gene classes in the CBF regulon. Proc Natl Acad Sci USA 104:21002–21007

Olien CR, Smith MN (1997) Ice adhesions in relation to freeze stress. Plant Physiol 60:499–503

Oono Y, Seki M, Najo T, Narusaka M, Fujita M, Satoh R, Satou M, Sakurai T, Ishida J, Kakiyama K, Iida K, Maruyama K, Satoh S, Yamaguchi-Shinozaki K, Shinozaki K (2003) Monitoring expression profiles of Arabidopsis gene expression during rehydration process after dehydration using a ca. 7000 full length cDNA microarray. Plant J 34:868–887

Pan A, Hayes PM, Chen F, Chen THH, Blake T, Wright S, Karsai I, Bedo Z (1994) Genetic analysis of the components of winter hardiness in barley (*Hordeum vulgare* L.). Theor Appl Genet 89(7–8):900–910

Pramanik MHR, Imai R (2005) Functional identification of a trehalose-6-phosphatase gene that is involved in transient induction of trehalose biosynthesis during chilling stress in rice. Plant Mol Biol 58:751–762

Pulli S (1989) Metabolic effects of flooding in red clover and bromegrass during growth and hardening. Iceland Agric Sci 2:75–85

Pulli S, Hjortsholm K, Larsen A, Gudleifsson B, Larsson S, Kristiansson B, Hommo L, Tronsmo AM, Ruuth P, Kristensson C (1996) Development and evaluation of laboratory testing methods for winterhardiness breeding. Sweden, Nordic Gene Bank, pp 1–68

Quarrie SA, Laurie DA, Zhu JH, Lebreton C, Semikhodskii A, Steed A, Witsenboer H, Calestani C (1997) QTL analysis to study the association between leaf size and abscisic acid accumulation in droughted rice leaves and comparisons across cereals. Plant Mol Biol 35(1–2):155–165

Quinn PJ (1985) A lipid phase separation model of low temperature damage to biological membranes. Cryobiology 22:28–46

Rapacz M (2002) Regulation of frost resistance during cold deacclimation and reacclimation in oilseed rape. A possible role of PSII redox state. Physiol Plant 115:236–243

Rapacz M, Gasior D, Zwierzykowski Z, Lesnieweska-Bocianowska A, Humphreys MW, Gay AP (2004) Changes in cold tolerance and the mechanisms of acclimation of photosystem II to cold hardening generated by another culture of *Festuca pratensis* and *Lolium multiflorum* cultivars. New Phytol 162:105–114

Rapacz M, Gasior D, Humphreys MW, Zwierzykowski Z, Plazek A, Lesniewska-Bocianowska A (2005) Variation for winter hardiness generated by androgenesis from *Festuca pratensis* × *Lolium multiflorum* amphidiploid cultivars with different winter susceptibility. Euphytica 142 (1–2):65–73

Roberts JKM, Wemmer D, Ray PM, Jardetzky O (1982) Regulation of cytoplasmic and vacuolar pH in maize root tips under different experimental conditions. Plant Physiol 69:1344–1347

Robson P, Mos M, Clifton-Brown J, Donnison I (2012) Phenotypic variation in senescence in *Miscanthus*: towards optimising biomass quality and quantity. Bioenergy Res 5:95–105

Roderick HW, Morgan WG, Harper JA, Thomas HM (2003) Introgression of crown rust (*Puccinia graminis*) from meadow fescue (*Festuca pratensis*) into Italian ryegrass (*Lolium multiflorum*) and physical mapping of the locus. Heredity 91:396–400

Sandve SR, Fjellheim S (2010) Did gene family expansions during the Eocene–Oligocene boundary climate cooling play a role in Pooideae adaptation to cool climates? Mol Ecol 19 (10):2075–2088

Sandve SR, Rudi H, Asp T, Rognli OA (2008) Tracking the evolution of a cold stress associated gene family in cold tolerant grasses. BMC Evol Biol 8:245

Sanghera GS, Wani SH, Hussain W, Singh NB (2011) Engineering cold stress tolerance in crop plants. Curr Genomics 12:30–43

Satoh R, Nakashima K, Seki M, Shinozaki K, Yamaguchi-Shinozaki K (2002) ACTCAT, a novel cis-acting element for praline and hypoosmolarity-responsive expression of the *ProDH* gene encoding proline dehydrogenase in Arabidopsis. Plant Physiol 130:709–719

Scarth R, Law CN (1984) The control of the daylength response in wheat by the group 2 chromosomes. Plant Breed 92:140–150

Senser M, Beck E (1982) Frost resistance in Spruce (*Picea abies* (L.) Karst): V. Influence of photoperiod and temperature on the membrane lipids of the needles. Z Pflanzenphysiol 108:71–85

Shima S, Matsui H, Tahara S, Imai R (2007) Biochemical characterization of rice trehalose-6-phosphate phosphatases supports distinctive functions of these plant enzymes. FEBS J 274:1192–1201

Shinozaki K, Yamaguchi-Shinozaki K (2000) Molecular responses to dehydration and low temperature: differences and cross-talk between two stress signaling pathways. Plant Biol 3:217–223

Shinozaki K, Yamaguchi-Shinozaki K, Seki M (2003) Regulatory network of gene expression in the drought and cold stress responses. Plant Biol 6:410–417

Shinwari ZK, Nakashima K, Miura S, Kasuga M, Seki M, Yamaguchi-Shinozaki K, Shinozaki K (1998) An Arabidopsis gene family encoding DRE/CRT binding proteins involved in low-temperature-responsive gene expression. Biochem Biophys Res Commun 250:161–170

Skinner J, Szucs P, von Zitzewitz J, Marquez-Cedillo L, Filichkin T, Stockinger EJ, Thomashow MF, Chen THH, Hayes PM (2006) Mapping of barley homologs to genes that regulate low temperature tolerance in Arabidopsis. Theor Appl Genet 112(5):832–842

Smallwood M, Bowles JD (2002) Plants in a cold climate. Philos Trans R Soc B Biol Sci 357 (1423):831–847

Sourdille P, Snape JW, Cadalen T, Charmet G, Nakata N, Bernard S, Bernard M (2000) Detection of QTLs for heading time and photoperiod response in wheat using a doubled-haploid population. Genome 43(3):487–494

Stamp P (1984) Chilling tolerance of young plants demonstrated on the example of maize (*Zea mays* L.). J Agron Crop Sci 7(suppl):1–83

Steponkus PL (1984) Role of the plasma membrane in freezing injury and cold acclimation. Annu Rev Plant Physiol 35:543–584

Steponkus PL, Uemura M, Webb MS (1993) A contrast of the cryostability of the plasma membrane of winter rye and spring oat – two species that widely differ in their freezing tolerance and plasma membrane lipid composition. In: Steponkus PL (ed) Advance in low-temperature biology, vol 2. JAI, London, pp 211–312

Stockinger EJ, Gilmour SJ, Thomashow MF (1997) *Arabidopsis thaliana* CBF1 encodes an AP2 domain-containing transcriptional activator that binds to the C-repeat/DRE, a cis-acting DNA regulatory element that stimulates transcription in response to low temperature and water deficit. Proc Natl Acad Sci USA 94(3):1035–1040

Storlie EW, Allan RE, Walker-Simmons MK (1998) Effect of the Vrn1-Fr1 interval on cold hardiness levels in near-isogenic wheat lines. Crop Sci 38:483–488

Sui N, Li M, Zhao SJ, Li F, Liang H, Meng QW (2007) Overexpression of glycerol-3-phosphate acyl transferase gene improves chilling tolerance in tomato. Planta 226:1097–1108

Sutka J, Snape JW (1989) Location of a gene for frost resistance on chromosome 5A of wheat. Euphytica 42:41–44

Suzuki N, Bajad S, Shuman J, Shulaev V, Mittler R (2008) The transcriptional co-activator MBF1c is a key regulator of thermotolerance in *Arabidopsis thaliana*. J Biol Chem 283:9269–9275

Tamura K, Yamada T (2007) A perennial ryegrass CBF gene cluster is located in a region predicted by conserved synteny between Poaceae species. Theor Appl Genet 114(2):273–283

Tao DL, Oquist G, Wingsle G (1998) Active oxygen scavengers during cold acclimation of Scots pine seedlings in relation to freezing tolerance. Cryobiology 37:38–45

Thomas H, James AR (1993) Freezing tolerance and solute changes in contrasting genotypes of *Lolium perenne* L. acclimated to cold and drought. Ann Bot 72:249–254

Thomashow MF (1990) Molecular genetics of cold acclimation in higher plants. In: Scandalios JG (ed) Advances in genetics, genomic responses to environmental stress, vol 28. Academic, New York, pp 99–131

Thomashow MF (1998) Role of cold-responsive genes in plant freezing tolerance. Plant Physiol 118:1–8

Thomashow MF (1999) Plant cold acclimation: freezing tolerance genes and regulatory mechanisms. Annu Rev Plant Physiol Plant Mol Biol 50:571–599

Tremblay K, Ouellet F, Fournier J, Danyluk J, Sarhan F (2005) Molecular characterization and origin of novel bipartite cold-regulated ice recrystallization inhibition proteins from cereals. Plant Cell Physiol 46(6):884–891

Uemura M, Steponkus PL (1994) A contrast of the plasma membrane lipid composition of oat and rye leaves in relation to freezing tolerance. Plant Physiol 104:479–496

Uemura M, Steponkus PL (1997) Artificial manipulation of the intracellular sucrose content alters the incidence of freeze-induced membrane lesions of isolated protoplasts of *Arabidopsis thaliana*. Cryobiology 35:336

Vagujfalvi A, Galiba G, Cattivelli L, Dubcovsky J (2003) The cold-regulated transcriptional activator Cbf3 is linked to the frost-tolerance locus Fr-A2 on wheat chromosome 5A. Mol Genet Genomics 269(1):60–67

Valluru R, Van den Ende W (2008) Plant fructans in stress environments: emerging concepts and future prospects. J Exp Bot 59:2905–2916

Valluru R, Lammens W, Claupein W, Van den Ende W (2008) Freezing tolerance by vesicle-mediated fructan transport. Trends Plant Sci 13:409–414

Vartapetian BB, Jackson B (1997) Plant adaptations to anaerobic stress. Ann Bot 79(Suppl A):3–20

Vogel JT, Zarka DG, van Anbuskirk HA, Fowler SG, Thomashow MF (2005) Roles of the CBF2 and ZAT12 transcription factors in configuring the low temperature transcriptome of Arabidopsis. Plant J 41:195–211

Wagner W, Wiemken A, Matile P (1986) Regulation of fructan metabolism in leaves of barley (*Hordeum vulgare* L. cv. Gerbel). Plant Physiol 81:444–447

Wang W, Vinocur B, Altman A (2003) Plant responses to drought, salinity and extreme temperatures: towards genetic engineering for stress tolerance. Planta 218:1–14

Wani SH, Sandhu JS, Gosal SS (2008) Genetic engineering of crop plants for abiotic stress tolerance. In: Malik CP, Kaur B, Wadhwani C (eds) Advanced topics in plant biotechnology and plant biology. MD Publications, New Delhi, pp 149–183

Watts WR (1972) Role of temperature in regulation of leaf extension in *Zea mays*. Nature 229:46–47

Welsh JR, Keim DL, Pirasteh B, Ricarhds RD (1973) Genetic control of photoperiod response in wheat. In: Sears ER, Sears LMS (eds) Proceedings of the 4th international wheat genetics symposium. University of Missouri, Columbia, MO, pp 879–884

Welti R, Li W, Li M, Sang Y, Biesiada H, Zhou H-E, Rajashekar CB, Williams TD, Wang X (2002) Profiling membrane lipids in plant stress responses. Role of phospholipase in freezing induced lipid changes in Arabidopsis. J Biol Chem 277:31994–32002

Williams WP (1990) Cold-induced lipid phase transitions. Philos Trans R Soc Lond 326:555–570

Worrall D, Elias L, Ashford D, Smallwood M, Sidebottom C, Lillford P, Telford J, Holt C, Bowles D (1998) A carrot leucine-rich-repeat protein that inhibits ice recrystallization. Science 282 (5386):115–117

Xin Z, Browse J (1998) Eskimo1 mutants of Arabidopsis are constitutively freezing-tolerant. Proc Natl Acad Sci USA 95:7799–7804

Xin Z, Li PH (1993) Relationship between proline and abscisic acid in the induction of chilling tolerance in maize suspension-cultured cells. Plant Physiol 103(2):607–613

Xin Z, Mandaokar A, Chen J, Last RL, Browse J (2007) *Arabidopsis ESK1* encodes a novel regulator of freezing tolerance. Plant J 49:786–799

Yamaguchi-Shinozaki K, Shinozaki K (2006) Transcriptional regulatory networks in cellular responses and tolerance to dehydration and cold stresses. Annu Rev Plant Biol 57:781–803

Zachariassen KE, Kristiansen E (2000) Ice nucleation and antinucleation in nature. Cryobiology 41(4):257–279

Zachos J, Pagani M, Sloan L, Thomas E, Billups K (2001) Trends, rhythms, and aberrations in global climate. Ma to present. Science 65:686–693

Zhang YJ, Yang JS, Guo SJ, Meng JJ, Zhang YL, Wan SB, He QW, Li XG (2011) Over-expression of the Arabidopsis CBF1 gene improves resistance of tomato leaves to low temperature under low irradiance. Plant Biol 13(2):362–367

Zhu J, Shi H, Lee BH, Damsz B, Cheng S, Strim V, Zhu JK, Hasegawa PM, Bressan RA (2004) An Arabidopsis homeodomain transcription factor gene, HOS9, mediates cold tolerance through a CBF-independent pathway. Proc Natl Acad Sci USA 101:9873–9878

Zhu J, Verslues PE, Zheng X, Lee BH, Zhan X, Manabe Y, Sokolchik I, Zhu Y, Dong CH, Zhu JK, Hasegawa PH, Bressan RA (2005) *HOS10* encodes an R2R3-type MYB transcription factor essential for cold acclimation in plants. Proc Natl Acad Sci USA 102:9966–9971

# Chapter 4
# Heat Tolerance

Timothy G. Porch and Anthony E. Hall

**Abstract** Predicted global warming would make it more difficult for farmers to achieve the increases in crop productivity needed to meet expected increases in demand for food during this century—because an increase in temperature of 1 °C has been shown to decrease grain production of some annual crop species by about 10 %. In considering strategies for breeding heat-resistant cultivars that have greater yield than current cultivars under hot conditions, high-temperature effects on germination, vegetative growth, reproductive development, and yield are reviewed. For several annual crop species, pollen development and seed or fruit set have been shown to be particularly sensitive to high temperatures occurring in the late-night to early-morning period. The few studies that have been conducted indicated that elevated atmospheric carbon dioxide concentration will not enable plants to overcome this problem. For a few crop species, heat-resistant cultivars have been bred by conventional hybridization and selection for heat tolerance during reproductive development and/or yield. The progress that has been made in breeding for heat resistance in cowpea, common bean, cotton, tomato, rice, and wheat are reviewed. The successes achieved in breeding with these crops using conventional hybridization and selection provide guidelines whereby further progress can be made in increasing the heat resistance of these and other crop species. For the future, DNA markers for what appear to be major genes conferring heat tolerance during reproductive development would be valuable because their use in selection could substantially enhance the efficiency whereby heat-resistant cultivars

T.G. Porch (✉)
USDA-ARS Tropical Agriculture Research Station, 2200 P.A. Campos Avenue, Suite 201, Mayaguez, PR 00680, USA
e-mail: Timothy.Porch@ars.usda.gov

A.E. Hall
Department of Botany and Plant Sciences, University of California, Riverside, CA 92521-0124, USA

2922 Lindsay Lane, Quincy, CA 95971-9602, USA

C. Kole (ed.), *Genomics and Breeding for Climate-Resilient Crops*, Vol. 2,     167
DOI 10.1007/978-3-642-37048-9_4, © Springer-Verlag Berlin Heidelberg 2013

are bred. More upstream research on the development of crops with facultative apomictic breeding systems is warranted. Cultivars with an appropriate type of apomixis could have tolerance to the many stresses that damage reproductive development including chilling and drought, in addition to heat, because these cultivars do not require pollen development to achieve seed production. Apomictic cultivars have additional values including the ability of hybrids to produce true-breeding seed permitting the development of hybrid cultivars for crop species where it currently is not economically feasible.

## 4.1  Introduction

Predicted global warming would make it more difficult for farmers to achieve the increases in crop productivity (yield per unit area) needed to meet expected increases in demand for food and animal feed during this century. Due to projected increases in human population and the need for people who currently are poorly fed to receive more, higher-quality food, farmers will need to produce twice as much food and feed crops on about the same area of cultivated land by the end of this century (World Bank 2010). It may be difficult to achieve this goal because the productivity of some crops has not increased in recent years.

Doubling productivity of food and feed crops in about 69 years would require yield increases of 1 % per year. During these 69 years it has been predicted that temperatures could increase several degrees celcius, and high temperatures already are reducing yields of some crops in some climatic zones. For example, rice (*Oryza sativa* L.) yields for crops grown under optimal management on the experimental farm of the International Rice Research Institute in the Philippines over a 12-year period from 1992 through 2003 were negatively correlated with temperatures during the growing seasons (Peng et al. 2004). Day temperatures had increased 0.35 °C while night temperatures had increased 1.13 °C from 1979 to 2003. Grain yields exhibited a 10 % decrease per °C increase in night temperature with no correlation with day temperature. This indicates that increases in temperature of 0.05 °C per year, equivalent to 3.5 °C over 69 years, could cause yield decreases of 0.5 % per year. Apparently, the development of heat-resistant cultivars that produce greater yields than current cultivars under hot conditions could contribute to maintaining or hopefully increasing productivity during a period of global warming.

As will be discussed later, heat-resistant cultivars of a few crop species have been developed by selecting for greater heat tolerance during reproductive development. But, in general, there has been little emphasis on selecting for heat tolerance in public plant breeding programs. We do not know the extent of selection for heat tolerance in commercial plant breeding programs, but it likely has not been done to a major extent, since virtually none of these programs advertise their cultivars as having heat resistance.

In this review we will examine plant responses to high temperature during germination, vegetative growth, and reproductive development to provide a physiological basis for choosing selection criteria in breeding for heat tolerance in annual crops. We will emphasize those crops that produce grain and/or fruit because most of the earlier research on heat tolerance has been conducted with these types of crops, and the cereals and grain legumes are major sources of food and feed. Studies of these types of crops provide an opportunity to determine whether heat stress is reducing reproductive yield through detrimental effects on the photosynthetic source of carbohydrate or through damage to reproductive development. Whether heat stress is mainly damaging the source or the sink or both processes will determine which heat tolerance selection strategies are most likely to enhance heat resistance.

We will review methods for breeding for heat tolerance building on general reviews of this topic by Singh et al. (2011) and Hall (1992), a web site on breeding for heat tolerance (http://www.plantstress.com), and reviews of specific crops: cotton (*Gossypium hirsutum* L.) by Singh et al. (2007), cowpea (*Vigna unguiculata* L. Walp.) by Hall (2011), rice by Lafarge et al. (2011), and wheat (*Triticum* spp.) by Trethowan and Mahmood (2011). Finally, we will describe those programs we are aware of where breeding for heat tolerance has made significant progress in developing heat-resistant cultivars.

## 4.2 Plant Responses to High Temperatures

Relatively hot temperatures can impair plant function or development through either direct effects of high tissue temperature or indirect effects of the high evaporative demand and water stress that accompany hot weather. Evaporative demand exhibits a near exponential increase with increase in temperature and can result in large increases in transpiration and substantial decreases in leaf water potential (Hall 2001). Through transpirational cooling and traits influencing radiation loading, plants can avoid high tissue temperatures. This review, however, will mainly consider the direct effects of high tissue temperature on plants.

Hot temperatures can have either reversible effects on plant function or irreversible damaging effects on plant development and/or function. This review will consider only the irreversible damaging effects of heat stress on plants. The magnitude of heat stress depends on intensity (temperature), duration of exposure, and rate of increase, because plants have some ability to acclimate and more rapid increases in temperature can be more damaging. Threshold temperatures where heat stress begins in natural environments are particularly relevant to understanding effects on crops under field conditions.

## 4.2.1   Germination and Seedling Survival Under Heat Stress

In the semiarid tropics, inadequate seedling establishment due to heat stress can reduce productivity and stability of production of sorghum (*Sorghum bicolor* L. Moench) (Peacock 1982) and pearl millet (*Pennisetum glaucum* L. R. Br.) (Peacock et al. 1993). Inadequate seedling establishment has occurred when soil seed-zone temperatures exceed 45 °C and soil surface temperatures exceed 55 °C. This could be due to failed germination, inhibited epicotyl emergence from the soil, or death of seedlings caused by heat girdling (Peacock et al. 1990).

Maximum threshold temperatures for germination and emergence are higher for warm-season annuals than for cool-season annuals (as defined in Hall 2001). For example, the maximum threshold seed-zone temperature for cowpea is about 37 °C (El-Kholy et al. 1997), compared with about 33 °C for lettuce (*Lactuca sativa* L.) (Argyris et al. 2005). Inadequate seed germination due to heat stress can be a problem for cool-season crops such as lettuce when they are sown in late summer to accommodate a fall harvest (Borthwick and Robbins 1928).

Maximum threshold temperatures at which high temperatures kill seedlings can depend on plant preconditioning. Seedlings subjected to high but sublethal temperatures for a few hours subsequently can survive higher temperatures than seedlings that have been maintained at moderate temperatures (Yarwood 1961). This acclimation to heat can be induced by the gradual diurnal increases in temperature that occur in hot natural environments (Vierling 1991). The "heat-shock" response involves repression of the synthesis of most normal proteins and mRNAs, and the initiation of transcription and translation of a small set of heat-shock proteins (Vierling 1991). Studies of loss-of-function mutants of *Arabidopsis thaliana* demonstrated that the enhanced thermotolerance can be associated with at least three independent effects: the synthesis of a novel set of proteins (specifically Hsp101), protection of membrane integrity, and recovery of protein activity/synthesis (Queitsch et al. 2000; Hong et al. 2003). The heat-shock protein Hsp101 likely functions as a molecular chaperone in the renaturation of cellular proteins that have a tendency to unfold and aggregate at very high temperatures. With increases in temperature, membranes become more fluid and electrolytes leak more readily from plant tissues. Heat acclimation reduces the tendency for membranes to leak under hot conditions.

Variation occurs among species in the maximum threshold temperatures that result in the death of seedlings. Among the cool-season annuals, pea (*Pisum sativum* L.) is very sensitive, dying when daytime temperatures exceed about 35 °C for sufficient duration, whereas barley (*Hordeum vulgare* L.) can withstand hotter temperatures. Warm-season annuals usually can withstand higher temperatures than cool-season annuals. For example, cowpea and cotton can survive about the highest temperatures experienced in crop production zones (maximum daytime air temperatures in a weather station shelter of 50 °C) and produce substantial amounts of vegetative biomass providing the crops have an adequate supply of water.

Germination, emergence, and seedling survival are relatively simple systems, and methods have been developed that can screen large numbers of plants to detect genotypic differences in heat tolerance (Wilson et al. 1982; Soman and Peacock 1985; Peacock et al. 1993). This type of selection, however, only will confer a useful level of heat resistance for those field conditions where heat stress has major detrimental effects on plant stands because many crops have some capacity to exhibit compensatory growth when the population of seedlings is low.

## 4.2.2  Effects of High Temperatures on Vegetative Growth and Development

The rate of vegetative growth decreases when canopy photosynthesis is reduced by heat stress. If a maximum threshold temperature is imposed on a plant for sufficient duration, the photosynthetic system is damaged such that the rate of $CO_2$ uptake is substantially reduced and the damage may be irreversible. Sensitivity of photosynthesis to heat may be primarily due to damage to components of photosystem II (PSII) located in the thylakoid membranes of the chloroplast (Al-Khatib and Paulsen 1999). The extreme sensitivity of PSII may be due to effects of temperature on the membranes in which it is embedded. Murakami et al. (2000) developed transgenic tobacco (*Nicotiana tabacum* L.) plants with altered chloroplast membranes by silencing the gene encoding chloroplast omega-3 fatty acid desaturase. The transgenic plants had less trienoic fatty acids and more dienoic fatty acids in their chloroplasts than the wild-type plants. The transgenic plants also had greater photosynthesis and growth rates than the wild-type plants in hot environments. In addition, there are indications that a small heat-shock protein in the chloroplast may protect PSII electron transport during heat stress (Heckathorn et al. 1989).

In comparisons of contrasting species, PSII of the cool-season species wheat was more sensitive to heat than PSII of either rice or pearl millet, which are warm-season species adapted to much hotter climates (Al-Khatib and Paulsen 1999). Portable instruments are available to rapidly measure chlorophyll fluorescence parameters of intact leaves that could be used to screen many plants to quantify the extent that PSII has been damaged and hopefully detect genotypic differences in heat tolerance. Selection of this type with grain crops only will enhance heat resistance where grain yield is being limited by heat-stress effects on the supply of carbohydrate through effects on PSII.

For spring wheat growing in hot, irrigated environments, cultivar differences in grain yield were positively correlated with photosynthetic rate per unit leaf area (Reynolds et al. 1994). Even stronger positive associations were observed between grain yield and stomatal conductance. This indicates that the more open stomata of the heat-resistant cultivars may be enhancing photosynthesis both by facilitating the diffusion of $CO_2$ into leaves and by enhancing transpirational cooling bringing leaf temperatures below damage thresholds. In addition, cultivar differences in grain yield of spring wheat growing in a hot, irrigated environment were positively correlated with the number of kernels per spike (Shpiler and Blum 1991). This could be explained by either cultivar differences in heat tolerance during reproductive development or the possibility that processes determining kernel number may be linked to photosynthesis. Cultivar variation in kernel number was correlated with spike dry weight at anthesis and the ratio of solar irradiance to air temperature for the 30-day period prior to anthesis (Fischer 1985). Consequently, there are circumstances where damaging effects of heat on photosynthesis can reduce both the photosynthetic source and the reproductive sink, making it difficult to determine which of them is most limiting to grain yield. Also, pollen development in wheat is very sensitive to heat stress (see for review, Dolferus et al. 2011; Farooq et al. 2011).

The source versus sink issue concerning the limiting effects of heat stress on yield is further complicated by the possibility that photosynthetic capacity and stomatal behavior may be influenced by the extent of the reproductive sink through complex, long-term feedback effects. Pima cotton (*Gossypium barbadense* L.) cultivars were bred that have greater boll yields in hot environments by selecting plants with the ability to set more bolls on low nodes under very hot irrigated conditions (Feaster and Turcotte 1985). Subsequent studies showed that these heat-resistant cultivars also had greater rates of photosynthesis and higher stomatal conductances in hot conditions (Cornish et al. 1991). Plants that have higher photosynthetic capacity often have higher maximal stomatal conductance, and the mechanism of this long-term regulation is not known (Hall 2001). The mechanisms whereby the cotton cultivars are heat-resistant are not known. The simplest explanation is that they have greater ability to set bolls under hot conditions, because this is what they were selected for. But then why do they also have higher photosynthetic rates? Is this a consequence of or a cause for the greater boll set?

When grains are developing, leaves begin to senesce and this can be accelerated by late-season heat stress. Delayed-leaf-senescence traits have been sought as a means to enhance grain filling and lengthen the reproductive period under late heat stress or drought. This trait can be easily screened for visually in field nurseries providing one only selects plants that have both delayed leaf senescence and abundant fruit and/or grain because plants that have low fruit and/or grain set typically also exhibit a type of delayed leaf senescence that has limited agronomic value. Cultivars and genetic lines with delayed-leaf-senescence traits have been bred for sorghum (Reddy et al. 2011), cowpea (Gwathmey and Hall 1992), and wheat (Farooq et al. 2011). The extent to which they enhance heat resistance has not been adequately quantified.

## 4.2.3 Reproductive Development Under Heat Stress

In very hot environments many crop plants produce significant amounts of biomass but few flowers, fruits, or seeds. In these cases, reproductive development clearly is more sensitive to heat than photosynthesis and biomass production.

### 4.2.3.1 Flowering Under Heat Stress

Some emphasis is given to cowpea because much work has been done on heat-stress effects on flowering with this crop and several other warm-season crops appear to have similar responses (Hall 2004). Under long-day very hot field conditions, some cowpea genotypes produced floral buds, but they did not produce any flowers (Ehlers and Hall 1996). Prior to discussing growth chamber studies of this heat-stress effect, it is important to note that it depends on light quality. Floral bud development of a heat-sensitive cowpea genotype was arrested by high night temperatures and long days only in growth chambers that had red (655–665 nm)/far-red (725–735) ratios of 1.3–1.6 but not with an R/FR ratio of 1.9 (Ahmed et al. 1993b). Note that sunlight has an R/FR ratio of 1.2 above the canopy and a lower ratio within dense canopies, and suppression of floral buds does occur with heat-sensitive genotypes in very hot long-day field conditions. Also note that growth chambers that mainly use either fluorescent or metal-halide lamps and have a high F/R ratio often have been used in plant growth studies and could have produced results that are not relevant to natural environments.

The suppression of floral buds under high night temperatures and long days appears to be a phytochrome-mediated effect, except that a night-break of red light during a long night did not result in floral bud suppression (Mutters et al. 1989b). Suppression of floral buds was greater under a combination of hot nights and very hot days than under only hot nights (Dow El-Madina and Hall 1986). Experiments involving the transfer of plants between growth chambers with optimal or hot night temperatures for different periods demonstrated that plants did not have a particular stage of development where they were sensitive to high temperature but that the duration of the heat experience may be critical for the suppression of floral bud development (Ahmed and Hall 1993). A period of 2 weeks or more of consecutive or interrupted hot nights during the first 4 weeks after germination caused complete suppression of the first five floral buds on the main stem. The minimum day length required to elicit heat-induced suppression of floral bud development may be as short as 13 h, including civil twilight. Days that are longer than this minimum occur in all subtropical and some tropical zones during the season when cowpeas are grown. We are using the definitions of subtropical and tropical zones of Hall (2001) that are based on temperature.

It should be noted that genotypes of cowpea with "classical" sensitivity to photoperiod do not produce floral buds under long days. Under inductive short days and high night temperature, a cultivar with "classical" sensitivity to photoperiod produced floral buds and the buds were not sensitive to heat and developed normally, producing flowers (Mutters et al. 1989b).

In genetic studies, segregation of cowpea plants for heat tolerance during floral bud development was consistent with the hypothesis that it is conferred by a single recessive nuclear gene (Hall 1993). The presence or absence of flowers is easy to screen for in field nurseries, and subsequent breeding demonstrated that this trait can be reliably selected in the $F_2$ generation (Hall 2011) supporting the argument that the inheritance is controlled by a single recessive gene. Multiple forms of phytochrome are present in plants that have different physiological roles (Smith and Whitelam 1990). The dominant nuclear gene involved in conferring sensitivity to heat during floral bud development in cowpea might be involved in the synthesis of one of these phytochromes (Hall 1992).

Some cultivars of common bean (*Phaseolus vulgaris* L.) have been shown to exhibit floral bud suppression under hot long days (Konsens et al. 1991; Shonnard and Gepts 1994). The sensitivity to heat during early floral bud development of common bean may only occur in long days and may be consistent with the action of a single dominant gene (White et al. 1996) as it is with cowpea. Pima cotton also has been shown to not produce flowers under very hot conditions (Reddy et al. 1992).

### 4.2.3.2   Fruit and/or Seed Set Under Heat Stress

Growth chamber studies of heat-stress effects on cowpea produced unexpected results. Pod set was reduced much more by subjecting shoots to high temperature at night than when a higher temperature was imposed on shoots during the day (Warrag and Hall 1984a, b). The sensitivity of pod set in cowpea to high night temperature was confirmed under field conditions (Nielsen and Hall 1985b). In this study a system was used in which plots of cowpea were enclosed in plastic sheets only during the nighttime hours and a fan, air distribution system, heater, and differential thermostat were used to increase air temperature in the enclosure a fixed number of degrees above ambient air temperature (Nielsen and Hall 1985a). Growth chamber studies demonstrated that the pod set of cowpea was sensitive to heat during the last 6 h but not the first 6 h of the night (Mutters and Hall 1992).

Fruit and/or seed set has been shown to be particularly sensitive to high night temperature in several other warm-season crops including cotton (Gipson and Joham 1968), sorghum (Eastin et al. 1983), rice (Peng et al. 2004; Mohammed and Tarpley 2010), and common bean (Konsens et al. 1991). For peanut (*Arachis hypogaea* L.), high temperatures during the morning reduced fruit set, whereas higher temperatures during the afternoon had no effect on fruit set (Prasad et al. 2000). Apparently, fruit and/or seed set can be damaged by high temperatures occurring during the late night or early morning while they are not sensitive to much higher temperatures occurring later in the day. This late-night or early-morning sensitivity to heat stress has implications for the mechanism of the effect, choice of screening environments, and predictions concerning how global warming might influence productivity under field conditions.

The lack of pod set in cowpea under high night temperature was due to impaired pollination and a lack of seed set (Warrag and Hall 1984b). Artificial pollination studies demonstrated that the female part of the flower, the pistil, was not damaged by high night temperature (Warrag and Hall 1983). Reciprocal transfers of plants between growth chambers with high or optimal night temperatures demonstrated that the stage of floral development most sensitive to heat stress occurred 9–7 days before anthesis (Ahmed et al. 1992). The sensitive stage is after meiosis, which occurs 11 days before anthesis in cowpea. Damage occurred at the time when the tetrads were being released from the microspore mother cell sac (Warrag and Hall 1984b; Ahmed et al. 1992; Mutters and Hall 1992). In common bean, microsporogenesis was also found to be the period during reproductive development most sensitive to high-temperature stress using scanning and transmission electron microscopy, pollen viability, and yield (Porch and Jahn 2001). Premature degeneration of the tapetal tissue and lack of endothecium formation were observed in cowpea (Ahmed et al. 1992) and common bean (Porch and Jahn 2001), which could have been responsible for the low pollen viability, low anther dehiscence, and low pod set under high night temperature (Ahmed et al. 1992). Tapetal tissue plays an important role in providing nutrients to developing pollen grains, and its premature degeneration could thereby stunt pollen development. Mutters et al. (1989a) hypothesized that heat injury during floral development of sensitive cowpea genotypes may be due to reduced translocation of proline from anther walls and tapetal tissue to developing pollen, based on studies of cowpea genotypes with contrasting heat tolerance in field environments with either very hot or more optimal thermal regimes. Large amounts of proline are required for pollen development, pollen germination, and pollen tube growth. Tapetal malfunction has been considered the causal mechanism of some of the genetic male sterility occurring in plant species (Dundas et al. 1981; Nakashima et al. 1984).

High night temperatures damaged pollination in other warm-season crops including rice (Mohammed and Tarpley 2009), common bean (Gross and Kigel 1994; Porch and Jahn 2001), pepper (*Capsicum annuum* L.) (Wien 1997), sorghum (Eastin et al. 1983; Prasad et al. 2008), and cotton (Singh et al. 2007). High night and/or day temperatures damaged pollination in peanut (Prasad et al. 1999a, b), maize (*Zea mays* L.) (Herrero and Johnson 1980), and tomato (*Lycopersicon esculentum* Mill.) (Peet et al. 1998). Pollination in the cool-season crop wheat is particularly sensitive to high temperatures (Saini et al. 1984) and also may involve tapetal malfunction (Dolferus et al. 2011). It is likely that pollen development of many crop species is sensitive to high temperatures.

High night temperatures can be more damaging to pod set of cowpeas under the long days typical of subtropical zones than under the short days that can occur in the tropics (Ehlers and Hall 1998). Responses to red light during long nights, far-red light at the end of long days, and far-red followed by red light at the end of long days indicate that the greater sensitivity of cowpea to high night temperature under long days is a phytochrome-mediated effect (Mutters et al. 1989b). For rice also the detrimental effects of high temperatures on pollination are enhanced under long days, and this effect provides the basis for the photoperiod-sensitive genic male sterile system that has been used in the production of hybrid seed (Ziguo and Hanlai 1992; Yuan et al. 1993).

Why is the seed set of many warm-season crops sensitive to high temperatures during the late night or early morning, while they are not sensitive to higher temperatures occurring during midday and afternoon? Mutters and Hall (1992) hypothesized that there is a heat-sensitive physiological or developmental process in pollen development that is under circadian control. Note that reproductive events, such as anthesis, meiosis, and flower opening, and phytochrome-mediated events, have a degree of circadian control occurring at a particular time in the 24-h cycle. They hypothesized that natural selection would have favored plants in which the heat-sensitive process in pollen development takes place in the coolest part of the 24-h cycle, which is the late night and early morning.

Genetic control of the timing of reproductive events that has relevance to heat tolerance has been demonstrated. The time of day for flowering of different genotypes within the genus *Oryza* ranges widely from early morning to evening with many cultivars of rice exhibiting anthesis between 10 A.M. and noon. An early-morning flowering trait was introgressed from a wild rice that begins flowering at 6 A.M. into a rice cultivar with later flowering. An early-morning flowering line was bred and found to flower earlier and exhibit less spikelet sterility than the parental cultivar, which flowered when temperatures were hotter (Ishimaru et al. 2010).

In hot screening environments, fruit set of grain legumes, tomatoes, and cotton can be scored visually. For example, the number of pods per peduncle is a useful criterion for pod set in cowpea. For cereals, seed set can be scored visually. Additional criteria may be used, such as the proportion of anthers that dehisce or the extent of pollen viability.

### 4.2.4   Influence of Elevated Atmospheric Carbon Dioxide Concentration on Plant Responses to High Temperature

Analyses of air trapped in polar ice indicate that, prior to the year 1800, the atmospheric $CO_2$ fluctuated between 180 and 290 ppm for at least 220,000 years (Hall and Allen 1993). Since 1800, ice core data indicate increases in $CO_2$ from 280 to 315 ppm by 1958. Direct measurements of $CO_2$ indicate accelerating increases since 1958 from 315 to 380 ppm by the early 2000s. It has been predicted that atmospheric $CO_2$ could exceed 600 ppm by the end of this century.

Plants with the $C_4$ photosynthetic system evolved during an early period after atmospheric $CO_2$ became low, and this system represents a specific adaptation to the low $CO_2$ of the last 200,000 years. The extent and nature of the evolution of plants with the $C_3$ photosynthetic system, with respect to the low $CO_2$, are not known. It is likely that low $CO_2$ over 220,000 years resulted in evolutionary modifications to whole plant processes, such as increases in the ratio of the photosynthetic source to carbohydrate sink tissues. Consequently, some $C_3$ plants may not be well adapted to future or even present day levels of $CO_2$ due to inadequate investment in sink tissues (Hall and Ziska 2000). Photosynthetic rates of these $C_3$ plants may increase as $CO_2$ increases, but the rates may not increase as much as they would if the plants were adapted to function optimally at elevated $CO_2$. Progress during the twentieth century in increasing the productivity of several $C_3$ crops through plant breeding was estimated as mainly (77 %) resulting from increases in harvest index (the ratio of grain yield to total shoot biomass) with only 23 % due to increases in total shoot biomass (Gifford 1986). This indicates that these crops had an inadequate reproductive sink, for agricultural purposes, and with further increases in atmospheric $CO_2$, the reproductive sink may become even more incapable of supporting the full photosynthetic potential.

In early studies using controlled-environment enclosures, doubling $CO_2$ increased grain yield of various cereal crops by 32 % and of various grain legumes by 54 % at intermediate temperatures (Kimball 1983). More recent studies with free-air $CO_2$ enrichment (FACE) experiments under field conditions, however, resulted in grain yield responses to elevated $CO_2$ that were about 50 % lower than those obtained using enclosures (Leakey et al. 2009). FACE experiments provide crop responses that are more relevant to farming because the crops are grown under natural open-air field conditions. Yield increases in the FACE studies were less than the increases in photosynthesis that occurred with short-term doubling of $CO_2$ at the same temperature (Poorter 1993; Allen 1994). Limitations in sink demand were apparent in the FACE experiments where only a small proportion of the increase in photosynthate supply was partitioned to grain (Leakey et al. 2009). The results of the FACE experiments support the hypothesis that crop plants are not well adapted to the higher $CO_2$ likely to occur by the end of this century.

The influence of elevated atmospheric $CO_2$ is examined for those crops whose reproductive development is sensitive to high temperature under the current $CO_2$ concentration. For soybean (*Glycine max* L. Merr.) growing under controlled-environment field conditions, harvest index progressively decreased with an increase in temperature under either 330 or 660 ppm $CO_2$. Harvest index was lower with elevated $CO_2$ indicating a more severe imbalance between the reproductive sink and the photosynthetic supply (Baker et al. 1989). Reproductive development of Pima cotton can be so sensitive to high temperatures that the plants do not produce either flowers or bolls (Reddy et al. 1992). Studies in naturally sunlit, controlled-environment chambers demonstrated that at elevated $CO_2$ of 700 ppm, Pima cotton still did not produce any flowers or bolls when subjected to high temperatures (Reddy et al. 1995, 1997). Controlled-environment field studies with rice demonstrated that grain yield decreased 10 % per °C increase in average temperature above 26 °C at $CO_2$ of either 330 or 660 ppm (Baker and Allen 1993). The decrease in grain yield was mainly due to fewer grains per panicle. High day and night temperatures can cause decreases in viability of pollen grains at anthesis and decreases in seed set in rice (Ziska and Manalo 1996). Elevated $CO_2$ aggravated these heat-stress effects, causing a 1 °C decrease in the maximum threshold canopy surface temperature after which the percentage of spikelets having ten or more germinated pollen grains exhibited a precipitous decline (Matsui et al. 1997). Heat-induced increases in floret sterility may have been responsible for the downregulation of photosynthesis observed in rice under high temperatures and elevated $CO_2$ through indirect effects associated with reductions in reproductive sink strength (Lin et al. 1997).

A unique insight into the interactive effects of high night temperature and elevated $CO_2$ was obtained from studies with cowpea genotypes that are either heat tolerant or heat sensitive during reproductive development. With high night temperatures, many cowpea genotypes are heat sensitive and do not produce any flowers, while others are partially heat tolerant and produce flowers but no pods and a few with complete heat tolerance produce both flowers and pods (Ehlers and Hall 1996). In growth chamber studies with three contrasting genotypes under high night temperature, a heat-sensitive genotype did not produce any flowers and a partially heat-tolerant genotype produced flowers but did not set any pods under either 350 or 700 ppm $CO_2$ (Ahmed et al. 1993a). Interestingly, a completely heat-tolerant genotype had greater pod production under elevated $CO_2$ at both high and more optimal night temperatures than a genetically similar cultivar that does not have the heat-tolerance genes (Ahmed et al. 1993a).

These studies indicate that elevated atmospheric $CO_2$ will not overcome heat-stress effects on reproductive development. This supports the argument that these effects are direct and are not mediated by heat-stress effects on the photosynthetic source. For those many crops that are sensitive to heat during reproductive development, incorporating heat tolerance may also enhance their yield responses to elevated atmospheric $CO_2$ over a range of temperatures (Hall and Allen 1993; Hall and Ziska 2000). This important hypothesis should be more completely tested using field conditions with additional cultivars of cowpea and other heat-tolerant crop species. Hall and Allen (1993) hypothesized that cultivars with heat tolerance during reproductive development, high harvest index, high photosynthetic capacity per unit leaf area, small leaves, and low leaf area per unit ground area will be most responsive to elevated $CO_2$ under both hot and intermediate temperatures.

### 4.2.5   Effects of High Temperature on Grain Yield

Populations of plants under field conditions must be used when seeking estimates of the effects of heat stress on grain yield that are relevant to agriculture. In studies that use year-to-year variation in temperature or different locations with contrasting thermal regimes, other aspects of the environment also vary. In the study of variation in rice yield over 12 years on the experimental farm of the International Rice Research Institute (Peng et al. 2004), night temperature, day temperature, and solar radiation level varied. The authors used partial correlation analysis to establish that grain yield decreased 10 % per °C increase in night temperature independent of the changes in solar radiation. Ismail and Hall (1998) evaluated grain yields of six pairs of cowpea lines, with differences in heat tolerance during reproductive development but with simliar genetic background, growing under optimal irrigated management in four field environments with contrasting thermal regimes over 2 years. The solar radiation levels in these eight field environments were high and similar. Heat-susceptible lines exhibited a 13.6 % decrease in grain yield per °C increase in average minimum night temperature between emergence and first flowering above a threshold of 16 °C. The reduction in grain yield was mainly due to decreases in pod set and harvest index. In the three environments with the highest night temperatures, the heat-resistant lines had 54 % higher grain yield than the heat-susceptible lines mainly due to greater pod set and harvest index.

A reliable method to determine the effects of increases in temperature on productivity is to subject different plots of plants growing in the field to increases in temperature without otherwise disturbing the environment. This can be more readily achieved for increasing night temperature than for increasing day temperature. In one of the first studies of this type, the night temperature of sorghum plots was increased 5 °C for 1 week during floret differentiation and there was a 28 % decrease in grain yield associated with a 30 % decrease in the number of grains (Eastin et al. 1983). Plots of cowpea plants were subjected to elevated night temperatures during early-flowering stages using enclosures that were only imposed at night as was described previously (Nielsen and Hall 1985a). The cowpea plants exhibited a 4.4 % decrease in grain yield per °C increase in nighttime temperature above a threshold daily minimum temperature of 15 °C (Nielsen and Hall 1985b). The reductions in grain yields were due to reductions in the proportions of flowers producing pods. When rice plots were subjected to a 4 °C increase in temperature, during both night and day using open-top chambers, there was a decrease in grain yield of 18 % (Moya et al. 1998) due to reductions in spikelet and pollen fertility (Matsui et al. 1997).

It is clear that heat stress especially at night can decrease grain and/or fruit yield. Reduced seed and/or fruit set due to a lack of pollination are often responsible for this effect. Experience with breeding for heat tolerance during reproductive development has shown that it can overcome this problem.

## 4.3   Breeding and Selection Methods

Development of heat-resistant cultivars has relied principally on classical breeding methods using available genetic variation. Although heat-tolerant genotypes have been identified in landrace and cultivar collections, the proportions of heat-tolerant accessions have been low, such as in tomato (Opeña et al. 1989), cowpea (Patel and Hall 1990), the Sudan core of the sorghum collection (Dahlberg et al. 2004), a Mexican wheat collection (Hede et al. 1999), wheat cultivars (Reynolds et al. 1994), cucumber (Staub and Krasowska 1990), chickpea (Krishnamurthy et al. 2011), and the common bean core collection (Porch TG, unpublished). Wild species of crop plants provide a mainly untapped source of additional genetic diversity. For example, collection of wild *Triticum* and *Aegilops* species led to the identification of heat-tolerant accessions originating from specific high-temperature agroecological zones, including eastern Israel, western Jordan, and southwestern Syria (Edhaie and Waines 1992). Ecogeographical approaches, based on the identification of landraces or wild species in carefully identified high-temperature agroecological zones, might lead to the discovery of germplasm with high levels of heat tolerance.

## 4.3.1   Genetics and Heritability of Heat Resistance and Heat Tolerance

For effective yield gain in breeding, adequate additive genetic variance and heritability are needed. However, heritability of yield generally decreases under greater and more complex stress conditions. In addition, $G \times E$ interaction in the high ambient temperature stress response can be significant, e.g., effects of photoperiod (see for review, Hall 1992 and Wallace 1980), while additional abiotic or biotic stressors in the environment also need to be considered. For example, insect-pest attacks can make it difficult to assess genotypic differences in pod set in cowpea under hot field conditions, especially in West Africa. Although variance increases under stress, resulting in high CVs, the advantages to selection for yield in complex field environments include improving adaptation to the target production environment (Ceccarelli et al. 1987).

Inheritance of heat resistance, i.e., high yield in hot environments compared with other cultivars, has been shown to be complex whereas heat tolerance of individual components that contribute to yield such as flower production and seed set has been shown to be simply inherited in some cases. Research in cowpea on the interaction of heat stress under variable photoperiods showed that germplasm can be divided into two groups, day-neutral accessions and short-day accessions, and several subgroups (Ehlers and Hall 1996). Genetic analysis of heat tolerance focused on the day-neutral group. Inheritance of heat tolerance for flower production was controlled by a single recessive gene (Hall 1993). In contrast, for pod set, two genes appeared to control heat tolerance with one of them characterized as having dominant gene action (Marfo and Hall 1992). Heat tolerance for flower production was highly heritable and could be effectively selected in the $F_2$ generation (Hall 2011), while narrow-sense heritability of heat tolerance for pod set in cowpea was only 26 % (Marfo and Hall 1992). In snap bean, genetic control of heat tolerance, measured as pod production (Bouwkamp and Summers 1982) or as abscission tolerance (Rainey and Griffiths 2005a, b), was found to be under simple genetic control. The narrow-sense heritability for pod production, however, was less than 10 %. In dry bean, heat resistance was found to be a heritable trait with breeding potential (Román-Avilés and Beaver 2001, 2003). Indeterminate growth habit and heat resistance were correlated (Shonnard 1991; Porch 2001), indicating possible simple genetic control for this trait or escape related to indeterminacy in common bean. In tomato, a few dominant genes were found to control fruit set under heat stress but narrow-sense heritability was low (Shelby et al. 1978), while a few recessive major genes were found by Opeňa et al. (1989). In rice, high broad-sense and narrow-sense heritabiltiy, with mostly additive genetic control, was found for heat resistance (Yoshida et al. 1981). Pollen grain number under heat stress was found to be correlated with yield under heat stress and controlled by several recessive genes (Mackill and Coffman 1983).

While the majority of studies have seemed to indicate simple inheritance for reproductive traits under heat stress, there is also evidence for complex inheritance of heat tolerance. For floral development under hot conditions in common bean, Shonnard (1991) found complex inheritance with additive gene action, dominance effects, and interaction with the environment. For floral bud abscission, realized heritability was 36 %, while for seed set it was 22 %. In tomato, diallel experiments indicated both specific and general combining abilities for fruit set (El Ahmadi and Stevens 1979; Opeňa et al. 1987).

## 4.3.2   Selection Techniques

A major constraint to progress in breeding for heat tolerance is the identification of and access to appropriate, consistent selection environments. The selection environment plays a key role in the efficiency of the selection process. The predominant approach to breeding for heat tolerance has involved selection for reproductive traits or yield under high ambient temperature conditions in the field. This cost-effective approach for increasing productivity and multiple-stress tolerance in high ambient temperature target production environments has been shown to be effective for a number of crops such as cotton, cowpea, dry bean, snap bean, and tomato.

Characterization of field production environments is an important aspect of the response to high-temperature stress. Each crop differs in its tolerance to high ambient temperatures, thus crop-specific threshold temperatures are important for assessing those daytime and nighttime temperatures that result in significant yield reductions. These temperatures vary based on the crop and developmental stage but can also depend on other environmental conditions.

Use of significantly hotter field environments than the typical production environment has resulted in the improvement of heat tolerance during reproductive development in cowpea (Ehlers et al. 2000; Hall 2011) and dry bean (Shonnard 1991; Rosas et al. 1999; Porch et al. 2012) but may result in the exclusion of moderately tolerant germplasm with other advantageous characteristics (Abdul-Baki 1991). Different field temperatures can be attained through selection of field sites at different altitudes and/or latitudes within a production region (e.g., Wallace 1980).

Confounding factors often exist in the field, which lead to more challenging identification of specific genetic factors related to heat tolerance. Root rot, low soil fertility, foliar diseases, insect pests, and drought often occur in high ambient temperature stress locations. However, screening plants in complex field environments can produce germplasm that is tolerant to multiple stresses (e.g., Porch et al. 2012), which is a decided advantage.

In addition to temperature, relative humidity is a major environmental factor that differs between semiarid, subhumid, and humid production zones. Research in wheat has shown that the greatest source of interaction between genotypes and sites across years was due to relative humidity (Reynolds et al. 1998; Vargas et al. 1998), resulting in the recommendation that breeding for high and low relative humidity environments should be conducted separately (Reynolds et al. 2001). These findings suggested that disease pressure may be a more important constraint under high relative humidity environments, while drought stress could be a greater constraint in low relative humidity environments. Thus relative humidity, as well as soil moisture conditions, should be considered in addition to temperature stress characteristics when choosing a selection environment.

Photoperiod can be a critical parameter in selecting a screening environment for some crops. Sensitivity of reproductive development to heat has been shown to be influenced by day length in cowpea and rice (as reviewed above).

The timing of screening is important. Studies on a number of crops (e.g., cotton, cowpea, rice, tomato, and common bean), reviewed above, have shown that the greatest sensitivity to high-temperature stress occurs during reproductive development (note: some crops have shown sensitivity to high soil temperatures at planting or during vegetative growth). Since the sensitive periods during the reproductive development period tend to be relatively short, weeks to a month long, planting can be timed so that reproductive development coincides with this high-temperature period during the warm season. Some species are more sensitive to high night temperature and locations should be chosen that provide sufficiently high night temperature during reproductive development to cause stress. Differences in plant phenology between genotypes should be considered in order to avoid escapes. Field selection environments typically do not facilitate temperature control; however, nighttime temperatures can be modified using small canopies that can be placed over the crop at night.

Controlled environment, high-temperature screening, and selection methods have been used for crop improvement. Modern, tightly controlled greenhouse environments effectively control temperature, in addition to some other aspects of the environment. Greenhouse selection for heat tolerance has been widely employed in snap bean (e.g., Dickson and Petzolt 1989; Wasonga et al. 2010, 2012) and tomato (Berry and Rafique-Uddin 1988) and is effective in cowpea (Thiaw and Hall 2004). Temperatures generally differ between the field and greenhouse screening environments (often higher in the greenhouse, especially at night), in order to ensure a similar selection index. A significant advantage of the greenhouse environment is that it affords the opportunity for the application of consistent high night temperatures that are often only available in lowland tropical and subtropical but not temperate field environments. In addition, greenhouse environments generally offer better control of other sources of stress, thus eliminating the effects of interaction with drought and biotic constraints. A major advantage of hot greenhouse environments is that they can provide more consistent evaluations of plant heat tolerance than hot field environments, but they are more expensive to operate per plant tested. However, greenhouse environments are often not representative of field environments.

Correlations between genotype responses in field and greenhouse trials or between different field environments have been shown in some circumstances. For example, cowpea selection for yield in high-temperature tropical field environments (Patel and Hall 1986; Ehlers and Hall 1998) has resulted in germplasm tolerant to high temperature in subtropical field and greenhouse environments. Snap bean selected for heat tolerance in greenhouses has exhibited high yields in high-temperature field sites in Kenya and Puerto Rico (Wasonga et al. 2010, 2012). In wheat, canopy temperature depression was found to be correlated between screen-house trials and field trials (Hede et al. 1999). These results indicate that in some cases, selection in one environment has resulted in broad adaptation of heat-tolerant germplasm which could simplify breeding approaches and make germplasm useful over broad geographical areas.

Indirect selection approaches can be useful where specific traits are correlated with heat tolerance. Maintenance of membrane function has been considered a general mechanism for heat tolerance. Based on earlier research it had been proposed that selection for slow leaf-electrolyte leakage under heat stress could provide a method to breed for increased heat tolerance (Blum 1988). Genetic selection studies with cowpea demonstrated that low leaf-electrolyte leakage under hot conditions is associated with heat tolerance during pod set in cowpea and can be used as a selection criteria. Lines of cowpea that were heat tolerant during pod set were shown to have less leaf-electrolyte leakage under heat stress than a set of heat-sensitive lines (Ismail and Hall 1999). Thiaw and Hall (2004) selected two of these lines, a heat-sensitive cultivar and a heat-tolerant genetic line with similar genetic background, and used them as parents in a breeding program. They crossed the parents and then divergently selected one population for high and low pod set under heat stress and another population for low and high leaf-electrolyte leakage under heat stress. Genetically stable, selected lines were developed and evaluated in an extremely hot long-day field environment and a long-day glasshouse environment with high night temperature. In both environments, 66 lines selected in one generation for high pod set and 16 lines selected in two generations for low leaf-electrolyte leakage had greater pod set and grain yield than either the 40 lines selected for low pod set or the 16 lines selected for high leaf-electrolyte leakage. In both environments, highly significant negative correlations were obtained between leaf-electrolyte leakage and number of pods per peduncle and grain yield using data from both populations. These results indicate there is a strong association between ability to set pods under heat and maintenance of membrane function that may be pleiotropic. The indirect realized heritability for heat tolerance during pod set resulting from selection for slow leaf-electrolyte leakage was only 10–12 %. This indirect screen has value, however, in that it can be conducted at times of the year and in locations where field screening for heat tolerance is not possible. Field screening for pod set, however, still would be necessary.

Blum et al. (2001) made progress in breeding heat-resistant spring wheat cultivars that produced more grain than other cultivars in hot environments by selecting for slow leaf-electrolyte leakage under hot conditions. They proposed that selecting for slow leaf-electrolyte leakage under hot conditions could not replace selecting for yield in the field but that it can be useful for reducing a large population into the most likely heat-tolerant core in the early stages of a breeding program.

Selection of heat-shock proteins (HSP) may be another avenue to pursue for the indirect selection of heat-tolerant germplasm; however, to date, correlations have not been shown between HSPs and heat resistance as expressed in seed yield. Genetic lines of cowpea with substantial differences in heat tolerance during reproductive development and with a similar genetic background have produced the same set of heat-shock proteins in their leaves when subjected to high temperatures. Heat-shock protein profiles were examined in six common bean lines that differ in heat acclimation potential with respect to differences in heat-killing time based on leaf-electrolyte leakage (Li and Udomprasert 1993). No relationship was observed between patterns of HPSs and heat acclimation potential.

While there is no published evidence to indicate that genetic engineering has been used for the development of improved heat tolerance in a cultivar, basic research has been completed indicating that transcription factors (Yoshida et al. 2012), micro RNAs (Yu et al. 2012), and heat-shock proteins (Queitsch et al. 2000) increase tolerance of specific physiological processes in the laboratory, but this does not necessarily translate to the field environment. Some recent research demonstrated that the expression of AtSAP5 in cotton increased expression of stress-response genes and may have increased drought and heat tolerance (Hozain et al. 2012).

### 4.3.3   Breeding Methods

Breeding methods for heat tolerance vary based on the genetic nature of the tolerance to high ambient temperature stress and on the crop. The pedigree method has perhaps been used most extensively for breeding for heat tolerance across crops. This breeding approach has been employed for the improvement of snap bean at Cornell; dry bean at the University of Puerto Rico and USDA-ARS in Mayagüez, Puerto Rico; cowpea at the University of California, Riverside (Hall 2011); and tomato and Chinese cabbage at AVRDC (Opeña et al. 1987). Selection is often conducted in early generations. In cowpea, for example, selection of single $F_2$ plants with high numbers of flowers, pods per peduncle, seeds per pod, and seed quality has been conducted under very high-temperature field conditions. Recessive genes in the $F_2$ were thus fixed, which include photoperiod insensitivity and resistance to floral bud abortion, but this selection only provided partial heat tolerance during pod set. Early generation selection can substantially increase the speed of the breeding process when the trait of interest is under relatively simple control and has adequate inheritance. Selection of more complex or dominant traits, such as pod set in cowpea, mainly is conducted in later generations while agronomic traits can be selected in concurrent trials under commercial conditions. Evaluations of advanced lines are then conducted in multilocation trials throughout the target production zone.

Due to the simple genetic nature of heat tolerance during reproductive development in a number of crops, the backcross breeding approach has also been employed as method for introgression of tolerance into elite material. In rice, a large backcross population was advanced using single-seed descent and then selected for heat tolerance during the milky stage of seed development using differential planting dates (Liao et al. 2011). In cotton, backcross breeding was used to transfer pollen thermotolerance through treating pollen before pollination for 15 h with 35 °C temperatures, resulting in higher levels of fertile pollen in subsequent generations (Rodriquez-Garay and Barrow 1988). The backcross method is ideal for relatively simply inherited traits with adequate inheritance and can be combined with single-seed descent or the pedigree approach to result in an efficient method for germplasm improvement.

In order to combine multiple sources of heat tolerance, recurrent selection has been employed in diploid potato resulting in a 27 % increase in yield in a single cycle of recurrent selection (Gautney and Haynes 1983) and is being employed to combine heat and drought tolerance in common bean (Porch TG, unpublished).

Although quantitative trait locus (QTL) approaches have not been extensively used in breeding for heat tolerance, some QTLs that may be relevant have been identified in crop species. QTLs controlling pollen germination and pollen tube growth were identified in maize (Frova and Sari-Gorla 1994), and grain-filling duration QTLs were identified in wheat (Yang et al. 2002). While some studies have shown that heat tolerance is under relatively simple control (Porch et al. 2004; Rainey and Griffiths 2005a, b; Hall 2011), few QTL analyses have been completed and molecular markers have yet to be developed for marker-assisted selection (MAS). Additional QTL studies are needed that may show similar genetic control and similar mechanisms for heat tolerance across some species.

### 4.3.4 Facultative Apomixis as an Approach for Achieving General Tolerance to Several Stresses During Reproduction

Production of pollen and pollination are sensitive to several other stresses in addition to high temperature, including chilling, drought, and frost. A potential general solution to reproductive sensitivity is to develop cultivars that do not require sexual processes when growing in farmers' fields, i.e., apomixis.

Crop plants with an appropriate type of apomixis would be able to produce viable seed from maternal tissue without requiring either meiosis of the embryo mother cell or pollen production and pollination of the embryo or endosperm. Advantages of cultivars with this type of apomixis were described by Jefferson (1993) and are discussed below.

Because they do not require sexual processes, apomictic cultivars would have tolerance to the many stresses that have the potential to damage pollination and other aspects of sexual reproduction. Since meiosis is eliminated, apomixis would fix hybridity because all cultivars including $F_1$ hybrids would have true-breeding seeds. This would make possible the use of hybrid vigor in the many crop species where it is currently either not economic (e.g., cowpea, common bean, and wheat) or difficult (e.g., rice) to generate hybrid seed by crossing. Note that cowpea exhibits substantial hybrid vigor. Farmers would be able to reuse seed produced by apomictic hybrid cultivars in that the seed would retain its hybrid vigor. This would be a significant advantage to poor farmers. Seed industries would still be needed because farmers would still need to purchase high quality seed after several years of using their own seed. The quality of seed declines from year to year due to the presence of seed-borne pathogens and contamination by off-types either in the farmers' fields or due to mechanical mixing during harvesting or seed processing. Crops currently propagated vegetatively, such as most Irish potatoes (*Solanum tuberosum* L.), would benefit from apomictic cultivars because they could be propagated by seed instead. Breeding progress would be accelerated by using apomictic lines because they would confer the ability to immediately fix heterozygous genotypes.

Normal reproduction would still be needed in breeding programs to permit the continual breeding of improved cultivars. The integration of breeding with apomictic cultivar release could be achieved by using facultative apomixis systems, where the default state is apomictic and the sexual state can be "switched on" by the breeder. Switching might be achieved either by sprays with specific chemicals or by growing the plants in a specific environment. For example, there is a male sterility in rice that can be switched off by growing the plants in shorter-day cooler environments (Yuan et al. 1993). In this system a warm thermal regime combined with long days induces male sterility. This represents an extreme sensitivity of the similar system that has been observed in some other species such as cowpea (Hall 2004).

Several plant species have the genes needed to develop facultative apomictic breeding systems in crop plants. Through genetic engineering it may be possible to create and transfer the "cassette" of genes needed for facultative apomixis into crop cultivars (Jefferson 1993). This would not be an easy task but crop cultivars with facultative apomixis would result in a revolution in plant biology, plant breeding, crop production, and agriculture.

## 4.4 Heat-Resistant Cultivars and Breeding Lines

Relatively little progress has been made in breeding cultivars with heat resistance, i.e., cultivars with greater yields than other cultivars in hot commercial production environments. In addition to heat resistance, cultivars should have similar or greater grain yields than current cultivars in more moderate temperature environments within the target production zone and other desirable agronomic traits. Documented cases of the development of heat-resistant cultivars or breeding lines of annual crops that produce grain and/or fruit are described.

## 4.4.1   Grain Cowpea and Snap Cowpea

"California Blackeye 27" (CB27) was bred using a pedigree breeding procedure involving crosses among two accessions that are heat tolerant during reproductive development and two California cultivars (Ehlers et al. 2000). Segregating lines were subjected to selection for flower production and pod set over several years in a summer field nursery in Imperial Valley, California, that has extremely high night and day temperatures during early flowering (average daily maximum and minimum air temperatures of 41 and 24 °C) and long days. In addition, resistances to root-knot nematode and Fusarium wilt were incorporated. This was followed by selection for yield and other agronomic traits in experiment-station sites and farmer-managed trials in commercial production locations in California where the temperatures ranged from being cool to very hot (average daily maximum and minimum air temperatures ranging from about 26/13 to 35/17 °C during the first 60 days after sowing). As a result of this work, CB27 was bred, which is heat resistant in that it produced higher yields than cultivar CB5 in hot environments (Ismail and Hall 1998) and it had similar yields as the cultivar CB46 in more optimal environments. CB27 is a semidwarf cultivar with a higher harvest index and a more compact growth habit than CB5 and has exhibited greater yield advantages when grown on narrow row spacing (Ismail and Hall 2000).

"Marfo-Tuya" was bred by crossing a landrace from Ghana "Sumbrisogla" and breeding line 518-2 from the University of California, Riverside (Padi et al. 2004b). Line 518-2 was bred by crossing CB5 with a Nigerian landrace that has heat tolerance during reproductive development and selecting for flower production and pod set in an extremely hot, long-day summer field nursery in Imperial Valley, California. Segregating progeny from the cross between "Sumbrisogla" and 518-2 were selected for flower production and pod set in an extremely hot, long-day summer field nursery in Imperial Valley, California, and then for grain yield and other agronomic attributes in northern Ghana. In experiment-station trials over 7 years at four locations in northern Ghana, "Marfo-Tuya" had grain yields that were 52 % greater than a check cultivar. In farmer-managed agronomic trials over 3 years at 52 farm sites across northern Ghana, "Marfo-Tuya" produced significantly more grain than local checks in 72 % of the test locations.

"Apagbaala" was bred by a three-way cross involving a breeding line from the International Institute of Tropical Agriculture (IITA), a heat-tolerant cultivar from Nigeria, and a heat-tolerant breeding line from the University of California, Riverside (Padi et al. 2004a). Segregating progeny were selected for flower production and pod set in an extremely hot, long-day summer field nursery in Imperial Valley, California, over 2 years and two generations. Lines were then selected for grain yield and other agronomic attributes over 3 years in northern Ghana. In experiment-station yield trials over 7 years at four locations in northern Ghana, "Apagbaala" had grain yields that were 41 % greater than the local check cultivar. In farmer-managed agronomic trials over 4 years at 66 farm sites across northern Ghana, "Apagbaala" produced significantly higher grain yields than the local check in 74 % of the test locations.

"Mouride" was bred by crossing a landrace from Senegal that is used as a cultivar, 58-57, with a breeding line from IITA (Cisse et al. 1995). Early generation selections were conducted for resistance to cowpea weevil, cowpea aphid-borne mosaic virus, and bacterial blight. Selection for yield and other agronomic traits was conducted in Senegal at four experiment-station locations and at 35 on-farm sites over 3 years. In these trials, "Mouride" produced 18 % more grain but 17 % less forage than 58-57, and in subsequent years it has produced very high grain yields in many trials in the hot Sahelian zone. At no time was "Mouride" selected for heat tolerance, but it was selected for yield in very hot environments in Senegal. Subsequent glasshouse studies in California under short days with high night temperatures (Ehlers and Hall 1998) showed that "Mouride" had some heat toler-ance during pod set, whereas another cultivar bred in Senegal using similar methods but different parents, "Melakh" (Cisse et al. 1997), was heat sensitive during pod set. Selection for grain yield in hot regions of tropical Africa does not necessarily incorporate heat tolerance, but it has produced some cultivars that may be useful donors of heat-tolerance traits.

Five vegetable (edible pod snap-type) cowpea breeding lines were developed at the Indian Agricultural Research Institute, New Delhi, India, by incorporating resistance to bacterial blight and selecting for high pod yield under very hot, long-day field conditions in northern India (Patel and Singh 1984). These lines were shown to have heat tolerance during flowering and pod set in an extremely hot, long-day summer field nursery in Imperial Valley, California (Patel and Hall 1986). Under hot long-day field conditions in Riverside, California, these breeding lines produced green pod yields between 25–28 ton/ha, whereas the US cultivar "Snapea" yielded only 13 ton/ha and a snap bean (*Phaseolus vulgaris* L.) cultivar, "Contender," produced only 6.7 ton/ha.

### 4.4.2  Common Bean, Snap Bean, and Tepary Bean

Current estimates predict that the area suitable for common bean production could increase by over 50 % if beans were able to tolerate a 2 °C increase in temperature (Beebe et al. 2012). Thus, there is significant justification for investing in breeding common bean for heat tolerance (Porch et al. 2007). Plant breeding efforts have increased tolerance to heat stress in germplasm, mainly in the Mesoamerican gene pool. Early efforts to improve bean golden mosaic virus (BGMV) resistance in the lowlands of Central America resulted in the generation of heat-tolerant germplasm at CIAT, including DOR 364 and DOR 557. Specific breeding for heat tolerance at

Escuela Agricola Panamericana (Zamorano) resulted in the generation of the first heat-tolerant common bean variety, Tio Canela, with a small red seed type (Rosas et al. 1999). Significant gains in heat tolerance were achieved with the release of two additional small red genotypes, Amadeus 77 (Rosas et al. 2004) and CENTA Pipil, which resulted in significant yield improvement in lowland production environments and also in Amadeus 77 becoming the most widespread common bean variety in Central America. In the Caribbean, selection for heat tolerance led to the release of "Verano" (Beaver et al. 2008), a small white variety from the University of Puerto Rico, with heat tolerance, common bacterial blight resistance, and virus resistance. Andean beans, TARS-HT1 and HT2 kidney bean germplasm (Porch et al. 2010), were developed for heat tolerance at USDA-ARS in Mayagüez, Puerto Rico, while Mesoamerican beans TARS-SR05 (Smith et al. 2007) and TARS-MST1 (Porch et al. 2012) have multiple-stress tolerance, including heat tolerance.

At Cornell University, the snap bean breeding program has improved reproductive heat tolerance in snap bean germplasm for the USA, resulting in the development of Cornell 502 and 503. Ongoing efforts are combining heat tolerance and rust resistance for snap bean production in East Africa (Wasonga et al. 2010, 2012).

The genetic limits of heat tolerance in existing crop plants may require exploration into the development and use of new species, or historically used native species, as future food security crops. One possible example is tepary bean (*Phaseolus acutifolius* A. Gray), a native species from the Sonoran desert of Mexico and southern USA (Freytag and Debouck 2002), which has drought adaptation and heat tolerance that is superior to that of common bean. Although not currently a commercial crop, at the peak of tepary production in the 1930s, the Tohono O'odham tribe in southern Arizona grew more than 750 tons of tepary beans annually, and the crop has traditionally been produced by Native American tribes in the region (Nabhan and Felger 1978). Current production levels are far lower and dispersed through Arizona, New Mexico, Mexico, and Central America. However, there is increasing interest in tepary in the USA and production has begun in Africa by small-holder farmers in arid regions (Shisanya 2002). Tepary exhibits multiple drought and heat tolerance related traits including: stomatal control (Markhart 1985), dehydration avoidance (Mohammed et al. 2002), root characteristics (Butare et al. 2011), and yield stability (Miklas et al. 1994; Porch 2006).

Tepary landraces have been used for mass selection, resulting in the release of a dark yellow tepary (cited by Pratt and Nabhan 1988). "Redfield" tepary was released in a published registration (Garver 1934) and was the result of a selection from T.S. 3306, a Texas landrace. Recent breeding efforts at ARS-TARS, Puerto Rico, have resulted in the development of improved tepary lines, TARS-Tep 22 and TARS-Tep 32, with heat, drought, common bacterial blight, and bruchid resistance (Porch et al. 2013, accepted). They also exhibit a larger seed size and improved agronomic characteristics, which have been constraints to tepary production and consumer acceptability in the past. Virus resistance and culinary traits may also need to be improved, but further research is needed to explore these characteristics and the potential for improvement.

### 4.4.3   Pima Cotton and Upland Cotton

Breeding heat-tolerant Pima cotton began in the late 1950s at Phoenix, Arizona. Feaster and Turcotte (1985) established that in environments where adaptation depends on tolerance to high night temperature, the extent of boll set at low nodal positions under very hot conditions is an effective indicator of heat tolerance. Several cultivars (Feaster and Turcotte 1976, 1984; Turcotte et al. 1992) and germplasm lines (Turcotte et al. 1991; Percy and Turcotte 1997) were bred by selecting for greater boll set on low nodal positions under very hot field conditions. The realized genetic gain in lint yield from "Pima S-1" to "Pima S-5" was 57 % in very hot conditions at low elevation and 30 % at higher elevation. "Pima S-6," which was released in 1983, was estimated to have 69 % greater yield over Pima S-1 (the dominant cultivar in southwestern USA from 1955 to 1961) in hot environments and a 27–43 % yield advantage in cooler environments (Feaster and Turcotte 1985).

Average yields of Pima and upland cotton (*Gossypium hirsutum* L.) were compared for six counties in Arizona over a 30-year period (Kittock et al. 1988). Pima cotton lint yields increased substantially more than upland cotton lint yields, particularly in hot environments, as improved Pima cultivars were released over the 30-year period. The authors concluded that about 50 % of the 30-year lint yield increase of Pima cotton in hot environments resulted from the increased heat tolerance (i.e., boll set) of the improved cultivars. A historical set of eight heat Pima cultivars, which had been selected for boll set in a very hot field environment, were compared in a very hot commercial production environment (Lu et al. 1998). Genotypic differences in lint yield were positively correlated with stomatal conductance measured in the middle of the afternoon during peak flowering and fruiting. Lu et al. (1998) hypothesized that the adaptive advantage of the higher stomatal conductance is independent of photosynthesis and is associated with leaf cooling providing a heat avoidance type of heat resistance. However, this hypothesis would not explain heat-stress effects on boll set that occur in the late night or early morning when stomata would have little influence on plant temperature.

In earlier years upland cotton had greater heat tolerance than Pima cotton. Since this time, the success of commercial upland cotton breeding programs for the extremely hot environments of southwestern USA probably was also partially due to incorporation of heat tolerance during boll set.

Upland cotton cultivar CRIS-134 was bred to tolerate the hot period of June–August at the Cotton Research Institute, Sakarand, Sindh, Pakistan (Soomro 1998). This cultivar is capable of producing 32 bolls after 75 days from sowing compared with 11–17 bolls in check cultivars.

Heat-tolerant upland cotton genotypes have been developed at the Indian Agricultural Research Institute (IARI), New Delhi, India, using a shuttle-breeding approach involving changing environments during generations of selection (Singh et al. 2004). Selection was applied for high numbers of fruiting structures and early maturity (Singh et al. 2003). According to Singh et al. (2007), heat-tolerant genotype Pusa 17-52-10 has been registered with NBPGR (INGR No 03073) and the heat-tolerant upland cotton genotypes bred in India typically have compact plant type, as also was observed in heat-tolerant cowpea genotypes (Ismail and Hall 1998), and low first fruiting node number, as was observed in heat-tolerant Pima cotton genotypes (Feaster and Turcotte 1985). IARI has bred an upland cotton cultivar, "Aurobindo PSS-2," that has tolerance to high temperatures during reproductive stages and tolerance to low temperatures during germination (Singh et al. 2011). Cowpea lines also have been bred that combine tolerance to high night temperature during reproductive stages with chilling tolerance during germination and emergence (Hall 2011). It is surprising that chilling tolerance and heat tolerance, at different stages of development, can be combined because they may depend on different membrane fluidity properties (Lyons 1973).

### 4.4.4   Tomato

The heat-tolerant tomato cultivar "Saladette" was bred by Paul W. Leeper by selecting for fruit set under the hot summer conditions of the Lower Rio Grande Valley in Texas. Since this time, Saladette and other heat-tolerant genotypes (Stevens 1979) have been used as parents in breeding programs in the USA. Several commercial breeding programs in California have developed heat-resistant tomato cultivars for processing and fresh market use by incorporating heat tolerance during reproductive development.

Fresh market tomato breeding began at the University of Florida in 1922. A hybrid cultivar was released in 1989, "Solar Set," that was described as having a commercially acceptable level of heat tolerance (Scott 1999). Heat tolerance involved the ability to set fruit in hot field conditions. Solar Set was compared with three other heat-tolerant cultivars, nine heat-tolerant lines, and four heat-sensitive cultivars in an extremely hot glasshouse (38–40 °C/29–31 °C day/night air temperatures) and a hot (28–29 °C/18–19 °C day/night average air temperatures) field environment (Abdul-Baki 1991). "Solar Set" had higher yields than the other heat-tolerant cultivars and lines in the field environment but produced fewer flowers and normal fruits and had lower yields than the other heat-tolerant cultivars and lines in the extremely hot glasshouse. An open-pollinated heat-tolerant cultivar, "Neptune," was released that had similar field yields as "Solar Set" (Scott et al. 1995a). A heat-tolerant hybrid, "Equinox," was released that was reported to have the same high-temperature fruit-setting ability and similar field yields as "Solar Set" (Scott et al. 1995b). A heat-tolerant hybrid, "Solar Fire," was developed that was reported to have superior fruit-setting ability under high temperatures (>32 °C day/>21 °C night) than most current cultivars and to have produced greater yields than "Solar Set" in field trials in Florida (Scott et al. 2006). Note that the various tomato cultivars that were released also had differences in resistances to diseases and other traits such as those involving fruit quality.

The Asian Vegetable Research and Development Center (AVRDC) in Taiwan began its tomato breeding program in 1972 focusing on producing cultivars for use in the tropics (where night temperatures are high). Many biotic stresses constrain tomato production in the tropics, but during the first stage of the AVRDC tomato breeding program (1973–1980), heat tolerance during reproductive development was incorporated by selecting for fruit set in field nurseries grown at the hottest time of the year (Villareal and Lai 1979). As of 1992 this breeding program had contributed to the development and release of 52 tropical tomato lines in 32 countries (Opeňa et al. 1993). The relationships between reproductive-stage heat tolerance of some AVRDC lines and fruit yield under hot tropical conditions are discussed by de la Peňa et al. (2011). They recommend that screening for pollen amount and pollen viability can be useful surrogate traits to selecting for fruit set under very hot field conditions.

### 4.4.5 Rice

Genotypic differences in heat tolerance have been detected in rice in which there was a highly significant correlation between the percentage of florets setting seed and the number of pollen grains per stigma (Yoshida et al. 1981; Mackill et al. 1982). Many heat-tolerant lines were also early flowering (Yoshida et al. 1981). A similar association between heat tolerance at flowering and earliness also has been observed in cowpea (Ehlers and Hall 1996). In genetic studies with rice, the number of pollen grains per stigma and percentage seed set had broad-sense heritabilities of 83.7 % and 69.2 %, respectively (Mackill and Coffman 1983). Phenotypic and genetic correlations between the two traits were 0.58 and 0.65, respectively. A narrow-sense heritability of 48 % indicated that it should be possible to select for seed set under high temperature.

A breeding project was initiated at the International Rice Research Institute in 2007 to incorporate heat tolerance during seed set into rice and to change the time of flowering from between 10 A.M. and noon to earlier in the morning when it is cooler (Redoňa et al. 2007). Advanced breeding lines have been developed, some of which were selected using a shuttle-breeding approach involving naturally hot locations in different countries. These heat-tolerant rice lines are now being evaluated for yield and other agronomic traits (E. D. Redoňa, personal communication).

### *4.4.6 Wheat*

Progress has been made in breeding heat-resistant wheat cultivars as was shown by evaluations of spring wheat cultivars (Shpiler and Blum 1991; Leithold et al. 1997) and winter wheat cultivars (Assad and Paulsen 2002; Tewolde et al. 2006). Heat-resistant cultivars usually produced more grain per spike and had greater individual grain weight. A major factor in gaining heat resistance may have been selection for grain yield in field environments where heat stress occurred. Through the use of appropriate experimental designs and field procedures and computerized combine harvesters, more extensive yield testing can be done with wheat than with most other crop species. For example, the spring wheat breeding program of E. A. Hurd in Saskatchewan, Canada, yield-tested about 1,000 lines per cross in early generations (Hurd 1971).

Spring wheat breeding programs of the International Maize and Wheat Improvement Center (CIMMYT) use selection in field nurseries with managed stress to incorporate stress tolerance (Trethowan and Mahmood 2011). For example, a late-planted field nursery in northwestern Mexico provides heat stress during grain development but is well watered to avoid confounding effects due to drought. Seed are bulk harvested, in some cases, and separated for seed size and density on a gravity table. Large well-filled grain are selected and then resown again in a late-planted field nursery where plants are subjected to heat stress during grain development.

Progress has been made at CYMMT in combining selection for physiological criteria with empirical selection for grain yield and individual seed weight. Grain yield of spring wheat cultivars in hot irrigated environments had been shown to be positively correlated with stomatal conductance, and it was shown that canopy temperature depression (CTD) as measured with an infrared thermometer is an effective method for screening for this trait (Amani et al. 1996). Based on studies with breeding lines, Reynolds et al. (1998) proposed that selection for CTD could complement empirical selection for heat resistance. Selection for CTD has now been used in the $F_4$ generation to favorably skew gene frequency, resulting in a higher proportion of drought-adapted materials with cooler canopies (Trethowan and Turner 2009). Blum et al. (2001) proposed that selection for slow leaf-electrolyte leakage under heat stress also might be useful for reducing a large population into the most likely heat-tolerant core in the early stages of a breeding program.

## 4.5   Future Directions and Conclusions

Due to global warming, more effort should be devoted to breeding for heat resistance in annual crops. Work by a few breeders with a few annual crop species has shown that this can be done by conventional hybridization with selection for heat tolerance during reproductive development. Pollen development of many species is particularly sensitive to high temperatures occurring in the late night or early morning. DNA markers should be developed for heat tolerance during pollen development since this could make breeding for heat resistance more effective and markers could be determined in plants grown in cooler as well as warmer environments. Collaboration in marker development is warranted since similar markers may be effective in several species and since comparative genomics can be employed for testing candidate genes. More consideration should be given to the development of facultative apomictic breeding systems since they could provide a general solution to the damage that many stresses cause to reproductive development in crop plants.

Future breeding methods should involve a greater integration of molecular breeding, whole genome selection, high throughput phenotyping, and reverse genetic approaches as compared to current improvement programs. Additional markers should be developed for different heat-tolerance traits. The complexity of target production environments, however, requires that phenotypic field selection likely will remain the foundation of crop improvement programs.

Where necessary, alternate crops can be considered where the reclamation of marginal high temperature agroecological zones is being pursued. For example, among the warm-season legumes, dry grain cowpea, snap cowpea, and tepary have much greater heat tolerance than common bean or snap bean.

# References

Abdul-Baki AA (1991) Tolerance of tomato cultivars and selected germplasm to heat stress. J Am Soc Hortic Sci 116:1113–1116

Ahmed FE, Hall AE (1993) Heat injury during early floral bud development in cowpea. Crop Sci 33:764–767

Ahmed FE, Hall AE, DeMason DA (1992) Heat injury during floral development in cowpea (*Vigna unguiculata*, Fabaceae). Am J Bot 79:784–791

Ahmed FE, Hall AE, Madore MA (1993a) Interactive effects of high temperature and elevated carbon dioxide concentration on cowpea (*Vigna unguiculata* (L.)Walp.). Plant Cell Environ 16:835–842

Ahmed FE, Mutters RG, Hall AE (1993b) Interactive effects of high temperature and light quality on floral bud development in cowpea. Aust J Plant Physiol 20:661–667

Al-Khatib K, Paulsen GM (1999) High-temperature effects on photosynthetic processes in temperate and tropical cereals. Crop Sci 39:119–125

Allen LH Jr (1994) Carbon dioxide increase: direct impacts on crops and indirect effects mediated through anticipated climate changes. In: Boote KJ, Bennett JM, Sinclair TR, Paulsen GM (eds) Physiology and determination of crop yield. Crop Science Society of America, Madison, WI, pp 425–459

Amani I, Fischer RA, Reynolds MP (1996) Canopy temperature depression associated with yield of irrigated spring wheat cultivars in a hot climate. J Agron Crop Sci 176:119–129

Argyris J, Truco MJ, Ochoa O, Knapp SJ, Still DW, Lenssen GM, Schut JW, Michelmore RW, Bradford KJ (2005) Quantitative trait loci associated with seed and seedling traits in *Lactuca*. Theor Appl Genet 111:1365–1376

Assad MT, Paulsen GM (2002) Genetic changes in resistance to environmental stresses by U.S. Great Plains wheat cultivars. Euphytica 128:87–96

Baker JT, Allen LH (1993) Contrasting species responses to $CO_2$ and temperature: rice, soybean and citrus. Vegetation 104(105):239–260

Baker JT, Allen LH, Boote KJ, Jones P, Jones JW (1989) Responses of soybean to air temperature and carbon dioxide concentration. Crop Sci 29:98–105

Beaver JS, Porch TG, Zapata M (2008) Registration of 'Verano' white bean. J Plant Regist 2:187–189

Beebe S, Ramirez J, Jarvis A, Rao IM, Mosquera G, Bueno JM, Blair MW (2012) Genetic improvement of common beans and the challenges of climate change. In: Yadav SS, Redden RJ, Hatfield JL, Lotze-Campen H, Hall AE (eds) Crop adaptation to climate change. Wiley-Blackwell, Chichester, pp 356–369

Berry SZ, Rafique-Uddin M (1988) Effect of high temperature on fruit set in tomato cultivars and selected germplasm. HortScience 23:606–608

Blum A (1988) Plant breeding for stress enviroments. CRC, Boca Raton, FL

Blum A, Klueva N, Nguyen HT (2001) Wheat thermotolerance is related to yield under heat stress. Euphytica 117:117–123

Borthwick HA, Robbins WW (1928) Lettuce seed and its germination. Hilgardia 3:275–305

Bouwkamp JC, Summers WL (1982) Inheritance of resistance to temperature-drought stress in snap bean. J Hered 73:385–386

Butare L, Rao IM, Lepoivre P, Polania J, Cajiao C, Cuasquer J, Beebe S (2011) New genetic sources of resistance in the genus Phaseolus to individual and combined aluminium toxicity and progressive soil drying stresses. Euphytica 181:385–404

Ceccarelli S (1987) Breeding strategies to improve barley yield and stability in drought prone environments. In: Monti L, Porceddu E (eds) Drought resistance in plants. Physiological and genetic aspects. pp 333–348

Cisse N, Ndiaye M, Thiaw S, Hall AE (1995) Registration of 'Mouride' cowpea. Crop Sci 35:1215–1216

Cisse N, Ndiaye M, Thiaw S, Hall AE (1997) Registration of 'Melakh' cowpea. Crop Sci 37:1978

Cornish K, Radin JW, Turcotte EL, Lu Z, Zeiger E (1991) Enhanced photosynthesis and stomatal conductance of Pima cotton (*Gossypium barbadense* L.) bred for increased yield. Plant Physiol 97:484–489

Dahlberg JA, Burke JJ, Rosenow DT (2004) Development of a sorghum core collection: refinement and evaluation of a subset from Sudan. Econ Bot 58:556–567

de la Peña RC, Ebert AW, Gniffke PA, Hanson P, Symonds RC (2011) Genetic adjustment to changing climates: vegetables. In: Yadav SS, Redden RJ, Hatfield JL, Lotze-Campen H, Hall AE (eds) Crop adaptation to climate change. Wiley-Blackwell, Chichester, pp 396–410

Dickson MH, Petzolt R (1989) Heat tolerance and pod set in green beans. J Am Soc Hortic Sci 114:833–836

Dolferus R, Ji X, Richard RA (2011) Abiotic stress and control of grain number in cereals. Plant Sci 181:331–341

Dow El-Madina IM, Hall AE (1986) Flowering of contrasting cowpea (*Vigna unguiculata* (L.) Walp.) genotypes under different temperatures and photoperiods. Field Crops Res 14:87–104

Dundas I, Saxena KB, Byth DE (1981) Microsporogenesis and anther wall development in male-sterile and fertile lines of pigeon pea (*Cajanuscajan* [L.] Millsp.). Euphytica 30:431–435

Eastin JD, Castleberry RM, Gerik TJ, Hutquist JH, Mahalakshmi V, Ogunela VB, Rice JR (1983) Physiological aspects of high temperature and water stress. In: Raper CD, Kramer PJ (eds) Crop reactions to water and temperature stresses in humid, temperate climates. Westview, Boulder, CO, pp 91–112

Edhaie B, Waines JG (1992) Heat resistance in wild *Triticum* and *Aegilops*. J Genet Breed 46:221–228

Ehlers JD, Hall AE (1996) Genotypic classification of cowpea based on responses to heat and photoperiod. Crop Sci 36:673–679

Ehlers JD, Hall AE (1998) Heat tolerance of contrasting cowpea lines in short and long days. Field Crops Res 55:11–21

Ehlers JD, Hall AE, Patel PN, Roberts PA, Matthews WC (2000) Registration of 'California Blackeye 27' cowpea. Crop Sci 40:854–855

El Ahmadi A, Stevens MA (1979) Reproductive responses of heat-tolerant tomatoes to high temperatures. J Am Soc Hortic Sci 104:686–691

El-Kholy AS, Hall AE, Mohsen AA (1997) Heat and chilling tolerance during germination and heat tolerance during flowering are not associated in cowpea. Crop Sci 37:456–463

Farooq M, Bramley H, Palta JA, Siddique KHM (2011) Heat stress in wheat during reproductive and grain-filling phases. Crit Rev Plant Sci 30:1–17

Feaster CV, Turcotte EL (1976) Registration of Pima S-2, S-3, S-4 and S-5 cotton. Crop Sci 16:603–604

Feaster CV, Turcotte EL (1984) Registration of Pima S-6 cotton. Crop Sci 24:382

Feaster CV, Turcotte EL (1985) Use of heat tolerance in cotton breeding. In: Proceedings of Beltwide cotton research conference, Phoenix, AZ, 9–10 Jan 1985. National Cotton Council, Memphis, pp 364–366

Fischer RA (1985) Number of kernels in wheat crops and the influence of solar radiation and temperature. J Agric Sci Camb 105:447–461

Freytag GF, Debouck DG (2002) Taxonomy, distribution and ecology of the genus *Phaseolus* (Leguminosae- Papilionoideae) in North America, Mexico and Central America. SIDA Bot Miscel 23:1–300

Frova C, Sari-Gorla M (1994) Quantitative trait loci (QTLs) for pollen thermotolerance detetected in maize. Mol Gen Genet 245:424–430

Garver S (1934) The Redfield tepary bean, an early maturing variety. J Am Soc Agron 3:397–403

Gautney TL, Haynes FL (1983) Recurrent selection of heat tolerance in diploid potatoes (*Solanum tuberosum* subsp. *phureja* and *stenotomum*). Am J Potato Res 60:537–542

Gifford RM (1986) Partitioning of photoassimilate in the development of crop yield. In: Lucas WJ, Cronshaw J (eds) Phloem transport. Alan R. Liss, New York, pp 535–549

Gipson JR, Joham HE (1968) Influence of night temperature on growth and development of cotton (*Gossypium hirsutum* L.). I. Fruiting and boll development. Agron J 60:292–295

Gross Y, Kigel J (1994) Differential sensitivity to high temperatures of stages in the reproductive development of common bean (*Phaseolus vulgaris* L.). Field Crops Res 36:201–212

Gwathmey CO, Hall AE (1992) Adaptation to midseason drought of cowpea genotypes with contrasting senescence traits. Crop Sci 32:773–778

Hall AE (1992) Breeding for heat tolerance. Plant Breed Rev 10:129–168

Hall AE (1993) Physiology and breeding for heat tolerance in cowpea and comparisons with other crops. In: Kuo CG (ed) Adaptation of food crops to temperature and water stress, Publ No 93-410. Asian Vegetable Research and Development Center, Shanua, pp 272–284

Hall AE (2001) Crop responses to environment. CRC, Boca Raton, FL

Hall AE (2004) Comparative ecophysiology of cowpea, common bean and peanut. In: Nguyen HT, Blum A (eds) Physiology and biotechnology integration for plant breeding. Marcel Dekker, New York, pp 271–325

Hall AE (2011) Breeding cowpea for future climates. In: Yadav SS, Redden RJ, Hatfield JL, Lotze-Campen H, Hall AE (eds) Crop adaptation to climate change. Wiley-Blackwell, Chichester, pp 340–355

Hall AE, Allen LH Jr (1993) Designing cultivars for the climatic conditions of the next century. In: Buxton DR, Allen LH (eds) International crop science I. Crop Science Society of America, Madison, WI, pp 291–297

Hall AE, Ziska LH (2000) Crop breeding strategies for the 21st century. In: Reddy KR, Hodges HF (eds) Climate change and global crop productivity. CABI Publishing, Oxon, pp 407–423

Heckathorn SA, Downs CA, Sharkey TD, Coleman JS (1989) The small methionine-rich chloroplast heat-shock protein protects photosystem II electron transport during heat stress. Plant Physiol 116:439–444

Hede AR, Skovmand B, Reynolds MP, Crossa J, Vilhelmsen AL, Stolen O (1999) Evaluating genetic diversity for heat tolerance traits in Mexican wheat landraces. Genet Resour Crop Evol 46:37–45

Herrero MP, Johnson RR (1980) High temperature stress and pollen viability in maize. Crop Sci 20:796–800

Hong S, Lee U, Vierling E (2003) Arabidopsis hot mutants define multiple functions required for acclimation to high temperatures. Plant Physiol 132:757–767

Hozain M, Abdelmageed H, Lee J, Kang M, Fokar M, Allen RD, Holaday AS (2012) Expression of AtSAP5 in cotton up-regulates putative stress-responsive genes and improves the tolerance to rapidly developing water deficit and moderate heat stress. J Plant Physiol 169 (13):1261–1270

Hurd EA (1971) Can we breed for drought resistance? In: Larson KL, Eastin JD (eds) Drought injury and resistance in crops, CSSA Special Publication Number 2. Crop Science Society of America, Madison, WI, pp 77–88

Ishimaru T, Hirabayashi H, Ida M, Takai T, San-Oh YA, Yoshinaga S, Ando I, Ogawa T, Kondo M (2010) A genetic resource for early-morning flowering trait of wild rice *Oryza officinalis* to mitigate high temperature-induced spikelet sterility at anthesis. Ann Bot 106:515–520

Ismail AM, Hall AE (1998) Positive and negative effects of heat-tolerance genes in cowpea lines. Crop Sci 38:381–390

Ismail AM, Hall AE (1999) Reproductive-stage heat tolerance, leaf membrane thermostability and plant morphology in cowpea. Crop Sci 39:1762–1768

Ismail AM, Hall AE (2000) Semidwarf and standard-height cowpea responses to row spacing in different environments. Crop Sci 40:1618–1623

Jefferson RA (1993) Beyond model systems-new strategies, methods and mechanisms for agricultural research. Ann NY Acad Sci 700:53–73

Kimball BA (1983) Carbon dioxide and agricultural yield: an assembly and analysis of 430 prior observations. Agron J 75:779–788

Kittock DL, Turcotte EL, Hofman WC (1988) Estimation of heat tolerance improvement in recent American Pima cotton cultivars. J Agron Crop Sci 161:305–309

Konsens I, Ofir M, Kigel J (1991) The effect of temperature on the production and abscission of flowers and pods in snap bean (*Phaseolus vulgaris* L.). Ann Bot 67:391–399

Krishnamurthy L, Gaur PM, Basu PS, Chaturvedi SK, Tripathi S, Vadez V, Rathore A, Varshney RK, Gowda CLL (2011) Large genetic variation for heat tolerance in the reference collection of chickpea (*Cicer arietinum* L.) germplasm. Plant Genet Resour 9:59–69

Lafarge T, Peng S, Hasegawa T, Quick WP, Jagadish SVK, Wassman R (2011) Genetic adjustment to changing climates: rice In: Yadav SS, Redden RJ, Hatfield JL, Lotze-Campen H, Hall AE (eds) Crop adaptation to climate change. Wiley-Blackwell, Chichester, pp 298–313

Leakey ADB, Ainsworth EA, Bernacchi CJ, Rogers A, Long SP, Ort DR (2009) Elevated $CO_2$ effects on plant carbon, nitrogen and water relations: six important lessons from FACE. J Exp Bot 60:2859–2876

Leithold B, Müller G, Weber WE, Westermann T (1997) Investigations of heat tolerance of spring wheat varieties of different origins under growth chamber conditions. J Agron Crop Sci 179:115–122

Li PH, Udomprasert N (1993) Improving performance of *Phaseolus vulgaris* in high-temperature environments by heat acclimation potential. In: Kuo CG (ed) Adaptation of food crops to temperature and water stress, Publ No 93-410. Asian Vegetable Research and Development Center, Shanua, pp 303–315

Liao J-L, Zhang H-Y, Shao X-L, Zhong P-A, Huang Y-J (2011) Identification for heat tolerance in backcross recombinant lines and screening of backcross introgression lines with heat tolerance at milky stage in rice. Rice Sci 18:279–286

Lin W, Ziska LH, Namuco OS, Bai K (1997) The interaction of high temperature and elevated CO2 on photosynthetic acclimation of single leaves of rice in situ. Physiol Plant 99:178–184

Lu Z, Percy RG, Qualset CO, Zeiger E (1998) Stomatal conductance predicts yields in irrigated Pima cotton and bread wheat grown at high temperatures. J Exp Bot 49:453–460

Lyons JM (1973) Chilling injury in plants. Annu Rev Plant Physiol 24:445–466

Mackill DJ, Coffman WR (1983) Inheritance of high temperature tolerance and pollen shedding in rice. Z Pflanzenzücht 91:61–69

Mackill DJ, Coffman WR, Rutger JN (1982) Pollen shedding and combining ability for high temperature tolerance in rice. Crop Sci 22:730–733

Marfo KO, Hall AE (1992) Inheritance of heat tolerance during pod set in cowpea. Crop Sci 32:912–918

Markhart AH (1985) Comparative water relation of *Phaseolus vulgaris* L. and *Phaseolus acutifolius* Gray. Plant Physiol 77:113–117

Matsui T, Namuco OS, Ziska LH, Horie T (1997) Effects of high temperature and CO2 concentration on spikelet sterility in indica rice. Field Crops Res 51:213–219

Miklas PN, Rosas JC, Beaver JS, Telek L, Freytag GF (1994) Field performance of select tepary bean germplasm in the tropics. Crop Sci 34:1639–1644

Mohamed MF, Keutgen N, Tawfik AA, Noga G (2002) Dehydration-avoidance response of tepary bean lines differing in drought resistance. J Plant Physiol 159:31–38

Mohammed AR, Tarpley L (2009) High night temperatures affect rice productivity through altered pollen germination and spikelet fertility. Agric For Meteorol 149:999–1008

Mohammed AR, Tarpley L (2010) Effects of high night temperature and spikelet position on yield-related parameters of rice (*Oryza sativa* L.) plants. Eur J Agron 33:117–123

Moya TB, Ziska LH, Namuco OS, Olsyk D (1998) Growth dynamics and genotypic variation in tropical, field-grown paddy rice (*Oryza sativa* L.) in response to increasing carbon dioxide and temperature. Glob Change Biol 4:645–656

Murakami Y, Tsuyama M, Kobayashi Y, Kodama H, Iba K (2000) Trienoic fatty acids and plant tolerance of high temperature. Science 287:476–479

Mutters RG, Hall AE (1992) Reproductive responses of cowpea to high temperatures during different night periods. Crop Sci 32:202–206

Mutters RG, Ferreira LGR, Hall AE (1989a) Proline content of the anthers and pollen of heat-tolerant and heat-sensitive cowpea subjected to different temperatures. Crop Sci 29:1497–1500

Mutters RG, Hall AE, Patel PN (1989b) Photoperiod and light quality effects on cowpea floral development at high temperatures. Crop Sci 29:1501–1505

Nabhan GP, Felger RS (1978) Teparies in the Southwestern North America. Econ Bot 3–19

Nakashima H, Horner HT, Palmer RG (1984) Histological features of anthers from normal and ms3 mutant soybean. Crop Sci 24:735–739

Nielsen CL, Hall AE (1985a) Responses of cowpea (*Vigna unguiculata* (L.) Walp.) in the field to high night temperatures during flowering. I. Thermal regimes of production zones and field experimental system. Field Crops Res 10:167–179

Nielsen CL, Hall AE (1985b) Responses of cowpea (*Vigna unguiculata* (L.) Walp.) in the field to high night temperatures during flowering. II. Plant responses. Field Crops Res 10:181–196

Opeňa RT, Kuo GC, Yoon JY (1987) Breeding for stress tolerance under tropical conditions in tomato and heading Chinese cabbage. In: Improved vegetable production in Asia, FFTC Book Series No 36. ASPAC-FFTC, Taipei, pp 88–109

Opeňa RT, Green SK, Talekar NS, Chen JT (1989) Genetic improvement of tomato adaptability to the tropics: progress and future prospects. In: Green SK (sced.), Griggs TD, Mc Lean BT (publeds) Tomato and pepper production in the tropics. Proceedings of the international symposium on integrated management practices. AVRDC, Shanhua, pp 71–85

Opeňa RT, Chen JT, Kuo CG, Chen HM (1993) Genetic and physiological aspects of tropical adaptation in tomato. In: Kuo CG (ed) Adaptation of food crops to temperature and water stress, Publ No 93-410. Asian Vegetable Research and Development Center, Shanua, pp 321–334

Padi FK, Denwar NN, Kaleem FZ, Salifu AB, Clottey VA, Kombiok J, Haruna M, Hall AE, Marfo KO (2004a) Registration of 'Apagbaala' cowpea. Crop Sci 44:1486

Padi FK, Denwar NN, Kaleem FZ, Salifu AB, Clottey VA, Kombiok J, Haruna M, Hall AE, Marfo KO (2004b) Registration of 'Marfo-Tuya' cowpea. Crop Sci 44:1486–1487

Patel PN, Hall AE (1986) Registration of snap-cowpea germplasms. Crop Sci 26:207–208

Patel PN, Hall AE (1990) Genotypic variation and classification of cowpea for reproductive responses to high temperature and long photoperiods. Crop Sci 30:614–621

Patel PN, Singh D (1984) New bacterial blight resistant vegetable cowpeas in India. Trop Grain Legume Bull 29:14–18

Peacock JM (1982) Response and tolerance of sorghum to temperature stress. In: House LR et al (eds) Sorghum in the Eighties. ICRISAT, Patencheru, pp 143–159

Peacock JM, Miller WB, Matsuda K, Robinson DL (1990) Role of heat girdling in early seedling death of sorghum. Crop Sci 30:138–143

Peacock JM, Soman P, Jayachandran R, Rani AU, Howarth CJ, Thomas A (1993) Effects of high soil surface temperature on seedling survival in pearl millet. Exp Agric 29:215–225

Peet MM, Sato S, Gardner R (1998) Comparing heat stress on male-fertile and male-sterile tomatoes. Plant Cell Environ 21:225–231

Peng S, Huang J, Sheehy JE, Laza RC, Visperas RM, Zhong X, Centeno GS, Kush GS, Cassman KG (2004) Rice yields decline with higher night temperature from global warming. Proc Natl Acad Sci USA 101:9971–9975

Percy RG, Turcotte EL (1997) Registration of 10 Pima cotton germplasm lines, P70 to P79. Crop Sci 37:632–633

Poorter H (1993) Interspecific variation in the growth responses of plants to an elevated ambient CO2 concentration. Vegetation 104(105):77–97

Porch TG (2001) Genetics and applications of heat tolerance in common bean. PhD Dissertation, Cornell University, Ithaca, NY

Porch TG (2006) Application of stress indices for heat tolerance screening of common bean. J Agron Crop Sci 192:390–394

Porch TG, Beaver JS, Brick MA (2013) Registration of tepary germplasm with multiple-stress tolerance, TARS-Tep 22 and TARS-Tep 32. J Plant Regist (accepted)

Porch TG, Jahn M (2001) Effects of high-temperature stress onmicrosporogenesis in heat-sensitive and heat-tolerant genotypes of *Phaseolus vulgaris*. Plant Cell Environ 24:723–731

Porch TG, Dickson MH, Long MC, Viands DR, Jahn M (2004) General combining ability effects for reproductive heat tolerance in snap bean. J Agric Univ Puerto Rico 88:161–164

Porch TG, Bernsten R, Rosas JC, Jahn M (2007) Climate change and the potential economic benefits of heat tolerant bean varieties for farmers in Atlántida, Honduras. J Agric Univ Puerto Rico 91:133–148

Porch TG, Smith JR, Beaver JS, Griffiths PD, Canaday CH (2010) TARS-HT1 and TARS-HT2 heat-tolerant dry bean germplasm. HortScience 45:1278–1280

Porch TG, Urrea CA, Beaver JS, Valentin S, Peña PA, Smith R (2012) Registration of TARS-MST1 and SB-DT1 multiple-stress tolerant black bean germplasm. J Plant Regist 6:75–80

Prasad PVV, Craufurd PQ, Summerfield RJ (1999a) Sensitivity of peanut to timing of heat stress during reproductive development. Crop Sci 39:1352–1357

Prasad PVV, Craufurd PQ, Summerfield RJ (1999b) Fruit number in relation to pollen production and viability in groundnut exposed to short periods of heat stress. Ann Bot 84:381–386

Prasad PVV, Craufurd PQ, Summerfield RJ, Wheeler TR (2000) Effects of short periods of heat stress on flower production, pod yield and yield components of groundnut (*Arachis hypogaea* L.). J Exp Bot 51:777–784

Prasad PVV, Pisipati SR, Mutava RN, Tuinistra MR (2008) Sensitivity of grain sorghum to high temperature stress during reproductive development. Crop Sci 48:1911–1917

Pratt RC, Nabhan GP (1988) Evolution and diversity of *Phaseolus acutifolius* genetic resources. In: Gepts P (ed) Genetic resources of *Phaseolus* beans. Kluwer Academic, Dordrecht, pp 409–440

Queitsch C, Hong S, Vierling E, Lindquist S (2000) Heat shock protein 101 plays a crucial role in thermotolerance in Arabidopsis. Plant Cell 12:479–492

Rainey KM, Griffiths PD (2005a) Inheritance of heat tolerance during reproductive development in snap bean (*Phaseolus vulgaris* L.). J Am Soc Hortic Sci 135:700–706

Rainey KM, Griffiths PD (2005b) Diallel analysis of yield components of snap bean exposed to two temperature stress environments. Euphytica 142:43–53

Reddy KR, Hodges HF, McKinion JM, Wall GW (1992) Temperature effects on Pima cotton growth and development. Agron J 84:237–243

Reddy KR, Hodges HF, McKinion JM (1995) Carbon dioxide and temperature effects on Pima cotton development. Agron J 87:820–826

Reddy KR, Hodges HF, McKinion JM (1997) A comparison of scenarios for the effect of global climate change on cotton growth and yield. Aust J Plant Physiol 24:707–713

Reddy BVS, Kumar AA, Ramesh S, Reddy PS (2011) Sorghum genetic enhancement for climate change adaptation. In: Yadav SS, Redden RJ, Hatfield JL, Lotze-Campen H, Hall AE (eds) Crop adaptation to climate change. Wiley-Blackwell, Chichester, pp 326–339

Redoña ED, Laza MA, Manigbas NL (2007) Breeding rice for adaptation and tolerance to high temperatures. Proceedings of the international workshop on cool rice for a warmer world, Huazhong Agricultural University, Wuhan, Hubei, 26–30 Mar 2007

Reynolds MP, Balota M, Delgado MIB, Amani I, Fischer RA (1994) Physiological and morphological traits associated with spring wheat yield under hot, irrigated conditions. Aust J Plant Physiol 21:717–730

Reynolds MP, Singh RP, Ibrahim A, Ageeb OAA, Larqué-Saavedra A, Quick JS (1998) Evaluating physiological traits to complement empirical selection for wheat in warm environments. Euphytica 100:85–94

Reynolds MP, Ortiz-Monasterio JI, Mcnab A (2001) Application of physiology in wheat breeding. CIMMYT, El Batan

Rodriquez-Garay B, Barrow JR (1988) Pollen selection for heat tolerance in cotton. Crop Sci 28:857–859

Román-Avilés B, Beaver J (2001) Heritability of heat tolerance of an Andean bean population. Annu Rep Bean Improv Coop 44:49–50

Román-Avilés B, Beaver J (2003) Inheritance of heat tolerance in common bean of Andean origin. J Agric Univ Puerto Rico 87:113–121

Rosas JC, Castro A, Beaver J, Perez CA, Morales A, Lepiz R (1999) Mejoramiento genético para tolerancia a altas temperaturas y resistencia a mosaico dorado en friojol común. XLV Reunion Anualdel PCCMCA

Rosas JC, Beaver JS, Escoto D, Pérez CA, Llano A, Hernández JC (2004) Registration of 'Amadeus 77' small red common bean. Crop Sci 44:1867–1868

Saini HS, Sedgley M, Aspinall D (1984) Developmental anatomy in wheat of male sterility induced by heat stress, water deficit or abscisic acid. Aust J Plant Physiol 11:243–253

Scott JW (1999) University of Florida tomato breeding accomplishments and future directions. Soil Crop Sci Soc FL 58:8–11

Scott JW, Jones JB, Somodi GC, Chellemi DO, Olson SM (1995a) 'Neptune', a heat-tolerant, bacterial-wilt-tolerant tomato. HortScience 30:641–642

Scott JW, Olson SM, Howe TK, Stoffella PJ, Bartz JA, Bryan HH (1995b) 'Equinox' heat-tolerant hybrid tomato. HortScience 30:647–648

Scott JW, Olson SM, Bryan HH, Bartz JA, Maynard DN, Stoffella PJ (2006) 'Solar Fire' hybrid tomato: Fla. 7776 tomato breeding line. HortScience 41:1504–1505

Shelby RA, Greenleaf WH, Peterson CM (1978) Comparative floral fertility in heat tolerant and heat sensitive tomatoes. J Am Soc Hortic Sci 103:778–780

Shisanya CA (2002) Improvement of drought adapted tepary bean (*Phaseolus acutifolius* A. Gray var. *latifolius*) yield through biological nitrogen fixation in semi-arid SE-Kenya. Eur J Agron 16:13–24

Shonnard GC (1991) Genetics of and selection for heat tolerance during reproductive development in common bean. PhD Dissertation, University of California, Davis, CA

Shonnard GC, Gepts P (1994) Genetics of heat tolerance during reproductive development in common bean. Crop Sci 34:1168–1175

Shpiler L, Blum A (1991) Heat tolerance for yield and its components in different wheat cultivars. Euphytica 51:257–263

Singh RP, Singh J, Lal CB, Sunita K, Elayaraja K (2003) Development of suitable cotton *G. hirsutum* L. plant type tolerant to high diurnal temperature with suitability to grow during spring/summer season. ICAR News 9:14–16

Singh RP, Singh J, Lal CB, Sunita K, Elayaraja K (2004) Evaluation of Punjab American cotton (*G. hirsutum* L.) genotypes tolerant to high temperature and their performance during spring summer season. Ann Agric Res 25:268–273

Singh RP, Prasad PVV, Sunita K, Giri SN, Reddy KR (2007) Influence of high temperature and breeding for heat tolerance in cotton: a review. Adv Agron 93:313–385

Singh RP, Prasad PVV, Sharma AK, Reddy KR (2011) Impacts of high-temperature stress and potential opportunities for breeding. In: Yadav SS, Redden RJ, Hatfield JL, Lotze-Campen H, Hall AE (eds) Crop adaptation to climate change. Wiley-Blackwell, Chichester, pp 166–185

Smith H, Whitelam GC (1990) Phytochrome, a family of photoreceptors with multiple physiological roles. Plant Cell Environ 13:695–707

Smith JR, Park SJ, Beaver JS, Miklas PN, Canaday CH, Zapata M (2007) Registration of TARS-SR05 multiple disease resistant dry bean germplasm. Crop Sci 47:457–458

Soman P, Peacock JM (1985) A laboratory technique to screen seedling emergence of sorghum and pearl millet at high soil temperature. Exp Agric 21:335–341

Soomro AR (1998) Central Cotton Research Institute, Pakistan, The Institute Profile. SAIC (SAARC Agricultural Information Center) Newsl 8:4

Staub JE, Krasowska A (1990) Screening of the U.S. germplasm collection for heat tolerance. Cucurbit Genet Coop Rep 13:4–7

Stevens MA (1979) Breeding tomatoes for processing. In: Cowell R (ed) Proceedings of the first international symposium on tropical tomato. Asian Vegetable Research and Development Center, Shanhua, pp 201–213

Tewolde H, Fernandez CJ, Erickson CA (2006) Wheat cultivars adapted to post-heading high temperature stress. J Agron Crop Sci 192:111–120

Thiaw S, Hall AE (2004) Comparison of selection for either leaf-electrolyte-leakage or pod set in enhancing heat tolerance and grain yield of cowpea. Field Crops Res 86:239–253

Trethowan RM, Mahmood T (2011) Genetics options for improving the productivity of wheat in water-limited and temperature-stressed environments. In: Yadav SS, Redden RJ, Hatfield JL, Lotze-Campen H, Hall AE (eds) Crop adaptation to climate change. Wiley-Blackwell, Chichester, pp 218–237

Trethowan RM, Turner MA (2009) New targets in wheat improvement. CAB reviews: perspectives in agriculture, veterinary science, nutrition and natural resources. CAB International, Wallingford

Turcotte EL, Feaster CV, Young EF (1991) Registration of six American Pima cotton germplasm lines. Crop Sci 31:495

Turcotte EL, Percy RG, Feaster CV (1992) Registration of Pima S-7 American Pima cotton. Crop Sci 32:1291

Vargas M, Crossa J, Sayre K, Reynolds M, Ramírez ME, Talbot M (1998) Interpreting genotype x environment interaction in wheat by partial least squares regression. Crop Sci 38:679–689

Vierling E (1991) The roles of heat-shock proteins in plants. Annu Rev Plant Physiol 42:579–620

Villareal RL, Lai SH (1979) Development of heat tolerant tomato varieties in the tropics. In: Cowell R (ed) Proceedings of the first international symposium on tropical tomato. Asian Vegetable Research and Development Center, Shanhua, pp 188–200

Wallace DH (1980) The genetics of photosynthesis and crop productivity: with emphasis on beans. In: Vartanian ME (ed) Proceedings of the 14th international congress of genetics, vol 1, book 2. MIR Publ, Moscow, pp 306–317

Warrag MOA, Hall AE (1983) Reproductive responses of cowpea to heat stress: genotypic differences in tolerance to heat at flowering. Crop Sci 23:1088–1092

Warrag MOA, Hall AE (1984a) Reproductive responses of cowpea (*Vigna unguiculata* (L.) Walp.) to heat stress. I. Responses to soil and day air temperatures. Field Crops Res 8:3–16

Warrag MOA, Hall AE (1984b) Reproductive responses of cowpea (*Vigna unguiculata* (L.) Walp.) to heat stress. II. Responses to night air temperature. Field Crops Res 8:17–33

Wasonga CJ, Pastor-Corrales MA, Porch TG, Griffiths PD (2010) Targeting gene combinations for broad spectrum rust resistance in heat tolerant snap beans developed for tropical environments. J Am Soc Hortic Sci 135:521–532

Wasonga CJ, Pastor-Corrales MA, Porch TG, Griffiths PD (2012) Predicting heat tolerance and rust resistance in snap bean genotypes targeted for East Africa through controlled environment selection. HortScience 47:1000–1006

White JW, Kornegay J, Cajiao C (1996) Inheritance of temperature sensitivity of the photoperiod response in common bean (*Phaseolus vulgaris* L.). Euphytica 91:5–8

Wien HC (1997) Pepper. In: Wien HC (ed) The physiology of vegetable crops. CAB International, Wallingford, pp 259–293

Wilson GL, Raju PS, Peacock JM (1982) Effect of soil temperature on seedling emergence in sorghum. Indian J Agric Sci 52:848–851

World Bank (2010) World development report 2010: development and climate change. The World Bank, Washington, DC, 439 p

Yang J, Sears RG, Gill BS, Paulsen GM (2002) Quantitative and molecular characterization of heat tolerance in hexaploid wheat. Euphytica 126:275–282

Yarwood CE (1961) Acquired tolerance of leaves to heat. Science 134:941–942

Yoshida S, Satake T, Mackill DJ (1981) High-temperature stress in rice, IRRI Research Paper Series Number 67. International Rice Research Institute, Manila, p 15

Yoshida T, Ohama N, Nakajima J, Kidokoro S, Mizoi J, Nakashima K, Maruyama K, Kim J-M, Seki M, Todaka D, Osakabe Y, Sakuma Y, Schöz F, Shinozaki K, Yamaguchi-Shinozaki K (2012) Arabidopsis HsfA1 transcription factors function as the main positive regulators in heat shock-responsive gene expression. Mol Genet Genomics 286:321–332

Yu X, Wang H, Lu Y, de Ruiter M, Cariaso M, Prins M, van Tunen A, He Y (2012) Identification of conserved and novel microRNAs that are responsive to heat stress in *Brassica rapa*. J Exp Bot 63:1025–1038

Yuan S-C, Zhang Z-G, He H-H, Zen H-L, Lu K-Y, Lian J-H, Wang B-X (1993) Two photoperiod-reactions in photoperiod-sensitive genic male-sterile rice. Crop Sci 33:651–660

Ziguo Z, Hanlai Z (1992) Fertility alteration in photoperiod-sensitive genic male sterile (PGMS) rice in response to photoperiod and temperature. Int Rice Res Newsl 17:7–8

Ziska LH, Manalo PA (1996) Increasing night temperature can reduce seed set and potential yield of tropical rice. Aust J Plant Physiol 23:791–794

# Chapter 5
# Drought Tolerance

Rodomiro Ortiz

**Abstract** Plant breeding has a limited success for developing new cultivars with enhanced adaptation to drought-prone environments, although it has been pursued for various decades. Water use efficiency and water productivity by crops are being sought by agricultural researchers to address water scarcity in drought-prone environments across the world. They may be improved through genetic enhancement. Research on the mechanisms underlying the efficient use of water by crops and water productivity remains essential for succeeding in this endeavor. Advances in genetics, "omics," precise phenotyping, and physiology coupled with new developments in bioinformatics and phenomics can provide new insights on traits that enhance adaptation to water scarcity. This chapter provides an update on research advances and breeding main grain crops for drought-prone environments.

## 5.1 Introduction

Water scarcity is among the key factors that limit crop production because drought stress affects yields. Global models predict a further increase of drought under climate change (Gornall et al. 2010; Turral et al. 2011). Moreover, irrigation will be important to ensure sustainable intensification of agriculture. Adapting crops to water-limited environments and improving their efficiency for water use, which have been always major objectives of plant breeding, will be key elements for developing climate-resilient cultivars that are capable of producing "more food per drop" (Ortiz et al. 2007).

Crop yields under water scarcity are a function of the amount of water used by the crop, how efficiently the crop uses this water for biomass growth (i.e., water use

R. Ortiz (✉)
Department of Plant Breeding and Biotechnology, Swedish University of Agricultural Sciences, Sundsvagen 14, P.O. Box 101, 23053 Alnarp, Sweden
e-mail: rodomiro.ortiz@slu.se

efficiency or aboveground biomass/water use), and the harvest index, which is the proportion of "edible yield" to aboveground biomass (Passioura 2004). The ratio of total dry matter accumulation to evapotranspiration and other water losses (or water entering and lost from the system that is not transpired through the plant) is known as water use efficiency (WUE). An increase in transpiration efficiency or a reduction in soil evaporation will therefore increase WUE. The concept of water productivity (WP) was recently defined as the net return for water use (Molden et al. 2010), i.e., the ratio of biomass with economic value (e.g., grain yield of cereals) compared to the amount of water transpired. Both WUE and WP may be improved through plant breeding (Farooq et al. 2009), as can biomass accumulation and harvest index. Understanding the mechanisms underlying the efficient use of water by crops is essential succeeding this endeavor (Chaves and Oliveira 2004). Such knowledge will lead to identifying key genes that will be further used in developing cultivars with enhanced adaptation to drought-prone environments.

Global spatial data on crop production, climate, and poverty were used to identify high-priority geographic areas to target international plant breeding undertakings for drought-prone environments (Hyman et al. 2008). These areas include 15 major farming systems, especially in South Asia, the Sahel, and Eastern and Southern Africa, where high diversity in drought frequency characterizes the environments. Furthermore, the decrease in perennial drainage will significantly affect present surface water access across 25 % of Africa by the end of this century (de Wit and Stankiewicz 2006). About 13 crops (especially cereals, legumes, and root crops) make up the bulk of food production in these drought-prone areas of Africa and Asia.

## 5.2 Getting Terms Right

Although the term "drought resistance" has been used in the literature (Blum 2005), this author, as noted above, prefers to use the concepts of WUE or WP when breeding crops for adapting to drought-prone environments. As indicated by Passoura (2007), the word "drought" has many meanings in relation to crop production, but plant breeders—together with agronomists and crop physiologists—should seek minimizing evaporative losses from the soil surface by better matching cultivar development to its target environment. There are of course constitutive traits that allow a crop to maintain a high plant water status (i.e., dehydration avoidance) under stress but they also contribute to WUE and WP.

Water productivity could be also regarded, in its broadest sense, as "productive" because transpiration is the only water flow in a field actually passing through the crop. WP can be therefore improved through crop breeding by reducing nontranspirational uses of water, lowering transpiration without jeopardizing production, increasing production without elevating transpiration, and enhancing crop adaptation to drought stress (Bennet 2003).

**Fig. 5.1** Total number of publications per year on plant breeding for drought-prone environments or water use efficiency from leading journals: *Crop Science, Euphytica, Molecular Breeding, Plant Breeding,* and *Theoretical and Applied Genetics* (1990–June 2012 including online ahead of press)

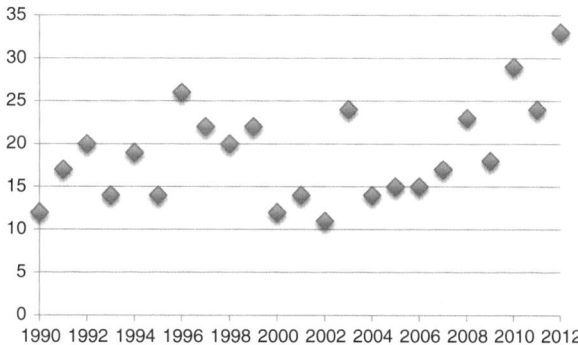

## 5.3   Crop Breeding for Drought-Prone Environments: Last Two Decades

Although, as acknowledged elsewhere (Tester and Bacic 2005), there has been a limited success in breeding new cultivars with enhanced adaptation to drought-prone environments, research on the subject has been in the agenda of plant breeding, crop physiology, and more recently molecular biology. The number of articles on this subject published since the 1990s (i.e., around the time that first-generation DNA markers began its use) in the five most cited international plant breeding journals (namely, *Crop Science, Euphytica, Molecular Breeding, Plant Breeding,* and *Theoretical and Applied Genetics*) illustrates these research investments (Fig. 5.1).

After lowering the number of journal articles towards the beginning of the last decade, the volume of publications on crop breeding for drought-prone environ-ment, and to a lesser extent on the application of DNA markers for assisting in this endeavor, increased and reached its peak in mid-2012. This finding may be associated to the launching in 2003 of the Generation Challenge Program (GCP) by the Consultative Group on International Agricultural Research (CGIAR). The GCP network (involving 200 partners worldwide) uses the rich pool of crop genetic diversity and advanced plant science (including genomic tools) to improve populations and cultivars for drought-prone environment. GCP facilitated plant breeding and research on 18 crops in its first phase I (2004–2008), while in its second phase (2009–2013), GCP selected 12 key crops (mainly cereals, legumes, and roots crops) for further enhancing their performance under drought stress.

Most of the 435 articles in the most cited plant breeding journals are in cereals (55.5 %) and legumes (19.4 %). Maize, rice, and wheat dominate the literature among cereals, while soybean and beans are the predominant for legumes. Such result was not surprising because it reflects the relative importance of each crop and the associated research investments. Several articles, especially in *Crop Science,* are related to cotton, forage species, sunflower, and turfgrass. Likewise, due to the scope of this journal, some articles are not solely on plant breeding but on related

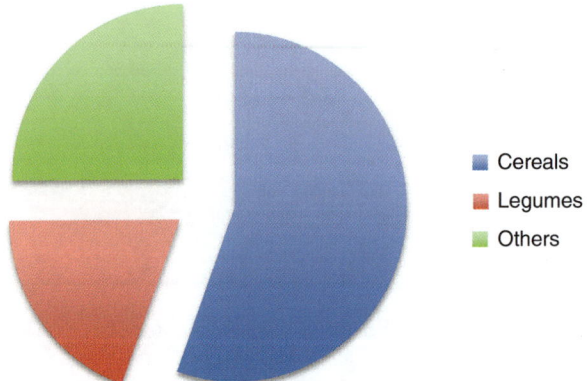

**Fig. 5.2** Percentage of publications according to crop groups per year on plant breeding for drought-prone environments or water use efficiency from leading journals: *Crop Science*, *Euphytica*, *Molecular Breeding*, *Plant Breeding*, and *Theoretical and Applied Genetics* (1990–June 2012 including online ahead of press). Cereals include barley, maize, pearl millet, rice, sorghum, and wheat, whereas bean, chickpea, cowpea, faba bean, groundnut (or peanut), lentil, pigeon pea, and soybean are among legumes. Other crops are cassava, canola (or oil seed rape), coconut, cotton, forages, Indian mustard, *Musa*, potato, *Salix*, sugar beet, sunflower, strawberry, and tomato. Review articles relevant to the subject (including screening methods or modeling) are also in this group

crop physiology research—especially for identifying traits or screening protocols (Fig. 5.2).

## 5.4   Advances in Plant Breeding Under Water Stress

Advances in genetics, "omics," precise phenotyping, and physiology coupled with new developments in bioinformatics and phenomics are or will be providing means for dissecting integrative traits that affect adaptation to stressful environments. In this regard, Tardieu and Tuberosa (2010) indicate that analyzing the effect of traits on crop yield with the aid of modeling and confirming through field experiment (and sound biometrics) will lead to identifying alleles with favorable alleles for enhancing adaptation to a stress-prone environment.

Some traits used as proxy for selecting germplasm with enhanced adaptation to drought-prone environments (especially among grain crops) are anthesis–silk interval, early flowering (that could provide partial relief to water shortage during grain filling), floral fertility (by minimizing severe water deficit-induced damage at flowering), early vigorous growth (which improves crop establishment and reduces soil evaporation), root architecture and size (for optimizing water and nutrient harvest), and tiller inhibition (that increases tiller survival rates and carbohydrate storage in stems for ensuring further grain filling), among others (Tuberosa et al.

2007a). Likewise, indirect selection has been used for improving WUE, e.g., through canopy temperature depression, carbon isotope discrimination ($\Delta^{13}C$) for $C_3$ crops (although both may differ across locations), and ear photosynthesis (Araus et al. 2007; Tambussi et al. 2007). Highlights of research advances in plant breeding for drought-prone environments are given below. They are drawn from the most important grain crops that feed the world: namely, maize, rice, wheat, and leading grain legumes.

### 5.4.1 Maize

Drought causes 17 % of annual average losses in maize (Edmeades et al. 1989), but it can be up to 60 % in Southern Africa (Rosen and Scott 1992). Drought at 1 or 2 weeks before anthesis is a particularly sensitive period to water stress for maize (Grant et al. 1989) due to an increased anthesis–silking interval (ASI) (Edmeades et al. 1989), delayed silk emergence, and rising grain abortion, which occurs 2–3 weeks after silking (Boyle et al. 1991). Drought-induced changes in ASI are related to silk expansion, while delayed silk emergence results from reduced tissue expansion and cell division rates under water deficit (Fuad-Hassan et al. 2008). Drought stress between tassel emergence and early grain filling affects significantly grain yield due to a reduction in kernel size (Bolaños and Edmeades 1993a, b).

Hund et al. (2008) found that high WUE and sufficient water acquisition by a deep root system can increase adaptation of tropical maize lines to drought-prone environments. Although maize inbred lines with adaptation to drought-prone environments showed significantly higher $\Delta$ than susceptible ones at flowering, Monneveux et al. (2008b) indicated that carbon isotope discrimination will be only accurate for initial screening of lines or hybrids but not advanced selection among hybrids.

Significant grain yield gains have been noted under drought in maize breeding due to a significant increase in numbers of ears per plant and grains per ear and significant reductions in ASI, ovule number, and abortion rate during grain filling (Monneveux et al. 2006). Recurrent selection under drought improved tropical maize source populations for performance under water deficits (Bolaños et al. 1993). Monneveux et al. (2006) indicated that this result could ensue from enhanced partitioning of assimilates to the ear at flowering, at the expense of tassel and stem growth, whereas Zaidi et al. (2008) found that reduced 5-day ASI, barrenness, delayed senescence, and minimum loss of leaf chlorophyll can account for improved performance of maize breeding lines in Asian drought-prone environments. For further gains, maize should show more ears, larger grains, and smaller tassels (Monneveux et al. 2008a).

Bänziger et al. (2004) highlight the importance of choosing the right selection environments for breeding maize when targeting the highly variable drought stressful environments, e.g., in Southern Africa. Farmers should be further involved in selecting maize bred germplasm to ensure their suitability for such risk-prone locations (Foti et al. 2008). Due to this innovative client-oriented breeding approach, which started in the late 1990s, more than 50 new maize cultivars were released and are now grown in several million ha in sub-Saharan Africa (Bänziger et al. 2006; Tollefson 2011). Farmers and researchers managed jointly "mother–baby" trials for evaluating breeding materials in drought-prone environments. The "mother" trial involved up to 12 breeding materials sown under varied researcher-designed treatments, was close to the community, and was managed by schools, colleges, or extension agencies. The "babies" were satellite subsets of the "mother" trial, comprising approximately 4–6 cultivars in the fields of participating farmers using their own inputs and equipment. In this system public researchers and private sector partners created a network of regional "stress breeding" sites that, for the first time, provide objective information on how maize cultivars perform under drought-prone infertile soils thereby meeting the needs of poor farmers who had not previously benefited from maize breeding programs. At the beginning of 2011, the US private seed sector announced the release of new maize hybrids—bred by crossbreeding—with a yield increase of 5–15 % in water-scarce environments (Tollefson 2011). They are also developing transgenic maize cultivars aiming to tap a US$ multibillion seed market (Gilbert 2010). These genetically enhanced maize seed-embedded technologies targeting drought-prone environments may reach farmers soon.

Molecular markers have been used for mapping traits affecting maize performance under drought stress and may provide tools for continuing breeding genotypes with stable grain yields in stressful environments (Bruce et al. 2002; Ribaut and Ragot 2007). DNA-aided analysis has also provided means for the genetic dissection of maize performance under drought (Ribaut et al. 1996, 1997; Quarrie et al. 1999; Tuberosa et al. 2002; Ribaut and Ragot 2007). Table 5.1 lists some of these recent research advances. Backcrossing has been used for introgressing a few quantitative trait loci (QTLs) into elite maize lines. This breeding method does not, however, appear to be very effective when many QTLs of small effect are involved. Furthermore, QTLs are often germplasm specific and the costs for marker-aided selection (MAS) for many QTLs of small effect may be higher than those from conventional crossbreeding of maize. Likewise, many putative QTLs could be likely of limited use in applied breeding because they depend on genetic background or to their sensitivity to the environment (Campos et al. 2004). Hence, as indicated by Ortiz et al. (2007), the challenge is to identify QTLs of major effect that are independent of genetic background, not affected by the genotype-by-environment interaction, and to devise more effective breeding approaches for the application of the resultant markers. In this regard, multitrait multi-environment (MTME) QTL models can help to identify genome regions responsible for genetic correlations (useful for indirect selection through trait components) and how genetic correlations depend on the environmental conditions (Malosetti et al. 2007a). For example, Malosetti et al. (2007b) detected 36 QTLs affecting maize yield, anthesis–silking interval, male flowering, ear number, and plant height in drought and nitrogen stress. Likewise, single nucleotide polymorphisms (SNP) can be used in routine large-scale genomics-assisted marker

**Table 5.1** Some recent research advances in mapping quantitative trait loci (QTLs) affecting maize performance under drought

| Trait and major finding(s) | Reference |
| --- | --- |
| Maximum leaf elongation rate per unit thermal time accounted for by five QTL, among which three co-localized with QTL for anthesis–silk interval (ASI) of well-watered plants. Alleles conferring high leaf elongation rate conferred a high silk elongation rate (i.e., low ASI) | Welcker et al. (2007) |
| 81, 57, 51, and 34 QTLs were noted for target traits (male flowering, anthesis-to-silking interval, grain yield, kernel number, 100-kernel fresh weight, and plant height) in water stress and well-watered trials in Mexico and Zimbabwe, respectively. About 80, 60, and 6 % of the QTLs did not show significant QTL-by-environment interactions (QTL × E) in the joint analyses per environment, per water regime, and across all experiments. QTL clusters for different traits were identified in chromosomes 1, 3, and 5 in water stress trials | Messmer et al. (2009) |
| Specific seedling QTLs water stress: root dry weight, shoot dry weight, and leaf area-to-root length ratio. Ear number QTLs collocated with QTLs for shoot-to-root dry weight and leaf area-to-root length ratios. ASI QTLs collocated with QTLs for the numbers of crown roots and seminal roots irrespective of water supply | Ruta et al. (2010) |

development and gene discovery. Very recently, Lu et al. (2010) used 2,052 SNPs (including 659 from drought-response candidate genes) in joint linkage–linkage disequilibrium mapping to detect QTLs underlying maize adaptation to drought-prone environments.

## 5.4.2 Rice

Rice production systems are very vulnerable to drought stress, which affects 10 million ha of upland rice and over 13 million ha of rainfed lowland rice in Asia alone (Wassmann et al. 2009 and references therein). Rainfall distribution seems to be more important than total seasonal rainfall. A short dry spell at flowering leads to a significant decrease of grain yield and harvest index (Serraj et al. 2008), because water stress at this stage reduces grain formation more significantly than at other reproductive stages (Boonjung and Fukai 1996). A reduction on spikelet fertility and panicle exertion due to drought affects significantly grain yield. Drought can also inhibit the development of reproductive organs (egg and pollen) and processes (anther dehiscence, pollen shedding and germination, and fertilization).

Although the physiological basis of genetic variation in drought response remains unclear (Laffite et al. 2004), research advances in physiology, molecular biology, and genetics have contributed tremendously to the understanding of rice performance under drought (Table 5.2). A deep root system seems to be the most important target trait for improving rice grain yield in drought-prone environments (Gowda et al. 2011; Kato et al. 2011). Near-isogenic lines (NILs) have been very important tools for dissecting traits accounting for adaptation to drought stress in rice.

**Table 5.2** A sample of progress of rice physiology, molecular biology, and genetic research under drought stress

| Ensuing knowledge | Reference |
| --- | --- |
| Grain percentage (spikelet fertility) is the most sensitive yield component. Spikelet fertility in the top four rachis branches was reduced due to decreasing anther dehiscence and lowering stigma pollen density | Liu et al. (2006) |
| There was a significant variation on water use efficiency among the five upland and three lowland cultivars used. Cultivars showing high photosynthesis and conductance were more productive across a soil moisture gradient | Centritto et al. (2009) |
| Near-isogenic lines (NILs) with broad leaves had higher biomass (and its individual components), less stomatal conductance, and higher transpiration efficiency under drought than NILs with narrow and short leaves, which explain their improved performance | Farooq et al. (2010) |
| Dehydration avoidance (characterized by significantly higher growth rate and biomass, efficient partitioning (measured by improved harvest index)) and drought escape by accelerated heading under water scarcity are the main mechanisms for adapting to drought-prone environments | Guan et al. (2010) |
| Leaf water status and leaf elongation are not particularly sensitive to water deficit but poor root system may account for drought sensitivity | Parent et al. (2010) |
| Adaptation to drought related to deep root growth and water uptake ability | Gowda et al. (2011) |
| Enhanced rooting depth enhances dehydration avoidance and rice adaptation to drought stress, but there are other mechanisms contributing to significant yield increase in lowland drought-prone environments | Venuprasad et al. (2011) |
| Cytokinin-mediated source–sink modifications delay stress-induced senescence under drought | Peleg et al. (2011) |
| High-yielding lines had maximum root length and root volume compared, thereby indicating positive influence of drought avoidance root traits contributing to adaptation to this stress | Gouda et al. (2012) |

Appropriate choice of parents, selection criteria, and robustness of the managed screening protocols will contribute to the success of breeding rice germplasm for drought-prone environments (Wassmann et al. 2009). Intermittent stress, which is imposed by withholding irrigation during the period bracketing the entire flowering and grain-filling stages, seems to be a reliable screening for ranking cultivars' performance under drought (Laffite and Courtois 2002). Significant progress has been achieved in breeding upland and aerobic cultivars for water-scarce environments (Bernier et al. 2008). The breeding gains on rainfed lowland germplasm have been however relatively slow until recently, when several lines were released in India and the Philippines (Verulkar et al. 2010). The use of drought-selected introgression lines for rainfed or water-scarce rice-growing regions (Laffite et al. 2006) contributed to this achievement.

Serraj et al. (2009) provide a recent overview on rice breeding for drought-prone environment. They found that increased crop yield and water productivity need optimizing physiological processes at the initial critical stages of plant response to soil drying, water use efficiency, and dehydration-avoidance mechanisms. They advocate a holistic interdisciplinary approach that integrates plant breeding with physiological dissection of resistance traits and molecular genetic tools together with agronomical practices. Such an approach, according to these authors, should lead to a better conservation and use of soil moisture and matching rice cultivars with the environment.

QTL analysis of rice traits in drought-prone environments has been widely used (Laffite et al. 2004). QTLs were noted for some secondary traits associated with drought response, e.g., rooting depth, membrane stability, and osmotic adjustment. Most QTL research has been useful for identifying promising genome regions for potential use in rice breeding for drought-prone environments. For example, seven QTLs for carbon isotope discrimination were mapped in five chromosomal regions through composite interval analysis (Xu et al. 2009). Likewise, Bernier et al. (2009) found one QTL improving grain yield under severe drought stress (*qtl12.1*) mainly through a slight improvement (7 %) in plant water uptake under water scarcity. Very recently, 23 QTLs linked to plant phenology and production traits under stress were identified with the aid of a meta-analysis (Sellamuthu et al. 2011). Chromosomes 1 and 9 seem to bear QTLs for reproductive-growth traits and grain yield under drought stress. Serraj et al. (2011) indicated further that high-throughput, high-precision phenotyping systems will allow to map genes for yield components under stress and to assess their effects on drought-related traits. Such undertaking will assist to simultaneously incorporate the most promising genes into widely grown rice cultivars and enhance gene detection and delivery for use in marker-aided breeding.

African rice (*Oryza glaberrima*) shows adaptation to harsh growing environments and has been used for breeding a new rice for Africa (Jones et al. 1997). Bimpong et al. (2011a) found that the fraction of transpirable soil water was higher in some African rice accessions than in Asian rice lines. Such finding shows the ability of these African rice accessions to close their stomata early in response to drought stress to keep transpiration rate similar to the rate of uptake of soil water, thereby resulting in maintenance of the water balance of the plant. Further research by Bimpong et al. (2011b) using $BC_2F_3$ offspring deriving from IR64 and African rice found one QTL on chromosome 2, which accounts for 22 % of the genetic variation, affecting positively grain yield under stress.

## 5.4.3  Wheat

Water scarcity could be the most important abiotic stress affecting grain yield and quality of wheat. Climate change may exacerbate this stress due to more random weather events or by extending drought, especially in rainfed environments. Early drought in the growing season affects seed germination and crop establishment in wheat. Leaf expansion and tillering are also very sensitive to water stress in the vegetative stage. Water deficit before flower initiation can decrease the number of spikelet primordia and grain number during the spike growth period. Moreover, water stress 10 days before spike emergence decreases spikelets per spike of fertile tillers (thereby maximizing grain yield loss), while water deficit prior to anthesis accelerates plant development. Furthermore, water stress during grain reduces grain weight because of the shortening of the grain-filling period due to accelerated plant senescence.

There are various traits that could improve wheat yield in dry environments. They may enhance plant establishment, early canopy development, root growth and depth, or soil water use (Richards et al. 2001). WUE can be improved in wheat by traits such as deep roots, plant phenology, seedling vigor, tiller inhibition, high transpiration efficiency, or osmotic adjustment. Furthermore, canopy temperature, stomatal conductance, stay-green intermediate, or leaf rolling may indicate deep roots in the crop. These drought-adaptive traits can be associated to distinct genes controlling early (pre-anthesis) growth, access to water, WUE, and photoprotection (Reynolds et al. 2005).

Mexico's diverse landraces, as per DNA fingerprinting, showed superior ability for water extraction from soil depth and an increased concentration of soluble carbohydrates in the stem shortly after anthesis, whereas resynthesized wheat lines, using wild ancestors, had increased partitioning of root mass to deep soil profiles (between 60 cm and 120 cm) and increased ability to extract moisture from these depths (Reynolds et al. 2007). Izanloo et al. (2008) noted that the capacity for osmotic adjustment was the main physiological trait related for adaptation to cyclic water stress in South Australian bread wheat cultivars, thereby enabling their plants to recover from water deficit. Carbon isotope discrimination ($\Delta$) could be another useful trait because it can indirectly measure yield potential, harvest index, and water status under irrigated and rainfed conditions, as shown by Zhu et al. (2008) when researching on wheat cultivars in northern China. Rebetzke et al. (2002) suggest that $\Delta$ could be used for indirect selection in water-limited environments because it may increase the probability of recovering high-yielding lines in a wheat breeding program. Hoffman et al. (2010) found that increasing the root/shoot ratio and decreasing the water potential and grain-filling duration enhanced adaptation to water deficiency in wheat–barley introgression lines.

Although wheat breeders may have interest in the above traits (especially those with high heritability and showing low genotype-by-environment interactions), a few of them have been used to develop cultivars because of the little capacity of breeding programs to screen for these traits, their lack of validation through relevant field experiments, and the limited knowledge regarding their economic benefit for selecting one trait against another. A reliable drought screening may need to use managed environment facilities (MEF) that control water availability and remove the impact of seasonal changes and timing of rainfall on the performance of trial crops. The accuracy of performance measurements can, therefore, increase, as well as the precision for attributing phenotypic effects to individual traits and their underlying genetics.

Progress in international wheat breeding aiming drought-affected environments was measured for a 20-year period using data from global semiarid wheat yield trials (Trethowan et al. 2002). The yearly progress was determined by measuring change in the percentage of trial mean (%TM) and change in trial mean (TM) between 1979 and 1998. The increases were 4.38 and 0.09 % per year for %TM and TM, respectively, in environments whose average grain yield was below 4 t ha$^{-1}$, while in those environments yielding 4 t ha$^{-1}$ or above, these rates were 0.85 and 2.87 % per year. Further analysis for the period 1994–2010 (Manès et al. 2012) shows a rate of 1 % per year (or 31 kg ha$^{-1}$ per year) when grain yield was expressed as a percentage of the long-term check cultivar "Dharwar Dry," being the yield increase rate twice in high-yielding environments than in low-yielding environments. "Attila" and "Pastor," which are key parents in international wheat breeding, and their derivatives were high-yielding lines in several environments. A resynthesized wheat derivative ("Vorobey"), which includes "Pastor" as parent, had outstanding yields in recent multi-environment trials. A data subset from these trials (122 locations) was also used for assessing the relationship among international drought-prone test sites and further clustering of environments (Trethowan et al. 2001). Some weak associations among locations could be attributed to their inherent variability, which is a major characteristic of most rainfed, drought-prone environments.

As noted above, several traits contribute to wheat performance under drought stress. Hence, none of them alone will be able to improve wheat yields in environments affected by water scarcity. Moreover, adaptation to drought seems to be a quantitative trait controlling a complex phenotype, which may be often confounded by plant phenology (Fleury et al. 2010). Breeding for drought-prone environments may be further confused due to interaction with other stress factors (e.g., high temperature and irradiance, nutrient toxicity, or deficiency), which explains so far the limited success using physiological and molecular breeding approaches. Nonetheless DNA markers can assist dissecting some of these traits (Table 5.3) and mapping QTLs, e.g., for plant height at distinct water regimes, flag leaf senescence, and grain yield under drought stress. An interdisciplinary approach that takes into account interactions among stress factors and plant phenology, integrates physiology and genetics of traits for enhancing adaptation to drought, and uses high-throughput MEF and genomic tools will facilitate wheat breeding under water scarcity.

## 5.4.4 Legumes

Adaptation of legumes to drought-prone environments remains challenging because of the complexity of the various traits involved as well as the broad variability in the environments where water scarcity occurs. This section includes research highlights for cowpea, chickpea, groundnut (or peanut), bean, and soybean.

**Table 5.3** QTL sample for wheat characters measured in drought-prone environments

| Trait | Finding | Reference |
|---|---|---|
| Flag leaf senescence | Complex genetic mechanism (in chromosomes 2B and 2D) of this trait involving the remobilization of resources from the source to the sink during senescence | Verma et al. (2004) |
| Grain yield | A grain yield QTL on the proximal region of chromosome 4AL (contributed by cv. "Dharwar Dry") with a significant effect on performance under reduced moisture | Kirigwi et al. (2007) |
| Adaptive QTL | QTL located on 4A accounts for 27 % of grain yield variation under drought | Pinto et al. (2010) |
| | QTL on 3B-b explains 14 % of canopy temperature variation | |
| | Common QTL for drought and heat stress traits on 1B-a, 2B-a, 3B-b, 4A-a, 4B-b, and 7A-a, thereby confirming their generic value across both stress factors | |
| Plant height (PH) | No single QTL continually active in all periods of PH growth | Wu et al. (2010) |
| | QTL with additive effects expressed in the period from the original point to the jointing stage formed a foundation for PH development | |
| | PH development under control of a network of genes with additive and epistatic effects, but the former are more important than the latter or QTL × environment interactions | |

Cowpea—the most important legume crop of the semiarid tropics of Africa—shows an inherent adaptation to drought but yet suffers significant damage leading to grain yield loss in the Sahel and dryland savannas, which are affected by unstable rainfall patterns and frequent droughts. Agbicodo et al. (2009) provide the most recent overview of cowpea breeding for drought-prone environments. Research on this crop under drought stress includes the analysis of the relationship between drought response and grain yield components, plus other morphological and physiological traits. The extensive research on drought screening was only successful for selecting cowpea parents showing distinct mechanisms to adapting to drought. There have been some recent advances in identifying DNA markers defining QTLs for traits related to adapting the crop to drought stress. Table 5.4 provides a summary of some research findings regarding the genetics of traits favoring adaptation of this crop to environments affected by water scarcity.

Chickpea could be the most investigated legume crop under drought stress (Dwivedi et al. 2013 and references therein). Its adaptation to drought seems to be related to drought avoidance root traits (root length density, root to total dry plant weight ratio, root depth, and root to shoot length density), transpiration efficiency (TE), carbon isotope discrimination, SPAD chlorophyll meter reading, and canopy temperature. Plant breeders have used some of these traits to enhance chickpea performance in drought-prone environments. Krishnamurthy et al. (2010) and Zaman-Allah et al. (2011) were able to identify various chickpea lines with enhanced adaptation to terminal drought.

**Table 5.4** Research advances on trait analysis, physiology, and genetics for adapting cowpea to drought-prone environments

| Findings | Reference |
| --- | --- |
| Water stress regulates the synthesis and the activity of superoxide dismutase isoforms, which contribute to protect photosynthesis against the damageable effects of superoxide radicals in cowpea | Brou et al. (2007) |
| Delayed leaf senescence associated with adaptation to drought | Muchero et al. (2008) |
| 10 QTLs associated with seedling drought-induced senescence on linkage groups 1, 2, 3, 5, 6, 7, 9, 10, each contributing between 5 and 12 % to the phenotypic variance | Muchero et al. (2009) |
| Germplasm with enhanced adaptation to drought had more pods and seeds per plant and grain yield per plant and show high harvest index | Hegde and Mishra (2009) |
| Genetic variation in water-saving traits such as lower plant transpiration rate (Tr) under no stress and restricted Tr under high vapor pressure deficit (VPD) | Belko et al. (2012) |

There are various *Arachis* germplasm sources for adapting groundnuts to mid- and end-of-season terminal drought (Cruickshank et al. 2003 and articles therein), e.g., subspecies *hypogaea* and *fastigiata* show variation for physiological traits such as specific leaf area (SLA), chlorophyll content, amount of water transpired ($T$), TE, WUE, and harvest index under drought stress (Dwivedi et al. 2013 and references therein). Most groundnut breeding programs follow an empirical approach (largely based on pod yield as selection criterion for adaptation to drought), which explains its slow genetic progress (Nigam et al. 2005). Recently Ravi et al. (2011) found a few major, many minor, and epistatic QTLs, thereby revealing that the groundnut adaptation to drought is a complex and multigenic trait.

Common beans show significant plasticity at the biochemical and cellular levels under drought, especially for stomatal conductance, photosynthetic rate, abscisic acid synthesis, and resistance to photoinhibition (Lizana et al. 2006). Deep rooting system, maximization of WUE, greater photosynthate transport to seed under stress through efficient remobilization, early maturity, and recovery from drought are among the traits that confer enhanced adaptation in common bean in water-limiting stressful environments (Dwivedi et al. 2013 and references therein). Several germplasm and breeding lines showing these traits have been selected worldwide. Some recently bred lines outyielded the commercial checks by 15–25 % under favorable environments or up to 36 % under drought stress (Beebe et al. 2008). This enhanced adaptation to drought seems to be multigenic (Beebe et al. 2008), and recurrent selection could be an effective breeding method. Few QTLs that enhance adaptation to drought have been noted (Blair et al. 2012). Asfaw et al. (2012) mapped nine QTLs for ten traits enhancing adaptation of common bean to drought-prone environments on 6 of the 11 linkage groups. Six of these QTLs had a significant interaction with the environment.

There has been a tremendous progress for identifying the physiological and genetic basis of traits for water-limited environments and quantifying their impact on soybean grain yield (Sadok and Sinclair 2011 and references therein). Such advances translated in the release of the first ever N-fixing soybean cultivars for drought stressful environments and the use of "slow-wilting" germplasm in soybean breeding programs. Their success ensued from an interdisciplinary approach relying on field-based phenotyping, dissecting traits using physiological tools and genetic analysis, and crop modeling.

## 5.5  Outlook

Water remains as one of the most serious limiting factors for world agriculture and it may likely worsen with the anticipated global climate change. Adaptation to drought is a complex, multigenic phenotype that may be affected by both the environment and other stress factors. Mir et al. (2012) suggest that genomics and genetics coupled with precise phenotyping methods could unravel the genes and metabolic pathways that confer crop adaptation to drought. "Omics" research, which emerged mostly in the last one and a half decades, provides further insights for breeding WUE in crops, while phenomics will assist to quantify precisely the traits enhancing crop adaptation to water scarcity. Phenotyping protocols (within a plant breeding strategy for improving crop yields under drought) should screen for water use, WUE, and harvest index (Salekdeh et al. 2009). Digital imaging could be used for accurately doing high-throughput phenotyping of ground cover and early vigor traits (Mullan and Reynolds 2010). It may replace the destructive sampling methods, which are regarded as time consuming by plant breeding programs.

Modeling offers means for simulating how virtual plants—bearing a diverse combination of alleles—could respond to distinct drought scenarios (Tardieu 2003). Chapman (2008) indicated that this approach has the additional advantage of manipulating separately "biological and experimental noise." Crop simulation models are, therefore, an important tool for genomics research and breeding crop adaptation to drought.

The focus of the recent years in crop-drought research has been on dissecting traits that enhance adaptation to this stress (Collins et al. 2009). QTL mapping or gene discovery through linkage and association mapping, QTL cloning, candidate gene identification, transcriptomics, and functional genomics has been used to understand crop responses to drought. In some crops several QTLs controlling characters that enhanced adaptation to drought have been detected. Dissecting complex phenotypes into their constituting QTLs offers a more direct access to tap valuable genetic diversity regulating the adaptive response to drought (Tuberosa et al. 2007b). As suggested by Courtouis et al. (2009) for rice roots traits, compiling these data could significantly assist on identifying candidate genes through positioning consensus QTLs with more precision through meta-QTL analysis.

Gene expression microarrays are useful for gaining insights into physiological and biochemical pathways of crop adaption to drought. This technology can also lead to identifying new candidate genes that can be used for enhancing plant breeding for environments affected by water scarcity. For example, Luo et al.

(2010) examined some gene expression patterns in several maize stress response-associated pathways and found that specific genes were responsive to drought stress positively, while Kathiresan et al. (2006) noted contrasting drought responses when comparing upland- and lowland-adapted rice cultivars and between traditional and improved upland rice types.

"Omics" tools can address the multi-genecity of plant responses under stress (Bohnert et al. 2006). The analysis of genomes, transcriptomes, protein dynamics, and metabolomes will provide means for comparing adaptation to drought across species. Advances in crop genome sequencing and resequencing will further provide means for dissecting genetically the basis for adapting crops to stressful environments (Roy et al. 2010). Likewise, molecular-aided breeding, which includes marker-assisted backcrossing, marker-assisted recurrent selection, and genome-wide selection, may offer means for developing new cultivars with enhanced adaptation to drought stressful environments. The judicious use of plant genetic resources, precise phenotyping, and knowledge-intensive crop breeding led by physiology and "omics" sciences should be also able to deliver germplasm with enhanced water productivity. The success in this endeavor will contribute towards ensuring food, feed, fiber, and other feedstocks under the likely unstable and more variable climate to be brought by global warming.

# References

Agbicodo EM, Fatokun CA, Muranaka S, Visser RGF, van der Linder CG (2009) Breeding drought tolerant cowpea: constraints, accomplishments, and future prospects. Euphytica 167:353–370

Araus JL, Slafer G, Reynolds MP, Royo C (2007) Plant breeding and drought in cereals: what we should breed for? Ann Bot 89:925–940

Asfaw A, Blair MW, Struik PC (2012) Multienvironment quantitative trait loci analysis for photosynthate acquisition, accumulation, and remobilization traits in common bean under drought stress. G3 2:579–595

Bänziger M, Setimela PS, Hodson D, Vivek B (2004) Breeding for improved drought tolerance in maize adapted to southern Africa. In: New directions for a diverse planet, Proceedings of the 4th international crop science congress, Brisbane, Australia, 26 Sep–1 Oct 2004 (Published on CDROM)

Bänziger M, Setimela PS, Hodson D, Vivek B (2006) Breeding for improved drought tolerance in maize adapted to southern Africa. Agric Water Manag 80:212–224

Beebe S, Rao IM, Cajiao C, Grajales M (2008) Selection for drought resistance in common bean also improves yield in phosphorus limited and favorable environments. Crop Sci 48:582–592

Belko N, Zaman-Allah M, Cisse N, Diop ND, Zombre G, Ehler JD, Vadez V (2012) Lower soil moisture threshold for transpiration decline under water deficit correlates with lower canopy conductance and higher transpiration efficiency in drought tolerant cowpea. Funct Plant Biol. doi:10.1007/FM11282

Bennet J (2003) Opportunities for increasing water productivity of CGIAR crops to plant breeding and molecular biology. In: Kijne JW, Barker R, Molden D (eds) Water productivity in agriculture: limits and opportunities from improvement. CAB International, Wallingford, Oxon, pp 103–126

Bernier J, Atlin GN, Serraj R, Kumar A, Spaner D (2008) Breeding upland rice for drought resistance (a review). Sci Food Agric 88:927–939

Bernier J, Serraj R, Kumar A, Venuprasad R, Impa S, Gowda V, Oane R, Spaner D, Atlin G (2009) Increased water uptake explains the effect of qtl12.1 a large-effect drought-resistance QTL in upland rice. Field Crop Res 110:139–146

Bimpong IK, Serraj R, Chin JH, Mendoza EMT, Hernandez J, Mendioro MS (2011a) Determination of genetic variability for physiological traits related to drought tolerance in African rice (*Oryza glaberrima*). J Plant Breed Crop Sci 3:60–67

Bimpong IK, Serraj R, Chin JH, Ramos J, Mendoza EMT, Hernandez J, Mendioro MS, Brar DS (2011b) Identification of QTLs for drought-related traits in alien introgression lines derived from crosses of rice (*Oryza sativa* cv IR64) × *O. glaberrima* under lowland moisture stress. J Plant Biol 54:237–250

Blair MW, Galeano CH, Tovar E, Torres MCM, Castrillón AV, Beebe SE, Rao IM (2012) Development of a Mesoamerican intra-genepool genetic map for quantitative trait loci detection in a drought tolerant × susceptible common bean (*Phaseolus vulgaris* L.) cross. Mol Breed 29:71–88

Blum A (2005) Drought resistance, water-use efficiency, and yield potential—are they compatible, dissonant, or mutually exclusive? Aust J Agric Res 56:1159–1168

Bohnert HJ, Gong Q, Li P, Ma S (2006) Unraveling abiotic stress tolerance mechanisms – getting genomics going. Curr Opin Plant Biol 9:180–188

Bolaños J, Edmeades GO (1993a) Eight cycles of selection for drought tolerance in lowland tropical maize. 1. Responses in grain yield, biomass, and radiation utilization. Field Crop Res 31:233–252

Bolaños J, Edmeades GO (1993b) Eight cycles of selection for drought tolerance intropical maize. II. Responses in reproductive behavior. Field Crops Res 31:253–268

Bolaños J, Edmeades GO, Martinez L (1993) Eight cycles of selection for drought tolerance in tropical maize. III. Responses in drought-adaptive physiological and morphological traits. Field Crops Res 31:269–286

Boonjung H, Fukai S (1996) Effects of soil water deficit at different growth stages on rice growth and yield under upland conditions. 2. Phenology, biomass production and yield. Field Crops Res 48:47–55

Boyle MG, Boyer JS, Morgan PW (1991) Stem infusion of liquid culture medium prevents reproductive failure of maize at low water potential. Crop Sci 31:1246–1252

Brou YC, Zézé A, Diouf O, Eyletters M (2007) Water stress induces overexpression of superoxide dismutases that contribute to the protection of cowpea plants against oxidative stress. Afr J Biotechnol 6:1982–1986

Bruce WB, Edmeades GO, Barker TC (2002) Molecular and physiological approaches for maize improvement for drought tolerance. J Exp Bot 53:13–25

Campos H, Cooper M, Habben JE, Edmeades GO, Schussler JR (2004) Improving drought tolerance in maize: a view from industry. Field Crops Res 90:19–34

Centritto M, Lauteri M, Monteverdi C, Serraj R (2009) Leaf gas exchange, carbon isotope discrimination and grain yield in contrasting rice genotypes subjected to water deficits during reproductive stage. J Exp Bot 60:2325–2339

Chapman SC (2008) Use of crop models to understand genotype by environment interactions for drought in real-world and simulated plant breeding trials. Euphytica 161:195–208

Chaves MM, Oliveira MM (2004) Mechanisms underlying plant resilience to water deficits: prospects for water-saving agriculture. J Exp Bot 55:2365–2384

Collins NC, Tardieu F, Tuberosa R (2009) Quantitative trait loci and crop performance under abiotic stress: where do we stand? Plant Physiol 147:469–486

Courtouis B, Ahmadi N, Khowaja F, Price AH, Rami J-F, Frouin J, Hamelim C, Ruiz M (2009) Rice root genetic architecture: meta-analysis from a drought QTL database. Rice 2:115–128

Cruickshank AW, Rachaputi NC, Wright GC, Nigam SN (eds) (2003) Breeding of drought-resistant peanuts. ACIAR Proceedings 112. Australian Centre for International Agricultural Research, Canberra

de Wit M, Stankiewicz J (2006) Changes in surface water supply across Africa with predicted climate change. Science 311:1917–1921

Dwivedi SL, Sahrawat K, Upadhyaya H, Ortiz R (2013) Food, nutrition and agro-biodiversity under global climate change. Adv Agron 120:1–128

Edmeades GO, Bolaños J, Lafitte HR, Rajaram S, Pfeiffer W, Fischer RA (1989) Traditional approaches to breeding for drought resistance in cereals. In: Baker FWG (ed) Drought resistance in cereals. ICSU–CABI, Wallingford, pp 27–52

Farooq M, Kobayashi N, Wahid A, Ito O, Basra SMA (2009) Strategies for producing more rice with less water. Adv Agron 101:351–388

Farooq M, Kobayashi N, Ito O, Wahid A, Serraj R (2010) Broader leaves result in better performance of indica rice under drought stress. J Plant Physiol 167:1066–1075

Fleury D, Jefferies S, Kuchel H, Langridge P (2010) Genetic and genomic tools to improve drought tolerance in wheat. J Exp Bot 61:3211–3222

Foti R, Mapiye C, Mutenje M, Mwale M, Mlambo N (2008) Farmer participatory screening of maize seed varieties for suitability in risk prone, resource-constrained smallholder farming systems of Zimbabwe. Afr J Agric Res 3:180–185

Fuad-Hassan A, Tardieu F, Turc O (2008) Drought-induced changes in anthesis-silking interval are related to silk expansion: a spatio-temporal growth analysis in maize plants subjected to soil water deficit. Plant Cell Environ 31:1349–1360

Gilbert N (2010) Food: inside the hothouse of industry. Nature 466:548–551

Gornall J, Betts R, Burke E, Clark R, Camp J, Willett K, Wiltshire A (2010) Implications of climate change for agricultural productivity in the early twenty-first century. Philos Trans R Soc B 365:2973–2989

Gouda PK, Varma CMK, Saikumar S, Kiran B, Shenoy V, Sashidhar HE (2012) Direct selection for grain yield under moisture stress in *Oryza sativa* cv. IR58025B × *O. meridionalis* population. Crop Sci 52:644–653

Gowda VRP, Henry A, Yamauchi A, Shashidhar HE, Serraj R (2011) Root biology and genetic improvement for drought avoidance in rice. Field Crops Res 122:1–13

Grant RF, Jackson BS, Kiniry JR, Arkin GF (1989) Water deficit timing effects on yield components in maize. Agron J 81:61–65

Guan YS, Serraj R, Liu SH, Xu JL, Ali J, Wang WS, Venus E, Zhu LH (2010) Simultaneously improving yield under drought stress and non-stress conditions: a case study of rice (*Oryza sativa* L.). J Exp Bot 61:4145–4156

Hegde VS, Mishra SK (2009) Landraces of cowpea, *Vigna unguiculata* (L.) Walp., as potential sources of genes for unique characters in breeding. Genet Resour Crop Evol 56:615–627

Hoffman B, Aranyi NR, Molnár-Lán R (2010) Characterization of wheat-barley introgression lines for drought tolerance. Acta Agron Hung 58:211–218

Hund A, Ruta N, Liedgens M (2008) Rooting depth and water use efficiency of tropical maize inbred lines, differing in drought tolerance. Plant Soil 318:311–325

Hyman G, Fujisaka S, Jones P, Wood S, de Vicente C, Dixon J (2008) Strategic approaches to targeting technology generation: assessing the coincidence of poverty and drought-prone crop production. Agric Syst. doi:10.1016/j.agsy.2008.04.001

Izanloo A, Condon AG, Langridge P, Tester M (2008) Different mechanisms of adaptation to cyclic water stress in two South Australian bread wheat cultivars. J Exp Bot 59:3327–3346

Jones MP, Dingkuhn M, Aluko GK, Monde S (1997) Interspecific *O. sativa* L. *O. glaberrima* Steud: progenies in upland rice improvement. Euphytica 92:237–246

Kathiresan A, Lafitte HR, Chen J, Mansueto L, Bruskiewich R, Bennett J (2006) Gene expression microarrays and their application in drought stress research. Field Crops Res 97:101–110

Kato Y, Henry A, Fujita D, Katsura K, Kobayashi N, Serraj R (2011) Physiological characterization of introgression lines derived from an indica rice cultivar, IR64, adapted to drought and water-saving agriculture. Field Crops Res 123:130–138

Kirigwi FM, van Ginkel M, Brown-Guedira G, Gill BS, Paulsen GM, Fritz AK (2007) Markers associated with a QTL for grain yield in wheat under drought. Mol Breed 20:401–413

Krishnamurthy L, Kashiwagi J, Gaur PM, Upadhyaya HD, Vadez V (2010) Sources of tolerance to terminal drought in the chickpea (*Cicer arietinum* L.) mini core germplasm. Field Crops Res 119:322–330

Laffite HR, Courtois B (2002) Interpreting cultivar × environment interactions for yield in upland rice: assigning value to drought-adaptive traits. Crop Sci 42:1409–1420

Laffite HR, Ismail A, Bennett J (2004) Abiotic stress tolerance in rice for Asia: progress and the future. In: New directions for a diverse planet. Proceedings of the 4th international crop science congress, Brisbane, Australia, 26 Sept–1 Oct 2004 (Published on CDROM)

Laffite HR, Li ZK, Vijayakumar CHM, Gao YM, Shi Y, Xu JL, Fu BY, Yu SB, Ali AJ, Domingo J, Maghirang R, Torres R, Mackill D (2006) Improvement of rice drought tolerance through backcross breeding: evaluation of donors and selection in drought nurseries. Field Crops Res 97:77–86

Liu JX, Liao DQ, Oane R, Estenor L, Yang XE, Li ZC, Bennet J (2006) Genetic variation in the sensitivity of anther dehiscence to drought stress in rice. Field Crops Res 97:87–100

Lizana C, Wentworth M, Martinez JP, Villegas D, Meneses R, Murchie EH, Pastenes C, Lercari B, Vernieri P, Horton P, Pinto M (2006) Differential adaptation of two varieties of common bean to abiotic stress. I. Effects of drought on yield and photosynthesis. J Exp Bot 57:685–697

Lu Y, Zhang S, Shah T, Xie C, Hao Z, Li X, Farkhari M, Ribaut J-M, Cao M, Rong T, Xu Y (2010) Joint linkage–linkage disequilibrium mapping is a powerful approach to detecting quantitative trait loci underlying drought tolerance in maize. Proc Natl Acad Sci USA 107:19585–19590

Luo M, Liu J, Lee RD, Scully BT, Guo B (2010) Monitoring the expression of maize genes in developing kernels under drought stress using oligo-microarray. J Integr Plant Biol 52:1059–1074

Malosetti M, Ribaut JM, Vargas M, Crossa J, Boer MP, van Eeuwijk FA (2007a) Multitrait multi-environment QTL modelling for drought-stress adaptation in maize. In: Spiertz JHC, Struik PC, van Laar HH (eds) Scale and complexity in plant systems research: gene-plant-crop relations. Springer, Dordrecht, pp 25–36

Malosetti M, Ribaut JM, Vargas M, Crossa J, van Eeuwijk FA (2007b) A multi-trait multi-environment QTL mixed model with an application to drought and nitrogen stress trials in maize (*Zea mays* L.). Euphytica 161:241–257

Manès Y, Gomez HF, Puhl L, Reynolds M, Braun HJ, Trethowan R (2012) Genetic yield gains of the CIMMYT international semi-arid wheat yield trials from 1994 to 2010. Crop Sci 52:1543–1552

Messmer R, Fracheboud Y, Bänziger M, Vargas M, Stamp P, Ribaut J-M (2009) Drought stress and tropical maize: QTL-by-environment interactions and stability of QTLs across environments for yield components and secondary traits. Theor Appl Genet 119:913–930

Mir RR, Zaman-Allah M, Sreenivasulu N, Trethowan R, Varshney RK (2012) Integrated genomics, physiology and breeding approaches for improving drought tolerance in crops. Theor Appl Genet. doi:10.1007/s00122-012-1904-9

Molden D, Oweis T, Steduto P, Bindraban P, Hanjra MA, Kijne J (2010) Improving agricultural water productivity: between optimism and caution. Agric Water Manag 97:528–535

Monneveux P, Sánchez C, Beck D, Edmeades GO (2006) Drought tolerance improvement in tropical maize source populations: evidence of progress. Crop Sci 46:180–191

Monneveux P, Sánchez C, Tiessen A (2008a) Future progress in drought tolerance in maize needs new secondary traits and cross combinations. J Agric Sci Camb 146:287–300

Monneveux P, Sheshshayee MS, Akther J, Ribaut J-M (2008b) Using carbon isotope discrimination to select maize (*Zea mays* L.) inbred lines and hybrids for drought tolerance. Plant Sci 173:390–396

Muchero W, Ehlers JD, Roberts PA (2008) Seedling stage drought-induced phenotypes and drought-responsive genes in diverse cowpea genotypes. Crop Sci 48:541–552

Muchero W, Ehlers JD, Close TJ, Roberts PA (2009) Mapping QTL for drought stress induced premature senescence and maturity in cowpea [*Vigna unguiculata* (L.) Walp.]. Theor Appl Genet 118:849–863

Mullan DJ, Reynolds MP (2010) Quantifying genetic effects of ground cover on soil water evaporation using digital imaging. Funct Plant Biol 37:703–712

Nigam SN, Chandra S, Sridevi KR, Bhukta M, Reddy AGS, Nageswara Rao RC, Wright GC, Reddy PV, Deshmukh MP, Mathur RK, Basu MS, Vasundhara S, Varman PV, Nagda AK (2005) Efficiency of physiological trait-based and empirical selection approaches for drought tolerance in groundnut. Ann Appl Biol 146:433–439

Ortiz R, Iwanaga M, Reynolds MP, Wu H, Crouch JH (2007) Overview on crop genetic engineering for drought-prone environments. J Semi-Arid Trop Agric Res 4. http://www.icrisat.org/journal/SpecialProject/sp3.pdf

Parent B, Suard B, Serraj R, Tardieu F (2010) Rice leaf growth and water potential are resilient to evaporative demand and soil water deficit once the effects of root system are neutralized. Plant Cell Environ 33:1256–1267

Passioura J (2004) Increasing crop productivity when water is scarce – from breeding to field management. In: New directions for a diverse planet. Proceedings of the 4th international crop science congress, Brisbane, Australia, 26 Sept–1 Oct 2004 (Published on CDROM)

Passoura J (2007) The drought environment: physical, biological and agricultural perspectives. J Exp Bot 58:113–117

Peleg Z, Reguera M, Tumimbang E, Walia H, Blumwald E (2011) Cytokinin-mediated source/sink modifications improve drought tolerance and increase grain yield in rice under water-stress. Plant Biotechnol J 9:747–758

Pinto RS, Reynolds MP, Mathews KL, McIntyre CL, Olivares-Villegas JJ, Chapman SC (2010) Heat and drought adaptive QTL in a wheat population designed to minimize confounding agronomic effects. Theor Appl Genet 121:1001–1021

Quarrie SA, Lazić-Jančić V, Kovačević D, Steed A, Pekić S (1999) Bulk segregant analysis with molecular markers and its use for improving drought resistance in maize. J Exp Bot 50:1299–1306

Ravi K, Vadez V, Isobe S, Mir RR, Guo Y, Nigam SN, Gowda MVC, Radhakrishnan T, Bertioli DJ, Knapp SJ, Varshney RK (2011) Identification of several small main-effect QTLs and a large number of epistatic QTLs for drought tolerance related traits in groundnut (*Arachis hypogaea* L.). Theor Appl Genet 122:1119–1132

Rebetzke GJ, Condon AG, Richards RA, Farquhar GD (2002) Selection for reduced carbon isotope discrimination increases aerial biomass and grain yield of rainfed bread wheat. Crop Sci 42:739–745

Reynolds MP, Mujeeb-Kazi A, Sawkins M (2005) Prospects for utilising plant-adaptive mechanisms to improve wheat and other crops in drought- and salinity-prone environments. Ann Appl Biol 146:239–259

Reynolds M, Dreccer F, Trethowan R (2007) Drought-adaptive traits derived from wheat wild relatives and landraces. J Exp Bot 58:177–186

Ribaut J-M, Ragot M (2007) Marker-assisted selection to improve drought adaptation in maize: the backcross approach, perspectives, limitations, and alternatives. J Exp Bot 58:351–360

Ribaut J-M, Hoisington DH, Deutsch JA, Jiang C, Gonzalez-de-Leon D (1996) Identification of quantitative trait loci under drought conditions in tropical maize. 1. Flowering parameters and the anthesis-silking interval. Theor Appl Genet 92:905–914

Ribaut J-M, Hoisington DH, Deutsch JA, Jiang C, Gonzalez-de-Leon D (1997) Identification of quantitative trait loci under drought conditions in tropical maize. 2. Yield components and marker-assisted selection strategies. Theor Appl Genet 94:887–896

Richards RA, Condon AG, Rebetzke GJ (2001) Traits to improve yield in dry environments. In: Reynolds MP, Ortiz-Monasterio JI, McNab A (eds) Application of physiology in wheat breeding. Centro Internacional de Mejoramiento de Maíz y Trigo, Mexico DF, pp 88–100

Rosen S, Scott L (1992) Famine grips sub-Saharan Africa. Outlook Agric 191:20–24

Roy SJ, Tucker EJ, Tester M (2010) Genetic analysis of abiotic stress tolerance in crops. Curr Opin Plant Biol 14:232–239

Ruta N, Stamp P, Liedgens M, Fracheboud Y, Hund A (2010) Traits and yield components of tropical maize under water stress conditions. Crop Sci 50:1385–1392

Sadok W, Sinclair TR (2011) Crops yield increase underwater-limited conditions: Review of recent physiological advances for soybean genetic improvement. Adv Agron 113:325–349

Salekdeh GH, Reynolds M, Bennett J, Boyer J (2009) Conceptual framework for drought phenotyping during molecular breeding. Trends Plant Sci 14:488–496

Sellamuthu R, Liu GF, Chandra Babu R, Serraj R (2011) Genetic analysis and validation of quantitative trait loci associated with reproductive-growth traits and grain yield under drought stress in a doubled haploid line population of rice (Oryza sativa L.). Field Crops Res 124:46–58

Serraj R, Dimayuga G, Gowda V, Guan Y, He H, Impa S, Liu DC, Mabesa RC, Sellamuthu R, Torres R (2008) Drought-resistant rice: physiological framework for an integrated research strategy. In: Serraj R, Bennett J, Hardy B (eds) Drought frontiers in rice – crop improvement for increased rainfed production. World Scientific, Singapore, pp 139–170

Serraj R, Kumar A, McNally KL, Slamet-Loedin I, Bruskiewich R, Mauleon R, Cairns J, Hijmans RJ (2009) Improvement of drought resistance in rice. Adv Agron 103:41–99

Serraj R, McNally KL, Slamet-Loedin I, Kohli A, Haefele SM, Atlin G, Kumar A (2011) Drought resistance improvement in rice: an integrated genetic and resource management strategy. Plant Prod Sci 14:1–14

Tambussi EA, Bort J, Araus JL (2007) Water use efficiency in C 3 cereals under Mediterranean conditions: a review of physiological aspects. Ann Appl Biol 150:307–321

Tardieu F (2003) Virtual plants: modelling as a tool for the genomics of tolerance to water deficit. Trends Plant Sci 8:9–14

Tardieu F, Tuberosa R (2010) Dissection and modelling of abiotic stress tolerance in plants. Curr Opin Plant Biol 13:206–212

Tester M, Bacic A (2005) Abiotic stress tolerance in grasses. From model plants to crop plants. Plant Physiol 137:791–793

Tollefson J (2011) Drought-tolerant maize gets US debut. Nature 469:144

Trethowan RM, Crossa J, van Ginkel M, Rajaram S (2001) Relationships among bread wheat international yield testing locations in dry areas. Crop Sci 41:1461–1469

Trethowan RM, van Ginkel M, Rajaram S (2002) Progress in breeding wheat for yield and adaptation in global drought affected environments. Crop Sci 42:1441–1446

Tuberosa R, Salvi S, Sanguineti MC, Landi P, Maccaferri M, Conti S (2002) Mapping QTLs regulating morphophysiological traits and yield in drought-stressed maize: case studies, shortcomings and perspectives. Ann Bot 89:941–963

Tuberosa R, Giuliani S, Parry MAJ, Araus JL (2007a) Improving water use efficiency in Mediterranean agriculture: what limits the adoption of new technologies? Ann Appl Biol 150:157–162

Tuberosa R, Salvi S, Giullani S, Sanguineti MC, Bellotti M, Conti S, Landi P (2007b) Genome-wide approaches to investigate and improve maize response to drought. Crop Sci 47: S120–S141

Turral H, Burke J, Faurès J-M (2011) Climate change, water and food security. Food and Agriculture Organization of the United Nations, Rome

Venuprasad R, Impa SM, Gowda RPV, Atlin GN, Serraj R (2011) Rice near-isogenic-lines (NILs) contrasting for grain yield under lowland drought stress. Field Crop Res 123:36–46

Verma V, Foulkes MJ, Worland AJ, Sylvester-Bradley R, Caligari PDS, Snape JW (2004) Mapping quantitative trait loci for flag leaf senescence as a yield determinant in winter wheat under optimal and drought-stressed environments. Euphytica 135:255–263

Verulkar SB, Mandal NP, Dwivedi JL, Singh BN, Sinha PK, Mahato RN, Dongre P, Singh ON, Bose LK, Swain P, Robin S, Chandrababu R, Senthil S, Jain A, Shashidhar HE, Hittalmani S, Vera Cruz C, Paris T, Raman A, Haefele S, Serraj R, Atlin G, Kumar A (2010) Breeding resilient and productive genotypes adapted to drought-prone rainfed ecosystem of India. Field Crops Res 117:197–208

Wassmann R, Jagadish SVKS, Heuer S, Ismail A, Redona E, Serraj R, Singh RK, Howell G, Pathak H, Sumfleth K (2009) Climate change affecting rice production: the physiological and agronomic basis for possible adaptation strategies. Adv Agron 101:59–121

Welcker C, Boussuge B, Bencivenni C, Ribaut JM, Tardieu F (2007) Are source and sink strengths genetically linked in maize plants subjected to water deficit? A QTL study of the responses of leaf growth and of anthesis-silking interval to water deficit. J Exp Bot 58:339–349

Wu X, Wang Z, Chang X, Jing R (2010) Genetic dissection of the developmental behaviours of plant height in wheat under diverse water regimes. J Exp Bot. doi:10.1093/jxb/erq117

Xu Y, This D, Pausch RC, Vonhof WM, Coburn JR, Comstock JP, McCouch SR (2009) Leaf-level water use efficiency determined by carbon isotope discrimination in rice seedlings. Theor Appl Genet 118:1065–1081

Zaidi PH, Yadav M, Singh DK, Singh RP (2008) Relationship between drought and excess moisture tolerance in tropical maize (*Zea mays* L.). Aust J Crop Sci 1:78–96

Zaman-Allah M, Jenkinson DM, Vadez V (2011) Chickpea genotypes contrasting for seed yield under terminal drought stress in the field differ for traits related to control of water use. Funct Plant Biol 38:270–281

Zhu L, Liang ZS, Xu X, Li SH, Jing JH, Monneveux P (2008) Relationships between carbon isotope discrimination and leaf morphophysiological traits in spring-planted spring wheat under drought and salinity stress in Northern China. Aust J Agric Res 59:941–949

# Chapter 6
# Water Use Efficiency

Helen Bramley, Neil C. Turner, and Kadambot H.M. Siddique

**Abstract** Water use efficiency (or transpiration efficiency) describes the intrinsic trade-off between carbon fixation and water loss that occurs in dryland plants because water evaporates from the interstitial tissues of leaves whenever stomata open for $CO_2$ acquisition. The transpiration efficiency of crop plants is generally low as they typically lose several 100-fold more water than the equivalent units of carbon fixed by photosynthesis. With the increasing demand for sustainable water use and increasing agricultural productivity, the need to improve transpiration efficiency (TE) of crops has received much attention, although this trait may not be beneficial in all water-limited environments. This chapter shows that TE is predominantly driven by hydraulic properties and genes that modulate TE are mostly involved in gas exchange. Genetic variation exists in crop plants for most of the respective traits, but more research is needed to determine their relative influence on TE under well-watered as well as water-limited conditions. Moreover, research is needed to demonstrate that improvements in TE will improve yields in different environments.

H. Bramley • K.H.M. Siddique (✉)
The UWA Institute of Agriculture, M082, The University of Western Australia, 35 Stirling Highway, Crawley, WA 6009, Australia
e-mail: kadambot.siddique@uwa.edu.au

N.C. Turner
The UWA Institute of Agriculture, M082, The University of Western Australia, 35 Stirling Highway, Crawley, WA 6009, Australia

Centre for Legumes in Mediterranean Agriculture, M080, The University of Western Australia, 35 Stirling Highway, Crawley, WA 6009, Australia

C. Kole (ed.), *Genomics and Breeding for Climate-Resilient Crops*, Vol. 2,                    225
DOI 10.1007/978-3-642-37048-9_6, © Springer-Verlag Berlin Heidelberg 2013

## 6.1 Introduction

$CO_2$ acquisition by dryland plants inherently results in water loss. Vast amounts of water are transpired in comparison with the small amounts of carbon that are fixed by photosynthesis. Typical crop plants transpire 200–1,000 g of water per g of assimilated carbon (Martin et al. 1976). With diminishing fresh water supplies, the threat of more frequent droughts due to climate change, and the increasing demand for more food crops, the possibility of increasing "agricultural water productivity" through agronomic and genetic means has received much attention (Araus et al. 2008; Reynolds and Tuberosa 2008; Passioura and Angus 2010). Although conservation agriculture, improved irrigation management, and other agronomic practices have made significant contributions to improving yields under water-limited conditions (Anderson et al. 2005; Turner and Asseng 2005), genotypes that are better matched to their target environments are also needed.

One theoretical avenue for improving yields with less water is through manipulation of the relationship between carbon gain (photosynthesis) and water loss (transpiration). The ratio between these two parameters, called water use efficiency (WUE and see below for full definition), essentially describes how "efficient" the plant is at optimizing carbon gain while minimizing water loss. Plants that are drought tolerant or have evolved in environments with limited available water tend to have higher WUE than plants adapted to conditions with freely available water (Smith et al. 1989). It is, therefore, not surprising that different plant species vary in their WUE (Briggs and Shantz 1913; Rawson and Begg 1977; Siddique et al. 2001). Many species also display plasticity in their WUE (Smith et al. 1989), acclimating to drier conditions, but caution needs to be taken in comparing studies from different environments as differences in transpiration associated with vapor pressure deficits need to be taken into account (Rawson and Begg 1977). Identification that there is intraspecific genetic variation for WUE has encouraged breeders to develop selection programs for improving WUE as a way to improve drought resistance (Farquhar and Richards 1984; Condon et al. 1990; Bonhomme et al. 2009; Barbour et al. 2010; Galmés et al. 2010; Devi et al. 2011). However, traits that are associated with high WUE are often associated with low yield potential, and therefore, there is debate whether selection for high WUE results in higher yield under water-limited conditions (Blum 2005, 2009). Selection for high WUE may only be appropriate for irrigated crops or crops that are grown on stored soil moisture (Blum 2005, 2009). With this caveat in mind, we will explore the traits that are associated with high WUE in this chapter and indicate their relationship to yield production.

Genetic variation in photosynthetic efficiency and transpiration has also driven research to identify mechanisms and genes controlling WUE in the hope of speeding up breeding efforts. Given the multifarious hydraulic and biochemical processes involved in controlling water flow through the plant and photosynthesis rates, many genes and their interaction are likely to be involved in determining the WUE of a plant during development and under water-limited conditions. The problem arises in identifying which of the plethora of genes will have the most impact if their activity is modified in such a complex system. The outlook however is promising, as a few signaling and transcription factors have been identified that control a cascade of subsequent events and development of relevant traits (Masle et al. 2005; Karaba et al. 2007), making these individual genes propitious candidates for genetic manipulation or targeting by breeders. Because these individual genes have the potential to most rapidly influence development of new crop varieties with modified WUE, we will focus on these genes in this chapter.

## 6.2 Definition of Water Use Efficiency

WUE is used in various ways depending on the level of the observations. At the level of the crop in the field, WUE is usually the ratio of aboveground biomass or economic yield to water use or evapotranspiration (ET):

$WUE_b$ = aboveground biomass/water use
$WUE_g$ = grain yield/water use

Water use or ET needs to be measured for the whole growing season. In Mediterranean climatic regions where crops grow on current rainfall and there is no runoff or deep drainage from the soil, growing-season rainfall may be a useful measure of crop water use (French and Schultz 1984a, b; Siddique et al. 1990a, 2001). However, in glasshouse experiments where pots can be weighed and roots extracted, WUE may be on a total biomass basis:

$WUE_t$ = total biomass/water use

One of the problems associated with the measurement of WUE in the field is that water use or ET includes water loss by soil evaporation as well as transpiration by the crop so that changes in WUE may reflect changes in soil evaporation rather than changes in plant production/transpiration. It is difficult to measure soil evaporation within the crop in the field as it will differ from bare soil evaporation even when measured nearby. In pots it is possible to cover the soil to reduce the soil evaporation, so that water use is now a measure of transpiration. Transpiration efficiency (TE) over the growing season or part of the growing season can be measured where

$TE_t$ = total biomass/transpiration
$TE_b$ = aboveground biomass/transpiration
$TE_g$ = grain yield or economic yield/transpiration

For comparisons between locations or seasons, differences in humidity or vapor pressure deficits will influence the value of the measured transpiration, so that WUE and TE are likewise affected. Average values of the vapor pressure deficit for the period of measurement of transpiration can be used to make the values of WUE and TE comparable.

At the level of the leaf, the instantaneous TE (sometimes incorrectly termed instantaneous WUE) can be measured by gas exchange equipment as the ratio of the rate of net leaf photosynthesis ($A_N$) to rate of leaf transpiration ($E$):

$$A_N/E = p_a(1 - p_i/p_a)/1.6(e_i - e_a)$$

where $p_a$ and $p_i$ are the partial pressures of carbon dioxide and $e_a$ and $e_i$ are the vapor pressure deficits of the air external ($a$) and internal ($i$) to the leaf. Integrated over the life of the plant, the losses of carbon by respiration ($\varphi_c$) and the losses of water through the cuticle and partly closed stomata ($\varphi_w$) must be also accounted for when determining TE (Hubick and Farquhar 1989; Turner 1997):

$$TE = (p_a(1 - p_i/p_a)(1 - \varphi_c))/(1.6(e_i - e_a)(1 - \varphi_w))$$

Thus TE is affected by both environmental factors, primarily vapor pressure deficit, and plant factors. It has long been known that some species such as maize and sorghum have much higher values of WUE, and particularly TE, than others (Briggs and Shantz 1913), and this was shown to be due to the higher rates of assimilation and higher ratios of internal to external partial pressures for $CO_2$ in $C_4$ species than $C_3$ species (Tanner and Sinclair 1983). Initially it was considered that the TE of a species did not vary among genotypes (Fischer 1981), but subsequent analysis and measurement demonstrated that TE did vary among genotypes and that it could be selected in breeding programs (Farquhar and Richards 1984; Condon et al. 1990; Richards 2006). Analysis has shown that variation among genotypes can arise from either or both differences in photosynthetic efficiency and differences in stomatal conductance ($g_s$) (Condon et al. 1990). Differences on the discrimination of $^{12}C$ to $^{13}C$ due to different diffusion rates through the stomatal pore and uptake within the stomatal cavity can be used to measure integrated TE. Carbon isotope discrimination (CID) also varies with drought, which generally increases TE. Thus, genetic differences in TE are usually identified in well-watered plants in the vegetative phase and not on the seed as differences in harvest index and CID in the conversion of carbon to starch or sugars can cloud differences in TE (Richards 2006).

## 6.3   Plant Mechanisms and Characteristics That Influence WUE

Figure 6.1 shows plant traits that influence water use and photosynthetic efficiency. Essentially, the plant is a hydraulic conduit between the soil and atmosphere. Canopy and root architecture, organ morphology, anatomy, and aquaporin activity influence a plant's water uptake capacity, transport efficiency, and water loss. These in turn influence plant water status (Fig. 6.2). A well-hydrated status is needed for growth and many biochemical processes. The trade-off between water loss (transpiration) and $CO_2$ assimilation exists because the atmosphere is almost always

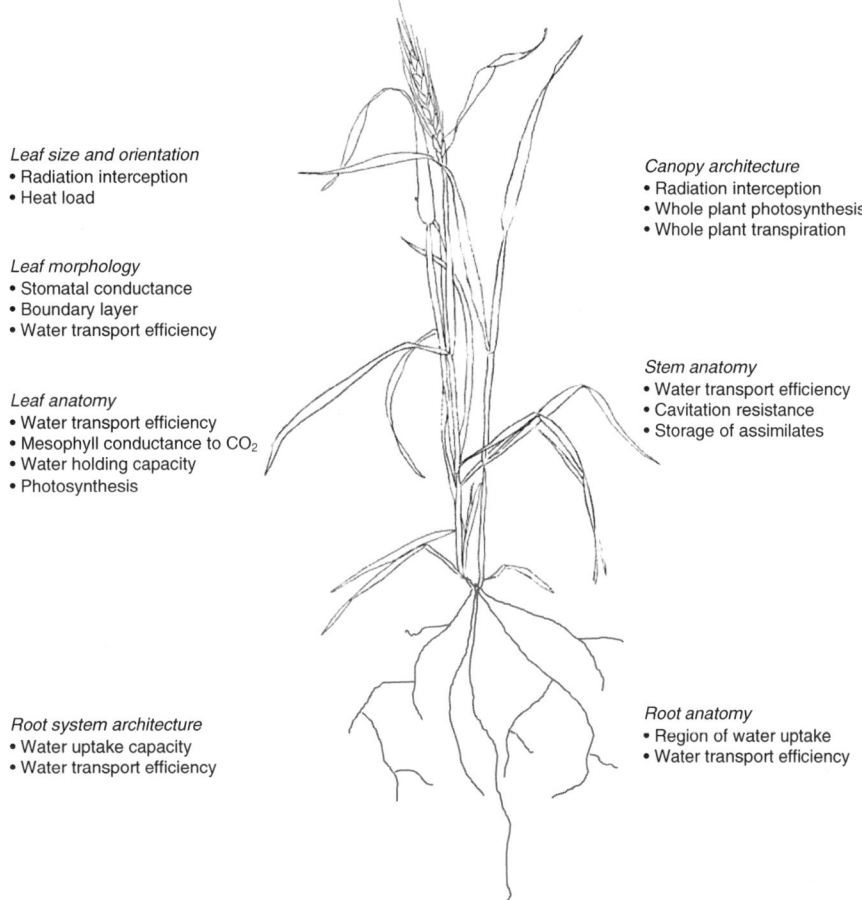

**Leaf size and orientation**
• Radiation interception
• Heat load

**Leaf morphology**
• Stomatal conductance
• Boundary layer
• Water transport efficiency

**Leaf anatomy**
• Water transport efficiency
• Mesophyll conductance to $CO_2$
• Water holding capacity
• Photosynthesis

**Root system architecture**
• Water uptake capacity
• Water transport efficiency

**Canopy architecture**
• Radiation interception
• Whole plant photosynthesis
• Whole plant transpiration

**Stem anatomy**
• Water transport efficiency
• Cavitation resistance
• Storage of assimilates

**Root anatomy**
• Region of water uptake
• Water transport efficiency

**Fig. 6.1** Attributes that influence water use and photosynthetic efficiency

drier than the substomatal cavities of leaves. Consequently, when the stomata open, water evaporates from the interstitial tissues into the atmosphere. The four attributes listed above also influence net assimilation rates, but photosynthetic efficiency is also influenced by leaf anatomy and biochemical pathways (Fig. 6.2).

The $C_4$ biochemical pathway, in which the first products of photosynthesis are $C_4$ carboxylic acids, and specific bundle sheath anatomy of leaves enable higher rates of photosynthesis than the $C_3$ biochemical pathway. As $C_4$ plants frequently, but not always, have lower stomatal conductances, the TE of $C_4$ species is considerably greater than that of $C_3$ species when directly compared in the same environment. However, $C_4$ species are often grown in warmer and drier environments than $C_3$ species, so that WUE may be similar because of the greater vapor pressure deficits in the warmer and drier environments or seasons. The highest values of TE are found in those plants that fix carbon via the carboxylic acid metabolism (CAM)

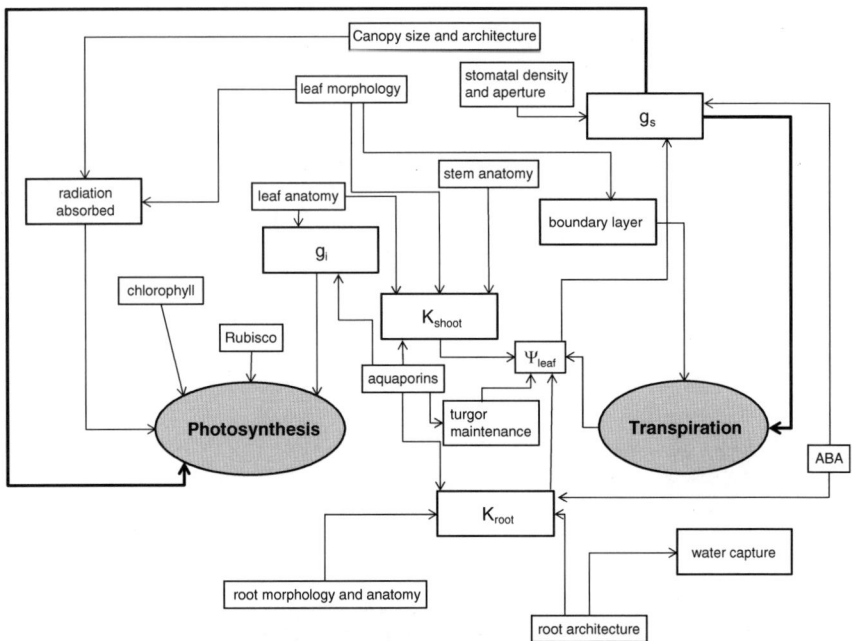

ABA = abscisic acid; $g_i$ = internal conductance to $CO_2$; $g_s$ = stomatal conductance to water vapour; $K_{root}$ = root hydraulic conductance; $K_{shoot}$ = shoot hydraulic conductance; $\Psi_{leaf}$ = leaf water potential

**Fig. 6.2** Schematic diagram of the complex interaction between hydraulic and biochemical components that influence transpiration efficiency

pathway, whereby the plant opens its stomata and takes up $CO_2$ in the dark when vapor pressure deficits and water loss are low. CAM plants store the carbon as the $C_4$ malate in mesophyll cells and use sunlight to convert the malate to carbohydrates behind closed stomata (Ting 1985). While the CAM biochemical pathway is very efficient in terms of water productivity, dry matter production is very low and is generally found in specialized plants living in extremely dry environments. However, a crop species that has the CAM pathway is pineapple (*Ananas comosus*), but as some species can switch between CAM and $C_3$ pathways depending on water supply (Dodd et al. 2002), it is likely that pineapple uses the $C_3$ pathway when irrigated for fruit production.

## 6.3.1   Physical Attributes That Influence Transpiration and Photosynthesis

### 6.3.1.1   Canopy Size and Architecture

Canopy size and architecture will influence WUE and TE through its effects on ET, $E$, and $A_N$. For example, more-closed canopies will have higher humidity and, hence, potentially lower rates of transpiration. The water-conserving benefits of a denser canopy may, however, be outweighed by reduced light penetration and lower rates of photosynthesis. Canopy architecture affects light interception and reflectance, such that erectophile canopies with leaves more vertically orientated allow greater light penetration and reflectance during the day due to the angle of the sun, compared with planophile canopies where leaves are orientated more in a horizontal plane (Pinter et al. 1985; Yunasa et al. 1993).

Canopy size and architecture in the field affects ET through soil moisture evaporation. While agronomic methods such as stubble retention and application of mulches can help reduce evaporation from the soil (Yunasa et al. 1994; Richards et al. 2002; Gregory 2004), soil moisture evaporation will also be ameliorated by a crop canopy that shades the soil surface (Siddique et al. 1990b). Soil moisture evaporation may be greater early in the season, before the canopy has fully established, depending on temperature and relative humidity. The trait for early vigor may be a way of establishing canopy cover quicker to minimize soil evapora-tion (Richards et al. 2002). Wheat genotypes with early vigor traits have been tested in a number of environments (Regan et al. 1992; Turner and Nicolas 1998; Botwright et al. 2002). However, although vigorous growth is supported by higher rates of photosynthesis in wheat, these genotypes appear to have much higher rates of transpiration resulting in lower instantaneous TE in comparison with commercial cultivars (Dias de Oliveira et al. 2013). These genotypes may, therefore, deplete the soil moisture more quickly, leaving insufficient water supply for the reproductive and grain filling stages, which would be a problem for crops growing on predomi-nantly stored soil moisture.

While early vigor may not be suitable where high TE is required, this does not exclude it as a suitable trait in other water-limited environments. For example, in southwestern Australia, winter rainfall during the growing season usually exceeds crop demand for water, but terminal drought during spring and early summer is a common occurrence (Turner and Asseng 2005). Terminal drought is the greatest limitation to crop yields (Boyer 1982; Loss and Siddique 1994; Dracup et al. 1998). Early vigor genotypes flower earlier, thereby avoiding terminal drought, and their high growth rates facilitated by high photosynthesis and transpiration rates mean that there is sufficient biomass to support grain production after flowering despite a shorter vegetative stage (Dias de Oliveira et al. 2012). In this type of environment, more effective use of available water such as early vigor rather than high TE may be the key to improving and sustaining crop yields (Siddique et al. 1989; Turner 1997). Blum (2005, 2009) has argued in support of this paradigm, and Sinclair (2012) has demonstrated that although improvements in TE are possible, it is unlikely to dramatically improve yields, but more effective water use may be a better focus for breeding efforts.

### 6.3.1.2 Leaf Morphology and Anatomy

The attributes of leaves are probably the greatest determinants of plant TE. Natural variation in the complexity of leaf shape is controlled in most eudicots by expression of a *KNOX1* gene (Bharathan et al. 2002; Kimura et al. 2008), but monocots have only simple leaves where *KNOX* is downregulated. Auxin is also involved in regulating whether leaf margins are serrated or lobed and many other genes are, of course, involved in regulating the length and width of leaf laminas (see review by Nicotra et al. 2011). The function of leaf shape (excepting leaf size) is still a matter of debate (Nicotra et al. 2011) and is mainly relevant in an ecological context, but it should not be dismissed completely from this discussion because the water relations of leaves are strongly influenced by leaf shape (Nicotra et al. 2011), and therefore, manipulation of leaf shape may be an avenue to improving TE in crop species with complex or lobed leaves.

The size of a leaf, its color, orientation, and its topology determine the amount of light intercepted and its radiation and heat load. Small and thick leaves are associated with drought tolerance and high TE (Abrams et al. 1990, 1994). In many crop species, leaf size decreases with water deficit, and in fact, leaf expansion is one of the most sensitive processes to water stress (Boyer 1970; Hsiao et al. 1985). Leaf thickness also tends to increase in many crop species in response to water deficit, particularly genotypes with greater drought tolerance and higher TE (Nobel 1980; Wright et al. 1994). Increasing leaf thickness is associated with water conservation. Increasing leaf thickness will increase the boundary layer thickness and, hence, decrease the rate of evaporation from the leaf (see below), but boundary layer thickness will be conversely reduced by a reduction in leaf size. Thicker cell walls, as well as increased suberization of bundle sheaths, help protect from dehydration but would probably also decrease leaf hydraulic conductance (see below). Leaf thickness also contributes to higher TE through an increase in $A_N$ (Dornhoff and Shibles 1976; Schulze et al. 2006) due to greater abundance of photosynthetic apparatus.

Topological features such as pubescence and glaucousness are associated with drought tolerance and high TE in xerophytes, and many plant species adapted to semiarid environments and environments with seasonal rainfall (Mulroy 1979; Ehleringer 1980; Ehleringer and Clark 1987; Rieger and Duemmel 1992; Abrams et al. 1994). Despite the limited research on pubescence and glaucousness in crop species providing no clear indication that they will lead to improvements in yield by increasing TE, there is increasing recent interest in selection of glaucousness and pubescence for drought tolerance by plant breeders (Reynolds et al. 2001).

Depending on their density and length, leaf trichomes help to keep the surface of the leaf more humid, thereby reducing transpiration, and trichomes may also help reduce the heat load of a leaf by reflecting light (Ehleringer 1980, 1984). Trichome density varies within and across species (Ehleringer 1984), and trichome density of some species and genotypes increases in response to water deficit (Ehleringer 1982). Varying trichome density has been observed in several crop species (Sharma and Waines 1994; Webster et al. 1994; Sagaram et al. 2007). Trichomes, like leaf thickness, increase the thickness of the undisturbed air surrounding the leaf, that is, the boundary layer (Schuepp 1993). The conductance of water vapor across the boundary layer decreases with increasing boundary layer thickness, and therefore, the rate of evaporation from the leaf surface will be lower (Nobel 2009). However, the conductance of $CO_2$ across the boundary layer will be similarly reduced, and so the presence of leaf trichomes may negatively impact on photosynthetic rate. It should be noted, however, that the boundary layer resistance to $CO_2$ diffusion will only impact photosynthetic rates if it is greater than stomatal and internal resistance to $CO_2$ (Nobel 1999). Selecting for leaf pubescence in crops would, therefore, only have a beneficial effect on TE if the increased resistance to water vapor diffusion out of the leaf was greater than the increased resistance to $CO_2$ diffusion into the leaf. Pubescence as a trait may be beneficial in the drier climates expected for the future because greater atmospheric $CO_2$ due to climate change will increase the gradient for $CO_2$ diffusion into the leaf. Leaf pubescence in *Brachypodium distachyon* has been reported to be under the control of a single dominant gene, *SPUB* (Garvin et al. 2008), which provides the opportunity to manipulate expression of this trait and evaluate its influence on the physiology of a species closely related to the major crops wheat and barley.

The aboveground parts of almost all land plants are covered by a waxy cuticle, which helps reduce evaporation from the nonstomatal epidermis. Transpiration across the leaf cuticle varies between species and across genotypes probably due to differences in thickness and hydrophilic properties of the epicuticular wax (Jordan et al. 1984; Larsson and Svenningsson 1986; Svenningsson and Liljenberg 1986; Clarke and Richards 1988; González and Ayerbe 2009). As cuticular transpiration is usually measured when the stomata are assumed to be closed in the dark, it needs to be recognized that some of the putative variation in cuticular transpiration may be the result of incomplete stomatal closure. High glaucousness and low cuticular transpiration are associated with drought tolerance, which can vary with environmental conditions depending on the species (Oppenheimer 1960). Glaucousness may also help in reflecting light to reduce radiation load. Evidence in support of this comes from a study on durum wheat lines isogenic for glaucousness (Richards et al. 1986). Glaucous lines had slightly cooler leaf temperatures under water deficit in the field, compared with nonglaucous wheat lines (Richards et al. 1986). In some species, transpiration through the stomata can increase with glaucousness (Kerstiens 1997, 2006), which would cool the leaf surface, but Richards et al. (1986) reported similar transpiration rates for glaucous and nonglaucous lines. Glaucousness of wheat flag leaves had no effect on TE, or photosynthetic and transpiration rates, but increased leaf duration under drought (Richards et al. 1986). Although glaucousness reduced photosynthesis and transpiration rates of wheat ears, the greater proportional reduction in transpiration resulted in increased TE of ears in comparison with nonglaucous ears (Richards et al. 1986).

Glaucousness did not correlate with yield or yield components in a doubled haploid population derived from a cross between a drought-tolerant and a relatively drought-intolerant bread wheat line (Bennett et al. 2012), despite glaucousness of parental material being associated with higher TE and yield under water deficit (Izanloo et al. 2008). The disparity between these two studies may be related to differences in behavior in glasshouse compared with field conditions, although Bennett et al. (2012) speculated that their method of visually scoring for glaucousness may not correlate with physical wax content, especially those lines that were given mid-high scores, as was observed by Clarke et al. (1993). Epicuticular wax content should also have been measured (Bennett et al. 2012). In barley breeding lines, increased drought resistance and greater yields compared with commercial cultivars were associated with greater epicuticular wax load (González and Ayerbe 2009). Water deficit increased the epicuticular wax load and reduced nonstomatal transpiration. Epicuticular wax load was negatively associated with nonstomatal transpiration and positively associated with grain yield (González and Ayerbe 2009). The negative association between epicuticular wax load and nonstomatal transpiration has also been observed for sorghum (Jordan et al. 1984), but the converse was observed in earlier studies on barley and oats (Larsson and Svenningsson 1986; Svenningsson and Liljenberg 1986).

Stomata are small pores on leaf and stem surfaces that facilitate the exchange of gases between the leaf interior and the atmosphere. They are an important determinant of TE, being physically located as the gateway controlling the trade-off between carbon capture and water loss, and, as such, have warranted the greatest amount of research. Stomatal conductance has a greater effect on transpiration than photosynthesis, so maximum TE will occur through coordination of stomatal opening and capability for $CO_2$ fixation (Nobel 2009; Yoo et al. 2009).

Stomatal density and size of their aperture determine a leaf's conductance to $CO_2$ and water vapor (assuming gas exchange across the nonstomatal epidermis is negligible). Stomatal density increases stomatal conductance to water vapor ($g_s$) of a leaf in a linear dependent manner, but conductance to water vapor through a single pore increases with radius of the aperture to the fourth power, i.e., doubling the number of stomata will double $g_s$, but doubling stomatal aperture (keeping density the same) will increase $g_s$ 16-fold. Hetherington and Woodward (2003) showed that there is a strong negative curvilinear relationship between stomatal density and stoma size across a range of plant groups and that the relationship between $g_s$ and $A_N$ approximately fit within one of two curvilinear relationships depending on whether the plants have a $C_3$ or $C_4$ photosynthetic pathway. Stoma aperture size is regulated by turgor pressure of the flanking guard cells. An uptake of ions ($K^+$ and $Cl^{-1}$) (Turner 1972) generates the osmotic gradient to induce water uptake, which probably occurs through aquaporins (Heinen et al. 2009). Guard cell volume increases with turgor pressure causing the stoma to open due to radial orientation of the cellulose microfibrils of the guard cells walls, causing the cells to swell more in one direction. The reverse happens to cause a stoma to close. Guard cell length, therefore, ultimately determines aperture size of the stoma. It has been suggested that small stomata can open and close more rapidly and by being generally associated with high densities provide the capacity for rapid changes in $g_s$ with fluctuating environmental conditions (Hetherington and Woodward 2003).

Stomatal density is under strict genetic control, but the number of stomata can also be influenced by environmental factors (see review by Casson and Gray (2008) and Sect. 6.4.1). Stomatal density tends to decrease when plants are grown under elevated atmospheric $CO_2$ and has been shown to be modulated by the *HIGH CARBON DIOXIDE* gene in *Arabidopsis*. Light intensity also induces changes in stomatal density in developing leaves, but the genes regulating the response are not yet known (Casson and Gray 2008). Drought can influence stomatal density of developing leaves, depending on the species (Quarrie and Jones 1977; Clifford et al. 1995; Xu and Zhou 2008; Hamanishi et al. 2012), but this may be confounded by inhibited leaf expansion at low soil water potentials (see above).

Stomatal opening depends on the water status of the plant and is influenced by irradiance, $CO_2$ concentration, relative humidity, phytohormones such as ABA, and stress (see reviews by Buckley 2005; Shimazaki et al. 2007). Evapotranspirational demand (temperature, relative humidity, and wind speed) also determines $g_s$ in field conditions. Variation in $g_s$ across species in response to soil water deficit has been observed (Henson et al. 1989a). Stomatal closure occurred in lupin, but not wheat, before changes in the leaf water potential were detected. The disparity in the responses was later attributed to differences in sensitivity to ABA (Henson et al. 1989b). However, field-grown wheat has also been observed closing stomata before leaf water potential decreased under drought (Ali et al. 1999), indicating that there may be genotypic variation in ABA production and/or stomatal sensitivity to soil water deficit. Genotypic variation in the relationship between $g_s$ and leaf water potential under water-limited conditions has been observed for a number of other crop species (Ackerson et al. 1980; González et al. 1999).

Modern phenotyping programs are using canopy temperature, as assessed through thermal imaging, as a surrogate for selecting genotypes with high $g_s$ (Jones et al. 2009; Munns et al. 2010). Lower canopy temperatures have correlated well with greater yields in genotypes of wheat (Reynolds et al. 2007) and rice (Garrity and O'Toole 1995), but not soybean (Harris et al. 1984). Low canopy temperature was also correlated with yield in a comparison between four legume species (Pandey et al. 1984), whereas lower canopy temperature was associated with higher susceptibility to drought in genotypes of potato (Stark et al. 1991). Lower canopy temperatures imply higher $g_s$ facilitating transpirational cooling and favoring net photosynthesis. Higher $g_s$ also indicates enhanced capacity for water uptake or better ability at maintaining plant water status. In these situations, low TE may in fact be selected since canopy temperature depression is related to transpirational cooling. Transpiration is linearly related to $g_s$, whereas photosynthesis saturates at high $g_s$ (Yoo et al. 2009) due to metabolic limitations of carbon assimilation and fixation (Farquhar and Sharkey 1982). The ideal then for improved TE is to select genotypes with less than maximum $g_s$, when the slope of the relationship between $A_N$ and $g_s$ is greater than the slope of the relationship between $E$ and $g_s$ (Yoo et al. 2009).

The internal anatomy of the leaf determines the pathway for water flow from the petiole to the substomatal cavities and the distance for diffusion of water vapor from the evaporation sites to the external atmosphere as well as the diffusion distance for $CO_2$ to reach the photosynthetic apparatus. These properties, therefore, determine leaf hydraulic conductance (water transport efficiency) and internal conductance to $CO_2$, respectively, although measurements of the internal conductance to $CO_2$ also include the permeability of mesophyll plasma membranes to $CO_2$.

The reader is referred to Sack and Holbrook (2006) for an in-depth review of leaf hydraulics, but we will highlight aspects relevant to TE in crop plants for the purpose of this chapter. Leaf hydraulic resistance accounts for up to a third of the total plant resistance to liquid water transport (Sack and Holbrook 2006), which is an overlooked factor potentially affecting TE of crops (see more detailed discussion below on roots, regarding constraints of hydraulic resistance to gas exchange). Leaf hydraulics has mainly been studied in tree leaves in an ecophysiological context, but many aspects would also be applicable to crop species. The hydraulic conductance of leaves from annual crop species is apparently two- to threefold greater than woody species, although annual crop species accounted for only 7 % of the species so far examined (Sack and Holbrook 2006). The high hydraulic conductance of crop leaves may account for the small differences in water potential observed between leaves and stems (Tsuda and Tyree 2000). Leaf hydraulic conductance is variable, changing with development and environmental conditions (Tsuda and Tyree 2000; Brodribb and Holbrook 2004; Nardini et al. 2005; Simonin et al. 2012). Irradiance can also stimulate increased hydraulic conductance (independently of stomatal opening) in some species, which has been attributed to increased aquaporin activity (Cochard et al. 2007; Baaziz et al. 2012). Although not yet shown to involve aquaporins, the hydraulic conductance of sunflower shoots was on average fourfold higher when the leaves were illuminated than under ambient light in the laboratory, but only when leaves were well hydrated (Guyot et al. 2012). The function of this irradiance-induced stimulation of leaf hydraulic conductance is not yet fully understood but when combined with stomatal regulation of $g_s$ provides flexible regulation of leaf water status (Guyot et al. 2012).

The myriad of leaf hydraulic properties and responses reflects differences in anatomy, principally vein density, and extravascular pathways for water flow (Cochard et al. 2004; Brodribb et al. 2007). When water enters the leaves, it travels through the veins, the density and pattern of which determine the distribution of water within the leaf. Water distribution (and assimilate translocation) within $C_4$ grass leaves is superior to $C_3$ grasses because of a more denser network of the smaller longitudinal and transverse veins (Ueno et al. 2006). Water exits the major or minor veins into the bundle sheath, but then its journey is less clear (Sack and Holbrook 2006). Water may travel apoplastically, around the cells, or transcellulary, through the cells of the mesophyll. Transcellular water transport may occur through the symplast via plasmodesmata or across aquaporin-embedded membranes. The contribution of these pathways is not yet known, but the rapid and reversible changes in leaf hydraulic conductance stimulated by temperature or irradiance indicate significant transcellular transport in those species. Moreover, suberin deposition in the walls of the bundle sheath, such as those found in the mestome sheath of wheat and oats (O'Brien and Carr 1970), may obstruct apoplastic water flow. The sites of water evaporation within the leaf are also not yet known but will likely be determined by anatomy and the pathway for water flow through the leaf.

Soybean is one of the few crop species where genotypic differences in leaf hydraulics have been examined. High diversity of leaf morphological features was observed in soybean cultivars, where $g_s$ was significantly correlated with stomatal density (Tanaka et al. 2010). The greater productivity of US cultivars compared with Asian cultivars of soybean was attributed to greater $g_s$ due to greater stomatal density. Interestingly, US cultivars had slightly smaller guard cells, smaller leaves, and higher vein density, which indicate that leaves of US cultivars also potentially had greater water transport efficiency. Conversely, a low leaf hydraulic conductance contributed to higher TE in a soybean genotype through water conservation when transpirational demands were high (Sinclair et al. 2008). Either high or low hydraulic conductance may, therefore, aid adaptation to drought, the significance of either being dependent on coordination with other leaf properties.

The internal conductance to $CO_2$ ($g_i$) is the remaining part of this discussion about leaf properties to be considered with regard to TE of crop plants. The greatest physical constraint to photosynthetic rates, excepting the biochemical process, was long considered to be the stomatal resistance to $CO_2$ diffusion into the leaf. However, $g_i$ is also a major constraint (Warren 2008). The internal conductance includes the diffusion of $CO_2$ once inside the leaf to the sites of carboxylation. The diffusion distance through the intercellular air spaces of the mesophyll is, therefore, a major determinant of $g_i$ as well as the permeability of the mesophyll membranes to $CO_2$. The dissolution of gaseous $CO_2$ to the liquid phase at the mesophyll cell wall and diffusion across the cytoplasm and chloroplast envelope should also not be discounted.

There is high variability among species in the relative limitation of $A_N$ due to $g_i$, but generally $A_N$ increases with increasing $g_i$ (see review by Warren 2008). Herbaceous species tend to have higher $g_i$, so the capacity for manipulating this property to improve TE, by reducing a constraint to $A_N$, in crop species may be limited, but $g_i$ will be influenced by anatomical changes in response to water stress. Moreover, aquaporin-mediated $CO_2$ diffusion and carbonic anhydrase activity can change $g_i$ rapidly (Flexas et al. 2007).

There are few genotypic comparisons of $g_i$ but we can make some inferences from anatomical studies. In flag leaves of diploid, tetraploid, and hexaploid wheat and *Aegilops* species, the surface area of the mesophyll cells per unit volume of mesophyll tissue was similar, but diploid genotypes had the highest photosynthetic rates at light saturation (Austin et al. 1982; Kaminski et al. 1990). This occurred despite thinner leaves and lower chlorophyll abundance. Diploid genotypes may have had a greater $g_i$ due to shorter diffusion distance between the leaf surface and carboxylation sites in thin leaves; diffusion across the cytoplasm would also have been shorter because diploid mesophyll cells were smaller. Interestingly, greater efficiency in the import of nutrients and export of assimilates was also related to improved productivity as vein density was also positively correlated with $A_N$ (Austin et al. 1982). Light-saturated $A_N$ was also negatively correlated with mesophyll cell size in ryegrass leaves, but this was dependent on growth temperatures (Wilson and Cooper 1969). Conversely, slower growth rates and lower photosynthetic rates were correlated with thinner leaves in bean genotypes (Sexton et al. 1997). However, RuBisCO may have been limiting in bean genotypes with lower photosynthetic rates, as indicated by the lower leaf nitrogen contents.

### 6.3.1.3 Stem Anatomy

Apart from the obvious structural role in supporting leaves above the ground, stems are the conveyors of water, nutrients, and photosynthetic assimilates between roots and leaves. Their anatomy, especially the vascular bundles, will determine transport efficiency which, similar to roots and leaves, will influence TE. The primary structure of stems is highly variable with the vascular tissue forming either a continuous cylinder sandwiched between the cortex and pith or a cylinder of discrete strands within the undifferentiated ground tissue, or in monocots such as maize, the vascular bundles may be spread throughout the ground tissue. Stems may also be involved in storage and remobilization of assimilates or water, particularly in response to abiotic stress. Remobilization of water-soluble carbohydrates is a major contributor to drought adaptation in a number of crop species (McIntyre et al. 2012), but it can confound measurements of TE if the whole plant biomass at maturity is considered, and therefore, TE is usually measured during the vegetative stages.

Within the vascular bundles, the number of xylem vessels varies. Xylem vessels are analogous to pipes and hence, their abundance and diameters largely determine the hydraulic conductance of a stem. Accordingly, the hydraulic conductance of stems is generally much greater than leaves or roots (Steudle 2000). Studies on the resistance to lodging indicate that there is high genetic variation for the number of vascular bundles in grass species, including the major cereal crops (Jellum 1962; Dunn and Briggs 1989; Khanna 1991; Kaack et al. 2003). Genetic variation in vascular bundles also exists in other crop species. In tomato, thicker stems with more conducting tissue (xylem vessels) were related to drought resistance (Kulkarni and Deshpande 2006).

Water flow through the stem, and for that matter roots and leaves, can be disrupted if tensions in the xylem become large and cavitations occur. The susceptibility of xylem to cavitation-induced hydraulic disruption depends on the species, but most research has examined this phenomenon in woody species. Xylem cavitation can occur in annual crop species at fairly high water potentials in comparison with woody species (Tyree et al. 1986). However, herbaceous species tend to generate high root pressures when they are not transpiring (Steudle and Jeschke 1983; Bramley et al. 2010), and so it is possible that embolized vessels may be refilled during the night. Although embolisms may be reversed, when cavitation occurs, water transport to the leaves will be disrupted, which will ultimately affect crop productivity. The occurrence and susceptibility of crop species to hydraulic disruption should, therefore, be investigated.

### 6.3.1.4   Root Architecture and Anatomy

The architecture of a root system, or spatial distribution within the soil, is described by its length and branching pattern, which in turn are determined by the rate, angle, and duration of root growth. The root system architecture, therefore, determines the capacity for water and nutrient uptake from the soil (Lynch 1995, 2007). Root systems of different species display a high degree of phenotypic plasticity, responding to environmental conditions, but developmental instability has also been proposed as another adaptive trait generating nongenetic variation (Forde 2009). Intraspecific variation in root system architecture of major crop species is receiving increasing interest as a potential means to improve water and nutrient uptake efficiency (Ratnakumar and Vadez 2011; Chen et al. 2012; Wasson et al. 2012), but the fact that roots are below ground and the soil environment is generally heterogeneous presents an enormous challenge in root research. Phenotyping is, thus, generally undertaken on roots growing in artificial medium (Chen et al. 2011; Clark et al. 2011) or in glass-walled chambers (Liao et al. 2006; Manschadi et al. 2006; Bramley et al. 2011). Moreover, breeding programs need to ensure that traits identified in the laboratory are translated into the field (Wasson et al. 2012).

In order to identify which root traits are relevant for improving TE, we need to consider the target environment because this will determine temporal and spatial variation in available water, as well as physical and chemical constraints to root growth. For example, crops growing on deep stored moisture may need long roots and high branch-root density at depth (Wasson et al. 2012), whereas high shallow-root density may be more advantageous for crops growing on sandy soils relying on intermittent in-season rainfall (Manschadi et al. 2006). There is considerable genetic variation for root length density that could be exploited (Siddique et al. 1990a; Mwanamwenge et al. 1998; Chen et al. 2011; Ratnakumar and Vadez 2011).

We also need to consider the particular species when selecting root traits for high TE because root hydraulic properties determine the region of root involved in water uptake and the transport efficiency of the root system (hydraulic conductance), which vary between species. For example, cereal roots absorb water preferentially in a region just behind the root tip, whereas legumes absorb water along the entire root length (Bramley et al. 2009). The hydraulic conductivity of cereal roots will, hence, be influenced by the number of branches, whereas the hydraulic conductivity of legume roots will be influenced by total root length.

The root system is a major constraint to gas exchange and, consequently, TE because it forms the largest resistance to liquid water flow in the plant (Steudle 2000). Root hydraulic conductivity needs to be high enough to supply the leaf with sufficient water to maintain leaf hydration so stomata can remain open. High root hydraulic conductivity will also minimize the drop in water potential needed to drive water uptake from the soil and transport it to the shoot. However, root hydraulic conductivity also needs to be variable and coordinated with stomatal opening and environmental conditions to maximize hydraulic efficiency; otherwise, the plant runs the risk of losing water to the soil when it becomes extremely dry or excessive water loss to the atmosphere if the vapor pressure deficit is high. While water transport is a passive process, roots themselves are not passive structures.

Hydraulic conductivity of a root system is determined by morphology, anatomy of individual roots, and aquaporin activity. All roots are cylindrical, comprising several concentric tissue layers: epidermis, exodermis (if present), cortex, and endodermis. The endodermis surrounds the stele, which contains the pericycle (origin of lateral roots), parenchyma, and vascular tissue. The outer surface of the epidermis, and walls of the exodermis and endodermis, can contain lignin or suberin, which reduces the permeability of the apoplast to water and nutrients (Steudle et al. 1993; Zimmermann et al. 2000). Development of these wall modifications depends on the species, developmental stage, and environmental conditions (Barrowclough et al. 2000; Zimmermann et al. 2000; Bramley et al. 2009; Vandeleur et al. 2009). An exodermis does not develop in wheat, barley, or perennial ryegrass (Perumalla et al. 1990) probably because the root cortex dies back in mature roots (Hamblin and Tennant 1987), revealing the heavily suberized endodermis. Lupin and chickpea roots also do not form an exodermis, but this has not been tested in response to soil water deficit (Hartung et al. 2002; Bramley et al. 2009). Increased suberization of the exodermis and endodermis has been observed in many crop species in response to water deficit, especially woody perennials, and was correlated with decreased root hydraulic conductivity (Lo Gullo et al. 1998; Rieger and Litvin 1999; Vandeleur et al. 2009). Maize is one of the few annual crop species where an exodermis has been reported, and it becomes more suberized under drought (Stasovski and Peterson 1991). The fact that increased exodermal suberization can also be stimulated by growing maize roots in a mist culture has demonstrated the reduction in root hydraulic conductance caused by this feature (Zimmermann et al. 2000).

Air-filled spaces called aerenchymas can also develop in the cortex, particularly in adventitious (nodal) roots, in response to waterlogging-induced oxygen deficiency, some nutrient deficiencies, and occasionally water deficit (Drew et al. 1989; Stasovski and Peterson 1991; Colmer 2003; Zhu et al. 2010). Aerenchyma development may be a more common root response to soil water deficit than has been reported because very few field studies have examined root anatomy, and cell plasmolysis or death occurs with dehydration. Aerenchymas also serve as radial barriers to water and nutrient transport (see review by Bramley and Tyerman 2010), but they also reduce the plant's respiratory costs (Zhu et al. 2010). Aerenchyma development, then, may be a beneficial trait for drought tolerance (Zhu et al. 2010) and even TE but only if water is preferentially absorbed in a region close to the root tip, where aerenchymas do not form.

Principal anatomical differences occur between monocotyledon and eudicotyledon roots, especially in vasculature development and because dicots have secondary growth (Klepper 1983). Monocot roots generally develop a ring of small xylem vessels surrounding a central pith or large xylem vessel, whereas xylem development in dicot roots initially occurs in poles and develops centripetally in a diarch or tetrarch pattern. Consequently, the axial hydraulic conductance of dicots is generally greater than monocots, and axial hydraulic conductance increases with root length in dicots because of increasing vessel development (Bramley et al. 2009).

Genetic variation in xylem vessel diameter has been exploited in wheat to reduce axial hydraulic conductance (Richards and Passioura 1989). The philosophy was based on the idea that reduced axial hydraulic conductance would reduce the rate of water uptake, conserving soil water for later in the season when it would be available for grain filling. Passioura (1972) first tested this theory by growing a wheat cultivar entirely on stored soil moisture and removing all but one seminal root to simulate reduced axial conductance, as all water uptake must flow through the one remaining central xylem vessel (discounting the ring of smaller vessels). Nodal roots were allowed to grow but were mostly restricted from accessing the soil moisture. Single-rooted plants compensated for the initially small root system with increased growth rate and branching, so that total root mass at maturity was the same as the control plants. Single-rooted plants initially maintained a similar water use as the control plants, but after a few weeks and until after anthesis, plant water use was lower than the control plants. Single-rooted plants ultimately had greater grain yield because control plants ran out of water for grain production. Passioura (1972) did not measure root hydraulic conductance or other physiological parameters, but he speculated that the estimated threefold reduction in axial conductance (three seminal roots reduced to one) would have caused stomatal closure due to much lower leaf water potentials, thereby conserving water. However, translocation of stored carbon in response to water deficit also probably contributed to grain yield (Passioura 1976). Remobilization of water-soluble carbohydrates is a well-recognized drought-adaptive trait in wheat (Reynolds et al. 2009). Moreover, partial excision of roots in young wheat plants increased hydraulic conductivity of the remaining root, probably by an ABA-induced change in aquaporin activity (Vysotskaya et al. 2004), so single-rooted wheat plants growing on stored moisture may be more efficient in water use through a number of mechanisms.

Reducing the diameter of the central xylem vessel by 10 μm through plant breeding increased wheat yield in the field by about 10 % in one of the selected genotypes, in comparison with unselected controls, but only in the low-yielding or drier environments (Richards and Passioura 1989). Theoretically this would reduce axial hydraulic conductance of the seminal root system by about half (vessel diameter decreased from 65 to 55 μm). It is not clear what effect the selection for narrower xylem vessels in seminal roots had on morphology and anatomy of nodal roots, anatomy of stems, or other physiological processes such as gas exchange and leaf water status. Moreover, the yield increase was not observed in an environment relying on in-season rainfall. Despite narrower xylem vessels being a potential drought-adaptive trait for environments relying on stored soil moisture, no current commercial wheat varieties appear to have been developed with narrower xylem vessels. Subsequent investigations of genetic differences in xylem anatomy in rice have led to the conclusion that manipulating vessel radii will not improve drought resistance, but in that study the aim was to increase axial hydraulic conductance (Yambao et al. 1992). Conservative water use has also been shown to be important in tolerance of chickpea genotypes to drought, which was correlated with lower $g_s$ but not root growth components (Zaman-Allah et al. 2011) and suggests that either greater sensitivity to ABA or lower root hydraulic conductance induced stomatal closure.

Surprisingly, studies on manipulating xylem anatomy to improve drought tolerance or TE have discounted the fact that it is the radial, not axial, hydraulic conductance that mostly constrains total root hydraulic conductance (Steudle and Peterson 1998). When water is taken up by the root, it has to cross the cylinder of living tissue described above, either by crossing membranes or the symplast (via plasmodesmata) or through cell walls (apoplast), and it is for this reason that the radial hydraulic conductance is several orders of magnitude less than the axial conductance. The radial pathway, though, is the main source attributing to variable root hydraulic conductance through anatomical changes (described above) and changes in aquaporin activity (see Sect. 6.4.4), depending on the pathway taken by water across the root radius. If water crosses membranes at some point in the radial pathway, aquaporins can facilitate rapid and reversible changes in hydraulic conductance. Of greatest significance to TE are the aquaporin-mediated changes in hydraulic conductance that can match transpirational demands (Henzler et al. 1999; Vandeleur et al. 2009), but also those that improve water transport efficiency in response to soil water deficit (Sade et al. 2010). However, more research is needed linking aquaporins to their physiological roles.

Differences in radial root anatomy and aquaporin activity have been found between genera, closely related species, and cultivars (Bramley et al. 2009; Vandeleur et al. 2009). Genotypic differences in root anatomy have also been observed in a number of crop species (Saliendra and Meinzer 1992; Zhu et al. 2010). However, the only study linking genotypic differences in aquaporin activity to root hydraulics within a species has been undertaken on accessions of *Arabidopsis* (Sutka et al. 2010).

## 6.3.2    Biochemicals and Enzymes

### 6.3.2.1    ABA and Sensitivity to ABA

The phytohormone abscisic acid (ABA) is an important stress hormone produced in the roots as the soil dries (Davies and Zhang 1991; Turner and Hartung 2012) and is transported in the xylem to the leaves where it closes the stomata (Henson et al. 1989b; Zhang and Davies 1989, 1990). It is, therefore, an important hormone in controlling water loss. While ABA triggers stomatal closure in the leaves, it also induces a reduction in leaf growth (Davies et al. 2005), an increase in root growth at low water potentials (Spollen et al. 2000), and an increase in the hydraulic conductivity of roots (Hose et al. 2000; Aroca et al. 2006), all of which provide an integrated response to water shortage. Although the influence of ABA on root hydraulic conductivity may be transient (Hose et al. 2000) or inhibitory (Fiscus 1981), ABA production has been shown to increase yields of wheat under moderate stress conditions (Travaglia et al. 2010) and improve TE in tree seedlings, particularly when subject to a water deficit (Duan et al. 2007). As there is genetic variation in the response of stomata to ABA concentration (Henson and Turner 1991), this provides another genetic mechanism for differences in TE to be expressed (see Sect. 6.4.4 for further details).

### 6.3.2.2  Isotopic Discrimination

Farquhar et al. (1982) first developed the theory that predicted that in $C_3$ plants the discrimination of the small amounts of the naturally occurring stable isotope of carbon, $^{13}C$, in photosynthesis will be least in those plants that fix the most carbon per unit amount of water transpired, i.e., in those that have the highest TE. This was confirmed by Farquhar and Richards (1984) who showed that the ratio of $^{13}C$ to $^{12}C$ in the leaves of wheat genotypes was correlated with the TE measured from changes in dry weight and water use for the period before sampling for CID. This relationship between CID and TE has been shown to exist in a range of crop species (Turner 1993), but not all crop species (Turner et al. 2007b). The relationship has been used to select wheat genotypes that yield better under drought (Richards 2006).

### 6.3.2.3  Maintenance of Turgor and Leaf Water Potential

Crop plants have been shown to maintain their turgor potential/pressure as the plant water status decreases by the accumulation of solutes (Jones et al. 1980; Turner and Jones 1980), a process known as osmotic adjustment or osmoregulation (Turner and Jones 1980; Morgan 1984). Wheat with genes for high osmotic adjustment has been shown to have increased yields in water-limited environments (Richards 2006). However, osmotic adjustment in chickpea was shown to be not highly inherited, and the degree of osmotic adjustment was not associated with yield under dryland conditions in the field (Turner et al. 2007a). Other crop species such as lupin show limited capacity for osmotic adjustment, and they are relatively intolerant to drought (see review by Palta et al. 2012).

Changes in tissue elasticity is another feature of adaptation to drought (Bowman and Roberts 1985; Lo Gullo and Salleo 1988). The properties of cell walls predominantly determine leaf elastic modulus and, therefore, water-holding capacity. Leaf density correlates strongly with bulk leaf elastic modulus (Niinemets 2001). The capacity to reduce leaf elastic modulus in response to soil moisture deficit enhances the water-holding capacity and when combined with osmotic adjustment increases turgor maintenance under lower relative water contents (Jensen and Henson 1990).

High water-holding capacity, therefore, aids in maintaining leaf water status. Conversely, with a high leaf elastic modulus (low elasticity), a leaf's water potential decreases more for a given change in tissue water content, which may aid in water uptake from dry soils (Bowman and Roberts 1985) or may induce stomatal closure. Measurement of leaf elastic modulus is laborious, which is probably the reason why few crop studies have investigated this attribute, but genetic diversity exists (Johnson et al. 1984; Rascio et al. 1988). Moreover, leaf elastic modulus and hydraulic conductance may not be mutually exclusive properties because a high water-holding capacity would buffer against water loss when the root or shoot resistance to water flow is high. Inferences about leaf hydraulic capacitance can be made by plotting leaf water potential against transpiration rates; hysteresis between dehydration and rehydration indicates that either the hydraulic conductance is variable or stored water is initially used in transpiration (Zhang and Davies 1989). Whether these relationships are relevant to crop species, particularly annual crops remains to be seen.

### 6.3.2.4 Photosynthesis and Respiration

Earlier, TE was shown to be a ratio between the rates of $A_N$ and $E$. While this is controlled at the level of the leaf by leaf conductance to $CO_2$ and water vapor, $A_N$ is also affected by other anatomical and biochemical factors in the carbon assimilation pathway. Measurement under similar light intensity reveals variation in the rate of leaf net photosynthesis among plant and crop species (e.g., Turner et al. 1984; Leport et al. 1998). The reasons for this variation can be several. The concentration of chlorophyll that captures the light energy may vary, but as the rate of photosynthesis is maintained over a relatively wide range of leaf chlorophyll concentrations, a low concentration may not necessarily result in lower rates of photosynthesis. The conversion of $CO_2$ to carbohydrates is not the subject of this chapter, but with the energy from sunlight captured in the thylakoid membranes of the chloroplasts, the $CO_2$ is converted into $C_3$ sugars via a series of reactions in the Calvin cycle. One of the key enzymes in the Calvin cycle is ribulose-1,5-bisphosphate carboxylase/oxygenase (RuBisCO), which comprises 50 % (20–30 % of the total leaf nitrogen) of the soluble protein in the leaf of $C_3$ plants and 30 % (5–9 % of the total leaf nitrogen) of the soluble leaf protein in $C_4$ plants. However, it is notoriously inefficient in converting ribulose-1,5-bisphosphate to glycerate-3-phosphate and glucose, and therefore, there has been considerable effort to increase the efficiency of the enzyme and, hence, the photosynthetic efficiency of crop plants (Spreitzer and Salvucci 2002; Parry et al. 2003). Although Spreitzer and Salvucci (2002) concluded that this will not be easy due to the range of atomic structures of the enzyme, attempts to genetically modify the enzyme and increase photosynthetic efficiency have continued (see Masle et al. (2005) and discussion in Sect. 6.4.1), but with limited success to date in crop plants. In the presence of high concentrations of $CO_2$ in the leaf, RuBisCO converts the $CO_2$ to sugars, but if the $CO_2$ concentration decreases, for example, when stomata close with water deficits, RuBisCO also has an affinity for oxygen and reduces ribulose-1,5-bisphosphate to glycerate-3-phosphate at a reduced rate and greater metabolic cost. This process is known as photorespiration and is estimated to reduce photosynthesis by as much as 25 % in $C_3$ plants (Sharkey 1988).

However, photosynthetic efficiency is much higher in $C_4$ plants because they are able to increase the $CO_2$ concentration in the leaf, thereby reducing the oxygenation of ribulose-1,5-bisphosphate and reducing the rate of photorespiration. In order to bypass the photorespiration pathway, $C_4$ plants have a mechanism to efficiently deliver $CO_2$ to the enzyme RuBisCO. They have a different anatomy from $C_3$ plants in that chloroplasts exist not only in the mesophyll cells but also in specialized bundle sheath cells that surround the leaf xylem. Instead of direct fixation to RuBisCO via the Calvin cycle, $CO_2$ is incorporated into the 4-carbon oxaloacetate by the enzyme PEP carboxylase and then converted to malate, which can be transferred to the bundle sheath cells that have the ability to regenerate $CO_2$ in the chloroplasts. Bundle sheath cells can then utilize this $CO_2$ to generate carbohydrates by the $C_3$ pathway (Slack and Hatch 1967). There have been many attempts to increase the photosynthetic efficiency of $C_3$ plants by the incorporation of the $C_4$ photosynthetic pathway, but to date all attempts have been unsuccessful.

Respiration, the conversion of carbohydrates back to $CO_2$ with the release of energy, is required for growth, metabolism, and nutrient uptake. It consumes between 25 and 70 % of the carbon fixed and also varies among species and genotypes (Lambers et al. 2005). Attempts to reduce carbon losses without affecting growth have met with little success.

Finally, as mentioned above, the specialized leaf anatomy of $C_4$ plants enables them to fix $CO_2$ more efficiently than $C_3$ plants. Additionally, leaf anatomy such as thicker leaves also affects the rate of photosynthesis. Schulze et al. (2006) compared 65 species of *Eucalyptus* at 73 sites in Australia and showed that thicker leaves were associated with higher nitrogen content (and presumably higher RuBisCO content) and higher TE as determined by CID.

## 6.4   Genes That Influence WUE

The interactive networks of traits and biochemical processes that influence or determine TE imply that TE is under the control of a whole suite of genes. The findings of Masle et al. (2005) suggested that TE may be under the control of a single gene. During the last decade a number of other individual genes have also been identified as having dominant roles in influencing WUE or TE (Table 6.1). These genes are predominantly involved in signaling, transcription, or water transport (Table 6.1). A summary of the genes that have been shown to influence WUE or TE follows in the next section.

**Table 6.1** Genes that modulate TE

| Gene | Plant species | Tissue location | Method of testing | Associated function | Reference |
|---|---|---|---|---|---|
| ERECTA Leucine-rich repeat receptor-like kinase (LRR-RLK) Signaling factor | Arabidopsis thaliana | Broad expression in shoot apical meristem and leaf primordia | Gene complementation in erecta null mutants | Regulation of stomatal density | Masle et al. (2005) |
| HARDY AP2-ERF-like transcription factor | A. thaliana, Oryza sativa, Trifolium alexandrinum | Inflorescence tissue | Overexpression | Stress response | Karaba et al. (2007), Abogadallah et al. (2011) |
| ESKIMO1 Unknown function DUF231 | A. thaliana | | Null mutation | Stress response, metabolic profile | Bouchabke-Coussa et al. (2008) |
| HVA1 ABA-inducible group 3 LEA gene | Triticum aestivum | | Overexpression | Stress response | Sivamani et al. (2000), Bahieldin et al. (2005) |
| ME NADP-malic enzyme | Nicotiana tabacum | Chloroplast | Overexpression | Malate metabolism | Laporte et al. (2002) |
| NCED1 9-cis-epoxycarotenoid dioxygenase | Solanum lycopersium | | Overexpression | ABA production | Thompson et al. (2007) |
| DREB1 Transcription factor | Arachis hypogaea | | Stress-induced expression | Stress response, antioxidant activities | Bhatnagar-Mathur et al. (2007), Devi et al. (2011) |
| NtAQP1 Aquaporin | S. lycopersium | Flowers, stems, leaves, and roots | Overexpression | Water and $CO_2$ transport | Sade et al. (2010) |
| SITIP2;2 | S. lycopersium | All organs | Overexpression | Water transport, osmoregulation? | Sade et al. (2009) |
| OsPIP1;3 | Oryza sativa | Roots, leaves, anthers | Overexpression | Water transport | Lian et al. (2004) |
| GTL1 Transcription factor | A. thaliana | Abaxial epidermal cells | Null mutant | Regulation of stomatal density | Yoo et al. (2010) |
| ABP9 Transcription factor | A. thaliana | Leaves | Overexpression | Stress response, ABA production, production of photosynthetic pigments | Zhang et al. (2008) |

## 6.4.1    Signaling Factors

One of the most promising discoveries for manipulation of TE arose in the last decade when a leucine-rich repeat receptor-like kinase (LRR-RLK) called *ERECTA* was identified as being a major contributor to the genetic locus for CID (Masle et al. 2005). Transmembrane receptor kinases, such as ERECTA, sense external signals via extracellular ligands, which activate the molecule's intracellular kinase domain resulting in downstream signal transduction and gene regulation (Torii 2004). *ERECTA*, known for its involvement in regulating organ shape (Torii et al. 1996), was found to regulate the coordination between photosynthesis and transpiration in *Arabidopsis thaliana* (Masle et al. 2005). Complementation of *erecta* mutants with the wild-type allele restored TE to the same levels as wild-type plants under both well-watered and water-deficient conditions. *ERECTA* coordinates the early phase of stomatal development that determines stomatal density (see for review Bergmann and Sack 2007), which subsequently influences stomatal conductance and transpiration. More densely packed spongy mesophyll cells, greater electron transport capacity, and higher RuBisCO carboxylation rates in complemented ERECTA lines contributed to their greater $CO_2$ assimilation rates (Masle et al. 2005). Overexpression of *ERECTA* from *Populus nigra* in *Arabidopsis* produced similar results and improved dry biomass (Xing et al. 2011). These results indicate that manipulation of *ERECTA* may be a way of potentially improving crop performance (Masle et al. 2005).

Despite the significance of the *ERECTA* discovery in regulating TE, it has not yet directly translated into crop varieties with greater yield under water-limited conditions due to improved TE, although the gene has presumably been selected in some crop species where breeding programs for drought tolerance used genotypic screening of CID to select parental material (Condon et al. 2002). Wheat genotypes with improved TE have already been produced from these programs (Condon et al. 2004). Homologs of the *A. thaliana ERECTA* locus have been identified as well as *ERECTA* genes isolated from the crop species rice, sorghum, maize, and wheat, which will aid marker-assisted selection and breeding programs (Masle et al. 2004). More recently, transgenic maize plants overexpressing *ERECTA* (*ZmERECTA A*) showed more vigorous growth in the field compared with nontransformed controls, but yield data was not provided (Guo et al. 2011). Interestingly, leaves on transgenic maize had not only decreased stomatal density but also increased pubescence (Guo et al. 2011).

*ERECTA*, along with the other members of its family (*ERECTA-LIKE1* and *ERECTA-LIKE2*), and the leucine-rich repeat receptor-like protein (LRR-RLP), *TOO MANY MOUTHS*, are all needed to coordinate stomatal production and spacing, but the interacting relationships of these genes and their involvement in signal transduction are not yet fully understood (Bergmann and Sack 2007; Rowe and Bergmann 2010). ERECTA-family members have subtly different functions, and TOO MANY MOUTHS modulates ERECTA signaling in a tissue-specific manner (Rowe and Bergmann 2010), possibly through formation of heterodimers

**Table 6.2** Genes involved in stomatal development

| Gene | Function |
|---|---|
| EPIDERMAL PATTERNING FACTORS (EPF1, EPF2, CHALLAH, STOMAGEN) | Signaling peptides |
| ERECTA-family | Leucine-rich repeat receptor-like kinase (LRR-RLK) |
| TOO MANY MOUTHS | Leucine-rich repeat receptor-like protein (LRR-RLP) |
| STOMATAL DENSITY AND DISTRIBUTION 1 | Subtilisin-like serine protease |
| YODA | Mitogen-activated protein kinase (MAPK) |
| FOUR LIPS | R2Y3 MYB transcription factor |
| MYB88 | Transcription factor |
| FAMA | Basic helix-loop-helix transcription factor |
| CYCLIN-DEPENDENT KINASE B1;1 | Cyclin-dependent kinase |
| SPEECHLESS | Basic helix-loop-helix transcription factor |
| SCREAM | Basic helix-loop-helix transcription factor |
| MUTE | Basic helix-loop-helix transcription factor |

Data summarized from Bergmann and Sack (2007), Gray (2007), and Torii (2012)

(Torii 2004). Rapid progress is being made in identifying the many other genes and signaling peptides that act upstream and downstream of the ERECTA-family and TOO MANY MOUTHS (Table 6.2) in the stomatal development pathway. While the specific details of stomatal development are not a primary focus of this chapter, the reader is referred to the excellent reviews of Bergmann and Sack (2007), Nadeau (2009), and Rowe and Bergmann (2010). Expression of any of these genes, in addition to *ERECTA*, could potentially influence TE by modulating stomatal conductance given their interdependent relationships and influence on stomatal density. Almost all of the stomatal development research has been undertaken on the model plant *Arabidopsis*, but to have greater relevance to agricultural crops, more research needs to be undertaken on other species to determine whether the mechanisms identified in *Arabidopsis* are conserved across species. *B. distachyon* would be an ideal model for cereals because of the close synteny between the genomes (Huo et al. 2009) and similar root and shoot development (Watt et al. 2009). Greater understanding of how the environment interacts with stomatal development is also needed as plasticity in TE may be a useful trait in different crops and different growing regions.

## 6.4.2 Transcription Factors Influencing WUE

Transcription factors control the expression of other genes, and therefore, those implicated in influencing stomatal density and aperture (ABA signaling) and root or shoot growth are likely to play a significant role in TE. Several transcription factors that affect TE have been identified (Table 6.1), but not all of the downstream events that they regulate. *GTL1* appears to be involved in the development of stomata as it interacts with the promoter of STOMATAL DENSITY AND DISTRIBUTION1 (Yoo et al. 2010) (Table 6.2), a subtilizing protease, which is believed to be one of the ligands that interacts with TOO MANY MOUTHS (Bergmann and Sack 2007). Lower transpiration of *gtl1*-mutant *Arabidopsis* plants was associated with upregulated STOMATAL DENSITY AND DISTRIBUTION1 and fewer stomates on the abaxial leaf surface. *GTL1* is one of the first genes to be shown to be involved in stomatal development in response to environmental conditions as it was shown to be downregulated with water stress (Yoo et al. 2010).

Several other transcription factors influencing TE independently from the genes involved in the stomata developmental pathway described in the section above have been identified (Table 6.1). *HARDY*, an AP2-ERF-like transcription factor, was first identified in phenotypic screening of *Arabidopsis* mutants (Karaba et al. 2007). Overexpression of *HARDY* in *Arabidopsis* induced thicker leaves due to extra palisade and mesophyll cells, and the root system was denser. Transformed plants survived longer periods when water was withheld and were more tolerant to salinity than wild-type plants. To test whether *HARDY* could contribute to improved TE in crop plants, Karaba et al. (2007) expressed the *Arabidopsis HARDY* gene in the rice cultivar Nipponbare. *HARDY* rice lines had greater TE, growth, and consequently biomass, but yield data was not reported. The root system contributed most to the greater biomass. In comparison with nontransformed plants, *HARDY* rice lines also had higher TE and instantaneous TE under well-watered and water-deficient conditions. Greater TE was facilitated by lower stomatal conductance and greater assimilation rates; however, the mechanism responsible for this is not known (Karaba et al. 2007). *HRD* overexpression induced clusters of genes that are normally expressed under drought stress, which probably contributed to the greater drought tolerance of the transgenic rice.

Overexpression of the *Arabidopsis HARDY* gene in *Trifolium alexandrinum* was consistent with the results in rice (Abogadallah et al. 2011). *HARDY-Trifolium alexandrinum* lines had greater instantaneous TE due to lower transpiration rates under water-deficient conditions. In the field, transgenic *T. alexandrinum* plants were more drought tolerant because of slower soil water depletion and had improved biomass in comparison with wild-type plants.

Other transcription factors that have been associated with TE include *DREB1* and *ABP9* (Table 6.1). Both of these genes are known for their association with drought tolerance (Lata and Prasad 2011). The *Arabidopsis* genes of *DREB1* were expressed in groundnut driven by a stress-responsive promoter (Bhatnagar-Mathur et al. 2007; Devi et al. 2011). Five transgenic lines were compared against wild-type parental material under well-watered and water-deficient conditions in the field. TE varied across the lines and was correlated with specific leaf area and leaf chlorophyll content, but only under water deficit and was not correlated with CID under either water treatment. High TE lines maintained gas exchange longer, until the soil water content was lower than low TE genotypes (Devi et al. 2011). *DREB1's* involvement in increased drought tolerance is associated with increased antioxidant activities (Li et al. 2011). *DREB1* is also reported to stimulate root growth (Vadez et al. 2007). *DREB* homologous genes have been isolated in a variety of plant species, and some are also responsive to other abiotic stresses such as heat and cold, but overexpression does not always result in phenotypic changes possibly because constitutive promoters are not always functional in other species (Nakashima et al. 2009).

*ABP9* is involved in the ABA-dependent signaling pathway. Transgenic *Arabidopsis* plants overexpressing *ABP9* were more tolerant to drought and high temperature compared with wild-type plants and maintained higher rates of $A_N$ and $g_s$ (Zhang et al. 2008). Increased ABA content, photosynthetic pigments, and stress-responsive genes were some of the positive changes associated with increased instantaneous TE of transgenic plants.

### 6.4.3 Aquaporins Influencing TE

Aquaporins are membrane intrinsic proteins that predominantly form water-conducting pores, although some isoforms appear to conduct a wide range of other neutral molecules (see for review Hachez and Chaumont 2010). Although aquaporins are found in every organism, their abundance and diversity is greatest in plants (Tyerman et al. 2002). More than 30 aquaporin genes have been identified in *Arabidopsis*, maize, rice, and tomato (Chaumont et al. 2001; Johanson et al. 2001; Sakurai et al. 2005; Sade et al. 2009); at least 23 aquaporins have been detected in grapevine (Shelden et al. 2009) and 55 in poplar (*Populus trichocarpa*) (Gupta and Sankararamakrishnan 2009). Wheat apparently has an even greater number of aquaporin genes, as 24 PIPs (plasma membrane intrinsic proteins) and 11 TIPs (tonoplast intrinsic proteins) have already been identified (Forrest and Bhave 2008) and more may be discovered because allohexaploid wheat has three diploid genomes.

Aquaporin gene expression varies in different parts the plant, but they tend to be concentrated in tissue and cells needing greatest osmotic regulation and regulation of water flows (Maurel 1997), with the majority of aquaporin isoforms being expressed in roots (Bramley et al. 2007). Aquaporins that are located in plasma membranes (PIPs) may be primarily responsible for controlling transcellular water flow (Javot and Maurel 2002). TIPs and NIPs (nodulin-like intrinsic proteins) may also be involved in osmoregulation and transport of some osmolytes (Maurel 1997; Maurel et al. 2009), processes that are also important in drought resistance. Aquaporins that belong to the other subgroups, SIPs (small basic intrinsic proteins), and XIPs (X-intrinsic proteins) are rarely expressed (Chaumont et al. 2001; Danielson and Johanson 2008). In *Arabidopsis*, SIPs have been observed in endoplasmic reticulum membranes of specific cell types (Maeshima and Ishikawa 2008), and XIPs of the Solanaceae family transport uncharged molecules, but not water (Bienert et al. 2011).

The hydraulic conductance of membranes and tissue, if water takes the transcellular pathway, can be controlled by the abundance and gating (opening/

closing) of aquaporins (Tyerman et al. 1999). The high diversity and complex expression patterns of aquaporins is one of the reasons why there is limited understanding of the role of individual isoforms. Nevertheless, pharmacological agents, that block aquaporin "pores" or change protein conformation, and reverse genetics have demonstrated the influence of aquaporins on controlling water flow across membranes and through roots or leaves of a number of species (Cochard et al. 2007; Bramley et al. 2009; Ehlert et al. 2009; Vandeleur et al. 2009; Sutka et al. 2010). Aquaporin-mediated regulation of hydraulic conductance endows the capability for rapid and reversible regulation of water flow through the plant to match transpiration and in response to the environment, unlike changes in morphology and anatomy, which are slow and growth dependent, i.e., efficiency in water transport can match the needs of the plant and environmental conditions (Tyerman et al. 1999). However, the majority of aquaporin studies have focused on details at the molecular level; a greater understanding of their physiology is needed.

A detailed discussion about the mechanisms controlling aquaporin activity is beyond the scope of this chapter, but it should be pointed out that aquaporin gating is directly or indirectly regulated by apoplastic water potential, free $Ca^{2+}$ concentration, cytosolic pH, $H_2O_2$, and phosphorylation (see for review Bramley and Tyerman 2010), all of which are pertinent to cellular responses to water stress. The expression of various aquaporin isoforms has also been shown to change in response to drought and osmotic stress (Jang et al. 2004; Lian et al. 2004; Parent et al. 2009; Vandeleur et al. 2009), and ABA has been implicated in either stimulating gating or changes in aquaporin expression (Wan et al. 2004; Aroca et al. 2006; Parent et al. 2009). There have been few studies demonstrating the role of aquaporins in maintaining leaf water status. Moreover, these proteins are prime candidates for modulating TE, particularly if plant hydraulic conductance is tightly coupled with $g_s$. Those aquaporins shown to be involved in TE (Table 6.1) or productivity will be discussed in the remainder of this chapter.

In contrast with the signaling and transcription factors described above, instantaneous TE of tomato plants was increased through overexpression of the tobacco aquaporin *NtAQP1* (Sade et al. 2010). This aquaporin is one of the most intensely studied isoforms, and while highly abundant in roots (see references in Bramley et al. 2007), expression has been detected in all plant organs, including leaves where it is concentrated in the spongy mesophyll (Biela et al. 1999; Siefritz et al. 2001). Antisense tobacco plants with inhibited *NtAQP1* expression demonstrated the importance of this aquaporin in maintaining plant water status, especially under water stress (Siefritz et al. 2002). Moreover, *NtAQP1* is induced by stress, possibly through the production of the phytohormones ABA or gibberellic acid (Siefritz et al. 2001). In addition to transporting water (Siefritz et al. 2002; Sade et al. 2010), the NtAQP1 pore is also permeable to $CO_2$ and influences $A_N$ through moderation of a leaf's internal conductance to $CO_2$ (or mesophyll conductance) (Uehlein et al. 2003, 2008). Interestingly, *NtAQP1*-transformed tomato plants had not only higher $A_N$ but also higher $g_s$ that was facilitated by larger stomatal apertures and not through increased stomatal density (Sade et al. 2010). The difference in instantaneous TE between transformed and wild-type tomato plants was even greater under salt stress.

Manipulation of aquaporin expression can cause changes in plant growth or hydraulic conductance (Martre et al. 2002; Siefritz et al. 2002), but overexpression of *NtAQP1* in tomato did not affect root hydraulic conductivity (normalized to xylem cross-sectional area). Indeed, overexpression of this aquaporin in tobacco increased leaf growth, but did not affect root mass or plant height (Uehlein et al. 2003), whereas plant size and root mass were affected by inhibited expression (Siefritz et al. 2002). Although *NtAQP1*-transformed and wild-type tomato plants had similar root hydraulic conductivities under control conditions, salinity reduced root hydraulic conductivity of *NtAQP1*-transformed plants considerably less than wild-type plants. The maintenance of root hydraulic conductivity in *NtAQP1*-transformed plants likely supported their higher daily rates of transpiration under salinity (Sade et al. 2010).

Rice has several aquaporin genes that respond to abiotic stresses. A plasma membrane aquaporin, *OsPIP1;3*, was upregulated in an upland subspecies that is more tolerant of drought than a lowland subspecies (Lian et al. 2004). When *OsPIP1;3* was overexpressed in transgenic lowland plants and exposed to osmotic stress (decreased water potential of the root solution culture), transgenic plants had higher root hydraulic conductivity, leaf water potential, and transpiration (Lian et al. 2004). In contrast, overexpression of *PIP1b* in *Arabidopsis* had an adverse effect on drought resistance, which caused premature wilting (Aharon et al. 2003). The productivity of transgenic plants was improved under well-watered conditions due to greater transpiration rates and photosynthetic efficiency, which was facilitated by increased stomatal density. These results indicate that selection of the relevant isoforms should be conducted under water-deficient conditions if the aim is to improve TE under drought.

Modulation of aquaporin activity to improve TE would need to lead to improved yields before this trait would be considered for selection in breeding programs. *NtAQP1* is a promising candidate because there was no yield penalty associated with overexpression of this isoform in tomato (Sade et al. 2010). Furthermore, individual fruit of *NtAQP1*-transformed tomato plants weighed more than wild-type plants under salt treatment (salt did not reduce individual fruit weight of transformed plants). Under control conditions the number of fruit per plant was higher on transformed plants and overall total fruit fresh weight was slightly higher in transformed plants under salinity compared with wild type (Sade et al. 2010), but whether this was due to higher water content or greater dry weight is not clear. Sade et al. (2010) also demonstrated that the beneficial effect of *NtAQP1* overexpression on plant production is not isolated to tomato because transformed *NtAQP1*-*Arabidopsis* plants produced more biomass under control and saline conditions compared with wild-type plants.

Sade et al. (2009) tested overexpression of a TIP in tomato, *SlTIP2;2*, on yield performance under salt and drought stress in the field. TOM-SlTIP2;2 plants had more biomass and fruit yields under drought stress compared with nontransgenic plants. Transgenic TOM-SlTIP2;2 plants had higher transpiration rates than controls when grown in pots and irrigated with saline or nonsaline solution. Higher and sustained rates of transpiration under drought led to lower relative water contents in leaves of transgenic plants. TIP aquaporins are expressed in the tonoplast, and it has been speculated that their function may be buffering of the cytoplasm against deleterious changes in volume (Maurel 1997; Tyerman et al. 1999) or they may function as osmotic and turgor pressure sensors (Hill et al. 2004). The novel results of Sade et al. (2009) demonstrate that SlTIP2;2 is involved in leaf water status, but rather than maintaining leaf water status, the leaves dehydrated to lower relative water contents suggesting that overexpression of this TIP served a protective role.

### 6.4.4  Other Regulatory Genes of TE

Several other direct or indirect regulators of TE have been identified (Table 6.1), the most studied of which is *ESKIMO1*, which encodes a protein with unknown function (Xin et al. 2007). *ESKIMO1* was first identified for its negative effect on freezing tolerance, which was associated with changes in expression of transcription factors, signaling components, and stress-responsive genes (Xin et al. 2007). The overlap between the genes regulated by *ESKIMO1* with those that respond to osmotic, salt, and ABA treatments suggested that *ESKIMO1* may be involved in drought tolerance, but *esk1*-mutants wilted earlier than wild-type plants (Xin et al. 2007). In contrast, Bouchabke-Coussa et al. (2008) observed improved drought tolerance in *esk1*-mutants through a lower transpiration rate and improved TE. *Esk1* mutants were smaller than wild-type plants and produced fewer seeds under well-watered conditions, but *esk1*-mutants stayed green longer and wilted later than wild-type plants when water was withheld. Interestingly, of the 135 genes that were differentially expressed between wild-type and *esk1*-mutants under water deficit, a TIP and a PIP aquaporin were downregulated in the mutant plants. The PIP aquaporin (*PIP1;5*) is predominantly expressed in roots and has consistently been shown to be downregulated in response to abiotic stress (Jang et al. 2004; Alexandersson et al. 2005). Downregulation of *PIP1;5* could have reduced the water transport capability of *esk1*-mutant roots and may explain why rosettes of *esk1*-mutants had lower relative water contents (Bouchabke-Coussa et al. 2008; Lugan et al. 2009). Lower root hydraulic conductance in *esk1*-mutants compared with wild-type plants was confirmed in a subsequent study, but this was correlated with xylem deformation (Lefebvre et al. 2011). Genes involved in cell wall synthesis (*CesA7*) have also been found to influence TE through xylem deformation (Liang et al. 2010). The rosettes of *Esk1*-mutants also had higher ABA contents, which may be a water-conserving mechanism inducing stomatal closure because of the decreased water uptake capability (Lefebvre et al. 2011). Changes in metabolite content and osmotic potential further indicate that *ESKIMO1* is involved in maintaining water balance (Lugan et al. 2009).

The relationship between ABA synthesis and $g_s$ has long been known, but recent discoveries about the genetic control of ABA biosynthesis and its role in TE provide new avenues for genetic manipulation of crop plants. Overexpression of NCED1, an enzyme (Table 6.1) that catalyzes a rate-limiting step in ABA biosynthesis, increased ABA content and reduced $g_s$ in transgenic tomato in comparison with wild-type plants (Thompson et al. 2007). Biomass of transgenic and wild-type plants was not significantly different, but one of the transgenic lines had the lowest total water use and highest TE. Drought reduced the $g_s$ of wild-type plants but not transgenic plants. The reduced $g_s$ were associated with greater ABA production in wild-type plants and not increased sensitivity to ABA. Because of the reduced rate of water loss, the decrease in leaf water potential of transgenic plants was slower than wild-type plants, and the transgenic plants retained their leaves longer. In a separate experiment, hydroponically grown transgenic plants tended to have greater root hydraulic conductivities, which were calculated from the exudation rate of detopped roots and the difference in osmotic pressure between the root exudates and the solution bathing the roots (Thompson et al. 2007). The authors speculated that ABA stimulated an increased solute flux into the xylem stream to drive the greater exudation rates, but the osmolarity of the exudates was lower than wild-type plants (Thompson et al. 2007). However, a decrease in the force driving water efflux is compensated by the increased hydraulic conductance. Exogenous application of ABA to root systems has been shown previously to stimulate increased root hydraulic conductance, in some species, probably via an aquaporin-mediated mechanism (Quintero et al. 1999; Aroca et al. 2006), and therefore, the overproduction of ABA in transgenic plants may have stimulated increased aquaporin activity. It should be noted that interpretations based on root pressure-driven exudation rates from detopped plants should be made with caution as the water flow rates and driving forces are not representative of transpiring plants.

Another gene associated with ABA signaling and shown to improve TE is the ABA-inducible HVA1 gene (Sivamani et al. 2000). Expression of the barley HVA1 in transgenic wheat improved TE in all transgenic lines except one. Transgenic plants had larger root systems under moderate water deficit compared with controls. Shoot dry weight was also greater, but seed weight was only greater in one homozygous transgenic line. In the field, some of the HVA1-transgenic wheat lines outyielded the wild-type plants but only in some seasons (Bahieldin et al. 2005).

Manipulation of stomatal aperture by perturbation of the enzyme that converts malic acid to pyruvate in guard cells can also influence TE (Laporte et al. 2002). Malate is one of the anions that balance the positive charge generated by the influx of potassium ions during turgor-driven stomatal opening. Expressing the maize NADP-malic enzyme (ME) in transgenic tobacco increased the malate content of leaves, decreased $g_s$, and improved TE compared with wild-type plants. Under drought conditions, the lower $g_s$ of ME-transformed plants drew soil water content down more slowly so that the wild type wilted earlier than transformed plants.

## 6.5   Conclusion and Future Directions

This chapter has demonstrated that there are many hydraulic and biochemical processes involved in controlling water flow through the plant and in controlling photosynthesis. As a consequence, there are potentially many genes and their interaction involved in determining the TE of a plant during development and under water-limited conditions. Nevertheless, TE is predominantly driven by hydraulic properties, such as stomatal conductance and anatomy of the water transport pathway, or factors that regulate those hydraulic properties, such as stomatal density (Fig. 6.2). Thus, most of the genes associated with modulation of TE are involved in regulating gas exchange, and although $g_s$ will also affect $A_N$, $A_N$ is further limited by the efficiency of the photosynthetic apparatus. In $C_3$ plants, the greatest limitation is RuBisCO, which provides few avenues for improvement. These results indicate that the main way of improving TE of agricultural crops will be through a reduction in water use rather than increases in photosynthetic efficiency. Many studies have observed that higher genotypic TE is generally due to lower transpiration rates (see reviews by Blum 2005, 2009).

Moreover, despite the extensive research on TE and aspirations to improve TE of crops, evidence demonstrating the benefits of improved TE in the field and for increasing yield is substantially deficient. Most of the studies discussed in this chapter indicate that higher TE is predominantly associated with slower growth. Most of the transgenic plants with improved TE were smaller than their related parental wild types. Higher TE was also related to more gradual depletion of the soil moisture, which suggests that high TE could be a suitable trait for improving drought tolerance under water-limited conditions, but when the water supply is not limiting, it may lead to yield penalties. In addition, slower growth is not a suitable trait in environments that experience terminal drought, but a more conservative water use would be beneficial in environments where crops are grown predominantly on stored soil moisture. Future research should, therefore, be directed towards examining the influence of high TE on yield production in targeted environments, as the research to date indicates that it is not a trait that would be beneficial in all growing regions.

## References

Abogadallah GM, Nada RM, Malinowski R, Quick P (2011) Overexpression of *HARDY*, an AP2/ERF gene from Arabidopsis, improves drought and salt tolerance by reducing transpiration and sodium uptake in transgenic *Trifolium alexandrinum* L. Planta 233:1265–1276

Abrams MD, Kubiske ME, Steiner KC (1990) Drought adaptations and responses in five genotypes of *Fraxinus pennsylvanica* Marsh: photosynthesis, water relations and leaf morphology. Tree Physiol 6:305–315

Abrams MD, Kubiske ME, Mostoller SA (1994) Relating wet and dry year ecophysiology to leaf structure in contrasting temperate tree species. Ecology 75:123–133

Ackerson RC, Krieg DR, Sung FJM (1980) Leaf conductance and osmoregulation of field-grown Sorghum genotypes. Crop Sci 20:10–14

Aharon R, Shahak Y, Winiger S, Bendov R, Kapulnik Y, Galili G (2003) Overexpression of a plasma membrane aquaporin in transgenic tobacco improves plant vigor under favorable growth conditions but not under drought or salt stress. Plant Cell 15:439–447

Alexandersson E, Fraysse L, Sjövall-Larsen S, Gustavsson S, Fellert M, Karlsson M, Johanson U, Kjellbom P (2005) Whole gene family expression and drought stress regulation of aquaporins. Plant Mol Biol 59:469–484

Ali M, Jensen CR, Mogensen VO, Bahrun A (1999) Drought adaptation of field grown wheat in relation to soil physical conditions. Plant Soil 208:149–159

Anderson WK, Hamza MA, Sharma DL, D'Antuono MF, Hoyle FC, Hill N, Shackley BJ, Amjad M, Zaicou-Kunesch C (2005) The role of management in yield improvement of the wheat crop- a review with special emphasis on Western Australia. Aust J Agric Res 56:1137–1149

Araus JL, Slafer GA, Royo C, Serret MD (2008) Breeding for yield potential and stress adaptation in cereals. Crit Rev Plant Sci 27:377–412

Aroca R, Ferrante A, Vernieri P, Chrispeels MJ (2006) Drought, abscisic acid and transpiration rate effects on the regulation of PIP aquaporin gene expression and abundance in *Phaseolus vulgaris* plants. Ann Bot 98:1301–1310

Austin RB, Morgan CL, Ford MA (1982) Flag leaf photosynthesis of *Triticum aestivum* and related diploid and tetraploid species. Ann Bot 49:177–189

Baaziz KB, Lopez D, Rabot A, Combes D, Gousset A, Bouzid S, Cochard H, Sakr S, Venisse J-S (2012) Light-mediated Kleaf induction and contribution of both the PIP1s and PIP2s aquaporins in five tree species: walnut (*Juglans regia*) case study. Tree Physiol 32:423–434

Bahieldin A, Hesham HT, Eissa HF, Saleh OM, Ramadan AM, Ahmed IA, Dyer WE, Madkour MA (2005) Field evaluation of transgenic wheat plants stably expressing the *HVA1* gene for drought tolerance. Physiol Plant 123:421–427

Barbour M, Warren C, Farquhar G, Forrester G, Brown H (2010) Variability in mesophyll conductance between barley genotypes, and effects on transpiration efficiency and carbon isotope discrimination. Plant Cell Environ 33:1176–1185

Barrowclough DE, Peterson CA, Steudle E (2000) Radial hydraulic conductivity along developing onion roots. J Exp Bot 51:547–557

Bennett D, Izanloo A, Reynolds MP, Kuchel H, Langridge P, Schnurbusch T (2012) Genetic dissection of grain yield and physical grain quality in bread wheat (*Triticum aestivum*) under water-limited environments. Theor Appl Genet 125:255–271

Bergmann DC, Sack FD (2007) Stomatal development. Annu Rev Plant Biol 58:163–181

Bharathan G, Goliber TE, Moore C, Kessler S, Pham T, Sinha NR (2002) Homologies in leaf form inferred from *KNOX1* gene expression during development. Science 296:1858–1860

Bhatnagar-Mathur P, Devi MH, Reddy DS, Lavanya M, Vadez V, Serraj R, Yamaguchi-Shinozaki K, Sharma KK (2007) Stress-inducible expression of *At*DREB1A in transgenic peanut (*Arachis hypogaea* L.) increases transpiration efficiency under water-limiting conditions. Plant Cell Rep 26:2071–2082

Biela A, Grote K, Otto B, Hoth S, Hedrich R, Kaldenhoff R (1999) The *Nicotiana tabacum* plasma membrane aquaporin NtAQP1 is mercury-insensitive and permeable for glycerol. Plant J 18:565–570

Bienert GP, Bienert MD, Jahn TP, Boutry M, Chaumont F (2011) Solanaceae XIPs are plasma membrane aquaporins that facilitate the transport of many uncharged substrates. Plant J 66:306–317

Blum A (2005) Drought resistance, water-use efficiency, and yield potential – are they compatible, dissonant, or mutually exclusive? Aust J Agric Res 56:1159–1168

Blum A (2009) Effective use of water (EUW) and not water-use efficiency (WUE) is the target of crop yield improvement under drought stress. Field Crops Res 112:119–123

Bonhomme L, Monclus R, Vincent D, Carpin S, Lomenech AM, Plomion C, Brignolas F, Morabito D (2009) Leaf proteome analysis of eight *Populus x euramericana* genotypes: genetic variation in drought response and in water-use efficiency involves photosynthesis-related proteins. Proteomics 9:4121–4142

Botwright TL, Condon AG, Rebetzke GJ, Richards RA (2002) Field evaluation of early vigour for genetic improvement of grain yield in wheat. Aust J Agric Res 53:1137–1145

Bouchabke-Coussa O, Quashie ML, Seoane-Redondo J, Fortabat MN, Gery C, Yu A, Linderme D, Trouverie J, Granier F, Téoulé E, Durand-Tardif M (2008) *ESKIMO1* is a key gene involved in water economy as well as cold acclimation and salt tolerance. BMC Plant Biol 8:125

Bowman WD, Roberts SW (1985) Seasonal changes in tissue elasticity in chaparral shrubs. Physiol Plant 65:233–236

Boyer JS (1970) Leaf enlargement and metabolic rates in corn, soybean and sunflower at various leaf water potentials. Plant Physiol 43:1056–1062

Boyer JS (1982) Plant productivity and environment. Science 218:543–548

Bramley H, Tyerman S (2010) Root water transport under waterlogged conditions and the roles of aquaporins. In: Mancuso S, Shabala S (eds) Waterlogging signalling and tolerance in plants. Springer, Berlin, pp 151–180

Bramley H, Turner DW, Tyerman SD, Turner NC (2007) Water flow in the roots of crop species: the influence of root structure, aquaporin activity, and waterlogging. Adv Agron 96:133–196

Bramley H, Turner NC, Turner DW, Tyerman SD (2009) Roles of morphology, anatomy, and aquaporins in determining contrasting hydraulic behavior of roots. Plant Physiol 150:348–364

Bramley H, Turner NC, Turner DW, Tyerman SD (2010) The contrasting influence of short-term hypoxia on the hydraulic properties of cells and roots of wheat and lupin. Funct Plant Biol 37:183–193

Bramley H, Tyerman SD, Turner DW, Turner NC (2011) Root growth of lupins is more sensitive to waterlogging than wheat. Funct Plant Biol 38:910–918

Briggs LJ, Shantz HL (1913) The water requirement of plants. Bureau of Plant Industry Bulletin. US Department of Agriculture, Washington, DC, pp 284–285

Brodribb T, Holbrook N (2004) Diurnal depression of leaf hydraulic conductance in a tropical tree species. Plant Cell Environ 27:820–827

Brodribb TJ, Feild TS, Jordan GJ (2007) Leaf maximum photosynthetic rate and venation are linked by hydraulics. Plant Physiol 144:1890–1898

Buckley TN (2005) The control of stomata by water balance. New Phytol 168:275–292

Casson S, Gray JE (2008) Influence of environmental factors on stomatal development. New Phytol 178:9–23

Chaumont F, Barrieu F, Wojcik E, Chrispeels MJ, Jung R (2001) Aquaporins constitute a large and highly divergent protein family in maize. Plant Physiol 125:1206–1215

Chen YL, Dunbabin VM, Postma J, Diggle AJ, Palta JA, Lynch JP, Siddique KHM, Rengel Z (2011) Phenotypic variability and modelling of root structure of wild *Lupinus angustifolius* genotypes. Plant Soil 348:345–364

Chen YL, Dunbabin VM, Diggle AJ, Siddique KHM, Rengel Z (2012) Assessing variability in root traits of wild *Lupinus angustifolius* germplasm: basis for modelling root system structure. Plant Soil 354:141–155

Clark RT, MacCurdy RB, Jung JK, Shaff JE, McCouch SR, Aneshansley DJ, Kochian LV (2011) Three-dimensional root phenotyping with a novel imaging and software platform. Plant Physiol 156:455–465

Clarke JM, Richards RA (1988) The effects of glaucousness, epicuticular wax, leaf age, plant height, and growth environment on water loss rates of excised wheat leaves. Can J Plant Sci 68:975–982

Clarke JM, McCaig TN, Depauw RM (1993) Relationship of glaucousness and epicuticular wax quantity of wheat. Can J Plant Sci 73:961–967

Clifford SC, Black CR, Roberts JA, Stronach IM, Singleton-Jones PR, Mohamed AD, Azam-Ali SN (1995) The effect of elevated atmospheric $CO_2$ and drought on stomatal frequency in groundnut (*Arachis hypogaea* (L.)). J Exp Bot 46:847–852

Cochard H, Nardini A, Coll L (2004) Hydraulic architecture of leaf blades: where is the main resistance? Plant Cell Environ 27:1257–1267

Cochard H, Venisse J-S, Barigah TS, Brunel N, Herbette S, Guilliot A, Tyree MT, Sakr S (2007) Putative role of aquaporins in variable hydraulic conductance of leaves in response to light. Plant Physiol 143:122–133

Colmer TD (2003) Long-distance transport of gases in plants: a perspective on internal aeration and radial oxygen loss from roots. Plant Cell Environ 26:17–36

Condon AG, Farquhar GD, Richards RA (1990) Genotypic variation in carbon isotope discrimination and transpiration efficiency in wheat. Leaf gas exchange and whole plant studies. Funct Plant Biol 17:9–22

Condon AG, Richards RA, Rebetzke GJ, Farquhar GD (2002) Improving intrinsic water-use efficiency and crop yield. Crop Sci 42:122–131

Condon AG, Richards RA, Rebetzke GJ, Farquhar GD (2004) Breeding for high water-use efficiency. J Exp Bot 55:2447–2460

Danielson J, Johanson U (2008) Unexpected complexity of the Aquaporin gene family in the moss *Physcomitrella patens*. BMC Plant Biol 8:45

Davies WJ, Zhang J (1991) Root signals and the regulation of growth and development of plants in drying soil. Annu Rev Plant Biol 42:55–76

Davies WJ, Kudoyarova G, Hartung W (2005) Long-distance ABA signalling and its relation to other signalling pathways in the detection of soil drying and the remediation of the plant's response to drought. J Plant Growth Regul 24:285–295

Devi MJ, Bhatnagar-Mathur P, Sharma KK, Serraj R, Anwar SY, Vadez V (2011) Relationships between transpiration efficiency and its surrogate traits in the rd29A:DREB1A transgenic lines of groundnut. J Agron Crop Sci 197:272–283

Dias de Oliveira E, Bramley H, Siddique KHM, Henty S, Berger J, Palta JA (2013) Can elevated $CO_2$ combined with high temperature ameliorate the effect of terminal drought in wheat? Funct Plant Biol 40:160–171

Dodd AN, Borland AM, Haslam RP, Griffiths H, Maxwell K (2002) Crassulacean acid metabolism: plastic, fantastic. J Exp Bot 53:569–580

Dornhoff GM, Shibles R (1976) Leaf morphology and anatomy in relation to $CO_2$-exchange rate of soybean leaves. Crop Sci 16:377–381

Dracup M, Reader M, Palta JA (1998) Timing of terminal drought is an important cause of yield variability in lupin grown on duplex soils in souther Australia. Aust J Agric Res 49:799–810

Drew MC, He C-J, Morgan PW (1989) Decreased ethylene biosynthesis, and induction of aerenchyma, by nitrogen- or phosphate-starvation in adventitious roots of *Zea mays* L. Plant Physiol 91:266–271

Duan B, Yang Y, Lu Y, Korpelainen H, Berninger F, Li C (2007) Interactions between water deficit, ABA, and provenances in *Picea asperata*. J Exp Bot 58:3025–3036

Dunn GJ, Briggs KG (1989) Variation in culm anatomy among barley cultivars differing in lodging resistance. Can J Bot 67:1838–1843

Ehleringer J (1980) Leaf morphology and reflectance in relation to water and temperature stress. In: Turner NC, Kramer PJ (eds) Adaptation of plants to water and high temperature stress. Wiley Interscience, New York, pp 295–308

Ehleringer J (1982) The influence of water stress and temperature on leaf pubescence development in *Encelia farinosa*. Am J Bot 69:670–675

Ehleringer J (1984) Ecology and ecophysiology of leaf pubescence in North American desert plants. In: Rodrigues E, Healy PL, Mehta I (eds) Biology and chemistry of plant trichomes. Plenum, New York, pp 113–132

Ehleringer JR, Clark C (1987) Evolution and adaptation in *Encelia* (Asteraceae). In: Gottlieb LD, Jain SK (eds) Plant evolutionary biology. Chapman and Hall, London, pp 221–248

Ehlert C, Maurel C, Tardieu F, Simonneau T (2009) Aquaporin-mediated reduction in maize root hydraulic conductivity impacts cell turgor and leaf elongation even without changing transpiration. Plant Physiol 150:1093–1104

Farquhar GD, Richards RA (1984) Isotopic composition of plant carbon correlates with water-use efficiency of wheat genotypes. Aust J Plant Physiol 11:539–552

Farquhar GD, Sharkey TD (1982) Stomatal conductance and photosynthesis. Annu Rev Plant Physiol 33:317–345

Farquhar GD, O'Leary MH, Berry JA (1982) On the relationship between carbon isotope discrimination and the intercellular carbon dioxide concentration in leaves. Aust J Plant Physiol 9:121–137

Fischer RA (1981) Optimizing the use of water and nitrogen through breeding of crops. Plant Soil 58:249–278

Fiscus EL (1981) Effects of abscisic acid on the hydraulic conductance and the total ion transport through *Phaseolus* root systems. Plant Physiol 68:169–174

Flexas J, Ribas-Carbo M, Diaz-Espejo A, Galmes J, Medrano H (2007) Mesophyll conductance to $CO_2$: current knowledge and future prospects. Plant Cell Environ 31:602–621

Forde BG (2009) Is it good noise? The role of developmental instability in the shaping of a root system. J Exp Bot 60:3989–4002

Forrest K, Bhave M (2008) The PIP and TIP aquaporins in wheat form a large and diverse family with unique gene structures and functionally important features. Funct Integr Genomics 8:115–133

French RJ, Schultz JE (1984a) Water use efficiency of wheat in a Mediterranean-type environment. 1. The relation between yield, water use and climate. Aust J Agric Res 35:743–764

French RJ, Schultz JE (1984b) Water use efficiency of wheat in a Mediterranean-type environment. 2. Some limitations to efficiency. Aust J Agric Res 35:765–775

Galmés J, Conesa MA, Ochogavía JM, Perdomo JA, Francis DM, Ribas-Carbó M, Savé R, Flexas J, Medrano H, Cifre J (2010) Physiological and morphological adaptations in relation to water use efficiency in Mediterranean accessions of *Solanum lycopersicum*. Plant Cell Environ 34:245–260

Garrity DP, O'Toole JC (1995) Selection for reproductive stage drought avoidance in rice, using infrared thermometry. Agron J 87:773–779

Garvin DF, Gu YQ, Hasterok R, Hazen SP, Jenkins G, Mockler TC, Mur LAJ, Vogel JP (2008) Development of genetic and genomic research resources for *Brachypodium distachyon*, a new model system for grass crop research. Crop Sci 48:S69–S84

González A, Ayerbe L (2009) Effect of terminal water stress on leaf epicuticular wax load, residual transpiration and grain yield in barley. Euphytica 172:341–349

González A, Martín I, Ayerbe L (1999) Barley yield in water-stress conditions: the influence of precocity, osmotic adjustment and stomatal conductance. Field Crops Res 62:23–34

Gray JE (2007) Plant development: three steps for stomata. Curr Biol 17:R213–R215

Gregory P (2004) Agronomic approaches to increasing water use efficiency. In: Bacon M (ed) Water use efficiency in plant biology. Blackwell, Oxford, pp 142–170

Guo M, Rupe M, Simmons C, Sivasankar S (2011) The maize *ERECTA* genes for improving plant growth, transpiration efficiency and drought tolerance in crop plants. United States Patent Application, US 2011/0035844 A1

Gupta A, Sankararamakrishnan R (2009) Genome-wide analysis of major intrinsic proteins in the tree plant Populus trichocarpa: characterization of XIP subfamily of aquaporins from evolutionary perspective. BMC Plant Biol 9:134

Guyot G, Scoffoni C, Sack L (2012) Combine impacts of irradiance and dehydration on leaf hydraulic conductance: insights into vulnerability and stomatal control. Plant Cell Environ 35:857–871

Hachez C, Chaumont F (2010) Aquaporins: a family of highly regulated multifunctional channels. In: Jahn TP, Bienert GP (eds) MIPs and their role in the exchange of metalloids. Springer, New York, pp 1–17

Hamanishi ET, Thomas BR, Campbell MM (2012) Drought induces alterations in the stomatal development program in *Populus*. J Exp Bot 63:4959–4971

260                                                                                    H. Bramley et al.

Hamblin A, Tennant D (1987) Root length density and water uptake in cereals and grain legumes: how well are they correlated? Aust J Agric Res 38:513–527

Harris DS, Schapaugh WT, Kanemasu ET (1984) Genetic diversity in soybeans for leaf canopy temperature and the association of leaf canopy temperature and yield. Crop Sci 24:839–842

Hartung W, Leport L, Ratcliffe RG, Sauter A, Duda R, Turner NC (2002) Abscisic acid concentration, root pH and anatomy do not explain growth differences of chickpea (*Cicer arietinum* L.) and lupin (*Lupinus angustifolius* L.) on acid and alkaline soils. Plant Soil 240:191–199

Heinen RB, Ye Q, Chaumont F (2009) Role of aquaporins in leaf physiology. J Exp Bot 60:2971–2985

Henson IE, Turner NC (1991) Stomatal responses to abscisic-acid in 3 lupin species. New Phytol 117:529–534

Henson IE, Jensen CR, Turner NC (1989a) Leaf gas exchange and water relations of lupins and wheat. I. Shoot responses to soil water deficits. Aust J Plant Physiol 16:401–413

Henson IE, Jensen CR, Turner NC (1989b) Leaf gas exchange and water relations of lupins and wheat. III. Abscisic acid and drought-induced stomatal closure. Aust J Plant Physiol 16:429–442

Henzler T, Waterhouse RN, Smyth AJ, Carvajal M, Cooke DT, Schäffner AR, Steudle E, Clarkson DT (1999) Diurnal variations in hydraulic conductivity and root pressure can be correlated with the expression of putative aquaporins in the roots of *Lotus japonicus*. Planta 210:50–60

Hetherington AM, Woodward FI (2003) The role of stomata in sensing and driving environmental change. Nature 424:901–908

Hill AE, Shachar-Hill B, Shachar-Hill Y (2004) What are aquaporins for? J Membr Biol 197:1–32

Hose E, Steudle E, Hartung W (2000) Abscisic acid and hydraulic conductivity of maize roots: a study using cell- and root-pressure probes. Planta 211:874–882

Hsiao TC, Silk WW, Jing J (1985) Leaf growth and water deficits: biophysical effects. In: Baker NR, Davies WJ, Ong CK (eds) Control of leaf growth. Cambridge University Press, Cambridge, pp 239–266

Hubick KT, Farquhar GD (1989) Genetic variation in carbon isotope discrimination and the ratio of carbon gained to water lost in barley. Plant Cell Environ 12:795–804

Huo N, Vogel JP, Lazo GR, You FM, Ma Y, McMahon S, Dvorak J, Anderson OD, Luo M-C, Gu YQ (2009) Structural characterization of *Brachypodium* genome and its syntenic relationship with rice and wheat. Plant Mol Biol 70:47–61

Izanloo A, Condon AG, Langridge P, Tester M, Schnurbusch T (2008) Different mechanisms of adaptation to cyclic water stress in two South Australian bread wheat cultivars. J Exp Bot 59:3327–3346

Jang JY, Kim DG, Kim YO, Kim JS, Kang H (2004) An expression analysis of a gene family encoding plasma membrane aquaporins in response to abiotic stresses in *Arabidopsis thaliana*. Plant Mol Biol 54:713–725

Javot H, Maurel C (2002) The role of aquaporins in root water uptake. Ann Bot 90:301–313

Jellum MD (1962) Relationships between lodging resistance and certain culm characters in oats. Crop Sci 2:263–267

Jensen CR, Henson IE (1990) Leaf water relations characteristics of *Lupinus angustifolius* and *L. cosentini*. Oecologia 82:114–121

Johanson U, Karlsson M, Johansson I, Gustavsson S, Sjovall S, Fraysse L, Weig AR, Kjellbom P (2001) The complete set of genes encoding major intrinsic proteins in Arabidopsis provides a framework for a new nomenclature for major intrinsic proteins in plants. Plant Physiol 126:1358–1369

Johnson RC, Nguyen HT, Croy LI (1984) Osmotic adjustment and solute accumulation in two wheat genotypes differing in drought resistance. Crop Sci 24:957–962

Jones MM, Osmond CB, Turner NC (1980) Accumulation of solutes in leaves of sorghum and sunflower in response to water deficits. Aust J Plant Physiol 7:193–205

Jones HG, Serraj R, Loveys BR, Xiong L, Wheaton A, Price AH (2009) Thermal infrared imaging of crop canopies for the remote diagnosis and quantification of plant responses to water stress in the field. Funct Plant Biol 36:978–989

Jordan WR, Shouse PJ, Blum A, Miller FR, Monk RL (1984) Environmental physiology of sorghum. II. Epicuticular wax load and cuticular transpiration. Crop Sci 24:1168–1173

Kaack K, Schwarz K-U, Brander PE (2003) Variation in morphology, anatomy and chemistry of stems of *Miscanthus* genotypes differing in mechanical properties. Ind Crops Prod 17:131–142

Kaminski A, Austin RB, Ford MA, Morgan CL (1990) Flag leaf anatomy of *Triticum* and *Aegilops* species in relation to photosynthetic rate. Ann Bot 66:359–365

Karaba A, Dixit S, Greco R, Aharoni A, Trijatmiko KR, Marsch-Martinez N, Krishnan A, Nataraja KN, Udayakumar M, Pereira A (2007) Improvement of water use efficiency in rice by expression of HARDY, an Arabidopsis drought and salt tolerance gene. Proc Natl Acad Sci USA 104:15270–15275

Kerstiens G (1997) *In vivo* manipulation of cuticular water permeance and its effect on stomatal response to air humidity. New Phytol 137:473–480

Kerstiens G (2006) Water transport in plant cuticles: an update. J Exp Bot 57:2493–2499

Khanna VK (1991) Relationship of lodging resistance yield to anatomical characters of stem in wheat [*Triticum* spp.], triticale and rye. Wheat Inform Serv 73:19–24

Kimura S, Koenig D, Kang J, Yoong FY, Sinha N (2008) Natural variation in leaf morphology results from mutation of a novel KNOX gene. Curr Biol 6:672–677

Klepper B (1983) Managing root systems for efficient water use: axial resistances to flow in root systems – anatomical considerations. In: Taylor HM, Jordan WR, Sinclair TR (eds) Limitations to efficient water use in crop production. American Society of Agronomy, Madison, WI, pp 115–125

Kulkarni M, Deshpande U (2006) Comparative studies in stem anatomy and morphology in relation to drought resistance in tomato (*Lycopersicon esculentum*). Am J Plant Pathol 1:82–88

Lambers H, Robinson SA, Ribos-Carbo M (2005) Regulation of respiration in vivo. In: Lambers H, Ribas-Carbo M (eds) Plant respiration. Springer, Dordhecht, pp 1–15

Laporte MM, Shen B, Tarczynski MC (2002) Engineering for drought avoidance: expression of maize NADP-malic enzyme in tobacco results in altered stomatal function. J Exp Bot 53:699–705

Larsson S, Svenningsson M (1986) Cuticular transpiration and epicuticular lipids of primary leaves of barley (*Hordeum vulgare*). Physiol Plant 68:13–19

Lata C, Prasad M (2011) Role of DREBs in regulation of abiotic stress responses in plants. J Exp Bot 62:4731–4748

Lefebvre V, Fortabat MN, Ducamp A, North HM, Maia-Grondard A, Trouverie J, Boursiac Y, Mouille G, Durand-Tardif M (2011) *ESKIMO1* disruption in Arabidopsis alters vascular tissue and impairs water transport. PLoS One 6:e16645

Leport L, Turner NC, French RJ, Tennant D, Thomson BD, Siddique KHM (1998) Water relations, gas exchange and growth of cool-season grain legumes in a Mediterranean-type environment. Eur J Agron 9:295–303

Li X, Cheng X, Liu J, Zeng H, Han L, Tang W (2011) Heterologous expression of the *Arabidopsis DREB1A/CBF3* gene enhances drought and freezing tolerance of transgenic *Lolium perenne* plants. Plant Biotechnol Rep 5:61–69

Lian H-L, Yu X, Ye Q, Ding X-S, Kitagawa Y, Kwak S-S, Su W-A, Tang Z-C (2004) The role of aquaporin RWC3 in drought avoidance in rice. Plant Cell Physiol 45:481–489

Liang YK, Xie X, Lindsay SE, Wang YB, Masle J, Williamson L, Leyser O, Hetherington AM (2010) Cell wall composition contributes to the control of transpiration efficiency in *Arabidopsis thaliana*. Plant J 64:679–686

Liao M, Palta JA, Fillery IRP (2006) Root characteristics of vigorous wheat improve early nitrogen uptake. Aust J Agric Res 57:1097–1107

Lo Gullo MA, Salleo S (1988) Different strategies of drought resistance in three Mediterranean sclerophyllous trees growing in the same environmental conditions. New Phytol 108:267–276

Lo Gullo MA, Nardini A, Salleo S, Tyree MT (1998) Changes in root hydraulic conductance ($K_R$) of *Olea oleaster* seedlings following drought stress and irrigation. New Phytol 140:25–31

Loss SP, Siddique KHM (1994) Morphological and physiological traits associated with wheat yield increases in Mediterranean environments. Adv Agron 52:229–276

Lugan R, Niogret M-F, Kervazo L, Larher FR, Kopka J, Bouchereau A (2009) Metabolome and water status phenotyping of *Arabidopsis* under abiotic stress cues reveals new insight into *ESK1* function. Plant Cell Environ 32:95–108

Lynch JP (1995) Root architecture and plant productivity. Plant Physiol 109:7–13

Lynch JP (2007) Roots of the second green revolution. Aust J Bot 55:494–512

Maeshima M, Ishikawa F (2008) ER membrane aquaporins in plants. Pflugers Arch Eur J Physiol 456:709–716

Manschadi AM, Christopher J, deVoil P, Hammer GL (2006) The role of root architectural traits in adaptation of wheat to water-limited environments. Funct Plant Biol 33:823–837

Martin JH, Leonard WH, Stamp DL (1976) Principles of field crop production. Macmillan, New York

Martre P, Morillon R, Barrieu F, North GB, Nobel PS, Chrispeels MJ (2002) Plasma membrane aquaporins play a significant role during recovery from water deficit. Plant Physiol 130:2101–2110

Masle J, Farquhar GD, Gilmore SR (2004) Method of producing plants having enhanced transpiration efficiency and plants produced there from. International application, Patent Cooperation Treaty, WO 2004/005555 A1

Masle J, Gilmore SR, Farquhar GD (2005) The ERECTA gene regulates plant transpiration efficiency in Arabidopsis. Nature 436:866–870

Maurel C (1997) Aquaporins and water permeability of plant membranes. Annu Rev Plant Physiol Plant Mol Biol 48:399–429

Maurel C, Santoni V, Luu D-T, Wudick MM, Verdoucq L (2009) The cellular dynamics of plant aquaporin expression and functions. Curr Opin Plant Biol 12:690–698

McIntyre CL, Seung D, Casu RE, Rebetzke GJ, Shorter R, Xue GP (2012) Genotypic variation in the accumulation of water soluble carbohydrates in wheat. Funct Plant Biol 39:560–568

Morgan JM (1984) Osmoregulation and water stress in higher plants. Annu Rev Plant Physiol 35:299–319

Mulroy TW (1979) Spectral properties of heavily glaucous and non-glaucous leaves of a succulent rosette-plant. Oecologia 38:349–357

Munns R, James RA, Sirault XRR, Furbank RT, Jones HG (2010) New phenotyping methods for screening wheat and barley for beneficial responses to water deficit. J Exp Bot 61:3499–3507

Mwanamwenge J, Loss SP, Siddique KHM, Cocks PS (1998) Growth, seed yield and water use of faba bean (*Vicia faba* L) in a short-season Mediterranean-type environment. Aust J Exp Agric 38:171–180

Nadeau JA (2009) Stomatal development: new signals and fate determinants. Curr Opin Plant Biol 12:29–35

Nakashima K, Ito Y, Yamaguchi-Shinozaki K (2009) Transcriptional regulatory networks in response to abiotic stresses in Arabidopsis and grasses. Plant Physiol 149:88–95

Nardini A, Salleo S, Andri S (2005) Circadian regulation of leaf hydraulic conductance in sunflower (*Helianthus annuus* L. cv Margot). Plant Cell Environ 28:750–759

Nicotra AB, Leigh A, Boyce CK, Jones CS, Niklas KJ, Royer DL, Tsukaya H (2011) The evolution and functional significance of leaf shape in the angiosperms. Funct Plant Biol 38:535–552

Niinemets Ü (2001) Global-scale climatic controls of leaf dry mass per area, density, and thickness in trees and shrubs. Ecology 82:453–469

Nobel PS (1980) Leaf anatomy and water use efficiency. In: Turner NC, Kramer PJ (eds) Adaptation of plants to water and high temperature stress. Wiley Interscience, New York, pp 43–55

Nobel PS (1999) Physicochemical and environmental plant physiology. Academic, San Diego, CA

Nobel PS (2009) Physicochemical and environmental plant physiology. Academic, Oxford

O'Brien TP, Carr DJ (1970) A suberized layer in the cell walls of the bundle sheath of grasses. Aust J Biol Sci 23:275–288

Oppenheimer HR (1960) Adaptation to drought, xerophytism. Arid zone research. XV. Plant water relationships in arid and semi-arid conditions. UNESCO, Paris, pp 105–138

Palta JAP, Berger J, Bramley H (2012) Physiology of the yield under drought: lessons from studies with lupin. In: Aroca R (ed) Plant responses to drought stress: from morphological to molecular features. Springer, Heidelberg, pp 417–440

Pandey RK, Herrera WAT, Pendleton JW (1984) Drought response of grain legumes under irrigation gradient. II. Plant water status and canopy temperature. Agron J 76:553–557

Parent B, Hachez C, Redondo E, Simonneau T, Chaumont F, Tardieu F (2009) Drought and abscisic acid effects on aquaporin content translate into changes in hydraulic conductivity and leaf growth rate: a trans-scale approach. Plant Physiol 149:2000–2012

Parry MAJ, Andralojc PJ, Mitchell RA, Madgwick PJ, Keys AJ (2003) Manipulation of RuBisCO: the amount, activity, function and regulation. J Exp Bot 54:1321–1333

Passioura JB (1972) The effect of root geometry on the yield of wheat growing on stored water. Aust J Agric Res 23:745–752

Passioura JB (1976) Physiology of grain yield in wheat growing on stored water. Aust J Plant Physiol 3:559–565

Passioura JB, Angus JF (2010) Improving productivity of crops in water-limited environments. Adv Agron 106:37–75

Perumalla CJ, Peterson CA, Enstone DE (1990) A survey of angiosperm species to detect hypodermal Casparian bands. I. Roots with a uniseriate hypodermis and epidermis. Bot J Linn Soc 103:93–112

Pinter PJ, Jackson RD, Ezra CE (1985) Sun-angle and canopy-architecture effects on the spectral reflectance of six wheat cultivars. Int J Remote Sens 6:1813–1825

Quarrie SA, Jones HG (1977) Effects of abscisic acid and water stress on development and morphology of wheat. J Exp Bot 28:192–203

Quintero JM, Fournier JM, Benlloch M (1999) Water transport in sunflower root systems: effects of ABA, $Ca^{2+}$ status and $HgCl_2$. J Exp Bot 50:1607–1612

Rascio A, Cedola MC, Sorrentino G, Pastore D, Wittmer G (1988) Pressure-volume curves and drought resistance in two wheat genotypes. Physiol Plant 73:122–127

Ratnakumar P, Vadez V (2011) Groundnut (*Arachis hypogaea*) genotypes tolerant to intermittent drought maintain a high harvest index and have small leaf canopy under stress. Funct Plant Biol 38:1016–1023

Rawson HM, Begg JE (1977) The effect of atmospheric humidity on photosynthesis, transpiration and water use efficiency of leaves of several plant species. Planta 134:5–10

Regan KL, Siddique KHM, Turner NC, Whan BR (1992) Potential for increasing early vigour and total biomass in spring wheat. II. Characteristics associated with early vigour. Aust J Agric Res 43:541–553

Reynolds MP, Tuberosa R (2008) Translational research impacting on crop productivity in drought-prone environments. Curr Opin Plant Biol 11:171–179

Reynolds MP, Ortiz-Monasterio JI, McNab A (eds) (2001) Application of physiology of wheat breeding. CIMMYT, DF Mexico

Reynolds MP, Pierre CS, Saad ASI, Vargas M, Condon AG (2007) Evaluating potential genetic gains in wheat associated with stress-adaptive trait expression in elite genetic resources under drought and heat stress. Crop Sci 47:S172–S189

Reynolds MP, Manes Y, Izanloo A, Langridge P (2009) Phenotyping approaches for physiological breeding and gene discovery in wheat. Ann Appl Biol 155:309–320

Richards RA (2006) Increasing crop productivity when water is scarce—from breeding to field management. Agric Water Manage 80:176–196

Richards RA, Passioura JB (1989) A breeding program to reduce the diameter of the major xylem vessel in the seminal roots of wheat and its effect on grain yield in rain-fed environments. Aust J Agric Res 40:943–950

Richards RA, Rawson HM, Johnson DA (1986) Glaucousness in wheat: its development and effect on water-use efficiency, gas exchange and photosynthetic tissue temperatures. Aust J Plant Physiol 13:465–473

Richards RA, Rebetzke GJ, Condon AG, van Herwaarden AF (2002) Breeding opportunities for increasing the efficiency of water use and crop yield in temperate cereals. Crop Sci 42:111–121

Rieger M, Duemmel MJ (1992) Comparison of drought resistance among *Prunus* species from divergent habitats. Tree Physiol 11:369–380

Rieger M, Litvin P (1999) Root system hydraulic conductivity in species with contrasting root anatomy. J Exp Bot 50:201–209

Rowe MH, Bergmann DC (2010) Complex signals for simple cells: the expanding ranks of signals and receptors guiding stomatal development. Curr Opin Plant Biol 13:548–555

Sack L, Holbrook NM (2006) Leaf hydraulics. Annu Rev Plant Biol 57:361–381

Sade N, Vinocur BJ, Diber A, Shatil A, Ronen G, Nissan H, Wallach R, Karchi H, Moshelion M (2009) Improving plant stress tolerance and yield production: is the tonoplast aquaporin SlTIP2;2 a key to isohydric to anisohydric conversion? New Phytol 181:651–661

Sade N, Gebretsadik M, Seligmann R, Schwartz A, Wallach R, Moshelion M (2010) The role of tobacco aquaporin1 in improving water use efficiency, hydraulic conductivity, and yield production under salt stress. Plant Physiol 152:245–254

Sagaram M, Lombardini L, Grauke LJ (2007) Variation in leaf anatomy of pecan cultivars from three ecogeographic locations. J Am Soc Hortic Sci 132:592–596

Sakurai J, Ishikawa F, Yamaguchi T, Uemura M, Maeshima M (2005) Identification of 33 rice aquaporin genes and analysis of their expression and function. Plant Cell Physiol 46:1568–1577

Saliendra NZ, Meinzer FC (1992) Genotypic, developmental and drought-induced differences in root hydraulic conductance of contrasting sugarcane cultivars. J Exp Bot 43:1209–1217

Schuepp PH (1993) Leaf boundary layers. New Phytol 125:477–507

Schulze E-D, Turner NC, Nicoll D, Schumacher J (2006) Leaf and wood carbon isotope ratios, specific leaf areas and wood growth of *Eucalyptus* species across a rainfall gradient in Australia. Tree Physiol 26:479–492

Sexton PJ, Peterson CM, Boote KJ, White JW (1997) Early-season growth in relation to region of domestication, seed size, and leaf traits in common bean. Field Crops Res 52:69–78

Sharkey TD (1988) Estimating the rate of photorespiration in leaves. Physiol Plant 63:147–152

Sharma HC, Waines JG (1994) Inheritance of leaf pubescence in diploid wheat. J Hered 85:286–288

Shelden MC, Howitt SM, Kaiser BN, Tyerman SD (2009) Identification and functional characterisation of aquaporins in the grapevine, *Vitis vinifera*. Funct Plant Biol 36:1065–1078

Shimazaki K-i, Doi M, Assmann SM, Kinoshita T (2007) Light regulation of stomatal movement. Annu Rev Plant Biol 58:219–247

Siddique KHM, Belford RK, Perry MW, Tennant D (1989) Growth, development and light interception of old and modern wheat cultivars in a Mediterranean-type environment. Aust J Agric Res 40:473–487

Siddique KHM, Belford RK, Tennant D (1990a) Root:shoot ratios of old and modern, tall and semi-dwarf wheats in a Mediterranean environment. Plant Soil 121:89–98

Siddique KHM, Tennant D, Perry MW, Belford RK (1990b) Water use and water use efficiency of old and modern wheat cultivars in a Mediterranean-type environment. Aust J Agric Res 41:431–447

Siddique KHM, Regan KL, Tennant D, Thomson BD (2001) Water use and water use efficiency of cool season grain legumes in low rainfall Mediterranean-type environments. Eur J Agron 15:267–280

Siefritz F, Biela A, Eckert M, Otto B, Uehlein N, Kaldenhoff R (2001) The tobacco plasma membrane aquaporin NtAQP1. J Exp Bot 52:1953–1957

Siefritz F, Tyree MT, Lovisolo C, Schubert A, Kaldenhoff R (2002) PIP1 plasma membrane aquaporins in tobacco: from cellular effects to function in plants. Plant Cell 14:869–876

Simonin KA, Limm EB, Dawson TE (2012) Hydraulic conductance of leaves correlates with leaf lifespan: implications for lifetime carbon gain. New Phytol 193:939–947

Sinclair TR, Zwieniecki MA, Holbrook NM (2008) Low leaf hydraulic conductance associated with drought tolerance in soybean. Physiol Plant 132:446–451

Sinclair TR (2012) Is transpiration efficiency a viable plant trait in breeding for crop improvement? Funct Plant Biol 39:359–365

Sivamani E, Bahieldin A, Wraith JM, Al-Niemi T, Dyer WE, Ho TD, Qu R (2000) Improved biomass productivity and water use efficiency under water deficit conditions in transgenic wheat constitutively expressing the barley *HVA1* gene. Plant Sci 155:1–9

Slack CR, Hatch MD (1967) Comparative studies on the activity of carboxylases and other enzymes in relation to the new pathway of photosynthetic carbon dioxide fixation in tropical grasses. Biochem J 103:660–665

Smith JAC, Popp M, Lüttge U, Cram WJ, Diaz M, Griffiths H, Lee HSJ, Medina E, Schäfer C, Stimmel KH, Thonke B (1989) Ecophysiology of xerophytic and halophytic vegetation of a coastal alluvial plain in northern Venezuela. VI. Water relations and gas exchange of mangroves. New Phytol 111:293–307

Spollen WG, LeNoble ME, Sammuels TD, Bernstein N, Sharp RE (2000) Abscisic acid accumulation maintains maize primary root elongation at low water potentials by restricting ethylene production. Plant Physiol 122:967–976

Spreitzer RJ, Salvucci ME (2002) RuBisCO: structure, regulatory interactions, and possibilities for a better enzyme. Annu Rev Plant Biol 53:449–475

Stark JC, Pavek JJ, McCann IR (1991) Using canopy temperature measurements to evaluate drought tolerance of potato genotypes. J Am Soc Hortic Sci 116:412–415

Stasovski E, Peterson CA (1991) The effects of drought and subsequent rehydration on the structure and vitality of *Zea mays* seedling roots. Can J Bot 69:1170–1178

Steudle E (2000) Water uptake by roots: effects of water deficit. J Exp Bot 51:1531–1542

Steudle E, Jeschke WD (1983) Water transport in barley roots. Planta 158:237–248

Steudle E, Peterson CA (1998) How does water get through roots? J Exp Bot 49:775–788

Steudle E, Murrmann M, Peterson CA (1993) Transport of water and solutes across maize roots modified by puncturing the endodermis. Plant Physiol 103:335–349

Sutka M, Li G, Boudet J, Boursiac Y, Doumas P, Maurel C (2010) Natural variation of root hydraulics in Arabidopsis grown in normal and salt-stressed conditions. Plant Physiol 155:1264–1276

Svenningsson M, Liljenberg C (1986) Changes in cuticular transpiration rate and cuticular lipids of oat (*Avena sativa*) seedlings induced by water stress. Physiol Plant 66:9–14

Tanaka Y, Fujii K, Shiraiwa T (2010) Variability of leaf morphology and stomatal conductance in soybean [*Glycine max* (L.) Merr.] cultivars. Crop Sci 50:2525–2532

Tanner CB, Sinclair TR (1983) Efficient water use in crop production: research and re-search? In: Taylor HM, Jordan WR, Sinclair TR (eds) Limitations to efficient water use in crop production. American Society of Agronomy, Madison, WI, pp 1–27

Thompson AJ, Andrews J, Mulholland BJ, McKee JMT, Hilton HW, Horridge JS, Farquhar GD, Smeeton RC, Smillie IRA, Black CR, Taylor IB (2007) Overproduction of abscisic acid in tomato increases transpiration efficiency and root hydraulic conductivity and influences leaf expansion. Plant Physiol 143:1905–1917

Ting IP (1985) Crassulacean acid metabolism. Annu Rev Plant Physiol 36:595–622

Torii KU (2004) Leucine-rich repeat receptor kinases in plants: structure, function, and signal transduction pathways. Int Rev Cytol 234:1–46

Torii KU (2012) Mix-and-match: ligand-receptor pairs in stomatal development and beyond. Trends Plant Sci 17(12):711–719

Torii KU, Mitsukawa N, Oosumi T, Matsuura Y, Yokoyama R, Whittier RF, Komeda Y (1996) The Arabidopsis ERECTA gene encodes a putative receptor protein kinase with extracellular leucine-rich repeats. Plant Cell 8:735–746

Travaglia C, Reinoso H, Cohen A, Luna C, Tommasino E, Castillo C, Bottina R (2010) Exogenous ABA increases yield in field-grown wheat with moderate water restriction. J Plant Growth Regul 29:366–374

Tsuda M, Tyree MT (2000) Plant hydraulic conductance measured by the high pressure flow meter in crop plants. J Exp Bot 51:823–828

Turner NC (1972) $K^+$ uptake of guard cells stimulated by fusicoccin. Nature 235:341–342

Turner NC (1993) Water use efficiency of crop plants: potential for improvement. In: Buxton DR, Shibles R, Forsberg RA, Blad BL, Asay KH, Paulsen GM, Wilson RF (eds) International crop science 1. Crop Science Society of America, Madison, WI, pp 75–82

Turner NC (1997) Further progress in crop water relations. Adv Agron 58:293–337

Turner NC, Asseng S (2005) Productivity, sustainability, and rainfall-use efficiency in Australian rainfed Mediterranean agricultural systems. Aust J Agric Res 56:1123–1136

Turner NC, Hartung W (2012) Dehydration of isolated roots of seven *Lupinus* species induces synthesis of different amounts of free, but not conjugated, abscisic acid. Plant Growth Regul 66:265–269

Turner NC, Jones MM (1980) Turgor maintenance by osmotic adjustment: a review and evaluation. In: Turner NC, Kramer PJ (eds) Adaptation of plants to water and high temperature stress. Wiley Interscience, New York, pp 87–103

Turner NC, Nicolas ME (1998) Early vigour: a yield-positive characteristic for wheat in drought-prone Mediterranean-type environments. In: Behl RK, Singh DP, Lodhi GP (eds) Crop improvement for stress tolerance. CCS Haryana Agricultural University/Max Mueller Bhawan, Hisar/New Delhi

Turner NC, Schulze E-D, Gollan T (1984) The responses of stomata and leaf gas exchange to vapour pressure deficits and soil water content. I. Species comparisons at high soil water contents. Oecologia 63:338–342

Turner NC, Abbo S, Berger JD, Chaturvedi SK, French RJ, Ludwig C, Mannur DM, Singh SJ, Yadava HS (2007a) Osmotic adjustment in chickpea (*Cicer arietinum* L.) results in no yield benefit under terminal drought. J Exp Bot 58:187–194

Turner NC, Palta JA, Shrestha R, Ludwig C, Siddique KHM, Turner DW (2007b) Carbon isotope discrimination is not correlated with transpiration efficiency in three cool-season grain legumes (pulses). J Integr Plant Biol 49:1478–1483

Tyerman SD, Bohnert HJ, Maurel C, Steudle E, Smith JAC (1999) Plant aquaporins: their molecular biology, biophysics and significance for plant water relations. J Exp Bot 50:1055–1071

Tyerman SD, Niemietz CM, Bramley H (2002) Plant aquaporins: multifunctional water and solute channels with expanding roles. Plant Cell Environ 25:173–194

Tyree MT, Fiscus EL, Wullschleger SD, Dixon MA (1986) Detection of xylem cavitation in corn under field conditions. Plant Physiol 82:597–599

Uehlein N, Lovisolo C, Siefritz F, Kaldenhoff R (2003) The tobacco aquaporin NtAQP1 is a membrane $CO_2$ pore with physiological functions. Nature 425:734–737

Uehlein N, Otto B, Hanson DT, Fischer M, McDowell N, Kaldenhoff R (2008) Function of *Nicotiana tabacum* aquaporins as chloroplast gas pores challenges the concept of membrane $CO_2$ permeability. Plant Cell 20:648–657

Ueno O, Kawano Y, Wakayama M, Takeda T (2006) Leaf vascular systems in $C_3$ and $C_4$ grasses: a two-dimensional analysis. Ann Bot 97:611–621

Vadez V, Krishnamurthy L, Kashiwagi J, Kholova J, Devi JM, Sharma KK, Bhatnagar-Mathur P, Hoisington DA, Hash CT, Bidinger FR, Keatinge JDH (2007) Exploiting the functionality of root systems for dry, saline, and nutrient deficient environments in a changing climate. SAT eJournal. ICRISAT, Pattancheru, pp 1–61

Vandeleur RK, Mayo G, Shelden MC, Gilliham M, Kaiser BN, Tyerman SD (2009) The role of plasma membrane intrinsic protein aquaporins in water transport through roots: diurnal and drought stress responses reveal different strategies between isohydric and anisohydric cultivars of grapevine. Plant Physiol 149:445–460

Vysotskaya LB, Kudoyarova GR, Jones HG (2004) Unusual stomatal behaviour on partial root excision in wheat seedlings. Plant Cell Environ 27:69–77

Wan X, Steudle E, Hartung W (2004) Gating of water channels (aquaporins) in cortical cells of young corn roots by mechanical stimuli (pressure pulses): effects of ABA and of $HgCl_2$. J Exp Bot 55:411–422

Warren CR (2008) Stand aside stomata, another actor deserves centre stage: the forgotten role of the internal conductance to $CO_2$ transfer. J Exp Bot 59:1475–1487

Wasson AP, Richards RA, Chatrath R, Misra SC, Prasad SVS, Rebetzke GJ, Kirkegaard JA, Christopher J, Watt M (2012) Traits and selection strategies to improve root systems and water uptake in water-limited wheat crops. J Exp Bot 63:3485–3498

Watt M, Schneebeli K, Dong P, Wilson IW (2009) The shoot and root growth of Brachypodium and its potential as a model for wheat and other cereal crops. Funct Plant Biol 36:960–969

Webster JA, Inayatullah C, Hamissou M, Mirkes KA (1994) Leaf pubescence effects in wheat on yellow sugarcane aphids and greenbugs (Homoptera: Aphididae). J Econ Entomol 87:231–240

Wilson D, Cooper JP (1969) Effect of temperature during growth on leaf anatomy and subsequent light-saturated photosynthesis among contrasting Lolium genotypes. New Phytol 68:1115–1123

Wright GC, Nageswara R, Farquhar GD (1994) Water-use efficiency and carbon isotope discrimination in peanut under water deficit conditions. Crop Sci 34:92–97

Xin Z, Mandaokar A, Chen J, Last RL, Browse J (2007) Arabidopsis ESK1 encodes a novel regulator of freezing tolerance. Plant J 49:786–799

Xing HT, Guo P, Xia XL, Yin WL (2011) PdERECTA, a leucine-rich repeat receptor-like kinase of poplar, confers enhanced water use efficiency in Arabidopsis. Planta 234:229–241

Xu Z, Zhou G (2008) Responses of leaf stomatal density to water status and its relationship with photosynthesis in a grass. J Exp Bot 59:3317–3325

Yambao EB, Ingram KT, Real JG (1992) Root xylem influence on the water relations and drought resistance of rice. J Exp Bot 43:925–932

Yoo CY, Pence HE, Hasegawa PM, Mickelbart MV (2009) Regulation of transpiration to improve crop water use. Crit Rev Plant Sci 28:410–431

Yoo CY, Pence HE, Jin JB, Miura K, Gosney MJ, Hasegawa PM, Mickelbart MV (2010) The Arabidopsis GTL1 transcription factor regulates water use efficiency and drought tolerance by modulating stomatal density via transrepression of SDD1. Plant Cell 22:4128–4141

Yunasa IAM, Siddique KHM, Belford RK, Karimi MM (1993) Effect of canopy structure on efficiency of radiation interception and use in spring wheat cultivars during the pre-anthesis period in a Mediterranean-type environment. Field Crops Res 35:113–122

Yunasa IAM, Sedgley RH, Siddique KHM (1994) Influence of mulching on the pattern of growth and water use by spring wheat and moisture storage on a fine textured soil. Plant Soil 160:119–130

Zaman-Allah M, Jenkinson DM, Vadez V (2011) A conservative pattern of water use, rather than deep or profuse rooting, is critical for the terminal drought tolerance of chickpea. J Exp Bot 62:4239–4252

Zhang J, Davies WJ (1989) Sequential response of whole plant water relations to prolonged soil drying and the involvement of xylem sap ABA in the regulation of stomatal behaviour of sunflower plants. New Phytol 113:167–174

Zhang J, Davies WJ (1990) Changes in the concentration of ABA in xylem sap as a function of changing soil water status can account for changes in leaf conductance and growth. Plant Cell Environ 13:277–285

Zhang X, Wollenweber B, Jiang D, Liu F, Zhao J (2008) Water deficits and heat shock effects on photosynthesis of a transgenic *Arabidopsis thaliana* constitutively expressing *ABP9*, a bZIP transcription factor. J Exp Bot 59:839–848

Zhu J, Brown KM, Lynch JP (2010) Root cortical aerenchyma improves the drought tolerance of maize (*Zea mays* L.). Plant Cell Environ 33:740–749

Zimmermann HM, Hartmann K, Schreiber L, Steudle E (2000) Chemical composition of apoplastic transport barriers in relation to radial hydraulic conductivity of corn roots (*Zea mays* L.). Planta 210:302–311

# Chapter 7
# Flooding and Submergence Tolerance

Abdelbagi M. Ismail

**Abstract** Rice is the major crop in most rainfed and flood-prone environments of South and Southeast Asia and it provides food for millions of subsistence farming families. The productivity of these environments is low because high-yielding varieties tolerant of prevailing abiotic stresses are lacking. The hydrologic conditions in these environments are particularly harsh and unpredictable, leading to recurring floods and sometimes drought and salinity. Early floods cause poor crop establishment, and plant survival and yield decline drastically when plants are partially or completely submerged during the season. Traditional rice landraces predominate in these areas because they acquired partial tolerance against most of these stresses, but they are low yielding. Transferring tolerance traits into high-yield backgrounds will help boost and sustain the productivity in these areas, and this is becoming more feasible with the recent availability of effective genomics and molecular breeding tools that are being used to dissect and transfer these favorable traits and genes. These tools are providing opportunities to develop resilient varieties for the less favorable environments, and varieties that can tolerate complete flooding during germination and vegetative stage (conferred by *SUB1*) as well as yield well under long-term stagnant floods are becoming available. These varieties will also provide opportunities for better land productivity by being more responsive to inputs and other adjustments in the system, helping farmers cope with the existing problems and future adversities of climate change.

A.M. Ismail (✉)
International Rice Research Institute, DAPO Box 7777, Metro Manila, Philippines
e-mail: a.ismail@irri.org

C. Kole (ed.), *Genomics and Breeding for Climate-Resilient Crops*, Vol. 2,
DOI 10.1007/978-3-642-37048-9_7, © Springer-Verlag Berlin Heidelberg 2013

## 7.1   Introduction

Rice is the major crop in most flood-prone environments of South and Southeast Asia and rice is the staple food for more than half of the world population and over 90 % of it is produced and consumed in Asia. Hence, this chapter on flooding and submergence tolerance deliberates only on rice. Temperature and water availability largely determine the ecological and geographic distribution of rice. Rice is now cultivated on more than 144 million hectares (ha) worldwide, from 50°N in northern China to 35°S in New South Wales, Australia, and in Argentina. Two broad categories are identified within *Oryza sativa* commonly grown in Asia: *japonica* varieties are mostly grown in temperate regions, while *indica* varieties are grown in tropical and subtropical areas. A third category—tropical japonica—is mostly grown in the uplands of the tropics and subtropics. Rice is grown on a variety of soils, and the physical ability of the soil to hold water is an important property, with medium- and heavy-textured soils typically preferred over light-textured sandy soils. It is also grown under variable water regimes and hydrologic conditions, from aerobic soils in uplands to flooded soils in irrigated and rainfed lowlands to long-duration inundated conditions in flood-prone areas (Fig. 7.1; Maclean et al. 2002).

Rainfed lowland and flood-prone rice areas in Asia cover about 47 million ha or about 35 % of the total global rice area. These areas usually experience variable types of excess water stress caused by various factors such as direct heavy rains and flooding from adjacent rivers because of much rain in upper catchments or during high tides. This excess water together with poor or nonexisting drainage facilities creates serious problems for rice and other crops. Rainfed lowland areas generally experience frequent flash floods for a relatively short duration, commonly ranging from a few days to 2 weeks. Longer-term stagnant flooding (SF) of 20–50-cm water depth can occur through most of the season, also called medium-deep or semi-deep areas. In deepwater areas, water at depth of over 50 cm to a few meters stagnates in the rice field for a long duration, even up to a few months. The depth of water in some of these areas can surpass 4 m as in floating-rice areas (Fig. 7.1). Areas affected by these types of floods are not strictly distinct, and in some cases, combinations of these conditions occur within the same season (Khush 1984; Lafitte et al. 2006; Sarkar et al. 2006).

Widespread and persistent rural poverty is a long-standing problem in rice-producing countries in Asia, particularly in rainfed and flood-prone areas. In these areas, rice is the predominant food and sometimes a cash crop because it thrives well under excess water during the wet season. It is the staple food of rural communities, providing up to 60 % of their energy intake and 50–70 % of the labor force employment. Average productivity of rice in these areas is very low, only 0.5–2.0 t ha$^{-1}$, and is unstable. The large gap in the productivity of rainfed ecosystems compared with the input-intensive irrigated system (about 5.0 t ha$^{-1}$) is caused by abiotic stresses such as drought, floods, and poor soils, coupled with the lack of well-adapted high-yielding varieties. These biophysical limitations in

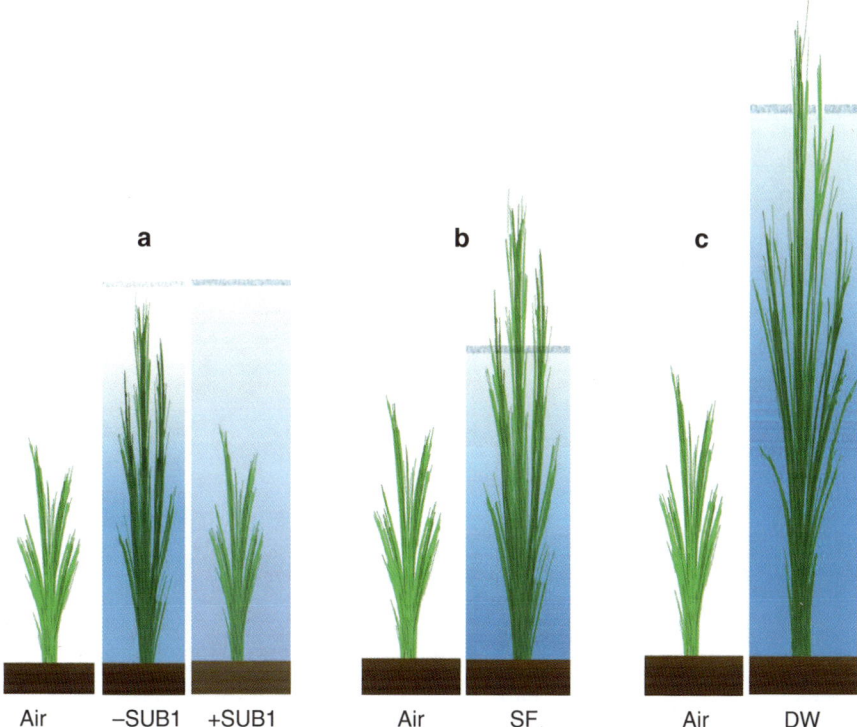

**Fig. 7.1** Diagram showing the types of floods affecting rice in rainfed lowlands and flood-prone areas. (**a**) Transient flash floods cause complete submergence for up to 2 weeks, intolerant genotypes (*–SUB1*) attempt to escape through partial elongation, which exhausts their energy reserves, while tolerant genotypes (*+SUB1*) stay dormant until water recedes. (**b**) In areas affected by stagnant floods (SF, medium deep), cultivars partially elongate in response to rising water to maintain contact with air. (**c**) In deepwater (DW) and floating-rice areas, adapted cultivars elongate at higher rates to keep up with the fast-rising water (drawings by Mohamed A. Ismail)

flood-prone areas are expected to become even worse with the alarming changes in climate. Moreover, access to new technologies and information in these areas is inadequate and marketing infrastructure and seed production systems are poorly developed. Obviously, improving the productivity in these systems is becoming even more essential with the ominous food crisis and the inability of most developing countries to produce sufficient food or meet their food demand even from international markets because of the high costs and limited supply. Knowledge about these systems and awareness of their constraints and potential have improved considerably over the recent past, and this suggests considerable potential for exploitation to contribute to and sustain food production and help eliminate poverty in these resource-poor communities (Alpuerto et al. 2009; Mackill et al. 2012).

## 7.2 Types of Floods Affecting Rice and Impacts of Climate Change

Uncontrolled floods annually affect more than 20 million ha of rice area in South and Southeast Asia. Yields are low, averaging 1.5 t ha$^{-1}$, and yet these areas support more than 100 million people. Traditional varieties still dominate in farmers' fields because they are more adapted to water fluctuations than modern varieties, and this predominant use of low-yielding landraces renders the productivity of flood-affected areas quite low. Modern high-yielding varieties with enhanced tolerance of partial and/or complete submergence are needed to stabilize rice production in these rainfed lowlands.

### 7.2.1 Flash Floods

Transient flash floods that result in complete inundation of farmers' fields for variable durations are highly unpredictable and can occur at any growth stage and sometimes more than once during the monsoon season. Yield losses due to these flash floods can range from 10 % to 100 % depending on water depth, duration of submergence, temperature, turbidity of floodwater, soil fertility, light intensity, and age of the crop (Setter et al. 1997; Das et al. 2009). In areas where direct seeding is practiced, farmers could encounter flooding or waterlogging if rain occurs directly after seeding, and this can lead to complete crop failure because of the high sensitivity of rice to the anaerobic conditions caused by flooding during germination (Yamauchi et al. 1993; Ismail et al. 2009). Flooding slows seed germination and delays seedling establishment in direct-seeded rice, causing poor crop establishment. Recently, progress was made in identifying rice landraces with better tolerance of submergence during germination, and these landraces were further studied for the genetics and physiological bases of tolerance (Ismail et al. 2009, 2012; Angaji et al. 2010; Ella et al. 2010).

The most common and seriously damaging type of flooding is short-term inundation, which is also called flash floods. More than 20 million ha of rainfed lowlands in Asia (excluding China) and significant areas of lowland rice production in Africa are affected by flash floods that cause complete inundation of rice for durations ranging from a few days to more than 2 weeks (Khush 1984; Bailey-Serres et al. 2010; Mackill et al. 2012). Significant areas are also damaged by flash floods in China (Dey and Upadhyaya 1996). In these areas, tolerant rice varieties are needed that can survive transient submergence through limited elongation. Recently, varieties that tolerate complete submergence have become available through the incorporation of the *SUB1A* gene. These varieties can survive 4–18 days of complete submergence, with considerable yield benefits in farmers' fields (Sarkar et al. 2009; Singh et al. 2009; Mackill et al. 2012).

## 7.2.2   Stagnant Longer-Term Floods

Long-term or stagnant flooding (SF) is common in low-lying areas where water accumulates through most of the growing season at various depths (Singh et al. 2011). Flooding in these areas is normally caused by direct rain, overflow of nearby rivers and canals, or sometimes tidal movements as in coastal areas. In some of these areas, rice is partially flooded with a water depth that normally does not exceed 50 cm but persists for a few weeks to several months, often referred to as stagnant or medium-deep flooding. Generally, SF of over 25-cm depth adversely affects the growth and survival of modern rainfed lowland rice varieties even though the plants are not fully submerged. In particular, it hinders tillering and increases lodging and, in some cases, causes a severe reduction in crop stand (Tuong et al. 2000; Singh et al. 2011). Prolonged SF sometimes occurs following short-term complete submergence, when it becomes even more devastating, especially for varieties that tolerate complete submergence through reduced elongation (the quiescent strategy; Bailey-Serres and Voesenek 2008). In areas that are prone to both submergence and stagnant floods, farmers still use traditional low-yielding landraces because of their partial tolerance to transient complete flooding and their ability to withstand stagnant floods through stem and leaf elongation, though they have lower yields and poor grain quality.

## 7.2.3   Deepwater and Floating Rice

Under more extreme conditions, water depths can reach from 50 cm to 100 cm in deepwater areas and can even surpass 4 m in floating-rice areas and persist for several weeks to several months. Elongation ability of leaves and internodes is essential under deepwater conditions to keep pace with the rising water and escape complete submergence. This will ensure oxygen supply through the continuum of aerenchyma tissue for aeration of lower plant parts, including roots. Elongating plants will also gain access to $CO_2$ and light and maintain energy supply for fast growth (Catling 1992; Hattori et al. 2009; Voesenek and Bailey-Serres 2009). Noticeably, deepwater areas have been declining in recent years because existing varieties have low yields as a consequence of the extensive energy spent during elongation. Better opportunities are now available for farmers to plant high-yielding, shorter maturing rice varieties after flood periods with the availability of supplementary irrigation systems as in Vietnam and shallow tube wells in Bangladesh and some parts of India.

## 7.3    Impacts of Climate Change on Rice in Rainfed and Flood-Prone Areas

Flooding costs farmers an estimated US$1 billion a year in rice losses in South and Southeast Asia. These losses are expected to increase substantially with the worsening flooding incidences and intensities caused by the mostly human-provoked global warming. This is recently being witnessed as heavier rainfall, sea-level rise, and an increase in frequencies of natural disasters such as coastal storms, cyclones, and typhoons. Over the past 100 years, average global sea level has risen by 10–15 cm (IPCC 2007), and a further increase of about 50 cm is anticipated by 2100 (IPCC 2001). This is mainly due to the expansion of ocean water and melting of glaciers caused by rising temperatures, and it will seriously threaten coastal areas with floods and erosion. Other factors could also contribute to the severity of regional floods such as deforestation causing excessive runoff from uplands and hilly areas, usually accompanied by landslides; draining of wetlands that act as reservoirs for excess water to make room for agricultural expansion and urban uses; and other development activities that interfere with natural drainage such as the building of roads, polders, and other structures.

Apparently, the areas already being affected by floods are expected to increase substantially as a consequence of sea-level rise and the predicted increases in frequencies and intensities of flooding (Bates et al. 2008). These areas include vast coastal rice production lands in the tropics and subtropics, especially the deltas along the coastlines of South and Southeast Asia. Effects of increased flooding incidences on land-use activities are already being witnessed in some of the highly vulnerable areas. In most cases, these areas are heavily populated with poor communities, with little opportunity for other livelihood options. Rice farming is the dominant agricultural activity along the tropical coasts of South and Southeast Asia, including these low-lying deltas, where it constitutes the main component of agricultural and food production, for both domestic supply and exports (Ismail and Tuong 2009). For example, in Vietnam, the deltaic areas provide more than 70 % of the national rice production, with the Mekong Delta providing about 54 % and the Red River Delta providing an additional 17 %. Similarly, the Ayeyarwady Delta of Myanmar provides 68 % of the national rice production, while the Ganges–Brahmaputra Delta of Bangladesh provides 34 % of the total rice production in Bangladesh (Wassmann et al. 2009).

These tropical deltas are more vulnerable to even a small rise in sea level because they are mostly low lying and are the first to be affected by any changes in the frequencies of coastal storms and other disasters, particularly after the removal of most of the protective mangrove belts as in southern Bangladesh. In recent years, increasing incidences of moderate to severe storms were witnessed in several countries, as in the Philippines and Bangladesh, with some devastating storms, such as cyclone Nargis in Myanmar in May 2008, with a surge of over 4 m, which damaged about 1.7 million ha of rice (United States Department of Agriculture-Foreign Agricultural Service (USDA-FAS 2008) http://www.pecad.fas.usda.gov/highlights/2008/

05/Burma_Cyclone_Nargis_Rice_Impact.htm). Cyclone Sidr in southern Bangladesh resulted in a loss of more than 640,000 ha of cropland in 2007, with rice as the main crop during the wet season and the most severely damaged (Hossain et al. 2008). Lands in most of these large coastal deltas in Asia are also subsiding because of sediment compaction caused by overpumping of water, gas, and oil in some areas, coupled with the trapping of sediments in newly constructed upstream reservoirs as in Bangladesh (Syvitski et al. 2009). Any decline in rice production in these coastal deltas will have considerable impacts on national food security as well as on international trade in rice, with countries such as Vietnam regarded as one of the most vulnerable, despite currently being the second-largest rice exporter in the world. In Bangladesh, the transplanted wet-season or "T. aman" rice provides over 50 % of the total national production, and in Myanmar, wet-season rice constitutes more than 85 % of the national production. However, in both countries, wet-season rice remains at substantial risk from any further increase in flooding incidence (Wassmann et al. 2004).

Efforts are, therefore, needed to halt any further decline in productivity and to maintain a steady increase in rice production in both coastal and inland areas that remain vulnerable to further deterioration. This will ultimately contribute to the global efforts for coping with the consequences of climate warming and food shortage. Developing submergence-tolerant rice varieties provides a step forward in these efforts (Ismail et al. 2008; Ismail and Tuong 2009; Mackill et al. 2012). This will involve integrating tolerance of submergence and other coexisting abiotic stresses into modern high-yielding rice varieties and elite lines. Breeding for tolerance to such abiotic stresses has been considered difficult, and despite the substantial attention by rice breeders to the various stresses in recent years, progress is still scant, with very few successes in which new tolerant varieties are adopted by farmers. This is often attributed to the genetic and physiological complexity of tolerance of these stresses. However, recent scientific discoveries and advances— particularly in genetics, genomics, and plant molecular physiology—have opened up new frontiers to cope with these adversities and to mitigate their negative impacts on food security. One of the most promising approaches is the use of molecular markers to identify and select for genes controlling tolerance traits (Thomson et al. 2010b; Ismail and Thomson 2011; Mackill et al. 2012).

## 7.4   Germplasm Improvement for Submergence- and Flood-Prone Areas and Target Traits

Various tolerance traits or mechanisms are necessary for high and stable productivity in flood-affected areas. Varieties are needed that can germinate and establish in flooded soils where direct seeding is practiced or is likely to dominate in the future because of lower costs and other benefits (Ismail et al. 2012). Transient flash floods are commonly experienced at any time, and mostly more than once during the season, in rainfed lowlands and flood-prone areas. Desired varieties should withstand complete submergence while being capable of fast regeneration and recovery after submergence. Other types of varieties are required for areas where water stagnates for longer durations—from a few weeks to several months during the season. These varieties should have the ability to produce higher yields under either stagnant floods, where water depth is 20–50 cm, or deepwater conditions, where water depth exceeds 50 cm. Combinations of several tolerance traits are often required for most areas as more than one type of flooding could occur within the season and depending on the crop stage of development (Mackill et al. 2012).

## 7.4.1 Germination and Crop Establishment

Farmers in rainfed areas and particularly in flood-prone areas often face difficulties in adopting a direct-seeding system for crop establishment because of the uncertain flooding after seeding, with the consequent partial to complete crop failure as rice is highly sensitive to flooding during germination (Yamauchi et al. 1993; Ismail et al. 2009, 2012; Angaji et al. 2010). Farmers are forced to replant their fields, incurring additional costs and running into other problems late in the season, retransplant older seedlings in standing water into heavily damaged areas, or leave their fields barren, with a consequent shortage of food and other uncertainties. Heavy rains or early floods also pose serious problems for transplanted rice in rainfed areas, and early crop establishment through direct seeding may help avoid further damage caused by floods just after transplanting. Breeding cultivars with higher tolerance of flooding during germination and early seedling establishment will help avoid these troubles. Benefits of this trait could also be extended to irrigated ecosystems where flooding after direct seeding can effectively suppress most weeds (Ismail et al. 2012).

Attempts to develop rice varieties appropriate for direct seeding began in the past but with limited success, mainly because donors with sufficient tolerance were not identified (Yamauchi et al. 1993; Yamauchi and Winn 1996; Biswas and Yamauchi 1997). However, substantial genetic variation in the ability to germinate and establish under flooding was found in rice after screening a large number of accessions and breeding lines (Yamauchi et al. 1993; Biswas and Yamauchi 1997; Jiang et al. 2004). We have recently screened more than 8,000 accessions from different sources and identified several lines that can germinate and establish in flooded soils. These lines were further studied to reveal the genetics and physiological bases of tolerance and they are also used in breeding. Several mapping populations were subsequently developed and used to identify quantitative trait loci (QTLs) associated with tolerance (Angaji et al. 2010; Septiningsih et al. 2013). Significant progress was also made in unraveling the physiological and molecular bases of tolerance (Ismail et al. 2009). Unlike most major cereal crops such as maize, wheat, barley, oats, and sorghum, which are extremely sensitive to low-oxygen stress during germination, rice can germinate in flooded soils even under anoxia or the absence of oxygen (Taylor 1942; Yamauchi et al. 1993;

Perata et al. 1997; Vartapetian and Jackson 1997; Ella and Setter 1999; Angaji et al. 2010). However, this is mostly limited to elongation of the coleoptiles with failure to form functional roots and leaves (Biswas and Yamauchi 1997; Ismail et al. 2009).

Germinating embryos in flooded soils can suffer from hypoxia or even anoxia in severe cases, and this prevents further growth because oxygen is necessary for the functioning of the enzymes needed for the breakdown and mobilization of stored carbohydrates and for the oxidative pathways to generate energy for growing embryos (Drew 1990; Greenway and Setter 1996). Apparently, germinating rice seeds acquired several mechanisms to cope with oxygen deficiency in flooded soils. One of these adaptive features is the ability of the coleoptile to grow fast, and it can even elongate faster under low $O_2$ than in air (Alpi and Beevers 1983); however, considerable genetic variation was observed in this elongation ability among rice genotypes (Turner et al. 1981). Elongating coleoptiles could facilitate contact with air in waterlogged or flooded soils and maintain adequate aeration of the growing embryo. Another well-characterized survival strategy of germinating seeds under anaerobic conditions is the primary shift from aerobic metabolism to alcoholic fermentation to generate the energy needed to sustain growth of the germinating embryo (Guglielminetti et al. 1995).

Unlike other cereals, rice seeds can mobilize stored starch into readily ferment-able simple sugars when germinating under hypoxia or anoxia (Atwell and Green-way 1987). The importance of α-amylases in starch degradation when $O_2$ is limiting was reported in several studies, and both expression and translation of relevant genes were observed under anoxia (Perata et al. 1993, 1997; Guglielminetti et al. 1995; Hwang et al. 1999). Recently, we reported substantial genetic variation within rice in the ability to break down starch reserves in germinating rice seeds, with genotypes tolerant to anoxia showing greater capacity to mobilize and use stored carbohydrates than the sensitive ones and with strong correlation of both survival and elongation rate of the coleoptile with total amylase activity (Ismail et al. 2009; Angaji et al. 2010). Germinating rice seeds of tolerant rice genotypes can, therefore, effectively degrade starch to soluble sugars under anaerobic conditions. The next sensitive process is the use of these sugars as substrates to generate ATP for the growing embryo, because aerobic respiration is inhibited when oxygen—the terminal electron acceptor during aerobic respiration—is low. Germinating rice seeds overcome this problem by switching to anaerobic metabo-lism (Avadhani et al. 1978; Jackson et al. 1982; Gibbs et al. 2000). The key enzymes for anaerobic fermentation—pyruvate decarboxylase (PDC), which catalyzes the decarboxylation of pyruvate to yield carbon dioxide and acetaldehyde, and alcohol dehydrogenase (ADH), which reduces acetaldehyde to ethanol, with NADH being oxidized in the process to maintain glycolysis—are both upregulated under anaerobic conditions. We also observed considerable genetic variation in the activity of PDC and ADH, with tolerant genotypes having much higher activity, supporting the perception that anaerobic respiration is important for the survival of rice seeds germinating under anaerobic conditions. We are now conducting further studies to develop a better understanding of the mechanisms associated with the tolerant phenotype.

Rice varieties appropriate for direct seeding need to incorporate tolerance of flooding during germination and higher vigor during seedling establishment. Breeding of such varieties began at IRRI in 2003, using tolerant landraces such as "Khaiyan," "Khao Hlan On," and "Ma-Zhan Red" identified during initial screening (Angaji et al. 2010). Crosses were made with several popular varieties adapted to both rainfed and irrigated ecosystems, and stable breeding lines were developed and are being evaluated in the field. Several QTLs associated with tolerance to flooding during earlier stages were reported before (Jiang et al. 2004, 2006). We developed several mapping populations using the recently identified tolerant genotypes. In the first backcross population, six QTLs derived from "Khao Hlan On," a tolerant donor from Myanmar, were identified (Angaji et al. 2010). The largest of these QTLs was mapped on the long arm of chromosome 9 (*qAG-9-2*), with an LOD score of 20.3 and contributing about 33.5 % of the phenotypic variance. This QTL was further fine-mapped to within a 58-kb region based on the "Nipponbare" sequence, and several candidate genes were annotated in this region and are being further analyzed (Septiningsih et al. unpublished). This QTL is now being incorporated into popular varieties and elite lines through marker-assisted backcrossing. Another large-effect QTL derived from the tolerant landrace "Ma-Zhan Red" was identified on the short arm of chromosome 7 and is being fine-mapped for use in breeding and for cloning (Septiningsih et al. 2013). Our aim is to combine these two QTLs into high-yielding varieties for direct seeding in both irrigated and rainfed ecosystems. These varieties will help in the large-scale adoption of direct-seeded rice, especially in flood-prone areas as flooding intensity, timing, and the extent of affected areas are expected to worsen with climate change. Additionally, these varieties will be important in irrigated systems for weed management using early flooding, which will considerably reduce the cost of hand weeding and hazardous herbicide use (Ismail et al. 2012).

### 7.4.2   Flooding During Vegetative Stage

All modern high-yielding rice varieties are sensitive to complete submergence or flash floods of just a few days. However, a few landraces were identified that can withstand inundation for up to 2 weeks (Mackill et al. 1993). The effects of flash flooding on rice as well as the physiological basis of tolerance were extensively studied in the past two decades, and significant understanding of the mechanisms of tolerance was achieved (Ram et al. 2002; Jackson and Ram 2003; Sarkar et al. 2006). Survival of flooding in rice is greatly affected by the genotype, stage of development, and characteristics of the floodwater, particularly dissolved gases, irradiance and water temperature, and turbidity (Das et al. 2009; Mackill et al. 2012). Among the important plant traits associated with the tolerant phenotype are the ability to maintain high nonstructural carbohydrate concentration in shoots before and following submergence, slower growth, and ability to retain high chlorophyll in leaves during submergence (Ella et al. 2003b; Jackson and Ram 2003; Sarkar et al. 2006). Retention of higher carbohydrates after submergence is necessary for survival and faster recovery and is better correlated with survival than carbohydrate concentration in shoots before submergence (Das et al. 2005). A rapid regeneration and recovery following submergence ensures the production of sufficient effective tillers and leaf area, and plants that exhibit only limited elongation during submergence are more tolerant of complete submergence.

Over the past few decades, extensive efforts were devoted to developing rice varieties tolerant of complete submergence, but with little progress. This is mainly because of the many undesirable traits associated with the tolerant landraces used as donors. A few varieties were released but they were not largely adopted by farmers because the yield of most of them was low and their grain quality was below the standards of farmers in these areas (Mackill et al. 1993, 2012). Substantial progress was accomplished only after the discovery of the major QTL, *Submergence 1* (*SUB1*), on chromosome 9 (Xu and Mackill 1996), which originated from the landrace FR13A from Odisha, India. This single QTL explained about 70 % of the phenotypic variance in the population used for mapping. The strong effect of *SUB1* as the major determinant of submergence tolerance in rice was further confirmed by other groups, and a few other minor QTLs were also identified (Nandi et al. 1997; Kamolsukyunyong et al. 2001; Toojinda et al. 2003). *SUB1* was later fine-mapped to a small interval of about 0.06 cM using an $F_2$ population (Xu et al. 2000, 2004). This also helped in the positional cloning of *SUB1*, and the locus was found to contain three ethylene-response-like transcription factors designated *SUB1A*, *SUB1B*, and *SUB1C* (Xu et al. 2006). Complementation studies determined that *SUB1A* was the gene largely responsible for tolerance, even though *SUB1C* showed a tolerance-specific allele and, like *SUB1A*, was induced by submergence (Fukao et al. 2006; Xu et al. 2006). These results were further confirmed using recombinants within the *SUB1* locus (Septiningsih et al. 2009). Strong expression of *SUB1* is required for submergence tolerance and the gene seems to have an additive effect. This is observed in $F_1$ hybrids of crosses between tolerant and sensitive parents showing an intermediate tolerance, suggesting that *SUB1* should be introduced in both parents to develop tolerant hybrid varieties (Septiningsih et al. 2009; Singh et al. 2010). Introgression of *SUB1* into varieties with intermediate tolerance such as IR64 also resulted in higher tolerance than in less tolerant recipients (Singh et al. 2009). *SUB1A* is absent in all *japonica* genotypes evaluated so far and is also absent in some *indica* genotypes, and the tolerance conferred by this gene is allele specific. Two different alleles of the *SUB1A* gene have been identified, *SUB1A-1* and *SUB1A-2*, and the *SUB1A-1* allele was found to be the major determinant of submergence tolerance (Xu et al. 2006).

The physiological and molecular bases of tolerance conferred by *SUB1* were extensively studied. The main effect of *SUB1A-1* is to suppress elongation of the shoot when the plants are submerged, thus limiting anaerobic catabolism and leading to the preservation of carbohydrate reserves. In contrast, the intolerant genotypes lacking the *SUB1A-1* allele tend to elongate when submerged, in an "attempted escape" strategy to reach the water surface and to maintain contact with oxygen and light (Das et al. 2005; Fukao et al. 2006; Xu et al. 2006). Our recent studies also suggested the involvement of *SUB1* in chlorophyll protection during flooding, with the consequent enhancement of underwater photosynthesis and maintenance of carbohydrate reserves during submergence (Winkel et al. 2012). This is also supported by our previous observations that blocking of the early ethylene-induced chlorophyll degradation rescues the phenotype and improves survival in both tolerant and sensitive genotypes (Ella et al. 2003b). Furthermore, survival correlates better with carbohydrate status at the time of de-submergence than with that before submergence (Das et al. 2005), supporting the importance of continuous carbohydrate synthesis during submergence. A third role potentially mediated through *SUB1A-1* is the regulation of reactive oxygen species generated immediately after submergence, when leaves were subjected to light. *Sub1* genotypes showed lower concentrations of hydrogen peroxide and malondialdehyde, the product of membrane breakdown under oxidative stress, but higher concentrations of reduced ascorbate, the scavenger of reactive oxygen species. In addition, gene expression as well as the activity of most of the enzymes involved in detoxification of reactive oxygen species such as superoxide dismutase, ascorbate peroxidase, and glutathione reductase was upregulated in tolerant genotypes upon de-submergence (Ella et al. 2003a). Both shoot elongation (Das et al. 2005; Bailey-Serres and Voesenek 2008) and chlorophyll degradation underwater (Ella et al. 2003b) are triggered by ethylene, which increases in concentration in plant tissue during submergence because of enhanced synthesis and entrapment. High ethylene concentration reduces ABA (abscisic acid) but enhances synthesis and sensitivity to gibberellins (GA) through two GA-signaling repressors, *SLENDER RICE-1* (*SLR1*) and *SLR1 LIKE-1* (*SLRL1*) (Fukao and Bailey-Serres 2008). These tolerance strategies—limiting elongation and maintaining underwater photosynthetic carbon fixation—prevent an energy crisis caused by carbohydrate starvation with a consequent increase in survival of submerged plants.

The discovery and cloning of *SUB1* helped in shortening the breeding cycle to 2–3 years and in overcoming the undesirable effects of additional unwanted introgressions or linkage drag normally associated with conventional backcrossing. This becomes feasible after the development and use of a marker-assisted backcrossing system that helped to incorporate *SUB1* into numerous popular rice varieties. A historical treatise of *SUB1* from discovery to deployment through breeding was recently provided (Bailey-Serres et al. 2010; Mackill et al. 2012). The presence of rice varieties that are well known to farmers and are already covering several millions of hectares provided an opportunity to use these varieties for deploying *SUB1* to farmers in flood-prone areas. *SUB1* has been incorporated into eight varieties that are popular among farmers in South and Southeast Asia using marker-assisted backcrossing, and each of these varieties covers between 1 million and more than 6 million ha of rainfed and irrigated rice areas (Neeraja et al. 2007; Septiningsih et al. 2009; Iftekharuddaula et al. 2011; Mackill et al. 2012).

These varieties, together with their counterparts lacking the *SUB1* introgression, were extensively evaluated both on research stations and in farmers' fields in several countries in Asia, and the evaluation was carried out both under control conditions to test whether this gene has any other effects and under flooded conditions to quantify its impact. Data from these trials showed that *SUB1* considerably improved plant survival during submergence and recovery afterward when plants were submerged for 4–18 days, and this is reflected in grain yield. On average, the yield advantage of Sub1 lines ranges from 1 to over 3.5 t ha$^{-1}$ depending on the duration of flooding and the conditions of the floodwater in a particular trial or farmer's field, with its effects being more pronounced with a longer duration of flooding. However, in the absence of stress, *SUB1* showed no effects on growth, yield, or grain quality. This can be explained by the fact that, being an ethylene-response factor (ERF), the gene is induced at the transcript level only during submergence (Xu et al. 2006). Other outcomes of these extensive trials are that *SUB1* reduces the delay in flowering and maturity caused by submergence, is effective from seedling establishment to about one week before heading, works well in all genetic backgrounds tested so far, and is effective under variable environments in both Asia and Africa (Sarkar et al. 2009; Singh et al. 2009, 2011; Manzanilla et al. 2011; Mackill et al. 2012).

Apparently, *SUB1* can provide sufficient protection during transient complete flooding; however, this tolerance is consistently below that of the original donor variety FR13A. Under certain conditions and years, higher tolerance is desired, particularly when flooding persists for a longer duration or the flooding conditions are conducive to greater damage (Das et al. 2009), which is expected in some flood-prone areas, and with the current trends in climate change. Efforts are, therefore, needed to enhance the current tolerance conferred by *SUB1*, and this could be addressed through various strategies (1) identifying and using additional QTLs from FR13A or other donors to be combined with *SUB1* to enhance its effectiveness under longer and more frequent submergence incidences, (2) developing genotypes with faster recovery ability to speed the regeneration of early tillers after flooding and to minimize the delay in flowering and maturity caused by submergence, and (3) developing management strategies that could further boost *SUB1* tolerance and promote faster recovery (Ella and Ismail 2006; Ella et al. 2011).

## 7.4.3  Longer-Term Partial Flooding

Prolonged partial flooding is common in vast areas of rainfed lowlands with water depths as low as 30–50 cm in stagnant flood (SF) areas, to more than 1 m in deepwater areas, and to even several meters in floating-rice areas (Fig. 7.1). Even though the plants are not completely submerged under these conditions, grain yield declines markedly for several reasons, the most important of which is the lack of suitable varieties. Still, old landraces predominate in most areas but are low yielding because of their poor tillering ability and susceptibility to lodging. Previously, efforts to develop varieties suitable for stagnant flooding and deepwater areas in Asia were made over several decades at IRRI but with relatively limited progress (HillcRisLambers and Vergara 1982; Catling 1992; Mackill et al. 2010).

### 7.4.3.1 Tolerance of Stagnant Flooding

Modern rice varieties are not adapted to stagnant flood conditions and their yields decrease markedly when water depth exceeds 30 cm in the field. The effects of SF are even more devastating when it follows short-term complete submergence, as commonly witnessed in flood-prone ecosystems (Singh et al. 2011). Breeding for SF areas received less attention in the past despite the enormous yield losses and the large areas affected each year. In recent years, we screened large sets of diverse rice germplasm accessions and breeding lines from the IRRI genebank and other sources, but mostly from flood-prone areas, to identify genotypes with reasonable tolerance. Less than one-third of the accessions tested showed high survival and produced filled grains when partially flooded with a water depth of about 50 cm through maturity. To our surprise, almost all of the modern varieties tested showed high mortality, fewer tillers, and poor yield under these partial-flood conditions. This is probably why farmers in affected areas still rely on growing low-yielding local landraces.

Our understanding of the basis of tolerance of these conditions is still inadequate; however, our initial data showed that traits such as low mortality, ability to tiller well and produce sufficient leaf area, and larger panicles with high fertility are some of the most important traits required in varieties for SF conditions. Variation in elongation of the internodes of the tolerant genotypes was also observed in a manner dependent on the water depth (facultative elongation), and survival and yield are largely dependent on the extent of this partial elongation, together with lesser carbohydrate depletion and thicker culms to avoid lodging and for better internal oxygen transport (Vergara and Ismail unpublished). More recently, breeding lines that are high yielding and perform well under SF have been developed at IRRI through conventional breeding (Mackill et al. 2010) and are being evaluated in target areas in South and Southeast Asia and also in flood-affected areas of sub-Saharan Africa. Tolerance of SF is apparently more complex than tolerance of submergence mediated by *SUB1*, and numerous QTLs/genes probably need to be combined for higher tolerance. Mapping populations are being developed to identify loci associated with various tolerance traits.

When we started evaluating Sub1 introgression lines in the field, we observed that these lines are not particularly successful when floodwater stagnates even at shallower depths of 15–25 cm following complete submergence, mainly because most of them are dwarf and their growth is further suppressed when *SUB1* is induced by complete submergence (Singh et al. 2011). We then conducted a series of experiments to test whether the poor performance under SF following submergence is related to *SUB1* or other factors. Pairs of lines with and without *SUB1* were completely submerged for 12 days and then subjected to three depths of stagnant flooding of 2–5 cm (control), 15–20 cm, and 25–30 cm through flowering. Another set of trials was conducted under 30 cm and 50 cm of partial flooding but without exposure to complete submergence. The data showed that survival of all genotypes decreased considerably when SF followed complete submergence, and the reduction was substantially greater when the genotypes were exposed to deeper stagnant water. The yield of Sub1 genotypes was much lower than that of non-Sub1 genotypes when SF followed complete submergence, but with no differences under SF alone. The taller and submergence-tolerant landrace FR13A included in these trials as a check exhibited a much lower reduction in grain yield under SF as well as under submergence followed by SF. Our recent studies using a larger set of contrasting lines showed that rice genotypes can tolerate SF when they are inherently tall or are capable of partial elongation in response to submergence (Kato and Ismail unpublished). These studies clearly showed that tolerance of SF is independent of the *SUB1* gene and that higher yields could be attained when *SUB1* is introgressed into taller varieties or varieties that are capable of partial elongation under SF for areas expected to experience both stresses.

The findings of these studies helped redesign the breeding strategies for flood-prone areas, and efforts have started to combine tolerance of stagnant flooding with submergence tolerance (based on *SUB1*). As *SUB1* promotes survival of submerged plants by hindering shoot elongation to conserve energy reserves, its incorporation into shorter varieties makes them more sensitive to partial stagnant flooding. However, the fact that response to SF is dependent on the background of the recipient genotype suggested greater opportunities for combining tolerance to both stresses, which will ensure broader adaptation in almost all flood-prone areas. Breeding lines combining SF and *SUB1* are now becoming available at IRRI and some of them are being tested in several countries in Asia.

### 7.4.3.2 Deepwater and Floating Rice

Under more extreme conditions, floodwater can surpass 100 cm and even reach a few meters and stagnate for several months in deepwater and floating-rice areas, depending on water depth (Catling 1992). Elongation ability of leaves, leaf sheaths, and internodes is an essential feature of the cultivars adapted to these conditions to keep pace with the fast-rising water in order to escape complete inundation. Sub1 types will not survive these conditions because of the extended durations of floods. This fast elongation will ensure an oxygen supply to roots via the well-developed continuum of aerenchyma tissue (Setter et al. 1997). Traditional cultivars adapted to these environments can elongate by more than 20 cm per day based on the depth and rate by which floodwater rises in the field. However, these cultivars are generally poor-yielding because of excessive vegetative growth, low tillering ability as tillering is usually suppressed under low oxygen, and susceptibility to lodging. Areas under deepwater rice are now shrinking because of low productivity, and in most countries, these areas are being replaced by dry-season rice with the use of shallow tube wells during the dry season as in Bangladesh and India and surface water with improved flood control and irrigation systems as in Vietnam and other countries.

The most important trait for the survival of deepwater and floating rice is rapid underwater internode elongation, to ensure that upper leaves maintain contact with air (Vergara et al. 1976; Catling 1992). This rapid elongation is induced by ethylene and GA in a manner opposite to the effects of *SUB1A*. Numerous studies detected QTLs associated with individual internode elongation and number of elongating internodes on chromosomes 1, 3, and 12, using different donors, and most of these QTLs were found to be common across several mapping populations (Sripongpangkul et al. 2000; Nemoto et al. 2004; Hattori et al. 2007, 2008; Kawano et al. 2008). Furthermore, the QTL on chromosome 12 was found to control most of the rapid internode elongation phenotype, and this QTL was recently cloned. Two genes, *SNORKEL1* (*SK1*) and *SNORKEL2* (*SK2*), were identified to be responsible for internode elongation in deepwater rice at this locus, and both are ERF genes of subgroup VII, which also includes *SUB1A* (Hattori et al. 2009). With the worsening of weather patterns in recent years and the increase in incidences of typhoons and coastal storms, more areas are expected to be affected by deeper and more prolonged floods. Renewed empahsis is needed to develop high-yielding varieties for this system, especially after cloning of these major genes.

## 7.5   Combining Adaptive Traits for Flooding- and Submergence-Prone Rice Environments

The extent of tolerance provided by the major QTLs currently being targeted for breeding such as *AG1* and *AG2* for tolerance of flooding during germination and *SUB1* for tolerance of submergence during the vegetative stage is mostly less than that of the original donor landraces, and in some cases, donors of novel and independent QTLs are being identified. Efforts should, therefore, continue to focus on mining new sources of tolerance through the collection and evaluation of native germplasm to identify new QTLs/genes. Combining multiple QTLs with either additive effects or synergistic interactions will help develop varieties that are highly tolerant for stable productivity in flood-affected areas, especially as flooding events become more frequent and severe. Recent studies showed that it is possible to combine tolerance of anaerobic conditions during germination and early growth with that during the vegetative stage conferred by the *SUB1* gene, with *SUB1* causing inactivation of shoot elongation when induced (Mackill et al. 2010). The regulatory mechanisms that mediate these opposing responses are not yet understood, but they seem to be stage specific and probably mediated by sugar and light signaling, as the seedlings shift from being dependent on stored reserves to carbohydrates supplied through photosynthesis. Both tolerance of longer-term (stagnant) flooding and tolerance of flooding during germination are independent of submergence tolerance based on *SUB1*. However, both tolerance traits are also compatible with *SUB1* (Mackill et al. 2010; Singh et al. 2011). Therefore, tolerances of the multiple types of flooding can be combined into a single variety, which would have major advantages for achieving higher and more stable yields in flood-affected areas.

In nearly all the unfavorable rice environments in South and Southeast Asia where submergence is a major constraint to rice production, incidences of other abiotic stresses are also common, such as drought in rainfed lowland areas and salinity in coastal zones and some inlands. Drought and submergence can occur even within the same season, and varieties that combine tolerance to both stresses would be extremely useful. Breeders are adopting different strategies to develop varieties that combine drought and submergence tolerance. In one approach, some of the major QTLs associated with drought tolerance identified recently (Bernier et al. 2007; Kumar et al. 2007; Venuprasad et al. 2009) are being transferred into a few Sub1 varieties that are popular in areas prone to both drought and submergence, such as Swarna-Sub1 and IR64-Sub1, using marker-assisted backcrossing. In a second approach, the *SUB1* gene is being backcrossed into some recently released drought-tolerant varieties (Verulkar et al. 2010).

In coastal areas, salinity is commonly experienced at the beginning and end of the rainy season, whereas submergence can occur at any time during the season (Ismail et al. 2008; Ismail and Tuong 2009). Salinity tolerance in rice is also under the control of multiple QTLs during both seedling and reproductive stages (Moradi et al. 2003; Ismail et al. 2007). For example, the *Saltol* locus, a major QTL on the short arm of chromosome 1, is known to confer significant tolerance during the seedling stage (Thomson et al. 2010a). Progress has recently been made in introducing the *SUB1* gene into salt-tolerant varieties, and work is ongoing to introduce both *SUB1* and *Saltol* into several popular varieties in South and Southeast Asia using marker-assisted backcrossing (Ismail et al. unpublished). Opportunities to combine adaptive complex traits and even QTLs with relatively smaller effects had recently become more feasible with the progress made in genomics and molecular breeding, especially with the use of whole-genome sequencing to clone multiple QTLs and to develop efficient single nucleotide polymorphism (SNP) and insertion-deletion (InDel) markers for use in breeding (Thomson et al. 2012; Platten and Ismail unpublished). Combining tolerance of multiple stresses into high-yielding varieties is now feasible and these varieties will be very useful for enhancing the productivity of areas unfavorably affected by environmental disturbances.

## 7.6  Summary and Outlook

Flood-prone areas hold enormous potential for food production and future food security. This is because these areas are currently underexploited and the yield gap is particularly large compared with that of the more favorable areas, despite their vast resources of good-quality water and fertile soils. Rice remains the most suitable crop for these mostly wetlands because it can grow well in flooded soils; however, its productivity decreases markedly when partially or totally flooded. Other challenges in these areas include periodic droughts in rainfed lowlands and salinity and other soil problems, particularly in coastal areas and in some salt-affected inlands. In addition, most rice production areas in the tropics are confined to low-lying deltas and coastal zones that are more vulnerable to weather disruptions and natural disasters. Recently, rice varieties with better adaptation to these conditions and with higher yields are being developed using genomic tools that helped shorten the breeding cycle. These varieties are expected to gradually replace the existing low-yielding landraces. The higher attainable yields of these varieties in farmers' fields and their greater responsiveness to inputs and other crop management options will ensure even higher and more stable productivity. The success of Sub1 varieties in flash flood-prone areas and salt- and drought-tolerant varieties in other areas provided good examples of such successes. Future efforts should focus on enhancing the tolerance of particular stresses in these varieties, together with combining tolerances to the major abiotic stresses prevailing in the main rice production areas. This is now becoming more feasible with the use of modern genomics and molecular breeding tools, with which relevant traits and genes are being identified and pyramided in suitable backgrounds. New varieties with stronger tolerance of multiple stresses will also open new opportunities for further increases in land productivity through the use of more inputs and increasing the intensity of land use through adjusting cropping calendars and types of crops and varieties to ensure higher and more stable benefits. The enormous challenges facing agricultural productivity require that these efforts be intensified to cope with unfavorable climate change and to keep up with the ever-rising global food demand.

**Acknowledgments** The research summarized in this chapter was funded by numerous donors; I would like to particularly acknowledge the support of the German Federal Ministry for Economic Cooperation and Development (BMZ) and the Bill and Melinda Gates Foundation for funding most of our recent work. The author is also indebted to the various colleagues who coauthored most of the papers cited in this review.

# References

Alpi A, Beevers H (1983) Effect of $O_2$ concentration on rice seedlings. Plant Physiol 71:30–34

Alpuerto V-LEB, Norton GW, Alwang J, Ismail AM (2009) Economic impact analysis of marker-assisted breeding for tolerance to salinity and phosphorous deficiency in rice. Rev Agric Econ 31:779–792

Angaji SA, Septiningsih EM, Mackill DJ, Ismail AM (2010) QTLs associated with tolerance of flooding during germination in rice (*Oryza sativa* L.). Euphytica 172:159–168

Atwell BJ, Greenway H (1987) Carbohydrate-metabolism of rice seedlings grown in oxygen deficient solution. J Exp Bot 38:466–478

Avadhani PN, Greenway H, Lefroy R, Prior L (1978) Alcoholic fermentation and malate metabolism in rice germinating at low oxygen concentrations. Aust J Plant Physiol 5:15–25

Bailey-Serres J, Voesenek LACJ (2008) Flooding stress: acclimations and genetic diversity. Annu Rev Plant Biol 59:313–339

Bailey-Serres J, Fukao T, Ronald P, Ismail AM, Heuer S, Mackill D (2010) Submergence tolerant rice: SUB1's journey from landrace to modern cultivar. Rice 3:138–147

Bates BC, Kundzewicz ZW, Wu S, Palutikof JP (eds) (2008) Climate change and water. IPCC Secretariat, Geneva, pp 1–210. http://www.ipcc.ch/ipccreports/tp-climate-change-water.htm

Bernier J, Kumar A, Ramaiah V, Spaner D, Atlin G (2007) A large-effect QTL for grain yield under reproductive-stage drought stress in upland rice. Crop Sci 47:507–518

Biswas JK, Yamauchi M (1997) Mechanism of seedling establishment of direct-seeded rice (*Oryza sativa* L.) under lowland conditions. Bot Bull Acad Sin 38:29–32

Catling D (1992) Rice in deep water. International Rice Research Institute, Manila, 542 p

Das KK, Sarkar RK, Ismail AM (2005) Elongation ability and non-structural carbohydrate levels in relation to submergence tolerance in rice. Plant Sci 168:131–136

Das KK, Panda D, Sarkar RK, Reddy JN, Ismail AM (2009) Submergence tolerance in relation to variable floodwater conditions in rice. Environ Exp Bot 66:425–434

Dey MM, Upadhyaya HK (1996) Yield loss due to drought, cold and submergence in Asia. In: Evenson RH, Herdt RW, Hossain M (eds) Rice research in Asia: progress and priorities. CAB International, Wallingford, pp 291–303

Drew MC (1990) Sensing soil oxygen. Plant Cell Environ 13:681–693

Ella ES, Ismail AM (2006) Seedling nutrient status before submergence affects survival after submergence in rice. Crop Sci 46:1673–1681

Ella ES, Setter TL (1999) Importance of seed carbohydrates in rice seedling establishment under anoxia. Acta Hortic 504:209–216

Ella ES, Kawano N, Osamu H (2003a) Importance of active oxygen scavenging system in the recovery of rice seedlings after submergence. Plant Sci 165:85–93

Ella E, Kawano N, Yamauchi Y, Tanaka K, Ismail AM (2003b) Blocking ethylene perception enhances flooding tolerance in rice seedlings. Funct Plant Biol 30:813–819

Ella ES, Dionisio-Sese ML, Ismail AM (2010) Proper management improves seedling survival and growth during early flooding in contrasting rice genotypes. Crop Sci 50:1997–2008

Ella ES, Dionisio-Sese ML, Ismail AM (2011) Application of silica at sowing affects growth and survival of rice following submergence. Philip J Crop Sci 36:1–11

Fukao T, Bailey-Serres J (2008) Submergence tolerance conferred by *Sub1A* is mediated by SLR1 and SLRL1 restriction of gibberellin responses in rice. Proc Natl Acad Sci USA 105:16814–16819

Fukao T, Xu K, Ronald PC, Bailey-Serres J (2006) A variable cluster of ethylene response factor-like genes regulates metabolic and developmental acclimation responses to submergence in rice. Plant Cell 18:2021–2034

Gibbs J, Morrell S, Valdez A, Setter TL, Greenway T (2000) Regulation of alcoholic fermentation in coleoptiles of two rice cultivars differing in tolerance to anoxia. J Exp Bot 51:785–796

Greenway H, Setter TL (1996) Is there anaerobic metabolism in submerged rice plants? A view point. In: Singh VP, Singh RK, Singh BB, Zeigler RS (eds) Physiology of stress tolerance in rice. Narendra Deva University of Agriculture and Technology/International Rice Research Institute, Faizabad/Manila, pp 11–30

Guglielminetti L, Yamaguchi J, Perata P, Alpi A (1995) Amylolytic activities in cereal seeds under aerobic and anaerobic conditions. Plant Physiol 109:1069–1076

Hattori Y, Miura K, Asano K, Yamamoto E, Mori H, Kitano H, Matsuoka M, Ashikari M (2007) A major QTL confers rapid internode elongation in response to water rise in deepwater rice. Breed Sci 57:305–314

Hattori Y, Nagai K, Mori H, Kitano H, Matsuoka M, Ashikari M (2008) Mapping of three QTLs that regulate internode elongation in deepwater rice. Breed Sci 58:39–46

Hattori Y, Nagai K, Furukawa S, Song XJ, Kawano R, Sakakibara H, Wu JZ, Matsumoto T, Yoshimura A, Kitano H, Matsuoka M, Mori H, Ashikari M (2009) The ethylene response factors *SNORKEL1* and *SNORKEL2* allow rice to adapt to deep water. Nature 460:1026–1031

HilleRisLambers D, Vergara BS (1982) Summary results of an international collaboration on screening methods for flood tolerance. In: Proceedings of the 1981 international deepwater rice workshop. International Rice Research Institute, Los Baños, pp 347–353

Hossain MZ, Islam MT, Sakai T, Ishida M (2008) Impact of tropical cyclones on rural infrastructures in Bangladesh. Invited overview. Agri EnggInt: CIGR JX April, 2008. http://www.cigrjournal.org/index.php/Ejounral/article/view/1036/1029

Hwang YS, Thomas BR, Rodriguez RL (1999) Differential expression of α-amylase genes during seedling development under anoxia. Plant Mol Biol 40:911–920

Iftekharuddaula K, Newaz M, Salam M, Ahmed H, Mahbub M, Septiningsih E, Collard B, Sanchez D, Pamplona A, Mackill D (2011) Rapid and high-precision marker assisted backcrossing to introgress the *SUB1* QTL into BR11, the rainfed lowland rice mega variety of Bangladesh. Euphytica 178:83–97

IPCC, IPoCC (eds) (2001) Climate change 2001: the scientific basis. Cambridge University Press, New York, pp 1–881

IPCC, IPoCC (eds) (2007) Fourth assessment report of the intergovernmental panel on climate change: the impacts, adaptation and vulnerability (Working Group III). Cambridge University Press, New York

Ismail AM, Thomson MJ (2011) Molecular breeding of rice for problem soils. In: de Costa Oliveira A, Varshney RK (eds) Root genomics, Chap 12. Springer, London, pp 289–312

Ismail AM, Tuong TP (2009) Brackish water coastal zones of monsoon tropics: challenges and opportunities. In: Haefele S, Ismail AM (eds) Natural resource management for poverty reduction and environmental sustainability in fragile rice-based systems. Limited Proc No 15, International Rice Research Institute, Manila, pp 113–121

Ismail AM, Heuer S, Thomson M, Wissuwa M (2007) Genetic and genomic approaches to develop rice germplasm for problem soils. Plant Mol Biol 65:547–570

Ismail AM, Thomson MJ, Singh RK, Gregorio GB, Mackill DJ (2008) Designing rice varieties adapted to coastal areas of South and Southeast Asia. J Indian Soc Coast Agric Res 26:69–73

Ismail AM, Ella ES, Vergara GV, Mackill DJ (2009) Mechanisms associated with tolerance of flooding during germination and early seedling growth in rice (*Oryza sativa*). Ann Bot 103:197–209

Ismail AM, Johnson D, Ella E, Vergara GV, Baltazar A (2012) Adaptation to flooding during emergence and seedling growth in rice and weeds, and implications for crop establishment. AoB Plants 2012. doi: 10.1093/aobpla/pls019

Jackson MB, Ram PC (2003) Physiological and molecular basis of susceptibility and tolerance of rice plants to complete submergence. Ann Bot 91:227–241

Jackson MB, Herman B, Goodenough A (1982) An examination of the importance of ethanol in causing injury to flooded plants. Plant Cell Environ 5:163–172

Jiang L, Ming-yu HW, Ming C, Jian-min W (2004) Quantitative trait loci and epistatic analysis of seed anoxia germinability in rice (*Oryza sativa* L.). Rice Sci 11:238–244

Jiang L, Liu S, Hou M, Tang J, Chen L, Zhai H, Wan J (2006) Analysis of QTLs for seed low temperature germinability and anoxia germinability in rice (*Oryza sativa* L.). Field Crops Res 98:68–75

Kamolsukyunyong W, Ruanjaichon V, Siangliw M, Kawasaki S, Sasaki T, Vanavichit A, Tragoonrung S (2001) Mapping of quantitative trait locus related to submergence tolerance in rice with aid of chromosome walking. DNA Res 8:163–171

Kawano R, Doi K, Yasui H, Mochizuki T, Yoshimura A (2008) Mapping of QTLs for floating ability in rice. Breed Sci 58:47–53

Khush GS (1984) Terminology of rice growing environments. International Rice Research Institute, Manila, pp 5–10

Kumar R, Venuprasad R, Atlin GN (2007) Genetic analysis of rainfed lowland rice drought tolerance under naturally-occurring stress in eastern India: heritability and QTL effects. Field Crops Res 103:42–52

Lafitte RH, Ismail AM, Bennett J (2006) Abiotic stress tolerance in tropical rice: progress and the future. Oryza 43:171–186

Mackill DJ, Amante MM, Vergara BS, Sarkarung S (1993) Improved semi-dwarf rice lines with tolerance to submergence of seedlings. Crop Sci 33:749–753

Mackill DJ, Ismail AM, Pamplona AM, Sanchez DL, Carandang JJ, Septiningsih EM (2010) Stress tolerant rice varieties for adaptation to a changing climate. Crop Environ Bioinforma 7:250–259

Mackill DJ, Ismail AM, Singh US, Labios RV, Paris TR (2012) Development and rapid adoption of submergence-tolerant (Sub1) rice varieties. Adv Agron 115:303–356

Maclean JL, Dawe DC, Hardy B, Hettel GP (eds) (2002) Rice almanac. International Rice Research Institute, Los Baños, pp 16–24

Manzanilla DO, Paris TR, Vergara GV, Ismail AM, Pandey S, Labios RV, Tatlonghari GT, Acda RD, Chi TTN, Duoangsila K, Siliphouthone I, Manikmas MOA, Mackill DJ (2011) Submergence risks and farmers' preferences: implications for breeding Sub1 rice in Southeast Asia. Agric Syst 104:335–347

Moradi F, Ismail AM, Gregorio G, Egdane J (2003) Salinity tolerance of rice during reproductive development and association with tolerance at seedling stage. Indian J Plant Physiol 8:105–116

Nandi S, Subudhi PK, Senadhira D, Manigbas NL, Sen-Mandi S, Huang N (1997) Mapping QTLs for submergence tolerance in rice by AFLP analysis and selective genotyping. Mol Gen Genet 255:1–8

Neeraja C, Maghirang-Rodriguez R, Pamplona A, Heuer S, Collard B, Septiningsih E, Vergara G, Sanchez D, Xu K, Ismail AM, Mackill D (2007) A marker-assisted backcross approach for developing submergence-tolerant rice varieties. Theor Appl Genet 115:767–776

Nemoto K, Ukai Y, Tang DQ, Kasai Y, Morita M (2004) Inheritance of early elongation ability in floating rice revealed by diallel and QTL analyses. Theor Appl Genet 109:42–47

Perata P, Geshi N, Yamaguchi J, Akazawa T (1993) Effect of anoxia on the induction of alpha-amylase in cereal seeds. Planta 191:402–408

Perata P, Guglielminetti L, Alpi A (1997) Mobilization of endosperm reserves in cereal seeds under anoxia. Ann Bot 79:49–56

Ram P, Singh B, Singh A, Ram P, Singh P, Singh H, Boamfa I, Harren SF, Jackson M, Setter T, Reuss J, Wade L, Singh V, Singh R (2002) Submergence tolerance in rainfed lowland rice: physiological basis and prospects for cultivar improvement through marker-aided breeding. Field Crops Res 76:131–152

Sarkar RK, Reddy JN, Sharma SG, Ismail AM (2006) Physiological basis of submergence tolerance in rice and implications for crop improvement. Curr Sci 91:899–906

Sarkar RK, Panda D, Reddy JN, Patnaik SSC, Mackill DJ, Ismail AM (2009) Performance of submergence tolerant rice (Oryza sativa) genotypes carrying the Sub1 quantitative trait locus under stressed and non-stressed natural field conditions. Indian J Agric Sci 79:876–883

Septiningsih EM, Ignacio JCI, Sendon PMD, Sanchez DL, Ismail AM, Mackill DJ (2013) QTL mapping and confirmation for tolerance of anaerobic conditions during germination derived from the rice landrace Ma-Zhan Red. Theor Appl Genet. doi:10.1007/s00122-013-2057-1

Septiningsih EM, Pamplona AM, Sanchez DL, Neeraja CN, Vergara GV, Heuer S, Ismail AM, Mackill DJ (2009) Development of submergence tolerant rice cultivars: the Sub1 locus and beyond. Ann Bot 103:151–160

Setter TL, Ellis M, Laureles EV, Ella ES, Senadhira D, Mishra SB, Sarkarung S, Datta S (1997) Physiology and genetics of submergence tolerance in rice. Ann Bot 79:67–77

Singh S, Mackill DJ, Ismail AM (2009) Responses of SUB1 rice introgression lines to submergence in the field: yield and grain quality. Field Crops Res 113:12–23

Singh N, Dang TTM, Vergara GV, Pandey DM, Sanchez D, Neeraja CN, Septiningsih EM, Mendioro M, Tecson-Mendoza EM, Ismail AM, Mackill DJ, Heuer S (2010) Molecular marker survey and expression analyses of the rice submergence-tolerance gene SUB1A. Theor Appl Genet 121:1441–1453

Singh S, Mackill DJ, Ismail AM (2011) Tolerance of longer-term partial stagnant flooding is independent of the SUB1 locus in rice. Field Crops Res 121:311–323

Sripongpangkul K, Posa GBT, Senadhira DW, Brar D, Huang N, Khush GS, Li ZK (2000) Genes/QTLs affecting flood tolerance in rice. Theor Appl Genet 101:1074–1081

Syvitski JPM, Kettner AJ, Overeem I, Hutton EWH, Hannon MT, Brakenridge GR, Day J, Vörösmarty C, Saito Y, Giosan L, Nicholls RJ (2009) Sinking deltas due to human activities. Nat Geosci 2:681–686

Taylor DL (1942) Influence of oxygen tension on respiration, fermentation and growth of wheat and rice. Am J Bot 29:721–738

Thomson MJ, de Ocampo M, Egdane J, Rahman MR, Sajise AG, Adorada DL, Tumimbang-Raiz E, Blumwald E, Seraj ZI, Singh RK, Gregorio GB, Ismail AM (2010a) Characterizing the *Saltol* quantitative trait locus for salinity tolerance in rice. Rice 3:148–160

Thomson MJ, Ismail AM, McCouch SR, Mackill MJ (2010b) Marker assisted breeding. In: Pareek A, Sopory SK, Bohnert HJ, Govindjee (eds) Abiotic stress adaptation in plants: physiological, molecular and genomic foundation. Springer, Dordrecht, pp 451–469

Thomson MJ, Zhao K, Wright M, McNally KL, Rey J, Tung CW, Reynolds A, Scheffler B, Eizenga G, McClung A, Kim H, Ismail AM, de Ocampo M, Mojica C, Reveche MY, Dilla CJ, Mauleon R, Leung H, Bustamante C, McCouch SR (2012) High-throughput SNP genotyping for breeding applications in rice using the BeadXpress platform. Mol Breed 29:875–886

Toojinda T, Siangliw M, Tragoonrung S, Vanavichit A (2003) Molecular genetics of submergence tolerance in rice: QTL analysis of key traits. Ann Bot 91:243–253

Tuong TP, Pablico PP, Yamauchi M, Confesor R, Moody K (2000) Increasing water productivity and weed suppression of wet seeded rice: effect of water management and rice genotypes. Exp Agric 36:71–89

Turner FT, Chen CC, McCauley GN (1981) Morphological development of rice seedlings in water at controlled oxygen levels. Agron J 73:566–570

Vartapetian BB, Jackson MB (1997) Plant adaptations to anaerobic stress. Ann Bot 79:3–20

Venuprasad R, Dalid CO, Del Valle M, Zhao D, Espiritu M, Cruz MTS, Amante M, Kumar A, Atlin GN (2009) Identification and characterization of large-effect quantitative trait loci for grain yield under lowland drought stress in rice using bulk-segregant analysis. Theor Appl Genet 120:177–190

Vergara BS, Jackson MB, De Datta SK (1976) Deepwater rice and its response to deep-water stress. In: Climate and rice. International Rice Research Institute, Los Baños, 301 p

Verulkar SB, Mandal NP, Dwivedi JL, Singh BN, Sinha PK, Mahato RN, Dongre P, Singh ON, Bose LK, Swain P, Robin S, Chandrababu R et al (2010) Breeding resilient and productive genotypes adapted to drought-prone rainfed ecosystem of India. Field Crops Res 117:197–208

Voesenek LACJ, Bailey-Serres J (2009) Genetics of high-rise rice. Nature 460:959–960

Wassmann R, Hien NX, Hoanh CT, Tuong TP (2004) Sea level rise affecting the Vietnamese Mekong Delta: water elevation in the flood season and implications for rice production. Clim Change 66:89–107

Wassmann R, Jagadish SVK, Sumfleth K, Pathak H, Howell G, Ismail A, Serraj R, Redoña E, Singh RK, Heuer S, Donald LS (2009) Regional vulnerability of climate change impacts on Asian rice production and scope for adaptation. Adv Agron 102:91–133

Winkel A, Colmer TD, Ismail AM, Pedersen O (2012) Internal aeration of paddy field rice (*Oryza sativa* L.) during complete submergence–importance of light and floodwater $O_2$. New Phytol 197(4):1193–1203

Xu K, Mackill DJ (1996) A major locus for submergence tolerance mapped on rice chromosome 9. Mol Breed 2:219–224

Xu K, Xu X, Ronald PC, Mackill DJ (2000) A high-resolution linkage map in the vicinity of the rice submergence tolerance locus *Sub1*. Mol Gen Genet 263:681–689

Xu KN, Deb R, Mackill DJ (2004) A microsatellite marker and a codominant PCR-based marker for marker-assisted selection of submergence tolerance in rice. Crop Sci 44:248–253

Xu K, Xu X, Fukao T, Canlas P, Maghirang-Rodriguez R, Heuer S, Ismail AM, Bailey-Serres J, Ronald RC, Mackill DJ (2006) *Sub1A* is an ethylene-response factor-like gene that confers submergence tolerance to rice. Nature 442:705–708

Yamauchi M, Winn T (1996) Rice seed vigor and seedling establishment in anaerobic soil. Crop Sci 36:680–686

Yamauchi M, Aguilar AM, Vaughan DA, Seshu DV (1993) Rice (*Oryza sativa* L.) germplasm suitable for direct sowing under flooded soil surface. Euphytica 67:177–184

# Chapter 8
# Disease Resistance

Harbans S. Bariana, Urmil K. Bansal, Daisy Basandrai, and Mumta Chhetri

**Abstract** All crop plants are affected by plant diseases caused by fungal, bacterial, nematode, and viral pathogens, and the research investment for a given crop disease is driven by its economic value. Disease resistance has always been a major objective in crop improvement programs. While plant breeders focus on deployment of diverse sources of resistance to achieve durable disease control, pathogen populations keep on evolving to produce new virulent races. A successful breeding for resistance program therefore relies on the knowledge of genetic variation for resistance, evolutionary potential of the pathogen, and choice of breeding methods in line with modern selection technologies. There have been enormous developments in genomic research and its application to plant breeding in the first decade of the twenty-first century. Genome sequence projects of brachypodium, barley, and rice have been completed, and physical maps are publicly available. Genomic developments in plant science have accelerated gene discovery, development of disease response-linked markers, and marker-assisted selection activities. High-throughput and cost-effective genotyping has facilitated identification of genomic regions that control economic traits in crops through association analysis. Such regions can be selected in breeding populations through the use of associated marker fingerprints. Genome-wide association selection (GWAS) and marker-assisted recurrent selection (MARS) schemes are being used in crop breeding programs to improve various traits including disease resistance. This chapter details the role of genomic developments in discovery, characterization, and deployment of diverse sources of resistance in new cultivars, and

H.S. Bariana (✉) • U.K. Bansal • M. Chhetri
Department of Plant and Food Sciences, Faculty of Agriculture, Food and Environment,
The University of Sydney Plant Breeding Institute-Cobbitty, PMB4011, Narellan, NSW2567,
Australia
e-mail: harbans.bariana@sydney.edu.au

D. Basandrai
Hill Agricultural Research and Extension Centre, CSK Himachal Pradesh Agricultural University,
Dhaula Kuan, Sirmour 173001, India

C. Kole (ed.), *Genomics and Breeding for Climate-Resilient Crops*, Vol. 2,
DOI 10.1007/978-3-642-37048-9_8, © Springer-Verlag Berlin Heidelberg 2013

more importantly the role of efficient information management in making informed decisions is also discussed.

## 8.1 Introduction

Global food consumption predominantly includes wheat, rice, maize, potato, pulses, and oilseed crops. Irrespective of the acreage, all crops are affected by plant diseases caused by fungal, bacterial, nematode, and viral pathogens. An extensive effort is being expended globally to produce disease-resistant crops. The International Rice Research Institute (IRRI) in the Philippines, International Wheat and Maize Improvement Center (CIMMYT) in Mexico, International Center for Agricultural Research in the Dry Areas (ICARDA) in Syria, International Crops Research Institute for the Semi-Arid Tropics (ICRSAT) in India, and many other international centers contribute effectively to breeding for disease resistance in their target crops. National programs in all nations also invest significantly in breeding disease-resistant crop cultivars.

Disease resistance is one of the most important objectives in crop improvement programs, and the research investment for a given crop disease is driven by its economic value. Murray and Brennan (2009) estimated current average loss due to wheat pathogens to be A$913 m annually in Australia. Similarly, losses valued at US$3.31 b across top 10 soybean-producing nations were reported in 1994 (Wrather et al. 1996). Economic losses due to geminivirus infections on different crops and geographic regions are summarized by Varma and Malathi (2003). Severe bacterial leaf blight (BLB) epidemic can cause 20–50 % yield losses in rice (Adhikari et al. 1995; Sonti 1998). Other major crops also suffer from such huge losses in the event of disease epidemics. Historically, epidemics of late blight of potato in Ireland (1845–1846), the Bengal famine (1943), and corn leaf blight (1970) represent examples that signify the need to control crop diseases. Economic losses due to crop diseases can be saved either by spraying chemicals or through genetic control. Genetic control is preferred over chemical control for its eco-friendliness. While scientists focus on their individual disciplines, plant breeders have to combine disease resistance with other desirable economic traits including high yield and adaptation to different climatic regions in a single genotype recurrently.

Developments in plant genomics in the last decade have added significantly to breeder's toolbox to perform marker-assisted selection (MAS) for disease resistance and other traits. This chapter summarizes the current status and future prospects of application of genomic tools in breeding disease-resistant crop cultivars with examples from cereal and pulse crops. The whole process can be divided into five key components:

• Identification of genetically diverse sources of resistance
• Development of closely linked markers/resistance gene cloning
• MAS for disease resistance

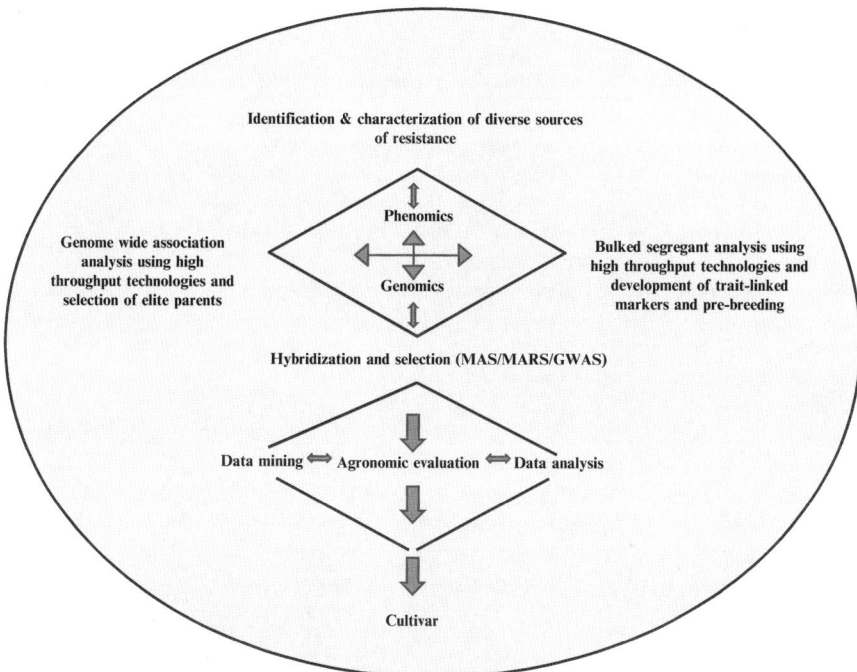

**Fig. 8.1** An integrated model for application of genomics to breeding for disease resistance

- Transgenic approach to disease resistance
- Information management

## 8.2 Identification of Genetically Diverse Sources of Resistance

Breeding for disease resistance relies on the availability of diverse genetic variation. This component relies on combination of phenomics and genomics (Fig. 8.1). Success of a disease resistance improvement program relies on detailed knowledge of pathotypic variation in target pathogen populations (Bariana 2003). Specialized national and international pathogen surveillance programs play vital role to collect and maintain pathogen cultures for gene discovery. Genomic advancements for important crop pathogens are still far behind. Two surveys by the journal of *Molecular Plant Pathology* listed top 10 fungal (Dean et al. 2012) and viral (Scholthof et al. 2011) pathogens that are attracting attention of molecular biologists. Some of these fungi and viruses are not economically important, but represent model systems that can add significantly to our understanding of host-pathogen interactions. There are still only a few examples of cloned avirulence genes in plant pathogens of global economic importance.

**Table 8.1** Examples of wild relatives that contributed towards disease resistance in different crops

| Donor species | Resistance to pathogens |
| --- | --- |
| Wheat | |
| *Triticum tauschii* | *Puccinia spp.*, cereal cyst nematode |
| *T. speltoides* | *Puccinia spp.* |
| *T. timopheevii* | *Puccinia spp.* and *Erysiphe graminis* |
| *Aegilops ventricosa* | *Puccinia spp.* and cereal cyst nematode |
| *A. geniculata* | *Puccinia spp.* |
| *Thinopyrum intermedium* | BYDV, Tanspot, and *Puccinia spp.* |
| *Th. junceum* | Tanspot |
| *Th. elongatum* | Tanspot |
| *Th. ponticum* | Tanspot, *Puccinia spp* |
| *Leymus racemosus* | Tanspot |
| Rice | |
| *Oryza rufipogon* | *Sclerotium oryzae* |
| *O. nivara* | Grassy stunt |
| *O. longistaminata* | Bacterial leaf blight |
| Maize | |
| *Tripsacum floridanum* | *Helminthosporium turcicum* |
| Tomato | |
| *Lycopersicon hirsutum* | *Septoria lycopersici*, TMV, *Odium lycopersici* |
| *L. chilense, L. peruvian* | *Heterodera maioni* |
| *L. peruvianum* | *Pyrenochaeta lycopersici, Meloidogyne spp.*, Heterodera maioni |
| Potato | |
| *Solanum demissum* | *Phytophthora infestans* |
| *Solanum stoloniferum* | *Phytophthora infestans*, PVA and PVY |
| *Solanum acaule* | Potato virus X |

Diversity is often contributed by cultivars grown in different geographical locations, unimproved landraces, and wild relatives of a given crop. In the case of wheat, almost 50 % named rust resistance genes have been contributed by its wild relatives. The 1BL/1RS translocation in "Veery" wheats in the last century is an example of contribution of resistance to three rust diseases and powdery mildew by rye. Resistances to tanspot, barley yellow dwarf virus, and cereal cyst nematode are among other examples of intergeneric and interspecific gene transfer in wheat. Examples of transfer of resistance to pathogens from related species in other crops are listed in Table 8.1. The transfer of BLB resistance gene *Xa21* from *Oryza longistaminata* was a significant achievement of disease resistance research in rice as this gene imparted broad-spectrum resistance to pathovars of *Xanthomonas oryzae* from both India and Philippines. Unlike other BLB genes, *Xa21* is seedling susceptible and expresses resistance at post-seedling growth stages.

In addition to wild relatives of crop species, landraces can be good sources of resistance. The "Green Revolution" in the last century resulted from the introduction of height-reducing genes in wheat. While this effort increased wheat yield, ample genetic variation for all traits including disease resistance in older tall wheat genotypes remained unexplored. Screening of an old collection of wheat genotypes collected by Arthur Watkins during 1920–1930 from 33 nations against three rust diseases and genotyping with molecular markers linked with known rust resistance genes facilitated the short listing of genotypes that were expected to carry yet uncharacterized rust resistance loci (Bansal et al. 2011; Bariana and coworkers unpublished data). Bhullar et al. (2009) isolated seven resistance alleles of powdery mildew resistance gene *Pm3* in wheat landraces from the ICARDA genebank. The bean golden mosaic virus gene, *bgm1*, was transferred from a Mexican landrace (Urrea and Miklas 1996). BLB resistance genes *Xa4*, *xa5*, *Xa7*, and *xa13* were also identified in basmati landraces using linked molecular markers (Jamil et al. 2012) for use as donor sources.

Scientists can become complacent and think that sustainable disease control has been achieved. The detection of virulence for stem rust resistance gene *Sr31* in the race Ug99 Pretorius et al. (2000) after more than 20 years of its deployment reminds us to be vigilant. While Ug99 possesses virulence for such a widely deployed gene *Sr31*, it was avirulent on *Sr28*, a widely ineffective gene in many parts of the world. The previously undetected presence of this gene in an Australian wheat cultivar Gabo and its derivatives was demonstrated by marker genotyping of Ug99 resistant Australian germplasm (Bansal et al. 2012).

Chromosomal location of diverse genetic variation for disease resistance traits in crops was based on cytogenetic techniques in the last century. Cytogenetic techniques were well established in tomato, rice, wheat, maize, and many other crops. This technology was very laborious and time consuming. Molecular mapping has replaced the cytogenetic technology in the twenty-first century for chromosomal location of resistance genes. Development of genetic maps in various crops has shortened the process of genomic location of resistance genes. For example, chromosomal location of resistance in wheat involved crossing of a resistance source with a complete set of 21 monosomics, identification of monosomic $F_1$ plants, and development of 21 monosomic $F_2$ populations prior to testing. Progeny testing of $21F_2$ populations is essential to confirm results based on single-plant tests. This process determines only the critical chromosome, and precise location relies on telocentric mapping and/or crosses with genotypes known to carry other genes on the target chromosome. Chromosomal location of quantitatively inherited traits using cytogenetic techniques was even more cumbersome. In contrast, molecular mapping can provide chromosome arm location of a given locus at a much faster rate. The available genomic resources also facilitate fine mapping and cloning of resistance genes. Genomic location of stripe rust resistance genes *Yr28* to *Yr54*, leaf rust resistance genes from *Lr46* to *Lr71*, and stem rust resistance genes *Sr46* to *Sr56* has been based on molecular mapping. Genomic locations of barley leaf rust resistance gene *Rph20* (Hickey et al. 2011) and *Rph21* (Sandhu et al. 2012) were determined through molecular mapping. Similar examples exist in rice, soybean, and other crops (Table 8.2).

**Table 8.2** Examples of trait/gene-marker associations for disease resistance in three cereal crops and five pulse crops

| Disease | Resistance locus | Molecular marker | Reference |
|---|---|---|---|
| **Wheat** | | | |
| Powdery mildew | Several | SSR, STS | Cowgar et al. (2012) |
| Rust diseases | >10 for each rust disease | STS, SSR, CAPS | Bariana et al. (2007), McCallum et al. (2012), Sambasivam et al. (2008), Wellings et al. (2012) |
| Septoria diseases | Several | SSR | Goodwin (2012) |
| Tanspot | *Tsn1* | STS | Faris et al. (2010) |
| Common bunt | *Bt10* | STS | Laroche et al. (2000) |
| Fusarium head blight | Several QTLs | SSR | Buerstmayar et al. (2012) |
| Flag smut | Four QTLs | SSR/DArT | Toor and Bariana (2012) |
| **Rice** | | | |
| Bacterial leaf blight | >10 genes | STS, SSR, CAPS | Chunwongse et al. (1993), Datta (2010), Yoshimura et al. (1995), Zhang et al. (1996) |
| Rice blast | >10 genes | STS, SSR, CAPS | Datta (2010), Fjellstrom et al. (2004), Hayashi et al. (2004), Jeon et al. (2003) |
| **Barley** | | | |
| Barley yellow mosaic virus | *rym4/rym5* | SSR | Graner et al. (1999) |
| | *rym4, rym9, rym11* | SSRs | Werner et al. (2003) |
| Rust diseases | *Rph7* | CAPS | Graner et al. (2000) |
| | *Rph14* | SSR | Golegaonkar et al. (2009) |
| | *Rph15* | AFLP-derived STS | Weerasena et al. (2004) |
| | *Rph20* | STS | Hickey et al. (2011) |
| | *Rpg1* | Gene based | Brueggemann et al. (2002) |
| | *Rpg4* | RAPD | Borovkova et al. (1995) |
| **Common bean** | | | |
| Anthracnose | *Co-1* | RAPD | Young and Kelly (1997) |
| | *Co-2* | RAPD | Young and Kelly (1996) |
| | *Are* | SCAR, RAPD, and RFLP | Adam-Blondon et al. (1994) |
| | *Co-4* | RAPD | Young et al. (1998) |
| | *Co-5* | RAPD | Young and Kelly (1997) |
| *Co-6* | RAPD | Young and Kelly (1997) | |
| BCMV | *I* | RAPD | Haley et al. (1994b), Mellotto and Kelly (1998) |
| | *Bc-3* | RAPD | Haley et al. (1994a), Johanson et al. (1997) |
| | *Bgm-1* | RAPD | Urrea and Miklas (1996) |
| *Macrophomina* | *Mp-1, Mp-2* | RAPD | Olaya et al. (1996) |
| Rust | *Ur-3* | RAPD | |

(continued)

**Table 8.2** (continued)

| Disease | Resistance locus | Molecular marker | Reference |
|---|---|---|---|
| | | | Haley et al. (1994c), Johnson et al. (1995) |
| | *Ur-4* | RAPD | Miklas et al. (1993) |
| | *Ur-5* | RAPD | Haley et al. (1993) |
| | *Ur-9* | RAPD | Jung et al. (1996a) |
| Soybean | | | |
| Sclerotinia stem rot | 16 QTLs | SSRs | Vuong et al. (2008), Guo et al. (2008), Huynh et al. (2010) |
| Halo blight | *Pse-1* | SCAR | Miklas et al. (2009) |
| *Phytophthora root rot* | 3–8 QTLs/genes | AFLPs, SSRs, RAPDs, SCARs | Han et al. (2008), Li et al. (2010), Wang et al. (2010) |
| Brown stem rot | 4 QTLs/genes | RFLPs, AFLPs, SSRs | Patzoldt et al. (2005) |
| Asian soybean rust | 5 loci/QTLs | SSRs, SNPs | Garcia et al. (2008), Silva et al. (2008), Hyten et al. (2009), Chakraborty et al. (2009) |
| Soybean mosaic virus | *Rsv1*, *Rsv3* | SSRs | Shi et al. (2008) |
| Sudden death syndrome | 4 QTLs | SSRs | Kazi et al. (2008) |
| Soybean cyst nematode | *rhg1* | SSR | Cregan et al. (1999) |
| Chickpea | | | |
| Ascochyta blight | *AR2, ar1, ar1a, ar1b, ar2a, ar2b, Ar19* | SSRs, RAPDs, DAF | Cho et al. (2004) |
| | $QTL_{AR1}$, $QTL_{AR2}$ | SCARs, SSRs, RAPDs | Iruela et al. (2006, 2007) |
| | 13 QTLs | SSRs, RAPDs | Kottapalli et al. (2009) |
| Fusarium wilt | *Foc-0, foc-1, foc-2, foc-3, foc-4, foc-5* | SSRs, STSs, RAPDs | Cobos et al. (2005), Iruela et al. (2007) |
| Rust | 1 QTL | SSR | Madrid et al. (2008) |
| Lentil | | | |
| Fusarium wilt | *Fw* | SSR | Eujayl et al. (1998), Hamwieh et al. (2005) |
| Anthracnose | *Lct-2* | OPE06, UBC704 | Tullu et al. (2003) |
| | *3 QTLs QTL1, QTL2 and QTL3* | | Nguyen et al. (2001), Tar'an et al. (2003) |
| Ascochyta blight | *AbR1A* | SCARs | Ford et al. (1999) |
| | *ral2* | UBC227, OPD 10 | Chowdhury et al. (2001) |
| | *C-TTA/M-AC QLT1 and QTL2, M20 QTL3* | | |
| Rust | QTL | SRAP F7XEM4a | Saha et al. (2010) |
| | | RAPD | Barilli et al. (2010) |
| Pea | | | |
| Powdery mildew | *Er3* | SCAR | Fondevilla et al. (2008) |

## 8.3 Development of Closely Linked Molecular Markers and Resistance Gene Cloning

Availability of effective selection technologies underpins disease resistance breeding. Traditionally, selection for disease resistance has been based on bioassays. The maintenance of different pathogen isolates has been recognized as a great challenge. Even if such resources are maintained, it is difficult to ascertain the presence of combinations of two or more highly effective resistance genes in a single genotype without formal genetics. Disease response-linked markers offer an alternative phenotype-neutral technology for selection of molecularly tagged genes either alone or in combinations. A successful marker development process is driven by reliable phenotyping and reproducible genotyping. The robustness of disease response-linked markers depends on the quality of phenotypic data used for its initial detection. Validation of linked markers across different genetic backgrounds ensures their successful implementation in breeding programs. Attributes of a good marker include reliability, level of polymorphism, and cost-effectiveness (Bariana 2003; Collard and Mackill 2008). Table 8.2 lists DNA markers that were developed in key crops related to this chapter. Molecular markers closely linked with rust resistance genes $Sr2$, $Sr15$, $Sr22$, $Sr24/Lr24$, $Sr25/Lr19$, $Sr26$, $Sr31/Lr26/Yr9$, $Sr33$, $Sr36$, $Sr38/Lr37/Yr17$, $Sr39/Lr35$, $Sr40$, $Sr45$, $Sr50$, $Lr17a$, $Lr21$, $Lr34/Yr18$, $Lr42$, $Lr47$, $Lr51$, $Lr57/Yr40$, $Lr58$, $Lr67/Yr46/Sr55$, $Yr4$, $Yr10$, $Yr15$, $Yr24$, $Yr32$, $Yr35$, $Yr36$, $Yr47$, and $Yr49$ have been reported. All, except five of these genes, are major genes and can be assayed reliably under controlled environment conditions, and hence, high-quality phenotypic data can be obtained. Many of these genes have not been used in practical wheat breeding yet. Similarly, markers for other diseases of wheat have been developed (Table 8.2).

Francia et al. (2005) listed disease response-linked markers for wheat, rice, barley, maize, sugar beet, and tomato. Markers linked with several BLB and blast resistance genes in rice have been developed and listed in Datta (2010). Kelly and Miklas (1998) reviewed the development of markers linked with resistance to anthracnose, bean common mosaic virus, bean golden mosaic virus, and bean rust (Table 8.2). Markers linked with resistance to common blight, web blight, and rust in common bean were reported by Jung et al. (1996a, b). Table 8.2 lists markers linked with resistance to lentil, pea, chickpea, and soybean diseases.

The availability of high-throughput DNA marker systems such as diversity arrays technology (DArT) and single nucleotide polymorphism (SNP) chips has revolutionized the marker development and quantitative trait loci (QTL) mapping in crop plants. SNPs are gaining predominance in crop species where whole genomes have been sequenced. Many SNPs have been identified between *indica* and *japonica* subspecies of rice, and information is freely available online (http://www.plantgenome.uga.edu/snp; http://www.ricesnp.org). Recently wheat genome has been sequenced using Illumina 454 platform, and SNPs are being identified under the NSF project (http://wheat.pw.usda.gov/SNP/new/index.shtml). With the availability of high-throughput platforms, the time taken to identify a closely linked marker has been significantly reduced in recent years. There were more than 25 publications on QTL mapping of stripe rust resistance in wheat in the last decade. This is a huge increase compared to less than five such publications in the previous decade. A majority of the studies on QTL analyses have not been followed up to isolate individual components (Mendelizing of QTL) for detailed mapping and formal naming.

Molecular cloning of several disease resistance genes has been achieved in commercial crop species (Table 8.3). Qualitatively inherited major genes mainly belong to NBS-LRR class, whereas genes with minor and post-seedling resistance belong to different categories. These studies have led to the development of gene-specific markers and will have a role in producing gene combination cassettes for crops where transformation technology is available. Individual trait-marker associations will continue to play a leading role in prebreeding (Fig. 8.1). Wide hybridization has been used in many crops to introgress genetic variation for different traits including disease resistance. One of the major issues was the linkage drag of undesirable genetic variation on the translocated chromosome segments. Precise molecular maps in different crops are enabling identification of the size of introduced segments. Reduced-sized introgressed segments have been identified using genetic distances between markers (Dundas et al. 2007; Niu et al. 2011), and closely linked markers for the ameliorated segments have been developed.

## 8.4 Marker-Assisted Selection for Disease Resistance

Breeding for resistance to crop diseases has been traditionally based on bioassay in greenhouse and artificially inoculated and/or disease-prone nurseries under field conditions. Selection for simply inherited resistance has been more effective in classical breeding; however, it was not possible to predict the presence of resistance gene combinations without formal genetic studies. Disease resistance-linked markers facilitate the selection of genotypes carrying combinations of resistance genes. MAS for disease resistance can be performed in different ways depending on breeding objectives.

### 8.4.1 Genotyping of Crossing Block/Diverse Germplasm

A majority of breeding research results often remain unpublished, and it is very difficult to assess the full picture unless personal communication is made with plant breeders. Many public plant breeding programs around the world have started to use gene-linked markers to screen their crossing blocks and advanced breeding lines. For example, Urrea and Miklas (1996) used random amplified polymorphic DNA (RAPD) marker linked with a recessive gene *bgm1* for detecting the presence of bean golden mosaic virus resistance in advanced bean breeding lines. Markers

**Table 8.3** Examples of cloned disease resistance genes

| Gene | Pathogen | Predicted features of R protein | Reference |
|------|----------|--------------------------------|-----------|
| Maize | | | |
| *Hm1* | *Helminthosporium maydis* | HC-toxin reductase | Johal and Briggs (1992) |
| *Rp1-D* | *Puccinia sorghi* | NBS-LRR | Collins et al. (1999) |
| *Rp3* | *Puccinia sorghi* | NBS-LRR | Webb et al. (2002) |
| Wheat | | | |
| *Lr1* | *P. triticina* | NBS-LRR | Cloutier et al. (2007) |
| *Lr10* | *P. triticina* | NBS-LRR | Feuillet et al. (2003) |
| *Lr21* | *P. triticina* | NBS-LRR | Huang et al. (2003) |
| *Pm3* | *Blumeria graminis* | NBS-LRR | Yahiaoui et al. (2004) |
| *Lr34* | *P. triticina* | ATP transporter | Krattinger et al. (2009) |
| *Yr36* | *P. striiformis* | Receptor kinase-like protein | Fu et al. (2009) |
| Rice | | | |
| *Xa-21* | *Xanthomonas oryzae* | LRR serine/threonine protein kinase | Song et al. (1995) |
| *Xa-26* | *Xanthomonas oryzae* | LRR receptor kinase-like protein | Sun et al. (2004) |
| *Pi-b* | *Magnaporthe grisea* | NBS-LRR | Wang et al. (1999) |
| Barley | | | |
| *mlo* | *Blumeria graminis* | Transmembrane protein | Buschges et al. (1997) |
| *Mla series* | *Blumeria graminis* | NBS-LRR | Wei et al. (1999) |
| *Rpg1* | *P. graminis* | Protein kinase | Brueggemann et al. (2002) |
| *Rpg5* | *P. graminis* | NBS-LRR | Brueggemann et al. (2008) |
| Tomato | | | |
| *Cf-9* | *Cladosporium fulvum* | NBS-LRR | Jones et al. (1994) |
| *Cf-2* | *C. fulvum* | NBS-LRR | Dixon et al. (1996) |
| *Cf-4* | *C. fulvum* | NBS-LRR | Thomas et al. (1997) |
| *Cf-5* | *C. fulvum* | NBS-LRR | Dixon et al. (1998) |
| *Pto* | *Pseudomonas syringae* pv. *tomato* | Intracellular serine/threonine protein kinase | Martin et al. (1993) |
| *Mi* | *Meloidogyne* spp. | NBS-LRR | Milligan et al. (1998) |

linked with known rust resistance genes are used to ascertain the presence of target genes in advanced breeding lines of wheat in Australia (Bariana unpublished data). Similar genotyping is also performed on international germplasm distributed by CIMMYT and ICARDA. Young et al. (1998) used markers tightly linked with bean anthracnose resistance genes, *Co-4* and *Co-5*, to identify a third genetically independent gene, *Co-7*, in natural gene pyramids in highly resistant cultivars G2332 and G2338.

Whole-genome scan of advanced and putatively homozygous breeding material is preferred over gene-linked marker approach due to its value in gene discovery for a range of traits that show variation among the set of genotypes used. Association analysis using whole-genome molecular scans has been used by various workers to identify new genetic variation for various traits. Bansal and Bariana (unpublished results) employed this technology to identify genomic regions that control resistance to rust diseases in a set of wheat landraces. Validation of association mapping results through formal genetic studies facilitated naming of rust resistance genes *Yr47* (Bansal et al. 2011), *Yr51*, and *Sr49* (Bansal and Bariana unpublished results). Huang et al. (2010) genotyped 517 rice landraces using the Illumina NGS (new generation sequence) technology to conduct association analysis for various agronomic traits. A total of 3.6 million SNPs were identified. Eighty associations were identified for 14 agronomic traits; some of them were located close to previously characterized genes. On an average, the strong associations explained 36 % of the phenotypic variation (6–68 % for different traits), which was much higher than that observed in human studies. Uncovering the genetic basis of agronomic traits in crop landraces that have adapted to various agroclimatic conditions is important for global food security and to identify genes for broader climatic adaptations.

## 8.4.2 Marker-Assisted Backcrossing of Resistance Genes

Backcross breeding method involves the transfer of resistance gene(s) from a donor source to agronomically superior but susceptible genetic background through repetitive crossing. This method is often used for parent building/prebreeding activities. Nevertheless, cultivars have been released from backcrossing programs. A majority of Australian prime hard-quality wheat cultivars released by the University Sydney program were backcross derived. In contrast to classical backcrossing, more than one gene with similar phenotype can be backcrossed using MAS. Two stripe rust QTLs were transferred to barley cultivar Streptoe through restriction fragment length polymorphism (RFLP) marker-assisted backcrossing (Toojinda et al. 1998). Various workers transferred BLB resistance genes *Xa4*, *xa5*, *Xa7*, *xa13*, *Xa17*, and *Xa21* through MAS in different combinations (Chen et al. 2000, 2001; Sanchez et al. 2000; Singh et al. 2001). Mago et al. (2011) performed marker-assisted pyramiding of stem rust resistance genes *Sr24*, *Sr26*, *SrR*, and *Sr31* in wheat. Bariana et al. (2007) selected combinations of rust resistance genes from parental cultivars Sunco and Kukri through a single backcross scheme followed by recurrent phenotypic selection, and the presence of resistance genes *Lr34/Yr18*, *Sr2*, *Sr24*, and *Sr36* was later confirmed through marker genotyping. A marker linked with barley yellow dwarf virus resistance gene *Yd2* was used in backcrossing by Jefferies et al. (2003). Kelly et al. (2003) summarized results on bean and cowpea mapping studies and the role of markers linked with bean anthracnose, bean common mosaic virus, bean common mosaic necrosis virus, bean golden mosaic virus, bean rust, and common bacterial blight in marker-assisted pyramiding. Chowdhury et al. (2001) reported the development of RAPD markers linked with resistance to Ascochyta blight of lentil and discussed their role in combining dominant and recessive genes through marker-assisted backcrossing. Shi et al. (2009) pyramided three soybean mosaic virus resistance genes (*Rsv1*, *Rsv3* and *Rsv4*) using linked molecular markers. Backcross breeding is analogous to human vaccination programs. It is essential to develop "vaccines" against different pathotypes of target pathogens.

### 8.4.3    MAS: $F_2$ Population

Information generated from characterization of crossing block genotypes with disease response linked markers/whole-genome scan provides information on genetic diversity of resistance genes present and facilitates a more targeted crossing scheme. The target genes can then be tracked among $F_2$ or more advanced segregating generations. The aim in this case is to enrich the frequency of positive alleles in early generations. Codominant markers can allow selection of genotypes homozygous for the target gene(s) at the $F_2$ stage. The parental knowledge about resistance genes present in two parents in a biparental cross also enables combining genes for resistance to different diseases and other agronomic traits. Genotyping more markers on the same set of DNA increases the cost-effectiveness of MAS.

MAS can play an important role in anticipatory breeding for resistance to exotic pathogens and/or races of pathogens that are not present in a given geographical location. For example, many wheat-producing nations are free of a highly virulent stem rust pathotype Ug99, and information on the effectiveness of molecularly tagged stem rust resistance genes is enabling anticipatory breeding in the absence of this pathotype in various parts of the world. Liu et al. (2000) produced dual combinations of powdery mildew resistance genes *Pm2*, *Pm4a*, and *Pm21* in a wheat cultivar Yang158. $F_2$-based MAS has been performed in barley for barley yellow mosaic virus gene *rym1* and *rym5* (Okada et al. 2004) and blast resistance genes *Pi-1*, *Piz-5*, and *Pi-ta* in rice (Hittalmani et al. 2000). Multinational breeding companies involved in plant breeding are using MAS on a much larger scale; the detailed information is however not available for general public.

MAS for rust resistance genes and other traits was performed on $F_2$ populations derived from several crosses, and doubled haploid plants were generated from genotypes carrying maximum number of positive alleles using wheat-maize system in an Indo-Australian program on implementation of molecular markers to enhance selection efficiency in wheat breeding programs (Trethowan and coworkers unpublished results). It involves the University of Sydney, Australia; Punjab Agricultural University, Ludhiana, India; and the Directorate of Wheat Research, Karnal, India. The resultant material is planted in both nations for evaluation. This program demonstrated that implementation of genomics in breeding programs will require fine-tuning of breeding methodologies.

### 8.4.4   Whole-Genome Analysis Using High-Throughput Marker Systems

High-throughput, efficient, and cost-effective whole-genome genotyping systems (DArT and SNP chips) are available for many crop species (Gupta et al. 2008), and fee-for-service systems are in operation. Whole-genome scan of advanced breeding lines, where a large volume of data for disease responses and other traits are available, could determine molecular diversity among those genotypes. It can also enable identification of genomic regions that control phenotypic variation for various traits through association analysis.

Marker-assisted recurrent selection (MARS) scheme (Bernardo and Charcosset 2006) is more commonly used in open-pollinated crops (Eathington et al. 2007). It has been successfully used in sweat corn (Edwards and Johnson 1994), soybean, and sunflower (Eathington et al. 2007). Trethowan and Mahmood (2011) outlined a MARS scheme for wheat to select for abiotic stresses. This scheme should work for other self-pollinated crops where high-throughput genotyping facilities are available. In this scheme parents and intermediate generations ($F_3$ or $F_4$) are genotyped. This scheme is devised for quantitatively inherited abiotic stresses; however, it can also be used for improvement of disease resistance traits that are genetically simple and complex. Trethowan and Mahmood (2011) in a flow diagram showed that 300 $F_3$ families are genotyped and progeny tested as $F_4$. The breeding value of each marker allele is then calculated. The MARS scheme combines the benefit of MAS with identification of currently unknown genomic regions that could be important in crop improvement. These workers emphasized the need to select adapted parents to avoid segregation for phenological traits that may affect comparison of growth stage-specific traits.

Genome-wide association selection (GWAS) approach is another way to develop superior germplasm. Genotypes from a phenotyped population ("training population") are subjected to genome-wide molecular marker analysis, and breeding values of alternative alleles of markers are fitted as random effects in a linear model (Mir et al. 2012). Selections in subsequent generations are based on sum of genetically estimated allelic breeding values (GEBV) (Meuwissen et al. 2001). The main advantage of this approach lies in reduction of phenotyping frequency and cycle time (Rutkoski et al. 2010). This scheme has been employed to select durable stem rust resistance in wheat and can be applied to different traits and crops.

## 8.5   Transgenic Approach to Disease Resistance in Crop Plants

While MAS allows selection of a target trait through linked marker genotyping, transgenic approaches are based on "tightly regulated transgenes" (Gurr and Rushton 2005). These authors predicted the development of "designer promoters" in which the expression pattern, strength, and inducibility of a promoter can be tailored to an individual strategy. Despite all the problems of expression, there have been some success stories. Punja (2001) listed examples of genetically engineered plant species with resistance to fungal diseases using expression of hydrolytic enzymes; expression of pathogenesis-related proteins; expression of antimicrobial proteins, peptides, or compounds; expression of phytoalexins; inhibition of pathogen virulence products; alteration of structural components; regulation of plant defense responses; and expression of combined gene products. The release of a zucchini yellow mosaic virus (ZYMV) and watermelon mosaic virus II (WMVII)-resistant yellow squash variety in the USA represents the use of coat protein-mediated resistance (Shah et al. 1995). These authors also reported the effectiveness of full-length PVY or PLRV replicase genes. Tu et al. (1998) reported transfer of BLB resistance gene *Xa21* into a rice cultivar IR72 through transformation. Since cloning of resistance genes has expedited recently (Table 8.3), it will be possible to combine more than one gene in a cassette to achieve durability. Despite significant recent progress, transgenic approach suffers from consumer preference.

## 8.6 Information Management

Plant breeders deal with a substantial volume of pedigrees and phenotypic information every crop season. The implementation of MAS in a breeding program adds significantly to data generation. The establishment of an efficient system that can manage pedigrees, phenotypes, and genotypic (background and foreground) data is critical to make more informed decisions. The International Crop Information System (ICIS) is publicly available database that integrates components for management of such datasets (DeLacy et al. 2009).

On the commercial side, different plant breeding software systems used by Monsanto group of seed companies have been replaced by a centralized "global all crop database system," which allows all their breeders to access genetic material inventory, pedigree information, and phenotypic data (Eathington et al. 2007). Monsanto also adopted a similar system for genotypic data that allows effective management of the whole genotyping process. These individual database developments were followed by an integrated molecular decision-making system that provides a conduit between breeder decisions and field and laboratory information technology systems. It allows a "seamless flow to the next stage of the breeding process" (Eathington et al. 2007). National funding bodies should invest in this "global all crop database approach" to increase efficiency of selection of various traits including disease resistance. Public and private partnerships should be encouraged to achieve best outcomes.

## 8.7 Concluding Remarks and Future Directions

Significant advances in the development and use of disease response-linked molecular markers have been made in the last decade. Low-throughput (gel-based) to high-throughput (array-based) systems are available to suit the needs of breeding programs. An integrated program for application of genomics in disease resistance breeding is depicted in Fig. 8.1. Although high-throughput systems will be the choice for future and are already being used for gene discovery and MAS in some breeding programs, individual gene-marker associations will remain crucial for targeted parent building/prebreeding activities (Fig. 8.1). Identification and characterization of new sources of resistance has been expedited through the evolution of genomic technologies. The fee-for-service options of high-throughput systems have increased efficiency of discovery of new loci that can be used for MAS. Genotype-by-sequencing technology (GBS) offered by DArT Pty Ltd, Australia, will further add to the arsenal to uncover uncharacterized genetic variation. Fig. 8.2 provides a comparison of different breeding strategies and emphasizes the fact that a holistic approach to plant breeding is necessary. The selection efficiency for a given trait will be higher in the case of MAS and GWS due to better parental choice and effective tracking of target genomic regions in segregating populations. Management of huge data sets generated in marker-assisted breeding programs is a major challenge. Joint mining of phenotypic and genotypic data for various economic traits and detailed analysis will ensure better breeding and gene discovery outcomes.

Genomic research on crop plants has facilitated a closer interaction among plant scientists to achieve common goals. For example, investment by the Australian Grains Research and Development Corporation in the last two decades through the Australian Winter Cereal Molecular Marker Program encouraged interaction among cereal scientists working on different aspects of cereal research. Two special issues of the *Australian Journal of Agricultural Research* were published to cover various aspects of wheat and barley research. Validation (Sharp et al. 2001) and implementation (Eagles et al. 2001) information on markers in relation to Australian wheat breeding programs was detailed. Generation Challenge Program (GCP) initiative to improve heat and drought tolerance in wheat through integrated breeding platforms such as MARS scheme involving research institutes from India, China, and Australia and similar MARS programs for drought tolerance in chickpea at ICRISAT, in sorghum at CIRAD, and in cowpea at the University of California, Riverside, USA, will be useful for uncovering genomic regions that control disease resistance in these crops, if phenotypic assessments for disease responses are also made on breeding populations segregating for abiotic stresses. There is a need for scientists working on different traits to work together. A holistic selection scheme to capture variation for additional traits on breeding populations developed for improving a particular trait through MARS and/or GWS schemes will result in cost-effectiveness of MAS. It will also aid in the understanding of relationship of underlying physiological processes that may be common for both biotic and abiotic stresses.

MAS is effectively used by multinational companies; however, the detailed information on the output is often confidential due to commercial reasons. Public plant breeding programs perform MAS for parent building through backcrossing and genotyping of advanced breeding lines with resistance gene-linked markers. Pathogens never sleep and therefore breeding for disease resistance remains a continuos journey. To tackle climate change and variation in pathogen populations,

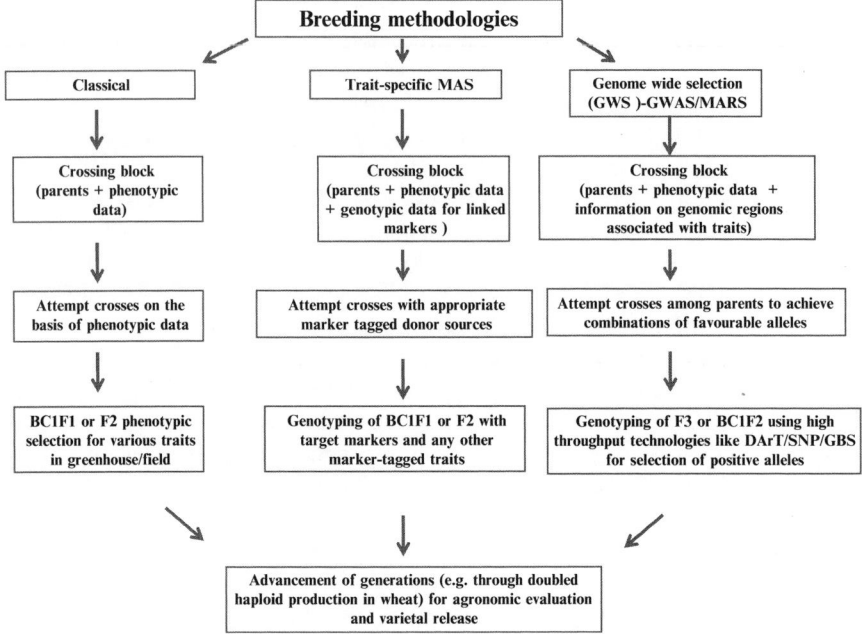

**Fig. 8.2** Classical and molecular breeding strategies

faster and reliable selection technologies are needed to shorten the breeding cycle. The whole-genome scan approach to selection in breeding programs will not only enable selection of target traits more effectively, but it will also facilitate discovery of unknown trait-genome associations. The success lies in understanding parental variation for key traits to make better crosses. Emphasis on effective information management will be important to harvest full potential of any MAS initiative.

**Acknowledgments** We thank Dr Hanif Miah for assistance in compiling the reference list. First and second authors are employed in the Australian Cereal Rust Control Program funded by GRDC Australia.

# References

Adam-Blondon AF, Sevignac M, Bannerot H, Dron M (1994) SCAR, RAPD and RFLP markers linked to a dominant gene (Are) conferring resistance to anthracnose in common bean. Theor Appl Genet 88:865–870

Adhikari TB, Cruz CMV, Zhang Q, Nelson RJ, Skinner DZ, Mew TW, Leach JE (1995) Genetic diversity of *Xanthomonas oryzae* pv. *oryzae* in Asia. Appl Environ Microbiol 61:966–971

Bansal UK, Forrest KL, Hayden MJ, Miah H, Singh H, Bariana HS (2011) Characterization of a new stripe rust resistance gene *Yr47* and its genetic association with the leaf rust resistance gene *Lr52*. Theor Appl Genet 122:1461–1466

Bansal UK, Zwart R, Bhavani S, Wanyera R, Gupta V, Braiana HS (2012) Microsatellite mapping identifies TTKST-effective stem rust resistance gene in wheat cultivar VL404 and Janz. Mol Breed 30(4):1757–1765. doi:10.1007/s11032-012-9759-y

Bariana HS (2003) Breeding for disease resistance. In: Thomas B (ed) Encyclopedia of applied plant sciences. Academic, Harcourt, pp 244–253

Bariana HS, Brown GN, Bansal UK, Miah H, Standen GE, Lu M (2007) Breeding triple rust resistant wheat cultivars for Australia using conventional and marker-assisted selection technologies. Aust J Agric Res 58:576–587

Barilli E, Satovic Z, Rubiales D, Torres AM (2010) Mapping of quantitative trait loci controlling partial resistance against rust incited by *Uromyces pisi* (Pers.) Wint. in a *Pisum fulvum* L. intraspecific cross. Euphytica 175:151–159

Bernardo R, Charcosset A (2006) Usefulness of gene information in marker-assisted recurrent selection: a simulation appraisal. Crop Sci 46:614–621

Bhullar NK, Street K, Mackay M, Yahiaoui N, Keller B (2009) Unlocking wheat genetic resources for molecular identification of previously undescribed functional alleles at the *Pm3* resistance locus. Proc Natl Acad Sci USA 106:9519–9524

Borovkova I, Steffenson BJ, Jin Y, Rasmussen JB, Kilian A, Kleinhofs A, Rossnagel BG, Kao KN (1995) Identification of molecular markers linked to the stem rust resistance gene *rpg4* in barley. Phytopathology 85:181–185

Brueggemann R, Rostoks N, Kudrna D, Killian A, Han F, Chen J, Druka A, Steffenson B, Klelnhofs A (2002) The barley stem rust-resistance gene *Rpg1* is a novel disease-resistance gene with homology to receptor kinases. Proc Natl Acad Sci USA 99:9328–9333

Brueggemann R, Druka A, Nirmala J, Cavileer T, Drader T, Rostoks N, Mirlohi A, Bennypaul H, Gill U, Kudrna D, Whitelaw C, Kilian A, Han F, Sun Y, Gill K, Steffenon B, Kleinhofs A (2008) The stem rust resistance gene *Rpg5* encodes a protein with nucleotide-binding-site, leucine-rich, and protein kinase domains. Proc Natl Acad Sci USA 105:14970–14975

Buerstmayar H, Adam G, Lemmens M (2012) Resistance to head blight caused by *Fusarium* spp. In: Sharma I (ed) Disease resistance in wheat, CABI Plant Protection Series. CAB International, Wallingford, pp 236–276

Buschges R, Hollricher K, Panstruga R, Simons G, Wolter M, Frijters A, van Daelen R, Lee TV, Diergaarde P, Groenendijk J, Topsch S, Vos P, Salamini F, Schulze-Lefert P (1997) The barley *mlo* gene: a novel control element of plant pathogen resistance. Cell 88:695–705

Chakraborty N, Curley J, Frederick RD, Hyten DL, Nelson RL, Hartman GL, Diers BW (2009) Mapping and confirmation of a new allele at *Rpp1* from soybean PI 594538A conferring RB lesion type resistance to soybean rust. Crop Sci 49:783–790

Chen S, Lin XH, Xu CG, Zhang QF (2000) Improvement of bacterial blight resistance of Minghui 63 an elite restorer line of hybrid rice, by molecular marker-assisted selection. Crop Sci 40:239–244

Chen S, Xu CG, Lin XH, Zhang Q (2001) Improving bacterial blight resistance of '6078', an elite restorer line of hybrid rice, by molecular marker-assisted selection. Plant Breed 120:133–137

Cho SH, Chen WD, Muehlbauer FJ (2004) Pathotype-specific genetic factors in chickpea (*Cicer arietinum* L.) for quantitative resistance to ascochyta blight. Theor Appl Genet 109:733–739

Chowdhury MA, Andrahennadi CP, Slinkard AE, Vandenberg A (2001) RAPD and SCAR markers for resistance to ascochyta blight in lentil. Euphytica 118:331–337

Chunwongse J, Martin GB, Tanksley SD (1993) Pregermination genotypic screening using PCR amplification of half seeds. Theor Appl Genet 86:694–698

Cloutier S, McCallum B, Loutre C, Banks TW, Wicker T, Feuillet C, Keller B, Jordan MC (2007) Leaf rust resistance gene Lr1, isolated from bread wheat (*Triticum aestivum* L.) is a member of the large psr567 gene family. Plant Mol Biol 65:93–106

Cobos MJ, Fernandez M, Rubio J, Kharrat M, Moreno MT, Gil J, Millan T (2005) A linkage map of chickpea (*Cicer arietinum* L.) based on populations from Kabuli x Desi crosses: location of genes for resistance to *Fusarium* wilt race 0. Theor Appl Genet 110:1347–1353

Collard BCY, Mackill DJ (2008) Marker-assisted selection: an approach for precision plant breeding in the twenty-first century. Philos Trans R Soc B 363:557–572

Collins NC, Drake J, Ayliffe M, Sun Q, Ellis J, Hulbert S, Pryor T (1999) Molecular characterisation of the maize *Rp1-D* rust resistance haplotype and its mutants. Plant Cell 11:1365–1376

Cowgar C, Miranda L, Griffey C, Hall M, Murphy JP, Judd M (2012) Wheat powdery mildew. In: Sharma I (ed) Disease resistance in wheat, CABI Plant Protection Series. CAB International, Wallingford, pp 84–119

Cregan PB, Mudge J, Fickus EW, Danesh D, Denny R, Young ND (1999) Two simple sequence repeat markers to select for soybean cyst nematode resistance conditioned by the *rhg1* locus. Theor Appl Genet 99:811–818

Datta SK (2010) Rice improvement through application of biotechnology tools. Rice Knowledge Management Portal (RKMP). http://www.rkmp.co.in

Dean R, Van Kan JAL, Pretorius ZA, Hammond-Kosack KE, Pietro AD, Spanu PD, Rudd JJ, Dickman M, Kahmann R, Ellis J, Foster GD (2012) The top 10 fungal pathogens in molecular plant pathology. Mol Plant Pathol 13:414–430

Delacy IH, Fox PN, McLaren G, Trethowan R, White JW (2009) A conceptual model for describing processes of crop improvement in database structures published in crop. Crop Sci 49:2100–2112

Dixon MS, Jones DA, Keddie JS, Thomas CM, Harrison K, Jones JDG (1996) The tomato *Cf-2* disease resistance locus comprises two functional genes encoding leucine-rich repeat proteins. Cell 84:451–459

Dixon MS, Hatzixanthis K, Jones DA, Harrison K, Jones JDG (1998) The Tomato *Cf-5* disease resistance gene and six homologs show pronounced allelic variation in leucine-rich repeats copy number. Plant Cell 10:1915–1925

Dundas IS, Anugrahwati DR, Verlin DC, Park RF, Bariana HS, Mago R, Islam AKMR (2007) New sources of rust resistance from alien species: meliorating linked defects and discovery. Aust J Agric Res 58:545–549

Eagles HA, Bariana HS, Ogbonnaya FC, REbetzke GJ, Hollamby GJ, Henry RJ, Henschke P, Carter M (2001) Implementation of markers in Australian wheat breeding. Aust J Agric Res 52:1349–1356

Eathington SR, Crosbie TM, Edwards MD, Reiters RS, Bull JK (2007) Molecular markers in a commercial breeding program. Crop Sci 47:S154–S163

Edwards MD, Johnson L (1994) RFLPs for rapid recurrent selection. In: Proceedings of the joint plant breeding symposium series of CSSA and ASHA, Corvallis, OR, 5–6 Aug 1994. American Society for Horticultural Science, Alexandria, VA, pp 33–40

Eujayl I, Baum M, Powell W, Erskine W, Pehu E (1998) A genetic linkage map of lentil (*Lens* sp.) based on RAPD and AFLP markers. Theor Appl Genet 97:83–89

Faris JD, Zhang Z, Lu H, Lu S, Reddy L, Cloutier S, Fellers JP, Meinhardt SW, Rasmussen JD, Xu SS, Oliver RP, Simons KJ, Friesen TL (2010) A unique wheat disease resistance-gene like gene governs effector-triggered susceptibility to necrotrophic pathogens. Proc Natl Acad Sci USA 107:13544–13549

Feuillet C, Travella S, Stein N, Albar L, Nublat KB (2003) Map-based isolation of the leaf rust disease resistance gene *Lr10* from the hexaploid wheat (*Triticum aestivum* L.) genome. Proc Natl Acad Sci USA 100:15253–15258

Fjellstrom R, Conaway-Bormans C, McClung AM, Marchetti MA, Shank AR, Park WD (2004) Development of DNA markers suitable for marker assisted selection of three *Pi* genes conferring resistance to multiple *Pyricularia grisea pathotypes*. Crop Sci 44:1790–1798

Fondevilla S, Rubiales D, Moreno MT, Torres AM (2008) Identification and validation of RAPD and SCAR markers linked to the gene *Er3* conferring resistance to *Erysiphe pisi* Dc in pea. Theor Appl Genet 22:193–200

Ford R, Pang ECK, Taylor PWJ (1999) Genetics of resistance to aschochyta blight (*Ascochyta lentis*) of lentil and the identification of closely linked RAPD markers. Theor Appl Genet 98:93–98

Francia E, Tacconi G, Crosatti C, Barabaschi D, Bulgarelli D, Dall'Aglio E, Valè G (2005) Marker assisted selection in crop plants. Plant Cell Tiss Org Cult 82:317–342

Fu DL, Uauy C, Distelfeld A, Blechl A, Epstein L, Chen XM, Sela H, Fahima T, Dubcovsky J (2009) A kinase-START gene confers temperature – dependent resistance to wheat stripe rust. Science 323:1357–1360

Garcia A, Calvo ES, de Souza Kiihl RA, Harada A, Hiromoto DM, Vieira LG (2008) Molecular mapping of soybean rust (*Phakopsora pachyrhizi*) resistance genes: discovery of a novel locus and alleles. Theor Appl Genet 117:545–553

Golegaonkar PG, Karaoglu H, Park RF (2009) Molecular mapping of leaf rust resistance gene *Rph14* in *Hordeum vulgare*. Theor Appl Genet 119:1281–1288

Goodwin SB (2012) Resistance in wheat to Septoria diseases caused by *Mycosphaerella graminicola* (*Septoria tritici*) and *Phaeosphaeria* (*Stagonospora nodorum*). In: Sharma I (ed) Disease resistance in wheat, CABI Plant Protection Series. CAB International, Wallingford, pp 151–159

Graner A, Streng S, Kellermann A, Schiemann A, Baner E, Waugh R, Pellio B, Ordon F (1999) Molecular mapping of the *rym5* locus encoding resistance to different strains of the barley yellow mosaic virus complex. Theor Appl Genet 98:285–290

Graner A, Streng S, Drescher A, Jin Y, Borovkova I, Steffenson B (2000) Molecular mapping of the leaf rust resistance gene *Rph7* in barley. Plant Breed 119:389–392

Guo X, Wang D, Gordon SG, Helliwell E, Smith T, Berry SA, St Martin SK, Dorrance AE (2008) Genetic mapping of QTLs underlying partial resistance to *Sclerotinia sclerotiorum* in soybean PI 391589A and PI 391589B. Crop Sci 48:1129–1139

Gupta PK, Rustgi S, Mir RR (2008) Array-based high throughput DNA markers for crop improvement. Heredity 101:5–18

Gurr SJ, Rushton PJ (2005) Engineering plants with increased disease resistance: how are we going to express it? Trends Biotechnol 23:283–290

Haley SD, Miklas PN, Stavely JR, Byrum J, Kelly JD (1993) Identification of RAPD markers linked to major rust resistance gene block in common bean. Theor Appl Genet 85:505–512

Haley SD, Afanador L, Kelly JD (1994a) Selection for monogenic pest resistance traits with coupling and repulsion phase RAPD markers. Crop Sci 34:1061–1066

Haley SD, Afanador L, Kelly JD (1994b) Identification and application of a random amplified polymorphic marker for the *I* gene (Potyvirus resistance) in common bean. Phytopathology 84:157–160

Haley SD, Afanador LK, Miklas PN, Stavely JR, Kelly J (1994c) Heterogeneous inbred populations are useful as sources of near-isogenic lines for RAPD marker localization. Theor Appl Genet 88:337–342

Hamwieh A, Udapa SM, Choumane W, Sarker A, Dreyer F, Jung C, Buam M (2005) A genetic linkage map of *Lens* sp based on microsatellite and AFLP markers and the localisation of fusarium vascular wilt. Theor Appl Genet 110:669–677

Han Y, Teng W, Yu K, Poysa V, Anderson T, Qiu L, Lightfoot DA, Li W (2008) Mapping QTL tolerance to *Phytophthora* root rot in soybean using microsatellite and RAPD/SCAR derived markers. Euphytica 162:231–239

Hayashi K, Hashimoto N, Daigen M, Ashikawa I (2004) Development of PCR-based SNP markers for rice blast resistance genes at the *Piz* locus. Theor Appl Genet 108:1212–1220

Hickey LT, Lawson W, Platz GJ, Arief VN, German S, Fletcher S, Park RF, Singh D, Pereyra S, Franckowiak J (2011) Mapping *Rph20*: a gene conferring adult plant resistance to *Puccinia hordei* in barley. Theor Appl Genet 123:55–68

Hittalmani S, Parco A, Mew TV, Zeigler RS, Huang N (2000) Fine mapping and DNA marker-assisted pyramiding of the three major genes for blast resistance in rice. Theor Appl Genet 100:1121–1128

Huang L, Brooks SA, Li W, Fellers JP, Trick HN, Gill BS (2003) Map-based cloning of leaf rust resistance gene *Lr21from* the large and polyploidy genome of bread wheat. Genetics 164:655–664

Huang X, Wei X, Sang T, Zhao Q, Feng Q, Zhao Y, Li C, Zhu C, Lu T, Zhang Z et al. (2010) Genome-wide association studies of 14 agronomic traits in rice landraces. Nat Genet 42:961–967

Huynh TT, Bastien M, Iquira E, Turcotte P, Belzile F (2010) Identification of QTLs associated with partial resistance to white mold in soybean using field-based inoculation. Crop Sci 50:969–979

Hyten DL, Smith JR, Frederick RD, Tucker ML, Song Q, Cregan PB (2009) Bulked segregant analysis using the GoldenGate assay to locate the *Rpp3* locus that confers resistance to soybean rust in soybean. Crop Sci 49:265–271

Iruela M, Rubio J, Barro F, Cubero JI, Millan T, Gil J (2006) Detection of two quantitative trait loci for resistance to ascochyta blight in an intra-specific cross of chickpea (*Cicer arietinum* L): development of SCAR markers associated with resistance. Theor Appl Genet 112:278–287

Iruela M, Castro P, Rubio J, Cubero JI, Jacinto C, Millan T, Gil J (2007) Validation of a QTL for resistance to ascochyta blight linked to resistance to *Fusarium* wilt race 5 in chickpea (*Cicer arietinum* L). Eur J Plant Pathol 199:29–37

Jamil US, Iqbal MZ, Shaheen HL, Hasni SM, Jabeen S, Mehmood A, Akhter M (2012) Detection of bacterial blight resistance genes in basmati rice landraces. Genet Mol Res 11:1960–1966

Jefferies SP, King BJ, Barr AR, Warner P, Logue SJ, Langridge P (2003) Marker-assisted backcross introgression of the *Yd2* gene conferring resistance to barley yellow dwarf virus in barley. Plant Breed 112:52–56

Jeon JS, Chen D, Yi GH, Wang GL, Ronald PC (2003) Genetic and physical mapping of *Pi5(t)*, a locus associated with broad-spectrum resistance to rice blast. Mol Genet Genomics 269:280–289

Johal GS, Briggs SP (1992) Reductase activity encoded by the *HM1* disease resistance gene in maize. Science 258:985–987

Johanson WC, Guzman P, Mandala D, Mkandawire D, Temple ABC, Gilbertson RS, Gepts RL, Gepts P (1997) Molecular tagging of the *bc3* gene for introgression into Andean common bean. Crop Sci 37:248–254

Johnson E, Miklas P, Stavely J, Martínez-Cruzado J (1995) Coupling- and repulsion-RAPDs for marker-assisted selection of PI 181996 rust resistance in common bean. Theor Appl Genet 90:659–664

Jones DA, Thomas CM, Hammond-Kosack KE, Balint-Kurti PJ, Jones JDG (1994) Isolation of the tomato *Cf-9*gene for resistance to *Cladosporium fulvum* by transposon tagging. Science 266:789–793

Jung G, Coyne DP, Skroch PW, Nienhuis J, Arnaud-Santana E, Bokos J, Steadman JR (1996a) RAPD marker liked to a gene for specific rust resistance in common bean. Annu Rep Bean Improv Coop 39:59–60

Jung G, Coyne DP, Skroch PW, Nienhuis J, Arnaud-Santana E, Bokosi J, Aruyarathne HM, Steadman JR, Beaver JS, Kaeppler SM (1996b) Molecular markers associated with plant architecture and resistance to common blight, web blight and rust in common bean (*Phaseolus vulgaris* L). J Am Soc Hortic Sci 121:794–803

Kazi S, Shultz J, Afzal J, Johnson J, Njiti VN, Lightfoot DA (2008) Separate loci underlie resistance to root infection and leaf scorch during soybean sudden death syndrome. Theor Appl Genet 116:967–977

Kelly JD, Miklas PN (1998) The role of RAPD markers in breeding for disease resistance in common bean. Mol Breed 4:1–11

Kelly JD, Gepts P, Miklas PN, Coyne DP (2003) Tagging and mapping of gene and QTL and molecular marker assisted selection for traits of economic importance in bean and cowpeas. Field Crops Res 82:135–154

Kottapalli P, Gaur PM, Katiyar SK, Crouch JH, Buhariwalla HK, Pande S, Gali KK (2009) Mapping and validation of QTLs for resistance to an Indian isolate of *Ascochyta* blight pathogen in chickpea. Euphytica 165:79–88

Krattinger SG, Lagudah ES, Spielmeyer W, Singh RP, Huerta-Espino J, McFadden H, Bossolini E, Selter LL, Keller B (2009) A putative ABC transporter confers durable resistance to multiple fungal pathogens in wheat. Science 323:1360–1363

Laroche A, Demeke T, Gaudet DA, Puchalski B, Frick M, McKenzie R (2000) Development of a PCR marker for rapid identification the *Bt-10* gene for common bunt resistance in wheat. Genome 43:217–223

Li X, Han Y, Teng W, Zhang S, Yu K, Poysa V, Anderson T, Ding J, Li W (2010) Pyramided QTL underlying tolerance to *Phytophthora* root rot in mega-environments from soybean cultivars Conrad and Hefeng 25. Theor Appl Genet 121:651–658

Liu J, Liu D, Tao W, Li W, Wang S, Chen P, Cheng S, Gao D (2000) Molecular marker-facilitated pyramiding of different genes for powdery mildew resistance in wheat. Plant Breed 119:21–24

Madrid E, Rubiales D, Moral A, Moreno MT, Millan T, Gil J, Rubio J (2008) Mechanism and molecular markers associated with rust resistance in a chickpea interspecific cross (*Cicer arietinum* × *Cicer reticulatum*). Eur J Plant Pathol 121:45–53

Mago R, Lawrence G, Ellis J (2011) Application of DNA markers and doubled-haploid for stacking multiple stem rust resistance genes in wheat. Mol Breed 27:329–335

Martin GB, Brommonschenkel SH, Chunwongse J, Frary A, Ganal MW, Spivey R, Wu T, Earle ED, Tanksley SD (1993) Map-based cloning of a protein kinase gene conferring disease resistance in tomato. Science 262:1432–1436

McCallum B, Hiebert C, Huerta-Espino J, Cloutier S (2012) Wheat leaf rust. In: Sharma I (ed) Disease resistance in wheat, CABI Plant Protection Series. CAB International, Wallingford, pp 33–62

Mellotto M, Kelly JD (1998) SCAR markers linked to major disease resistance genes in common bean. Annu Rept Bean Improv Coop 41:64–65

Meuwissen T, Hayes BJ, Goddard ME (2001) Prediction of total genetic value using genome-wide dense marker maps. Genetics 157:1819–1829

Miklas PN, Stavely JR, Kelly JD (1993) Identification and potential use of a molecular marker for rust resistance in common bean. Theor Appl Genet 85:745–749

Miklas PN, Fourie D, Wagner J, Larsen RC, Mienie CMS (2009) Tagging and mapping *Pse-1* gene for resistance to halo blight in common bean differential cultivar UI-3. Crop Sci 49:41–48

Milligan SB, Bodeau J, Yaghoobi J, Kaloshian I, Zabel P (1998) The root knot nematode resistance gene Mi from tomato is a member of the leucine zipper, nucleotide binding, leucine-rich repeat family of plant genes. Plant Cell 10:1307–1319

Mir RR, Zaman-Allah M, Sreenivasulu N, Trethowan R, Varshney RK (2012) Integrated genomics, physiology and breeding for improving drought tolerance in crops. Theor Appl Genet 125:625–645. doi:101007/s00122-012-1904-9

Murray GM, Brennan JP (2009) Estimating disease losses to the Australian wheat industry. Aust Plant Pathol 38:558–570

Nguyen TT, Taylor PWJ, Brouwer JB, Pang ECK, Ford R (2001) A novel source of resistance in lentil (*Lens culinaris* ssp culinaris) to ascochyta blight cased by *Ascochyta pisi*. Aust Plant Pathol 30:211–215

Niu Z, Klindworth DL, Friesen TL, Chao S, Jin Y, Cai X, Xu SS (2011) Targeted introgression of a wheat stem rust resistance gene by DNA marker-assisted chromosome engineering. Genetics 187:1011–1021

Okada Y, Kanatani R, Arai S, Ito K (2004) Interaction between barley yellow mosaic disease resistance genes *rym1* and *rym5*, in the response to BaYMV. Breed Sci 54:319–325

Olaya G, Awabi GS, Weeden NF (1996) Inheritance of the resistance to *Macrophomina phaseolina* and identification of RAPD markers linked to the resistance genes in beans. Phytopathology 86:674–679

Patzoldt ME, Grau R, Stephens PA, Kurtzweil NC, Carlson SR, Diers BW (2005) Localization of a quantitative trait locus providing brown stem rot resistance in the soybean cultivar Bell. Crop Sci 45:1241–1248

Pretorius ZA, Singh RP, Wagoire WW, Payne TS (2000) Detection of virulence to wheat stem rust resistance gene *Sr31* in *Puccinia graminis* f. sp. *tritici* in Uganda. Plant Dis 84:203

Punja ZK (2001) Genetic engineering of plants to enhance resistance to fungal pathogens – a review of progress and future prospects. Can J Plant Pathol 23:216–235

Rutkoski JE, Heffner EL, Sorrells ME (2010) Genomic selection for durable stem rust resistance in wheat. Euphytica 179:161–173

Saha G, Sarker A, Chen W, Vandemark GJ, Muehlbauer FJ (2010) Identification of markers associated with genes for rust resistance in *Lens culinaris* medic. Euphytica 175:261–265

Sambasivam PK, Bansal UK, Hayden MJ, Dvorak J, Lagudah ES, Bariana HS (2008) Identification of markers linked with stem rust resistance genes *Sr33* and *Sr45*. In: Appels R, Eastwood R, Lagudah E, Langridge P, Mackay M, McIntyre L, Sharp P (eds) Proceedings of the 11th international wheat genetics symposium. Sydney University Press, Sydney, pp 351–353

Sanchez AC, Brar DS, Huang N, Li Z, Khush GS (2000) Sequence tagged site marker-assisted selection for three bacterial blight resistance genes in rice. Crop Sci 40:792–797

Sandhu KS, Forrest KL, Kong S, Bansal UK, Singh D, Hayden MJ, Park RF (2012) Inheritance and molecular mapping of gene conferring seedling resistance against *Puccinia hordei* in the barley cultivar Ricardo. Theor Appl Genet. doi:10.1007/s00122-012-1921-8

Scholthof KG, Adkins S, Czosnek H, Palukaitis P, Jacquot E, Hohn T, Hohn B, Saunders K, Candresse T, Ahlquist P, Hemenway C, Forster GD (2011) Top ten plant viruses in molecular plant pathology. Mol Plant Pathol 12:938–954

Shah DM, Rommens CMT, Beachy RN (1995) Resistance to diseases and insects in transgenic plants: progress and applications to agriculture. Trends Biotechnol 13:362–368

Sharp PJ, Johnston S, Brown G, Pallotta M, Bariana H, Khatkar S, Singh R, Khairallah M, Potter R, Jones M (2001) Validating molecular markers for wheat breeding. Aust J Agric Res 52:1357–1366

Shi A, Pengyin C, Cuiming Z, Hou A, Zhang B (2008) A PCR based marker for the *Rsv1* locus conferring resistance to soybean mosaic virus. Crop Sci 48:262–268

Shi A, Chen P, Li D, Cuiming Z, Zhang B, Hou A (2009) Pyramiding of multiple genes for resistance to soybean mosaic virus in soybean using molecular markers. Mol Breed 23:113–124

Silva DCG, Yamanaka N, Brogin RL, Arias CAA, Nepomuceno AL, Di Mauro AO, Pereira SS, Nogueira LM, Passianotto ALL, Abdelnoor RV (2008) Molecular mapping of two loci that confer resistance to Asian rust in soybean. Theor Appl Genet 117:57–63

Singh S, Sidhu JS, Huang N, Vikal Y, Li Z, Brar DS, Dhaliwal HS, Khush GS (2001) Pyramiding three bacterial blight resistance genes (*xa5, xa13* and *Xa21*) using marker-assisted selection into *indica* rice cultivar PR106. Theor Appl Genet 102:1011–1015

Song W-Y, Wang G-L, Chen L-L, Kim HS, Pi L-Y et al. (1995) A receptor kinase like protein encoded by the rice disease resistance gene, *Xa21*. Science 270:1804–1806

Sonti RV (1998) Bacterial leaf blight of rice: new insights from molecular genetics. Curr Sci 74:206–212

Sun X, Cao Y, Yang Z, Xu C, Li X, Wang S, Zhang Q (2004) *Xa26*, a gene conferring resistance to *Xanthomonas oryzae* pv. *Oryzae* in rice, encodes an LRR receptor kinase-like protein. Plant J 37:517–527

Tar'an B, Warkentin T, Somers DJ, Miranda D, Vandenberg A, Blade S et al. (2003) Quantitative trait loci for lodging resistance, plant height and partial resistance to *mycosphaerella* blight in field pea (*Pisum sativum* L). Theor Appl Genet 107(8):1482–1491

Thomas CM, Jones DA, Parniske M, Harrison K, Balint-Kurti PJ, Hatzixanthis K, Jones JDG (1997) Characterization of the tomato *Cf-4* gene for resistance to *Cladosporium fulvum* identifies sequences that determine recognitional specificity in *Cf-4* and *Cf-9*. Plant Cell 9:2209–2224

Toojinda T, Baird E, Booth A, Broers L, Hayes P, Powell W, Thomas W, Vivar H, Young G (1998) Introgression of quantitative trait loci (QTLs) determining stripe rust resistance in barley: an example of marker-assisted line development. Theor Appl Genet 96:123–131

Toor AK, Bariana HS (2012) Flag smut of wheat-pathogen biology and host resistance. In: Sharma I (ed) Disease resistance in wheat, CABI plant protection series. CAB International, Wallingford, pp 295–303

Trethowan R, Mahmood T (2011) Genetic options for improving the productivity of wheat in water-limited and temperature-stressed environments. In: Yadav S, Redden RJ, Hatfield J, Lotze-Campen H, Hall AE (eds) Crop adaptation to climate change. Willey Blackwell, West Sussex, pp 218–237

Tu J, Ona I, Zhang Q, Mew TW, Khush GS, Datta SK (1998) Transgenic rice variety 'IR72' with *Xa21* is resistant to bacterial blight. Theor Appl Genet 97:31–36

Tullu A, Buckwaldt I, Warkentin T, Tarn B, Vandenberg A (2003) Genetics of resistance to anthracnose and identification of AFLP and RAPD markers linked to the resistance gene in PI 320937 germplasm of lentil (*Lens culinaris* Medikus). Theor Appl Genet 106:428–434

Urrea CA, Miklas PN, Beaver JS, Riley H (1996) A codominant randomly amplified DNA marker useful for indirect selection of bean golden mosaic virus resistance in common bean. J Am Soc Hortic Sci 121:1035–1039

Varma A, Malathi VG (2003) Emerging geminivirus problems: a serious threat to crop production. Ann Appl Biol 142:145–164

Vuong TD, Diers BW, Hartman GL (2008) Identification of QTL for resistance to Sclerotinia stem rot (*Sclerotinia sclerotiorum*) in soybean plant introduction 194639. Crop Sci 48:2209–2214

Wang ZX, Yano M, Yamanouchi U, Iwamoto M, Monna L, Hayasaka H, Katayose Y, Sasaki T (1999) The *Pib* gene for rice blast resistance belongs to the nucleotide binding and leucine-rich repeat class of plant disease resistance genes. Plant J 19:55–64

Wang H, Waller L, Tripathy S, St Martin SK, Zhou L, Krampis K, Tucker DM, Mao Y, Hoeschele I, Saghai Maroof MA, Tyler BM, Dorrance AE (2010) Analysis of genes underlying soybean quantitative trait loci conferring partial resistance to *Phytophthora sojae*. Plant Genome 3:23–40

Webb CA, Richter TE, Collins NC, Nicolas M, Trick HN, Pryor T, Hulbert SH (2002) Genetic and molecular characterization of the maize *rp3* rust resistance locus. Genetics 162:381–394

Weerasena JS, Steffenson BJ, Falk AB (2004) Conversion of an amplified fragment length polymorphism marker into a co-dominant marker in the mapping of the *Rph15* gene conferring resistance to barley leaf rust, *Puccinia hordei* Otth. Theor Appl Genet 108:712–719

Wei F, Gobelman-Werner K, Morroll SM, Kurth J, Mao L, Wing R, Leister D, Schulze-Lefert P, Wise RP (1999) The *Mla* (powdery mildew) resistance cluster is associated with three NBS-LRR gene families and suppressed recombination within a 240-kb DNA interval on chromosome 5S (1HS) of barley. Genetics 153:1929–1948

Wellings CR, Boyd LA, Chen XM (2012) Resistance to stripe rust in wheat: pathogen biology driving resistance breeding. In: Sharma I (ed) Disease resistance in wheat, CABI plant protection series. CAB International, Wallingford, pp 63–83

Werner K, Friedt W, Laubach E, Waugh R, Ordon F (2003) Dissection of resistance to soil-borne yellow-mosaic-inducing viruses of barley (BaMMV, BaYMV, BAYMV-2) in a complex breeders' cross by means of SSR and simultaneous mapping of BaYMV/BaYMV-2 resistance of var "Chikurin Ibaraki 1". Theor Appl Genet 106:1425–1432

Wrather JA, Stienstra WC, Koenning SR (1996) Soybean disease loss estimates for the United States from 1996 to 1998. Can J Plant Pathol 23:122–131

Yahiaoui N, Srichumpa P, Dudler R, Keller B (2004) Genome analysis at different ploidy levels allow cloning of the powdery mildew resistance gene *Pm3b* from hexaploid wheat. Plant J 37:528–538

Yoshimura S, Yoshimura A, Iwata N, McCouch SR, Abenes ML, Baraoidan MR, Mew TW, Nelson RJ (1995) Tagging and combining bacterial blight resistance genes in rice using RAPD and RFLP markers. Mol Breed 1:375–387

Young RA, Kelly JD (1996) RAPD markers flanking the *Are* gene for anthracnose resistance in common bean. J Am Soc Hortic Sci 121:37–41

Young RA, Kelly JD (1997) RAPD markers linked to three major anthracnose resistance genes in common bean. Crop Sci 37:940–946

Young RA, Melotto AR, Nodari AJ, Kelly JD (1998) Marker-assisted dissection of oligogenic anthracnose resistance in the common bean cultivar, 'G2333'. Theor Appl Genet 96:87–94

Zhang G, Angeles ER, Abenes ML, Khush GS, Huang N (1996) RAPD and RFLP mapping for bacterial blight resistance gene *xa-13* in rice. Theor Appl Genet 93:65–70

# Chapter 9
# Insect Resistance

Chandrakanth Emani and Wayne Hunter

**Abstract** Insect pests exhibit a diverse array of genetic-based responses when interacting with crop systems; these changes can be in response to pathogens, symbiotic microbes, host plants, chemicals, and the environment. Agricultural research has for decades focused on gathering crucial information on the biochemical, genetic, and molecular realms that deal with plant–insect interactions in changing ecosystems. Environmental conditions, which include the overall conditions of climate change, are a reality that needs to be considered as one of the crucial phenomena of changing ecosystems when planning future crop security and/or pest management strategies. Focused research in documenting the interactions that occur between crop systems and insect pests under changing climates will be a needed addition to current research efforts whose aim is to define future strategies that crop scientists relate to the broader societal concerns of food security in an ever-changing environment. This chapter attempts to integrate the past and present research in classical and molecular breeding, transgenic technology, and pest management within the context of climate change. The integrated approach will direct present research efforts that aim at creating plant–insect pest interaction–climate change models that reliably advise future strategies to develop improved insect-resistant, climate-resilient plant varieties.

C. Emani (✉)
Department of Biology, Western Kentucky University-Owensboro, 4821 New Hartford Road, Owensboro, KY 42303, USA
e-mail: chandrakanth.emani@wku.edu

W. Hunter
US Horticultural Research Laboratory, 2001 South Rock Road, Fort Pierce, FL 34945, USA

C. Kole (ed.), *Genomics and Breeding for Climate-Resilient Crops*, Vol. 2,
DOI 10.1007/978-3-642-37048-9_9, © Springer-Verlag Berlin Heidelberg 2013

## 9.1   Introduction

The Nobel Peace Prize-winning efforts of the Intergovernmental Panel on Climate Change (IPCC) produced important documentation relevant to climate change as vital information for policy makers and the general public (IPCC 2001). In their later reports, they laid emphasis on the continual environmental changes related to increasing carbon dioxide levels, temperatures, and ultraviolet levels (IPCC 2007). Agriculture and its vital role in creating the present array of food systems across the world were witnesses to transforming factors such as population increase, changing income patterns, urbanization and globalization that affect food production, food marketing, and food consumption (Von Braun 2007). The new driving force in the form of climate change adds to the existing transforming factors which threatens the very subsistence of food production and food security due to the dual effects of climate variability and climate change on agricultural systems (Parry et al. 2004). The effects of climate change on plant disease epidemiology were examined around the same time as the IPCC report was being prepared (Coakley et al. 1999). Combined with this, the differences in pathosystems in different locations result in the inability of researchers to generalize their conclusions on the effects of climate change on plant disease (Coakley et al. 1999). Overwhelming evidence, however, suggested that climate change has a definite effect on the dynamics of plant disease epidemiology as it involves alterations in the developmental stages of pathogens, modifying host resistance and changes in the physiology of host–pathogen interactions (Garrett et al. 2006). Insects and pathogens form vital components of plant disease epidemiology, and recent reviews were an effort on integrating the pests and pathogens into the scenario of climate change (Gregory et al. 2009; Trumble and Butler 2009). The chapter aims to supplement such efforts in a more comprehensive manner by integrating the past and present research in classical and molecular breeding, transgenic technology, and pest management within the context of climate change. The integrated approach will direct present research efforts that aim at creating plant–insect pest interaction–climate change models that provide reliably model parameters, which will aid the development of future strategies to improved insect-resistant, climate-resilient plant varieties.

## 9.2   Integrating Plant–Insect Pest Interactions and Climate Change

The epochal natural event of colonization of plants 510 million years ago was soon followed by the interactions between plants and insects, and over one million or more insect pests have been at the forefront of the spread of plant diseases as the causative agents as they consume plants as herbivores (Howe and Jander 2008). In an evolutionary struggle that is still evidenced by present research, plants continually have evolved mechanisms to minimize the insect consumption, while insects evolved mechanisms to overcome these plant defense strategies (Wu and Baldwin 2010; Smith and Clement 2012). This evolutionary relationship is now being examined closer under the context of climate change, which is seeing elevated $CO_2$ and temperature levels that are altering the interactions between plants and their insect pests (De Lucia et al. 2012).

The broader context of plant disease epidemiology examines the broader focus of the disease triangle that includes the triumvirate interacting factors of host plant, pathogen, and insect vector, all being influenced by environment, and none of these factors can ever be excluded in researching or discussing plant diseases (Petzoldt and Seaman 2007). The major factors of climate change, namely, temperature, moisture, and $CO_2$, were shown to have an effect on all the interacting disease factors (Petzoldt and Seaman 2007; Gregory et al. 2009; Trumble and Butler 2009), and any plant disease–climate change model that has to be developed will be a complex process that takes into account all the intervening aspects of climate and interacting aspects of disease. Diseases caused by plant insect pests when viewed in the context of climate change will include a more mechanistic inclusion of the pests into plant disease–climate change models and recent research has started documenting such evidences (Gregory et al. 2009; De Lucia et al. 2012).

## 9.2.1  Molecular Events of Plant–Insect Pest Interactions

A general molecular model that summarizes all the key host–pathogen interaction events in a plant cell attacked by an insect pest is shown in Fig. 9.1. The model essentially accounts for many early signaling events in a plant cell induced by an insect pest attack that includes eight essential steps (Wu and Baldwin 2010): Step 1 involves the insect pest-derived elicitors being perceived by certain unidentified plasma membrane receptors. In step 2, these perception events trigger the activation of $Ca^{2+}$ channels that result in $Ca^{2+}$ influxes. In step 3, the binding of $Ca^{2+}$ to NADPH oxidase and its phosphorylation by calcium-dependent protein kinases (CDPKs) leads to step 4, where the production of reactive oxygen species (ROS) modifies ROS-dependent transcription factors that results in an array of plant defense responses (Lamb and Dixon 1997). In step 5, mitogen-activated protein kinases (MAPKs) are activated, among which salicylic acid-induced protein kinases (SIPKs) and wound-induced protein kinases (WIPKs) trigger the synthesis of jasmonates. In step 6, the jasmonates induce an array of molecular responses that inhibit the myelocytomatosis (MYC) transcription factors that in step 7 increases ethylene synthesis. The final step 8 is a series of signaling events induced by elevated ethylene production leading to increased activities of ethylene-responsive transcription factors that are translated into accumulation of metabolites that function as defensives in plants. Recent studies have indicated that the molecular events summarized above (for a detailed review of the events, see Wu and Baldwin 2010; Smith and Clement 2012) can be significantly affected by climate change factors especially in altering the levels of defense hormones such as jasmonic acid and

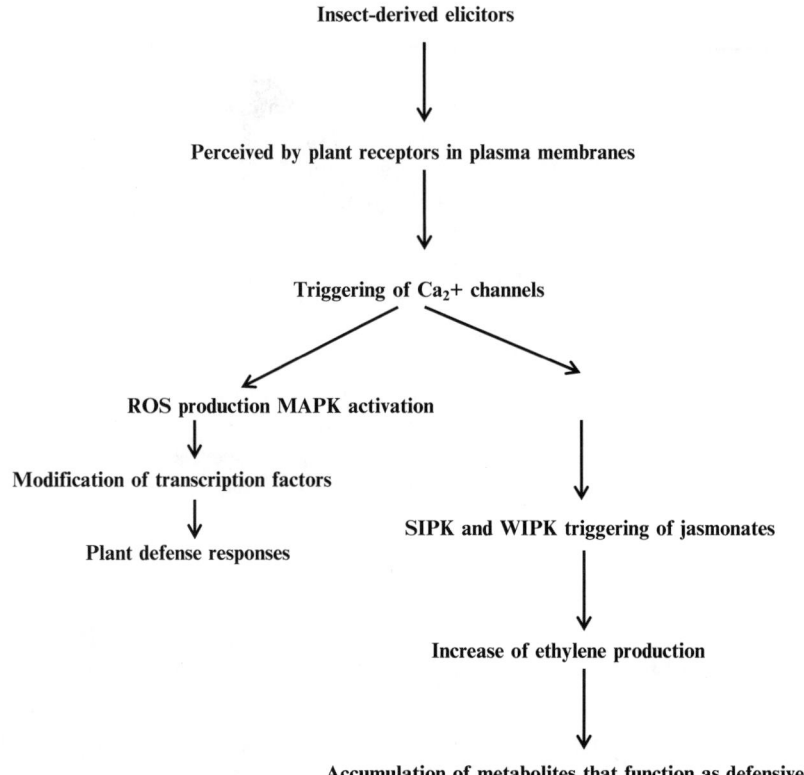

**Fig. 9.1** Plant–insect pest interactions emphasizing those that can be affected by climate change

salicylic acid (De Lucia et al. 2012). The differential effect on defense compounds and the subsequent changes in plant hormones potentially increases susceptibility to the attacking pests and enhances the plant's innate resistance to pathogens (De Lucia et al. 2012). A deeper understanding of the effect of individual factors of climate change and their role in specifically affecting the interactions between plants and insects will enable researchers to design new strategies to develop climate-resilient pest-resistant plants.

## 9.2.2 Effects of Rising Carbon Dioxide Levels on Insect Pests

Carbon is a primary constituent of plant tissues, and elevated $CO_2$ levels aid rapid plant growth due to increased assimilation of carbon, a fact that has been success-fully adapted by greenhouse growers who add $CO_2$ to promote rapid plant growth (Trumble and Butler 2009). Increased $CO_2$ was falsely thought to promote increased photosynthesis and improved crop yields (LaMarche et al. 1984), but this was shown to be an inaccurate hypothesis with the main reason for the absence of potential high yields due to increased feeding by insect herbivores. The increased $CO_2$ levels caused lowered nitrogen content in the leaves, and insects fed more on such leaves to obtain sufficient nitrogen for their metabolism (Hunter 2001). The increased consumption of low-quality food by insect pests to meet nutrient limitations is termed as "compensatory feeding," and this seems to damage plants in both managed and natural ecosystems with rising $CO_2$ levels (Cornelissen 2011).

Studies dating back to more than a decade have examined the effects of enhanced $CO_2$ levels on the plant–insect pest interactions (Bezemer and Jones 1998) using growth chambers, open-top chambers, and, more recently, free air carbon dioxide enrichment (FACE) approaches that create an atmosphere with $CO_2$ and $O_2$ concentrations that mirror climate change models for the middle of twenty-first century (Petzoldt and Seaman 2007). The interactions between insect pests and plants grown under conditions of ambient and elevated $CO_2$ levels were investigated by collecting data on 43 pests that represent 61 plant–insect pest interactions (Bezemer and Jones 1998). The $CO_2$ environmental changes on host plant quality in terms of their nitrogen content, water content, carbohydrate content, and secondary plant compounds were correlated with feeding habits of the pests that were leaf chewers, leaf miners, phloem feeders (root and shoot), xylem feeders, whole cell feeders, and seed eaters as larvae and pupae. Elevated $CO_2$ levels were shown to increase carbon-to-nitrogen ratios in plant tissues that slow insect development and increase the length of life stages that are vulnerable to parasitoid attack (Coviella and Trumble 1999). Hamilton et al. (2005) in a FACE technology mediated study showed that soybeans grown in elevated $CO_2$ atmosphere had a 57 % increase in insect pest damage and attributed the measured increases with levels of simple sugars in soybean leaves to have promoted the increased insect feeding.

The clear correlations seen between $CO_2$ as it affects the food consumption of insect herbivores at various developmental stages was later extended to more realistic climate change scenarios where it was observed that the effects of elevated $CO_2$ levels affecting the defense-related phenolic levels in leaves can be mitigated by elevated temperatures (Zvereva and Kozlov 2006). The $CO_2$-temperature interplay in affecting the insect populations and in turn the spread of certain viruses for which the insects act as vectors was illustrated in a crop model for cereal aphid (*Rhopalosiphum padi*) where an increase in winged forms could result in greater spread and incidence of barley yellow dwarf virus for which this aphid is a vector (Newman 2004). The phenolic and tannin contents of plants that act as a defense to reduce insect feeding by binding to proteins that interfere with the insects' digestive tracts were seen to increase due to elevated $CO_2$ levels (Trumble and Butler 2009).

Plant–insect interactions affected by elevated $CO_2$ levels can have a tangential effect on crop yields as well as soil fertility as shown in legumes (Gregory et al. 2009). Elevated $CO_2$ promotes larger $N_2$-fixing nodules housing $N_2$-fixing bacteria (Soussana and Hartwig 1996). In white clover, the elevated $CO_2$ levels were shown to increase legume nodule size and number, while simultaneously promoting the growth of soil-dwelling *Sitona* spp. weevil larvae (Staley and Johnson 2008) that are known to target legume root nodules (Murray et al. 1996). The parallel events involving the nodule and the nodule-damaging larval development resulted in the declining nitrogen concentrations in the soil, thus demonstrating one case, which would negate the agricultural benefits of legume arable crop rotations due to a changed climate scenario (Gregory et al. 2009).

Examining the effects of elevated $CO_2$ on the molecular pathways and molecular elements of the plant–pest interactions showcased certain interesting results such as in soybean where the elevated $CO_2$ caused downregulation of genes encoding cysteine protease inhibitors that have specific effects on coleopteran pests of corn, *Diabrotica virgifera virgifera* (western corn rootworm) and *Popillia japonica* (Japanese beetle) (Zavala et al. 2008; O'Neill et al. 2010). In tomato, elevated $CO_2$ enhances induced defenses derived from salicylic acid-induced pathogenesis-related (PR) proteins and reduces the jasmonate-signaling defense (Sun et al. 2011; Guo et al. 2012).

## 9.2.3 Effects of Increased Temperature on Insect Pests

The dramatic increases of temperatures around the world that are results of both warmer summer days and less number of winter days (IPCC 2007) have affected insect pests in terms of increased migration patterns (Epstein et al. 1998; Parmesan 2006), developmental rates and oviposition (Regniere 1983), sudden insect outbreaks (Bale et al. 2002), invasive species introductions (Dukes and Mooney 1999), and insect extinctions (Thomas et al. 2004). Temperature increases also were cited as the main reason for decreased effectiveness of insect biocontrol by fungi (Stacey and Fellowes 2002), reduced insect diversity in ecosystems (Erasmus et al. 2002), and decreasing parasitism (Fleming and Volney 1995; Hance et al. 2007).

Some of the best known studies that illustrate the effects of climate change variables on pests to be able to predict the impacts on plant pathology and pest management have been referred to as climate change fingerprinting (Scherm 2004). One of the detailed studies undertaken was the use of historical flight phenology records of aphids (Fleming and Tatchell 1995). This study had important ramifications in suggesting that climate change can potentially increase the spread of plant pathogens by aphid vectors in diverse crops that might result in a dramatic reduction of yields (Malloch et al. 2006; Harrington et al. 2007; Gregory et al. 2009). A potential insect pest outbreak that is expected due to increased temperatures can result in ecological changes related to carbon–nitrogen cycles, biomass decomposition, and energy flows as instances such as premature leaf drops will change the leaf litter composition affecting the success of organisms that decompose biomass (Haack and Byler 1993; Trumble and Butler 2009). Another devastating effect seen in Southern California was the sudden unpredictable intensity in increase of forest fires that was attributed to increased temperatures that led to greater susceptibility of trees to bark beetle attacks and the dead trees fueling huge forest fires (Trumble and Butler 2009). In combination with elevated $CO_2$

levels, increased temperature decreases the nitrogen content of plants, thus increasing insect foliage consumption, and also stimulates defense-related compounds such as jasmonic acid, ethylene, and salicylic acid, thus enhancing the defense mechanisms of plants against insect pest attacks (De Lucia et al. 2012). Elevated temperature increases the production of biogenic volatile compounds (BVOCs) that include plant defense-aiding, pollination-inducing, and communication-aiding chemicals that will have a direct effect on the complex plant–insect pest interactions (Arimura et al. 2009; Penuelas and Staudt 2010).

### 9.2.4 Effects of Rainfall, Precipitation Patterns, and Moisture on Insect Pests

Scientific studies on the effect of rainfall, precipitation patterns, and moisture on insect pests are fewer compared to temperature-related studies (Petzoldt and Seaman 2007). The general practice of farmers has been to note the insect pests that are sensitive to precipitation and choose effective pest management strategies by planting crops in appropriate water conditions as seen in the methods employed to control onion thrips by planting crops during heavy rains (Reiners and Petzoldt 2005) or controlling cranberry insect pests by flooding the soil (Vincent et al. 2003). Unseasonal increase in rainfall has significant effects on both the population and migratory patterns of insect pests (Gregory et al. 2009). Staley et al. (2007) observed a rapid increase of the potato wireworm during a spell of enhanced summer rainfall when compared to ambient and drought conditions confirming earlier speculations of crop researchers about the pest's increased effects on potato crop due to climate change (Parker and Howard 2001; Johnson et al. 2008). Migratory patterns of the desert locust *Schistocerca gregaria* that causes serious crop damage in the semiarid tropics were seen to be increasingly driven by changes in rainfall patterns (Cheke and Tratalos 2007), with decreases in the winter rainfall and increases in the summer and seasonal rainfall being shown most beneficial for the spread of this pest (Hulme et al. 2001). Certain pests such as the pea aphid, *Acyrthosiphon pisum*, decrease in populations during drought conditions (Mcvean and Dixon 2001).

## 9.3 Approaches to Develop Insect-Resistant Climate-Resilient Crops

Insect pests account for approximately 15 % of the world's agricultural produce losses annually making them a major constraint in agricultural production food security (Christou 2005). The widespread use of insecticides that are environmental pollutants and human-livestock toxins also increases the possibility of evolving resistance in insect pests while targeting nontarget insects, thus offsetting the ecological balance. One of the solutions to offset this environmental disaster situation was the use of biological control as illustrated by the use of microbes as an important component of what is known as integrated pest management (IPM, Trumble 1998), but these have resulted in limited successes in large-scale agricultural practices especially if viewed in the perspective of sustainable food production (Bale et al. 2008). Breeding resistant cultivars through both conventional and molecular breeding strategies was another viable option that aims at exploiting the plant's inherent resistance and the genetic and molecular components of the plant–pest interaction mechanisms (Wu and Baldwin 2010). Transgenic technology to genetically improve pest resistance in economically important plants is a complementary alternative strategy to supplement the efforts of plant breeders (Gatehouse 2008; Gatehouse and Gatehouse 1998) and is considered a vital component of IPM (Kos et al. 2009). These and other biotechnological approaches have now come under major reassessment of strategies in the contextual global scenario of climate change. Farmers and agricultural researchers across the world are now experiencing major impacts on insect management control strategies linked with changes in climate, thus present research calls for integrating the pests and pathogens into the climate change/food security debate (Gregory et al. 2009).

## 9.3.1 Classical and Molecular Breeding

Breeding for insect resistance has been a field of research with a momentous history in terms of both gene discovery and generation of cultivars that have fed the world during the hard times that galvanized agricultural efforts such as the Green Revolution. After the rediscovery of Mendel's laws, the breeding of arthropod-resistant plants blossomed as a significant agricultural research endeavor (Painter 1951) resulting in spectacular success in both public and private sectors with notable classical examples such as the Green Revolution's multiple insect-resistant rice cultivar IR36 that catapulted agricultural yields in Asia by $1 billion in 11 million hectares in a period of just 11 years (Panda and Khush 1995; Smith 2005). Classical breeding in the 1970s alone oversaw an effort that resulted in over 500 arthropod-resistant food and fiber cultivars in the USA (Smith 1989), and the advent of molecular tools in the past three decades has showcased a diverse array of developing new cultivars with durable insect pest resistance (Smith 2005) especially with the methodologies of tagging and mapping insect-resistant loci in plants (Yencho et al. 2000).

Plant breeders developing insect-resistant cultivars have a three-pronged approach in studying the mechanisms of insect pest resistance based on a plant-focused classification, namely, an antixenosis framework that examines the insects' non-preference to accepting the plant as a host due to the plant's morphological or chemical factors, an antibiosis framework that examines the resistant plant adversely affecting the life history of the insect, and a tolerance framework where the plant is naturally resistant to the pest (Painter 1941). Based on these considerations, classical breeders have identified categories of resistance documenting 40 arthropod resistance genes characterized by molecular mapping (Smith and Clement 2012), of which 90 % are cases of antibiosis alone or in combination with other factors, with plant tolerance to pests being in less than 10 % of cases. The wealth of information has been used by classical breeders working with cereal, food and forage legumes, fruit, and vegetable cultivars, and resistant genes have been introgressed into breeding lines and cultivars with recent notable successes such as in wheat, the Hessian fly-resistant genes (Berzonsky et al. 2010); in raspberry, the snow drop lectin *GNA* gene for resistance to aphids (Birch et al. 2002); and in apple, the *Er* resistance genes for resistance to aphids (Bus et al. 2008). The efforts by classical breeders are continually being complemented by molecular breeders armed with molecular biological techniques such as DNA markers to develop genetic linkage maps in important crops such as apple, barrel medic, raspberry, and wheat (Smith and Clement 2012). An efficient technology that oversaw significant successes in the last decade is that of quantitative trait loci (QTL) mapping that lists about 10 crop genera covering resistance to 21 arthropod species from the orders of Coleoptera, Hemiptera, and Lepidoptera (Yencho et al. 2000; Smith and Clement 2012). Taking into consideration the three-pronged approach of antixenosis–antibiosis–tolerance, QTLs related to both antixenotic and antibiotic resistance were mapped in a diverse array of plants such as for aphid resistance in wheat (Castro et al. 2005), midge resistance in sorghum (Tao et al. 2003), small brown plant hopper resistance in rice (Duan et al. 2010), corn earworm resistance in soybean (Rector et al. 2000), aphid and whitefly resistance in melon (Boissot et al. 2010), and aphid resistance in apple (Stoeckli et al. 2008).

More recently, QTL mapping accounted for the 2,4-dihydroxy-7-methoxy-2H-1,4-benzoxazin-3-one (DIMBOA) metabolic pathway *bx* genes (Butron et al. 2010) and the *p1* locus as a key to maysin biosynthesis that regulates flavone synthesis in maize (Zhang et al. 2003).

Plant and arthropod genomic researchers complementing their efforts of classical and molecular breeding techniques have successfully bred cultivars with arthropod-resistant genes, and if we look at these success stories in the context of climate change, almost all the characteristics listed above are affected by elevated $CO_2$ levels, temperature, and precipitation patterns (Gregory et al. 2009). Recent research that shows elevated $CO_2$ and temperature affecting plant-defensive compounds as well as insect development and migration patterns (De Lucia et al. 2012) calls to attention to the fact that research into plant–insect pest interactions needs to be seen in the light of long-term climate change factors. This calls for renewed and focused efforts of both classical and molecular breeders to revisit their gene databases and update the genetic and molecular data in relation to climate change factors. Decisions based upon well-documented pest and crop management records collected over time, along with evolving economics and environmental impacts of pest control, will determine the feasibility of redesigning pest management strategies, or growing certain crops in a particular climate will showcase the efforts of developing pest-resistant, climate-resilient crops (Petzoldt and Seaman 2007; Gregory et al. 2009).

## 9.3.2   Transgenic Approaches

Transgenic technology has been at the forefront of developing insect pest-resistant plants with the first known commercial success story of insect-resistant *Bacillus thuringiensis* (*Bt*) plants in corn that also served as a sounding board for the public and political debate on genetically modified plants' acceptance. Transgenic *Bt* cotton, maize, and rice resistant to a range of lepidopteran pests are now major components of agriculture with well-documented insect virulence with an extreme degree of antibiosis (Tabashnik et al. 2003). *Bt* crops are also model systems to discuss fitness costs in the fields (Gassmann et al. 2009), and many of the findings from well-documented studies related to area-wide suppression of insect pest control using *Bt* plants (Hutchison et al. 2010) can be extended to include climate change parameters. One significant study examined the interactions of elevated $CO_2$ and nitrogen fertilization effects on the production of *Bt* toxins in transgenic cotton, and it was observed that $CO_2$ level elevation from ambient 370 ppm to 900 ppm indicated reduced *Bt* toxin production and the increased leaf consumption was not compensated even if the plants were treated with increased contents of nitrogen fertilizers (Coviella et al. 2000). Recent studies focused on improving the target insect spectrum of *Bt* crops by extending the transgene pyramiding by fusing nontoxic ricin B-chain molecule to *Bt* transgene constructs (Mehlo et al. 2005) for a sustainable and durable insect pest resistance; for wider acceptance of this technology in the farming community, see Christou (2005).

Insect-resistant plants comprised 15 % of global transgenic plants (that were 8 % of total agricultural crops) in 2008 (James 2008), and they were also seen as vital components of integrated pest management due to their potential qualities of being environmentally benign yet durable pest management systems (Kos et al. 2009). Table 9.1 lists important milestones achieved in generating transgenic plants resistant to insect pests. A majority of the examples are related to identifying proteins with insecticidal activities that were genetically engineered into economically important crops (Gatehouse and Gatehouse 1998). The practical success of a limited number of transgenic insect pest-resistant crops in some countries has also been parallel to a not so successful public and political acceptance of this technology, and recent efforts are focusing on educating the farmers and general public about both economic and environmental benefits of this technology (Gatehouse 2008). Integrating these efforts into the realm of climate change debate that is also facing the same consequences in terms of public acceptance cannot come at a more appropriate time.

**Table 9.1** Transgenic plants expressing insecticidal plant genes

| Plant | Genes | Insect target | Reference |
|---|---|---|---|
| Cotton | Bt Cry1Ac + Cry2Ab | Helicoverpa zea | Stewart et al. (2001) |
| | | Spodoptera frugiperda | Chitkowski et al. (2003) |
| | | Heliothis virescens | Gahan et al. (2005) |
| Tobacco | CpTI | Heliothis virescens | Hilder et al. (1987) |
| | Pot PI II | Lepidoptera | Johnson et al. (1989) |
| | GNA | Myzus persicae | Hilder et al. (1995) |
| | p-lec | H. virescens | Boulter et al. (1990) |
| | CpTI + p-lec | H. virescens | Boulter et al. (1990) |
| | Na PI | Helicoverpa punctigera | Heath et al. (1997) |
| | BtCry2Aa2 | Helicoverpa armigera | Kota et al. (1999) |
| | Bt Cry2Aa2 | Helicoverpa armigera | De Cosa et al. (2001) |
| Potato | CpTI | Lacanobia oleracea | Gatehouse et al. (1997) |
| | GNA | L. oleracea | Gatehouse et al. (1997) |
| | | M. persicae | Gatehouse et al. (1996) |
| | | Aulacorthum solani | Down et al. (1996) |
| | GNA + BCH | M. persicae | Gatehouse et al. (1996) |
| | Bt Cry1Ba + Cry IIa | Phthomera opercula | Naimov et al. (2003) |
| | | Leptinotarsa decemlineata | |
| Rice | Pot PI II | Sesamia inferens | Duan et al. (1996) |
| | | Chilo suppressalis | Duan et al. (1996) |
| | CpTI | Sesamia inferens | Xu et al. (1996) |
| | | Chilo suppressalis | Xu et al. (1996) |
| | GNA | Nilaparvata lugens | Rao et al. (1998) |
| | | Nephotettix virescens | Foissac et al. (2000) |
| | ASA-L | Hemiptera | Saha et al. (2006) |
| Strawberry | CpTI | Otiorhynchus sulcatus | Graham et al. (1995) |
| Pea | α-AI | Zabrotes subfasciatus | Shade et al. (1994) |
| | α-AI | Bruchus pisorum | Morton et al. (2000) |
| Azuki bean | α-AI | Callosobruchus chinensis | Ishimoto et al. (1996) |
| Poplar | OC-I | Chrysomela tremulae | Leple et al. (1995) |
| Soybean | Bt Cry1Ab | Lepidoptera | Dufourmantel et al. (2005) |
| Maize | Bt Cry34/35 | Diabrotica virgifera | Moellenbeck et al. (2001) |
| | Bt Cry genes | Lepidoptera | Grainnet (2007) |
| | Bt Cry3Bb1 | Rootworm | Vaughn et al. (2005) |
| | Avidin | Coleoptera | Kramer et al. (2000) |
| | RNAi ATPase | Coleoptera | Baum et al. (2007) |
| Broccoli | Bt Cry1Ac/1C | Plutella xylostella | Zhao et al. (2005) |

## 9.3.3 Modifying Integrated Pest Management

Integrated pest management integrates biological controls such as the predators, parasites, and pathogens; chemical controls in the form of pesticides; and cultural controls such as resistant crop varieties and planting times in focused attempts to reduce insect populations in the field to circumvent economic losses (Trumble and

Butler 2009). If we closely look at each of the factors in the management strategy, each of them is directly affected by climate changes. Recent studies have shown that modest increases in temperature reduce effectiveness of insect pathogens (Stacey and Fellowes 2002) and also the pest suppression provided by parasites (Hance et al. 2007). Increasing temperatures also were found to favor insects with multiple generations in each farming season as opposed to those with a single generation (Bale et al. 2002). Lessons can also be learned from the research done in insects that are vectors of human pathogens and how their life cycles and spread are interlinked with climate change factors such as temperature and flooding (Trumble and Butler 2009) to redesign the integrated pest management strategies currently being employed by farmers and agricultural researchers.

## 9.4   Climate, Insect Resistance, and Food Security

The impact of insect pests on crop yields and as a global limiting factor in terms of food security is limited, but research studies have attested the role of pests and diseases as both yield limitation factors and early indicators of environmental changes due to their testable variables such as short generation times, high reproductive rates, and geographical migration patterns (Scherm et al. 2000). Insects were valuable in developing models of current and future climate changes by documenting their distribution using climatic mapping in ecological niches (Baker et al. 2000). Insect pests such as aphids were also seen to show varied appearance and outbreaks with changing seasons (Gregory et al. 2009), and this unpredictable change interrelated to climate change factors indicates greater consequences for agrosystems and agricultural yields. This calls for a reassessment in terms of research studies to integrate insect pest–plant interaction research into crop productivity in the context of global climate change. In a model that considers agronomic research in a climate change context linked to food security policy (Ingram et al. 2008; Gregory et al. 2009), it has been emphasized that the mechanistic inclusion of pests and disease effects on crop systems will enable researchers to make realistic predictions of geographically specific crop yields, thus assisting the development of robust regional food security policies. A systematic monitoring of plant–insect pest interactions to include climate change as an important determinant with a changed research emphasis in both the breeding and transgenic areas of developing insect-resistant plants and revisiting the fundamental concepts of the genetic and molecular bases of plant–insect pest interactions in the light of the effect of climate change will be a new focus for research that will determine future policies of food security.

# References

Arimura G-I, Matsui K, Takabayashi J (2009) Chemical and molecular ecology of herbivore induced plant volatiles: proximate factors and their ultimate functions. Plant Cell Physiol 50:911–923

Baker RHA, Sansford CE, Jarvis CH, Canon RJC, MacLeod A, Walters KFA (2000) The role of climatic mapping in predicting the potential geographical distribution of non-indigenous pests under current and future climates. Agric Ecosyst Environ 82:57–71

Bale JS, Masters GJ, Hodkinson ID, Awmack C, Bezemer M et al (2002) Herbivory in global climate change research: direct effects of rising temperature on insect herbivores. Glob Change Biol 8:1–16

Bale JS, van Lanteren JC, Bigler F (2008) Biological control and sustainable food production. Phil Trans R Soc Lond Sr B 363:761–776

Baum JA, Bogaert T, Clinton W, Heck GR, Feldmann P, Ilagan O, Johnson S, Plaetinck G, Munyikwa T, Pleau M et al (2007) Control of coleopteran insect pests through RNA interference. Nat Biotechnol 25:1322–1326

Berzonsky WA, Ding H, Haley SD, Harris MO, Lamb RJ et al (2010) Breeding wheat for resistance to insects. Plant Breed Rev 22. doi:10.1002/9780470650202.ch5

Bezemer TM, Jones TH (1998) Plant–insect herbivore interactions in elevated atmospheric $CO_2$: quantitative analyses and guild effects. Oikos 82:212–222

Birch ANE, Jones AT, Fenton B, Malloch G, Geoghegan I et al (2002) Resistance-breaking raspberry aphid biotypes: constraints to sustainable control through plant breeding. Acta Hortic 585:315–317

Boissot N, Thomas S, Sauvion N, Marchal C, Pavis C, Dogimont C (2010) Mapping and validation of QTLs for resistance to aphids and whiteflies in melon. Theor Appl Genet 121:9–20

Boulter D, Edwards GA, Gatehouse AMR, Gatehouse JA, Hilder VA (1990) Additive protective effects of different plant-derived insect resistance genes in transgenic tobacco plants. Crop Prot 9:351–354

Bus VGM, Chagne D, Bassett HCM, Bowatte D, Calenge F et al (2008) Genome mapping of three major resistance genes to woolly apple aphid (*Eriosoma lanigerum* Hausm.). Tree Genet Genomes 4:233–236

Butron A, Chen YC, Rottinghaus GE, McMullen MD (2010) Genetic variation at *bx1* controls DIMBOA content in maize. Theor Appl Genet 120:721–734

Castro AM, Vasicek A, Manifiesto M, Gimenez DO, Tacaliti MS et al (2005) Mapping antixenosis genes on chromosome 6A of wheat to greenbug and to a new biotype of Russian wheat aphid. Plant Breed 124:229–233

Cheke RA, Tratalos JA (2007) Migration, patchiness, and population processes illustrated by two migrant pests. Bioscience 57:145–153

Chitkowski RL, Turnipseed SG, Sullivan MJ, Bridges WC (2003) Field and laboratory evaluations of transgenic cottons expressing one or two *Bacillus thuringiensis* var. kurstaki Berliner proteins for management of noctuid (Lepidoptera) pests. J Econ Entomol 96:755–762

Christou P (2005) Sustainable and durable insect pest resistance in transgenic crops. http://www.isb.vt.edu/news/2005/artspdf/aug0503.pdf

Coakley SM, Scherm H, Chakraborty S (1999) Climate change and disease management. Annu Rev Phytopathol 37:399–426

Cornelissen T (2011) Climate change and its effects on terrestrial insects and herbivory patterns. Neotrop Entomol 40:155–163

Coviella C, Trumble J (1999) Effects of elevated atmospheric carbon dioxide on insect plant interactions. Conserv Biol 13:700–712

Coviella CE, Morgan DJ, Trumble JT (2000) Interactions of elevated CO2 and nitrogen fertilization: effects on production of *Bacillus thuringiensis* toxins in transgenic plants. Environ Entomol 29:781–787

De Cosa B, Moar W, Lee SB, Miller M, Daniell H (2001) Overexpression of the Bt cry2Aa2 operon in chloroplasts leads to formation of insecticidal crystals. Nat Biotechnol 19:71–74

De Lucia E, Nabity P, Zavala J, Berenbaum M (2012) Climate change: resetting plant-insect interactions. Plant Physiol. doi:10.1104/pp. 112.204750

Down RE, Gatehouse AMR, Hamilton DO, Gatehouse JA (1996) Snowdrop lectin inhibits development and decreases fecundity of the Glasshouse Potato Aphid (Aulacorthum solani) when administered in vitro and via transgenic plants both in laboratory and glasshouse trials. J Insect Physiol 42:1035–1045

Duan X, Li X, Xue Q, Abo-El-Saad M, Xu D, Wu R (1996) Transgenic rice plants harboring an introduced potato proteinase inhibitor II gene are insect resistant. Nat Biotechnol 14:494–498

Duan CX, Su N, Cheng ZJ, Lei CL, Wang JL et al (2010) QTL analysis for the resistance to small brown planthopper (*Laodelphax striatellus* Fallen) in rice using backcross inbred lines. Plant Breed 129:63–67

Dufourmantel N, Tissot G, Goutorbe F, Garcon F, Muhr C, Jansens S, Pelissier B, Peltier G, Dubald M (2005) Generation and analysis of soybean plastid transformants expressing *Bacillus thuringiensis* Cry1Ab protoxin. Plant Mol Biol 58:659–668

Dukes JS, Mooney HA (1999) Does global change increase the success of biological invaders? Trends Ecol Evol 14:135–139

Epstein PR, Diaz HF, Elias S, Grabherr G, Graham NE et al (1998) Biological and physical signs of climate change: focus on mosquito borne diseases. Bull Am Meteorol Soc 79:409–417

Erasmus BFN, van Jaarsveld AS, Chown S, Kshatriya M, Wessels KJ (2002) Vulnerability of South African animal taxa to climate change. Glob Change Biol 8:679–693

Fleming RA, Tatchell GM (1995) Shifts in the flight season of British aphids: a response to climate warming? In: Harrington R, Stork NE (eds) Insects in a changing environment. Academic, London, pp 505–508

Fleming RA, Volney WJA (1995) Effects of climate change on insect defoliator population processes in Canada boreal forest—some plausible scenarios. Water Air Soil Poll 82:445–454

Foissac X, Loc NT, Christou P, Gatehouse AMR, Gatehouse JA (2000) Resistance to green leafhopper (*Nephotettix virescens*) and brown planthopper (*Nilaparvata lugens*) in transgenic rice expressing snowdrop lectin (*Galanthus nivalis* agglutinin; GNA). J Insect Physiol 46:573–583

Gahan LJ, Ma YT, Coble MLM, Gould F, Moar WJ, Heckel DG (2005) Genetic basis of resistance to Cry1Ac and Cry2Aa in *Heliothis virescens* (Lepidoptera: Noctuidae). J Econ Entomol 98:1357–1368

Garrett KA, Dendy SP, Frank EE, Rouse MN, Travers SE (2006) Climate change effects on plant disease: genomes to ecosystems. Annu Rev Phytopathol 44:489–509

Gassmann AJ, Carriere Y, Tabashnik BE (2009) Fitness costs of insect resistance to *Bacillus thuringiensis*. Annu Rev Entomol 54:147–163

Gatehouse JA (2008) Biotechnological prospects for engineering insect-resistant plants. Plant Physiol 146:881–887

Gatehouse AMR, Gatehouse JA (1998) Identifying proteins with insecticidal activity: use of encoding genes to produce insect-resistant transgenic crops. Pestic Sci 52:165–175

Gatehouse AMR, Down RE, Powell KS, Sauvion N, Rahbe Y, Newell CA, Merryweather A, Hamilton WDO, Gatehouse JA (1996) Transgenic potato plants with enhanced resistance to the peach-potato aphid *Myzus persicae*. Entomol Exp Appl 34:295–307

Gatehouse AMR, Davison GM, Newell CA, Merryweather A, Hamilton WDO, Burgess EPJ, Gilbert RJC, Gatehouse JA (1997) Transgenic potato plants with enhanced resistance to the tomato moth, *Lacanobia oleracea*: growth room trials. Mol Breed 3:49–63

Graham J, McNicol RJ, Greig K (1995) Towards genetic based insect resistance in strawberry using the cowpea trypsin inhibitor. Ann Appl Biol 127:163–173

Grainnet (2007) Monsanto and Dow agrosciences launch "SmartStax", industry's first-ever eight-gene stacked combination in corn. http://www.grainnet.com/

Gregory RD, Willis SG, Jiguet F, Voříšek P, Klvaňová A et al (2009) An indicator of the impact of climatic change on European bird populations. PLoS One 4:e4678

Guo H, Sun Y, Ren Q, Zhu-Salzman K, Kang L, Wang C, Li C, Ge F (2012) Elevated $CO_2$ reduces the resistance and tolerance of tomato plants to *Helicoverpa armigera* by suppressing the JA signaling pathway. PLoS One 7(7):e41426

Haack RA, Byler JW (1993) Insects and pathogens: regulators of forest ecosystems. J For 1:32–37

Hamilton JM, Maddison DJ, Tol RSJ (2005) Climate change and international tourism: a simulation study. Global Environ Change 15:253–266

Hance T, van Baaren J, Vernon P, Boivin G (2007) Impact of extreme temperatures on parasitoids in a climate change perspective. Annu Rev Entomol 52:107–126

Harrington R, Clark SJ, Welham SJ, Verrier PJ, Denholm CH, Hulle M, Maurice D, Rounsevell MD, Cocu N (2007) Environmental change and the phenology of European aphids. Glob Change Biol 13:1550–1564

Heath R, McDonald G, Christeller JT, Lee M, Bateman K, West J, van Heeswijck R, Anderson MA (1997) Proteinase inhibitors from *Nicotiana alata* enhance plant resistance to insect pests. J Insect Physiol 43:833–842

Hilder VA, Gatehouse AMR, Sherman SE, Barker RF, Boulter D (1987) A novel mechanism for insect resistance engineered into tobacco. Nature 330:160–163

Hilder VA, Powell KS, Gatehouse AMR, Gatehouse JA, Gatehouse LN, Shi Y, Hamilton WDO, Merryweather A, Newell CA, Timans JC, Peumans WJ, van Damme E, Boulter D (1995) Expression of snowdrop lectin in transgenic tobacco plants results in added protection against aphids. Transgenic Res 4:18–25

Howe GA, Jander G (2008) Plant immunity to insect herbivores. Annu Rev Plant Biol 59:41–66

Hulme M, Dohrety R, Ngara T, New M, Lister D (2001) African climate change 1900–2100. Clim Res 17:145–168

Hunter MD (2001) Insect population dynamics meets ecosystem ecology: effects of herbivory on soil nutrient dynamics. Agric For Entomol 3:77–84

Hutchison WD, Burkness EC, Mitchell PD, Moon RD, Leslie TW et al (2010) Areawide suppression of European corn borer with *Bt* maize reaps savings to non-*Bt* maize growers. Science 330:222–225

Ingram JSI, Gregory PJ, Izac A-M (2008) The role of agronomic research in climate change and food security policy. Agric Ecosyst Environ 126:4–12

Intergovernmental Panel on Climate Change (2007) Climate change 2007: the physical science basis. Summary for policymakers. Contribution of working group I to the 3rd assessment report of the IPCC. IPCC Secretariat, Geneva

IPCC (2001) In: Houghton JT, Ding Y, Griggs DJ, Noguer M, van der Linden PJ et al (eds) Climate change 2001: the scientific basis. Contribution of working group to the third assessment report of the intergovernmental panel on climate change. Cambridge University Press, Cambridge, 94 p

Ishimoto M, Sato T, Chrispeels MJ, Kitamura K (1996) Bruchid resistance of transgenic azuki bean expressing seed a-amylase inhibitor of common bean. Entomol Exp Appl 79:309–315

James C (2008) Global status of commercialized biotech/GM crops: 2008 (ISAAA Brief 39), International Service for the Acquisition of Agri-biotech Applications. http://isaaa.org/resources/publications/briefs/39/download/isaaa-brief-39-2008.pdf

Johnson R, Narvaez J, An G, Ryan C (1989) Expression of proteinase inhibitors I and II in transgenic tobacco plants: effects on natural defense against *Manduca sexta* larvae. Proc Natl Acad Sci USA 86:9871–9875

Johnson SN, Anderson A, Dawson G, Griffiths DW (2008) Varietal susceptibility of potatoes to wireworm herbivory. Agric For Entomol 10:167–174

Kos M, van Loon JJA, Dicke M, Vet EM (2009) Transgenic plants as vital components of integrated pest management. Trends Biotechnol 27:621–627

Kota M, Daniell H, Varma S, Garczynski SF, Gould F, Moar WJ (1999) Overexpression of the *Bacillus thuringiensis* (*Bt*) Cry2Aa2 protein in chloroplasts confers resistance to plants against susceptible and Bt-resistant insects. Proc Natl Acad Sci USA 96:1840–1845

Kramer KJ, Morgan TD, Throne JE, Dowell FE, Bailey M, Howard JA (2000) Transgenic avidin maize is resistant to storage insect pests. Nat Biotechnol 18:670–674

LaMarche VC, Grabyll DA, Fritts HC, Rose MR (1984) Increasing atmospheric carbon dioxide: tree ring evidence for growth enhancement in natural vegetation. Science 225:1019–1021

Lamb C, Dixon RA (1997) The oxidative burst in plant disease resistance. Annu Rev Plant Physiol Plant Mol Biol 48:251–275

Leple JC, Bonade-Bottino M, Augustin S, Pilate G, Dumanois G, Le Tan VD, Delplanque A, Cornu D, Jouanin L (1995) Toxicity to *Chrysomela tremulae* (Coleoptera: *Chrysomelidae*) of transgenic poplars expressing a cysteine proteinase inhibitor. Mol Breed 1:319–328

Malloch G, Highet F, Kasprowicz L, Pickup J, Neilson R, Fenton B (2006) Microsatellite marker analysis of peach-potato aphids (*Myzus persicae*, Homoptera: *Aphididae*) from Scottish suction traps. Bull Entomol Res 96:573–582

Mcvean R, Dixon AFG (2001) The effect of plant drought-stress on populations of the pea aphid *Acyrthosiphon pisum*. Ecol Entomol 26:440–443

Mehlo L, Gahakwa D, Nghia P-T, Loc N-T, Capell T, Gatehouse J, Gatehouse A, Christou P (2005) An alternative strategy for sustainable pest resistance in genetically enhanced crops. Proc Natl Acad Sci USA 102:7812–7816

Moellenbeck DJ, Peters ML, Bing JW, Rouse JR, Higgins LS, Sims L, Nevshemal T, Marshall L, Ellis RT, Bystrak PG et al (2001) Insecticidal proteins from *Bacillus thuringiensis* protect corn from corn rootworms. Nat Biotechnol 19:668–672

Morton RL, Schroeder HE, Bateman KS, Chrispeels MJ, Armstrong E, Higgins TJV (2000) Bean alpha-amylase inhibitor 1 in transgenic peas (*Pisum sativum*) provides complete protection from pea weevil (*Bruchus pisorum*) under field conditions. Proc Natl Acad Sci USA 97:3820–3825

Murray PJ, Hatch DJ, Cliquet JB (1996) Impact of insect root herbivory on the growth and nitrogen and carbon contents of white clover (*Trifolium repens*) seedlings. Can J Bot 74:1591–1595

Naimov S, Dukiandjiev S, de Maagd RA (2003) A hybrid *Bacillus thuringiensis* delta-endotoxin gives resistance against a coleopteran and a lepidopteran pest in transgenic potato. Plant Biotechnol J 1:51–57

Newman JA (2004) Climate change and cereal aphids: the relative effects of increasing $CO_2$ and temperature on aphid population dynamics. Glob Change Biol 10:5–15

O'Neill BF, Zangerl AR, Dermody O, Bilgin DD, Casteel CL, Zavala JA, DeLucia EH, Berenbaum MR (2010) Impact of elevated levels of atmospheric CO2 and herbivory on flavonoids of soybean (*Glycine max* L.). J Chem Ecol 36:35–45

Painter RH (1941) The economic value and biologic significance of insect resistance in plants. J Econ Entomol 34:358–367

Painter RH (1951) Insect resistance in crop plants. University of Kansas Press, Lawrence, KS, 520 p

Panda N, Khush GS (1995) Host plant resistance to insects. CABI/IRRI, Wallingford/Manila, 431 p

Parker WE, Howard J (2001) The biology and management of wireworms (*Agriotes* spp.) on potato with particular reference to the UK. Agric For Entomol 3:85–98

Parmesan C (2006) Ecological and evolutionary responses to recent climate change. Annu Rev Ecol Evol Syst 37:637–669

Parry ML, Rosenzweig C, Iglesias A, Livermore M, Fischer G (2004) Effects of climate change on global food production under SRES emissions and socio-economic scenarios. Glob Environ Change 14:53–67

Penuelas J, Staudt M (2010) BVOCs and global change. Trends Plant Sci 15:133–144

Petzoldt C, Seaman A (2007) Climate change effects on insects and pathogens. Fact Sheet. http://www.climateandfarming.org/pdfs/FactSheets/III.2Insects.Pathogens.pdf

Rao KV, Rathore KS, Hodges TK, Fu X, Stoger E, Sudhakar D, Williams S, Christou P, Bharathi M, Bown DP et al (1998) Expression of snowdrop lectin (GNA) in transgenic rice plants confers resistance to rice brown planthopper. Plant J 15:469–477

Rector BG, All JN, Parrott WA, Boerma HR (2000) Quantitative trait loci for antibiosis resistance to corn earworm in soybean. Crop Sci 40:233–238

Regniere J (1983) An oviposition model of the spruce budworm, *Choristoneura fumiferana* (Lepidoptera: Tortricidae). Can Entomol 115:1371–1382

Reiners S, Petzoldt C (2005) Integrated crop and pest management guidelines for commercial vegetable production. Cornell Cooperative Extension publication #124VG. http://www.nysaes. cornell.edu/recommends/

Saha P, Majumder P, Dutta I, Ray T, Roy SC, Das S (2006) Transgenic rice expressing *Allium sativum* leaf lectin with enhanced resistance against sap-sucking insect pests. Planta 223:1329–1343

Scherm H (2004) Climate change: can we predict the impacts on plant pathology and pest management? Can J Plant Pathol 26:267–273

Scherm H, Sutherst RW, Harrington R, Ingram JSI (2000) Global networking for assessment of impacts of global change on plant pests. Environ Poll 108:333–341

Shade RE, Schroeder HE, Pueyo JJ, Tabe LM, Murdock LL, Higgins TJV, Chrispeels MJ (1994) Transgenic peas expressing the α-amylase inhibitor of the common bean are resistant to bruchid beetles. Biotechnology 12:793–796

Smith CM (1989) Plant resistance to insects: a fundamental approach. Wiley, New York, 286 p

Smith CM (2005) Plant resistance to arthropods: molecular and conventional approaches. Springer, Dordrecht, 423 p

Smith CM, Clement SL (2012) Molecular bases of plant resistance to arthropods. Annu Rev Entomol 57:309–328

Soussana JF, Hartwig UA (1996) The effects of elevated $CO_2$ on symbiotic $N_2$ fixation: a link between the carbon and nitrogen cycles in grassland ecosystems. Plant Soil 187:321–332

Stacey DA, Fellowes MDE (2002) Influence of temperature on pea aphid *Acyrthosiphon pisum* (Hemiptera: Aphididae) resistance to natural enemy attack. Bull Entomol Res 92:351–357

Staley JT, Johnson SN (2008) Climate change impacts on root herbivores. In: Johnson SN, Murray PJ (eds) Root feeders: an ecosystem perspective. CABI, Wallingford, UK, pp 192–213

Staley JT, Hodgson CJ, Mortimer SR, Morecroft MD, Masters GJ, Brown VK, Taylor ME (2007) Effects of summer rainfall manipulations on the abundance and vertical distribution of herbivorous soil macro-invertebrates. Eur J Soil Biol 43:189–198

Stewart SD, Adamczyk JJ, Knighten KS, Davis FM (2001) Impact of Bt cottons expressing one or two insecticidal proteins of *Bacillus thuringiensis* Berliner on growth and survival of noctuid (Lepidoptera) larvae. J Econ Entomol 94:752–760

Stoeckli S, Mody K, Gessler C, Patocchi A, Jermini M, Dorn S (2008) QTL analysis for aphid resistance and growth traits in apple. Tree Genet Genomes 4:833–847

Sun Y, Yin J, Cao H, Li C, Kang L, Ge F (2011) Elevated $CO_2$ influences nematode-induced defense responses of tomato genotypes differing in the JA pathway. PLoS One 6(5):e19751

Tabashnik BE, Carriere Y, Dennehy TJ, Morin S, Sisterson MS et al (2003) Insect resistance to transgenic *Bt* crops: lessons from the laboratory and field. J Econ Entomol 96:1031–1038

Tao YZ, Hardy A, Drenth J, Henzell RG, Franzmann BA et al (2003) Identifications of two different mechanisms for sorghum midge resistance through QTL mapping. Theor Appl Genet 107:116–122

Thomas CD, Cameron A, Green RE, Bakkenes M, Beaumont LJ et al (2004) Extinction risk from climate change. Nature 427:145–148

Trumble JT (1998) IPM: overcoming conflicts in adoption. Integr Pest Manage Rev 3:195–207

Trumble JT, Butler CD (2009) Climate change will exacerbate California's insect pest problems. Cal Agric 63(2):73–78

Vaughn T, Cavato T, Brar G, Coombe T, DeGooyer T, Ford S, Groth M, Howe A, Johnson S, Kolacz K et al (2005) A method of controlling corn rootworm feeding using a *Bacillus thuringiensis* protein expressed in transgenic maize. Crop Sci 45:931–938

Vincent C, Hallman G, Panneton B, Fleurat-Lessardú F (2003) Management of agricultural insects with physical control methods. Annu Rev Entomol 48:261–281

Von Braun J (2007) The world food situation: new driving forces and required actions. International Food Policy Research Institute, Washington, DC

Wu J, Baldwin IT (2010) New insights into plant responses to the attack from insect herbivores. Annu Rev Genet 44:1–24

Xu DP, Xue QZ, McElroy D, Mawal Y, Hilder VA, Wu R (1996) Constitutive expression of a cowpea trypsin-inhibitor gene, CpTI, in transgenic rice plants confers resistance of two major rice insect pests. Mol Breed 2:167–173

Yencho GC, Cohen MB, Byrne PF (2000) Applications of tagging and mapping insect resistance loci in plants. Annu Rev Entomol 45:393–422

Zavala JA, Casteel CL, DeLucia EH, Berenbaum MR (2008) Anthropogenic increase in carbon dioxide compromises plant defense against invasive insects. Proc Natl Acad Sci USA 105:5129–5133

Zhang P, Wang Y, Zhang J, Maddock S, Snook M, Peterson T (2003) A maize QTL for silk maysin levels contains duplicated Myb-homologous genes which jointly regulate flavone biosynthesis. Plant Mol Biol 52:1–15

Zhao JZ, Cao J, Collins HL, Bates SL, Roush RT, Earle ED, Shelton AM (2005) Concurrent use of transgenic plants expressing a single and two *Bacillus thuringiensis* genes speeds insect adaptation to pyramided plants. Proc Natl Acad Sci USA 102:8426–8430

Zvereva EL, Kozlov MV (2006) Consequences of simultaneous elevation of carbon dioxide and temperature for plant-herbivore interactions: a metaanalysis. Glob Change Biol 12:27–41

# Chapter 10
# Nutrient Use Efficiency

Glenn McDonald, William Bovill, Chunyuan Huang, and David Lightfoot

**Abstract** Much of the recent gains in global crop production have been underpinned by greater use of fertilizer, especially nitrogen and phosphorus, and continued improvements in plant nutrition will be needed to meet the increasing demands for food and fiber from a growing world population. Climate change presents many challenges to improvements in nutrient use efficiency by its direct effects on the growth and yield of plants, and hence on nutrient demand, and by its influence on soil nutrient cycling, nutrient availability, and uptake. However, the consequences of climate change on plant nutrition are difficult to predict because of the complexity of the soil–plant–atmosphere system. Empirical data suggests that enhanced as well as reduced nutrient availability and uptake may occur as a result of climate change, depending on the nutrient in question and the component of the climate that changes. Notwithstanding the uncertainty of the effects of climate change on soil nutrient availability and plant nutrient uptake, improvements in nutrient use efficiency will be required to sustain productivity into the future.

Over significant areas of the world's arable land, high inputs of nutrients have increased soil nutrient reserves and fertilizer use efficiency is low, while in other

G. McDonald (✉)
School of Agriculture, Food and Wine, The University of Adelaide; Waite Campus, PMB 1 Glen Osmond, South Australia 5064, Australia
e-mail: glenn.mcdonald@adelaide.edu.au

W. Bovill
CSIRO Plant Industry, CSIRO, Canberra, ACT, Australia
e-mail: william.bovill@csiro.au

C. Huang
Australian Centre for Plant Functional Genomics, The University of Adelaide, SA, Australia
e-mail: chunyuan.huang@acpfg.com.au

D. Lightfoot
Department of Plant, Soil and General Agriculture, Mailcode 4415, Southern Illinois University-Carbondale, Carbondale, IL 62901-4415, USA
e-mail: ga4082@siu.edu

C. Kole (ed.), *Genomics and Breeding for Climate-Resilient Crops*, Vol. 2,        333
DOI 10.1007/978-3-642-37048-9_10, © Springer-Verlag Berlin Heidelberg 2013

regions, impoverished soils and low rates of fertilizer use have limited the capacity of farmers to provide adequate amounts of nutritious food. Developing varieties with enhanced nutrient use efficiency provides a way of improving productivity in both situations, although the traits that are targeted may differ. The two pathways by which nutrient use efficiency can be improved are by better uptake efficiency or by enhanced utilization efficiency. The relative importance of these strategies will reflect the amount and availability of nutrients in the soil. Genetic variation in nutrient use efficiency in plants is well documented, but improvements in nutrient use efficiency in the major food crops so far have been modest. Reasons why progress has been limited include inconsistent and sometimes confusing definitions of nutrient use efficiency, incomplete understanding of the genetic and physiological bases of differences in nutrient use efficiency, lack of field validation of assays, and little consideration of genotype × environment interactions in the expression of nutrient use efficiency. However, currently a powerful array of molecular and genomic techniques promises considerable advances in understanding nutrient use efficiency and developing varieties that are more nutrient efficient. Combined with traditional disciplines of plant breeding, crop physiology, and agronomy, new opportunities are developing to study genetic differences in nutrient use efficiency and to allow agriculture to meet the challenges of increased production of quality grain in a variable environment.

## 10.1   Introduction

The changes in climate that are predicted to occur during the next century present many challenges to sustainable crop production and food security. A burgeoning world population accompanied by increasing standards of living will require unprecedented levels of production of food, fiber, and industrial crops. This needs to be achieved with little further increase in the area of arable land and with finite and increasingly expensive supplies of fertilizer. Greater productivity needs to occur at a time when large areas of the world's agricultural land will experience increases in the frequency and severity of heat stress and drought (IPCC 2007; Handmer et al. 2012). Improvements are also needed in the nutrient content of the major stable food crops to alleviate the chronic nutritional problems that occur in many countries, but particularly in developing countries (Graham et al. 2001, 2007).

These challenges have brought the importance of plant nutrition to sustainable agricultural production into sharp focus and have highlighted the need to improve nutrient use efficiency (NTUE). The higher yields that will be required to maintain (or improve) food security will require increased uptake of most of the essential nutrients at a time when shortages of some fertilizers are being predicted (Cordell et al. 2009).

Most agricultural soils are deficient in one or more essential nutrients or have other nutritional constraints to yield (Lynch and St. Clair 2004). The substantial increases in global grain production that have occurred up to now have been based, in part, on improvements in crop nutrition. However, successes of the past are no guarantee for future improvements. It is acknowledged that the yield improvement that was associated with using increasingly high rates of fertilizer has led to rates of nutrient input in excess of crop requirements, leading to low NTUE, a waste of input of nutrients, and reductions in soil and water quality in many regions of the world (Vitousek et al. 2009). On the other hand, there are still regions of the world where chronically low soil fertility is limiting agricultural production. Improvements in NTUE will rely on identifying weaknesses in current production practices and correcting past failures, as well as developing novel approaches to improve nutrient supply and nutrient efficiency.

Despite the central role of plant nutrition in sustaining the productivity of agricultural systems, the effect of climate change on nutrient availability and uptake has received little consideration. Attention has focused on breeding for tolerance to heat and drought resistance and only brief mention is made about crop nutrition, and then comments are confined largely to nitrogen nutrition (Reddy and Hodges 2000; Semenov and Halford 2009; Reynolds 2010; McClean et al. 2011; Olesen et al. 2011). Balanced nutrition of crops is not only important in its own right, but maintaining an adequate level of crop nutrition is also important to help plants cope with biotic and abiotic stress (Huber and Graham 1999; Cakmak 2000; Walters and Bingham 2007; Cakmak and Kirkby 2008). Consequently, the productivity and resilience of crop production in the face of changes in climate will be linked to the ability to maintain the nutritional health of crops and to enhance the NTUE of the cropping system.

There are two approaches to improving NTUE: using crop management to improve the supply and efficiency of nutrient uptake and its conversion to a harvestable product, and improving the ability of crop plants to take up and use nutrients from the soil and fertilizer. The two approaches are complementary and substantial gains in the NTUE of a farming system are likely to come when both strategies are used. We have witnessed the effect of combining variety improvement with fertilizer use in the past when high yielding varieties of crops allowed higher rates of fertilizer (and other inputs) to be applied, leading to large increases in productivity, which also resulted in an increase in NTUE (Ortiz-Monasterio et al. 1997).

The effects of climate change on productivity will be variable and will be influenced by the ability of farmers to adapt to a changing environment. This will be affected not only by their financial resources, and access to information and technology but also their perception of risk and the foibles of human decision making (Lobell and Burke 2010; Hayman et al. 2011). Growing a new variety with superior traits is frequently a low-risk investment and farmers readily adopt new varieties if they perceive a benefit in doing so. Modern high-yielding varieties have been widely adopted in developing countries because they have increased yields and yield stability (Maredia et al. 2000; Renkow and Byerlee 2010). Genetic improvement in NTUE has the potential to make an important contribution to overcoming the challenges of climate change, improving productivity and moderating the adverse effects of climate change on the sustainability of agricultural systems. It may be especially important in developing countries where poverty and lack of infrastructure limit the options of farmers to respond to climate change and who, as a consequence, are the most vulnerable to climate change.

In this chapter, the role of breeding for improved NTUE as one response to climate change will be examined. The review will focus on genetic improvement to overcome nutrient deficiencies and to enhance NTUE, yield, and grain quality. Some consideration will be given to the effects of climate change on nutrient supply from the soil and nutrient demand by crops as this will influence the nutrient balance of cropping systems and NTUE. The focus of the chapter will be on nitrogen (N) and phosphorus (P) nutrition and improvements in the efficient use of these nutrients.

## 10.2  Current Patterns of Nutrient Use and Nutrient Use Efficiency

The past 50 years has witnessed a large increase in the use of fertilizer, which has occurred at a faster rate than grain production (Hinsinger et al. 2011). While global consumption of N and P fertilizer has increased, average rates of nutrient application and trends in fertilizer consumption vary considerably between regions (Fig. 10.1). The largest and most consistent increases in fertilizer consumption have occurred in Asia, while Africa has shown only modest increases in the rate of N fertilizer and a decline in the rates of P application since the 1980s. Average application rates of N and P in Africa are the lowest of all the regions. Fertilizer rates in Europe and the Americas increased until the 1990s after which time rates either stabilized (in the Americas) or declined (Europe) as a nutrient replacement strategy was developed and stringent nutrient management policies were introduced (Vitousek et al. 2009; Ott and Rechberger 2012). Globally, the increases in the rates of N application have generally occurred at a faster rate than P and consequently the N: P balance of applied fertilizer has widened. The consequence of this shift in fertilizer nutrient balance to adaptation to climate change is not known, although responses to N and to elevated $CO_2$ can be limited if P nutrition is suboptimal (Conroy 1992; Edwards et al. 2005; Gentile et al. 2012).

The effects of these trends in fertilizer consumption on fertilizer use efficiency are illustrated in Fig. 10.2. The apparent fertilizer use efficiency has generally declined in Asia, Europe, and the Americas as the rates of application increased, whereas in Africa, the substantial increases in N and P use efficiency since the 1980s reflect the low rates used and, with P use efficiency (PUE), the decline in P application over recent times. Rather than indicating efficient use of nutrients, the high fertilizer use efficiency in Africa is symptomatic of the gradual impoverishment of the soil (Edmonds et al. 2009). Europe has also seen an increase in P efficiency corresponding to the reduction in P application rates.

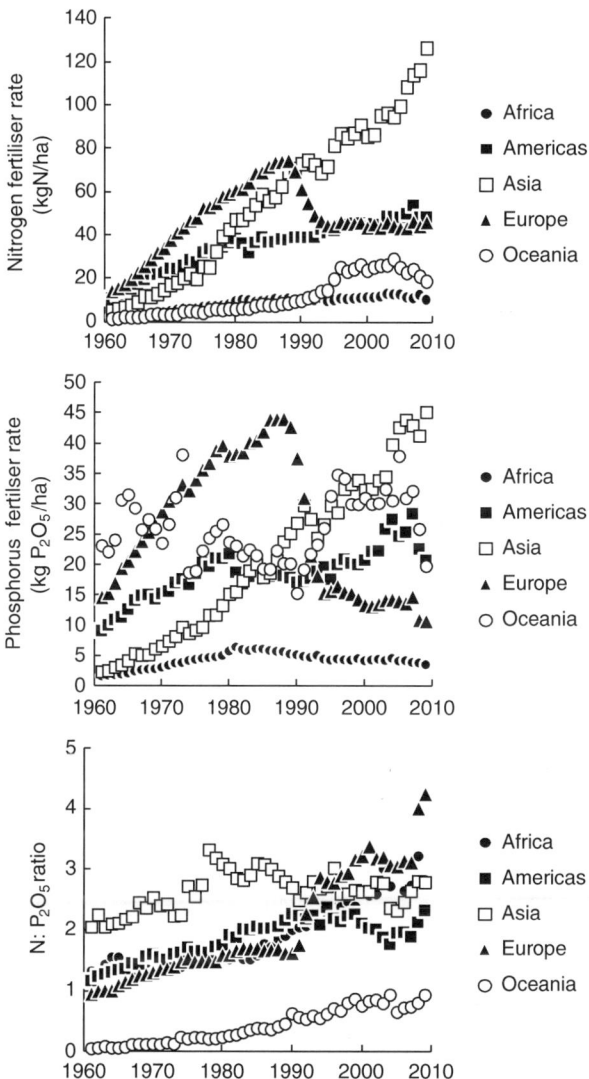

**Fig. 10.1** Changes in the nitrogen (N) and phosphorus ($P_2O_5$) fertilizer application rates and the ratio of $N:P_2O_5$ applied between 1960 and 2010. Fertilizer rates were estimated from the total amount of fertilizer consumed in each year and the total area of arable land and permanent crops (*Source*: FAOStat accessed April 2012)

The changes in fertilizer use over the last 50 years have resulted in marked regional differences in soil fertility that have important implications for future efforts to improve NTUE. The level of soil fertility can influence the physiological basis of NTUE, which will influence breeding objectives and selection criteria. When soil nutrient availability is low, traits associated with acquisition and uptake

**Fig. 10.2** Estimates of the N
and P use efficiency between
1960 and 2010. The fertilizer
use efficiency was based on
the total production of
cereals, coarse grains, pulses,
oilseeds, fiber crops, and
tuber and root crops in each
year and the total N and $P_2O_5$
consumed in each year
(*Source*: FAOstat accessed
April 2012)

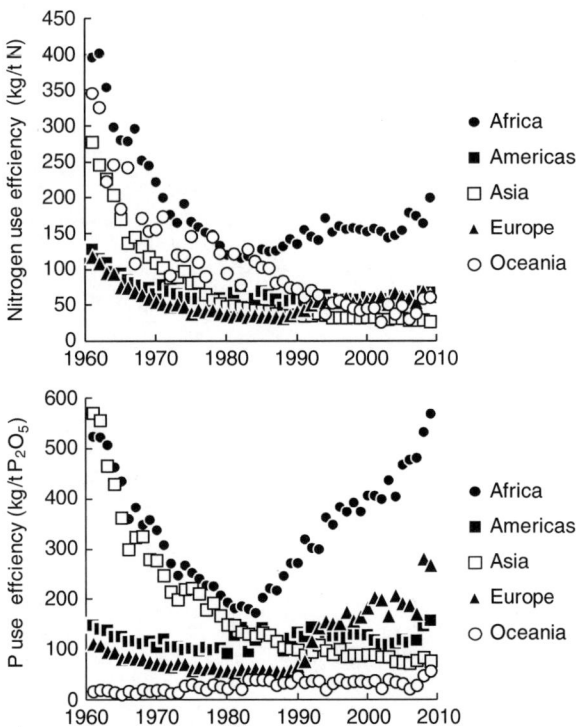

of nutrients tend to be important, while at high levels of fertility, when nutrient availability is less limiting to yield, traits related to utilization of the nutrient may become relatively more important (Ortiz-Monasterio et al. 1997; Wang et al. 2010).

In parts of Western Europe, China, and the USA, N and P are applied in excess of the crops' requirements, and this overfertilization has resulted in low NTUE and loss of nutrients from the agricultural system. As a result, degradation of soil and water systems is an issue in these regions (Khan et al. 2007; Vitousek et al. 2009; Wenqi et al. 2009; Wenqi et al. 2011; Ott and Rechberger 2012). In Australia, where soils over much of the agricultural areas are naturally low in P, farm gate P balances are positive with a high proportion of soil tests showing values well above the critical value (Weaver and Wong 2011). Detailed nutrient budgets for N and P illustrate the low recovery and poor efficiency of nutrient use. In China, only 4–13 % of the N fertilizer and 1–3 % of the P fertilizer applied to agricultural land are recovered in food (Wenqi et al. 2009, 2011). In the European Union, it has been estimated that net per capita consumption of P by agriculture is 4.7 kg P per year of which only 1.2 kg P per year is recovered in food (Ott and Rechberger 2012). In this instance, the P is estimated to be accumulating in agricultural soil at a rate of 2.9 kg P/ha per year. Breeding programs to improve NTUE in these regions will be conducted against a background of high levels of soil fertility and static or falling inputs of fertilizer.

In stark contrast, crops over large areas of sub-Saharan Africa receive suboptimal applications of fertilizer and suffer from chronic nutrient stress. This has limited productivity and has led to nutrient mining of soil and a drawing down of the soils' nutrient reserves (Edmonds et al. 2009; Vitousek et al. 2009). Improving NTUE therefore needs to be conducted in parallel with efforts to increase the soil nutrient base by the use of increased rates of fertilizer and inputs of organic matter and N.

## 10.3 Nutrient Use Efficiency

### 10.3.1 Definitions of Nutrient Use Efficiency and Implications for Breeding

NTUE is a deceptively simple term, which is used inconsistently (Table 10.1). There is a variety of terms used to describe how plants respond to nutrient supply. In some cases, the same measure of efficiency is called by different names and in others the same term is defined in different ways. Agronomic efficiency, in which the emphasis is placed on recovery of and response to fertilizer by a crop, may not be appropriate for genetic efficiency, where it may be better to exploit native soil reserves more effectively to reduce reliance of mineral fertilizer. There is little consensus on which definition is the most appropriate in breeding for improved nutrient efficiency.

Breeding for improved NTUE will be influenced in part by how NTUE is defined because it will affect the screening method used, including the measurements that need to be taken, whether fertilizer treatments need to be imposed and the selection of sites for assessment. More importantly, the type of germplasm that is developed can differ depending on which definition of NTUE is used to guide selection. The problem associated with the definition of NTUE in breeding for improved nutrient efficiency has been recognized for some time (Blair 1993; Gourley et al. 1994) but is far from being resolved.

One commonly used definition is that proposed by Moll et al. (1982), which was originally used for N use efficiency (NUE) but which has been subsequently extended to other nutrients (Ortiz-Monasterio et al. 2001). It is defined as the yield per unit of nutrient supplied (Table 10.1) and it has two components: the ability to extract nutrients from the soil (uptake efficiency) and the ability to convert the nutrients absorbed by the crop into grain (utilization or physiological efficiency). While plants derive their nutrients from soil and fertilizer, nutrient supply is often considered to be the nutrients supplied as fertilizer and thus NTUE is often the yield per unit of fertilizer applied. A problem with using this definition to identify more efficient genotypes is that one is essentially selecting for yield potential. If it is used to assess NTUE of a diverse range of genetic material, lower yielding varieties (such as landraces or old varieties) will have low NTUE

**Table 10.1** Some terms used to assess efficiency in nitrogen and phosphorus studies

| Efficiency term | Description | References |
|---|---|---|
| Nutrient use efficiency (I) | Shoot biomass per unit nutrient supplied | Steenbjerg and Jakobsen (1963) |
| Nutrient use efficiency (II) | Shoot biomass per unit nutrient uptake | Wissuwa et al. (1998) |
| Shoot nutrient utilization efficiency | Shoot biomass per unit nutrient uptake | Su et al. (2006) |
| Biomass utilization efficiency | Biomass yield per unit nutrient uptake | Su et al. (2009) |
| Nutrient use efficiency (grain) | Grain yield per unit nutrient supplied | Moll et al. (1982), Manske et al. (2002) |
| Nutrient uptake efficiency (I) | Total nutrient uptake per unit nutrient supplied | Moll et al. (1982), Osborne and Rengel (2002) |
| Nutrient uptake efficiency (II) | Total nutrient accumulated per unit root weight or length | Liao et al. (2008) |
| Nutrient acquisition efficiency | Total nutrient uptake per unit nutrient applied | Osborne and Rengel (2002) |
| Nutrient efficiency ratio (I) | Grain yield per unit nutrient uptake | Jones et al. (1989) |
| Nutrient efficiency ratio (II) | Shoot growth at low nutrient relative to shoot growth at high nutrient | Ozturk et al. (2005) |
| Nutrient utilization efficiency | Grain yield per unit nutrient uptake | Moll et al. (1982), Manske et al. (2002) |
| Shoot nutrient utilization efficiency | Shoot biomass per unit P uptake (shoots and roots minus seed P reserve) | Osborne and Rengel (2002) |
| Relative grain yield | Grain yield at low nutrient supply relative to grain yield at high nutrient supply | Graham (1984) |
| Apparent recovery | Net uptake of nutrient per unit nutrient applied | Crasswell and Godwin (1984) |
| Agronomic efficiency | Net increase in grain yield per unit nutrient applied | Crasswell and Godwin (1984), Hammond et al. (2009) |
| Physiological efficiency | Net increase in grain yield per unit net increase in nutrient uptake | Crasswell and Godwin (1984) |

even if they possess traits that may enhance nutrient uptake and use. Clearly, differences in yield potential confound assessment of efficiency when this definition is used.

Another definition, which is used frequently, is based on the relative yield at low and adequate levels of nutrient supply. Graham (1984) defined nutrient efficiency as the ability of a variety to produce a high yield in soil that is limiting in the particular nutrient. Consequently an efficient variety is one that has a high relative yield. Nutrient efficiency is estimated by growing plants at two rates of nutrient supply and calculating the ratio of biomass or yield at low (or zero) and high rates of

**Table 10.2** Nitrogen use efficiency of a range of maize genotypes when classified using different definitions of efficiency

| Hybrid | Grain yield Low N g per plant | High N | NUE Low N (g/gN) | High N | Agronomic eff. (g/gN) | Relative yield (%) |
|--------|------|------|------|------|------|------|
| 1 | 223 | 243 | 90.3 | 24.6 | 2.7 | 91.8 |
| 2 | 218 | 275 | 88.3 | 27.8 | **7.7** | 79.3 |
| 3 | 185 | 217 | 74.9 | 21.9 | 4.3 | 85.3 |
| 4 | 270 | 310 | 109.3 | 31.3 | 5.4 | 87.1 |
| 5 | 180 | 195 | 72.9 | 19.7 | 2.0 | 92.3 |
| 6 | 264 | 319 | 106.9 | **32.3** | 7.4 | 82.8 |
| 7 | 297 | 276 | **120.2** | 27.9 | -2.8 | **107.6** |
| 8 | 254 | 257 | 102.8 | 26.0 | 0.4 | 98.8 |

The hybrids were grown at low (2.47 gN per plant) and high (9.88 gN per plant). The most efficient hybrid is highlighted in bold for each definition. NUE is the grain yield divided by the N supply, agronomic efficiency is the increase in yield per unit of additional N, and relative yield is the yield at low N divided by the yield at high N (Adapted from Moll et al. 1982)

fertilizer. However, differences in yield potential can again affect the interpretation of results. Variation in biomass and yield is generally less under nutrient stress than under a nonlimiting supply of nutrients when differences in yield potential are expressed more strongly. Consequently a variety with a low yield potential may show a higher nutrient efficiency compared to a variety with a high yield potential (Gourley et al. 1994).

The problem of yield potential confounding interpretation of nutrient efficiency can be addressed by considering the two aspects separately. The responsiveness of a variety and its yield potential can be viewed as separate traits, and so varieties can be classified into four groups: low yielding and responsive, low yielding and nonresponsive, high yielding and responsive, and high yielding and nonresponsive (Blair 1993). This approach is useful when genotypes that differ widely in the yield potential are examined.

Different definitions of NTUE target different aspects of nutrient uptake and utilization, which creates the dilemma that genotypes may vary in their efficiency rankings depending on the definition used (Blair 1993; Gourley et al. 1994). The problem is illustrated using data from Moll et al. (1982) in Table 10.2, in which the ranking of the maize hybrids changes according to the criterion used to identify NTUE. The most appropriate criterion will most likely depend on the environment and the farming system that is being targeted.

## 10.3.2   Crop Responses to Nutrients and Nutrient Use Efficiency

Crops show a diminishing response to increasing supplies of a nutrient (Fig. 10.3a), which is most commonly described by the Mitscherlich curve. The $Y$-intercept represents the ability of crops to utilize native soil nutrients and the slope of the curve represents the responsiveness to each increment of additional nutrient and is related to the efficiency of nutrient use. Efficiency is greatest at the lowest rates of fertilizer use when the crops are most responsive and diminishes as the fertilizer rates approach that required for maximum yields.

Genetic improvement in NTUE can increase the ability to use native soil nutrients without altering the yield potential of the crop (Variety B), which will reduce the amount of fertilizer required to reach maximum yield. Increasing the yield potential alone (Variety C) may not alter the amount of fertilizer required to reach maximum yield but will increase the yield achieved for a given rate of a fertilizer below the optimum rate and thus may improve the profitability of fertilizer use. Increasing both the ability to exploit the native soil nutrients as well as increase the yield potential (Variety D) will enable lower fertilizer rates to be used as well as provide higher yield for a given rate below the optimum.

These differences in the ability to acquire nutrients and in yield potential highlight the problems of defining NTUE (Fig. 10.3b–d). Using the definition of Moll et al. (1982), sees relatively little variation in NTUE and the NTUE of the four genotypes quickly converge, reflecting the trends in grain yield (Fig. 10.3b). This is hardly surprising given this definition of NTUE means that efficiency is inversely related to nutrient supply. Using relative grain yield as the definition of NTUE sees a greater separation between varieties based on their ability to utilize native soil nutrients (Fig. 10.3c). If there are significant differences in grain yield at low levels of nutrient availability, differences in relative yield should be able to separate varieties. Agronomic efficiency (Fig. 10.3d) is also influenced by a variety's ability to exploit native soil nutrients; however, the important difference to note is that varieties with a high relative grain yield ("efficient" varieties) have a low agronomic efficiency, and thus, selection for genotypes with high relative yield may lead to varieties with low agronomic efficiency. The other point to note is that all measures of nutrient efficiency decline as fertilizer rates increase and genetic differences tend to be greater at lower levels of nutrient supply.

### 10.3.3   Breeding for Nutrient Use Efficiency

The case for breeding for greater nutrient efficiency has been argued strongly in the past (Graham 1984; Graham et al. 1992; Graham and Rengel 1993; Rengel 1999; Fageria et al. 2008; Khoshgoftarmanesh et al. 2011). If breeding for improved NTUE is to be successful, a number of conditions need to be met: (a) there needs to be useful genetic variation in NTUE; (b) the genetic basis of the trait needs to be understood; and (c) appropriate selection criteria need to be defined, which often will require an understanding of the important physiological determinants of nutrient efficiency. There also needs to be no yield penalty associated with improvements in nutrient efficiency.

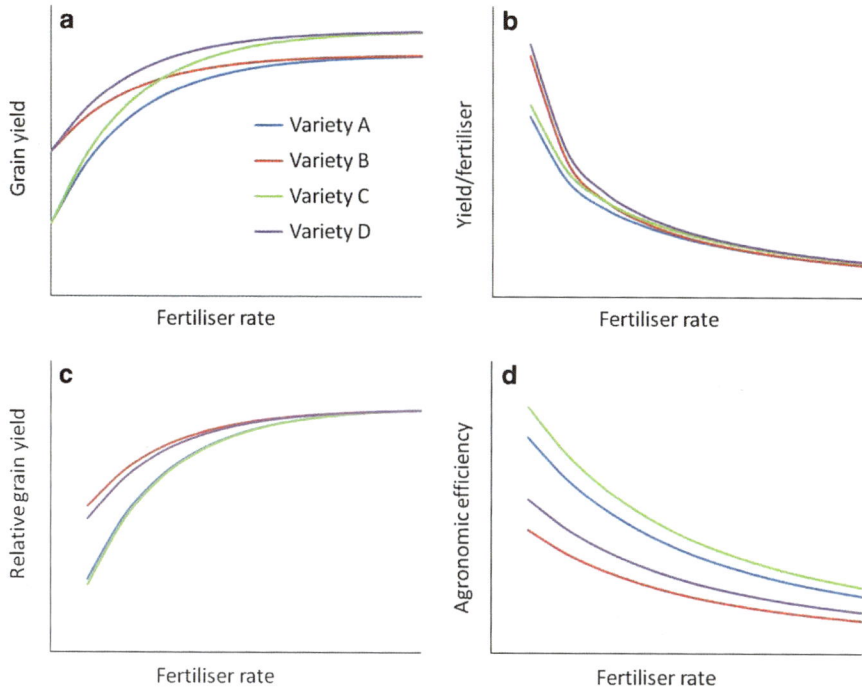

**Fig. 10.3** Possible effects of genetic differences in nutrient acquisition and response to fertilizer on different measures of nutrient use efficiency. Variety A is the standard variety. Variety B has the same yield potential but is able to extract soil nutrients more effectively than Variety A. Variety C has a higher yield potential than Variety A, while Variety D has both a higher yield potential and an improved ability to exploit soil nutrients. The graphs show (**a**) the grain yield response to fertilizer inputs and the differences in nutrient use efficiency when defined as (**b**) yield divided by fertilizer applied ($Y/F$; Moll et al. 1982), (**c**) relative grain yield ($Y_f/Y_{max}$; Graham 1984), or (**d**) agronomic efficiency ($(Y_f-Y_0)/F$). $Y_0$ is the grain yield with no fertilizer, $Y_f$ is the grain yield at a given fertilizer rate, $Y_{max}$ is the grain yield at a nonlimiting rate of fertilizer, and $F$ is the fertilizer rate

### 10.3.3.1   Genetic Variation in Nutrient Use Efficiency

Over the past 30 years, there has been a considerable amount of work that has characterized genetic variation in nutrient efficiency among the major food crops (Graham and Nambiar 1981; Graham 1984; Marcar and Graham 1987; Graham and Rengel 1993; Grewal et al. 1997; Ortiz-Monasterio et al. 1997; Cianzio 1999; Rengel 1999; Stangoulis et al. 2000; Baligar et al. 2001; Fageria et al. 2002; Torun et al. 2002; Hacisalihoglu et al. 2004; Yan et al. 2006; Hirel et al. 2007; Genc and McDonald 2008; Rengel and Damon 2008; Balint and Rengel 2009). This has demonstrated that there are significant levels of genetic variation in nutrient efficiency among genotypes of staple food crops, which can be exploited in breeding programs. Despite this, progress in developing nutrient efficient crop

**Table 10.3** Examples of the release of nutrient use efficient varieties or of germplasm development programs targeted specifically for improved nutrient use efficiency

| Nutrient | Crop | Region | Reference |
|---|---|---|---|
| Nitrogen | Maize | Southern and eastern Africa | Bänziger et al. (2006) |
| | Sugar cane | Brazil | Baldani et al. (2002) |
| Phosphorus | Soybean | China | Yan et al. (2006) |
| | Rice | | Chin et al. (2011) |
| | Wheat | China | Yan et al. (2006) |
| | Common bean | Central America | Lynch (2011) |
| | | Mozambique | McClean et al. (2011) |
| Manganese | Barley | South Australia | Jennings (2004), McDonald et al. (2001) |
| Iron | Soybean | USA | Wiersma (2010), Rodriguez de Cianzio (1991) |
| | Oat | USA | Rodriguez de Cianzio (1991) |
| | Sorghum | | |
| | Common bean | | |
| | Chickpea | WANA | Saxena et al. (1990) |

varieties has been slow (Fageria et al. 2008; Wissuwa et al. 2009), and there are few varieties that have been released specifically for improved NTUE (Table 10.3). Apart from conventional breeding methods, a transgenic approach has recently been used to develop lines of canola and rice requiring lower inputs of N fertilizer (Good et al. 2007; Shrawat et al. 2008). The slow progress is due to many factors: nutrient efficiency is a complex trait subject to considerable environmental variation, appropriate screening methods have been slow to be developed because of a poor understanding of the most limiting physiological processes, and the genetic control of nutrient efficiency is not well understood. At a pragmatic level, it is likely that many commercial plant-breeding programs in developed countries have viewed improved NTUE as peripheral to their major breeding objectives compared to characters such as disease resistance, quality, drought tolerance, and yield per se.

### 10.3.3.2 Mechanisms of Nutrient Use Efficiency

Several reviews have described the mechanisms of efficiency for a number of nutrients (Cianzio 1999; Rengel 1999; Cakmak and Braun 2001; Fageria et al. 2008; Khoshgoftarmanesh et al. 2011). The strategies used by plants to promote uptake and enhance yield can be considered in terms of two fundamental processes (a) the ability to acquire nutrients from the soil and (b) efficiency with which nutrients taken up by plants are used to produce biomass and grain. Mechanisms of nutrient acquisition include alterations to the chemical and biological properties of the rhizosphere to increase nutrient availability, increases in the volume of soil

**Fig. 10.4** The contributions to grain yield of phosphorus uptake efficiency (*filled circle*) and phosphorus utilization efficiency (*open circle*) among genotypes of bread wheat grown on a phosphorus deficient acid soil over 3 years (From Manske et al. 2001)

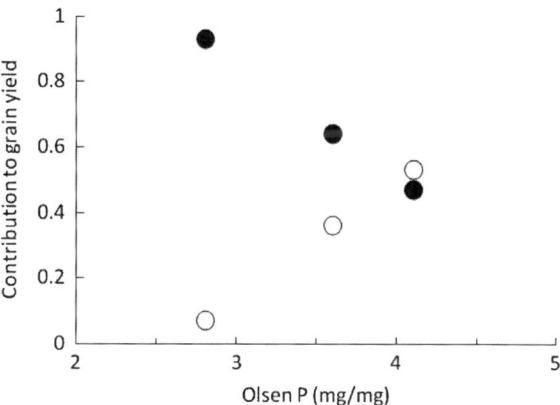

explored by increased root growth and changed root architecture, interactions with microbial populations in the rhizosphere, and changes in the expression of ion transporters in the roots to enhance uptake. Efficiency of utilization may include greater root to shoot translocation of nutrients, compartmentation of nutrients and partitioning within the plant, metabolic efficiencies and greater remobilization. The relative importance of different mechanisms is likely to vary with the severity of nutrient stress. Under severe deficiency, the ability to take up nutrients may be critical to meet the demands of the crop, particularly for nutrients that have low mobility (e.g., P) or which are present in very low concentrations, such as the micronutrients. When the concentration of available nutrients in the soil is high and uptake is less of a limitation, utilization efficiency may become more important to nutrient efficiency (Manske et al. 2002; Hirel et al. 2007; Wang et al. 2010; Rose and Wissuwa 2012). Figure 10.4 illustrates this effect in a study on the genetic variation in PUE: the importance of uptake efficiency is equal to or greater than that of utilization efficiency in each year, but its importance increases as the severity of the deficiency increases. There is some evidence to suggest that modern cultivars of crop plants developed for high input and intensive systems have lost the capacity to acquire some soil nutrients when availability is low (Wissuwa et al. 2009), which likely reflects the change in efficiency mechanisms as soil nutrient availability increases (Fig. 10.4).

While there can be some debate whether selection for nutrient efficiency should be based on traits associated with uptake or utilization efficiency (e.g., Wang et al. 2010; Rose and Wissuwa 2012), the natural variation in nutrient availability across sites and seasons may mean that both uptake and utilization will contribute to nutrient efficiency and their relative importance will vary with the availability of soil nutrients. If breeding for nutrient efficiency is targeting a region where soil nutrient availability is variable, both uptake efficiency and utilization efficiency are useful and should be combined.

## 10.4   Predicted Impacts of Climate Change and Implications for Nutrient Use Efficiency

The predicted effects on climate of changes in atmospheric concentrations of greenhouse gases and the consequences for local agricultural production are variable, with marked differences between regions (IPCC 2007; Jarvis et al. 2010; Olesen et al. 2011). Analysis of recent trends in rainfall between 1900 and 2005, for example, has shown that the eastern parts of North and South America, northern Europe, and north and central Asia have experienced increased rainfall, while the Sahel, the Mediterranean Basin, South Africa, and parts of south Asia have shown a drying trend (IPCC 2007). Despite the high degree of variability in regional and temporal trends, there is consistency in a number of long-term observations, which is concerning for agricultural production in the future. It is predicted that the world will face a generally warming climate with a greater frequency of warm spells and heat stress. Precipitation will be more variable and there will be a greater frequency of heavy precipitation and of drought, even in regions where annual precipitation is predicted to increase. For many regions of the world, nutrient management needs to be done in an environment of elevated $CO_2$, increased rates of N deposition, increasing temperatures, and more variable precipitation with a higher frequency of drought.

It is difficult to predict the consequences of these changes to nutrient use and NTUE because there are potential effects on soil properties and soil biological activity, on the growth of plants and their ability to take up nutrients, and on the partitioning of nutrients within the plant. Our understanding of the influence of many of the effects is poor, especially when they change together. Many past studies have been conducted in single factor experiments and have speculated on changes into the future without regard to the interactions that may occur. For example, rising atmospheric $CO_2$ levels can enhance plant biomass production (Long et al. 2004; Mittler and Blumwald 2010) and thus increase the demand for many nutrients, but the increase in temperatures that will accompany increases in $CO_2$ will hasten development, shorten the length of the growing season, and may reduce the demand for and uptake of nutrients (Nord and Lynch 2009). Other effects of individual components of climate change may influence nutrient uptake in different ways. Higher temperatures may increase transpiration rates by increasing atmospheric vapor pressure deficit, which may increase water flow to the roots and increase transport of some nutrients by mass flow, but may also lead to more rapid soil drying, which can reduce nutrient uptake.

Much of the past work on the effects of climate change on crop nutrition has focused on N nutrition (Lynch and St. Clair 2004). This is not surprising given the importance of N to crop growth and yield and the large quantities of N applied to crops (Fig. 10.1). However it tends to distort our understanding of the impact of climate change and underestimates the importance of nutrient balance in plants to productivity and NTUE.

Another aspect that is sometimes overlooked in discussions on plant responses to changing climate is the plasticity in growth and development that plants possess and, in the long term, the ability to respond to and adapt to changes in climate. Root systems respond to local variation in moisture and nutrients in the soil and are effective in integrating nutrient uptake in a heterogeneous soil environment. Long-term adaptation may also have profound influences on how plant nutrient demand changes in the future. For example, a community of spruce trees that had been growing under elevated levels of $CO_2$ for over 100 years because of their proximity to springs that emitted $CO_2$ had a lower rate of photosynthesis than nearby tress that had grown under normal $CO_2$ levels (Cook et al. 1998). The lower photosynthesis rate was associated with lower concentrations of chlorophyll in the leaves and a higher photosynthetic efficiency. It has been suggested that if this was a common response in other species, the greater photosynthetic efficiency would lead to a lower demand for nutrients involved in photosynthesis—N, sulfur (S), magnesium (Mg), and iron (Fe) (Brouder and Volenec 2008).

In its simplest terms, the nutrient status of a crop reflects the balance between the ability of the soil to provide nutrients to the crop and of the root system to take up nutrients (supply) and the nutrient required by the crop for optimum growth (demand). The two processes are not independent and there will be feedback between plant growth and aspects of soil nutrient availability. A study by Patil et al. (2010) illustrates this. Climate change is predicted to cause increases in rainfall and soil temperature in some regions of the world. In a lysimeter study, higher rainfall increased nitrate leaching during winter potentially reducing supply of N, but this was counteracted by warmer soils increasing crop growth rates, which increased demand and N uptake. Notwithstanding soil–plant interactions such as this, the net effect of climate change on crop nutrition will be a consequence of its relative effects on nutrient supply and demand. The success of genetic improvement to respond to changes in climate will depend on whether changes in availability and supply of nutrients or changes in the ability to acquire nutrients by changes in root growth and physiological efficiency will be more influential in meeting changes in plant demand for nutrients.

## 10.4.1   Soil Properties, Nutrient Availability, and Supply to the Roots

At present the impact of climate change on soil properties and its subsequent effect on mineral nutrient supply is largely conjectural because of the complexity of the soil system and the lack of long-term empirical data from agricultural systems. There are a number of ways that changes in climate can affect the availability of soil nutrients and their movement to the roots (Table 10.4). While the effects of these

**Table 10.4** Possible effects of changes in climate on availability of nutrient in soil (Adapted from St. Clair and Lynch 2010)

| Process | Influential climate change variables | Mineral nutrients affected |
|---|---|---|
| Transpiration-driven mass flow | $CO_2$, temperature, rainfall, vapor pressure deficit | Nitrate, sulfate, calcium, magnesium, silicon |
| Diffusion | Rainfall, temperature | Phosphorus, potassium zinc, iron |
| Soil C content and C cycling | $CO_2$, temperature, rainfall | Many nutrients |
| Leaching | Rainfall | Nitrate, sulfate, calcium, magnesium |
| Arbuscular mycorrhizae | $CO_2$, soil moisture | Phosphorus, zinc |
| Soil erosion | Rainfall | All nutrients |
| Soil pH | Rainfall, $CO_2$ | Aluminum, manganese, copper, manganese, zinc, iron |

individual factors have been the subject of much research, it is still unclear how they will interact to determine the effects of climate change on nutrient availability and how it will influence NTUE. When reviewing the influence of climate change on soil N cycling, Bijay-Singh (2011) observed:

> ... effects of climate change on soil N transformations can be complex, and the long-term implications on N retention and N use efficiency are unclear.

This statement summarizes our current state of understanding of the influence of climate change on the availability of soil N, a nutrient that has been the subject of intense research over many years: our understanding of the effects of climate change on the availability of other nutrients is even less. In part this is because the drivers of climate change can affect a number of different soil processes, and sometimes in different ways; however in much of the previous research into the effects of climate change, the response to one factor is often viewed in isolation from other climatic factors that may change at the same time. For example, rising atmospheric $CO_2$ levels induces stomatal closure and can reduce transpiration (Long et al. 2004), which has been suggested to reduce nutrient movement to roots by mass flow (St. Clair and Lynch 2010). However mean temperatures will also rise with increases in atmospheric $CO_2$, which will tend to increase transpiration by increasing atmospheric vapor pressure deficits. The very complexity of the soil–plant interaction in determining nutrient uptake and plant nutrient status requires multifactor experiments be conducted and that field data from long-term $CO_2$ enrichment experiments be collected. Alternatively, modeling approaches can be used to study the effects of climate variability on soil–plant–atmosphere interactions and investigate the impacts of climate change on nutrient cycling, plant growth, and nutrient uptake. Combining genetic and genomic information with crop simulation modeling can potentially provide a powerful tool for future studies on the influence of climate change on crop production.

### 10.4.1.1  Effects of Elevated $CO_2$

Soil organic matter has a central role in determining soil nutrient availability. Consequently, the potential influence of climate change on nutrient availability in the soil will be influenced by the changes in organic matter and the rates of organic matter cycling. Carbon and N are linked in the soil organic matter and changes in carbon cycling will also be reflected in changes in N cycling (Bijay-Singh 2011). Recent reviews strongly suggest that climate change will affect carbon and N cycling in soils (Bijay-Singh 2011), which in turn can alter other soil properties such as pH (Rengel 2011) that can also determine soil nutrient availability.

In one of the few long-term studies of the effects of $CO_2$ enrichment, it was found that soil under a pasture from an 11.5-year free air $CO_2$ enrichment experiment had lower available P (Olsen P 18 µg/L cf 25 µg/L) but similar available N (7.61 mg/kg cf 7.86 mg/kg) compared to pasture grown under ambient $CO_2$ (Gentile et al. 2012). This was associated with lower concentrations and content of P in the leaves of ryegrass and a lower response to N fertilizer. The lower available P was attributed to increased sequestration of P in the soil organic matter. Reductions in the availability of P associated with elevated $CO_2$ have also been reported in heathland soils (Andreson et al. 2010). If a long-term effect of elevated $CO_2$ is to reduce soil available P, the requirements for P fertilizer may increase and the need for more P efficient genotypes and farming systems will become greater.

### 10.4.1.2  Temperature

The predicted increases in atmospheric temperatures are likely to increase mean soil temperatures and there is evidence that some soils have warmed during the twentieth century (Qian et al. 2011). Increased soil temperature can increase nutrient uptake by plants (Singh and Subramaniam 1997; Bassirirad 2000), which is associated with changes in soil nutrient availability as well as changes in root growth.

Temperature is a major influence of organic matter turnover (Sanderman et al. 2010; Baldock et al. 2012), although its effect on decomposition of soil organic carbon will be influenced by other environmental factors that control the soil's biological activity and which determine the accessibility of soil organic carbon to degradative enzymes (Baldock et al. 2012). Modeling has suggested that global soil organic matter levels will decline as temperature increases, although local responses may vary (Jones et al. 2005), and there is considerable uncertainty about the temperature sensitivity of soil organic matter decomposition (Davidson and Janssens 2006). However, accelerated loss of soil organic matter will reduce available soil nutrients and increase the risk of nutrient stress over the long term unless inputs of organic carbon in farming systems can increase. Greater inputs of N fertilizer to boost crop production may not be successful in halting the decline (Khan et al. 2007).

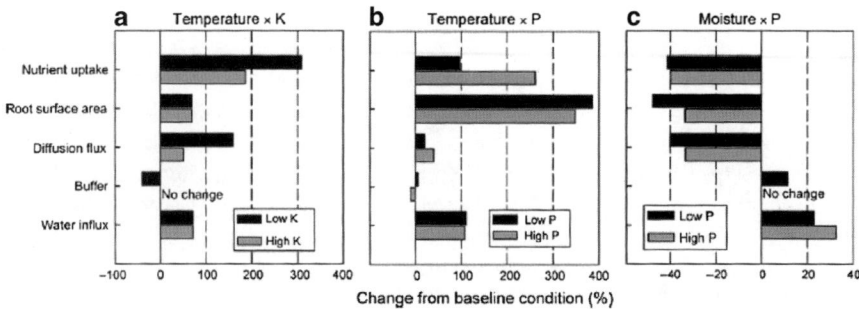

**Fig. 10.5** The effect of temperature or moisture on nutrient uptake by maize roots and on parameters of soil nutrient availability and root growth. The responses are shown as the percentage change from the baseline or control conditions and are based on experiments where (**a**) root zone temperature was increased from 15 °C to 29 °C at two rates of potassium, (**b**) root zone temperature was increased from 18 °C to 25 °C in low and high P-fertility soils, and (**c**) the root zone moisture potential was reduced from −33 kPa to −170 kPa in soils differing in P fertility (Brouder and Volenec 2008) (with permission)

Mineralization rates are higher in warmer soils (Bijay-Singh 2011; Lal 2011), and rates of nutrient adsorption and desorption may change (Barrow and Shaw 1975), which will alter nutrient availability. As well, warmer soils may increase rates of diffusion and water uptake (Fig. 10.5a, b) thereby increasing nutrient delivery to the roots, although any benefits of these responses to soil temperature will clearly depend on soil moisture (Brouder and Volenec 2008; St. Clair and Lynch 2010). Therefore, if there are no major limitations to nutrient uptake, such as intermittent drought, nutrient uptake by plants may be enhanced and this may help overcome some of the nutrient dilution effects often reported in plants grown at elevated $CO_2$. However, caution needs to be exercised when making generalized statements about the effects of temperature on soil nutrient availability and especially if extrapolating from short-term experiments. Bassirirad (2000) highlights the fact that the effect of temperature may not be similar over all temperature ranges and that there is evidence of differences among species in their sensitivity to rising soil temperatures. Short-term improvements in nutrient delivery from greater rates of mineralization may not be sustainable in the longer term if soil organic matter declines.

### 10.4.1.3 Water Availability

Changes in climate will see local variation in the distribution and intensity of rainfall. In many regions, more frequent periods of dry weather are predicted, and this may reduce nutrient availability and movement to the roots due to lower rates of mass flow and diffusion (Dunham and Nye 1976; Mengel and Kirkby 2001). However, Brouder and Volenec (2008) suggest the influence of changes in mass flow for immobile soil nutrients will be small because simulation modeling suggests total nutrient uptake is insensitive to changes in water flows to the root. They also argue that reducing soil moisture to the point that delivery of mobile nutrients to the roots by mass flow limits nutrient uptake would more than likely limit root and shoot growth by greater plant water deficits. Mass flow and diffusion are overlapping mechanisms of nutrient delivery to the root, and their relative importance varies with root zone conditions rather than with a specific nutrient. Consequently, there may be relatively little change in nutrient uptake as soil dries and as the balance of delivery changes from mass flow to diffusion. At some point however diffusive transport will be insufficient to meet the plants' demand and nutrient uptake will be reduced (Fig. 10.5c).

#### 10.4.1.4  Implications for Nutrient Use Efficiency

While climate change will alter a number of soil processes that affect soil nutrient availability and potentially affect the delivery of nutrients to the roots, the consequences of these changes to nutrient availability and the subsequent influences on nutrient uptake and NTUE are unclear. It is likely that accelerated loss of soil organic matter and greater periods of dry weather as a result of climate change will reduce the availability of nutrients over the long term and increase the risk of nutrient stress developing. Reductions in supply of nutrients will require crop varieties that are better at exploiting supplies of nutrients to maintain nutrient uptake and that make more efficient use of fertilizer. While the magnitude of the effect is uncertain, any decreases in soil nutrient availability as a result of climate change will increase the need to improve crop NTUE in the face of declining high-quality nutrient reserves and increasing costs of fertilizer.

### 10.4.2  Growth, Yield, and Nutrient Demand

There is still some uncertainty about the consequences of climate change on crop yields (Jarvis et al. 2010). There will be large regional variation in the responses and the effects on productivity will also depend on the capacity of farmers to adapt to changes in their environment (Lobell and Burke 2010). Nevertheless, to meet the predicted demands for food in the future, increases in yield need to be sustained in the face of increased frequency of heat and drought stress, and this will require commensurate changes in nutrient uptake. The concentrations of some nutrients such as potassium and zinc (Zn) may also help plants cope with the predicted increases in abiotic and biotic stresses that will be associated with climate change (Cakmak 2000; Walters and Bingham 2007; Amtmann et al. 2008; Peck and McDonald 2010). Acquisition of nutrients, therefore, becomes an important component in the capacity of plants to adapt to the changes in growth and nutrient availability that may result from climate change.

### 10.4.2.1 Increases in Shoot Growth and Yield

Shoot biomass in $C_3$ and $C_4$ crop species increases under elevated $CO_2$ concentrations, although increases are much smaller in $C_4$ crop plants (Kimball et al. 2002; Long et al. 2004). Increases in yield are more variable (Long et al. 2004); however, yield increases in $C_3$ crops under elevated $CO_2$ are frequently reported (Kimball et al. 2002). The relative increases in biomass production and yield from elevated $CO_2$ under water stress have been found to be equal to or greater than those under well-watered conditions, which contrast to the effect of nutrient stress: the ability of crops to respond to elevated $CO_2$ will depend on access to and uptake of available nutrients (Poorter 1998; Kimball et al. 2002; Luo et al. 2004; Edwards et al. 2005). Nutrient dilution at elevated $CO_2$ has often been reported (see below) but in many cases there is still an increased uptake of nutrients suggesting the increased demand for nutrients is often not met by increased uptake. Therefore, improvements in nutrient supply and in NTUE will be important to allow crops to take advantage of any benefits to growth and yield afforded by elevated $CO_2$. Genetic improvement in nutrient efficiency will make a valuable contribution to this goal.

### 10.4.2.2 Root Growth and Function

Modeling of nutrient uptake by plants suggests that changes in root length and surface area are important determinants of the responses of plants to changes in soil nutrient supply, soil temperature, and moisture (Barber and Mackay 1985; Mackay and Barber 1985; Brouder and Volenec 2008). The effects of changes in climate on root growth, therefore, will influence how plants can respond to changes to nutrient availability.

Changes in the partitioning of growth between root and shoot will alter nutrient balances within the plant and the nutrient composition of the shoot. Brouder and Volenec (2008) concluded that there was no significant change in the root to shoot ratio under elevated $CO_2$ and that changes in shoot biomass would most likely drive changes in nutrient demand. In contrast, the review of Kimball et al. (2002) found that root growth generally responded more strongly to elevated $CO_2$ than shoot growth and their data would suggest an average increase in root to shoot ratios of about 18 % among $C_3$ grasses and woody perennial crops. However, the often-reported reduction in shoot nutrient concentration under elevated $CO_2$ would suggest that greater partitioning to root growth may not result in uptake of nutrients sufficient to meet the additional nutrient demands of the crop.

The frequency of drought will increase as a consequence of climate change and this will reduce root growth (Weir and Barraclough 1986; Asseng et al. 1998; Buljovcic and Engels 2001). The decline may be most marked in the surface layers of soil although there may be proliferation of roots in the moist subsoil (Asseng et al. 1998). Nutrient uptake from the drying parts of the soil profile will be reduced and this is likely to lower total nutrient uptake if other parts of the profile are unable to maintain the supply of nutrients (Barber and Mackay 1985; Jupp and Newman 1987; Buljovcic and Engels 2001). The ability of roots to recover from periods of dry weather and resume nutrient uptake is also important for the nutrition of crops. Rewetting of the soil profile can lead to renewed root growth and water uptake, but there may be a considerable delay in the time of recovery. Studies on P uptake by Jupp and Newman (1987) found that uptake to the shoot was low following rewatering and there was some evidence of recovery only after 2 or 3 weeks. In the field, recovery of water use after a period of water deficit occurred approximately 10 days after rewatering (Asseng et al. 1998).

Root growth is a key factor in nutrient absorption from the soil and this is not likely to diminish under a changing climate. Root traits are an important aspect of nutrient efficiency and breeding for root characteristics that will enhance nutrient uptake will contribute to greater NTUE. Much of the past work on root growth has examined traits related to root architecture, but differences in recovery of root growth after rewetting may also be useful traits to examine.

### 10.4.2.3   Phenology and Nutrient Use Efficiency

The rate of crop development influences the duration of crop growth, biomass production, and partitioning, all of which influence nutrient uptake and demand. Phenology is sensitive to temperature and an important consequence of rising global temperatures is to hasten crop development in many regions (Sadras and Monzon 2006; Grab and Craparo 2011; Webb et al. 2012), while there has been a lengthening of the growing season in high latitudes associated with an earlier onset of the growing season (Jarvis et al. 2010; Olesen et al. 2011). There is some evidence that altering plant phenology may influence nutrient acquisition (Nord and Lynch 2009). Hastened development restricts P uptake; genotypes that flower quickly and have a short vegetative phase have low biomass production and P uptake. The delay in development that is characteristic of P deficiency is considered to be an adaptation to low P as it allows a longer period of P uptake (Nord and Lynch 2008). Increased global temperatures may hasten development and reduce uptake of P, leading to a reduction in PUE. However, this prediction overlooks the effect of rising soil temperatures on P uptake. Sowing wheat early into warmer soils can result in substantial reductions in the optimum rate of P, which is most likely to be due to the effects of soil temperature on root growth and P availability in the soil (Barrow and Shaw 1975; Batten et al. 1993, 1999).

### 10.4.2.4   Nutrient Toxicities and Nutrient Use Efficiency

Production over large areas of the world's agricultural land is affected by a number of soil toxicities such as aluminum toxicity, salinity, and boron toxicity, which restrict root growth and limit grain yield (Lynch and St. Clair 2004; Rengasamy

2006; Yau and Ryan 2008). These chemical constraints to root growth can exacerbate the effects of drought and heat stress by reducing the ability of crops to exploit soil moisture reserves, making crops more vulnerable to reductions in rainfall and increases in temperature induced by climate change. By restricting root growth and water use, soil toxicities not only reduce grain yield but may also reduce nutrient efficiency. Apart from influencing how crops may cope with an increasingly variable climate, soil constraints such as acidity and alkalinity can have a direct effect on nutrient uptake. For example, P deficiency occurs frequently in acid soils where aluminum toxicity occurs. Breeding to alleviate the effects of soil toxicities can enhance crop nutrient efficiency indirectly (Wissuwa et al. 2009).

### 10.4.3  Nutrient Concentrations and Grain Quality

Changes in the frequency and severity of soil water deficits and the effects of warmer growing seasons can influence plant nutrient concentrations by altering the balance between nutrient supply and crop requirement. However, these effects will be variable as they will largely reflect the regional and seasonal variations in the patterns of rainfall and temperature. Nutrient concentrations in plants are also responsive to elevated $CO_2$. Plant growth can be enhanced at high $CO_2$, the magnitude of the effect depending on the other nutritional limitations (Poorter 1998), which can lead to nutrient dilution if the rate of nutrient uptake does not increase at the same rate as biomass production.

Nitrogen concentrations are consistently lower in leaves and grain of plants grown under elevated $CO_2$ (Conroy 1992; Wu et al. 2004; Taub et al. 2008; Wieser et al. 2008; Pleijel and Danielsson 2009; Erbs et al. 2010; Pleijel and Uddling 2012). Reductions in grain N concentrations have also been associated with reductions in the protein fractions that are important for bread-making quality. Flour from wheat grown under elevated $CO_2$ has up to 20 % less gliadin and 15 % less glutenin (Wieser et al. 2008).

The effects on other mineral nutrients are more variable (Duval et al. 2012). Among crop plants, elevated $CO_2$ has been reported to lower the concentrations of S (Fangmeier et al. 1997; Erbs et al. 2010; Duval et al. 2012), Zn, and Fe (Fangmeier et al. 1997; Wu et al. 2004) in the grain and foliar Mg and Zn (Duval et al. 2012), but there is considerable variation in the effect. Despite the lower concentrations of nutrients in leaves and grain, total nutrient uptake is often greater under elevated $CO_2$, suggesting that the availability and/or uptake of nutrients may not be impaired but that there is proportionately greater production of biomass. In some cases, elevated $CO_2$ can increase nutrient uptake more than the increase in shoot biomass leading to increased concentrations. Increases in shoot Fe concentrations have been reported in tomato (Jin et al. 2009) and grasses (Duval et al. 2012) and the concentrations of cadmium, Zn, manganese, and Mg increased in the shoots of the cadmium/zinc hyperaccumulator *Sedum alfredii* under elevated $CO_2$ (Li et al. 2012).

The often-reported reductions in the concentrations of grain N, S, and Zn at high $CO_2$ levels have important implications for the quality and nutritive value of the food. Low concentrations of Zn and Fe in cereal grains is a major cause of the poor dietary intake of these nutrients and the attendant chronic health problems (Graham et al. 2012). Concentrations of Zn and Fe are often closely linked to the concentrations of N and S (Uauy et al. 2006; Morgounov et al. 2007; Gomez-Becerra et al. 2010), which can affect processing quality in wheat by altering the composition of storage protein in the grain (Shewry et al. 1995; Peck et al. 2008). Most of the experiments, which have reported a decline in nutrient concentrations at elevated levels of $CO_2$, have been conducted in soils with high fertility and in which total nutrient uptake increased. The impact when available soil nutrients are low is not known but will probably depend on the response in crop biomass and yield at elevated $CO_2$. Irrespective of the effect, a consequence of rising $CO_2$ may be to counteract recent efforts to increase grain Zn and Fe concentrations by plant breeding.

## 10.5   Genomic Approaches to Improving Nutrient Use Efficiency

While it is acknowledged that there is a need to increase NTUE in the major crops, improvement has been slow. There are many reasons for the slow rate of progress. NTUE is a complex, quantitative trait that is influenced greatly by environmental factors. The physiological and molecular bases of the use efficiency of many nutrients are, in general, poorly understood and often only defined in broad terms.

As with NTUE, there has been relatively little progress in improving drought tolerance based on physiological and molecular approaches despite the considerable amount of resources devoted to the task. Drought tolerance shares many of the features of NTUE, and the past approaches of aiming to improve drought tolerance by focusing on a narrow array of traits have been questioned: instead a broader multidisciplinary approach that integrates genomic and transgenic approaches with physiological and phenological dissection of responses has been proposed (Fleury et al. 2010). Selection and verification of the value of traits associated with NTUE under field conditions in the target environments is also a critical, and arguably the most important, step in the process (Wissuwa et al. 2009).

### 10.5.1   General Concepts

Genetic improvement of NTUE needs to take a multidisciplinary approach, which will integrate a number of different technologies and methods that span laboratory and field-based studies. Critical steps will include (a) identifying useful sources of genetic variation for the trait in question to allow introgression of desirable characteristics or to develop appropriate populations for genetic analysis and marker development, (b) developing a reliable phenotyping system that is predictive of field performance, and (c) developing efficient and rapid selection methods based on phenotypic or genetic screening to identify superior lines in the breeding program. In a forward genetic approach, undertaking detailed studies of the physiological and molecular responses to nutrient stress will allow improved understanding of how plants sense and respond to nutrient stress, thereby allowing improved understanding of the roles of specific genes in NTUE.

### 10.5.1.1   Sources of Genetic Variation

While genetic variation in NTUE has been reported for many crops, improvements in NTUE may be sometimes limited by a narrow level of genetic variation within modern germ plasm (e.g., Wissuwa et al. 2009). Novel sources of genetic variation can be derived from landraces and wild relatives (e.g., Cakmak et al. 1999; Manske et al. 2001; Genc and McDonald 2008; Gomez-Becerra et al. 2010). In wheat, the greater level of genetic variation found in wild relatives can be exploited by developing synthetic hexaploids (Ogbonnaya et al. 2007; Kishii et al. 2008). Using a transgenic approach to introduce specific genes or to manipulate gene expression to improve NTUE is another potential way of introducing new levels of genetic variation (Wissuwa et al. 2009; Mittler and Blumwald 2010). Apart from its role in introducing new levels of genetic variation for NTUE, using genetically modified plants should be seen as necessary to provide functional validation of any genes identified from approaches such as quantitative trait loci (QTL) analysis.

### 10.5.1.2   Using Marker-Assisted Selection to Introgress Specific Traits

Phenotypic selection by conventional plant breeding is time consuming and often depends on environmental conditions. Marker-assisted selection (MAS) is a technique that involves the selection of plants carrying genomic regions associated with a particular trait of interest by using molecular markers (Babu et al. 2004). It can be based on populations derived from a biparental cross or on a defined genetic population comprised of unrelated genotypes (association mapping) to identify marker-trait associations. The potential advantages of MAS to breeding programs include:

- Increased efficiency of backcross breeding strategies
- Combining (pyramiding) genes for traits of interest
- Incorporating target QTL into breeding programs (Collard et al. 2005; Francia et al. 2005)

The success of MAS will depend on the location of the marker(s) with respect to the gene contributing to the quantitative trait. Markers located within the gene of interest are the most sought after but these usually require the target gene to be cloned (Francia et al. 2005). Generally, markers are not located within the target gene and tightly linked flanking markers are required to locate accurately the QTL controlling a trait of interest. Markers located closely either side of a QTL are used to minimize the chance of double recombination events between the QTL and both flanking markers (Doerge 2002).

Marker-assisted selection for quantitative traits is often unsuccessful (Langridge and Chalmers 2005; Schuster 2011) and despite the large number of QTLs that have been mapped for a range of nutritional traits, few have been used in practical plant breeding. There are a number of reasons for this (Francia et al. 2005) which include the uncertainty of the QTL position, deficiencies in QTL analysis leading to underestimation or overestimation of the number and magnitude of effects of QTL, an inability to detect a QTL-marker association in divergent backgrounds, and the possibility of losing target QTL due to recombination between the marker and QTL. There may also be difficulty in evaluating epistatic effects and evaluating QTL × environment interactions (Francia et al. 2005).

The recent application of array-based methods for SNP genotyping is a valuable resource for genetic improvement by MAS (Akhunov 2011; McClean et al. 2011). Phenotyping a large association mapping panel under field conditions will help to overcome some problems associated with more traditional biparental QTL mapping. In addition, the possibility exists for combining the association mapping approach with a candidate gene approach to identify genes that could be targeted for transgenic manipulation. The advantage of this approach is that the identified genes would have a clear impact for breeding as they would have been shown to be important under commercial growing conditions. Both association and QTL mapping are essential steps in improving the efficiency of selecting for improved NTUE by breeding programs.

### 10.5.1.3   Gene Expression Studies

Important genes and key pathways involved in NTUE can be identified by combining genetic analysis with gene expression profiles under different levels of nutrient stress. These studies may integrate genomic, transcriptomic, proteomic, and metabolomic methods to elucidate the mechanisms of the perception of nutrient stress and the subsequent regulation of the physiological responses. This integrated approach, when applied to nutritional stresses, has been termed nutrinomics (Yan et al. 2006).

### 10.5.1.4   A Need for Field Validation

Despite the significant advances in application of molecular and genomic methods in the analysis of nutrient stress in recent years, there is still a need to validate results under conditions that are representative of the field environments where the crops are grown. There are a number of examples in the past where promising results obtained under highly controlled conditions either failed to show any advantage or showed much smaller benefits in the more complex field environment (Wissuwa et al. 2009). It is now being increasingly recognized that genomic approaches to improvements in abiotic stress need to integrate laboratory and field studies in a holistic approach to breeding (Mittler and Blumwald 2010).

## 10.5.2 Genomic Approaches to Breeding to Improve Nitrogen Use Efficiency

### 10.5.2.1 General Concepts

Nitrogen is a major factor in plant growth and crop yield (Marschner 1995). The growth and development of plants are often profoundly affected by the form and abundance of the N supply because the form of N significantly alters intracellular metabolism (Andrews et al. 2004; Forde and Lea 2007). Nitrogen also serves as a signaling molecule, with glutamate in particular being closely regulated within very limited concentration ranges (Forde and Lea 2007). Restricted or inappropriate N supply or form of N alters development, including shoot to root ratio, root development, seed development, and the rate of senescence. Activities of enzymes of primary metabolism respond to N supply, but so do the enzymes of photosynthesis, secondary metabolism, and metabolic control.

Since N fertilizers are rapidly depleted from most soil types and symbiotic $N_2$ fixation in many legumes ceases in mid-season, all field crops have some degree of dependence on applications of nitrogenous fertilizer (Marschner 1995), which is supplied to the soils surrounding growing crops before or during the growing season. Plants tend to absorb N released by soil microbes in the rhizosphere, not directly from fertilizer. One estimate had the plant as the seventh organism to assimilate the average N molecule applied to the soil! Nitrogen supply clearly alters the microbiome of the rhizosphere (Garcia-Teijeiro et al. 2009). As one of the cheapest agricultural inputs ($US200–400 per ha) in many countries, N fertilizer is often applied in excess of crop needs. Since many crops are heavily fertilized directly or from the rotational crop residue, high NUE during high soil concentrations is the metric commonly often targeted for improvement by breeders and biotechnologists in this field (Lightfoot et al. 2001, 2007; Lightfoot 2008, 2009). Enhanced NUE by plants should enable crops to be cultivated under reduced N availability, with slow release fertilizers, under water and biotic stress conditions, or poor soil quality. Improving NUE in crops could enable practices directed toward reducing groundwater (Lee and Nielsen 1987; David et al. 1997) and coastal water contamination (Fig. 10.6; Cherfas 1990; Burkholder et al. 1992) by nitrates. The decrease in undesirable environmental effects and reduced dietary nitrate

# Fertilizer derived Dead Zone in Gulf of Mexico

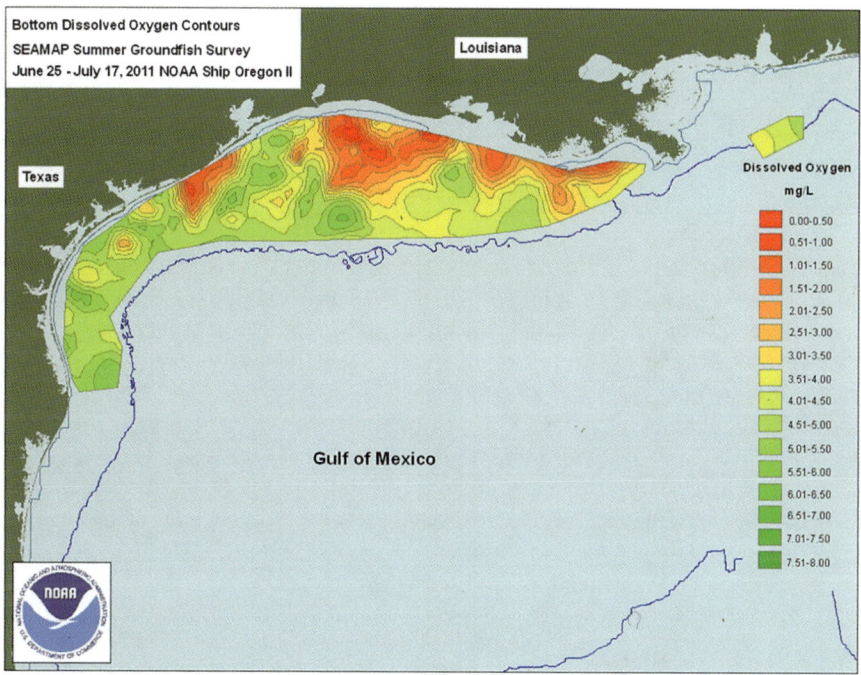

**Fig. 10.6** The effect of fertilizer runoff on coastal waters (From NAOA Web site)

concentrations could decrease several human and animal health problems (Tannenbaum et al. 1978; Mirvish 1985; Moller et al. 1989).

NUE is the product of many components. At a coarse scale, NUE is an expression of:

(a) Soil supply of N
(b) N uptake efficiency
(c) Developmental influence on the size of the N sinks
(d) Remobilization and translocation within the growing plant
(e) The fraction of N translocated to the seed at harvest

Consequently on a fine genomic scale, thousands of genes and hundreds of regulatory networks contribute to NUE in plants, from seed germination to final harvest and hundreds more to the microbial activity in the soil. For brevity this section focuses on genes and regulatory networks with major effects on NUE with an emphasis on patented technologies near commercialization and pending patents.

The Two Forms of Nitrogen Use Efficiency: Regulation of Nitrogen Partitioning and Yield in Crops

Ironically, despite intensive research, the biochemical bases of the regulation of N uptake, partitioning, and tolerance to high concentrations of N are not well understood in crops or model systems (Marschner 1995; Limami et al. 1999; Specht et al. 1999; von Wiren et al. 2000; Coruzzi and Bush 2001; Seebauer et al. 2004; Terce-Laforgue et al. 2004a, b; Century et al. 2008; Coque et al. 2008; Cañas et al. 2009, 2010; Seebauer et al. 2010; Vidal et al. 2010). However, genetically and phenotypically, carbon and N partitioning and yield are clearly interrelated. NUE is one expression of this coordinated regulation. The level of available N will influence the characteristics that will contribute to NUE. Under N limitation, NUE is the ability to yield well with the available resources, while under excess fertilizer N, it is the ability to assimilate all available N. The genes underlying both forms of NUE are very different: on the one hand, assimilation will be cardinal and on the other regulation will be key. The lack of satisfactory cell-free assays and easily quantifiable substrate changes has hindered progress in understanding the molecular biology of N regulation, in contrast to N assimilation where these assays are available. However, the whole plant responses to N limitation and excess are clear and easy to score. These phenotypic changes include several characteristics that are easy to assess, such as plant size, leaf chlorosis, and early senescence, and can be used in a genetic method to isolate the important genes underlying NUE and derive an understanding on their function and biochemistry.

Microbial Activity

It has been known for many years that each molecule of fertilizer N applied to a field will be metabolized by 6–7 different microbes before it is assimilated by the target crop plant (Marschner 1995). Consequently crop plants have been bred to optimize the NUE given current fertilization practices and soil microbial compositions (Specht et al. 1999; Duvick 2005). The area of nitrification and denitrification was recently revolutionized by the identification of slow-growing Archaea as major factors (Cabello et al. 2004). Many studies of plant NUE will have to be revised in view of these unrecognized variables in experiments.

New enhanced-release fertilizers, also known as temperature-dependent slow release N fertilizers, or intelligent fertilizers, are starting to be used to increase NUE and extend N availability over a longer part of the growing season (Trenkel 1997; Garcia-Teijeiro et al. 2009). These fertilizers also change the microbial populations in soil (Fig. 10.7). Their use will provide new opportunities for the genetic improvement of crops. Methylene urea, one type of slow release N, may play a major role as an environmentally safe source of N fertilizer because of its low-leaching potential. Methylene urea has been used widely in industrialized countries at a rate of over 220,000 Mt per year. However, the NUE of plants depends on the

**Fig. 10.7** Microbial populations are altered by N fertilization. *Panel* **a**. Shannon indexes (*H*) for characterizing soil bacterial diversity in soil treated with different N sources [urea and methylene urea (MU)] show differences develop over time. *Panel* **b**. The alterations contribute to differences in the height of corn plants (*Source*: Garcia-Teijeiro et al. 2009)

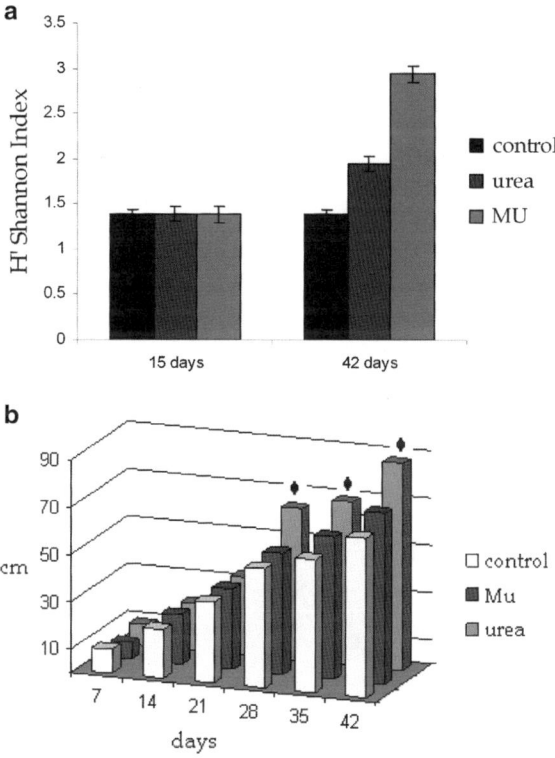

biological activity of microbes in the soil and their capacity to convert organic N into ammonium and nitrate available to the plant (Alexander and Helm 1990).

The decomposition of the methylene urea in soil is caused by both biological and abiotic factors (Koivunen and Horwath 2004; Koivunen et al. 2004b). Only a few species of bacteria are able to degrade methylene urea in soil. The bacterial species were isolated using traditional microbiological techniques, and almost all of them are potential plant and animal pathogenic bacteria, such as *Ochrobactrum anthropi* (Jahns et al. 1997), *Ralstonia paucula* (Jahns et al. 1999; Jahns and Kaltwasser 2000), and *Agrobacterium tumefaciens* (Koivunen and Horwarth 2004). These bacteria synthesize enzymes capable of hydrolyzing methylene urea to a form of N available to the plants. The enzymes are all classified as MDUase (EC 3.5.3.21; methylenediurea deaminase) but each appears different for each bacterial species. In *Ralstonia* sp., the enzyme is a gene fusion between a urease (EC 3.5.1.5) and a protein family of unknown function. Nevertheless, the sequences of the chemical reactions are the same in all cases, and methylene urea hydrolysis leads to three different products, formaldehyde, urea, and ammonium.

To improve the NUE of plants and diminish N losses by leaching and run off, it is important to understand the process responsible for degradation of methylene urea (Koivunen et al. 2004a; Koivunen and Horwath 2004). There is little information about the microbial populations involved in biodegradation of methylene urea in soil, although some microorganisms that degrade dimethylurea have been isolated. There is no information concerning the conditions that influence the transfer of mixed microbial populations that enhance the degradation of dimethylureas from one soil to another. Since the microorganisms isolated were all potential plant pathogens, it is important to elucidate the long-term abiotic and biotic mechanism triggered by this type of fertilizer (Garcia-Teijeiro et al. 2009).

Soil quality before and after the use of these N fertilizers should be compared by physical, chemical, and biological methods (Garcia-Teijeiro et al. 2009). All three approaches detect specific soil characteristics as well as possible interactions and thus can reflect changes in soil quality. Biomass, community structures, and specific functions of soil microorganisms appear to be of major importance for general soil functions and could serve as sensitive soil quality indicators. The microbiological characteristics of a soil reflect and integrate chemical, physical, and biological soil properties over time, since microbial soil communities strongly depend on the conditions of the habitat they colonize. Therefore, microbiological characteristics of a soil may provide indicators, which integrate short-, middle-, and long-term changes in soil quality. As soils display a multitude of biological characteristics and many of them may not be accessible, specific indicators have to be chosen (Zhao et al. 2005).

Nodule Effects

A surprising observation has been that nodules and $N_2$ fixation reduce legume yield and water use efficiency (WUE) in some genotypes (Sinclair et al. 2007; Valentine et al. 2011). Consequently, as breeders pursue high grain yields, they may have been inadvertently selecting against nodulation and limiting $N_2$ fixation. In some cases, N fertilizer is applied to legume crops, which will inhibit nodule formation because nodulation and nodule development are heavily suppressed when nitrate is present. New genetic resources have been found that provide for yield and $N_2$ fixation during drought stress (Sinclair et al. 2007). Breakthroughs in understanding the molecular control of nodule formation (Indrasumunar et al. 2011; Indrasumunar et al. 2012) have also shown that these genes are special derivatives of genes altering the microbial community around roots. Future studies matching root genotypes to microbial populations can potentially provide significant improvements in NUE in future.

Drought Tolerance and Nitrogen Use Efficiency

The importance of drought tolerance and WUE to adaptation to climate change is discussed in detail by Ortiz and Siddique, respectively, in this volume. However N and water interact strongly to determine crop productivity (Sadras 2005) and water availability and the severity of drought can affect NUE. Drought tolerance genes contribute to greater NUE because they improve biomass production over an extended range of soil moisture availability and weather conditions (Pennisi 2008; Harrigan et al. 2009). Effective genes, including *NF-YB1* (Nelson et al. 2007), *CspB* (Castiglioni et al. 2008), and *gdhA* (Mungur et al. 2006; Lightfoot et al. 2007), signal the cell and plant to maintain photosynthesis when water first becomes limiting and help photosynthesis recover quickly when water supply improves. Glutamate is an amino acid implicated in signaling and homeostasis because of the close regulation of cellular concentrations (Forde and Lea 2007). Using it, drought-tolerant crops were produced first in tobacco (Ameziane et al. 2000; Mungur et al. 2006) then in maize (Lightfoot et al. 2007). What was most interesting about the plants was that although total free amino acid concentrations doubled, glutamate did not, leading to the hypothesis that the drought tolerance was caused by signaling of sufficiency that caused an increase in compatible solutes. Metabolite analyses (Nolte et al. 2004; Mungur et al. 2005; Nolte 2009) showed that many hundreds of metabolites changed in abundance in shoots and roots. In addition, changes in crops can have beneficial side effects on plant pathogen interactions (Lightfoot and Fakhoury 2010). Therefore, the NUE/WUE technologies have promise for use during dry spells in agricultural production.

### 10.5.2.2  Genomic Approaches

Use of *Arabidopsis* Mutants

Gene function identification by mutagenesis of *A. thaliana* is an established protocol, and hundreds of mutant genes in a variety of plant processes have been defined, mapped, and some isolated (Huala et al. 2001). The recent development of TILLING programs for many crops (Cooper et al. 2008) promises an abundant supply of new mutants. Mutants in mineral nutrition, root development, and disease resistance have been isolated and mapped in *Arabidopsis*. Gene isolation by positional cloning has been reported for many mutated genes. However, mutants of relevance to NUE are rare (see Table 11.1); only the assimilation enzymes, gluR, and gsr1 lesions have been reported. Reviewing these will be informative.

Mutants in GS and GOGAT are lethal because of the fluxes through photorespiration (see review by Forde and Lea 2007). The two mutants in the mitochondria-located NADH-dependent GDH paralogs tend to be less resistant to abiotic stresses in both *A. thaliana* and *Zea mays* (Lea and Miflin 2011). Mutants in the NADPH GDH enzymes found in plant plastids and cytosol do not have clear phenotypes (Frank Turano, personal communication 2011). Mutants in aspartate and asparagine synthesis and transport are deleterious, underlining their role in transport and storage (Vidal et al. 2010). Double mutants in nitrite reductase are lethal because nitrite is very toxic whereas single mutants are phenotypic. In contrast, single mutants in nitrate reductase have either a neutral effect or can increase NUE. Kleinhofs et al. (1980) showed that barley with 10 % of the usual nitrate reductase activity was just as productive as the wild type. This phenomenon has been repeated in other plants. Nitrate reductase production by plants far exceeds the needs of a well-fertilized plant, which is probably a holdover from the predomestication era. Reducing the metabolic load of high nitrate reductase production may increase NUE.

Two regulatory mutants were well characterized. Gsr1 was a lesion that caused increased susceptibility to methylamine (Meyer 1997; Meyer et al. 2006). Different N sources provided different degrees of protection from the toxicity of methylamine. Glutamine was more protective than glutamate, nitrate, and nitrate mixed with ammonium more than ammonium. The lesion in gsr1 interfered with photosynthesis, emphasizing the control of C metabolism by the plant's N status. The lesion in gsr1 appears to map to a region of chromosome 5 encoding an amine oxidase, suggesting a mechanism for the lesion in detoxification. Lesions in putative glutamate receptors are known (Lam et al. 1998; Dennison and Spalding 2000; Coruzzi and Zhou 2001). These lesions provide alterations in growth and NUE as well as tolerance of second messaging inhibitors like okadaic acid. However, some mutants have unusual genes underlying N-regulatory like phenotypes (Godon et al. 1996) suggesting the field will continue to be recalcitrant to analysis.

Several transporters in plants and their microbial symbionts provide for alterations in NUE because all forms of N are accumulated against a steep concentration gradient (Godon et al. 1996; Kaiser et al. 1998; Gupta et al. 2011). Ammonium in particular is hard to accumulate and may pass in and out of cells several times before assimilation is successful. Consequently lesions in ammonium and nitrate transport are deleterious and do not improve NUE. The lesion in proton-dependent peptide transport is different. Resistance to herbicides targeted to the enzyme GS like MSX is provided by the lack of these transporters and NUE can be increased. These transporters will be particularly important to assimilating fertilizer N from root-associated microbes.

Microarrays and Transcript Analysis

Large numbers of genes are being discovered to be involved in processes like NUE by clustering during microarrays in crop plants. To obtain functional information on these genes, efficient expression monitoring methods have been developed (Wang et al. 2000). Rapid and simultaneous differential expression analysis of independent biological samples indicates activity. Using expression profiles, gene regulation perturbations in transgenic and mutant plants can be monitored and function inferred providing a central platform for plant functional genomics (Schenk et al. 2000; Wang et al. 2000). However, microarrays of relevance to NUE are rare: only gsr1 and a few enzyme lesions have been reported.

Microarrays have led to the patenting of sets of protein families implicated in NUE (Table 10.5). For example, Goldman et al. (2009) have applied for a patent for 157 protein families implicated in NUE by microarray and then tested for activity in transgenics which resulted in one or more positive results from assays of NUE,

**Table 10.5** Partial list of protein families implicated in nitrogen use efficiency by "-omics" approaches

| ADH_N | ADH_zinc_N | AP2 | AUX_IAA |
|---|---|---|---|
| Aa_trans | Acyl_transf. 1 | Aldedh | Aldo_ket_red |
| Alpha-amylase | Aminotran1-3 | Ammonium_transp | Arm |
| Asn_synthase | BAG | BSD | Beta_elim_lyase |
| Biotin_lipoyl | Brix | Bromodomain | $CI_4$ |
| CTP_transf2 | Catalase | CcmH | Chal_sti_synt_C |
| Cyclin_C | Cyclin_N | Cys_Met_Meta_PP | DAO |
| DIM1 | DPBB1 | DRMBL | DUF167 |
| DUF231 | DUF250 | DUF6 | DUF783 |
| DUF962 | E2F_TDP | E3_binding | EBP |
| Enolase_C | Enolase_N | | |

WUE and yield, in enhanced tolerance to salt, cold and heat, and in enhanced concentrations of oil and/or protein in seed.

Metanomic Tools for Extending Functional Genomics

In the postgenome sequence era, the determination of gene function(s) and relationships to pathways will be the focus. Multiparallel analysis of mRNA abundance and their protein products will suggest functions but not direct information on biological function. The multiplicity of gene interactions and metabolic network changes engineered by mutation are not always predictable, and many changes are cryptic. Metabolic profiling can link phenotype to biochemistry (Mungur et al. 2005; Nolte 2009). The methods are fast, reliable, sensitive, and automated due to improvements in mass spectrometry (MS). Libraries of compound identities have been developed for plants at mass accuracy of 0.01d, often by MS–MS fragmentation. However, the mass accuracy of ion cyclotron MS in a Fourier-transformed MS format provides for mass accuracy to 0.0001d. This accuracy allows for unequivocal identification of a larger number of compounds in fewer analyses.

Figure 10.8 shows the effects of GDH on pathways related to NUE in tobacco roots. FT-ICR-MS detected 2,012 ions reproducible in 2–4 ionization protocols. There were 283 ions in roots and 98 ions in leaves that appeared to significantly change abundance due to the measured GDH activity. About 58 % of ions could not be used to infer a corresponding metabolite. From the 42 % of ions that inferred known metabolites, many amino acids, organic acids, amines, and sugars increased and many fatty acids and amines decreased. These changes were profound and underlay the ability of the GDH transgene to increase NUE, ammonium assimilation, and nutritional value. The changes in core metabolism looked very similar to the changes reported for N-sink altered opaque mutants of *Z. mays*. The C skeleton map of Fig. 10.8 may not be the best way to look at the interactome. Two other views, the N fate map for N assimilation and the protein–protein interaction, are

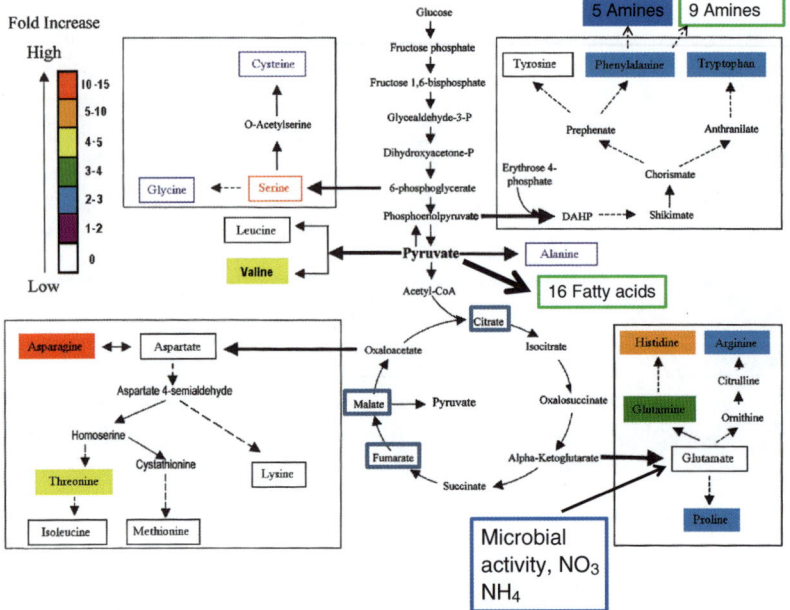

**Amino Acid Changes in Leaves without (boxes) and with (ovals) gluphosinate**

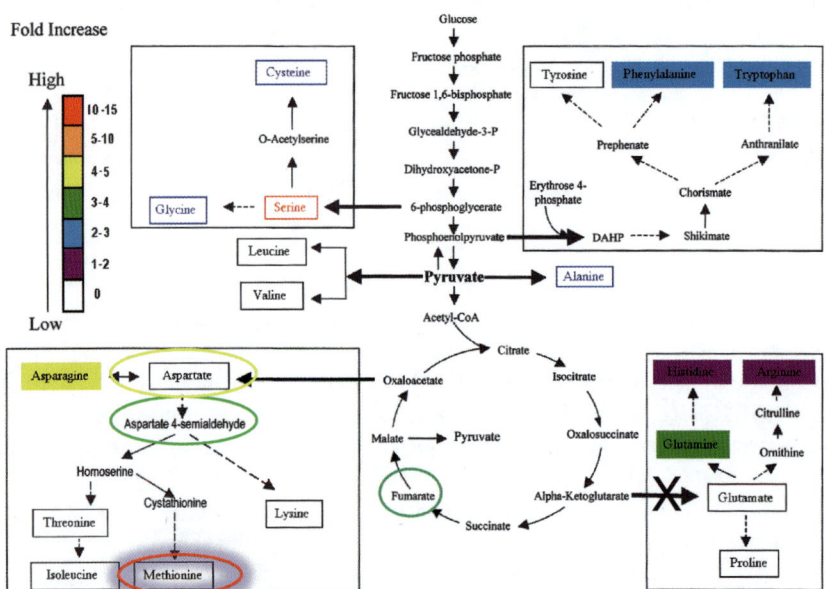

Metabolites in blue boxes were not detected. Metabolites in red boxes were used as internal standards and therefore not measured. Metabolites in black boxes were detected and not changed. Glow indicated increased by gluphosinate

**Fig. 10.8** Metabolite changes related to NUE in GDH transgenic roots (**a**) and leaves (**b**). Metabolites in *blue boxes* were not detected. Metabolites in *red boxes* were used as internal

also useful. In all three views, it is clear that N assimilation is at the nexus of many pathways. In fact, pyruvate and glutamate are at the center of metabolism, being related by 2.7 metabolisms on average to the metabolites in the cell. Still it is surprising that modest changes in glutamate concentrations in the cytoplasm had so many effects. Equally surprising was the observation that the transcriptome was not altered (Mungur et al. 2005). It should be expected that analysis of the proteome and metabolome promises to identify new genes useful for altering NUE that are missed by mutation, overexpression, and TA screens.

Metabolic analysis was also applied to the alanine aminotransferase (AlaAT) transgenic plants with improved NUE (Good et al. 2004). Rice (*O. sativa* L.) was genetically engineered by introducing a barley AlaAT cDNA driven by a rice tissue-specific promoter (OsAnt1; Shrawat et al. 2008). This modification increased the biomass and grain yield significantly in comparison with control plants when plants were well supplied with N. Compared with controls, transgenic rice plants also demonstrated significant changes in key metabolites and total N content, indicating increased N uptake efficiency. Metabolites included many of those reported for GDH. Goldman et al. (2009) report similar effects from an alanine decarboxylase in transgenic plants.

Transgenics Lacking A Priori Evidence for Nitrogen Use Efficiency

Table 11.1 (Chap. 11) summarizes some recent examples of transgenic approaches lacking a priori evidence of involvement in NUE that have been used to improve NUE. These approaches also showed NUE and WUE are closely interrelated. The connection between C metabolism and N metabolism underlies this association. These protein family and gene lists are very interesting, being mixtures of assimilation enzyme transport factors and esoteric proteins. Proteins like gluconases, catalases, and lyases are hard to place in relation to known pathways underlying NUE. However, the preponderance of proteins involved in TA control and protein degradation provides clues to the effect. Regulatory pathways in plants are hierarchical with about three layers. The enzymes in the basal layer are each controlled by several middle layer regulators. In turn, these are regulated by a few proteins sensitive to key environmental cues. The relationships are reciprocal to a change in a lyase that might alter the activity of a protein in a well-characterized NUE pathway. However, caution is warranted since even well-defined enzymes like GS (Kichey et al. 2005; Kichey et al. 2006) and their alleles (Ortega et al. 2006) have

---

**Fig. 10.8** (continued) standards and therefore detected. Metabolites in *blue boxes* were increased 2–3 fold. Metabolites in *black boxes* were detected and not changed. Metabolites in *green boxes* were decreased. Metabolite circled were altered by gluphosinate treatments. (adapted from Mungur et al. 2005)

complex relationships with plant growth and development (Hemon et al. 1990). Interactome analysis promises much in this area in the next few years.

## 10.6 Conclusions

Understanding of the control of NUE is at a new beginning (Krouk et al. 2010). Surprising effects of individual proteins abound showing our understanding of the processes behind NUE is preliminary. The potential for gain remains large. Most of the gradual increase in crop yield in major well-fertilized crops is likely an expression of NUE and WUE. However, to date the improvements in NUE provided by genes like GDH and AlaAT remain to be commercialized. The integrations of techniques made by systems biology hold promise. Can crop yields be moved on in quantum leaps by manipulating NUE? The very large yield gains from weed control provided by herbicide resistance technologies suggest NUE is improved in those crops. Can NUE be directly manipulated? If the soil's microbial community is considered and properly measured, that may finally be possible.

At the same time, as the need to increase in food production becomes greater, crops for biofuels are needed that do not compete with crops for food. Growth on marginal lands requires the second form of NUE—growth with low or no inputs. Here we must reverse centuries of crop breeding. The directed manipulation of NUE must be applied to both fields if this planet is to sustain a human population of nine billion by 2050.

## 10.6.1 Breeding for Improved Phosphorus Use Efficiency

After N, P is the second most important nutrient in crop production. Over half of the world's agricultural land has soils low in plant available P (Lynch 2011) and production relies on regular application of P fertilizer. In intensive systems in which regular applications of P occur, recovery of fertilizer P is low in the year of its application resulting in a gradual increase in soil P levels (Vance et al. 2003). The finite reserves of the world's supplies of high-quality rock phosphate have prompted concerns about the reliability of supplies in the future and have raised the specter of "peak phosphorus" (Cordell et al. 2009). There is a general consensus that improvements PUE are required and there is a need to explore the ability to improve crop PUE to limit further inputs of P (Hinsinger et al. 2011). In regions where soil P reserves have built up over time, improving the ability of plants to exploit soil reserves has been emphasized (Vance et al. 2003), although there is growing interest in improving the efficiency of utilization of P taken up by plants (Wang et al. 2010; Rose and Wissuwa 2012). However in regions where soil P levels are severely depleted, strategies to improve soil P will also need to be adopted in concert with genetic improvement.

Phosphorus in soil is poorly available to plants because of reactions with minerals such as aluminum, Fe, and calcium as well as being bound in organic matter. It also moves slowly to the root surface by diffusion. A low supply of P can induce a P starvation response to either enhance uptake from the soil or to maintain P homeostasis in actively growing tissue. There are many mechanisms that plants employ to increase uptake of P, including changes in root architecture; secretion of organic acids, protons, and phosphatases to increase the availability of phosphate in the rhizosphere; interactions with microorganisms; and symbiotic relationships with mycorrhizal fungi. These processes and their potential value to improvements in P nutrition have been extensively reviewed recently (Gahoonia and Nielsen 2004; Rengel and Marschner 2005; George and Richardson 2008; Kirkby and Johnston 2008; Richardson et al. 2009a, b). Low supplies of P may also lead to greater recycling of P within the plant from intracellular organic P compounds by increasing activities of APase PAPs, RNase, and scavenging Pi by replacing membrane phospholipids with galacto- and sulpho-lipids, which will contribute to greater PUE. However, under P starvation, it is important to distinguish between a general stress response and traits associated with improved tolerance to P stress to enable gains in PUE to be achieved (Pariasca-Tanaka et al. 2009).

The complexity of P nutrition of plants is because the availability and uptake of P depends on the interaction between soil, plant, and microbial processes and the plasticity afforded by increased recycling of P. Consequently, the relative importance of different processes and the effectiveness of different plant characteristics are likely to vary according to soil type, cropping history, and climate and weather. Therefore, targeting one specific mechanism to improve P uptake by plants may yield limited success. This is not to say that gains cannot be made as there have been reported improvements in PUE in a number of crops in China (Yan et al. 2006). However, a common thread among many studies is that the ability to take up P from small pools of available P in the soil is crucial to the P nutrition of crops. Therefore much of the interest in genetic improvement in PUE has centered on differences in genotypes being able to increase the availability P in soil and its uptake. Root traits have figured prominently in studies in PUE (Vance et al. 2003; Wissuwa et al. 2009).

At present conventional approaches to genetic improvement are being used to improve PUE, with QTL mapping and marker-aided selection being a particular focus. As our understanding of the physiological and molecular bases of P starvation responses and PUE improve, it is likely that genetic engineering will figure more prominently in future gains in PUE, although the success of a transgenic strategy will depend on public acceptance.

### 10.6.1.1 Current Challenges

While there has been a long history of characterizing genetic differences in NTUE among crop plants, breeding for improved NTUE is still in its infancy. The development of molecular technologies to understand the genetic basis of the trait and mechanisms of nutrient efficiency has provided new possibilities for significant advances to be made. Despite exciting opportunities, there are a number of challenges to improving PUE, whether based upon conventional genetic or transgenic approaches.

## Incomplete Understanding of Controls of P Uptake

The growth and P uptake of a crop relies on a complex interaction between the plant, the chemical and physical characteristics of the soil, and the biological properties of the rhizosphere. While there have been significant advances in our understanding of specific components of this system and of their genetic basis, time and again the comment is made in reviews of P nutrition that our understanding of the operation of the system is incomplete. There are many examples of traits that result in substantial improvements in P nutrition under controlled conditions, which fail to show similar advantages in field soil. This is one of the compelling reasons for rigorous testing in soils that are representative of commercial practice.

## Appropriate Phenotyping

Much of the reported work comes from glasshouse, controlled environment and hydroponic studies. This is out of convenience as well as necessity in the case of genetic modification (GM). There have been few studies in which results from controlled environment experiments have been compared with responses in field trials. Soil properties will influence the form and the availability of P but our understanding of the importance of specific mechanisms of PUE in different soils is poor. High-throughput screening methods are desirable, but if they cannot be demonstrated to correlate with results obtained from commercial growing situations, then their application may be limited.

## Environmental Variability in Expression of Phosphorus Use Efficiency

Grain yields of rainfed crops show large environmental variation, which reflects differences in seasonal conditions. Variation in soil moisture is likely to play an influential role in the availability of soil P, P uptake, and the consequent expression of PUE. Such considerations have not been addressed adequately as there have been few long-term assessments of genetic diversity for PUE under field conditions. Identifying genotypes, which show consistent PUE over a range of environments, is a key element to identify traits to improve PUE.

Limited Population-Based Studies

A review of the studies presented in Table 10.6 suggests there are a limited number of population-based mapping studies. A larger number of mapping studies within species will provide greater certainty of the value of QTL regions detected, and common QTLs could then be targeted for map-based cloning and potential transgenic approaches to improve PUE.

### 10.6.1.2   Genomic Approaches

QTL and Marker-Aided Selection

PUE is a complex trait subject to considerable environmental influence. Identifying QTL for PUE and subsequently developing molecular markers have the potential to improve selection for high PUE in crops. However, despite the large number of QTLs that have been identified for a variety of traits, including PUE, few have been actively deployed in plant-breeding programs and are used routinely by plant breeders (Collins et al. 2008). Lack of field validation in a range of genetic backgrounds and over environments as well as the small effects of many QTLs are contributing factors for the lack of success. These are important issues that need to be addressed if marker-aided selection is to be used to improve PUE in crop and pasture plants. However increasingly, marker-assisted selection is being used routinely and this trend will continue into the future. For example, marker-assisted selection has contributed to the release of more than 90 % of the varieties of common beans released in the USA over the past decade (McClean et al. 2011).

Phosphorus Uptake and Phosphorus Use Efficiency

There are relatively few studies that have identified QTLs for PUE in the major crop species (Table 10.6). Generally, biomass production, shoot P concentration, and uptake are the traits of interest, and these are often assessed under both limiting and nonlimiting P conditions. Given the central role of biomass and yield in most definitions of PUE, it is not surprising that the QTLs for biomass and yield often collocate with QTLs for P uptake and/or P utilization efficiency. For example, in wheat (Su et al. 2006, 2009), *Brassica* sp. (Hammond et al. 2009; Yang et al. 2010, 2011), soybean (Zhang et al. 2009), and rice (Wissuwa et al. 1998), QTLs for P uptake efficiency collocated with QTL for biomass production. This is because the correlation between biomass production and shoot P uptake is often high and shows that biomass production drives P uptake. Many studies show that QTLs detected for P uptake are linked in repulsion with QTLs for PUE (Wissuwa et al. 1998; Su et al. 2006, 2009; Zhang et al. 2009) with the authors suggesting that it will therefore be difficult to improve both traits simultaneously. It is not clear if this negative correlation is an artifact of the definitions that are used for P uptake and PUE.

**Table 10.6** Summary of studies that have identified QTL associated with phosphorus use efficiency

| Population | Environment(s) | Population size | Marker no | Traits[a] | QTL no | Variation explained (%) | Reference |
|---|---|---|---|---|---|---|---|
| Wheat (*Triticum aestivum*) | | | | | | | |
| Lovrin10/Chinese spring | Glasshouse | 92 DH lines | 253 | P uptake<br>P utilization efficiency | 39 | 5.7–34.6 | Su et al. (2006) |
| Hanxuan10/Lumai14 | Glasshouse and field | 120 DH lines | 395 | P uptake efficiency<br>P utilization efficiency | 195 | 4.1–38.8 | Su et al. (2009) |
| Barley (*Hordeum vulgare*) | | | | | | | |
| Association mapping panel | Field | 120 (56 winter types; 64 spring types) | 921 (winter)<br>843 (spring) | Shoot P concentration | 8 | Not provided | George et al. (2011) |
| *Brassica sp./Brassica napus* | | | | | | | |
| AD12DHd/GDDH33 | Glasshouse | 90 DH lines | Not provided (~90 % genome coverage) | Agronomic efficiency<br>P uptake efficiency<br>P utilization efficiency<br>Physiological P use efficiency<br>P efficiency ratio | 37 | 5.3–52. 6 | Hammond et al. (2009) |
| Eyou Changjia/B104-2 | Greenhouse | 124 RILs | 503 | P uptake | 62 | 8.1–17.1 | Yang et al. (2010) |
| Eyou Changjia/B104-2 | Greenhouse | 124 RILs | 553 | P uptake<br>P use efficiency | 71 | Not provided | Yang et al. (2011) |
| Common bean (*Phaseolus* sp.) | | | | | | | |
| DOR364/G19833 | Growth pouches and field | 86 RILs (growth pouches)<br>71 RILs (field) | 236 | P acquisition efficiency | 22 | 9.3–20.3 | Liao et al. (2004) |

| | | | | | | | |
|---|---|---|---|---|---|---|---|
| DOR364/G19833 | Field and greenhouse (hydroponic system) | 71 RILs (field) 86 RILs (greenhouse) | 236 | P acquisition efficiency | 26 | 9.4–51.3 | Beebe et al. (2006) |
| G19833/AND696 | Field | 75 RILs | 167 | P uptake<br>P use efficiency | 29 | 9.0–33 | Cichy et al. (2009) |
| Soybean (*Glycine max*) | | | | | | | |
| BD2/BX10 | Field | 106 RILs | 296 | Root and shoot P concentration | 31 | 9.1–31.3 | Liang et al. (2010) |
| Nannong94-156/Bogao | Greenhouse | 152 RILs | 371 | P acquisition efficiency<br>P use efficiency | 34 | 6.6–19.3 | Zhang et al. (2009) |
| Rice (*Oryza sativa*) | | | | | | | |
| Nipponbare/Kasalath | Glasshouse | 98 BILs | 245 | P uptake<br>P use efficiency | 7 | 5.8–27.9 | Wissuwa et al. (1998) |

[a]For clarity, only phosphorus use efficiency-related traits (and not each of the component traits) are reported in the table

If a genotype has a high P uptake efficiency (driven by biomass production), its P utilization efficiency (calculated as biomass production per unit P uptake) will be low. This further highlights the problem with commonly used definitions of NTUE.

An alternative, but infrequently used, approach is to examine QTLs for relative yield under low and high supplies of P to overcome the potential confounding effect of high biomass or yield on P uptake and PUE. Yang et al. (2010) assessed the relationship between QTLs for root traits and P uptake in *Brassica napus* and found that QTLs for P uptake and biomass production were linked. Later, they used relative yield and state that "these QTL were demonstrated to represent the true QTL for P efficiency" (Yang et al. 2011). Unfortunately the two sets of results cannot be directly compared due to the different format in which the maps were presented.

Currently some of the most promising works on breeding for improved PUE is in rice, where a QTL associated with P uptake (*Pup 1*) has been identified and found to confer significant improvements in growth and yield in upland rice under low soil P (Wissuwa et al. 1998; Chin et al. 2010, 2011). The identification of the QTL was based on growth under severely P-deficient conditions. Subsequent germ plasm surveys have shown that the P efficiency allele at *Pup 1* was most commonly found among genotypes developed for drought-prone, hostile upland environments (Chin et al. 2010). Interestingly, it seems that the value of *Pup 1* to PUE is not associated with genes directly related to P uptake and the underlying mode of action of *Pup 1* is yet to be described. The candidate genes associated with the QTL include a protein kinase gene and a dirigent-like gene that have been reported to be involved in root-specific growth functions and in tolerance to abiotic stress and a hypothetical gene that also has a root-specific function (Chin et al. 2011).

Root Traits

Another area of work that has shown promise in a number of crops is root architecture. To maximize P acquisition in low P conditions, plants change root growth and development by promoting the formation of a shallow, highly branched root system through a reduction of primary root growth and an increase in adventitious roots and lateral root density as well as the development of more and longer root hairs. Root architectural traits associated with enhanced topsoil foraging include shallower growth angles of axial roots, a greater number of adventitious axial roots, and greater dispersion of lateral roots (Péret et al. 2011). A number of genes controlling lateral root development have been identified in *Arabidopsis* (Péret et al. 2011) and rice (Coudert et al. 2010). At least six root QTL have been identified in maize and are good candidates for further evaluation (Hund et al. 2011). Genetic variation in root hair length and loci-controlling barley root hair formation has also been identified (Gahoonia and Nielsen 1997; Szarejko et al. 2005).

There have been a number of attempts to link QTLs for root development and architectural traits with QTLs for P uptake or utilization efficiency (Liao et al. 2004;

Beebe et al. 2006; Ochoa et al. 2006; Cichy et al. 2009; Hammond et al. 2009; Zhang et al. 2009; Liang et al. 2010; Yang et al. 2010, 2011). Common bean (*Phaseolus vulgaris* L.) has been the most widely studied species in this respect. When P is concentrated in the topsoil, lines of beans with shallow rooting were more productive than lines with deep roots (Rubio et al. 2003). Subsequent work used a population developed from a cross between Andean and Mesoamerican parents in which Beebe et al. (2006) and Liao et al. (2004) analyzed the same field data set either with hydroponic screening (Beebe et al. 2006) or growth pouch results (Liao et al. 2004). Three of the QTLs that contributed to P acquisition efficiency (*Pup4.1*, *Pup7.1*, and *Pup11.1*) in the field were linked to some of the QTLs associated with root gravitropism in growth pouches (Liao et al. 2004), suggesting root gravitropism contributes to PUE. Beebe et al. (2006) reported fewer QTLs for P acquisition efficiency in the field, but nevertheless, some of the QTLs for root architectural traits identified in the field and in hydroponics were associated with P acquisition QTLs (Beebe et al. 2006). Ochoa et al. (2006) examined adventitious root formation in a related population but the QTLs for adventitious root formation were not located in the same region as the QTLs for P acquisition efficiency as reported in Beebe et al. (2006) and Liao et al. (2004). Cichy et al. (2009) also could not find any relationship between root traits and P uptake in an Andean/Andean bean population. The contrasting results for the relationship between root architecture and PUE in these studies reflect the difficulty in phenotyping root architectural traits and the small effect of the QTLs that have been identified.

The Influence of Phenology

Crop phenology is an important influence on nutrient uptake and allocation and may play an important role in adaptation to low availability of soil P (Nord and Lynch 2009; Nord et al. 2011). The importance of phenology is also highlighted in a number of genetic studies, although it is unclear whether these associations are a result of the effect of P deficiency on development or play a role in PUE. In wheat, Su et al. (2006) identified a range of QTLs associated with P-deficiency tolerance, with three main clusters located on chromosome 4B, 5A, and 5D. Interestingly, the 5A and 5D QTLs were associated with the major vernalization genes, *Vrn-A1* and *Vrn-D1*. In a later study on a different wheat population, the effect of *Vrn-A1* and *Vrn-D1* were not detected, but it is unclear if this population was segregating at these loci (Su et al. 2006). In barley, George et al. (2011) attempted to identify QTLs for shoot P concentration in an association mapping panel of 120 barley genotypes, composed of 56 winter and 64 spring types. No common QTLs could be detected between winter and spring types. However, the associations only just exceeded the threshold for detection, and George et al. (2011) attribute this to the relatively small population sizes that were used for the study and/or limited genetic variation. QTLs for PUE in common bean were linked with the *fin* gene, which regulates determinism in this species (Cichy et al. 2009).

There are not enough studies of different populations within a species to make a strong case for the likely effectiveness of any particular QTLs that have been detected. A further issue arises from the population sizes that are used in many studies. The effect of population size on the accuracy of QTL mapping is well known; in small populations, only QTLs with large effect are likely to be identified, and their effect can be overinflated (Tanksley 1993). The small population sizes that are generally used probably arise from difficulty in phenotyping for PUE since the need to screen at both limiting compared to nonlimiting conditions doubles the amount of phenotyping that needs to be done.

A review of the studies presented in Table 10.6 suggests clear directions that should be taken for future QTL-mapping studies. Identification of QTLs under controlled conditions does not necessarily mean that the QTLs will be of practical value to plant breeding. A larger number of mapping studies within species will provide greater certainty of the value of QTL regions detected, and common QTLs could then be targeted for map-based cloning and potential transgenic approaches to improve PUE. Use of near isogenic lines to assess the effect of the QTLs will also be valuable. The effect of the QTLs also needs to be demonstrated under field conditions to examine the influence of the QTLs, and ideally, screening for PUE should be conducted on a range of different soil types and environments. Problems associated with definition of PUE need to be overcome—a comparison of results obtained using different definitions will help. Finally, the phenomenon of linkage of developmental genes with PUE should be investigated further to determine if certain combinations, particularly in wheat, can lead to improved PUE.

## Transgenic Approaches for Improving Phosphorus Use Efficiency

The use of molecular breeding to improve PUE is in its infancy. However, in the long-term, engineering key components in the regulatory network of the P starvation response represent a useful approach for molecular breeding of plants toward more efficient Pi uptake and use. This has been shown in *Arabidopsis* and rice using overexpression and gene knockdown or knockout. Artificial target mimics of miRNAs can also be used for functional studies on PSR and has potential to contribute to molecular breeding. Table 10.7 provides an overview of genes that have been assessed using transgenic approaches and Table 10.8 provides detail of the promoters that have been used to control gene expression.

### Overexpression

Overexpression of the high-affinity OsPHT1;8 increases Pi uptake and translocation from roots to shoots in rice (Jia et al. 2011) although no increase in Pi uptake has been seen in barley (Rae et al. 2004). Transgenic rice lines overexpressing OsPHT1;8 increased maximum influx by 3–5-fold, indicating the transgenic approach can enhance Pi uptake from soil in this crop (Jia et al. 2011).

**Table 10.7** Potential genes for enhanced phosphorus uptake and remobilization

| | Function | Source | Background | Promoter | Phenotype | Reference |
|---|---|---|---|---|---|---|
| P uptake | | | | | | |
| OsPHT1;8 | High-affinity Pi transporter | Rice | Rice | Maize ubiquitin | P uptake, P translocation to grain | Jia et al. (2011) |
| ALMT1 | Malate transporter | Wheat | Barley | Ubiquitin | P uptake | Delhaize et al. (2009) |
| AVP1 | Root proliferation, apoplast acidification | Arabidopsis | Arabidopsis Tomato Rice | AVP1 | P uptake, root size | Yang et al. (2007) |
| P translocation | | | | | | |
| OsPHT1;2 | Low-affinity Pi transporter | Rice | Rice | Ubiquitin | P translocation | Liu et al. (2010) |
| AtPHT1;5 | Pi transporter | Arabidopsis | Arabidopsis | Actin | P translocation | Nagarajan et al. (2011) |
| Transcription factors | | | | | | |
| PHR1 | MYB transcription factor | Arabidopsis | Arabidopsis | CaMV 35S | P uptake, translocation Root proliferation | Nilsson et al. (2007) |
| PHR2 | MYB transcription factor | Rice | Rice | CaMV 35S | P uptake, translocation, root length | Zhou et al. (2008) |
| miRNAs and other genes | | | | | | |
| *Target gene* | | | | | | |
| miR399 | Ubiquitin conjugase E2 | Arabidopsis Rice | Arabidopsis Rice | | P uptake, root to shoot transfer | Aung et al. (2006), Hu et al. (2011) |
| MiR399d | Ubiquitin conjugase E2 | Arabidopsis | Tomato | CaMV 35S | P uptake Acid phosphatases | Gao et al. (2010) |
| PHO2 | Unknown | Rice | Rice | mutation | P translocation | Hu et al. (2011) |

**Table 10.8** Tissue-specific and P-inducible promoters in plants

| Gene | Source | Expression patterns | Tissue specificity | Reference |
|------|--------|---------------------|--------------------|-----------|
| *AtPHT1;5* | *Arabidopsis* | P deficiency | Source to sink | Nagarajan et al. (2011) |
| *HvPHT1;1* | Barley | P deficiency | Epidermal, cortex, and stele of roots | Schunmann et al. (2004) |
| *HvPHT1;2* | Barley | P deficiency | Epidermal, cortex, and stele of roots | Schunmann et al. (2004) |
| *OsPHT1;8* | Rice | Constitutive | Roots and shoots | Jia et al. (2011) |
| *OsIPS1* | Rice | P deficiency | Phloem | Hou et al. (2005) |
| *MiR399* | *Arabidopsis* | P deficiency | Phloem | Aung et al. (2006) |

Overexpression of miR399s leads to a reduction in remobilization of Pi in *Arabidopsis* (Aung et al. 2006) and rice (Hu et al. 2011). This approach has also been used to modify secretion of acid phosphatase and protons in the roots of tomato, which facilitated the hydrolysis of soil organic P and dissolution of Pi (Gao et al. 2010).

## Gene Knockdown and Knockout

Knockdown OsPHT1;8 reduced Pi uptake and translocation (Jia et al. 2011). Knockout of *Osphf1* reduced Pi uptake and translocation from roots to shoots in rice as well as arsenate (Wu et al. 2011). The knockout mutant of *ltn1,* an ortholog of AtPHO2, shows several typical Pi-starvation responses, such as stimulation of phosphatase and RNase activities, lipid composition alteration, and N assimilation repression (Hu et al. 2011). The elongation of primary and adventitious roots is also enhanced in the *ltn1* mutant, suggesting that the modification of LTN1 expression may be able to enhance morphological, physiological, and biochemical responses to Pi starvation.

## Cell-Specific Expression in Roots

Specific expression driven by cell-specific promoters instead of constitutive promoters will be preferred in some cases. Phosphatases that release P from organic compounds would be more useful if produced by shallow roots than by deep roots, since soil organic matter typically decreases with depth (Lynch 2011). In contrast, carboxylates capable of releasing P from Fe and Al oxides may be more useful when released into deeper soil horizons where these forms of P predominate (Lynch 2011).

### 10.6.1.3  Mycorrhizal Associations

The reduction in concentration of shoot P at elevated $CO_2$ concentrations can often be alleviated by the formation of AM symbioses (Cavagnaro et al. 2011). Therefore, the importance of mycorrhizal fungi may increase in the future in attempts to reduce the rates of P fertilizer applied to crops. Genetic variation for AM associations has been demonstrated in a number of plant species (Baon et al. 1993; Zhu et al. 2001; Jakobsen et al. 2005; An et al. 2010), which offers the potential to select for AM responsiveness. However, the positive effect of mycorrhizal colonization decreases as soil P levels increase, and further work would benefit from assessment of colonization at P levels that are representative of agricultural soils. Understanding the physiological and genetic controls of the AM–plant interaction may enable root infection to occur even at high soil P concentrations, which may enhance P uptake over a wider range of soil P concentrations than occurs currently. Also a better understanding of the genetic controls of AM infection and P uptake and nutrition of the host plant is required to allow the synergistic relationship to be manipulated.

## 10.7  General Conclusions

An adequate and balanced supply of essential nutrients is a cornerstone of improvements in crop productivity. Nutrient efficiency will become increasingly important in the future as farmers strive to achieve higher levels of productivity and maintain profitable enterprises in the face of increasing fertilizer prices and under the influence of a changing climate. The effects of climate change on soil nutrient cycling, nutrient availability, and crop nutrient requirements are difficult to predict. However, the principles that have governed efficient nutrient management in the past are not going to alter with future changes in climate. Nutrient management under a changing climate is likely to operate within the same boundaries as current practices.

The strategies to improve NTUE will differ depending on past nutrient management practices. In regions of the world where crops are chronically malnourished, increases in soil fertility through soil improvement and fertilizer use will underpin increases in productivity, while in areas where fertilizer has been applied in excess of the crops' requirements, better use of the soil nutrient bank and a more sustainable use of fertilizer will be needed. In both cases, breeding for improved NTUE can play an important role in increasing the NTUE of the system, although the specific breeding objectives to achieve this may differ.

There is considerable variability in how crop species and varieties exploit soil nutrients and respond to fertilizer. The improvements in yield potential that have been achieved by breeding have resulted in a passive improvement in NTUE, but there are few examples of commercial varieties being developed for their high

NTUE. The complexity of the soil and plant processes that influence the nutrient status of crops, the incomplete understanding of the genetic control of NTUE and its underlying physiological and molecular basis, and a consistent conceptual understanding of NTUE have limited progress.

The rapid development of an array of molecular and genomic techniques provides an opportunity to overcome many of the hurdles that have hindered progress so far. Plant scientists are at the cusp of making considerable advances in understanding NTUE and developing varieties that are more nutrient efficient. However, an important aspect of the use and implementation of this approach is that material needs to be tested under realistic field conditions. Marshaling new methods and technologies with the traditional disciplines of plant breeding, crop physiology, and agronomy provides expanded opportunities to study genetic differences in NTUE and to link genotype to phenotype (Andrade et al. 2009; Messina et al. 2009).

# References

Akhunov E (2011) Next-generation tools for wheat genetics and breeding: high-throughput SNP genotyping assays and sequence-based genotyping. In: Dreisigacker S, Singh S (eds) 21st international Triticeae mapping initiative workshop, Mexico City, Mexico, 50 p

Alexander A, Helm HU (1990) Ureaform as a slow release fertilizer - a review. Z Pflanzenernaehr Bodenk 153:249–255

Ameziane R, Bernhard K, Lightfoot DA (2000) Expression of the bacterial gdhA encoding glutamate dehydrogenase in tobacco affects plant growth and development. Plant Soil 221:47–57

Amtmann A, Troufflard S, Armengaud P (2008) The effect of potassium nutrition on pest and disease resistance in plants. Physiol Plant 133:682–691

An GH, Kobayashi S, Enoki H, Sonobe K, Muraki M, Karasawa T, Ezawa T (2010) How does arbuscular mycorrhizal colonization vary with host plant genotype? An example based on maize (Zea mays) germplasm. Plant Soil 327:441–453

Andrade FH, Sala RG, Pontaroli AC, León A (2009) Integration of biotechology, plant breeding and crop physiology: dealing with complex interactions from a physiological perspective. In: Sadras VO, Calderini DF (eds) Crop physiology. Applications for genetic improvement and agronomy. Academic, San Diego, CA, pp 267–276

Andreson LC, Michelson A, Jonasson S, Schmidt IK, Mikkelsen TN, Ambus P, Beier C (2010) Plant nutrient mobilization in temperate heathland responds to elevated $CO_2$, temperature and drought. Plant Soil 328:381–396

Andrews M, Lea P, Raven J, Lindsey K (2004) Can genetic manipulation of plant nitrogen assimilation enzymes result in increased crop yield and greater N-use efficiency? An assessment. Ann Appl Biol 145:25–35

Asseng S, Ritchie JT, Smucker AJM, Robertson MJ (1998) Root growth and water uptake during water deficit and recovering in wheat. Plant Soil 201:265–273

Aung K, Lin SI, Wu CC, Huang YT, Su CL, Chiou TJ (2006) pho2, a phosphate overaccumulator, is caused by a nonsense mutation in a MicroRNA399 target gene. Plant Physiol Biochem 141:1000–1011

Babu R, Nair SK, Prasanna BM, Gupta HS (2004) Integrating marker-assisted selection in crop breeding - prospects and challenges. Curr Sci 87:607–619

Baldani JL, Reis VM, Baldani VLD, Dobereiner J (2002) A brief story of nitrogen fixation in sugarcane-reasons for success in Brazil. Funct Plant Biol 29:417–423

Baldock JA, Wheeler I, McKenzie N, McBrateny A (2012) Soils and climate change: potential impacts on carbon stocks and greenhouse gas emissions, and future research for Australian agriculture. Crop Pasture Sci 63:269–283

Baligar VC, Fageria NK, He ZL (2001) Nutrient use efficiency in plants. Commun Soil Sci Plant Anal 32:921–950

Balint T, Rengel Z (2009) Differential sulfur efficiency in canola genotypes at vegetative and grain maturity stage. Crop Pasture Sci 60:262–270

Bänziger M, Setimela PS, Hodson D, Vivek B (2006) Breeding for improved abiotic stress tolerance in maize adapted to southern Africa. Agric Water Manag 80:212–224

Baon J, Smith S, Alston A (1993) Mycorrhizal responses of barley cultivars differing in P efficiency. Plant Soil 157:97–105

Barber SA, Mackay AD (1985) Sensitivity analysis of the parameters of a mechanistic mathematical model affected by changing soil moisture. Agron J 77:528–531

Barrow NJ, Shaw TC (1975) The slow reactions between soil and anions: 2. Effect of time and temperature on the decrease in phosphate concentration in the soil solution. Soil Sci 119:167–177

Bassirirad H (2000) Kinetics of nutrient uptake by roots: responses to global change. New Phytol 147:155–169

Batten GD, Fettell NA, Mead JA, Khan MA (1993) Sowing date and phosphorus utilisation by wheat. Plant Soil 155(156):197–300

Batten GD, Fettell NA, Mead JA, Khan MA (1999) Effect of sowing date on the uptake and utilisation of phosphorus by wheat (cv Osprey) grown in central New South Wales. Aust J Exp Agric 39:161–170

Beebe SE, Rojas-Pierce M, Yan X, Blair MW, Pedraza F, Munoz F, Tohme J, Lynch JP (2006) Quantitative trait loci for root architecture traits correlated with phosphorus acquisition in common bean. Crop Sci 46:413–423

Bijay-Singh (2011) The nitrogen cycle: implications for management, soil health and climate change. In: Singh BP, Cowie AL, Chan KY (eds) Soil health and climate change, vol 29. Springer, Berlin, pp 107–129

Blair G (1993) Nutrient efficiency - what do we really mean? In: Randall PJ, Delhaize E, Richards RA, Munns R (eds) Genetic aspects of plant mineral nutrition. Kluwer Academic, Dordrecht, pp 205–213

Brouder SM, Volenec JJ (2008) Impact of climate change on crop nutrient and water use efficiencies. Physiol Plant 133:705–724

Buljovcic Z, Engels C (2001) Nitrate uptake ability by maize roots during and after drought stress. Plant Soil 229:125–135

Burkholder JM, Noga EJ, Hobbs CH, Glasgow HB Jr (1992) New 'phantom' dinoflagellate is the causative agent of major estuarine fish kills. Nat Aust 358:407–410

Cabello P, Roldan MD, Moreno-Vivián C (2004) Nitrate reduction and the nitrogen cycle in archaea. Microbiology 150:3527–3546

Cakmak I (2000) Tansley Review No. 111. Possible roles of zinc in protecting plant cells from damage by reactive oxygen species. New Phytol 146:185–205

Cakmak I, Braun H-J (2001) Genotypic variation for zinc efficiency. In: Reynolds MP, Ortiz-Monasterio JI, McNab A (eds) Applications of physiology in wheat breeding. CIMMYT, Mexico DF, pp 183–199

Cakmak I, Kirkby EA (2008) Role of magnesium in carbon partitioning and alleviating photo-oxidative damage. Physiol Plant 133:692–704

Cakmak I, Tolay I, Ozkan H, Ozdemir A, Braun HJ (1999) Variation in zinc efficiency among and within Aegilops species. J Plant Nutr Soil Sci 162:257–262

Cañas RA, Quilleré I, Christ A, Hirel B (2009) Nitrogen metabolism in the developing ear of maize (*Zea mays*): analysis of two lines contrasting in their mode of nitrogen management. New Phytol 184:340–352

Cañas RA, Quilleré I, Lea PJ, Hirel B (2010) Analysis of amino acid metabolism in the ear of maize mutants deficient in two cytosolic glutamine synthetase isoenzymes highlights the importance of asparagine for nitrogen translocation within sink organs. Plant Biotechnol J 8:966–978

Castiglioni P, Warner D, Bensen RJ, Anstrom DC, Harrison J, Stoecker M, Abad M, Kumar G, Salvador S, D'Ordine R, Navarro S, Back S, Fernandes M, Targolli J, Dasgupta S, Bonin C, Luethy MH, Heard JE (2008) Bacterial RNA chaperones confer abiotic stress tolerance in plants and improved grain yield in maize under water-limited conditions. Plant Physiol 147:446–455

Cavagnaro TR, Gleadow RM, Miller RE (2011) Plant nutrient acquisition and utilisation in a high carbon dioxide world. Funct Plant Biol 38:87–96

Century K, Reuber TL, Ratcliffe OJ (2008) Regulating the regulators: the future prospects for transcription-factor-based agricultural biotechnology products. Plant Physiol 147:20–29

Cherfas J (1990) The fringe of the ocean - under siege from land. Science 248:163–165

Chin JH, Lu XC, Haefele SM, Gamuyao R, Ismail A, Wissuwa M, Heuer S (2010) Development and application of gene-based markers for the major rice QTL Phosphorus uptake 1. Theor Appl Genet 120:1073–1086

Chin JH, Gamuyao R, Dalid C, Bustamam M, Prasetiyono J, Moeljopawiro S, Wissuwa M, Heuer S (2011) Developing rice with high yield under phosphorus deficiency: *Pup1* sequence to application. Plant Physiol 156:1202–1216

Cianzio S (1999) Breeding crops for improve nutrient use efficiency: soybean and wheat as case studies. In: Rengel Z (ed) Mineral nutrition of crops: fundamental mechanisms and implications. Haworth, Binghamton, NY, pp 227–265

Cichy KA, Blair MW, Galeano Mendoza CH, Snapp SS, Kelly JD (2009) QTL analysis or root architecture traits and low phosphorus tolerance in an Andean bean population. Crop Sci 49:49–68

Collard BCY, Jahufer MZZ, Brouwer JB, Pang ECK (2005) An introduction to markers, quantitative trait loci (QTL) mapping and marker-assisted selection for crop improvement: the basic concepts. Euphytica 142:169–196

Collins NC, Tardieu F, Tuberosa R (2008) Quantitative trait loci and crop performance under abiotic stress: where do we stand? Plant Physiol 147:469–486

Conroy JP (1992) Influence of elevated atmospheric $CO_2$ concentrations on plant nutrition. Aust J Bot 40:445–456

Cook AC, Tissue DT, Roberts SW, Oechel WC (1998) Effects of long-term elevated [$CO_2$] from springs on *Nardus stricta*: photosynthesis, biochemistry, growth and phenology. Plant Cell Environ 221:417–425

Cooper JL, Till BJ, Laport RG, Darlow MC, Kleffner JM, Jamai A, El-Mellouki T, Liu S, Ritchie R, Nielsen N, Bilyeu KD, Meksem K, Comai L, Henikoff S (2008) TILLING to detect induced mutations in soybean. BMC Plant Biol 8:9. doi:10.1186/1471-2229-8-9

Coque M, Martin A, Veyrieras JB, Hirel B, Gallais A (2008) Genetic variation for N-remobilization and postsilking N-uptake in a set of maize recombinant inbred lines. 3. QTL detection and coincidences. Theor Appl Genet 117:729–747

Cordell D, Drangert J-O, White S (2009) The story of phosphorus: global food security and food for thought. Glob Environ Change 19:292–305

Coruzzi G, Bush DR (2001) Nitrogen and carbon nutrient and metabolite signaling in plants. Plant Physiol 125:61–64

Coruzzi GM, Zhou L (2001) Carbon and nitrogen sensing and signaling in plants: emerging 'matrix effects'. Curr Opin Plant Biol 4:247–253

Coudert Y, Périn C, Courtois B, Khong NG, Gantet P (2010) Genetic control of root development in rice, the model cereal. Trends Plant Sci 15:219–226

Crasswell ET, Godwin DC (1984) The efficiency of nitrogen fertilizers applied to cereals grown in different climates. In: Tinker PB, Lauchli A (eds) Advances in plant nutrition, vol 1. Praeger, New York, pp 1–55

David MB, Gentry LE, Kovacic DA, S KM (1997) Nitrogen balance in and export from an agricultural watershed. J Environ Quart 26:1038–1048

Davidson EA, Janssens IA (2006) Temperature sensitivity of carbon decomposition and feedbacks to climate change. Nature 440:165–173

Delhaize E, Taylor P, Hocking PJ, Simpson RJ, Ryan PR, Richardson AE (2009) Transgenic barley (*Hordeum vulgare* L.) expressing the wheat aluminium resistance gene (*TaALMT1*) shows enhanced phosphorus nutrition and grain production when grown on an acid soil. Plant Biotechnol J 7:391–400

Dennison KL, Spalding EP (2000) Glutamate-gated calcium fluxes in Arabidopsis. Plant Physiol 124:1511–1514

Doerge RW (2002) Mapping and analysis of quantitative trait loci in experimental populations. Nat Rev Genet 3:43–52

Dunham RJ, Nye PH (1976) The influence of water content on the uptake of ions by roots. III. Phosphate, potassium, calcium, and magnesium uptake and concentration gradients in soil. J Appl Ecol 13:967–984

Duval BD, Blankinship JC, Dijkstra P, Hungate BA (2012) $CO_2$ effects on plant nutrient concentration depend on plant functional group and available nitrogen: a meta-analysis. Plant Ecol 213:505–521

Duvick DN (2005) The contribution of breeding to yield advances in maize (*Zea mays* L.). Adv Agron 86:83–145

Edmonds DE, Abreu SL, West A, Caasi DR, Conley TO, Daft MC, Desta B, England BB, Farris CD, Nobles TJ, Patel NK, Rounds EW, Sanders BH, Shawaqfeh SS, Lokuralalage L, Manandhar R, Raun WR (2009) Cereal nitrogen use efficiency in sub Saharan Africa. J Plant Nutr 32:2107–2122

Edwards EJ, McCafferty S, Evans JR (2005) Phosphorus status determines biomass response to elevated $CO_2$ in a legume:$C_4$ grass community. Glob Change Biol 11:1968–1981

Erbs M, Manderscheid R, Jansen G, Seddig S, Pacholski A, Weigel HJ (2010) Effects of free-air $CO_2$ enrichment and nitrogen supply on grain quality parameters and elemental composition of wheat and barley grown in a crop rotation. Agric Ecosyst Environ 136:59–68

Fageria NK, Baligar VC, Clark RB (2002) Micronutrients in crop production. Adv Agron 77:185–268

Fageria NK, Baligar VC, Li YC (2008) The role of nutrient efficient plants in improving crop yields in the twenty first century. J Plant Nutr 31:1121–1157

Fangmeier A, Grüters U, Högy P, Vermehren B, Jäger HJ (1997) Effects of elevated CO2, nitrogen supply and tropospheric ozone on spring wheat—II. Nutrients (N, P, K, S, Ca, Mg, Fe, Mn, Zn). Environ Pollut 96:43–59

Fleury D, Jefferies S, Kuchel H, Langridge P (2010) Genetic and genomic tools to improve drought tolerance in wheat. J Exp Bot 61:3211–3222

Forde BG, Lea PJ (2007) Glutamate in plants: metabolism, regulation, and signalling. J Exp Bot 58:2339–2358

Francia E, Tacconi G, Crosatti C, Barabaschi D, Bulgarelli D, Dall'Aglio E, Vale G (2005) Marker assisted selection in crop plants. Plant Cell Tiss Org Cult 82:317–342

Gahoonia TS, Nielsen NE (1997) Variation in root hairs of barley cultivars doubled soil phosphorus uptake. Euphytica 98:177–182

Gahoonia TS, Nielsen NE (2004) Root traits as tools for creating phosphorus efficient crop varieties. Plant Soil 260:47–57

Gao N, Su Y, Min J, Shen W, Shi W (2010) Transgenic tomato overexpressing ath-miR399d has enhanced phosphorus accumulation through increased acid phosphatase and proton secretion as well as phosphate transporters. Plant Soil 334:123–136

Garcia-Teijeiro R, Lightfoot DA, Hernandez JD (2009) Effect of a chemical modified urea fertilizer on soil quality: soil microbial populations around corn roots. Commun Soil Sci Plant Anal 40:2152–2168

Genc Y, McDonald GK (2008) Domesticated emmer wheat (*T. turgidum* L. subsp *dicoccon* (Schrank) Thell.) as a source for improvement of zinc efficiency in durum wheat. Plant Soil 310:67–75

Gentile R, Dodd M, Lieffering M, Brock SC, Theobald PW, Newton PCD (2012) Effects of long-term exposure to enriched $CO_2$ on the nutrient-supplying capacity of a grassland soil. Biol Fertil Soils 48:357–362

George TS, Richardson AE (2008) Potential and limitations to improving crops for enhanced phosphorus utilization. In: White PJ, Hammond JP (eds) The ecophysiology of plant-phosphorus interactions, vol 7. Springer, Amsterdam, pp 247–270

George T, Brown L, Newton A, Hallett P, Sun B, Thomas W, White PJ (2011) Impact of soil tillage on the robustness of the genetic component of variation in phosphorus (P) use efficiency in barley (*Hordeum vulgare* L.). Plant Soil 339:113–123

Godon C, Krapp A, Leydecker MT, DanielVedele F, Caboche M (1996) Methylammonium resistant mutants of Nicotiana plumbaginifolia are affected in nitrate transport. Mol Gen Genet 250:357–366

Goldman BS, Darveaux B, Cleveland J, Abad MS, Mahmood S (2009) Genes and uses for plant improvement. US Patent Application 2009/0,070,897

Gomez-Becerra HF, Yazici A, Ozturk L, Budak H, Peleg Z, Morgounov A, Fahima T, Saranga Y, Cakmak I (2010) Genetic variation and environmental stability of grain mineral nutrient concentrations in *Triticum dicoccoides* under five environments. Euphytica 171:39–52

Good AE, Shrawat AK, Muench DG (2004) Can less yield more? Is reducing nutrient input into the environment compatible with maintaining crop production? Trends Plant Sci 9:597–605

Good AG, Johnson SJ, De Pauw M, Carroll RT, Savidov N (2007) Engineering nitrogen use efficiency with alanine aminotransferase. Can J Bot 85:252–262

Gourley CJP, Allan DL, Russelle MP (1994) Plant nutrient use efficiency - a comparison of definitions and suggested improvement. Plant Soil 158:29–37

Grab S, Craparo A (2011) Advance of apple and pear tree full bloom dates in response to climate change in the southwestern Cape, South Africa: 1973–2009. Agric For Meteorol 151:406–413

Graham RD (1984) Breeding for nutritional characteristics in cereals. Adv Plant Nutr 1:57–102

Graham RD, Nambiar EKS (1981) Advances in research on copper deficiency in cereals. Aust J Agric Res 32:1009–1037

Graham RD, Rengel Z (1993) Genotypic variation in zinc uptake and utilization by plants. In: Robson AD (ed) Zinc in soils and plants. Kluwer Academic, Dordrecht, pp 107–118

Graham RD, Ascher JS, Hynes SC (1992) Selecting zinc-efficient cereal genotypes for soils of low zinc status. Plant Soil 146:241–250

Graham RD, Welch RM, Bouis HE (2001) Addressing micronutrient malnutrition through enhancing the nutritional quality of staple foods: principles, perspectives and knowledge gaps. Adv Agron 70:77–142

Graham RD, Welch RM, Saunders DA, Ortiz-Monasterio I, Bouis HE, Bonierbale M, de Haan S, Burgos G, Thiele G, Liria R, Meisner CA, Beebe SE, Potts MJ, Kadian M, Hobbs PR, Gupta RK, Twomlow S (2007) Nutritious subsistence food systems. Adv Agron 92:1–74

Graham RD, Knez M, Welch RM (2012) How much nutritional iron deficiency in humans globally is due to an underlying zinc deficiency? Adv Agron 115:1–40

Grewal HS, Stangoulis JCR, Potter TD, Graham RD (1997) Zinc efficiency of oilseed rape (Brassica napus and B-juncea) genotypes. Plant Soil 191:123–132

Gupta R, Liu J, Dhugga KS, Simmons CR (2011) Manipulation of ammonium transporters (AMTs) to improve nitrogen use efficiency in higher plants. US Patent Application 13/043,109

Hacisalihoglu G, Ozturk L, Cakmak I, Welch RM, Kochian L (2004) Genotypic variation in common bean in response to zinc deficiency in calcareous soil. Plant Soil 259:71–83

Hammond JP, Broadley MR, White PJ, King GJ, Bowen HC, Hayden R, Meacham MC, Mead A, Overs T, Spracklen WP, Greenwood DJ (2009) Shoot yield drives phosphorus use efficiency in *Brassica oleracea* and correlates with root architecture traits. J Exp Bot 60:1953–1968

Handmer J, Honda Y, Kundzewicz ZW, Arnell N, Benito G, Hatfield J, Mohamed IF, Peduzzi P, Wu S, Sherstyukov B, Takahashi K, Yan Z (2012) Changes in impacts of climate extremes: human systems and ecosystems. In: Field CB, Barros V, Stocker TF, Qin D, Dokken DJ, Ebi KL, Mastrandrea MD, Mach KJ, Plattner G-K, Allen SK, Tignor M, Midgley PM (eds), Managing the risks of extreme events and disasters to advance climate change adaptation. A special report of Working Groups I and II of the Intergovernmental Panel on Climate Change (IPCC). Cambridge University Press, Cambridge, pp 231–290

Harrigan GG, Ridley WP, Miller KD, Sorbet R, Riordan SG, Nemeth MA, Reeves W, Pester TA (2009) The forage and grain of MON 87460, a drought-tolerant corn hybrid, are compositionally equivalent to that of conventional corn. J Agric Food Chem 57:9754–9763

Hayman P, Crean J, Predo C (2011) A systems approach to climate risk in rainfed farming systems. In: Tow P, Cooper I, Partridge I, Birch C (eds) Rainfed farming systems. Springer Science + Business Media B.V, New York, NY, pp 75–100

Hemon P, Robbins MP, Cullimore JV (1990) Targeting of glutamine-synthetase to the mitochondria of transgenic tobacco. Plant Mol Biol 15:895–904

Hinsinger P, Betencourt E, Bernard L, Brauman A, Plassard C, Shen J, Tang X, Zhang F (2011) P for two, sharing a scarce resource: soil phosphorus acquisition in the rhizosphere of intercropped species. Plant Physiol 156:1078–1086

Hirel B, Le Gouis J, Ney B, Gallais A (2007) The challenge of improving nitrogen use efficiency in crop plants: towards a more central role for genetic variability and quantitative genetics within integrated approaches. J Exp Bot 58:2369–2387

Hou XL, Wu P, Jiao FC, Jia QJ, Chen HM, Yu J, Song XW, Yi KK (2005) Regulation of the expression of OsIPS1 and OsIPS2 in rice via systemic and local Pi signalling and hormones Plant. Cell Environ 28:353–364

Hu B, Zhu C, Li F, Tang J, Wang Y, Lin A, Liu L, Che R, Chu C (2011) LEAF TIP NECROSIS1 plays a pivotal role in the regulation of multiple phosphate starvation responses in rice. Plant Physiol Biochem 156:1101–1115

Huala E, Dickerman AW, Garcia-Hernandez M, Weems D, Reiser L, LaFond F, Hanley D, Kiphart D, Zhuang M, Huang W, Mueller LA, Bhattacharyya D, Bhaya D, Sobral BW, Beavis W, Meinke DW, Town CD, Somerville C, Rhee SY (2001) The Arabidopsis information resource (TAIR): a comprehensive database and web-based information retrieval, analysis, and visualization system for a model plant. Nucleic Acids Res 29:102–105

Huber DM, Graham RD (1999) The role of nutrition in crop resistance and tolerance to diseases. In: Rengal Z (ed) Mineral nutrition of crops: fundamental mechanisms and implications. Haworth, Binghamton, NY, pp 169–204

Hund A, Reimer R, Messmer R (2011) A consensus map of QTLs controlling the root length of maize. Plant Soil 344:143–158

Indrasumunar A, Searle I, Lin M-H, Kereszt A, Men A, Carroll BJ, Gresshoff PM (2011) Nodulation factor receptor kinase 1a controls nodule organ number in soybean (*Glycine max* L. Merr). Plant J 65:39–50

Indrasumunar A, Kereszt A, Gresshoff PM (2012) Soybean nodulation factor receptor proteins, encoding nucleic acids and uses thereof. US Patent Application 2011/0,231,952

IPCC (2007) In: Parry ML, Canziani OF, Palutikof JP, van der Linden PJ, Hanson CE (eds) Climate change 2007: impacts, adaptation and vulnerability. Contribution of Working Group II to the Fourth Assessment Report of the Intergovernmental Panel on Climate Change. Cambridge University Press, Cambridge, 976 p

Jahns T, Kaltwasser H (2000) Mechanism of microbial degradation of slow release fertilizers. J Polym Environ 8:11–16

Jahns T, Schepp R, Kaltwasser H (1997) Purification and characterization of an enzyme from a strain of Ochrobactrum anthropi that degrades condensation products of urea and formaldehyde (ureaform). Can J Microbiol 43:1111–1117

Jahns T, Schepp R, Siersdorfer C, Kaltwasser H (1999) Biodegradation of slow-release fertilizers (methyleneureas) in soil. J Environ Polym Degrad 7:75–82

Jakobsen I, Chen B, Munkvold L, Lundsgaard T, Zhu Y-G (2005) Contrasting phosphate acquisition of mycorrhizal fungi with that of root hairs using the root hairless barley mutant. Plant Cell Environ 28:928–938

Jarvis A, Ramirez J, Anderson B, Leibing C, Aggarwal P (2010) Scenarios of climate change within the context of agriculture. In: Reynolds MP (ed) Climate change and crop production. CABI, Wallingford, pp 9–37

Jennings G (2004) New varieties target specific needs Farming Ahead (Kondinin Group) 55: 41-42

Jia H, Ren H, Gu M, Zhao J, Sun S, Zhang X, Chen J, Wu P, Xu G (2011) The phosphate transporter gene OsPht1;8 is involved in phosphate homeostasis in rice. Plant Physiol Biochem 156:1164–1175

Jin CW, Du ST, Chen WW, Li GX, Zhang YS, Zheng SJ (2009) Elevated carbon dioxide improves plant iron nutrition through enhancing the iron-deficiency-induced responses under iron-limited conditions in tomato. Plant Physiol 150:272–280

Jones GPD, Blair GJ, Jessop RS (1989) Phosphorus efficiency in wheat – a useful selection criterion? Field Crops Res 21:257–264

Jones C, McConnell C, Coleman K, Cox P, Falloon P, Jenkinson D, Powlson D (2005) Global climate change and soil carbon stocks; predictions from two contrasting models for the turnover of organic carbon in soil. Glob Change Biol 11:155–166

Jupp AP, Newman EI (1987) Phosphorus uptake from soil by *Lolium perenne* during and after severe drought. J Appl Ecol 24:979–990

Kaiser BN, Finnegan PM, Tyerman SD, Whitehead LF, Bergersen FJ, Day DA, Udvardi MK (1998) Characterization of an ammonium transport protein from the peribacteroid membrane of soybean nodules. Science 281:1202–1206

Khan SA, Mulvaney RL, Ellsworth TR, Boast CW (2007) The myth of nitrogen fertilization for soil carbon sequestration. J Environ Qual 36:1821–1823

Khoshgoftarmanesh AH, Schulin R, Chaney RL, Daneshbakhsh B, Afyuni M (2011) Micronutrient-efficient genotypes for crop yield and nutritional quality in sustainable agriculture. In: Lichtfouse E, Hamelin M, Navarrete M, Debaeke P (eds) Sustainable agriculture, vol 2. Springer Science + Business Media, Amsterdam, pp 219–249

Kichey T, Le Gouis J, Sangwan B, Hirel B, Dubois F (2005) Changes in the cellular and subcellular localization of glutamine synthetase and glutamate dehydrogenase during flag leaf senescence in wheat (*Triticum aestivum* L.). Plant Cell Physiol 46:964–974

Kichey T, Heumez E, Pocholle D, Pageau K, Vanacker H, Dubois F, Le Gouis J, Hirel B (2006) Combined agronomic and physiological aspects of nitrogen management in wheat highlight a central role for glutamine synthetase. New Phytol 169:265–278

Kimball BA, Kobayashi K, Bindi M (2002) Responses of agricultural crops to free-air $CO_2$ enrichment. Adv Agron 77:293–368

Kirkby EA, Johnston AE (2008) Soil and fertilizer phosphorus in relation to crop nutrition. In: White PJ, Hammond JP (eds) The ecophysiology of plant-phosphorus interactions, vol 7. Springer, Dordrecht, pp 177–223

Kishii M, Delgado R, Rosas V, Cortes A, Cano S, Sanchez J, Mujeeb-Kazi A (2008) Exploitation of genetic resources through wide crosses In: Reynolds MP, Pietragalla J, Braun H-J (eds) International symposium on wheat yield potential: challenges to international wheat breeding, Mexico DF, pp 120–125

Kleinhofs A, Kuo T, Warner RL (1980) Characterization of nitrate reductase-deficient barley mutants. Mol Gen Genet 177:421–425

Koivunen M, Horwarth WR (2004) Effect of management history and temperature on the mineralization of methylene urea in soil. Nutr Cycl Agroecosyst 68:25–35

Koivunen ME, Horwath WR (2004) Effect of management history and temperature on the mineralization of methylene urea in soil. Nutr Cycl Agroecosyst 68:25–35

Koivunen M, Morisseau C, Horwarth WR, Hammock BD (2004a) Isolation of a strain of *Agrobacterium tumefaciens* (*Rhizobium radiobacter*) utilizing methylene urea (ureaformaldehyde) as nitrogen source. Can J Microbiol/Rev Can Microbiol 50:167–174

Koivunen ME, Morisseau C, Horwath WR, Hammock BD (2004b) Isolation of a strain of *Agrobacterium tumefaciens* (*Rhizobium radiobacter*) utilizing methylene urea (ureaformaldehyde) as nitrogen source. Can J Microbiol 50:167–174

Krouk G, Crawford NM, Coruzzi GM, Tsay Y-F (2010) Nitrate signaling: adaptation to fluctuating environments. Curr Opin Plant Biol 13:266–273

Lal R (2011) Soil health and climate change: an overview. In: Singh BP, Cowie AL, Chan KY (eds) Soil health and climate change, vol 29. Springer, Berlin, pp 3–24

Lam HM, Chiu J, Hsieh MH, Meisel L, Oliveira IC, Shin M, Coruzzi G (1998) Glutamate-receptor genes in plants. Nature 396:125–126

Langridge P, Chalmers K (2005) The principle: identification and application of molecular markers. In: Lorz H, Wenzel G (eds) Molecular marker systems in plant breeding and crop improvement, vol 55. Springer, Berlin, pp 3–22

Lea PJ, Miflin BJ (2011) Nitrogen metabolism in plants in the post-genomic era. In: Foyer C, Zhang H (eds) Annual plant reviews, vol 42, Nitrogen assimilation and its relevance to crop improvement. Wiley Blackwell, New Delhi, pp 1–40

Lee LK, Nielsen EG (1987) The extent and costs of groundwater contamination by agriculture. J Soil Water Conserv 42:243–248

Li T, Di Z, Han X (2012) Elevated $CO_2$ improves root growth and cadmium accumulation in the hyperaccumulator *Sedum alfredii*. Plant Soil 354:325–334

Liang Q, Cheng X, Mei M, Yan X, Liao H (2010) QTL analysis of root traits as related to phosphorus efficiency in soybean. Ann Bot 106:223–234

Liao H, Yan X, Rubio G, Beebe SE, Blair MW, Lynch JP (2004) Genetic mapping of basal root gravitropism and phosphorus acquisition efficiency in common bean. Funct Plant Biol 31:959–970

Liao M, Hocking PJ, Dong B, Delhaize E, Richardson AE, Ryan PR (2008) Variation in early phosphorus-uptake efficiency among wheat genotypes grown on two contrasting Australian soils. Aust J Agric Res 59:157–166

Lightfoot DA (2008) Blue revolution brings risks and rewards. Science 321:771–772

Lightfoot DA (2009) Genes for use in improving nitrate use efficiency in crops. In: Woods AJ, Jenks MA (eds) Genes for plant abiotic stress. Wiley-Blackwell, New York, NY, pp 167–182

Lightfoot DA, Fakhoury A (2010) Methods of using plants containing the *gdhA* gene: aflatoxin reduction and fungal root rot resistance Patent pending 2012/708,174

Lightfoot DA, Long LM, Vidal ME (2001) Plants containing the *gdhA* gene and methods of use thereof. US Patent # 6,329,573

Lightfoot DA, Mungur R, Ameziane R, Nolte S, Long L, Bernhard K, Colter A, Jones K, Iqbal MJ, Varsa E, Young B (2007) Improved drought tolerance of transgenic *Zea mays* plants that express the glutamate dehydrogenase gene (*gdhA*) of *E. coli*. Euphytica 156:103–116

Limami A, Phillipson B, Ameziane R, Pernollet N, Jiang QJ, Poy R, Deleens E, Chaumont-Bonnet M, Gresshoff PM, Hirel B (1999) Does root glutamine synthetase control plant biomass production in *Lotus japonicus* L.? Planta 209:495–502

Liu F, Wang ZY, Ren HY, Shen CJ, Li Y, Ling HQ, Wu CY, Lian XM, Wu P (2010) OsSPX1 suppresses the function of OsPHR2 in the regulation of expression of OsPT2 and phosphate homeostasis in shoots of rice. Plant J 62:508–517

Lobell D, Burke M (2010) Economic impacts of climate change on agriculture to 2030. In: Reynolds MP (ed) Climate change and crop production. CABI, Wallingford, pp 38–49

Long SP, Ainsworth EA, Rogers A, Ort DR (2004) Rising atmospheric carbon dioxide: plants FACE the future. Annu Rev Plant Biol 55:591–628

Luo Y, Su B, Currie WS, Dukes JS, Finzi A, Hartwig U, Hungate B, McMurtrie RE, Oren R, Parton WJ, Pataki DE, Shaw MR, Zak DR, Field CB (2004) Progressive nitrogen limitation of ecosystem responses to rising atmospheric carbon dioxide. BioScience 54:731–739

Lynch JP (2011) Root phenes for enhanced soil exploration and phosphorus acquisition: tools for future crops. Plant Physiol 156:1041–1049

Lynch JP, St. Clair SB (2004) Mineral stress: the missing link in understanding how global climate change will affect plants in real world soils. Field Crops Res 90:101–115

Mackay AD, Barber SA (1985) Soil moisture effect on potassium uptake by corn. Agron J 77:524–527

Manske GGB, Ortiz-Monasterio JI, van Ginkel M, Gonzalez M, Rajaram R, Molina E, Vlek PLG (2001) Importance of P uptake efficiency versus P utilization for wheat yield in acid and calcareous soils in Mexico. Eur J Agron 14:261–274

Manske GGB, Ortiz-Monasterio JI, van Ginkel RM, Rajaram S, Vlek PLG (2002) Phosphorus use efficiency in tall, semi-dwarf and dwarf near-isogenic lines of spring wheat. Euphytica 125:113–119

Marcar NE, Graham RD (1987) Genotypic variation for manganese efficiency in wheat. J Plant Nutr 10:2049–2055

Maredia MK, Byerlee D, Pee P (2000) Impacts of food crop improvement research: evidence from sub-Saharan Africa. Food Policy 25:531–559

Marschner H (1995) Mineral nutrition of higher plants. Academic, London, 889 p

McClean PE, Burridge J, Beebe S, Rao IM, Porch TG (2011) Crop improvement in the era of climate change: an integrated, multi-disciplinary approach for common bean (Phaseolus vulgaris). Funct Plant Biol 38:927–933

McDonald GK, Graham RD, Lloyd J, Lewis J, Lonergan P, Khabas-Saberi H (2001) Breeding for improved zinc and manganese efficiency in wheat and barley. In: Rowe B, Donaghy D, Mendham N (eds) Science and technology: delivering results for agriculture? Proceedings of the 10th Australian agronomy conference, Hobart, Tasmania. http://regional.org.au/au/asa/2001/6/a/mcdonald.htm. Accessed July 2012

Mengel K, Kirkby EA (2001) Principles of plant nutrition. Kluwer Academic, Dordrecht, 849 p

Messina C, Hammer G, Dong Z, Podlich D, Cooper M (2009) Modelling crop improvement in a GxExM framework via gene-trait-phenotype relationships. In: Sadras VO, Calderini DF (eds) Crop physiology. Applications for genetic improvement and agronomy. Academic, San Diego, CA, pp 235–265

Meyer R (1997) A genetic analysis of response to ammonium and methylammonium in Arabidopsis thaliana. MSc thesis, Southern Illinois University College, Carbondale, 118p

Meyer R, Yuan J, Afzal J, Iqbal MJ, Zhu M, Garvey G, Lightfoot DA (2006) Identification of Gsr1 in Arabidopsis thaliana: a locus inferred to regulate gene expression in response to exogenous glutamine. Euphytica 151:291–302

Mirvish SS (1985) Gastric-cancer and salivary nitrate and nitrite. Nature 315:461–462

Mittler R, Blumwald E (2010) Genetic engineering for modern agriculture: challenges and perspectives. Annu Rev Plant Biol 61:443–462

Moll RH, Kamprath EJ, Jackson WA (1982) Analysis and interpretation of factors which contribute to efficiency of nitrogen utilization. Agron J 74:562–564

Moller H, Landt J, Pedersen E, Jensen P, Autrup H, Jensen OM (1989) Endogenous nitrosation in relation to nitrate exposure from drinking water and diet in a Danish rural population. Cancer Res 49:3117–3121

Morgounov A, Gomez-Becerra HF, Abugalieva A, Dzhunusova M, Yessimbekova M, Muminjanov H, Zelenskiy Y, Ozturk L, Cakmak I (2007) Iron and zinc grain density in common wheat grown in Central Asia. Euphytica 155:193–203

Mungur R, Glass ADM, Goodenow DB, Lightfoot DA (2005) Metabolite fingerprinting in transgenic Nicotiana tabacum altered by the Escherichia coli glutamate dehydrogenase gene. J Biomed Biotechnol 2005(2):198–214

Mungur R, Wood AJ, Lightfoot DA (2006) Water potential is maintained during water deficit in Nicotiana tabacum expressing the Escherichia coli glutamate dehydrogenase gene. Plant Growth Regul 50:231–238

Nagarajan VK, Jain A, Poling MD, Lewis AJ, Raghothama KG, Smith AP (2011) Arabidopsis Pht1;5 mobilizes phosphate between source and sink organs and influences the interaction between phosphate homeostasis and ethylene signaling. Plant Physiol Biochem 156:1149–1163

Nelson DE, Repetti PP, Adams TR, Creelman RA, Wu J, Warner DC, Anstrom DC, Bensen RJ, Castiglioni PP, Donnarummo MG (2007) Plant nuclear factor Y (NF-Y) B subunits confer drought tolerance and lead to improved corn yields on water-limited acres. Proc Natl Acad Sci USA 104:16450–16455

Nilsson L, Muller R, Nielsen TH (2007) Increased expression of the MYB-related transcription factor, PHR1, leads to enhanced phosphate uptake in Arabidopsis thaliana. Plant Cell Environ 30:1499–1512

Nolte SA (2009) Metabolic analysis of resistance to glufosinate in gdhA transgenic tobacco. PhD thesis, Southern Illinois University, Carbondale, 244 p

Nolte SA, Young BG, Mungur R, Lightfoot DA (2004) The glutamate dehydrogenase gene gdhA increased the resistance of tobacco to glufosinate. Weed Res 44:335–339

Nord EA, Lynch JP (2008) Delayed reproduction in Arabidopsis thaliana improves fitness in soil with suboptimal phosphorus availability. Plant Cell Environ 31:1432–1441

Nord EA, Lynch JP (2009) Plant phenology: a critical controller of soil resource acquisition. J Exp Bot 60:1927–1937

Nord EA, Shea K, Lynch JP (2011) Optimizing reproductive phenology in a two-resource world: a dynamic allocation model of plant growth predicts later reproduction in phosphorus-limited plants. Ann Bot 108:391–404

Ochoa IE, Blair MW, Lynch JP (2006) QTL analysis of adventitious root formation in common bean under contrasting phosphorus availability. Crop Sci 46:1609–1621

Ogbonnaya F, Ye G, Trethowan R, Dreccer F, Lush D, Shepperd J, van Ginkel M (2007) Yield of synthetic backcross-derived lines in rainfed environments of Australia. Euphytica 157:321–336

Olesen JE, Trnka M, Kersebaum KC, Skjelvåg AO, Seguin B, Peltonen-Sainio P, Rossi F, Kozyra J, Micale F (2011) Impacts and adaptation of European crop production systems to climate change. Eur J Agron 34:96–112

Ortega JL, Moguel-Esponda S, Potenza C, Conklin CF, Quintana A, Sengupta-Gopalan C (2006) The 3' untranslated region of a soybean cytosolic glutamine synthetase (GS(1)) affects transcript stability and protein accumulation in transgenic alfalfa. Plant J 45:832–846

Ortiz-Monasterio JI, Sayre KD, Rajaram S, McMahon M (1997) Genetic progress in wheat yield and nitrogen use efficiency under four nitrogen rates. Crop Sci 37:898–904

Ortiz-Monasterio JI, Manske GGB, van Ginkel M (2001) Nitrogen and phosphorus use efficiency. In: Reynolds MP, Ortiz-Monasterio JI, McNab A (eds) Applications of physiology in wheat breeding. CIMMYT, Mexico DF, pp 200–207

Osborne LD, Rengel Z (2002) Growth and P uptake by wheat genotypes supplied with phytate as the only P source. Aust J Agric Res 53:845–850

Ott C, Rechberger H (2012) The European phosphorus balance. Resour Conserv Recycl 60:159–172

Ozturk L, Eker S, Torun B, Cakmak I (2005) Variation in phosphorus efficiency among 73 bread and durum wheat genotypes grown in a phosphorus-deficient calcareous soil. Plant and Soil 269:69–80

Pariasca-Tanaka J, Satoh K, Rose T, Mauleon R, Wissuwa M (2009) Stress response versus stress tolerance: a transcriptome analysis of two rice lines contrasting in tolerance to phosphorus deficiency. Rice 2:167–185

Patil RH, Laegdsmand M, Olesen JE, Porter JR (2010) Effect of soil warming and rainfall patterns on soil N cycling in Northern Europe. Agric Ecosyst Environ 139:195–205

Peck AW, McDonald GK (2010) Adequate zinc nutrition alleviates the adverse effects of heat stress in bread wheat. Plant Soil 337:355–374

Peck AW, McDonald GK, Graham RD (2008) Zinc nutrition influences the protein composition of flour in bread wheat (*Triticum aestivum* L.). J Cereal Sci 47:266–274

Pennisi E (2008) Plant genetics: the blue revolution, drop by drop, gene by gene. Science 320:171–173

Péret B, Clément M, Nussaume L, Desnos T (2011) Root developmental adaptation to phosphate starvation: better safe than sorry. Trends Plant Sci 16:442–450

Pleijel H, Danielsson H (2009) Yield dilution of grain Zn in wheat grown in open-top chamber experiments with elevated $CO_2$ and $O_3$ exposure. J Cereal Sci 50:278–282

Pleijel H, Uddling J (2012) Yield vs. quality trade-offs for wheat in response to carbon dioxide and ozone. Glob Change Biol 18:596–605

Poorter H (1998) Do slow-growing species and nutrient stressed plants respond relatively strongly to elevated $CO_2$? Glob Change Biol 4:681–697

Qian B, Gregorich EG, Gameda S, Hopkins DW, Wang X (2011) Observed soil temperature trends associated with climate change in Canada. J Geophys Res 116:D02106

Rae AL, Jarmey JM, Mudge SR, Smith FW (2004) Over-expression of a high-affinity phosphate transporter in transgenic barley plants does not enhance phosphate uptake rates. Funct Plant Biol 31:141–148

Reddy KR, Hodges HF (2000) Climate change and global crop productivity. CABI, Wallingford, 472 p

Rengasamy P (2006) World salinization with emphasis on Australia. J Exp Bot 57:1017–1023

Rengel Z (1999) Physiological mechanisms underlying differential nutrient efficiency of crop genotypes. In: Rengel Z (ed) Mineral nutrition of crops: fundamental mechanisms and implications. Haworth, Binghamton, NY, pp 227–265

Rengel Z (2011) Soil pH, soil health and climate change. In: Singh BP, Cowie AL, Chan KY (eds) Soil health and climate change, vol 29. Springer, Berlin, pp 69–85

Rengel Z, Damon PM (2008) Crops and genotypes differ in efficiency of potassium uptake and use. Physiol Plant 133:624–636

Rengel Z, Marschner P (2005) Nutrient availability and management in the rhizosphere: exploiting genotypic differences. New Phytol 168:305–312

Renkow M, Byerlee D (2010) The impacts of CGIAR research: a review of recent evidence. Food Policy 35:391–402

Reynolds MP (ed) (2010) Climate change and crop production. CAB International, Wallingford, 310p

Richardson A, Barea J-M, McNeill A, Prigent-Combaret C (2009a) Acquisition of phosphorus and nitrogen in the rhizosphere and plant growth promotion by microorganisms. Plant Soil 321:305–339

Richardson AE, Hocking PJ, Simpson RJ, George TS (2009b) Plant mechanisms to optimise access to soil phosphorus. Crop Pasture Sci 60:124–143

Rodriguez de Cianzio SR (1991) Recent advances in breeding for improving iron utilization by plants. Plant Soil 130:63–68

Rose TJ, Wissuwa M (2012) Rethinking internal phosphorus use efficiency: a new approach is needed to improve PUE in grain crops. Adv Agron 116:185–217

Rubio G, Liao H, Yan XL, Lynch JP (2003) Topsoil foraging and its role in plant competitiveness for phosphorus in common bean. Crop Sci 43:598–607

Sadras VO (2005) A quantitative top-down view of interactions between stresses: theory and analysis of nitrogen–water co-limitation in Mediterranean agro-ecosystems. Aust J Agric Sci 56:1151–1157

Sadras VO, Monzon JP (2006) Modelling wheat phenology captures rising temperature trends: shortened time to flowering and maturity in Australia and Argentina. Field Crops Res 99:136–146

Sanderman J, Farquharson R, Baldock J (2010) Soil carbon sequestration potential: a review for Australian agriculture. CSIRO, Canberra, 80 p

Saxena MC, Malhotra RS, Singh KB (1990) Iron deficiency in chickpea in the Mediterranean region and its control through resistant genotypes and nutrient application. Plant Soil 123:251–254

Schenk PM, Kazan K, Wilson I, Anderson JP, Richmond T, Somerville SC, Manners JM (2000) Coordinated plant defense responses in Arabidopsis revealed by microarray analysis. Proc Natl Acad Sci USA 97:11655–11660

Schunmann PHD, Richardson AE, Smith FW, Delhaize E (2004) Characterization of promoter expression patterns derived from the Pht1 phosphate transporter genes of barley (*Hordeum vulgare* L.). J Exp Bot 55:855–865

Schuster I (2011) Marker-assisted selection for quantitative traits. Crop Breed Appl Biotechnol Suppl 1:50–55

Seebauer JR, Moose SP, Fabbri BJ, Crossland LD, Below FE (2004) Amino acid metabolism in maize earshoots. Implications for assimilate preconditioning and nitrogen signaling. Plant Physiol 136:4326–4334

Seebauer JR, Singletary GW, Krumpelman PM, Ruffo ML, Below FE (2010) Relationship of source and sink in determining kernel composition of maize. J Exp Bot 61:511–519

Semenov MA, Halford NG (2009) Identifying target traits and molecular mechanisms for wheat breeding under a changing climate. J Exp Bot 60:2791–2804

Shewry PR, Napier JA, Tatham AS (1995) Seed storage proteins, structures and biosynthesis. Plant Cell Environ 7:945–956

Shrawat AK, Carroll RT, De Pauw M, Taylor GJ, Good AG (2008) Genetic engineering of improved nitrogen use efficiency in rice by the tissue-specific expression of alanine amino-transferase. Plant Biotechnol J 6:722–732

Sinclair TR, Purcell LC, King CA, Sneller CH, Chen P, Vadez V (2007) Drought tolerance and yield increase of soybean resulting from improved symbiotic N-2 fixation. Field Crops Res 101:68–71

Singh BR, Subramaniam V (1997) Phosphorus supplying capacity of heavily fertilized soils. II. Dry matter yield of successive crops and phosphorus uptake at different temperatures. Nutr Cycl Agroecosyst 47:123–134

Specht JE, Hume DJ, Kumudini SV (1999) Soybean yield potential - a genetic and physiological perspective. Crop Sci 39:1560–1570

St. Clair SB, Lynch JP (2010) The opening of Pandora's box: climate change impacts on soil fertility and crop nutrition in developing countries. Plant Soil 335:101–115

Stangoulis JCR, Grewal HS, Bell RW, Graham RD (2000) Boron efficiency in oilseed rape: I. Genotypic variation demonstrated in field and pot grown *Brassica napus* L. and *Brassica juncea* L. Plant Soil 225:243–251

Steenbjerg F, Jakobsen ST (1963) Plant nutrition and yield. Soil Sci 95:69–90

Su J, Xiao Y, Li M, Liu Q, Li B, Tong Y, Jia J, Li Z (2006) Mapping QTLs for phosphorus-deficiency tolerance at wheat seedling stage. Plant Soil 281:25–36

Su JY, Zheng Q, Li HW, Li B, Jing RL, Tong YP, Li ZS (2009) Detection of QTLs for phosphorus use efficiency in relation to agronomic performance of wheat grown under phosphorus sufficient and limited conditions. Plant Sci 176:824–836

Szarejko I, Janiak A, Chmilelwska B, Nawrot M (2005) Genetic analysis of several root hair mutants of barley. Barley Genet Newsl 35:36–38

Tanksley SD (1993) Mapping polygenes. Annu Rev Genet 27:205–233

Tannenbaum SR, Fett D, Young VR, Land PD, Bruce WR (1978) Nitrate and nitrite are formed by endogenous synthesis in human intestine. Science 200:1487–1489

Taub DR, Miller B, Allen H (2008) Effects of elevated $CO_2$ on the protein concentration of food crops: a meta-analysis. Glob Change Biol 14:565–575

Terce-Laforgue T, Dubois F, Ferrario-Mery S, de Crecenzo MAP, Sangwan R, Hirel B (2004a) Glutamate dehydrogenase of tobacco is mainly induced in the cytosol of phloem companion cells when ammonia is provided either externally or released during photorespiration. Plant Physiol 136:4308–4317

Terce-Laforgue T, Mack G, Hirel B (2004b) New insights towards the function of glutamate dehydrogenase revealed during source-sink transition of tobacco (Nicotiana tabacum) plants grown under different nitrogen regimes. Physiol Plant 120:220–228

Torun B, Kalayci M, Ozturk L, Torun A, Aydin M, Cakmak I (2002) Differences in shoot boron concentrations, leaf symptoms, and yield of Turkish barley cultivars grown on boron-toxic soil in field. J Plant Nutr 26:1735–1747

Trenkel ME (1997) Improving fertilizer use efficiency. Controlled release and stabilised fertilisers in agriculture. International Fertilizer Industry Association, Paris

Uauy C, Distelfeld A, Fahima T, Blechl A, Dubcovsky J (2006) A NAC gene regulating senescence improves grain protein, zinc, and iron content in wheat. Science 314:1298–1301

Valentine V, Benedito VA, Kang Y (2011) Legume nitrogen fixation and soil abiotic stress. In: Foyer C, Zhang H (eds) Nitrogen metabolism in plants in the post-genomic era, vol 42. Wiley Blackwell, New Delhi, pp 207–248

Vance CP, Uhde-Stone C, Allan DL (2003) Phosphorus acquisition and use: critical adaptations by plants for securing a nonrenewable resource. New Phytol 157:423–447

Vidal EA, Araus V, Lu C, Parry G, Green PJ, Coruzzi GM, Gutierrez RA (2010) Nitrate-responsive miR393/AFB3 regulatory module controls root system architecture in Arabidopsis thaliana. Proc Natl Acad Sci USA 107:4477–4482

Vitousek PM, Naylor R, Crews T, David MB, Drinkwater LE, Holland E, Johnes PJ, Katzenberger J, Martinelli LA, Matson PA, Nziguheba G, Ojima D, Palm CA, Robertson GP, Sanchez PA, Townsend AR, Zhang FS (2009) Nutrient imbalances in agricultural development. Science 324:1519–1520

von Wiren N, Gazzarrini S, Gojon A, Frommer WB (2000) The molecular physiology of ammonium uptake and retrieval. Curr Opin Plant Biol 3:254–261

Walters DR, Bingham IJ (2007) Influence of nutrition on disease development caused by fungal pathogens: implications for plant disease control. Ann Appl Biol 151:307–324

Wang RC, Guegler K, LaBrie ST, Crawford NM (2000) Genomic analysis of a nutrient response in arabidopsis reveals diverse expression patterns and novel metabolic and potential regulatory genes induced by nitrate. Plant Cell 12:1491–1509

Wang X, Shen J, Liao H (2010) Acquisition or utilisation, which is more critical for enhancing phosphorus efficiency in modern crops? Plant Sci 179:302–306

Weaver DM, Wong MTF (2011) Scope to improve phosphorus (P) management and balance efficiency of crop and pasture soils with contrasting P status and buffering indices. Plant Soil 349:37–54

Webb LB, Whetton PH, Bhend J, Darbyshire R, Briggs PR, Barlow EWR (2012) Earlier wine-grape ripening driven by climatic warming and drying and management practices. Nat Clim Change 2:259–264

Weir AH, Barraclough PB (1986) The effect of drought on the root growth of winter wheat and on its water uptake from a deep loam. Soil Use Manage 2:91–96

Wenqi M, Jianhu IL, Lin M, Fanghao W, Sisak I, Cushman G, Fusuo Z (2009) Nitrogen flow and use efficiency in production and utilisation of wheat, rice and maize in China. Agric Syst 99:53–63

Wenqi M, Li M, Jianhui L, Fanghao W, Sisak I, Fusuo Z (2011) Phosphorus balance and use efficiencies in production and consumption of wheat, rice and maize in China. Chemosphere 84:814–821

Wiersma JV (2010) Nitrate induced iron deficiency in soybean varieties with varying iron-stress responses. Agron J 102:1738–1744

Wieser H, Manderscheid R, Erbs M, Weigel HJ (2008) Effects of elevated atmospheric $CO_2$ concentrations on the quantitative protein composition of wheat grain. J Agric Food Chem 56:6531–6535

Wissuwa M, Yano M, Ae N (1998) Mapping of QTLs for phosphorus-deficiency tolerance in rice (*Oryza sativa* L.). Theor Appl Genet 97:777–783

Wissuwa M, Mazzola M, Picard C (2009) Novel approaches in plant breeding for rhizosphere-related traits. Plant Soil 321:409–430

Wu DX, Wang GX, Bai YF, Liao JX (2004) Effects of elevated $CO_2$ concentration on growth, water use, yield and grain quality of wheat under two soil water levels. Agric Ecosyst Environ 104:493–507

Wu Z, Ren H, McGrath SP, Wu P, Zhao FJ (2011) Investigating the contribution of the phosphate transport pathway to arsenic accumulation in rice. Plant Physiol Biochem 157:498–508

Yan X, Wu P, Ling H, Xu G, Xu F, Zhang Q (2006) Plant nutriomics in China: an overview. Ann Bot 98:473–482

Yang H, Knapp J, Koirala P, Rajagopal D, Peer WA, Silbart LK, Murphy A, Gaxiola RA (2007) Enhanced phosphorus nutrition in monocots and dicots over-expressing a phosphorus-responsive type IH+-pyrophosphatase. Plant Biotechnol J 5:735–745

Yang M, Ding G, Shi L, Feng J, Xu F, Meng J (2010) Quantitative trait loci for root morphology in response to low phosphorus stress in *Brassica napus*. Theor Appl Genet 121:181–193

Yang M, Ding G, Shi L, Xu F, Meng J (2011) Detection of QTL for phosphorus efficiency at vegetative stage in *Brassica napus*. Plant Soil 339:97–111

Yau SK, Ryan J (2008) Boron toxicity tolerance in crops: a viable alternative to soil amelioration. Crop Sci 48:854–865

Zhang D, Cheng H, Geng L, Kan G, Cui S, Meng Q, Gai J, Yu D (2009) Detection of quantitative trait loci for phosphorus deficiency tolerance at soybean seedling stage. Euphytica 167:313–322

Zhao Y, Li W, Zhou ZH, Wang LH, Pan YJ, Zhao LP (2005) Dynamics of microbial community structure and cellulolytic activity in agricultural soil amended with two biofertilizers. Eur J Soil Biol 41:21–29

Zhou J, Jiao FC, Wu ZC, Li YY, Wang XM, He XW, Zhong WQ, Wu P (2008) OsPHR2 is involved in phosphate-starvation signaling and excessive phosphate accumulation in shoots of plants. Plant Physiol Biochem 146:1673–1686

Zhu YG, Smith SE, Barritt AR, Smith FA (2001) Phosphorus (P) efficiencies and mycorrhizal responsiveness of old and modern wheat cultivars. Plant Soil 237:249–255

# Chapter 11
# Nitrogen Fixation and Assimilation

David A. Lightfoot

**Abstract** New crop plants suited to grow in semiarid environments will be fundamental to the future of agriculture. The interactions between nitrogen supply and
water availability that determine yield and quality in crops grown in semiarid
environments are being elucidated. Tools for analyzing the metabolic changes
associated with enhanced nitrogen assimilation under drought have been generated.
Here is summarized the crop and other plants that have altered NUE, yield
performances, and metabolic profiles caused by *in planta* expression of 31 different
transgenes generated in the past two decades. The change in nitrogen concentration
has profound effects on plant metabolisms. The metabolic changes resulted in
phenotypic changes that included increases in mean plant biomass production in
dry soils, tolerance to the herbicide phosphinothricin, tolerance to both severe and
mild water deficit, and resistance to rotting necrotrophs. Leaves, and grain had
higher nutritional value and higher yield, indicating improved NUE and WUE by
some of the transgenes. Therefore, in view of global climate change, continued
efforts to alter nitrogen fixation and assimilation by transgenesis and mutation
should be pursued through technology stacking.

## 11.1 Introduction

Nitrogen is an essential component in cellular physiology with only oxygen,
carbon, and hydrogen being more abundant (Marchner 1995; Andrews et al.
2004). Nitrogen is present in numerous essential compounds including nucleoside

D.A. Lightfoot (✉)
Genomics Core Facility, Department of Plant, Soil and Agricultural Systems, Southern Illinois
University at Carbondale, Carbondale, IL 62901-4415, USA

Genomics Core Facility, Center for Excellence, The Illinois Soybean Center, Southern Illinois
University at Carbondale, Carbondale, IL 62901-4415, USA
e-mail: ga4082@siu.edu

C. Kole (ed.), *Genomics and Breeding for Climate-Resilient Crops*, Vol. 2,
DOI 10.1007/978-3-642-37048-9_11, © Springer-Verlag Berlin Heidelberg 2013

phosphates and amino acids that form the building blocks of nucleic acids and proteins, respectively. In plants, nitrogen is used in large amounts in photosynthetic pigments, defense chemicals, and structural compounds. However, inorganic N is difficult to assimilate. Dinitrogen in the atmosphere is very inert. Reduction to ammonium requires the energy of a lightning bolt and two adenosine triphosphates (ATPs) or energy from petrochemicals or 12 ATP dephosphorylations per molecule within a nodule or other anaerobic environment. However, these ammonium fertilizers are prone to escape from the cell as ammonia gas. Photorespiration releases tenfold more ammonium than is assimilated from the environment and plants only re-assimilate 98 % of it. Consequently, a haze of ammonia gas is found floating above a photosynthetic canopy. That ammonia may be lost on the wind or returned to the plant or soil by rains or dew falls. Any improvements to these nitrogen cycles (Table 11.1; Tercé-Laforgue et al. 2004a,b; Seebauer et al. 2010; de Carvalho et al. 2011) can have a massive positive impact on the efficiency of agriculture, reduce its carbon footprint, and over geological time scales reverse some of the anthropogenic contributors to global warming.

The assimilation of ammonium has a second major problem associated with it. Ammonia is assimilated releasing one acidic proton per molecule (Marchner 1995). There is enough flux to reduce the pH of even well-buffered soils to concentrations that inhibit plant growth, both directly and by the release of toxic concentrations of micronutrients (Al and Mn in particular). Reduction within a nodule or other anaerobic environment compounds this problem by releasing two protons per ammonium produced (Indrasumunar et al. 2011, 2012). Soil acidification is a worldwide problem on a massive scale.

Nitrates and nitrites provide a solution to the acidification problem, as their reduction to ammonium absorbs 3–4 protons (Marchner 1995). So a pH-balanced fertilizer should theoretically be a 4 to 1 mixture of ammonium and nitrates. Nitrates and nitrites are the ions produced by those lightning bolts that provide about 10 % of the world's reduced nitrogen a year. However, they are not without costs and problems. Nitrite is highly toxic to photosynthesis and respiration and so must be immediately reduced to ammonium. Plants produce massive amounts of nitrite reductase for this purpose. Nitrate is benign, easy to store and transport, and consequently is the major form of inorganic N found in plants. However, plants still produce tenfold more nitrate reductase than is absolutely needed for assimilation, growth, and yield (Kleinhofs et al. 1980; Wang et al. 2012). Why? That is still unclear.

The major problem with nitrates in the environment is that they are water soluble and so rapidly leached from soils (Lee and Nielsen 1987; David et al. 1997). So much is lost from agricultural soils, industrial activity, and human waste treatments that the world's rivers, lakes, and oceans are significantly fertilized (Cherfas 1990; Burkholder et al. 1992). The algae are the microorganisms that benefit the most. Unfortunately, they run low on other nutrients (P, K) and so produce toxins to kill other organisms to obtain the limiting nutrients by their decomposition. In addition, they absorb much of the water's oxygen (at night) killing even toxin-resistant

**Table 11.1** Selected transgenic and mutant lines with effects on N fixation, N transport, primary N assimilating genes, and secondary N metabolism (adapted from Pathak et al. 2008; Lightfoot 2009)

| Gene and paralog/mutant allele number | Gene source | Promoter(s) | Target plant | Phenotype observed | |
|---|---|---|---|---|---|
| Nrt1.1—high-affinity nitrate transporter 1 | A. thaliana | CaMV 35S | A. thaliana | Increase in constitutive nitrate uptake but not the induced form | Liu et al. (1999) |
| Nrt2.1—high-affinity nitrate transporter 2 | N. plumbaginifolia | CaMV 35S, rolD | N. tabacum | Increased nitrate influx under low N conditions | Fraisier et al. (2000) |
| | | | | Three- to fourfold drop in NR protein and activity, no change in NR transcript | Vincentz and Caboche (1991) |
| NR—nitrate reductase isoform 1 | N. plumbaginifolia | CaMV 35S | N. tabacum | Increased NR activity, biomass, drought stress | Ferrario-Mery et al. (1998) |
| | N. tabacum | CaMV 35S | L. sativa | Reduced nitrate content, chlorate sensitivity | Curtis et al. (1999) |
| | N. tabacum | CaMV 35S | N. plumbaginifolia | Nitrite accumulation in high nitrate supply | Lea et al. (2004) |
| | H. vulgaris | NMS mutation | H. vulgaris | No phenotype, with just 10 % of NR activity | Kleinhofs et al. (1980), Oh et al. (1980) |
| | Porphyra sp. | RAB17 | Z. mays | Improved NUE, yield with N limitation | Loussaert et al. (2008) |
| NR—nitrate reductase isoform 2 | N. tabacum | CaMV 35S | S. tuberosum | Reduced nitrate levels | Djannane et al. (2002) |
| NiR—nitrite reductase | N. tabacum | CaMV 35S | N. plumbaginifolia | Increased NiR activity, no phenotypic difference | Crete et al. (1997) |
| | N. tabacum | CaMV 35S | A. thaliana | Increased NiR activity, no phenotypic difference | Crete et al. (1997) |
| | S. oleracea | CaMV 35S | A. thaliana | Higher NiR activity, higher nitrite accumulation | Takahashi et al. (2001) |
| AMT—ammonium transporters | A. thaliana | RAB17 | Z. mays | Increased NUE | Gupta et al. (2008, 2011) |
| | Z. mays | UBI | G. max | Increased NUE | Gupta et al. (2008, 2011) |

(continued)

**Table 11.1** (continued)

| Gene and paralog/mutant allele number | Gene source | Promoter(s) | Target plant | Phenotype observed | Reference |
|---|---|---|---|---|---|
| GS2—chloroplastic glutamine synthetase | O. sativa | CaMV 35S | N. tabacum | Improved photorespiration capacity and increased resistance to photooxidation | Kozaki and Takeba (1996) |
| | O. sativa | CaMV 35S | O. sativa | Enhanced photorespiration, salt tolerance | Hoshida et al. (2000) |
| | N. tabacum | Rbc SSU1 | N. tabacum | Enhanced growth rate | Migge et al. (2000) |
| | P. sativa | CaMV 35S | N. tabacum | Co-suppression reduced growth rate | Migge et al. (2000) |
| Fd-GOGAT—ferredoxin-dependent glutamate synthase | N. tabacum | CaMV 35S | N. tabacum | Diurnal changes in NH$_3$ assimilation | Ferrario-Mery et al. (2002) |
| GS1—cytosolic glutamine synthetase | P. vulgaris | CaMV 35S:: MITATPase | N. tabacum | Growth inhibited, insoluble GS in mitochondria | Hemon et al. (1990) |
| | P. vulgaris | CAMV 35S | N. tabacum | Increased herbicide tolerance | Hemon et al. (1990) |
| | G. max | CaMV 35S | L. corniculatus | Accelerated senescence | Vincent et al. (1997) |
| | G. max | rolD | L. japonicus | Decrease in biomass | Limami et al. (1999) |
| | P. vulgaris | Rbc SSU1 | T. aestivum | Enhanced capacity to accumulate nitrogen | Habash et al. (2001) |
| | M. sativa | CaMV 35S | N. tabacum | Enhanced growth under N starvation | Fuentes et al. (2001) |
| | M. sativa | CaMV 35S | N. tabacum | Herbicide tolerance | Donn et al. (1984) |
| | G. max | CaMV 35S | M. sativa | No increase in GS activity | Ortega et al. (2001) |
| | P. sativa | CaMV 35S | N. tabacum | Enhanced growth, leaf-soluble protein, ammonia levels | Oliveira et al. (2002) |
| | G. max | CaMV 35S | P. sativum | No change in whole-plant N | Fei et al. (2003) |
| | M. sativa | CaMV 35S | L. japonicas | Higher biomass and leaf proteins | Suarez et al. (2003) |
| | P. sylvestris | CaMV 35S | Populus sp. | Enhanced growth rate, leaf chlorophyll, total soluble protein | Gallardo et al. (1999) |

| Enzyme/gene | Source | Promoter | Plant | Effect | Reference |
|---|---|---|---|---|---|
| | *O. sativa* | CaMV 35S | *O. sativa* | Increased protein, amino acids, and nitrogen content | Cai et al. (2009) |
| | *E. coli* | CaMV 35S | *O. sativa* | Decreased salt, cold and drought tolerance; seed yield and amino acid content | Cai et al. (2009) |
| NADH-GOGAT–NADH-dependent glutamate synthase | *O. sativa* | CaMV 35S | *O. sativa* | Enhanced grain filling, increased grain weight | Yamaya et al. (2002) |
| | *M. sativa* | CaMV 35S | *M. sativa* | Higher total C and N content, increased dry weight | Schoenbeck et al. (2000) |
| GDH—glutamate dehydrogenases *gdhA* microbial NADPH dependent | *E. coli* | CaMV 35S | *N. tabacum* | Increased biomass, nutritional value | Ameziane et al. (2000) |
| | | | | N assimilation, NUE, WUE, herbicide tolerance | Mungur et al. (2005, 2006) |
| | | | | Amino acid and sugar content | Lightfoot et al. (1999) |
| | *E. coli* | OsUBI | *Z. mays* | Increased N assimilation, biomass, and sugar content | Lightfoot et al. (2007) |
| | | | | Herbicide tolerance, grain biomass, amino acids | Lightfoot et al. (1999) |
| | *E. coli* | CaMV 35S | *A. thaliana* | Increased biomass and herbicide tolerance | Lightfoot (unpublished) |
| | *C. sorokiniana* | CaMV 35S | *T. aestivum* | Increased biomass | Schmidt and Miller (2009) |
| | *C. sorokiniana* | CaMV 35S | *Z. mays* | Increased biomass | Schmidt and Miller (2009) |
| | *A. nidulans* | CaMV 35S | *L. esculentum* | Higher amino acid concentrations | Kisaka and Kida (2005, 2007) |
| | *A. niger* | CaMV 35S | *O. sativa* | Increase in dry weight, nitrogen content, and yield with high N | Abiko et al. (2010) |

(continued)

**Table 11.1** (continued)

| Gene and paralog/mutant allele number | Gene source | Promoter(s) | Target plant | Phenotype observed | |
|---|---|---|---|---|---|
| GDH1–nonmicrobial NADH dependent | A. thaliana | CaMV35S | A. thaliana | Increased abiotic stress tolerance | Coruzzi and Brears (1999) |
| | L. esculentum | CaMV 35S:: MIT | L. esculentum | Twice GDH activity, higher mRNA abundance, and twice | Kisaka et al. (2007) |
| AS1–glutamine-dependent asparagine synthetase | A. thaliana | CaMV 35S | A. thaliana | Enhanced seed protein | Lam et al. (2003) |
| | P. sativum | CaMV 35S | N. tabacum | Slightly increased biomass and increased level of free asparagines, plants required N source | Harrison et al. (2003) |
| | E. coli | CaMV 35S | L. sativa | Increased leaf mass, amino acids, proteins lower $NO_3$ and organic acids | Sobolev et al. (2010) |
| AspAT–mitochondrial aspartate aminotransferase | S. bicolor | CaMV 35S | N. tabacum | Increased AspAT, PEPCase activity | Sentoku et al. (2000) |
| AlaAT–alanine aminotransferase | H. vulgaris | btg26 | B. napus S. bicolor | Yields constant to 50% less N fertilizer Increased NUE | Good et al. (2007) |
| | | OsANT1 | O. sativa | Excess N increased biomass, grain yield metabolites, and total nitrogen content, indicating increased nitrogen uptake efficiency | |
| iGluR–glutamate receptors | A. thaliana | Mutants | A. thaliana | Growth inhibited or enhanced, kainate resistant | Coruzzi et al. (1998) |
| | A. thaliana | RAB17 | A. thaliana | Increased NUE | Dotson et al. (2009) |
| | Z. mays | RAB17 | Z. mays | Increased NUE and yield | |
| MMP1–S-methylmethionine permease1 | S. cerevisiae | AAP1 | P. sativum | Increased plant growth and seed number, S, N, and protein content | Tan et al. (2010) |
| POT–proton-dependent oligopeptide transport (PTR class) | A. thaliana | CaMV 35S | A. thaliana | Increased NUE, biomass and protein in seed, MSX resistance in mutants | Schneeberger et al. (2008) |
| | | Null mutant | | | |

| Gene/protein | Source organism | Promoter | Target organism | Phenotype/result | Reference |
|---|---|---|---|---|---|
| HMP–flavohemoglobin | E. coli | OsACT1 | Z. mays | Decreased NO, increased NUE, biomass, chlorophyll, and yield in the field | Basra et al. (2007) |
| ANR1–MADS transcription factor | A. thaliana | AtACT7 | G. max | Lateral root induction and elongation | Zhang and Forde (1998) |
|  | A. thaliana | CaMV 35S | A. thaliana |  |  |
| GLB1–PII regulatory protein | A. thaliana | CaMV 35S | A. thaliana | Growth rate, increased anthocyanin production in low N | Zhang et al. (2003) |
| Dof1–transcription factor | Z. mays | 35S C4PDK | A. thaliana | Enhanced growth rate under N-limited conditions, increase | Yanagisawa et al. (2004) |
|  | Z. mays | 35S C4PDK | O. sativa | In amino acid content | Kurai et al. (2011) |
| bZIP–transcription factor, MYB–transcription factor, glycosyl hydrolase family 9 | A. thaliana | CaMV 35S | A. thaliana | Improved NUE, plant size, vegetative growth, growth rate, light-inducible seedling vigor, and/or biomass | Nadzan et al. (2007) |
| Zinc finger C3HC4, NF-YB–plant nuclear factor Y | A. thaliana | CaMV 35S | A. thaliana | Improved NUE and WUE | Nelson et al. (2007) |
|  | Z. mays | OsRACT | Z. mays | Improved NUE and WUE |  |
| HAP3–transcription factor and interacting 14:3:3 proteins | A. thaliana | CaMV 35S | Z. mays | Improved NUE and seed yield | Dotson et al. (2009) |
|  |  |  | G. max | Improved NUE and WUE |  |
|  |  |  | G. hirsutum | Improved NUE, WUE, and yield |  |
| SnRK1–transcription factor | M. hupehensis | CaMV 35S | L. esculentum | Increased nitrogen uptake | Wang et al. (2012) |
| FUM2–fumarase | A. thaliana | Mutant | A. thaliana | Enhanced growth on high N | Pracharoenwattana et al. (2010) |
| IPT–cytokinin synthesis | A. tumefaciens | SARK | N. tabacum | Increased assimilation on low N | Rubio-Wilhelmi Mdel et al. (2012) |
| GCL–glyoxylate carboligase (EC 4.1.1.47) | E. coli | CaMV 35S | N. tabacum | Increased amino acid content | de Carvalho et al. (2011) |

aerobes. Finally, they bloom, blocking the light needed for photosynthesis by submerged organisms. Millions of acres of oceans are affected.

The major problem with nitrates in the human diet (water or food) is that they are metabolized to potent carcinogens (nitrosamines) in the acid of the human stomach (Tannenbaum et al. 1978; Moller et al. 1990; Mirvish 1985; Duncan et al. 1998). High nitrate and so nitrosamine amounts in human diets are associated with many different cancers as well as fertility problems. However, nitrates are naturally excreted in human and animal saliva for the purpose of producing some nitrosamines in the gut. This is because the combination of acid and nitrosamine effectively kills many human and animal pathogens resistant to one or the other. *Helicobacter pylori* is one example. This microbe causes stomach ulcers that left untreated often become cancerous. *H. pylori* is endemic and became more abundant as lifestyles became more stressful. Consequently, several epidemiological studies found diets high in nitrate to be healthful in the 1990s and beyond, whereas before that they were significantly unhealthful. Clearly, then the healthiest option is a low-nitrate diet and low-stress lifestyle. However where lifestyle change is not an option, for *H. pylori* and like pathogens, the lesions they cause are better treated with drugs than nitrosamines.

However produced and applied, microbes in the soil take up the bulk of all fertilizers before the plant can (Jahns and Kaltwasser 2000; Jahns et al. 1999; Trenkel 1997; Cabello et al. 2004; Garcia-Teijeiro et al. 2009; Koivunen et al. 2004). Microbes pass the N molecules through about seven microbial cells before plants absorb them. Ammonium can be assimilated or oxidized to nitrite, nitrate, nitrous oxide, or dinitrogen by microbial activities. Plants have to absorb N from microbes by force using highly efficient enzymes or by trade through symbiosis [reviewed by Ferguson and Indrasumunar (2011)]. In symbiosis the microbes are provided with sugars in return for ammonium. The microbes may be free living in the rhizosphere or housed in specialized structures like nodules. Symbiotic microbes produce a variety of chemical signals to elicit the delivery of sugars from the plants. These systems are ripe for manipulation by biotechnology approaches.

## 11.2   Plant Assimilations

Because soil particles do not naturally have many N-containing minerals, and because N can be readily lost from the rooting environment, it is the nutrient element that most often limits plant growth and so agricultural yields (Marchner 1995; Specht et al. 1999; Duvick 2005). As noted above, nitrogen is found in the environment in many forms and comprises about 80 % of the earth's atmosphere as triple-bonded nitrogen gas ($N_2$). However, this large fraction of N is not directly accessible by plants and must be bonded to one or more of three other essential nutrient elements including oxygen and/or hydrogen through N-fixation processes and carbon through N-assimilation processes. Plants are able to absorb a little $NH_3$

from the atmosphere through stomata in leaves, but this is dependent upon air concentrations. The ions $NO_3^-$ and $NH_4^+$ are the primary forms for uptake in by plants. The most abundant form that is available to the plant roots is $NO_3^-$, and the most abundant form in leaves is $NH_4^+$. The process of nitrification by soil bacteria readily converts fertilizer $NH_4^+$ to $NO_3^-$ (Trenkel 1997; Zhao et al. 2005). Relative nitrogen uptake is also dependent on soil conditions. Ammonium uptake is favored by a neutral pH and $NO_3^-$ uptake is favored by low pH. Nitrate also does not bind to the negatively charged soil particles; therefore, it is more freely available to plant roots especially through mass flow of soil water than is $NH_4^+$, which binds to negatively charged soil particles and so moves primarily by diffusion. As noted above, the assimilation of $NH_4^+$ by roots causes the rhizosphere to become acidic, while $NO_3^-$ causes the rhizosphere to become more basic.

### 11.2.1 Uptake of Nitrogen

Nitrogen uptake and assimilation summates a series of vital processes controlling plant growth and development (Godon et al 1996; Krouk et al 2010; Meyer et al. 2006). Nitrate, nitrite, and ammonium uptakes (and reuptakes following losses) occur against massive concentration gradients that require lots of energy to generate and maintain. Happily in agriculture, plants are spaced sufficiently that they have an excess of captured light energy relative to the N and C supplies. Transgenic plants overexpressing low-affinity nitrate uptake transporter Nrt1 increased the constitutive but not the induced nitrate uptake (Table 11.1; Liu et al. 1999). Equally, plants transgenic with Nrt2.1 the high-affinity nitrate transporter 2 increased nitrate influx under low-N conditions (Fraisier et al. 2000). Transgenic plants expressing an ammonium transporter increased nitrogen use efficiency (NUE; Gupta et al. 2008). Glutamate receptors in transgenic plants provided better growth. Equally the uptake of short peptides had positive effects. All these transport-associated phenotypes would be desirable in agricultural production systems directed toward greater efficiency and lower environmental impacts. A stack of the three transgenes would be of interest.

### 11.2.2 Nitrate Reduction

Nitrate acquired in the roots can be reduced in the shoot or the root or even stored in vacuoles in the root or shoot for later assimilation. However, nitrate must be reduced to a useable form. This occurs via a two-step process catalyzed by the enzymes nitrate reductase and nitrite reductase to form $NH_4^+$. Both enzymes are produced in massive excess compared to flux needed through the pathway so that mutants that reduce their amounts by 90 % do not have phenotypes (Table 10.1; Kleinhofs et al. 1980). Equally some transgenic plants overexpressing nitrate reductase (NR) increased nitrate reduction but were not altered in phenotypes (Crete et al. 1997; Curtis et al. 1999; Djannane et al. 2002; Lea et al. 2004). However, two studies of NR overexpressing transgenic plants did record altered phenotypes including increased biomass, drought stress (Ferrario-Mery et al. 1998), and improved NUE and yield during N limitation (Loussaert et al. 2008). These phenotypes would be desirable in agricultural crops. Coupling of NR to photosynthesis should be possible by the transformation of plants with a ferredoxin-dependent NR from cyanobacteria.

## 11.2.3   Dinitrogen Fixation

The ability to fix dinitrogen is restricted to the bacterial world but widespread among microbes (Ferguson and Indrasumunar 2011; Valentine et al. 2011; Reid et al. 2011). Many different *nif* gene families exist suggesting selections for variation have been favorable for species. The need for an anaerobic environment for nif activity means that transferring the enzymes to plants will be difficult. To date, transgenics in this field are bacterial, as in hydrogenase-enhanced microbes, or if plant they are designed to improve the chances of nodule occupancy by improved bacterial strains. Strains that are most likely to set up nodule occupancy are rarely the most efficient nitrogen fixers. Plants also often fail to maintain effective nodules through flowering and pod set (Sinclair et al. 2007). Soybean and common bean, for example, have senescent nodules by flowering. Some species do have indeterminate nodules, and it would be a valid goal of biotechnology to transfer this trait to major legume crops.

## 11.2.4   Ammonium Assimilation

The N acquired as $NH_4^+$ does not require reduction upon uptake into the root, thus providing some energy savings to the plant over that of the $NO_3^-$ form reviewed by Marchner (1995). However, it does require assimilation to avoid loss and at high concentrations ($>10$ mM) toxicity to the plant. Various studies have shown that under conditions of excessive $NH_4^+$ uptake, most plant species will transport this N source to the shoot, which is more sensitive to ammonium ions (Marchner 1995).

One important process to build key macromolecules in any living organism is the acquisition and utilization of inorganic forms of nitrogen during metabolism (Lea and Miflin 2011). Plants use amino acids and their precursors and catabolic products for important metabolic activities. Various other roles of amino acids include nitrogen storage and transport and the production of a very large number of secondary compounds including structural lignin compounds, light-absorbing pigments, phenolics, and plant hormones. Plants convert the available inorganic nitrogen into organic compounds through the process of ammonium assimilation, which occurs in plants by two main pathways. The first and primary pathway involves a reaction with glutamate to form glutamine, which is catalyzed by glutamine synthetase (GS, EC 6.3.1.2) and requires an energy source of ATP. There are two isoenzymes of GS based on their location in the plant, either in the cytosol (GS1) or in the root plastids or shoot chloroplasts (GS2). Expressed in germinating seeds or in the vascular bundles of roots and shoots, the cytosolic form (GS1) produces glutamine for intracellular nitrogen transport. GS2 located in root plastids produces amide nitrogen for local consumption, while GS2 in the shoot chloroplasts re-assimilates photorespiratory ammonium (Lam et al. 1996). GS1 is encoded by a set of 3–6 paralogs in different crop species, so hetero-hexamers can form. However, the affinity for the substrates hardly differs. Amino acid identity is very high even to GS2. GS2 has a short peptide extension at the C terminus that might be involved in regulation by phosphorylation. Alleles of the GS1 and GS2 encoding genes do exist that differ in their regulation. Alleles of GS appear to underlie quantitative trait loci (QTL) determining NUE and seed yield (Cañas et al. 2009, 2010; Coque et al. 2008). Transgenic analyses have been made of GS2 but not GS1 (Table 11.1). Among the 12 studies in nine plant species, the phenotypes reported included enhanced accumulation of N, growth under N starvation, herbicide (PPT) tolerance, leaf-soluble protein, ammonia, amino acids, and chlorophyll. Some genes and constructs though decreased growth; salt, cold, and drought tolerance; seed yield; and amino acid content. Therefore, the use of GS transgenics in agriculture will be useful and desirable but only with careful attention to regulation and expression.

## 11.2.5 Transaminases

The glutamine molecules produced by GS are used by a whole series of transaminases to produce the 20 protein amino acids and some nonprotein amino acids. Cardinal among the transaminases is the reaction catalyzed by glutamate synthase (GOGAT, EC 1.4.14) to form glutamate. There are two common isoenzymes of GOGAT including a ferredoxin-dependent GOGAT (Fdx-GOGAT) and an NADH-dependent GOGAT (NADH-GOGAT). While both forms are plastidic, the Fdx-GOGAT enzyme is predominately found in photosynthetic organs, and the NADH-GOGAT enzyme is found more in non-photosynthetic tissues, such as in roots and the vascular bundles of developing leaves (Schoenbeck et al. 2000: Yamaya et al. 2002). An NADPH-dependent GOGAT can be found in certain organs and in many bacteria. Plants transgenic with the NADH-dependent plant GOGAT have been reported. Phenotypes included enhanced grain filling, grain weight, total C and N content, and dry weight (Table 10.1). Phenotypes were very similar to the benefits reported from alanine dehydrogenase and asparagines synthase suggesting that transaminases are acting on a common pathway.

## 11.2.6   Glutamate Dehydrogenases

The second pathway for ammonium assimilation also results in the formation of glutamate through a reversible reaction catalyzed by glutamate dehydrogenase (GDH, EC 1.4.1.2), with a lower energy requirement than GS/GOGAT. There are also at least two forms of GDH that occur in plants that include an NADH-dependent form found in the mitochondria and an NADPH-dependent form localized in the chloroplasts of photosynthetic organs. In addition, there are enzymes capable of aminating reactions that resemble GDH (Turano Personal communication). GDHs present in plants serve as a link between carbon and nitrogen metabolism due to the ability to assimilate ammonium into glutamate or deaminate glutamate into 2-oxoglutarate and ammonium (Forde and Lea 2007). However, due to the reversibility of this reaction, the assimilatory role of GDH is severely inhibited at low concentrations of ammonium. Additionally, GDH enzymes have a low affinity for ammonium compared with GS, which further limits their assimilatory effectiveness. It has been suggested that the NAD-requiring form of GDH may be involved in carbon rather than nitrogen metabolism, (Coruzzi and Brears 1999; Kisaki et al. 2007; Nadzan et al. 2007) with glutamate catabolism providing carbon skeletons both for the TCA cycle and energy production during carbon or energy deficit. Alternate functions for GDH have also been proposed in which it has been assigned the role of re-assimilating excess ammonium, due to the limited ability of the GS/GOGAT cycle, during specific developmental stages (Loulakakis et al. 2002), such as during germination, seed set, and leaf senescence (Coruzzi and Brears 1999; Kisaka et al. 2007).

In contrast to plant GDHs, those found in microbes are very active in the assimilation of ammonium (Lightfoot et al. 1999; 2001). Plants did not have the opportunity to incorporate this type of NADPH-dependent GDH because the bacterial lines that gave rise to chloroplasts do not contain *gdhA* genes. The few cyanobacteria with GDH activity have acquired genes by transgenesis or cellular fusions. Transgenic plants in six crop species have been produced that express *gdhA* genes from three microbes (Ameziane et al. 2000; Lightfoot et al. 2007). Phenotypes in plants include increased biomass, water deficit tolerance, nutritional value, herbicide resistance, N assimilation, NUE, water use efficiency (WUE), amino acid, and sugar content (Mungur et al. 2005, 2006; Lightfoot 2008; Lightfoot and Fakhoury 2010; Nolte et al. 2004; Nolte 2009; Table 11.1). GDH genes used in this way are being evaluated for commercialization.

One problem faced by this and the alanine dehydrogenase transgenics (Good et al. 2005; Shrawat et al. 2008) is a dependence on soil type for some of the beneficial effects. GDH seems to provide a growth advantage on silty-loam clay soils common in the southern Midwest. In contrast, the alanine dehydrogenase transgenics seem to work best on sandy soils. Combining the technologies or altering their regulation might provide stable beneficial effects in many soil types and locations.

## 11.2.7  Other Aminases

A variety of other enzymes exist that are capable of aminating reactions. Each will be a candidate for overexpression in transgenic plants. Phenylalanine ammonia lyase has been used in many transgenic plants. Equally, the enzymes of cyanide assimilations (cysteine metabolism) might be more active than previously thought and could be manipulated. Alteration of the enzymes of heme and chlorophyll biosynthesis might be tried again. The *Escherichia coli hemA* gene was functional, but *hemB* became insoluble in plant chloroplasts (Denhart, Lightfoot and Gupta Unpublished). In this same pathway, the protoporphyrinogen oxidases are targets of increasingly used and useful selective herbicides. Another major sink of amines are the lignins and lignols. Emerging research suggests that transgenic manipulation of these pathways will alter NUE and therefore WUE (Castiglioni et al. 2008; Century et al. 2008; Goldman 2009; Vidal 2010; Jung et al. 2011). These enzymes might also be usefully manipulated in stacks with transgenes to improve NUE, WUE, and other traits including herbicide tolerance.

## 11.3  Conclusions

The assimilation of inorganic nitrogen is a key process in the productivity of crop plants. There are many steps at which metabolic improvements can be made. In the future, the chance to provide active nodules to nonlegumes will provide an impetus for biotechnology. In addition, the combination of existing transgenes and new promoter for their regulation will provide for new avenues in crop improvement.

## References

Abiko T, Wakayama M, Kawakami A, Obara M, Kisaka H, Miwa T, Aoki N, Ohsugi R (2010) Changes in nitrogen assimilation, metabolism, and growth in transgenic rice plants expressing a fungal NADP(H)-dependent glutamate dehydrogenase (*gdhA*). Planta 232(2):299–311

Ameziane R, Bernhard K, Lightfoot DA (2000) Expression of the bacterial *gdhA* encoding glutamate dehydrogenase in tobacco affects plant growth and development. Plant Soil 221:47–57

Andrews M, Lea PJ, Raven JA, Lindsey K (2004) Can genetic manipulation of plant nitrogen assimilation enzymes result in increased crop yield and greater N-use efficiency? An assessment. Ann Appl Biol 145:25–35

Basra A, Edgerton M, Lee GJ, Lu M, Lutfiyya LL, Wu W, Wu X (2007) Plants containing a heterologous flavohemoglobin gene and methods of use thereof. US Patent 20,070,074,312

Burkholder JM, Noga EJ, Hobbs CH, Glasgow HB Jr (1992) New 'phantom' dinoflagellate is the causative agent of major estuarine fish kills. Nature 358:407–410

Cabello P, Roldán MD, Moreno-Vivián C (2004) Nitrate reduction and the nitrogen cycle in archaea? Microbiology 150:3527–3546

Cai H, Zhou Y, Xiao J, Li X, Zhang Q, Lian X (2009) Overexpressed glutamine synthetase gene modifies nitrogen metabolism and abiotic stress responses in rice. Plant Cell Rep 28 (3):527–537

Cañas RA, Quilleré I, Christ A, Hirel B (2009) Nitrogen metabolism in the developing ear of maize (Zea mays): analysis of two lines contrasting in their mode of nitrogen management. New Phytol 184(2):340–352

Cañas RA, Quilleré I, Lea PJ, Hirel B (2010) Analysis of amino acid metabolism in the ear of maize mutants deficient in two cytosolic glutamine synthetase isoenzymes highlights the importance of asparagine for nitrogen translocation within sink organs. Plant Biotechnol J 8:966–978

Castiglioni P, Warner D, Bensen RJ, Anstrom DC, Harrison J, Stoeker M et al (2008) Bacterial RNA chaperones confer abiotic stress tolerance in plants and improved grain yield in maize under water-limited conditions. Plant Physiol 147:446–455

Century K, Reuber TL, Ratcliffe OJ (2008) Regulating the regulators: the future prospects for transcription-factor-based agricultural biotechnology products. Plant Physiol 147(1):20–29

Cherfas J (1990) The fringe of the ocean – under siege from land. Science 248:163–165

Coque M, Martin A, Veyrieras JB, Hirel B, Gallais A (2008) Genetic variation for N-remobilization and post-silking N-uptake in a set of maize recombinant inbred lines. 3. QTL detection and coincidences. Theor Appl Genet 117(5):729–747

Coruzzi G, Brears M (1999) Transgenic plants that exhibit enhanced nitrogen assimilation. US Patent 5,955,651

Coruzzi G, Oliveira I, Lam HM, Hsieh MH (1998) Plant glutamate receptors. US Patent 5, 824,867

Creelman RA, Ratcliffe O, Reuber TL, Zhang J, Nadzan G (2009) Polynucleotides and polypeptides in plants. US Patent 7,511,190

Crete P, Caboche M, Meyer C (1997) Nitrite reductase expression is regulated at the post-transcriptional level by the nitrogen source in Nicotiana plumbaginifolia and Arabidopsis thaliana. Plant J 11:625–634

Curtis IS, Power JB, de Llat AAM (1999) Expression of chimeric nitrate reductase gene in transgenic lettuce reduces nitrate in leaves. Plant Cell Rep 18:889–896

David MB, Gentry LE, Kovacic DA, Smith KM (1997) Nitrogen balance in and export from an agricultural watershed. J Environ Qual 26:1038–1048

de Carvalho JF, Madgwick PJ, Powers SJ, Keys AJ, Lea PJ, Parry MA (2011) An engineered pathway for glyoxylate metabolism in tobacco plants aimed to avoid the release of ammonia in photorespiration. BMC Biotechnol 11:111–121

Djannane S, Chauvin JE, Meyer C (2002) Glasshouse behavior of eight transgenic potato clones with a modified nitrate reductase expression under two fertilization regimes. J Exp Bot 53:1037–1045

Donn G, Tischer J, Smith A, Goodman HM (1984) Herbicide-resistant alfalfa cells: an example of gene amplification in plants. J Mol Appl Genet 2:621–635

Dotson SB, Edgerton MD, Wu J, Xie Z, Lee GJ, Adams TR, Nelson, DE (2009) Transgenic plants with enhanced agronomic traits. US Patent 20,090,049,573, Feb 19

Duncan D, Dykhuisen R, Frazer A, MacKenzie H, Golden M, Benjamin N, Leifert C (1998) Human health effects of nitrate. Gut 42:334–340

Duvick DN (2005) The contribution of breeding to yield advances in maize (Zea mays L.). Adv Agron 86:83–145

Fei H, Chaillou S, Mahon JD, Vessey JK (2003) Overexpression of a soyabean cytosolic glutamine synthetase gene linked to organ - specific promoters in pea plants grown in different concentrations of nitrate. Planta 216:467–474

Ferrario-Méry S, Valadier MH, Foyer C (1998) Overexpression of nitrate reductase in tobacco delays drought-induced decreases in nitrate reductase activity and mRNA. Plant Physiol 117:293–302

Ferrario-Méry S, Valadier MH, Godefroy N, Miallier D, Hirel B, Foyer CH, Suzuki A (2002) Diurnal changes in ammonia assimilation in transformed tobacco plants expressing ferredoxin-dependent glutamate synthase mRNA in the antisense orientation. Plant Sci 163:59–67

Ferguson BJ, Indrasumunar A (2011) Soybean nodulation and nitrogen fixation, Chap 4. In: Hendricks BP (ed) Agricultural research updates. Nova Science, Hauppauge, NY, pp 103–119

Forde BG, Lea PJ (2007) Glutamate in plants: metabolism, regulation, and signaling. J Exp Bot 58 (9):2339–2358

Fraisier V, Gojon A, Tillard P, Daniel-Vedele F (2000) Constitutive expression of a putative high affinity nitrate transporter in *Nicotiana plumbaginifolia*: evidence for post transcriptional regulation by a reduced nitrogen source. Plant J 23:489–496

Fuentes SI, Allen DJ, Ortiz-Lopez A, Hernández G (2001) Over-expression of cytosolic glutamine synthetase increases photosynthesis and growth at low nitrogen concentrations. J Exp Bot 52:1071–1081

Gallardo F, Fu J, Canton FR, Garcia-Gutierrez A, Canovas FM, Kirby EG (1999) Expression of a conifer glutamine synthetase in transgenic poplar. Planta 210:19–26

Garcia-Teijeiro R, Lightfoot DA, Hernandez JD (2009) Effect of a modified urea fertilizer on soil microbial populations around maize (*Zea mays* L.) roots. Commun Soil Sci Plant Anal 40:2152–2168

Godon C, Krapp A, Leydecker MT, Daniel-Vedele F, Caboche M (1996) Methylammonium-resistant mutants of *Nicotiana plumbaginifolia* are affected in nitrate transport. Mol Gen Genet 250(3):357–366

Goldman BS, Darveaux, B, Cleveland J, Abad MS, Sayed M (2009) Genes and uses for plant improvement. US Patent Application 2009/0,070,897

Good AG, Johnson SJ, DePauw MD, Carroll RT, Savidov N, Vidamir J, Lu Z, Taylor G, Stroeher V (2007) Engineering nitrogen use efficiency with alanine aminotransferase. Can J Bot 85:252–262

Gupta R, Liu J, Dhugga KS, Simmons CR (2008) Manipulation of ammonium transporters (AMTs) to improve nitrogen use efficiency in higher plants. US Patent Application 12/045,098

Gupta R, Liu J, Dhugga KS, Simmons CR (2011) Manipulation of ammonium transporters (AMTs) to improve nitrogen use efficiency in higher plants. US Patent Application 13/043,109

Habash DZ, Massiah AJ, Rong HL, Wallsgrove RM, Leigh RA (2001) The role of cytosolic glutamine synthetase in wheat. Ann Appl Biol 2001(138):83–89

Harrigan GG, Ridley WP, Miller KD, Sorbet R, Riordan SG, Nemeth MA, Reeves W, Pester TA (2010) The forage and grain of MON 87460, a drought-tolerant corn hybrid, are composition-ally equivalent to that of conventional corn. J Agric Food Chem 57:9754–9763

Harrison J, Crescenzo MP, Hirel B (2003) Does lowering glutamine synthetase activity in nodules modify nitrogen metabolism and growth of *Lotus japonicus* L. Plant Physiol 133:253–262

Hemon P, Robbins MP, Cullimore JV (1990) Targeting of glutamine synthetase to the mitochondria of transgenic tobacco. Plant Mol Biol 15:895–904

Hoshida H, Tanaka Y, Hibino T, Hayashi Y, Tanaka A (2000) Enhanced tolerance to salt stress in transgenic rice that overexpresses chloroplast glutamine synthetase. Plant Mol Biol 43:103–111

Indrasumunar A, Searle I, Lin M-H, Kereszt A, Men A, Carroll BJ, Gresshoff PM (2011) Nodulation factor receptor kinase 1a controls nodule organ number in soybean (*Glycine max* L. Merr). Plant J 59:61–68

Indrasumunar A, Kereszt A, Gresshoff PM (2012) Soybean nodulation factor receptor proteins, encoding nucleic acids and uses therefore. US Patent Application 2011/0,231,952

Jahns T, Kaltwasser H (2000) Mechanism of microbial degradation of slow release fertilizers. J Polym Environ 8:11–16

Jahns T, Schepp R, Siersdorfer C, Kaltwasser H (1999) Biodegradation of slow-release fertilizers (methyleneureas) in soil. J Environ Polym Degrad 7:75–82

Jung HG, Mertens DR, Phillips RL (2011) Effect of reduced ferulate-mediated lignin/arabinoxylan cross-linking in corn silage on feed intake, digestibility, and milk production. J Dairy Sci 94:5124–5137

Kisaka H, Kida T (2005) Method of producing transgenic plants having improved amino acid composition. US Patent 6, 969, 782

Kisaka H, Kida T, Miwa T (2007) Transgenic tomato plants that overexpress a gene for NADH-dependent glutamate dehydrogenase (legdh1). Breed Sci 57:101–106

Kleinhofs A, Kuo T, Warner RL (1980) Characterization of nitrate reductase-deficient barley mutants. Mol Gen Genet 177:421–425

Koivunen M, Morisseau C, Horwarth WR, Hammock BD (2004) Isolation of a strain of *Agrobacterium tumefaciens* (*Rhizobium radiobacter*) utilizing methylene urea (ureaformaldehyde) as nitrogen source. Can J Microbiol 50(3):167–174

Krouk G, Crawford NM, Coruzzi GM, Tsay YF (2010) Nitrate signaling: adaptation to fluctuating environments. Curr Opin Plant Biol 13:266–273

Kozaki A, Takeba G (1996) Photorespiration protects C3 plants from photooxidation. Nature 384:557–560

Kurai T, Wakayama M, Abiko T, Yanagisawa S, Aoki N, Ohsugi R (2011) Introduction of the *ZmDof1* gene into rice enhances carbon and nitrogen assimilation under low-nitrogen conditions. Plant Biotechnol J 9:826–837

Lam HM, Wong P, Chan HK, Yam KM, Chow CM, Corruzi GM (2003) Overexpression of the *ASN1* gene enhances nitrogen status in seeds of *Arabidopsis*. Plant Physiol 132:926–935

Lea PJ, Miflin BJ (2011) Nitrogen assimilation and its relevance to crop improvement, Chapter 1. In: Foyer C, Zhang H (eds) Nitrogen metabolism in plants in the post-genomic era. Annu Plant Rev 42:1–40

Lea US, ten Hoopen F, Provan F, Kaiser WM, Meyer C, Lillo C (2004) Mutation of the regulatory phosphorylation site of tobacco nitrate reductase results in constitutive activation of the enzyme in vivo and nitrite accumulation. Plant J 35:566–573

Lee LK, Nielsen EG (1987) The extent and costs of groundwater contamination by agriculture. J Soil Water Conserv 42:243–248

Lightfoot DA (2008) Blue revolution brings risks and rewards. Science 321:771–772

Lightfoot DA (2009) Genes for use in improving nitrate use efficiency in crops. In: Wood AJ, Jenks MA (eds) Genes for plant abiotic stress. Wiley Blackwell, New York, pp 167–182

Lightfoot DA, Fakhoury A (2010) Methods of using plants containing the *gdhA* gene: aflatoxin reduction and fungal rot resistance. US Patent 8,383,887

Lightfoot DA, Long LM, Vidal ME (1999) Plants containing the *gdhA* gene and methods of use thereof. US Patent 5,998,700

Lightfoot DA, Long LM, Vidal ME (2001) Plants containing the *gdhA* gene and methods of use thereof. US Patent 6,329,573

Lightfoot DA, Bernhardt K, Mungur R, Nolte S, Ameziane R, Colter A, Jones K, Iqbal MJ, Varsa EC, Young BG (2007) Improved drought tolerance of transgenic *Zea mays* plants that express the glutamate dehydrogenase gene (*gdhA*) of *E. coli*. Euphytica 156:106–115

Limami A, Phillipson B, Ameziane R, Pernollet N, Jiang Q, Roy R, Deleens E, Chaumont-Bonnet M. Gresshoff PM, Hirel B (1999) Does root glutamine synthetase control plant biomass production in lotus japonicus L.? Planta. 209, 495–502

Liu KH, Huang CY, Tsay YF (1999) CHL1 is a dual-affinity nitrate transporter of *Arabidopsis* involved in multiple phases of nitrate uptake. Plant Cell 11:865–874

Loulakakis KA, Primikirios NI, Nikolantonakis MA, Roubelakis-Angelakis KA (2002) Immunocharacterization of *Vitis vinifera* L. ferredoxin-dependent glutamate synthase, and its spatial and temporal changes during leaf development. Planta 215:630–638

Loussaert DF, O' Neill D, Simmons CR, Wang H (2008) Nitrate reductases from red algae, compositions and methods of use thereof. US Patent 20,080,313,775, Dec 18

Marchner H (1995) Mineral nutrition of higher plants. Academic, London, 889 p

Meyer R, Yuan Z, Afzal J, Iqbal J, Zhu M, Garvey G, Lightfoot d (2006) Identification of *Gsr1*: a locus inferred to regulate gene expression in response to exogenous glutamine. Euphytica 151:291–302

Migge A, Carrayol E, Hirel B, Becker TW (2000) Leaf specific overexpression of plastidic glutamine synthetase stimulates the growth of transgenic tobacco seedlings. Planta 2:252–260

Mirvish SS (1985) Gastric cancer and salivary nitrate and nitrite. Nature 315:461–462

Moller H, Landt J, Pederson E, Jensen P, Autrup H, Jensen OM (1990) Endogenous nitrosation in relation to nitrate exposure from drinking water and diet in a danish rural population. Cancer Res 49:3117–3121

Mungur R, Glass ADM, Goodenow DB, Lightfoot DA (2005) Metabolic fingerprinting in transgenic *Nicotiana tabacum* altered by the *Escherichia coli* glutamate dehydrogenase gene. J Biomed Biotechnol 2:198–214

Mungur R, Wood AJ, Lightfoot DA (2006) Water potential during water deficit in transgenic plants: *Nicotiana tabacum* altered by the *Escherichia coli* glutamate dehydrogenase gene. Plant Growth Regul 50:231–238

Nadzan G, Schneeberger R, Feldmann K (2007) Nucleotide sequences and corresponding polypeptides conferring improved nitrogen use efficiency characteristics in plants. US Patent 20,070,169,219

Nelson DE, Repetti PP, Adams TR, Creelman RA, Wu J, Warner DC et al (2007) Plant nuclear factor Y (NF-Y) B subunits confer drought tolerance and lead to improved corn yields on water-limited acres. Proc Natl Acad Sci USA 104:16450–16455

Nolte SA (2009) Metabolic analysis of resistance to glufosinate in gdhA transgenic tobacco. PhD Dissertation, Plant Biology, Southern Illinois University, Carbondale, IL, 244 p

Nolte SA, Young BG, Mungur R, Lightfoot DA (2004) The glutamate dehydrogenase gene *gdhA* increased the resistance of tobacco to glufosinate. Weed Res 44:335–339

Oh JY, Warner RL, Kleinhofs A (1980) Effect of nitrate reductase-deficiency upon growth, yield and protein in barley. Crop Sci 20:487–490

Oliveira IC, Brears T, Knight TJ, Clark A, Coruzzi GM (2002) Overexpression of cytosolic glutamate synthetase. Relation to nitrogen, light, and photorespiration. Plant Physiol 129:1170–1180

Ortega JL, Temple SJ, Sengupta-Gopalan C (2001) Constitutive overexpression of cytosolic glutamine synthetase (GS1) gene in transgenic alfalfa demonstrates that GS1 be regulated at the level of RNA stability and protein turnover. Plant Physiol 126:109–121

Pathak RR, Ahmad A, Lochab S, Raghuram N (2008) Molecular physiology of plant nitrogen use efficiency and biotechnological options for its enhancement. Curr Sci 94:1394–1403

Pracharoenwattana I, Zhou W, Keech O, Francisco PB, Udomchalothorn T, Tschoep H, Stitt M, Gibon Y, Smith SM (2010) Arabidopsis has a cytosolic fumarase required for the massive allocation of photosynthate into fumaric acid and for rapid plant growth on high nitrogen. Plant J 62:785–795

Reid DE, Ferguson BJ, Hayashi S, Lin Y-H, Gresshoff PM (2011) Molecular mechanisms controlling legume autoregulation of nodulation. Ann Bot 108:789–795

Rubio-Wilhelmi Mdel M, Sanchez-Rodriguez E, Rosales MA, Blasco B, Rios JJ, Romero L, Blumwald E, Ruiz JM (2012) Ammonium formation and assimilation in P(SARK):IPT tobacco transgenic plants under low N. J Plant Physiol 169:157–162

Schmidt RR, Miller P (2009) Polypeptides and polynucleotides relating to the α- and β-subunits of glutamate dehydrogenases and methods of use. US Patent 7,485,771

Schneeberger R, Margolles-Clark E, Park J-H, Jankowski B, Bobzin SC (2008) Modulating plant nitrogen levels. US Patent 7335510-A 1, 26 Feb 2008

Schoenbeck MA et al (2000) Decreased NADH-glutamate synthase activity in nodules and flowers of alfalfa (*Medicago sativa* L.) transformed with an antisense glutamate synthase transgene. J Exp Bot 51:29–39

Seebauer JR, Singletary GW, Krumpelman PM, Ruffo ML, Below FE (2010) Relationship of source and sink in determining kernel composition of maize. J Exp Bot 61:511–519

Sentoku N, Tanignchi M, Sugiyama T, Ishimaru K, Ohsugi R, Takaiwa F, Toki S (2000) Analysis of the transgenic tobacco plants expressing *Panicum miliaceum* aspartate aminotransferase genes. Plant Cell Rep 19:598–603

Shrawat AK, Carroll RT, DePauw M, Taylor GJ, Good AG (2008) Genetic engineering of improved nitrogen use efficiency in rice by the tissue-specific expression of alanine amino-transferase. Plant Biotechnol J 6:722–732

Sinclair TR, Purcell LC, King CA, Sneller CH, Chen P, Vadez V (2007) Drought tolerance and yield increase of soybean resulting from improved symbiotic N fixation. Field Crops Res 101:68–71

Sobolev AP, Testone G, Santoro F, Nicolodi C, Iannelli MA, Amato ME, Ianniello A, Brosio E, Giannino D, Mannina L (2010) Quality traits of conventional and transgenic lettuce (*Lactuca sativa* L.) at harvesting by NMR metabolic profiling. J Agric Food Chem 58(11):6928–6936

Specht JE, Hume DJ, Kumudini SV (1999) Soybean yield potential – a genetic and physiological perspective. Crop Sci 39:1560–1571

Suarez R, Márquez J, Shishkova S, Hernández G (2003) Overexpression of alfalfa cytosolic glutamine synthetase in nodules and flowers of transgenic Lotus japonicus plants. Physiol Plant 117:326–336

Takahashi M, Sasaki Y, Ida S, Morikawa H (2001) Nitrite reductase gene enrichment improves assimilation of NO2 in *Arabidopsis*. Plant Physiol 126:731–741

Tan Q, Zhang L, Grant J, Cooper P, Tegeder M (2010) Increased phloem transport of S-methyl-methionine positively affects sulfur and nitrogen metabolism and seed development in pea plants. Plant Physiol 154(4):1886–1896

Tannenbaum SR, Fett D, Young VR, Land PD, Bruce WR (1978) Nitrite and nitrate are formed by endogenous synthesis in the human intestine. Science 200:1487–1491

Tercé-Laforgue T, Dubois F, Ferrario-Méry S, Pou de Crecenzo M-A, Sangwan R, Hirel B (2004a) Glutamate dehydrogenase of tobacco is mainly induced in the cytosol of phloem companion cells when ammonia is provided either externally or released during photorespiration. Plant Physiol 136:4308–4317

Tercé-Laforgue T, Mäck G, Hirel B (2004b) New insights towards the function of glutamate dehydrogenase revealed during source-sink transition of tobacco (*Nicotiana tabacum*) plants grown under different nitrogen regimes. Physiol Plant 120:220–228

Trenkel ME (1997) Improving fertilizer use efficiency. Controlled-release and stabilized fertilizers in agriculture. International Fertilizer Industry Association, Paris, 100 p

Valentine V, Benedito VA, Kang Y (2011) Legume nitrogen fixation and soil abiotic stress, Chap 9. In: Foyer C, Zhang H (eds) Nitrogen metabolism in plants in the post-genomic era. Annu Plant Rev 42:207–248

Vidal EA, Araus V, Lu C, Parry G, Green PJ, Coruzzi GM, Gutiérrez RA (2010) Nitrate-responsive miR393/AFB3 regulatory module controls root system architecture in *Arabidopsis thaliana*. Proc Natl Acad Sci USA 107:4477–4482

Vincent R et al (1997) Over expression of a soybean gene encoding cytosolic glutamine synthetase in shoots of transgenic *Lotus cornicultus* L. plants triggers changes in ammonium and plant development. Planta 201:424–433

Vincentz M, Caboche M (1991) Constitutive expression of nitrate reductase allows normal growth and development of *Nicotiana plumbaginifolia* plants. EMBO J 10:1027–1035

Wang X, Peng F, Li M, Yang L, Li GJ (2012) Expression of a heterologous SnRK1 in tomato increases carbon assimilation, nitrogen uptake and modifies fruit development. Plant Physiol 169:1173–1182

Yamaya T, Obara M, Nakajima H, Sasaki S, Hayakawa T, Sato T (2002) Genetic manipulation and quantitative-trait loci mapping for nitrogen recycling in rice. J Exp Bot 53:917–925

Yanagisawa S, Akiyama A, Kisaka H, Uchimiya H, Miwa T (2004) Metabolic engineering with Dof1 transcription factor in plants: improved nitrogen assimilation and growth under low-nitrogen conditions. Proc Natl Acad Sci USA 101:7833–7838

Zhang H, Forde BG (1998) An *Arabidopsis* MADS box gene that controls nutrient-induced changes in root architecture. Science 279:407–409

Zhang Y, Dickinson JR, Paul MJ, Halford NJ (2003) Molecular cloning of an *Arabidopsis* homologue of GCN2, a protein kinase involved in co-ordinated response to amino acid starvation. Planta 217:668–675

Zhao Y, Li W, Zhou Z, Wang L, Pan Y, Zhao L (2005) Dynamics of microbial community structure and cellulolytic activity in agricultural soil amended with two biofertilizers. Eur J Soil Biol 41:21–29

# Chapter 12
# Carbon Sequestration

**Leland J. Cseke, Stan D. Wullschleger, Avinash Sreedasyam, Geetika Trivedi, Peter E. Larsen, and Frank R. Collart**

**Abstract** With the rising levels of atmospheric carbon dioxide ($CO_2$) threatening to alter global climate, carbon sequestration in plants has been proposed as a possible moderator or solution to the problem. This chapter examines the different mechanisms through which carbon sequestration can take place within the Earth's natural carbon cycle with special focus on events surrounding plant development. Unfortunately, endeavors that have purposefully and successfully altered plant traits to improve carbon sequestration are currently quite few. Consequently, we delve deeply into the specific biological processes that allow plants to capture, allocate, and store $CO_2$ long term in the form of both above-ground and below-ground biomass. Distinctions are made between the differing molecular mechanisms of $C_3$, $C_4$, and CAM plants, and we point out the importance of mycorrhizal and other soil community level interactions as an important reminder that healthy soils are required for the uptake of nutrients needed for efficient carbon sequestration. We suggest that, due the complexity of the biological interactions, modeling approaches designed to network multiple types of data input may provide

L.J. Cseke (✉)
Department of Biological Science, 302L Shelby Center for Science and Technology, The University of Alabama in Huntsville, Huntsville, AL 35899, USA
e-mail: csekel@uah.edu

S.D. Wullschleger
Environmental Sciences Division, Climate Change Science Institute (CCSI), Oak Ridge National Laboratory, Building 2040, Room E212, MS 6301, Oak Ridge, TN 37831, USA
e-mail: wullschlegsd@ornl.gov

A. Sreedasyam • G. Trivedi
Department of Biological Science, 302J Shelby Center for Science and Technology, The University of Alabama in Huntsville, Huntsville, AL 35899, USA
e-mail: as0005@uah.edu; gt0001@uah.edu

P.E. Larsen • F.R. Collart
Biosciences Division, Argonne National Lab, 9700 South Cass Ave., Lemont, IL 60439, USA
e-mail: plarsen@anl.gov; fcollart@anl.gov

C. Kole (ed.), *Genomics and Breeding for Climate-Resilient Crops*, Vol. 2,
DOI 10.1007/978-3-642-37048-9_12, © Springer-Verlag Berlin Heidelberg 2013

the best means for generating more useful hypotheses that can target specific traits for improved carbon sequestration.

## 12.1 Introduction

Carbon sequestration in the most general sense refers to the storage of elemental carbon within a material or reservoir. Examples include man-made processes where carbon dioxide ($CO_2$) is captured from flue gases at industrial plants and stored permanently in underground reservoirs, such as deep saline aquifers or depleted gas fields. Natural mineralization over long periods of time or during volcanic activity can sequester carbon. Carbon derived from organic matter can be trapped through the process of pyrolysis, the high-temperature degradation of organic material. This can generate biochar or charcoal, which can then be deposited into landfills or made into useful industrial products (Warnock et al. 2007). However, the most common example of carbon biosequestration occurs during the processes of photosynthesis in plants, where atmospheric $CO_2$ is captured (or fixed) with the help of water and light energy and stored in the form of sugar molecules that are used for subsequent metabolism and growth. Consequently, such carbon sequestration transfers $CO_2$ from the atmosphere to the leaves, stems, roots, and even surrounding soils of trees, plants, and crops, thus storing the carbon long term away from the atmosphere in the form of biomass.

The Earth has a natural carbon cycle. Carbon circulates in an endless cycle between the Earth's atmosphere, the oceans, the plants, and the soil primarily through the processes of photosynthesis and respiration of living organisms, although abiotic factors such as volcanoes, forest fires, and other forms of disturbance or land use also play a major role (Fig. 12.1). The major components or reservoirs of carbon include the oceans, phytoplankton, terrestrial vegetation and soils, and the Earth's atmosphere. The term "carbon sinks" is also used to refer these natural forms of carbon reservoirs when carbon sequestration is greater than carbon released over some time period. There is, however, only a fixed amount of carbon in Earth's system, and it is sequestered in and exchanged among the various sinks over time periods ranging from days to millennia, depending on the sink being considered. This carbon cycle helps to regulate the amount of $CO_2$ present in our atmosphere and is therefore a major component of the Earth's climate system.

### 12.1.1 Why Is Carbon Sequestration Important?

Carbon dioxide is a natural component of the atmosphere that helps reflect the Earth's infrared radiation back to its own surface. This reflection causes heat to be retained in an effect similar to that of a greenhouse, hence the reason that $CO_2$ is referred to as a greenhouse gas. While greenhouse gases are currently classified by

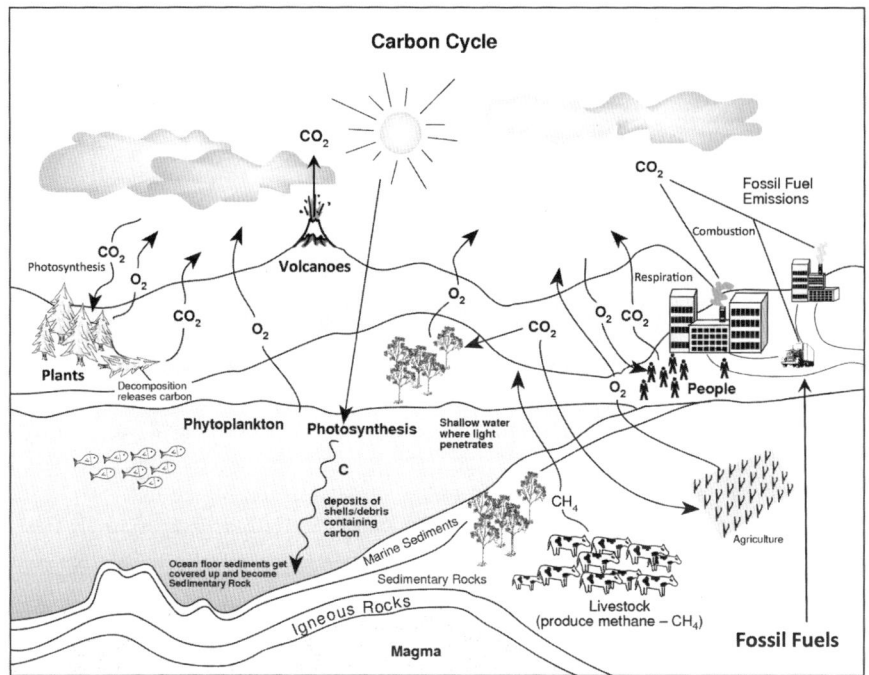

**Fig. 12.1** An overview of factors that play an important role in the cycling of $CO_2$ in the Earth's atmosphere (modified from http://kids.earth.nasa.gov/guide/earth_glossary.pdf)

the US Environmental Protection Agency (EPA) as air pollutants (allowing regulation of emissions of such gases), the fact that $CO_2$ can act as a temperature buffer in our atmosphere has played a key role in the ability of life to develop and survive on this planet. Without $CO_2$, the planet would be inhospitable, with daily surface temperatures varying by hundreds of degrees. Because they capture the $CO_2$ that would otherwise rise up and trap heat in the atmosphere, trees and plants are important players in offsetting the effects of temperature fluctuation and global warming (Haynes 1995). In the process, plants and trees make use of the captured carbon to generate an abundance of products that humans have used throughout the recorded history. These products come in the form of food, shelter, clothing, wood for fire, tools for hunting, and more modern items like paper, pharmaceutical agents, and biofuels.

The balance of just the right amount of $CO_2$ in the atmosphere has been key to the stability of our climate. Over the millennium prior to the Industrial Revolution, atmospheric concentrations of $CO_2$ were relatively stable. This is because the major carbon fluxes between terrestrial vegetation, the atmosphere, and the oceans were generally in equilibrium. However, since the 1800s, the world's population has grown tremendously with a corresponding increase in the use of coal, oil, and natural gas. Average $CO_2$ concentrations in the atmosphere, which were approximately 280 parts per million (ppm) in preindustrial times, have risen to approximately 390 ppm in 2010 (Barnola et al. 1995; NRC 2010; Stott et al. 2000). Such an unprecedented rate of change has been attributed by many as a direct result of the corresponding increase in environmentally detrimental human activities.

The burning of fossil fuels and deforestation has introduced an additional flux into the natural carbon cycle (Fig. 12.1). Scientists have estimated that these two activities currently reintroduce almost eight billion metric tons of carbon to the atmosphere every year, and about 20 % of this is the result of land-use change such as tropical deforestation (IPCC Special Report on LULUCF 2000). Roughly half of these human-induced carbon emissions remain in the atmosphere for as long as a century or more, while the remainder is taken up in nearly equal portions by the oceans and land vegetation (http://www.epa.gov/sequestration/ccyle.html). As more fossil fuels are burned, the concentration of $CO_2$ in the atmosphere rises, and more heat is trapped through the greenhouse effect. Indeed, the average temperatures throughout much of the world have been increasing, and while many factors have an impact on the climate, most scientists agree that the current rise in average global temperatures is due to our extensive use of fossil fuels. Whether caused by humans or not, the reason that carbon sequestration is important is that it can help slow down atmospheric $CO_2$ buildup, thus tempering subsequent climactic changes.

## 12.1.2   Approaches to the Solution of Rising Carbon Dioxide Levels

Because $CO_2$ accumulates in the atmosphere before being removed by natural processes, slowing and ultimately reversing atmospheric $CO_2$ buildup will require deep reductions in $CO_2$ emissions. Environmentalists and scientists tend to agree that aggressive afforestation is the best natural means for minimizing the impact of rising $CO_2$ levels and their subsequent relation to global warming. Land management practices can be altered to sustainably increase live biomass and store more carbon. For example, sustainable forestry practices can increase the ability of forests to sequester atmospheric carbon while enhancing other ecosystem processes through the improvement of soil and water quality. Reforesting cleared or mined lands, conserving stands of large trees, and improving forest health through thinning and prescribed burning are just a few ways to increase forest carbon sequestration in the long run. Harvested trees can be used to produce long-lasting products, including high-quality building materials and furniture that can keep captured $CO_2$ from reentering the atmosphere via decomposition for many years.

Soil carbon sequestration is another potential tool for combating climate change because soils offer a large carbon sink. Plants facilitate the soil sequestration process by removing carbon dioxide from the atmosphere during photosynthesis and storing it as biomass. However, this biomass later decomposes to organic matter in the soil, and over many years that organic material can add up to a tremendous amount of stored carbon. It is estimated that the Earth's soils contain roughly three times as much carbon as all plant biomass, and researchers have suggested that, globally, soils have the potential to sequester ~24 % of the total emissions from fossil fuel combustion (approximately 1.9 GtC potential stored out of an 8 GtC emission). In addition, sequestration of carbon in the soil also promotes soil quality and health, which subsequently improves the health of the plants that contribute the biomass to begin with.

Carbon sequestration is unlikely to be a stand-alone solution to rising atmospheric carbon dioxide. Multiple complementary solutions will likely be needed to meet the challenge of stabilizing climate change, including more efficient energy use, alternative fuels, alternative transportation technology, electricity from non-$CO_2$-emitting sources, and improved agricultural and forestry practices combined with natural carbon sequestration. Much attention is also focused on the potential to develop plants and trees that have an enhanced ability to sequester carbon in the form of biomass. To appreciate how researchers are approaching the improvement of carbon sequestration, one must first understand the rather complex mechanisms that plants, including trees, use to capture and distribute carbon to their leaves, stems, roots, and surrounding soil. The remainder of this chapter focuses on these aspects, the traits that have been singled out to help improve carbon sequestration, and the novel approaches to the discovery of new traits involved in improved biomass production.

## 12.2   Photosynthesis and Carbon Capture

The recent rapid rise in atmospheric $CO_2$ will likely continue in the future and will affect climate and biogeochemical cycles. In today's world, most of the anthropogenic $CO_2$ emissions result from the combustion of fossil fuels for energy production. The increasing demand for energy, particularly in developing countries, underlies the projected increase in $CO_2$ emissions. Addressing this energy demand without increasing $CO_2$ emissions requires more than merely increasing the efficiency of energy production. Carbon sequestration, capturing and storing carbon emitted from the global energy system, will likely be a major tool for reducing atmospheric $CO_2$ emissions from fossil fuel usage and store it in the form of biomass. Photosynthesis has long been recognized as a means to sequester anthropogenic $CO_2$. It is the first in the sequence of reactions that constitute global carbon cycle. As such, photosynthesis provides the primary route for energy to enter into the global ecosystem, and carbon assimilation produces most of the global biomass.

Carbon sequestration occurs in crops, forests, and soils primarily through the natural process of photosynthesis. Atmospheric $CO_2$ is taken up through tiny openings in leaves, called stomata, and transferred to the chloroplasts where it is incorporated into sugars (primarily glucose, fructose, and sucrose) with oxygen produced as a byproduct. The carbon within the sugars is subsequently used along with other nutrients to generate the plant body (leaves, stems, roots, flower, and fruits), including the woody biomass of trees and leaves of promising bioenergy crops. Around 90 % of the dry weight of a given plant consists of carbon and oxygen obtained from the atmosphere via photosynthetic carbon assimilation within the leaves. A typical leaf is made up of an upper and lower epidermis, the mesophyll, the vascular bundles, and a varying number of stomata. The mesophyll cells have chloroplasts and are the prime site of photosynthesis. The upper and lower epidermal cells do not have chloroplasts and serve primarily as protection for the leaf. The stomata are pores that occur primarily in the lower epidermis and allow for gaseous exchange (i.e., $CO_2$, $O_2$, and $H_2O$) for photosynthesis and respiration. The vascular bundles or veins in a leaf are part of the plant's transport system, conducting water, nutrients, and photosynthetic products throughout the plant.

The process of carbon assimilation starts with a light-dependent reaction that converts solar energy into chemical energy. The energy of light captured by chlorophyll pigment molecules, present in thylakoid membranes of chloroplasts, is used to release high-energy electrons from molecules of $H_2O$. These electrons are used in a series of electron transfers to produce NADPH while at the same time generating a proton (H+) gradient across the thylakoid membranes. This electrochemical gradient is then used by ATP synthase to generate ATP. The NADPH and ATP formed by the light reactions enable the reduction of carbon dioxide through a universal pathway called the Calvin cycle, often termed the "dark reaction" due to the fact that it does not use photons of light. Not only does the Calvin cycle require ATP and NADPH generated in light reaction but many of the enzymes involved in carbon assimilation are active only in the light. The high energy conversion efficiency of the Calvin cycle (approximately 90 %) is derived from reactions that involve rearrangement of chemical energy rather than energy transduction. Two molecules of NADPH and three molecules of ATP are required to reduce each molecule of $CO_2$ to a sugar $[CH_2O]n$. The first step in this cycle is the addition of $CO_2$ to a five-carbon compound, ribulose 1,5-bisphosphate (RuBP). The six-carbon compound formed is split, giving two molecules of a three-carbon compound called 3-phosphoglycerate (PGA). Since the first stable organic compound formed contains three carbon atoms, these plants are called $C_3$ plants. $C_3$ plants include most temperate plants (excluding many grasses) and represent more than 95 % of all Earth's plants. Some of the most common examples of $C_3$ plants are wheat, rice, soybean, barley, tobacco, carrot, potato, tea, and coffee.

The process for assimilation of $CO_2$ and replenishment of RuBP in the Calvin cycle is mediated by 13 enzymes located in the chloroplast. Three of these enzymes, ribulose bisphosphate carboxylase/oxygenase (RuBisCO), sedoheptulose 1,7-bisphosphatase, and phosphoribulokinase, are unique to the Calvin cycle. RuBisCO is the key enzyme responsible for photosynthetic carbon assimilation catalyzing the carboxylation of RuBP to form two molecules of 3-PGA. As its name suggests, RuBisCO is a bifunctional enzyme capable of catalyzing two distinct reactions, acting both as a carboxylase and as an oxygenase. It uses the carboxylation substrate, RuBP, in a parallel reaction with $O_2$ to form one molecule of 3-PGA and a two-carbon product, 2-phosphoglycolate, which is of no immediate use in the Calvin cycle. Interestingly, this reaction is catalyzed on the same active site as carboxylase reaction; hence, $CO_2$ and $O_2$ are competitive substrates; $O_2$ inhibits carboxylase reaction and $CO_2$ inhibits oxygenase reaction. Each time RuBisCO catalyzes reaction of RuBP with $O_2$ instead of $CO_2$, the plant makes 50 % less 3-PGA, thus reducing the net gain in photosynthetic carbon. Not only does RuBisCO lack specificity for $CO_2$, but it has very low specific activity and for this reason is produced in massive quantity in leaf. It has been estimated that RuBisCO constitutes up to half of the soluble protein in the plant leaf (Ellis 1979). In $C_3$ plants almost 25 % of the nitrogen is invested in RuBisCO alone, which accounts for the considerable interest in the relationship of this enzyme to the nitrogen nutrition of plants (Leegood et al. 2000).

In the oxygenase reaction of RuBisCO, the 2-phosphoglycolate product enters the photorespiratory pathway, which eventually returns some 3-PGA to the Calvin cycle. In converting 2-phosphoglycolate back to 3-PGA, there is loss of carbon as $CO_2$ and consumption of ATP, but this pathway provides a mechanism to restore carbon to the Calvin cycle that would otherwise have been lost. Through photorespiratory pathway as much as 75 % of the carbon initially lost from the Calvin cycle is thus returned. Although there is inefficient use of carbon, energy, and reductant in photorespiration, the enzymatic pathway has been maintained throughout evolution. One of the most compelling arguments for this retention is the advantage of its scavenging role, i.e., returning carbon to the Calvin cycle that would otherwise have been lost. It has also been suggested that photorespiration helps plants to withstand environmental stress.

Some plants like corn, sugarcane, and many other tropical grasses have a special mechanism to overcome the tendency of RuBisCO to wastefully fix $O_2$ rather than $CO_2$ by photorespiration. These plants use a supplementary method of $CO_2$ uptake in which a 4-carbon compound, oxaloacetate (OAA), is formed in place of one of the 3-carbon compound of the Calvin cycle. Therefore, these plants are called $C_4$ plants. These plants initially bind $CO_2$ using an enzyme called phosphoenolpyruvate carboxylase (PEP carboxylase). This helps in a more efficient harvest of $CO_2$, allowing the plant to capture sufficient $CO_2$ without opening its stomates too often. $C_4$ plants utilize their specific leaf anatomy (kranz anatomy), where mesophyll and bundle sheath cells cooperate to fix $CO_2$. In $C_4$ photosynthesis, $CO_2$ is first incorporated into a 3-carbon compound, phosphoenolpyruvate (PEP), by PEP carboxylase, forming OAA in mesophyll cells which is then pumped to the bundle sheath cells. There, it releases the $CO_2$ for carbon fixation by RuBisCO. The decarboxylation reaction also produces three-carbon organic acids ($C_3$) that return to the mesophyll cells and regenerate as PEP in a reaction catalyzed by the enzyme pyruvate orthophosphate dikinase (PPDK). By concentrating $CO_2$ in the bundle sheath cells, $C_4$ plants promote the efficient operation of the Calvin cycle and minimize photorespiration. This $CO_2$ concentrating mechanism makes $C_4$ plants more nitrogen efficient such that RuBisCO constitutes only 10–15 % of leaf nitrogen (Sage et al. 1987). However, the $C_4$ photosynthetic pathway requires two more moles of ATP than $C_3$ pathway per mole of $CO_2$. This additional ATP requirement means that the $C_4$ pathway needs more light energy than the $C_3$ pathway for the assimilation of same quantity of $CO_2$. Therefore, most of $C_4$ plants are native to the tropics and warm temperate zones with high temperature and light intensity. They exhibit higher photosynthetic and growth rates under these conditions because of the efficient use of carbon, water, and nitrogen. $C_4$ plants are among some of the world's most productive crops and pasture, for example, maize (*Zea mays*), sugar cane (*Saccharum officinarum*), sorghum (*Sorghum bicolor*), amaranth (*Amaranthus* spp.), bermuda grass (*Cynodon dactylon*), blue grama (*Bouteloua gracilis*), and rhodes grass (*Chloris gayana*) are $C_4$ plants. Also some of the most dangerous and damaging weeds like crabgrass, nut grass, and barnyard are also $C_4$ species. Although $C_4$ plants represent only a small fraction of the world's plant species, which is only 3 % of the vascular plants, they contribute about 20 % to the global primary productivity because of highly productive $C_4$ grasslands (Ehleringer et al. 1997).

Other than $C_3$ and $C_4$ photosynthetic pathways, there is yet another strategy used by plants in arid regions to cope with extreme hot and dry environments like desert. Plants like cacti and pineapple that live in extremely hot, dry areas can only safely open their stomates at night when the weather is cool to avoid dehydration. These plants open their stomates at night to take in $CO_2$, which is then incorporated into various organic compounds and stored in vacuoles. In the daytime, when the light reaction occurs and ATP is available, these plants extract $CO_2$ from the organic compounds for use in the Calvin cycle. These "CAM" plants (named for crassulacean acid metabolism after the plant family Crassulaceae where this process was first discovered) also initially attach $CO_2$ to PEP and form OAA similar to the process used in $C_4$ plants. However, CAM plants fix carbon at night and store the OAA in large vacuoles within the cell instead of fixing carbon during the day and pumping the OAA to other cells. In CAM photosynthesis there is temporal separation of two carboxylating enzymes, PEP carboxylase and RuBisCO. This allows CAM plants to avoid risk of dehydration and use the $CO_2$ for the Calvin cycle during the day, when it can be driven by the solar energy. Although primarily a means of surviving in arid conditions, CAM also enables plant to photosynthesize at low $CO_2$ because, like the $C_4$ pathway, it works by concentrating $CO_2$ for RuBisCO to assimilate into the Calvin cycle. There are many more species of CAM relative to $C_4$ species; however, only two CAM plants are of commercial importance, the pineapple (*Ananas comosus*) and the cactus (*Agave tequilana*). CAM plants are slow growing compared to $C_4$ and $C_3$ plants which have higher growth rates in their natural environment. The extra ATP requirement for CAM photosynthesis explains to some extent the slow growth rates relative to $C_3$ and $C_4$ plants. Also CAM plants undergo CAM idling, i.e., closing of stomata both day and night, which leads to complete retardation of growth in very arid environments. CAM species have not been exploited for carbon sequestration because of the diffusive (stomatal plus internal) constraints imposed by succulent CAM tissues on $CO_2$ supply to the cellular sites of carbon assimilation. However, areas of current and future research include elucidating the causes and consequences of CAM and providing a knowledge base that might inform and improve the potential of CAM plants for carbon sequestration and bioenergy production on marginal and degraded lands.

Overall carbon assimilation by $C_3$, $C_4$, and CAM photosynthesis each has distinct advantages and disadvantages in particular environment. $C_3$ plants are more abundant in temperate climates, $C_4$ plants in tropical climates, and CAM plants in arid regions with the distribution strongly influenced by water, nutrient, and light requirements. $C_4$ and CAM plants are more nitrogen and water efficient than $C_3$ plants. The $CO_2$ concentrating mechanisms of the $C_4$ and CAM pathways enable RuBisCO to function more efficiently than in $C_3$ plants, and therefore, these plants produce reduced amounts of RuBisCO protein relative to $C_3$ plants. $C_4$ plants also have the tendency to compete with $C_3$ plants in nitrogen-poor soils, where production of RuBisCO protein by leaves is limited by nitrogen availability. The $C_4$ and CAM pathways require more light energy than $C_3$ photosynthesis because of differences in ATP requirement due to additional reactions of the $C_4$ and CAM pathways. This gives $C_4$ and CAM plant an advantage in bright light and in warm temperatures.

## 12.2.1  Poplar as an Example of $C_3$ Plants

Poplar species, such as trembling aspen (*Populus tremuloides*), are the most widespread tree species throughout the world and have an important role in many different ecosystems. Poplar species have commercial utility for the production of wood products partly due to their ability to regenerate from their roots after cutting (Frey et al. 2003). Recent interest in terrestrial carbon sequestration is motivating efforts to explore opportunities for climate change mitigation and future energy resources. Since forests make a significant part of the global carbon cycle constant, efforts are being made to increase the sequestration of carbon in forests. Trees, through their growth process, act as a sink for atmospheric carbon. *Populus* ($C_3$ species) has emerged as a model tree system among all forest trees, which together comprise the largest fraction of the living terrestrial carbon reservoir (Dixon et al. 1994). The availability of extensive genetic resources like breeding populations, genetic maps, large expressed sequence tag (EST), and bacterial artificial chromosome (BAC) libraries and ease of transformation uniquely provide molecular tools and approaches to understand the mechanisms that control carbon allocation and partitioning (Cseke et al. 2007; Cheng and Tuskan 2009). Several studies on *Populus* have demonstrated that there is significant variation in biomass distribution among different clones and there exists a genetic basis for above-ground and below-ground dry mass distribution (Pregitzer et al. 1990; Heilman et al. 1994; Scarascia-Mugnozza et al. 1997; Dickson et al. 1998; Karim and Hawkins 1999; Cseke et al. 2009). Progeny from crosses within and between species has also shown considerable variation for the above- and below-ground distribution of biomass and tissue chemistry (Driebe et al. 2000; Schweitzer et al. 2004; Block et al. 2006; Fischer et al. 2006). Genomics and transcriptomics approaches are being used to identify potential genes that provide key control points for the flow and chemical transformations of carbon in roots. Particular focus is on genes that increase sink activity (greater root mass) and are involved in the synthesis of chemical forms of carbon that result in slower turnover rates of soil organic matter. Quantitative trait loci (QTL) analysis approach has been used to identify traits associated with soil carbon sequestration like productivity, biomass distribution to roots, and fine root/ coarse root ratios and regions of the genome linked to these traits (Wullschleger et al. 2005). Current strategies are directed to modification of genes that have been shown to play significant role in carbon partitioning, including the invertase family, which controls sucrose catabolism. Past studies have indicated that sink strength has a dominant influence on source photosynthesis and carbon partitioning (Paul et al. 2001; McCormick et al. 2006). Sink strength is essentially regulated by sucrose metabolism channeling carbon into storage or structural components. Metabolic engineering has been employed to target the activity of enzymes of sucrose metabolism like sucrose synthase, invertase, and ADP glucose pyrophosphorylase to increase sink strength (Capell and Christou 2004; Roitsch and Gonzalez 2004; Bieniawska et al. 2007; Coleman et al. 2007; Smidansky et al. 2007). Future studies are likely to focus on employing combinations of advanced breeding techniques and targeted genetic manipulations to select or develop hybrid poplars with traits favoring the enhanced carbon sequestration and storage capacity in long-lived soil pools.

## 12.2.2   Miscanthus *as an Example of* $C_4$ *Grasses*

Perennial $C_4$ grasses have highly efficient $C_4$ photosynthesis that often yields more biomass than $C_3$ species. These grasses are promising candidates as energy crops as they have the potential for increased soil carbon sequestration. In addition, they have a low demand for nutrient inputs and higher biomass yields on relatively poor-quality land. An advantage of perennial $C_4$ grasses compared with trees is that they can be established more quickly and produce an annual harvest with low moisture content. Perennial crops accumulate and sequester carbon in the soil as well as produce combustible material that substitutes for fossil fuels. Different perennial grasses differ in their potential productivity, chemical and physical biomass properties, environmental demands, and crop management requirements (Lewandowski et al. 2003). Among the candidate perennial $C_4$ grasses, one of the most intensively investigated potential new energy crops is *Miscanthus* species native to subtropical and tropical regions of Asia. The rapid growth, low mineral content, and high biomass yield of *Miscanthus* make it one of the most preferred choices for bioenergy production. *M. sacchariflorus*, *M. sinensis*, *M. floridulus*, and *M. giganteus* have caught particular interest in recent years for biomass production (Deuter 2000). *M. giganteus* (Giant Miscanthus), the triploid hybrid species, shows superior characteristics, such as high biomass yield, and has been considered as the most promising *Miscanthus* species for bioenergy production (Lewandowski et al.

2000; Pyter et al. 2007). In the United States, Giant Miscanthus has been proposed for use in combined heat and power generation and it is also a leading candidate feedstock for the production of cellulosic ethanol (Heaton et al. 2004; Khanna et al. 2008). The occurrence of this hybrid shows promise for interspecific hybridization within *Miscanthus* because the genus has substantial genetic diversity within and between species.

Most research in the past decade was focused on enhancing the productivity and economic potential of *Miscanthus* species (Jones and Walsh 2001; Khanna et al. 2008). Essential to this process is the development of efficient tissue culture methods and transformation tools that will allow introduction of desirable traits into this. Although the biotechnological approaches for molecular breeding of these plants are still in immature stages, a number of important genetic resources such as genetic maps, cross-species markers, molecular markers, and physical and comparative maps have been developed (Jiang et al. 2012; Kim et al. 2012; Swaminathan et al. 2012). Recently, studies involving micropropagation and plant regeneration from embryogenic callus of *Miscanthus sinensis* have been reported (Głowacka and Jeżowski 2009a, b; Głowacka et al. 2010; Zhang et al. 2012). Progress has been made in the development of transformation techniques; particle bombardment and *Agrobacterium*-mediated transformation have been established using embryogenic callus of *M. sinensis* (Wang et al. 2011). Transgenic approach is being used to target genes for value-added traits, such as enhanced biomass and fermentation efficiency. The target traits that are important and will make large impacts in the molecular breeding include herbicide resistance, biotic and abiotic tolerance, low-fertilizer needs, high-efficiency photosynthesis for high productivity, promoted and synchronized flowering for hybridization, and organellar transformation for the effective accumulation of high-value chemicals and proteins.

## 12.3  Allocation of Carbon to Leaves, Stems, and Roots

Perhaps the best-known commodity produced by a tree's vascular system is "wood"—a valuable, renewable resource for lumber, paper, and energy production. What we call "wood" is actually a complex vascular tissue, formed by many cell types organized within an extracellular matrix composed primarily of cellulose and lignin (Plomion et al. 2001; Li et al. 2003). The secondary thickening that results from an active vascular cambium in stems and roots increases the girth of the plant and is especially important to the production of biomass and the ability of trees to store water and to transport carbon and nutrients. The vascular system of all terrestrial plants is key to assimilating the carbon that is fixed during photosynthesis because it provides a mechanism for carbon-laden sugars to move throughout the plant body to the location of growth and development. This makes the understanding of above- and below-ground secondary development one of the more important aspects of developing tree-based technologies to sequester $CO_2$.

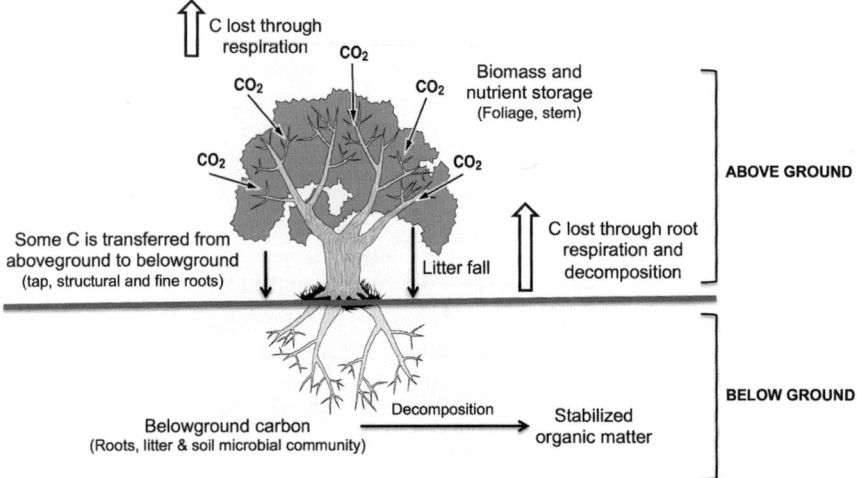

**Fig. 12.2** Carbon sequestration—above- and below- ground. *Arrows* pointed *downward* indicate carbon capture. *Arrows* pointed *upward* indicate emissions of $CO_2$ into the atmosphere

## 12.3.1 Above-Ground Carbon Allocation

Carbon compounds produced through photosynthesis, along with minerals from the soil, are transported from "sources" (photosynthetically active leaves) to "sinks" to support plant growth, fruit development, and maintenance of the plant's permanent structure (branches, stems, and roots) (Fig. 12.2). In plants, sucrose is the end product of photosynthesis and is converted to a wide variety of storage compounds in different tissues such as seeds, shoots, roots, and woody parts. The process of storage of carbon in different parts of plant including reproductive sinks (fruit and seeds), temporary storage sinks (tubers), shoots, roots, and woody parts is termed as carbon allocation. The relative amount of carbon allocated in the various organs, which is also called as biomass allocation, is not fixed but may vary over time, across environments and among species. The question of how plants allocate carbon among different organs has long been a topic of ecological interest.

Fundamentally, the assimilation of carbon by leaves, and acquisition of mineral nutrients by roots, must be in balance with the utilization of carbon and mineral nutrients for plant growth. This functional balance at whole-plant level can be represented in terms of carbon produced by assimilation and carbon consumed in growth, enabling root–shoot responses to nutrients. In the end, the plant has to balance the carbon allocation to leaves, stems, roots, and other storage organs in a way that matches the physiological activities and functions performed by these organs. The source organ (leaves) maintain high concentration of carbon assimilates in the phloem at the points of loading. The transportation of carbon substrates at the points of loading (source) and unloading (sinks) occurs either by symplast or apoplast. The root, shoot, and cambial sink (utilization sinks) appear to possess symplastic connections to phloem, with no active phloem unloading. In this scenario substrate gradient is maintained between phloem and sinks cells, across plasmodesmata, by the utilization for growth and associated respiration. However, in reproductive sinks like fruit and seeds, carbon substrates are transported by apoplast from phloem to sink, possessing active phloem-unloading mechanism. This is the reason reproductive sinks have large sink strength compared to utilization sinks. In many crop plants this loading is through apoplast involving active transport, whereas in some woody plants less efficient symplastic loading mechanisms are utilized (Gamalei 1991; Gamalei et al. 1992). In past research focus has been on understanding the sink properties of seeds, fruits, and specialized storage organs that constitute yield of crop plants compared to shoot and cambial sinks.

Trees store a large amount of carbohydrates in the parenchymatous tissues of their wood and bark, mainly as starch. These stored carbohydrates play an important role in tree functioning; they can be used when current photosynthesis is not enough to meet the carbon needs for maintenance and growth. Carbohydrate storage in tree wood parenchyma has been considered as only a passive accumulation process, but this view is being challenged recently. It has been demonstrated that an increased C demand does not necessarily result in a depletion of carbohydrate concentration in wood (Silpi et al. 2007). It has been suggested that trees tend to adapt the level of stored carbohydrate to current metabolic demand, at the possible expense of other sinks (Silpi et al. 2007). In order to better understand carbon allocation among functional sinks—growth and secondary metabolites—long-term studies that enable the comparison of contrasting levels of assimilate availability are required.

Forest C allocation has drawn particular interest due to its responsiveness and potential effect on carbon sequestration and the global carbon balance. The differences in lifespan and decomposition rates among tree organs suggest that C allocation in trees strongly influences forest carbon cycling rates. Owing to the importance of forest C allocation, a number of contrasting approaches exist to model forest C allocation. There are five main categories of approaches to allocation modeling that have been identified, based on the key principles used to predict allocation: empirical, allometric scaling, functional balance, evolution based, and entropy based (Purves and Pacala 2008; Ostle et al. 2009; Ise et al. 2010; Franklin et al. 2012). Although there are guidelines for identifying approaches that are appropriate to predict allocation for different purposes, more research is required to further increase an understanding of allocation and how it can best be modeled.

## 12.3.2 Below-Ground Carbon Allocation

Soils are the primary carbon repository for three-fourths of the terrestrial carbon with 4.5 times more than the biotic pool (Lal 2004). The total quantity of below-ground carbon allocated by plants is enormous and is in the form of living plant roots and root exudates. Below-ground carbon allocation (BCA) allows flow of organic carbon to the soil from photosynthetically fixed $CO_2$ and makes a significant impact on the global carbon cycle (Fig. 12.2). It is estimated that terrestrial plants allocate nearly 60 Pg (petagram) of carbon below-ground out of the 120 Pg fixed through photosynthesis (Schimel et al. 1994; Grace and Rayment 2000). The large amount of BCA is essential for plants to obtain mineral nutrients and water in resource-limited terrestrial ecosystems. Furthermore, BCA regulates soil organic matter formation and influences bulk density, water-holding capacity, and cation-exchange capacity of soil. Overall, BCA is profoundly important to ecosystem carbon budgets playing a dominant role in whole ecosystem carbon exchange (Valentini et al. 2000; Giardina et al. 2005). Although the relevance of BCA to the functioning of forest ecosystems and global carbon budget is far recognized, controls on BCA are not completely coherent (Ryan et al. 1996; Giardina and Ryan 2002). The above-ground plant carbon allocation is, however, precisely captured in leaf-based physiological process models (Landsberg and Gower 1997), whereas below-ground processes are poorly captured in process models. The complexity of the global changes in environmental factors and above-ground and below-ground interactions compounded by the soil matrix further hinders efforts to validate below-ground models. Often the ecosystem models describing BCA response to global environmental changes rely on the assumption that the functioning and dynamics of above-ground processes sufficiently describe those of below-ground processes (Binkley and Menyailo 2005).

### 12.3.3  How $CO_2$ Is Assimilated into Roots, Soil, and Soil Communities

Perennial and herbaceous plants capture atmospheric $CO_2$ through photosynthesis and store large amounts of organic carbon in above-ground structures followed by its translocation below-ground into plant roots (Hinsinger et al. 2009). Soil carbon is also accumulated through deposition of canopy litter and via rhizodeposition. The flow of organic matter from roots into the soil, which includes shedding of root cap and cortical cells, fine roots formation, and root exudation of simple sugars, polysaccharides, organic acids, amino acids, and proteins, is called rhizodeposition (Pritchard 2011). The rooting process itself, however, is the primary means for most carbon entering below-ground soil carbon pool that act as a means for substantial long-term carbon sequestration. Overall, the root turnover, a metabolically expensive process, accounts for about 30 % of global terrestrial net plant productivity (NPP) (Jackson et al. 1997), while fine root exudates comprise extra 0.5–20 % of net ecosystem C assimilation (Farrar et al. 2003; Frank and Groffman 2009; Pritchard 2011). It has been predicted that plants with deeper roots contribute more towards increase in the global carbon sequestration, as loss of carbon due to microbial decomposition is mainly high in the upper soil strata (Pritchard 2011).

Good examples are perennial grasses such as *Miscanthus* and switchgrass that exhibit increased (5–25 %) soil organic carbon deposition through carbon sequestration in deep roots and reduced net $CO_2$ emissions (Anderson-Teixeira et al. 2009). In temperate and boreal forests, soil carbon accounts for four times more carbon stored in vegetation, and it is almost 33 % higher than the total carbon sequestered in tropical forests. On the other hand in grasslands, 98 % of total sequestered carbon is stored below-ground, which in total adds to 8 % of global soil carbon (Lal 2004; Jansson et al. 2010).

The coevolution of plant roots along with soil inhabitants like fungi and micro- and macrofauna has resulted in the plant carbon allocation strategies, root physiological behaviors, and root structural properties (Johnston et al. 2004). Association of plant roots with soil microbes like mycorrhizal fungi increases total plant carbohydrate budget by 10–20 % to support and maintain these fungal symbiotic partners (Johnson and Gehring 2007) in return for mineral nutrients and water. The plant organic carbon transfer to extraradical mycelia is considered a vital feature of soil carbon processes as fungal mycelia constitutes 20–30 % of soil microbial biomass and it is estimated to be 15 % of the total soil carbon in certain ecosystems (Leake et al. 2004). And the soil carbon transfer in the form of root exudates stimulates soil food web, resulting in greater microbial biomass and activity, which in turn increases soil organic nitrogen turnover C (Carney et al. 2007). Therefore, the association of fine roots with mycorrhizal symbionts and the necessity of plants to feed the soil community for improved nitrogen mineralization are responsible for significant amount of organic carbon deposition into the soil.

The impact of rapidly changing climate, particularly increase in the atmospheric $CO_2$ concentration, on ecosystem carbon cycling and below-ground carbon deposition has gained much attention in the recent past. Plants exposed to increased $CO_2$ concentrations respond with a significant increase in photosynthesis and growth. Often, carbon allocation was high in below-ground processes as compared to that of the above-ground processes. This uneven distribution of carbon contributes to increased root growth as well as the root-to-shoot ratio (Rogers et al. 1996; Pritchard and Rogers 2000). A substantial increase in the root production in grasses (Rillig and Allen 1999; Milchunas et al. 2005), coniferous trees (Prichard et al. 2008), and field crops (Wechsung et al. 1995; Pritchard et al. 2006) was demonstrated under elevated $CO_2$ concentration. However, the reports of decreased or unchanged below-ground processes such as root production in $CO_2$-enriched environments also exist but are not common (Arnone et al. 2000; Johnson et al. 2006; Brown et al. 2007). Furthermore, thorough understanding of the consequences of global climatic change on the functioning of ecosystem also requires consideration of relationship between plants and their below-ground microbial communities (Bardgett and Wardle 2010). Drigo et al. (2008) reported that under elevated $CO_2$ atmospheres along with the changes in root developmental patterns, plants released more organic carbon compounds into the rhizosphere, the portion of soil where microorganism-mediated processes are under the influence of the root system stimulating greater microbial biomass (Carney et al. 2007). And an increase in the $CO_2$ concentration by double the ambient level exhibited a 47 %

increase in the mycorrhizal abundance (Treseder 2004). In a different study, 34 % increase in the mass of ectomycorrhizal fungi and that of arbuscular mycorrhizal fungi showed an increase of 21 % in $CO_2$-enriched environments (Alberton et al. 2005), demonstrating that the effects of atmospheric $CO_2$ enrichment on mycorrhizal fungi are profound. Therefore, an increase in the atmospheric $CO_2$ concentration has considerable positive impact on the downward flow of carbon and thus its influence on soil food webs, i.e., the relationship between plant roots and other soil inhabitants (Drigo et al. 2010). Root carbon transfer to the soil consequently controls the impact of climate change on carbon cycling and climate mitigation by reducing atmospheric $CO_2$. Studies designed to explore approaches to increase plant productivity and biomass distribution to roots and soil communities that contribute to carbon cycle processes under field environments could help in understanding of mechanisms to enhance soil carbon sequestration.

## 12.4 The Benefits of Mycorrhizal Interactions and Soil Communities

Plants constantly interact with a wide range of microbes in their environments. The beneficial microorganisms contribute to plant health by secreting plant hormones like auxins and cytokinins that positively affect plant growth and increase the availability of nitrogen, phosphorus, and other nutrients and provide protection against disease-causing microorganisms (Pritchard 2011). The majority of plants from terrestrial ecosystems form the mutually beneficial, long-term, symbiotic association between their roots and fungi to develop into functional structures called mycorrhizae. The functional basis of this relationship is the reciprocal transfer of nutrients and minerals, N and P, from the fungus to the plant and plant-derived carbohydrates to the fungus (Smith and Read 1997) (Fig. 12.3). Mycorrhizae are broadly categorized into two types, ectomycorrhizae (ECM) and arbuscular mycorrhizae (AM), particularly relying on the pattern of fungal colonization in plant roots (Bonfante and Genre 2010). AM fungi are widespread obligate biotrophs predominant in herbaceous plants. They form specialized organs called arbuscules within plant cells that aid in nutrient exchange with plant partners (Ferrol et al. 2002). Ectomycorrhizal fungi form symbiotic relationships with almost 90 % of the major forest trees belonging to temperate, boreal, and montane regions; thus, they dominate and can be said to have shaped the world's forests (Martin et al. 2008; Podila et al. 2009). The ECM fungal partner grows as a thick cover, termed the mantle, around the fine roots of the plant and can comprise up to 40 % of the colonized root biomass (Johansson et al. 2004). The ECM association increases the plant's fitness in many ways, beyond increasing the mineral N and P nutrition. It is known to increase the plant's water uptake efficiency and the plant's tolerance of various abiotic stresses, including salt, drought, and nutritional stress, allowing the plant to endure harsh climatic conditions (Larsen et al. 2011a). They

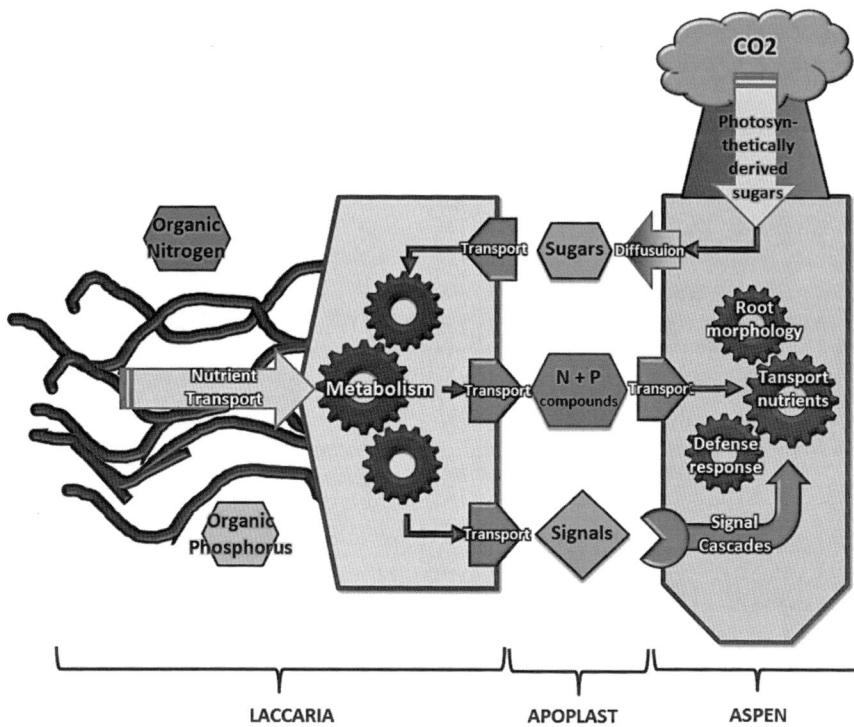

**Fig. 12.3** Schematic depicting the transport of nutrients between interacting soil mycorrhizal fungi (such as *Laccaria bicolor*) and tree roots (such as those from *Populus tremuloides* or quaking aspen)

protect trees against heavy metal toxicity by reducing the translocation of heavy metals to the host (Shetty et al. 1994; Jentschke and Godbold 2000; Schützendübel and Polle 2002; Langenfeld-Heyser et al. 2007). ECM symbiosis, thus, improves overall tree health and acts as the key component for the stability of forest ecosystems that positively impacts the global environment. Additionally, the fungal mycelium associated with root tips forms carbohydrate sinks for the tree, and up to 18 times more photosynthates are relocated to the mycorrhizal structures relative to non-mycorrhizal roots, thus contributing to carbon sequestration (Cairney et al. 1989; Smith and Read 1997). Mycorrhizal fungi transform plant-derived carbon into trehalose. Several studies have indicated the possible role of trehalose in fungal interaction with certain bacteria (Izumi et al. 2006; Uroz et al. 2007). Other microbes such as rhizobia, phosphorus-solubilizing bacteria, and free-living $N_2$-fixing organisms are considered to exert strong influence on the plant physiology and growth, hence the total ecosystem productivity.

## 12.4.1   The Importance of Soil Nutrients to Carbon Sequestration

Plant growth and productivity greatly depends upon the availability of water, accessible mineral nutrients, temperature, and light intensity. An imbalance of mineral nutrients in the soil leads to nutritional stress, which compromises overall plant growth. Nevertheless, plants sense soil nutrient limitations and acclimate to available nutrients. They do that primarily by diverting their carbohydrate resources and altering patterns of carbon allocation and partitioning among different organs, allowing optimal growth in a specific soil type (Marschner 1995; Aerts and Chapin 2000; Glynn et al. 2007). According to optimal allocation theory (OA), in nutrient-rich soils plants direct greater proportion of biomass towards above-ground tissues as mainly light limits the growth processes (Poorter and Nagel 2000). On the contrary, OA predicts that plants allocate more biomass to roots in nutrient-limiting soils in order to acquire scarce minerals (Ibrahim et al. 1997; Shipley and Meziane 2002). Nitrogen and phosphorus are major essential plant nutrients needed in large quantities, and their deficiency limits plant growth significantly. Amending the soil with fertilizers is routinely used to provide these essential minerals to enhance plant growth. Though plants gain from fertilization in severe nutrient-limiting soil types, excessive fertilization can have negative influence on plant health and physiology (Herms and Mattson 1997). Fertilization can alter patterns of carbohydrate partitioning in roots and reduce the content and concentration of nonstructural carbohydrates (e.g., simple sugars, starch) and defense-related secondary metabolites resulting in predisposing seedlings to injury from drought and pathogens (Bennett and Wallsgrove 1994; Pearce 1996; Kleczewski et al. 2010, 2012), having significant impact on plant growth and therefore carbon sequestration.

## 12.4.2   Dependence of Plant Carbon Sequestration on the Access to Nutrients Such as Nitrogen and Phosphorus

Nitrogen is present in various forms, including ammonium, nitrate, amino acids, peptides, and other complex insoluble nitrogen compounds in the soil. Though nitrate and ammonium are preferred nitrogen sources for plants, they can also uptake nitrogen in the form of amino acids that are available in abundance in certain soils, but the complex organic nitrogen compounds are not accessible (Williams and Miller 2001). Low soil nitrogen availability limits leaf nitrogen concentration, which in turn affect carbon assimilation and plant growth (Field and Mooney 1986). Many studies have suggested that carbon sequestration increases with the increasing nitrogen supply (King et al. 2005; Magnani et al. 2007; LeBauer and Treseder 2008; Xia and Wan 2008). These conclusions were, however, based on the observed changes in above-ground biomass production that do not always reflect the below-ground carbon deposition (Giardina et al. 2004; Lichter et al. 2005). Nitrogen enrichment can alter carbon allocation to below-ground processes. Specifically, high input of nitrogen increases fine root mortality while decreasing the fine root production and longevity (Gower et al. 1992; Haynes and Gower 1995). In contrast, addition of nitrogen-free fertilizer was reported to extend the fine root longevity (Majdi and Kangas 1997). Furthermore, addition of nitrogen also alters microbial community composition and reduces microbial activity significantly (Treseder 2004, 2008).

Likewise, low phosphorus availability is also a major limiting factor for plant growth (Lynch and Deikman 1998). Plants grown under phosphorus deficiency exhibit retarded growth and their roots are thinner with higher specific root length leading to increased root-to-shoot ratio (Bougher et al. 1990; Nielsen et al. 1998). An increase in length and density of root hairs is associated with adaptation to phosphorus deficiency, which potentially contributes to phosphorus acquisition (Gahoonia et al. 1999; Bates and Lynch 2000). The observed increase in carbon allocation to below-ground structures under limiting phosphorus conditions is attributed to decreased plant productivity (Qiu and Israel 1992). Consequently, increased carbon allocation to roots is a primary constraint to total carbon sequestration under phosphorus-limiting conditions. Plants with better phosphorus uptake efficiency or with greater ability to utilize acquired phosphorus, therefore, perform better under low phosphorus availability (Nielsen et al. 1998). Mack et al. (2004) showed that in a natural ecosystem, large declines in total soil organic carbon occur with experimental addition of nitrogen when phosphorus was added simultaneously. Studies also indicated that changes in soil phosphorus availability could alter ecosystem responses to nitrogen deposition and deposition of phosphorus alone could also potentially modify soil organic carbon dynamics (Wassen et al. 2005; Cleveland and Townsend 2006). Additionally, Bradford et al. (2008) suggest that not only the availability of nitrogen and phosphorus influences soil organic carbon sink strengths but also the rate of their deposition will be the critical determinant of whether these macronutrients increase or decrease the long-term sequestration of plant carbon inputs to soils. Therefore, an emphasis on studies that assess responses to multifactor (nitrogen and phosphorus) resource manipulations is crucial to gain better understanding of global carbon budgets.

## 12.4.3  How the Health of Soil Bacteria and Fungi Influences Carbon Sequestration

The stability and productivity of terrestrial ecosystems greatly depend on soil quality, requiring the management of soil–plant systems and the sustainability of soil resources (Altieri 1994). Diverse chemical, physical, and biological factors and their interactions determine the soil quality. For proper management of soil–plant systems, understanding how the physicochemical and the biological or microbial components of soil function and interact is critical (Kennedy and Smith 1995). While many studies have investigated the physicochemical properties of soil quality (Parr et al. 1992), the biological components did not gain much attention. However, microbial activities are particularly relevant at the root–soil interface, the rhizosphere, where microorganisms, plant roots, and soil constituents interact (Werner 1998; Bowen and Rovira 1999).

Plant roots continuously produce and release diverse array of carbon-containing primary and secondary metabolites into the rhizosphere (Rovira 2005), which acts as a key factor for the enrichment of specific microbial populations in the rhizo-sphere. And the rhizosphere is important for improving soil quality and microorganism-driven carbon sequestration and nutrient cycling in terrestrial ecosystems. The soil microflora consisting of rhizobia, phosphorus-solubilizing bacteria, free-living nitrogen-fixing organisms, and mycorrhizal fungi exert great influence on plant health. Numerous abiotic and biotic factors influence the struc-tural and functional diversity of the microbial community in the rhizosphere. However, in rhizosphere ecology not only the microbial and the physicochemical components interact, but different soil microflora also interact with each other (Berg and Smalla 2009).

Among the different organisms in soil microbial communities, mycorrhizal fungi are vital in regulating transfer of essential nutrients between the plants and the soil through widespread mycelial networks (Rooney et al. 2009). They also promote soil biological diversity through symbiotic interactions with other soil organisms. Some interactions between certain bacteria and mycorrhizal fungi are relevant to benefit plant fitness and sustainability of natural ecosystems (Jeffries and Barea 2001; Barea et al. 2002). For example, mycorrhiza helper bacteria (MHB) are associated with AM and ECM fungi and facilitate their root colonization capacity and suppress soil pathogens (Bending et al. 2006). Quoreshi and Khasa (2008) demonstrated the effectiveness of simultaneous inoculation of selected mycorrhizal fungal and bacterial species to poplar seedlings at the nursery stage with the perceived increase in plant nutrient status and mycorrhization. Therefore, the maintenance of diverse and active soil microbial communities is essential for improving soil quality (Kennedy and Smith 1995), carbon sequestration, ecosystem functioning, and nutrient cycling in natural ecosystems.

## 12.5 Approaches to Enhance Carbon Sequestration

One question that arises in discussions of carbon sequestration is how much carbon can forestry and agricultural practices actually sequester? To address such questions, it is important to remember that carbon sequestered in plants and soils can be released back to the atmosphere and there is a finite amount of carbon that can ultimately be sequestered within the sinks of Earth's carbon cycle. Carbon sequestration rates vary by plant species, soil type, regional climate, topography, and land management practices. Even with the advent of modern statistical sampling, computer modeling, and remote sensing, current estimates of carbon sequestration and emission sources vary over time despite the fact that they are more accurate and easier to generate than when rising $CO_2$ levels were first recognized.

In the USA, fairly well-established values for carbon sequestration rates are available for most tree species. Pine plantations in the Southeast can accumulate almost 100 metric tons of carbon per acre after 90 years, or roughly 1 metric ton of carbon per acre per year (Birdsey 1996). Interestingly, changes in forest management, such as lengthening the harvest-regeneration cycle, appear to result in less carbon sequestration on a per acre basis possibly because carbon accumulation in forests and their soils eventually reaches a saturation point, beyond which additional sequestration is no longer possible (Lal et al. 1999; West and Post 2002). This happens, for example, when trees reach maturity or when the organic matter in soils reaches equilibrium with the local plant life. Thus, the plant traits that are involved in carbon sequestration are complex and intimately connected with the surrounding ecology. A deep understanding of the molecular and cellular processes and how these processes interact is critical, which makes the targeting and manipulation of such traits all the more challenging.

Although the post-genomics era provides a unique opportunity to identify biochemical pathways and gene regulatory networks that underlie rate-limiting steps in carbon acquisition, transport, and fate, few studies have assessed the consequence of breeding for enhanced carbon uptake, allocation, or storage in perennial grass or woody crop systems. Investments in plant genomics could harness new approaches to increase biomass production and the distribution of that biomass to roots and recalcitrant pools of soil carbon in fast-growing trees and grasses grown in managed plantations (Jansson et al. 2010; Zhu et al. 2010; Garten et al. 2011). Research could focus on targeted improvements in light-use efficiency and photosynthesis (Long et al. 2006), in root growth and nutrient uptake (Hirel et al. 2007), and on overcoming constraints imposed on plant productivity by temperature and drought (Tuberosa and Salvi 2006). Genome-enabled increases in the production of plant biomass across a range of environments would, all else being equal, translate to enhanced input of carbon to soils via shoot and root litter, increasing the storage of carbon in terrestrial ecosystems. Microbial studies could target the interaction between plants and beneficial soil microorganisms as recently shown for important components of biogeochemical cycling in soils beneath *Miscanthus* (Mao et al. 2011) and for biomass production following inoculation of hybrid poplar with an endophytic, growth-promoting bacterium (Rogers et al. 2012). Gains in carbon sequestration might also be realized by understanding how genes and proteins that control the chemical composition of litter could impact the rates and magnitudes of carbon sequestration.

## 12.5.1   Genetic Manipulation to Improve Carbon Sequestration

### 12.5.1.1   Above-Ground Process

Attempts to increase the rates of photosynthesis through genetic engineering have focused on accomplishing this goal by increasing the total amount of RuBisCO in leaves (Suzuki et al. 2007). Contrary to original expectations, this research has been met with mixed success. Since RuBisCO is often rate-limiting for photosynthesis in plants, strategies have been employed to improve photosynthetic efficiency of plants by modifying RuBisCO (Spreitzer and Salvucci 2002). Additional research seeks to increase photosynthesis and plant productivity not by modifying the amount but by optimizing the distribution of resources between enzymes of carbon metabolism and/or by altering the kinetic properties of the RuBisCO enzyme itself. Theoretical analyses suggest that expressing RuBisCO as having either a higher specificity for $CO_2$ relative to $O_2$ or a higher maximum catalytic rate of carboxylation per active site could increase photosynthetic carbon gain by 25 % or more in $C_3$ plants (Zhu et al. 2004). Some have questioned whether substrate specificity of this enzyme can be improved (Tcherkez et al. 2006). Significant progress has been made in identifying natural variation in the catalytic properties of RuBisCO from different species. Also the development of the molecular tools for introduction of both novel and foreign RuBisCO genes into plants is in advance stages. The three-dimensional structure of RuBisCO has been determined for RuBisCO isolated from many organisms including tobacco, spinach, cyanobacterium (*Synechococcus*), purple bacterium (*Rhodospirillum rubrum*), and green sulfur bacterium (*Chlorobium tepidum*) (Schneider et al. 1990; Schreuder et al. 1993; Newman and Gutteridge 1993; Andersson 1996; Hanson and Tabita 2001).

One of the major challenges with the manipulation of RuBisCO in higher plants is that it is composed of eight large and eight small polypeptide subunits and that the genes for the small subunit are in the nuclear genome but those for the large subunit are encoded in the chloroplast genome (Chan and Wildman 1972; Kawashima and Wildman 1972; Smith and Ellis 1981). Therefore, the manipulation of the large subunit requires the gene (*rbc*L) to be introduced into the chloroplast genome. Though recent advances in chloroplast transformation have allowed experiments to be carried out to produce mutation and deletion in the *rbc*L gene, still there are very limited species for which chloroplast transformation has been successful. The other challenging issue has been the proper assembly of large and small subunits into the hexadecameric holoenzyme following the genetic manipulation, which requires sufficient expression, posttranslational modification, interaction with chaperonins, and interaction with RuBisCO activase (Gutteridge and Gatenby 1995).

One approach to increase RuBisCO efficiency is to produce hybrid enzymes, with the large subunit from one species and the small subunit from another species. One of the examples of hybrid RuBisCO is the replacement of the tobacco *rbc*L gene with the *rbc*L gene from sunflower by means of chloroplast transformation, which produced a catalytically active enzyme (Kanevski et al. 1999). Though the specificity factor of this hybrid enzyme was similar to that of wild-type tobacco, it had substantially reduced catalytic activity, achieving only 25 % of the carboxylation activity of either parent holoenzyme. One scenario is to introduce RuBisCO variants with naturally high specificity values, like the ones from the red alga *Galdieria partita* and purple photosynthetic bacterium *Rhodospirillum rubrum* into plants. This approach appears to hold promise in improving the photosynthetic efficiency of crop plants significantly; however, possible negative impacts have yet to be studied. Some progress has been made in this area including the replacement of the native *rbc*L of tobacco with *rbc*M from *Rhodospirillum rubrum* by chloroplast transformation (Andrews and Whitney 2003). A consistent finding from the above-mentioned studies and also from other genetic engineering experiments is that there is an inverse correlation between specificity and catalytic activity; with increased specificity for $CO_2$, the velocity of the carboxylase reaction decreases. This relationship has significantly affected the attempts to improve RuBisCO through genetic engineering, as each time the specificity is increased, the resulting modified enzyme has a reduced carboxylase activity. Another approach used is to increase RuBisCO content by overexpressing small subunit of RuBisCO gene, *rbc*S gene, but these have failed, often resulting in decreased RuBisCO content by cosuppression. Interestingly, increases in RuBisCO content have been observed in plants transformed with transgenes aimed at other targets (Pellny et al. 2002). Though considerable progress has been made in this area of research, but still a lot needs to be explored about the requirements for RuBisCO expression and its assembly in higher plants.

As an alternative, increased gains in carbon acquisition could be achieved by altering resource allocation to each of the enzymes involved in the Calvin cycle, photorespiratory metabolism, and sucrose and starch synthesis (Zhu et al. 2007). In this case, numerical simulations suggest that optimized allocation of resources to specific enzymes could greatly increase carbon gain without an increase in the total nitrogen invested in proteins involved in photosynthetic carbon metabolism. This process illustrates a potential win–win situation as even small gains in plant productivity distributed across a large land area could contribute to meaningful enhancements to carbon sequestration. Miguez et al. (2009) developed a process model to better understand the physiological controls on biomass production in *Miscanthus* × *giganteus* and concluded that harvestable yield (and presumably soil carbon storage) could be enhanced in this bioenergy crop through various mechanisms. Although none of these were explored in any detail, the approach illustrates how models and field studies could complement one another in support of breeding for traits of interest to carbon sequestration.

Since under optimum conditions of light and temperature, $C_4$ plants photosynthesize and grow at faster rates than $C_3$ plants, attempts have also been made to incorporate components of the $C_4$ pathway into $C_3$ species (Leegood 2002). Both conventional breeding and transgenic approaches have been used to achieve this goal (Haeusler et al. 2001). However, because of the independent inheritance of genes for morphological features, such as kranz anatomy, and of the $C_4$ pathway enzymes, breeding attempts have failed. Therefore, recent efforts are mainly focused on overexpressing the key $C_4$ cycle enzymes in $C_3$ plants, though there are pitfalls encountered when $C_3$ metabolism is perturbed by the overexpression of individual $C_4$ genes. Successful overexpression of $C_4$ PEP carboxylase and maize PEP carboxylase in $C_3$ plants like tobacco, rice, and potato has been reported (Izui et al. 1986; Ku et al. 1999; Hausler et al. 2002). One major issue with this approach is the differences in the regulation of $C_4$ cycle enzymes like PEP carboxylase in $C_3$ and $C_4$ plants. Similar to $C_4$ plants, PEP carboxylase in $C_3$ plants is subject to complex regulation by metabolites and covalent modification by reversible phosphorylation. In all of these experiments host plant regulatory mechanism, rather than the $C_4$ one, seemed to be operating in the transgenic plants. Therefore, though $C_4$ enzyme was overexpressed in $C_3$ plants, it was in an inactive state.

This issue has been tackled by introducing modified PEP carboxylase enzyme that lacks phosphorylation sites into $C_3$ plants. These studies reported high activity of PEP carboxylase in transgenic plants, but the plants were stunted with increasing PEP carboxylase activity (Haeusler et al. 2001). This is because $C_4$ enzymes are also present in $C_3$ plants and overexpression of individual $C_4$ enzymes interferes with $C_3$ metabolism. Other than PEP carboxylase, pyruvate orthophosphate dikinase (PPDK) and phosphoenolpyruvate carboxykinase (PEPCK) have been targeted for transgenic approaches (Ishimaru et al. 1998; Suzuki et al. 2000). However, overproduction of PEP carboxylase, PEPCK, and PPDK, by whatever means, does not improve photosynthesis in $C_3$ plants. A moderate overexpression of PEP carboxylase combined with PPDK or PEPCK also had no significant effect on photosynthetic efficiency of the plants (Haeusler et al. 2001). A more current approach is to introduce groups of $C_4$ genes in $C_3$ plants but it remains to be seen whether this strategy will be any more successful. One of the major challenges of this approach is distribution of these enzymes so that they provide a $CO_2$-rich environment around RuBisCO. While most of the $C_4$ plants express PEP carboxylase and RuBisCO in different cells, the discovery of $C_4$ photosynthesis in single cells of *Borszczowia aralocaspica* and *Bienertia cycloptera* leaves indicates that kranz anatomy is not essential for $C_4$ photosynthesis (Sage 2002). This discovery might lead to simpler approaches toward engineering $C_4$ photosynthesis.

### 12.5.1.2  Below-Ground Process

In addition to targeting traits specific to $CO_2$ acquisition and uptake, carbon could be sequestered in soils if genome-enabled modification of leaf or root turnover times could be achieved. Driebe and Whitham (2000) collected senesced leaves from trees along a hybridization zone near the Weber River in Utah and showed that genotypic variation in foliar condensed tannin concentrations could largely explain variation in rates of litter decomposition for $F_1$ and backcross hybrids of cottonwood (*Populus* spp.). Although these authors did not make an explicit connection between rates of litter decomposition and soil carbon sequestration, the implications of their work clearly suggest that genotypic variation in traits related to litter quality and decomposition may be sufficient to impact carbon and nitrogen cycles in terrestrial ecosystems. Wullschleger et al. (2005) demonstrated in a large field study that significant phenotypic variation exists in poplar for a number of traits related to carbon storage including biomass production and the allocation of carbon to below-ground roots. Quantitative trait loci (QTL) analysis revealed that above-ground and below-ground patterns of biomass distribution were under strong genetic control and that such a genetic underpinning could be leveraged to potentially enhance carbon sequestration in managed bioenergy plantations (Bradshaw and Stettler 1995; Rae et al. 2009).

Although promising, few studies have demonstrated that long-term plantings of poplar will enhance soil carbon stocks; at best the results for stands ranging in age from 3 to 12 years are mixed (Hansen 1993; Grigal and Berguson 1998; Coleman et al. 2004; Sanchez et al. 2007; Satori et al. 2007; Gupta et al. 2009). Lignin biosynthesis has also been shown to be under strong genetic control and poplar species have been modified to possess reduced lignin content. Hancock et al. (2008) showed in short-term studies that aspen (*P. tremuloides*) expressing high syringyl/guaiacyl (S/G) lignin accumulated less total plant carbon and subsequently accumulated less plant-derived carbon in soil. Furthermore, Garten et al. (2011), using a modeling study, demonstrated that breeding for specific traits could enhance carbon sequestration in managed plantations by focusing on improvements to above-ground production, below-ground carbon allocation, and root decomposition.

## 12.6 Modeling Approaches to Select Likely Targets for Eco-engineering Plant Systems for Increased Carbon Sequestration

Controlling the flow of carbon through system to favor synthesis of materials recalcitrant to easy degradation carbon (e.g., lignin) over biomass that will more quickly return to the carbon cycle is the goal of eco-engineering for increased carbon sequestration. One significant carbon sink in the plant ecosystems is the community of root-associated microorganisms. Terrestrial plants process about 15 % of the total atmospheric carbon dioxide each year, drawing about 450 billion tons of carbon dioxide from the atmosphere (Beer et al. 2010). Depending on conditions and on ecosystems, between 20 % (Gamper et al. 2005) and as much as 40 % (Drigo et al. 2010) of that fixed atmospheric carbon is incorporated directly into the subsurface ecosystems, making subsurface microorganisms a sink for potentially billions of tons of atmospheric carbon annually. Modifying systems for increased carbon sequestration involves engineering at the ecosystem level, incorporating plants and their subsurface community, with various bacterial and fungal components of soil community structure. The metabolic mechanisms for carbon sequestration in these systems are already in place. The most profitable targets to engineer carbon sequestration are regulatory, not metabolic. The goal then of eco-engineering is to determine how does the community interact with its environment and other members of its community and how does that interaction direct carbon sequestration. This analysis will yield specific molecular targets such as transmembrane sensor proteins, regulatory elements, and intracellular signaling compounds. While the leaves are the most obvious surface that the plant exposes to the atmosphere and the woody tissues are the evident locations for carbon storage, to find the plant tissues most amenable to eco-engineering, we turn to the plant root. The "root-brain hypothesis" was first proposed by Charles Darwin well over a century ago, but recent advances in investigation techniques have begun to identify the specific molecular mechanisms of this interaction (Baluška et al. 2009). In the root-brain hypothesis, plants are recognized as organisms capable of active behavior in response to changes in their environment, and the mechanisms for regulation of plant behavior reside in the root apex.

The following is a selection of computational tools for identifying likely targets for eco-engineering.

### 12.6.1  Ecological Systems as Artificial Neural Networks

Artificial neural networks (ANNs) are nonlinear computational modeling tools comprised of a network of interrelated nodes and have direct applications to modeling certain biological interactions (Thivierge et al. 2012). Two recent approaches that use ANNs to describe ecological relationships and relevant to identifying targets for eco-engineering systems are microbial assemblage prediction (MAP) (Larsen et al. 2012) and expression interaction networks (EINs) (Larsen and Dai 2009).

MAP generates ANNs that represent microbial community structure in terms of mathematical equations that best explain the data and uses them to predict the relative abundance of taxa in time or space as functions of environmental conditions. These ANNs capture potentially causal relationships between the changing abundances of different taxa, although relationships between taxa could arise through taxon proxies for changes in environmental parameters. This approach requires data for population structures, such as 16s/18s sequence data or shotgun metagenomics, and environmental parameter data corresponding to the time and place from which the populations were samples. This approach can be used, for example, to model the subsurface microbial populations as they change in response to surface conditions or changes in plant phenotype/genotype.

Similar to MAP models, expression interaction network (EIN) construct networks of gene regulation interaction and models these reactions as an ANN. EIN can use networks of gene expression interactions to link environmental conditions to measured environmental phenotypes such as biomass or nutrient concentrations. The model constructed by EIN can be used to predict the measureable phenotypic response to previously unobserved or novel environmental conditions. Genes whose expression is identified to be associated with a measured phenotype in an EIN are excellent gene/protein candidates for targeted eco-engineering plant systems.

## 12.6.2    Sensory Components as Protein–Protein Interaction Complexes

To identify possible protein–protein interaction (PPI) networks that are likely to be associated with environmental sensing or community signaling, tools such as likelihood of interaction (LOI) (Larsen et al. 2007) or function restricted value neighborhood (FRV-N) (Larsen et al. 2010) are useful computational tools.

LOI uses collected transcriptomic data to identify likely PPI complexes. Transcriptomic data collected from multiple experimental conditions is preferred for this approach, and archived transcriptomic data for many systems is available in public repositories like Gene Expression Omnibus (http://www.ncbi.nlm.nih.gov/geo/). LOI also requires a large set of previously identified PPI network and a set of relevant annotations for all proteins in the set of interactions. A number of such publicly available resources for experimentally observed PPIs are available for fungal, plant, or bacterial systems, such as the Saccharomyces Genome Database (http://www.yeastgenome.org/), the Arabidopsis Information Resource (http://www.arabidopsis.org/), and the Bacterial Protein Interaction Database (http://www.compsysbio.org/bacteriome/). The last input required by LOI is a discrete ontology for annotation of function assigned to predicted gene. The selected annotation needs to be uniform between genes in transcriptomic data and in databases of previously observed. Some possible sources of annotation include Gene Ontology (GO) annotation or descriptions of predicted subcellular location as in WoLF PSoRT (Horton et al. 2007). LOI used these inputs to identify pairs of function annotations enriched in previously observed PPI pairs and uses this information to propose likely PPI interactions from co-expression data in transcriptomic analysis. A significant advantage to LOI is its application of patterns identifies in model organism experimental data to organisms for which littler previous experimental data is available.

In addition to pairwise identification of pairwise PPIs as in LOI, the topology of entire PPI networks can be considered (Chen et al. 2006). A rank-based method function restricted value neighborhood (FRV-N) uses the network topology of previously experimentally observed PPI to impose biologically relevant network structures to the sets of PPIs predicted by methods like LOI.

As in LOI, FRV-N uses available large databases of known PPIs and a set of gene annotations. By identifying a characteristic neighborhood size associated with particular protein annotations, FRV-N applies a biologically relevant network topology to a set of predicted PPIs that provides increased accuracy to predicted PPI networks.

### 12.6.3 Metabolic Modeling of Systems

Predicted relative metabolic turnover (PRMT) is a method that infers metabolic data from metagenomic, transcriptomic, or proteomic data (Larsen et al. 2011b). Also, a database of known enzyme-mediated biochemical reactions, such as Kyoto Encyclopedia of Genes and Genomes (http://www.genome.jp/kegg/) (Ogata et al. 1999) or BioCyc (http://biocyc.org/), is required. To map detected gene or protein expression, an ontology describing the unique enzyme function of a gene product or detected protein, such as enzyme commission (EC) number (International Union of Biochemistry and Molecular Biology. Nomenclature Committee. and Webb 1992), is also needed. PRMT enables exploration of metabolite space inferred from the metagenome by associating changes in abundance of genes for enzyme activities with predicted changes in relative metabolic turnover for metabolites in a predicted metabolome. The complete output of PRMT analysis provides a numerical score for all thousands of predicted metabolites in a metabolome. An advantage of PRMT is its ability to make predictions for thousands of metabolites in complex systems from more easily obtainable sources of information like high-throughput sequencing data. The results of PRMT analysis can be used to direct more detailed biochemical analysis of systems for sets of metabolites predicted to be relevant to system interactions.

In the case of root community metabolome, significant carbon sequestration metabolism may not be discovered in root metabolism, but identification of the synthesis plant signaling compounds that regulate growth in more distant parts of the plant could be (e.g., auxin, gibberellin, or brassinosteroids). For example, in plants the hormone jasmonate regulates plant defense response, prioritizing defense over growth by blocking the gibberellin signaling (Yang et al. 2012). Sensor or signaling proteins in the plant root community might prove valuable targets for manipulating this interaction in plants, either by suppressing a defense response or by reducing jasmonate's ability to interfere with gibberellin-signaling cascade and promoting plant growth and carbon sequestration.

### 12.6.4 Assembling the Interactome

Only by understanding the interrelationships between plant and subsurface community, however, can one begin the process of rational eco-engineering of complex systems. Identifying the complete network of interactions between and within carbon-sequestering communities requires a "multi-omic" approach. While each of the above analytical tools can provide valuable information about interacting systems of themselves, no single investigative technique will uncover all the components of ecosystem interactions. The synthesis of the results of some or all of these tools into a single predictive model of regulatory and metabolic factors in carbon sequestration can be used to identify likely targets for manipulation and

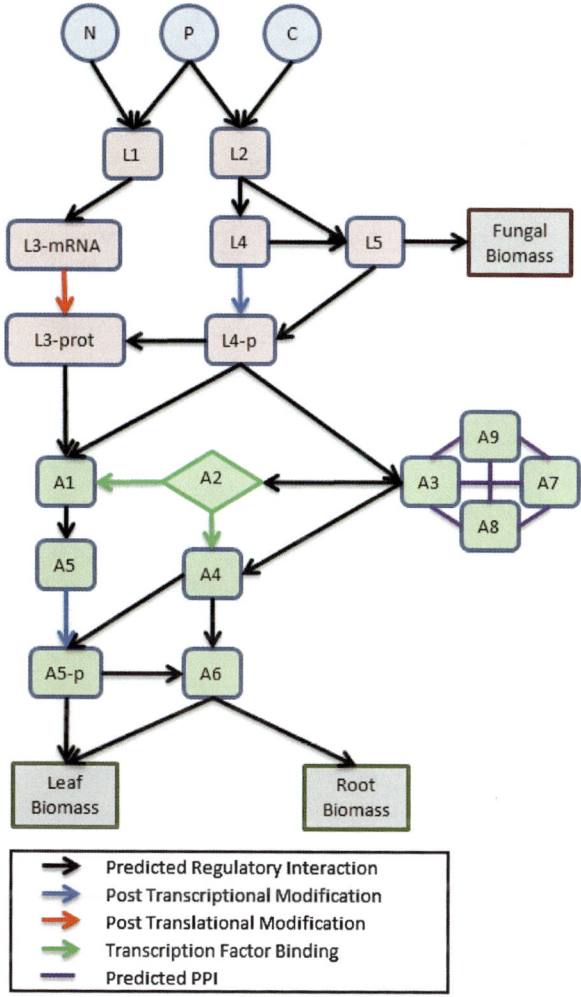

**Fig. 12.4** Example of a system-scale mycorrhizal regulatory network. This simple, hypothetical interaction network provides an example of all possible interaction mechanisms that go into the regulation of mycorrhizal interactions as they relate to soil nitrogen and phosphorus levels using *Laccaria bicolor* and *Populus tremuloides* as partners. In this network, the topmost nodes represent nitrogen (N)-, phosphorus (P)-, and carbon (C)-containing components. *Pink* nodes L1–L4 represent *Laccaria* genes or proteins, and *green* nodes A1–A6 represent aspen genes and proteins. *Gray* nodes are relative measures of fungal, leaf, and root biomass. *Black-directed edges* are DBN-predicted regulatory interactions. *Blue-directed edges* indicate a posttranslational modification, identified by the difference between transcript and protein relative abundance. *Red-directed edge* is a posttranslational regulation indicated by proteomic analysis. *Green edges* are direct transcription factor binding interactions identified by ChIP-seq. *Purple edges* are predicted PPI interactions

generate testable hypotheses for modification of carbon storage. This set of interactions is not best interpreted as a simple linear path of sequential regulatory steps but as an interconnected network of biological processes that occur in parallel (Fig. 12.4).

When choosing targets for increased carbon sequestration, root-associated microbes are an appealing target for eco-engineering. Microbes are far more tractable than genetic engineering plants and trees. Also, while the prospect of replanting the landscape with modified or engineered plants is the daunting work of many years for plants to reach maturity and fix useful amounts of carbon, engineering the subsurface microbial population by the addition of selected or modified microorganisms offers an opportunity to eco-engineer existing plant ecosystems with a more rapid response.

## 12.7 Conclusion

Greenhouse gases are accumulating in the Earth's atmosphere at least in part due to human activities. Over time, this is causing surface air temperatures and subsurface ocean temperatures to rise, which in turn affects growing seasons, precipitation, and the frequency and severity of extreme weather events and forest fires. Humans have the unique ability to be aware of such potential threats, and more importantly we have the capacity to alter our environment and ward off pending hardship. Practices that aim to increase carbon sequestration generally enhance the quality of soil, water, air, and wildlife habitat. Tree planting that restores forest cover not only may sequester carbon but also improves habitat suitability for wildlife. Likewise, preserving threatened tropical forests would avoid losses in both stored carbon and biodiversity. Similarly, reducing soil erosion through tree planting or soil conservation measures not only sequesters carbon but also improves water quality by reducing nutrient runoff.

Combined with improved land and crop management practices, researchers have been developing a deep understanding of the molecular and cellular processes that orchestrate the conversion of carbon captured from the atmosphere into plant biomass that can sequester the carbon for many years. The plant traits behind these biological processes are complex and intimately connected with the surrounding ecology, which makes the targeting and manipulating of such traits very challenging. However, progress is being made.

Several paradigms that could potentially lead to breakthroughs might be pursued. The first builds from a focus on specific pathways for altering processes of interest to carbon sequestration including those linked to the photosynthetic uptake of $CO_2$ from the atmosphere, litter chemistry, and modification of leaf and root turnover. Some of the more promising research targets include improvements in light-use efficiency and photosynthesis (Long et al. 2006), improvements in overcoming the constraints imposed on plant productivity by temperature and drought (Tuberosa and Salvi 2006), and improvements in root growth and nutrient

**Table 12.1** Summary of approaches used to enhance carbon sequestration

| Approaches | Implementation | Goal achieved | References |
|---|---|---|---|
| *Transgenic manipulation of candidate genes/mutagenesis* | | | |
| (a) Rubisco | Done | Partial success | Kanevski et al. (1999), Spreitzer and Salvucci (2002), Andrews and Whitney (2003), Suzuki et al. (2007) |
| (b) Phosphoenolpyruvate carboxylase (PEP carboxylase) | Done | No | Izui et al. (1986), Haeusler et al. (2001) |
| (c) Pyruvate orthophosphate dikinase (PPDK) | Done | No | Ishimaru et al. (1998), Suzuki et al. (2000) |
| (d) Phosphoenolpyruvate carboxykinase (PEPCK) | Done | No | Suzuki et al. (2006) |
| *Marker based* | | | |
| Quantitative trait loci (QTL) | Done | Successful | Bradshaw and Stettler (1995), Wullschleger et al. (2005), Rae et al. (2009) |
| *Breeding* | | | |
| | Done | Successful | Deuter (2000), Driebe and Whitham (2000) |
| *Modelling* | | | |
| | Done | Partial success | Purves and Pacala (2008), Ostle et al. (2009), Ise et al. (2010), Garten et al. (2011) |
| *Computational* | | | |
| (a) Artificial neural networks (ANNs) | Potential | Promising | Larsen and Dai (2009), Thivierge et al. (2012) |
| (b) Protein–protein interaction (PPI) networks | Potential | Promising | Chen et al. (2006), Larsen et al. (2007, 2010) |
| (c) Metabolic modeling of systems | Potential | Promising | Larsen et al. (2011b) |
| (d) Assembling the interactome | Potential | Promising | |

uptake (Hirel et al. 2007). The efficiency of the plants to access and uptake soil nutrients in particular is key to their ability to capture carbon and generate biomass. An improved understanding of such processes across a range of environments will improve the chances that manipulation of specific genes and proteins will translate to enhanced input of carbon to plant biomass and subsequently to soils via shoot and root litter.

The genetic and genomic resources are in place for the above studies, but the modification of traits for carbon sequestration is still in infancy (Table 12.1). In a second paradigm, focus is placed strictly on biomass production, knowing that increased biomass, through enhanced production of leaf and root litter, would lead to the storage of carbon away from the atmosphere as well as increases in carbon input to soils. Both of the above approaches would yield insights. However, it should be remembered that the process of carbon sequestration plays out over decades and verification for enhanced sequestration of carbon in soils can be exceedingly difficult.

In addition, other approaches could be used to focus on plant–microbe interactions and the role these play in shaping nutrient uptake, plant biomass, and soil carbon sequestration processes. When choosing targets for increased carbon sequestration, root-associated microbes, including bacteria and mycorrhizal fungi, are appealing targets for eco-engineering. Microbes are typically much more tractable than genetically engineering plants. Many tree species tend to have very long juvenile periods that inhibit breeding programs which are dependent on sexual maturity. Also, replanting the landscape with genetically modified plants is likely to be a daunting work, requiring many years for the plants to reach maturity and fix useful amounts of carbon. Thus, engineering the subsurface microbial population by addition of selected or modified microorganisms offers a promising alternative to alter existing plant ecosystems with a more rapid impact on stored carbon.

Investments in plant genomics, transcriptomics, proteomics, and metabolomics promise to identify and harness new approaches to increase biomass production. Modeling, based on the mining of such large datasets, is a largely untapped opportunity to translate our understanding of basic plant biochemistry and physiology into actions that might lead to enhanced carbon storage in plant biomass (i.e., wood) or soils. Garten et al. (2011) used this to some success in their model-based assessment of traits involved in carbon sequestration, but this need not be the only approach. Using new innovations in modeling, it is likely that we will identify novel traits that can be manipulated to improve carbon sequestration. Application of such modern system biology approaches and other advanced methodologies to improve fundamental understanding of soil microbial communities and their habitats is another approach that could further the potential for enhanced terrestrial sequestration. Such information could lead to new management practices and production of organic matter materials that optimize microbial activities for the transformation of residues specifically to enhance sequestration.

**Acknowledgments** This research was supported by the Office of Science (BER), U.S. Department of Energy.

# References

Aerts R, Chapin FS (2000) The mineral nutrition of wild plants revisited: a re-evaluation of processes and patterns. Adv Ecol Res 30:1–65

Alberton O, Kuyper TW, Gorissen A (2005) Taking mycocentrism seriously: mycorrhizal fungal and plant responses to elevated CO2. New Phytol 167:859–868

Altieri MA (1994) Sustainable agriculture. Encyclo Agric Sci 4:239–247

Anderson-Teixeira KJ, Davis SC, Masters MD, Delucia EH (2009) Changes in soil organic carbon under biofuel crops. Glob Change Biol Bioenergy 1:75–96

Andersson I (1996) Large structures at high resolution: the 1.6 Å crystal structure of spinach ribulose-1,5-bisphosphate carboxylase/oxygenase complexed with 2-carboxyarabinitol bisphosphate. J Mol Biol 259:160–174

Andrews TJ, Whitney SM (2003) Manipulating ribulose bisphosphate carboxylase/oxygenase in the chloroplasts of higher plants. Arch Biochem Biophys 414(2):159–169

Arnone JA, Zaller JG, Spehn JG, Niklaus PA, Wells CE, Korner C (2000) Dynamics of root systems in native grasslands: effects of elevated atmospheric CO2. New Phytol 147:73–85

Baluška F, Mancuso S, Volkmann D, Barlow P (2009) The 'root-brain' hypothesis of Charles and Francis Darwin: revival after more than 125 years. Plant Signal Behav 4(12):1121–1127

Bardgett RD, Wardle DA (2010) Aboveground-belowground linkages: biotic interactions, ecosystem processes, and global change. Oxford University Press, Oxford

Barea JM, Azcón R, Azcón-Aguilar C (2002) Mycorrhizosphere interactions to improve plant fitness and soil quality. Antonie Van Leeuwenhoek 81:343–351

Barnola JM, Anklin M, Porheron J, Raynaud D, Schwander J, Stauffer BTI (1995) CO2 evolution during the last millennium as recorded by Antarctic and Greenland ice. Tellus B 47:264–272

Bates T, Lynch J (2000) Plant growth and phosphorus accumulation of wild type and two root hair mutants of *Arabidopsis thaliana* (Brassicaceae). Am J Bot 87:958–963

Beer C, Reichstein M, Tomelleri E, Ciais P, Jung M et al (2010) Terrestrial gross carbon dioxide uptake: global distribution and covariation with climate. Science 329(5993):834–838

Bending GD, Aspray TJ, Whipps JM (2006) Significance of microbial interactions in the mycorrhizosphere. Adv Appl Microbiol 60:97–132

Bennett RN, Wallsgrove RM (1994) Secondary metabolites in plant defense mechanisms. New Phytol 127:617–633

Berg G, Smalla K (2009) Plant species and soil type cooperatively shape the structure and function of microbial communities in the rhizosphere. FEMS Microbiol Ecol 68(1):1–13

Bieniawska Z, Barratt DHP, Garlick AP, Thole V, Kruger NJ, Martin C, Zrenner R, Smith AM (2007) Analysis of the sucrose synthase gene family in Arabidopsis. Plant J 49:810–828

Binkley D, Menyailo O (eds) (2005) Tree species effects on soils: implications for global change. NATO Science Series. Kluwer Academic, Dordrecht

Birdsey RA (1996) Regional estimates of timber volume and forest carbon for fully stocked timberland, average management after final clearcut harvest. In: Sampson RN, Hair D (eds) Forests and global change, vol 2, Forest management opportunities for mitigating carbon emissions. American Forests, Washington, DC, pp 309–334

Block RMA, Van Rees KCJ, Knight JD (2006) A review of fine root dynamics in *Populus* plantations. Agrofor Syst 67:73–84

Bonfante P, Genre A (2010) Mechanisms underlying beneficial plant-fungus interactions in mycorrhizal symbiosis. Nat Commun 1:48

Bougher NL, Grove TS, Malajczuk N (1990) Growth and phosphorus acquisition of karri (*Eucalyptus diversicolor* F. Muell.) seedlings inoculated with ectomycorrhizal fungi in relation to phosphorus supply. New Phytol 114:77–85

Bowen GD, Rovira AD (1999) The rhizosphere and its management to improve plant growth. Adv Agron 66:1–102

Bradford MA, Fierer N, Jackson RB, Maddox TR, Reynolds JF (2008) Nonlinear root-derived carbon sequestration across a gradient of nitrogen and phosphorous deposition in experimental mesocosms. Glob Change Biol Bioenergy 14:1113–1124

Bradshaw HD, Stettler RF (1995) Molecular genetics of growth and development in Populus. IV. Mapping QTLs with large effects on growth, form, and phenology traits in a forest tree. Genetics 139:963–973

Brown ALP, Day FP, Hungate BA, Drake BG, Hinkle CR (2007) Root biomass and nutrient dynamics in a scrub-oak ecosystem under the influence of elevated atmospheric CO2. Plant Soil 292:219–232

Cairney JWG, Ashford AE, Allaway WG (1989) Distribution of photosynthetically fixed carbon within root systems of *Eucalyptus-Pilularis* plants ectomycorrhizal with *Pisolithus-Tinctorius*. New Phytol 112:495–500

Capell T, Christou P (2004) Progress in plant metabolic engineering. Curr Opin Biotechnol 15:148–154

Carney KM, Hungate BA, Drake BG, Megonigal JP (2007) Altered soil microbial community at elevated CO2 leads to loss of soil carbon. Proc Natl Acad Sci USA 104:4990–4995

Chan PH, Wildman SG (1972) Chloroplast DNA codes for the primary structure of the large subunit of fraction I protein. Biochim Biophys Acta 14:677–680

Chen G, Larsen P, Almasri E, Dai Y (2006) Sample scale-free gene regulatory network using gene ontology. Proceedings of the annual international conference of the IEEE Engineering in Medicine and Biology Society. IEEE Eng Med Biol Soc Conf 1:5523–5526

Cheng ZM, Tuskan GA (2009) *Populus* community mega-genomics: coming of age. Crit Rev Plant Sci 28:282–284

Cleveland CC, Townsend AR (2006) Nutrient additions to a tropical rain forest drive substantial soil carbon dioxide losses to the atmosphere. Proc Natl Acad Sci USA 103:10316–10321

Coleman MD, Isebrands JG, Tolsted DN, Tolbert VR (2004) Comparing soil carbon of short rotation poplar plantations with agricultural crops and woodlands in North Central United States. Environ Manag 33:299–308

Coleman HD, Canam T, Kang KY, Ellis DD, Mansfield SD (2007) Overexpression of UDP-glucose pyrophosphorylase in hybrid poplar affects carbon allocation. J Exp Bot 58:4257–4268

Cseke L, Cseke S, Podila G (2007) High efficiency poplar transformation. Plant Cell Rep 26:1529–1538

Deuter M (2000) Breeding approaches to improvement of yield and quality in *Miscanthus* grown in Europe. In: Lewandowski I, Clifton-Brown JC (eds) European Miscanthus improvement—Final report, Sept 2000. Institute of Crop Production and Grassland Research, University of Hohenheim, Stuttgart, pp 28–52

Dixon RK, Brown S, Houghton RA, Solomon AM, Trexler MC, Wisniewski J (1994) Carbon pools and flux of global forest ecosystems. Science 263:185–190

Driebe EM, Whitham TG (2000) Cottonwood hybridization affects tannin and nitrogen content of leaf litter and alters decomposition. Oecologia 123:99–107

Drigo B, Kowalchuk GA, van Veen JA (2008) Climate change goes underground: effects of elevated atmospheric $CO_2$ on microbial community structure and activities in the rhizosphere. Biol Fertil Soil 44:667–679

Drigo B, Pijl AS, Duytsc H, Kielaka AM, Gamper HA et al (2010) Shifting carbon flow from roots into associated microbial communities in response to elevated atmospheric $CO_2$. Proc Natl Acad Sci USA 107(24):10938–10942

Ehleringer JR, Cerling TE, Helliker BR (1997) C4 photosynthesis, atmospheric $CO_2$ and climate. Oecologia 112:285–299

Ellis RJ (1979) Most abundant protein in the world. Trends Biochem Sci 4:241–244

Farrar JF, Hawes M, Jones D, Lindow S (2003) How roots control the flux of carbon to the rhizosphere. Ecology 84:827–837

Ferrol N, Barea JM, Azcon-Aguilar C (2002) Mechanisms of nutrient transport across interfaces in arbuscular mycorrhizas. Plant Soil 244:231–237

Field C, Mooney HA (1986) The photosynthesis–nitrogen relationship in wild plants. In: Givnish TJ (ed) On the economy of plant form and function. Cambridge University Press, Cambridge, pp 22–55

Fischer DG, Hart SC, Rehill BJ, Lindroth RL, Keim P, Whitham TG (2006) Do high-tannin leaves require more roots? Oecologia 149:668–675

Frank DA, Groffman PM (2009) Plant rhizospheric N processes: what we don't know and why we should care. Ecology 90:1512–1519

Franklin O, Johansson J, Dewar RC, Dieckmann U, McMurtrie RE, Brännström A, Dybzinski R (2012) Modeling carbon allocation in trees: a search for principles. Tree Physiol 32 (6):648–666

Frey BR, Lieffers VJ, Landhäusser SM, Comeau PG, Greenway KJ (2003) An analysis of sucker regeneration of trembling aspen. Can J For Res 33(7):1169–1179

Gahoonia TS, Nielsen NE, Lyshede OB (1999) Phosphorus (P) acquisition of cereal cultivars in the field at three levels of P fertilization. Plant Soil 211:269–281

Gamalei Y (1991) Phloem loading and its development related to plant evolution from trees to herbs. Trees 5:50–64

Gamalei YV, Pakhomova MV, Sjutkina AV (1992) Ecological aspects of assimilate transport. I. Temperature. Fiziol Rast 39:1068–1078

Gamper H, Hartwig UA, Leuchtmann A (2005) Mycorrhizas improve nitrogen nutrition of Trifolium repens after 8 yr of selection under elevated atmospheric $CO_2$ partial pressure. New Phytol 167(2):531–542

Garten CT Jr, Wullschleger SD, Classen AT (2011) Review and model-based analysis of factors influencing soil carbon sequestration under hybrid poplar. Biomass Bioenergy 35:214–226

Giardina CP, Ryan MG (2002) Total belowground carbon allocation in a fast growing Eucalyptus plantation estimated using a carbon balance approach. Ecosystems 5:487–499

Giardina CP, Binkley D, Ryan MG, Fownes JH, Senock RS (2004) Belowground carbon cycling in a humid tropical forest decreases with fertilization. Oecologia 139:545–550

Giardina CP, Coleman MD, Binkley D, Hancock JE, King JS, Lilleskov EA, Loya WM, Pregitzer KS, Ryan MG, Trettin CC (2005) The response of belowground carbon allocation in forests to global change. In: Binkley D, Menyailo O (eds) Tree species effects on soils: implications for global change. Springer, Dordrecht, pp 119–154

Głowacka K, Jeżowski S (2009a) Genetic and nongenetic factors influencing callus induction in *Miscanthus sinensis* (Anderss.) anther cultures. J Appl Genet 50(4):341–345

Głowacka K, Jeżowski S (2009b) In vitro culture of *Miscanthus sinensis* anthers. Acta Biol Cracov Sr Bot 51(1):17

Głowacka K, Jeżowski S, Kaczmarek Z (2010) The effects of genotype, inflorescence developmental stage and induction medium on callus induction and plant regeneration in two *Miscanthus* species. Plant Cell Tiss Org Cult 102(1):79–86

Glynn C, Herms DA, Orians CM, Hansen RC, Larsson S (2007) Testing the growth-differentiation balance hypothesis: dynamic responses of willows to nutrient availability. New Phytol 176:623–634

Gower ST, Vogt KA, Grier CC (1992) Carbon dynamics of Rocky Mountain Douglas-fir: influence of water and nutrient availability. Ecol Monogr 62:43–65

Grace J, Rayment M (2000) Respiration in the balance. Nature 404:819–820

Grigal DF, Berguson WE (1998) Soil carbon changes associated with short-rotation systems. Biomass Bioenergy 14:371–377

Gupta N, Kukal SS, Bawa SS, Dhaliwal GS (2009) Soil organic carbon and aggregation under poplar based agroforestry system in relation to tree age and soil type. Agrofor Syst 76:27–35

Gutteridge S, Gatenby AA (1995) Rubisco synthesis, assembly, mechanism and regulation. Plant Cell 7:809–819

Haeusler RE, Rademacher T, Li J, Lipka V, Fischer KL, Schubert S, Kreuzaler F, Hirsch HJ (2001) Single and double overexpression of C4-cycle genes had differential effects on the pattern of endogenous enzymes, attenuation of photorespiration and on contents of UV protectants in transgenic potato and tobacco plants. J Exp Bot 52:1785–1803

Hancock JE, Bradley KL, Giardina CP, Pregitzer KS (2008) The influence of soil type and altered lignin biosynthesis on the growth and above and belowground biomass allocation in *Populus tremuloides*. Plant Soil 308:239–253

Hansen EA (1993) Soil carbon sequestration beneath hybrid poplar plantations in the north central United States. Biomass Bioenergy 5:431–436

Hanson TE, Tabita FR (2001) A ribulose-1,5-bisphosphate carboxylase/oxygenase (RuBisCO)-like protein from *Chlorobium tepidum* that is involved with sulfur metabolism and the response to oxidative stress. Proc Natl Acad Sci USA 98:4397–4402

Haynes RW (1995) The 1993 RPA timber assessment update. USDA Forest Service general technical Report RM-259. Rocky Mountain Forest Experiment Station, Fort Collins, CO

Haynes BE, Gower ST (1995) Belowground carbon allocation in unfertilized and fertilized red pine plantations in northern Wisconsin. Tree Physiol 15:317–325

Heaton EA, Clifton-Brown J, Voigt TB, Jones MB, Long SP (2004) *Miscanthus* for renewable energy generation: European Union experience and projections for Illinois. Mitig Adapt Strat Glob Change 9:433–451

Herms DA, Mattson WJ (1997) Trees, stress, and pests. In: Lloyd JE (ed) Plant health care for woody ornamentals. International Society for Arboriculture, Champaign, IL, pp 3–25

Hinsinger P, Bengough AG, Vetterlein D, Young IM (2009) Rhizosphere: biophysics, biogeochemistry and ecological relevance. Plant Soil 321:117–152

Hirel B, Le Gouis J, Ney B, Gallais A (2007) The challenge of improving nitrogen use efficiency in crop plants: towards a more central role for genetic variability and quantitative genetics within integrated approaches. J Exp Bot 58:2369–2387

Horton P, Park K-J, Obayashi T, Fujita N, Harada H, Adams-Collier CJ, Nakai K (2007) WoLF PSORT: protein localization predictor. Nucleic Acids Res 35:W585–W587

Ibrahim ML, Proe F, Cameron AD (1997) Main effects of nitrogen supply, and drought stress upon whole plant carbon allocation in poplar. Can J For Res 27:1413–1419

Ise T, Litton CM, Giardina CP, Ito A (2010) Comparison of modeling approaches for carbon partitioning: impact on estimates of global net primary production and equilibrium biomass of woody vegetation from MODIS GPP. J Geophys Res 115:G04025

Ishimaru K, Ohkawa Y, Ishige T, Tobias DJ, Ohsugi R (1998) Elevated pyruvate, orthophosphate dikinase (PPDK) activity alters carbon metabolism in C3 transgenic potatoes with a C4 maize PPDK gene. Physiol Plant 103:340–346

Izui K, Ishijima S, Yamaguchi Y, Katagiri F, Murata T, Shigesada K, Sugiyama T, Katsuki H (1986) Cloning and sequence analysis of a cDNA encoding active phosphoenolpyruvate carboxylase of the C4-pathway from maize. Nucleic Acids Res 14:1615–1628

Izumi H, Anderson IC, Alexander IJ, Killham K, Moore ER (2006) Endobacteria in some ectomycorrhiza of Scots pine (Pinus sylvestris). FEMS Microbiol Ecol 56:34–43

Jackson RB, Mooney HA, Schulze ED (1997) A global budget for fine root biomass, surface area, and nutrient contents. Proc Natl Acad Sci USA 94:7362–7366

Jansson C, Wullschleger SD, Kalluri UC, Tuskan GA (2010) Phytosequestration: carbon biosequestration by plants and the prospects of genetic engineering. Bioscience 60:685–696

Jeffries P, Barea JM (2001) Arbuscular mycorrhiza – a key component of sustainable plant–soil ecosystems. In: Hock B (ed) The Mycota, vol 9, Fungal associations. Springer, Berlin, pp 95–113

Jentschke G, Godbold DL (2000) Metal toxicity and ectomycorrhizas. Physiol Plant 109:107–116

Jiang JX, Wang ZH, Tang BR, Xiao L, Ai X, Yi ZL (2012) Development of novel chloroplast microsatellite markers for *Miscanthus* species (Poaceae). Am J Bot 99(6):230–233

Johansson JF, Paul LR, Finlay RD (2004) Microbial interactions in the mycorrhizosphere and their significance for sustainable agriculture. FEMS Microbiol Ecol 48:1–13

Johnson NC, Gehring CA (2007) Mycorrhizas: symbiotic mediators of rhizosphere and ecosystem processes. In: Cardon ZG, Whitbeck JL (eds) The rhizosphere. Elsevier Academic, Oxford, pp 73–100

Johnson MG, Rygiewicz PT, Tingey DT, Phillips DL (2006) Elevated CO2 and elevated temperature have no effect on Douglas-fir fine-root dynamics in nitrogen-poor soil. New Phytol 170:345–356

Johnston CA, Groffman P, Breshears DD, Cardon ZG, Currie W, Emanuel W, Gaudinski J, Jackson RB, Lajtha K, Nadelhoffer K, Nelson D Jr, Post WM, Retallack G, Wielopolski L (2004) Carbon cycling in soil. Front Ecol Environ 2:522–528

Jones MB, Walsh M (eds) (2001) Miscanthus for energy and fibre. James and James, London

Kanevski I, Maliga P, Rhoades DF, Gutteridge S (1999) Plastome engineering of ribulose-1,5 bisphosphate carboxylase/oxygenase in tobacco to form a sunflower large subunit and tobacco small subunit hybrid. Plant Physiol 119:133–141

Kawashima N, Wildman SG (1972) Studies on fraction I protein. IV. Mode of inheritance of primary structure in relation to whether chloroplast or nuclear DNA contains the code for a chloroplast protein. Biochim Biophys Acta 262:42–49

Kennedy AC, Smith KL (1995) Soil microbial diversity and the sustainability of agricultural soils. Plant Soil 170:75–86

Khanna M, Dhungana B, Clifton-Brown J (2008) Costs of producing miscanthus and switchgrass for bioenergy in Illinois. Biomass Bioenergy 32:482–493

Kim C, Zhang D, Auckland SA, Rainville LK, Jakob K, Kronmiller B, Sacks EJ, Deuter M, Paterson AH (2012) SSR-based genetic maps of *Miscanthus sinensis* and *M. sacchariflorus*, and their comparison to sorghum. Theor Appl Genet 124(7):1325–1338

King JS, Kubiske ME, Pregitzer KS, Hendrey GR, McDonald EP, Giardina CP et al (2005) Tropospheric O-3 compromises net primary production in young stands of trembling aspen, paper birch and sugar maple in response to elevated atmospheric $CO_2$. New Phytol 168:623–635

Kleczewski NM, Herms DA, Bonello P (2010) Effects of soil type, fertilization and drought on carbon allocation to root growth and partitioning between secondary metabolism and ectomycorrhizae of *Betula papyrifera*. Tree Physiol 30:807–813

Kleczewski NM, Herms DA, Bonello P (2012) Nutrient and water availability alter belowground patterns of biomass allocation, carbon partitioning, and ectomycorrhizal abundance in *Betula nigra*. Trees 26:525–533

Ku MSB, Agarie S, Nomura M, Fukayama H, Tsuchida H, Ono K, Hirose S, Toki S, Miyao M, Matsuoka M (1999) High-level expression of maize phosphoenolpyruvate carboxylase in transgenic rice plants. Nat Biotechnol 17:76–80

Lal R (2004) Soil carbon sequestration impacts on global climate change and food security. Science 304:1623–1627

Lal R, Kimble JM, Follett RF, Cole CV (eds) (1999) The potential of U.S. cropland to sequester carbon and mitigate the greenhouse effect. Lewis, Boca Raton, FL

Landsberg JJ, Gower ST (eds) (1997) Applications of physiological ecology to forest management. Academic, New York

Langenfeld-Heyser R, Gao J, Ducic T, Tachd P, Lu CF, Fritz E, Gafur A, Polle A (2007) *Paxillus involutus mycorrhiza* attenuate NaCl-stress responses in the salt-sensitive hybrid poplar *Populus canescens*. Mycorrhiza 17:121–131

Larsen P, Dai Y (2009) Using gene expression modeling to determine biological relevance of putative regulatory networks. Lect Notes Bioinformatics 5542:40–51

Larsen P, Almasri E, Chen G, Dai Y (2007) A statistical method to incorporate biological knowledge for generating testable novel gene regulatory interactions from microarray experiments. BMC Bioinformatics 8:317

Larsen P, Collart F, Dai Y (2010) Incorporating network topology improves prediction of protein interaction networks from transcriptomic data. Int J Knowl Discov Bioinformatics 1(3):1–17

Larsen P, Sreedasyam A, Trivedi G, Podila GK, Cseke LJ, Collart FR (2011a) Using next generation transcriptome sequencing to predict an ectomycorrhizal metabolome. BMC Syst Biol 5:70

Larsen PE, Collart F, Field D, Meyer F, Keegan KP et al (2011b) Predicted relative metabolomic turnover (PRMT): determining metabolic turnover from a coastal marine metagenomic dataset. Microb Informatics Exp 1(4)

Larsen P, Field D, Gilbert JA (2012) Predicting bacterial community assemblages using an artificial neural network approach. Nat Methods 9:621–625

Leake J, Johnson D, Donnely D, Muckle G, Boddy L, Read D (2004) Networks of power and influence: the role of mycorrhizal mycelium in controlling plant communities and agroecosystem functioning. Can J Bot 82:1016–1045

LeBauer DS, Treseder KK (2008) Nitrogen limitation of net primary productivity in terrestrial ecosystems is globally distributed. Ecology 89:371–379

Leegood C, Sharkey TD, Caemmerer SV (2000) Photosynthesis: physiology and metabolism, vol 9, Advances in photosynthesis and respiration. Kluwer Academic, Netherlands

Leegood RC (2002) C(4) photosynthesis: principles of CO(2) concentration and prospects for its introduction into C(3) plants. J Exp Bot 53:581–590

Lewandowski I, Clifton-Brown JC, Scurlock JMO, Huisman W (2000) Miscanthus: European experience with a novel energy crop. Biomass Bioenergy 19(4):209–227

Lewandowski I, Scurlock JMO, Lindvall E, Christou M (2003) The development and current status of perennial rhizomatous grasses as energy crops in the US and Europe. Biomass Bioenergy 25(4):335–361

Li L, Zhou Y, Cheng X, Sun J, Marita JM, Ralph J, Chiang VL (2003) Combinatorial modification of multiple lignin traits in trees through multigene cotransformation. Proc Natl Acad Sci USA 100:4939–4944

Lichter J, Barron SH, Bevacqua CE, Finzli AC, Irving KE, Stemmler EA et al (2005) Soil carbon sequestration and turnover in a pine forest after six years of atmospheric $CO_2$ enrichment. Ecology 86:1835–1847

Long SP, Zhu XG, Naidu SL, Ort DR (2006) Can improvement in photosynthesis increase crop yields? Plant Cell Environ 29:315–330

Lynch J, Deikman J (1998) Phosphorus in plant biology: regulatory roles in molecular, cellular, organismal, and ecosystem processes. American Society of Plant Physiologists, Rockville, MD

Mack MC, Schuur EAG, Bret-Harte MS, Shaver GR, Chapin FS III (2004) Ecosystem carbon storage in arctic tundra reduced by long-term nutrient fertilization. Nature 431:440–443

Magnani F, Mencuccini M, Borghetti M, Berbigier P, Berninger F, Delzon S et al (2007) The human footprint in the carbon cycle of temperate and boreal forests. Nature 447:848–850

Majdi H, Kangas P (1997) Demography of fine roots in response to nutrient applications in a Norway spruce stand in south-western Sweden. Ecoscience 4:199–205

Mao YJ, Yannarell AC, Mackie RI (2011) Changes in N-transforming Archaea and bacteria in soil during establishment of bioenergy crops. PLoS One 6:24750

Marschner H (1995) Mineral nutrition of higher plants, 2nd edn. Academic, San Diego, CA

Martin F, Aerts A, Ahren D, Brun A, Duchaussoy F, Kohler A, Lindquist E, Salamov A, Shapiro HJ, Wuyts J et al (2008) The genome sequence of the basidiomycete fungus *Laccaria bicolor* provides insights into the mycorrhizal symbiosis. Nature 452:88–92

McCormick AJ, Cramer MD, Watt DA (2006) Sink strength regulates photosynthesis in sugarcane. New Phytol 171:759–770

Miguez FE, Zhu XG, Humphries S, Bollero GA, Long SP (2009) A semimechanistic model predicting the growth and production of the bioenergy crop *Miscanthus* x *giganteus*: description, parameterization and validation. Glob Change Biol Bioenergy 1:282–296

Milchunas DG, Morgan JA, Mosier AR, LeCain DR (2005) Root dynamics and demography in shortgrass steppe under elevated $CO_2$, and comments on minirhizotron technology. Glob Change Biol 11:1837–1855

Newman J, Gutteridge S (1993) The X-ray structure of Synechococcus ribulose-bisphosphate carboxylase/oxygenase-activated quaternary complex at 2.2 Å resolution. J Biol Chem 268:25876–25886

Nielsen KL, Bouma TJ, Lynch JP, Eissenstat DM (1998) Effects of phosphorus availability and vesicular–arbuscular mycorrhizas on the carbon budget of common bean (*Phaseolus vulgaris*). New Phytol 139:647–656

NRC (2010) Advancing the science of climate change. National Research Council. National Academies, Washington, DC

Ogata H, Goto S, Sato K, Fujibuchi W, Bono H, Kanehisa M (1999) KEGG: Kyoto encyclopedia of genes and genomes. Nucleic Acids Res 27(1):29–34

Ostle NJ, Smith P, Fisher R, Woodward FI, Fisher JB et al (2009) Integrating plant-soil interactions into global carbon cycle models. J Ecol 97:851–863

Parr JF, Papendick RI, Hornick SB, Meyer RE (1992) Soil quality: attributes and relationship to alternative and sustainable agriculture. Am J Altern Agric 7:2–3

Paul MJ, Pellny TK, Goddijn O (2001) Enhancing photosynthesis with sugar signals. Trends Plant Sci 6:197–200

Pearce RB (1996) Antimicrobial defenses in the wood of living trees. New Phytol 132:203–233

Pellny T, Paul MJ, Goddijn JM (2002) Unravelling the role of trehalose-6-phosphate in metabolic signalling in photosynthetic tissue. Comp Biochem Physiol 132:102

Plomion C, Leprovost G, Stokes A (2001) Wood formation in trees. Plant Physiol 127 (4):1513–1523

Podila GK, Sreedasyam A, Muratet MA (2009) *Populus* rhizosphere and the ectomycorrhizal interactome. Crit Rev Plant Sci 28:359–367

Poorter H, Nagel O (2000) The role of biomass allocation in the growth response of plants to different levels of light, $CO_2$, nutrients and water: a quantitative review. Aust J Plant Physiol 27:1191

Pritchard SG (2011) Soil organisms and global climate change. Plant Pathol 60:82–99

Pritchard SG, Rogers HH (2000) Spatial and temporal deployment of crop roots in $CO_2$-enriched environments. New Phytol 147:55–71

Pritchard SG, Prior SA, Rogers HH, Davis MA, Runion GB, Popham TW (2006) The effects of elevated atmospheric $CO_2$ concentration on root dynamics of sorghum grown under sustainable and conventional agricultural management systems. Agric Ecosyst Environ 113:175–183

Purves D, Pacala S (2008) Predictive models of forest dynamics. Science 320:1452–1453

Pyter R, Voigt T, Heaton E, Dohleman F, Long S (2007) Giant miscanthus: biomass crop for Illinois. Proceedings of the 6th national new crops symposium, San Diego, CA, pp 39–42

Qiu J, Israel DW (1992) Diurnal starch accumulation and utilization in phosphorus-deficient soybean plants. Plant Physiol 98:316–323

Quoreshi AM, Khasa DP (2008) Effectiveness of mycorrhizal inoculation in the nursery on root colonisation, growth and nutrient uptake of aspen and balsam poplar. Biomass Bioenergy 32:381–391

Rae AM, Street NR, Robinson KM, Harris N, Taylor G (2009) Five QTL hotspots for yield in short rotation coppice bioenergy poplar: the poplar biomass loci. BMC Plant Biol 9:23

Rillig MC, Allen MF (1999) What is the role of arbuscular mycorrhizal fungi in plant-to-ecosystem responses to elevated atmospheric $CO_2$? Mycorrhiza 9:1–8

Rogers HH, Prior SA, Runion GB, Mitchell RJ (1996) Root to shoot ratio of crops as influenced by $CO_2$. Plant Soil 187:229–248

Rogers A, McDonald K, Muehlbauer MF, Hoffman A, Koenig K, Newman L, Taghavi S, van der Lelie LD (2012) Inoculation of hybrid poplar with the endophytic bacterium Enterobacter sp 638 increases biomass but does not impact leaf level physiology. Glob Change Biol Bioenergy 4:364–370

Roitsch T, Gonzalez MC (2004) Function and regulation of plant invertases: sweet sensations. Trends Plant Sci 9:606–613

Rooney DC, Killham K, Bending GD, Baggs E, Weih M, Hodge A (2009) Mycorrhizas and biomass crops: opportunities for future sustainable development. Trends Plant Sci 14 (10):542–549

Rovira AD (2005) Plant root excretions in relation to the rhizosphere effect. Plant Soil 7:178–194

Ryan MG, Hubbard RM, Pongracic S, Raison RJ, McMurtrie RE (1996) Autotrophic respiration in *Pinus radiata* in relation to nutrient status. Tree Physiol 16:333–343

Sage RF (2002) C4 photosynthesis in terrestrial plants does not require Kranz anatomy. Trends Plant Sci 7(7):1360–1385

Sage RF, Pearcy RW, Seemann JR (1987) The nitrogen use efficiency of C3 and C4 plants. III. Leaf nitrogen effects on the activity of carboxylating enzymes in *Chenopodium album* (L.) and *Amaranthus retroflexus* (L.). Plant Physiol 85:355–359

Sanchez FG, Coleman M, Garten CT, Luxmoore RJ, Stanturf JA, Trettin C, Wullschleger SD (2007) Soil carbon, after 3 years, under short-rotation woody crops grown under varying nutrient and water availability. Biomass Bioenergy 31:793–801

Satori F, Lal R, Ebinger MH, Eaton JA (2007) Changes in soil carbon and nutrient pools along a chronosequence of poplar plantations in the Columbia Plateau, Oregon, USA. Agric Ecosyst Environ 122:325–339

Schimel DS, Braswell BH, Holland EA, McKeown R, Ojima DS, Painter TH, Parton WJ, Townsend AR (1994) Climatic, edaphic, and biotic controls over storage and turnover of carbon in soils. Glob Biogeochem Cycles 8:279–293

Schneider G, Lindqvist Y, Lundqvist T (1990) Crystallographic refinement and structure of ribulose-1,5-bisphosphate carboxylase from *Rhodospirillum rubrum* at 1.7 Å resolution. J Mol Biol 211:989–1008

Schreuder HA, Knight S, Curmi PMG, Andersson I, Cascio D, Sweet RM, Brändén CI, Eisenberg D (1993) Crystal structure of activated tobacco rubisco complexes with the reaction-intermediate analog 2-caroxy-arabinitol 1.5-bisphosphate. Prot Struct 2:1136–1146

Schützendübel A, Polle A (2002) Plant responses to abiotic stresses: heavy metal- induced oxidative stress and protection by mycorrhization. J Exp Bot 53:1351–1365

Schweitzer JA, Bailey JK, Rehill BJ, Martinsen SD, Hart SC, Lindroth RL et al (2004) Genetically based trait in a dominant tree affects ecosystem processes. Ecol Lett 7:127–134

Shetty KG, Hetrick BAD, Figge DAH, Schwab AP (1994) Effects of mycorrhizae and other soil microbes on revegetation of heavy-metal contaminated mine spoil. Environ Pollut 86:181–188

Shipley B, Meziane D (2002) The balanced-growth hypothesis and the allometry of leaf and root biomass. Funct Ecol 16:326–331

Silpi U, Lacointe A, Kasempsap P, Thanysawanyangkura S, Chantuma P et al (2007) Carbohydrate reserves as competing sink: evidence from tapping rubber trees. Tree Physiol 27:881–889

Smidansky ED, Meyer FD, Blakeslee B, Weglarz TE, Greene TW, Giroux MJ (2007) Expression of a modified ADP-glucose pyrophosphorylase large subunit in wheat seeds stimulates photosynthesis and carbon metabolism. Planta 225:965–976

Smith SE, Read DJ (1997) Mycorrhizal symbiosis, 2nd edn. Academic, London

Smith SM, Ellis RJ (1981) Light-stimulated accumulation of transcripts of nuclear and chloroplast genes for ribulosebisphosphate carboxylase. J Mol Appl Genet 1:127–137

Spreitzer RJ, Salvucci ME (2002) Rubisco: structure, regulatory interactions, and possibilities for a better enzyme. Annu Rev Plant Biol 53:449–475. doi:10.1146/annurev.arplant.53.100301.135233

Stott PA, Tett SFB, Jones GS, Allen MR, Mitchell JFB, Jenkins GJ (2000) External control of 20th century temperature by natural and anthropogenic forcing. Science 290:2133–2137

Suzuki S, Murai N, Burnell JN, Arai M (2000) Changes in photosynthetic carbon Flow in transgenic rice plants that express $C_4$-type phosphoenolpyruvate carboxykinase from *Urochloa panicoides*. Plant Physiol 124:163–172

Suzuki S, Murai N, Kasaoka K, Hiyoshi T, Imaseki H, Burnell JN, Arai M (2006) Carbon metabolism in transgenic rice plants that express phosphoenolpyruvate carboxylase and/or phosphoenolpyruvate carboxykinase. Plant Sci 170(5):1010–1019

Suzuki Y, Ohkubo M, Hatakeyama H, Ohashi K, Yoshizawa R, Kojima S, Hayakawa T, Yamaya T, Mae T, Makino A (2007) Increased Rubisco content in transgenic rice transformed with the 'sense' rbcS gene. Plant Cell Physiol 48:626–637

Swaminathan K, Chae WB, Mitros T, Varala K, Xie L, Barling A, Glowacka K, Hall M, Jezowski S, Ming R et al (2012) A framework genetic map for *Miscanthus sinensis* from RNAseq-based markers shows recent tetraploidy. BMC Genomics 13(1):142

Tcherkez GG, Farquhar GD, Andrews TJ (2006) Despite slow catalysis and confused substrate specificity, all ribulose bisphosphate carboxylases may be nearly perfectly optimized. Proc Natl Acad Sci USA 103:7246–7251

Thivierge JP, Minai A, Siegelmann H, Alippi C, Geourgiopoulos M (2012) A year of neural network research: special issue on the 2011 international joint conference on neural networks. Neural Netw 32:1–2

Treseder KK (2004) A meta-analysis of mycorrhizal responses to nitrogen, phosphorus, and atmospheric CO2 in field studies. New Phytol 164:347–355

Treseder KK (2008) Nitrogen additions and microbial biomass: a meta-analysis of ecosystem studies. Ecol Lett 11:1111–1120

Tuberosa R, Salvi S (2006) Genomics-based approaches to improve drought tolerance of crops. Trends Plant Sci 11:405–412

Uroz S, Calvaruso C, Turpault MP, Pierrat JC, Mustin C, Frey-Klett P (2007) Effect of the mycorrhizosphere on the genotypic and metabolic diversity of the soil bacterial communities involved in mineral weathering in a forest soil. Appl Environ Microbiol 73:3019–3027

Valentini R, Matteucchi G, Dolman H, Schulze E-D, Rebmann C et al (2000) Respiration as the main determinant of carbon balance in European forests. Nature 404:861–865

Wang X, Yamada T, Kong FJ, Abe Y, Hoshino Y, Sato H, Takamizo T, Kanazawa A, Yamada T (2011) Establishment of an efficient in vitro culture and particle bombardment-mediated transformation systems in *Miscanthus sinensis* Anderss., a potential bioenergy crop. GCB Bioenergy 3:322–332

Warnock DD, Lehmann J, Kuyper TW, Rillig MC (2007) Mycorrhizal responses to biochar in soil – concepts and mechanisms. Plant Soil 300:9–20

Wassen MJ, Venterink HO, Lapshina ED, Tanneberger F (2005) Endangered plants persist under phosphorus limitation. Nature 437:547–550

Webb EC (1992) Enzyme nomenclature 1992: recommendations of the Nomenclature Committee of the International Union of Biochemistry and Molecular Biology on the nomenclature and classification of enzymes. Academic, New York

Wechsung G, Wechsung F, Wall GW, Adamsen FJ, Kimball BA et al (1995) Biomass and growth rate of a spring wheat root system grown in free-air $CO_2$ enrichment (FACE) and ample soil moisture. J Biogeogr 22:623–634

Werner D (1998) Organic signals between plants and microorganisms. In: Pinton R, Varanini Z, Nannipieri P (eds) The rhizosphere: biochemistry and organic substances at the soil–plant interfaces. MarcelDekker, New York

West TO, Post WM (2002) Soil carbon sequestration by tillage and crop rotation: a global data analysis. Soil Sci Soc Am J 66:1930–1946, Available at DOE CDIAC site

Williams L, Miller A (2001) Transporters responsible for the uptake and partitioning of nitrogenous solutes. Annu Rev Plant Physiol Plant Mol Biol 52:659–688

Wullschleger SD, Yin TM, DiFazio SP, Tschaplinski TJ, Gunter LE, Davis MF, Tuskan GA (2005) Phenotypic variation in growth and biomass distribution for two advanced-generation pedigrees of hybrid poplar. Can J For Res 35:1779–1789

Xia JY, Wan SQ (2008) Global response patterns of terrestrial plant species to nitrogen addition. New Phytol 179:428–439

Yang D-L, Yao J, Mei C-S, Tong X-H, Zeng L-J et al (2012) Plant hormone jasmonate prioritizes defense over growth by interfering with gibberellin signaling cascade. Proc Natl Acad Sci USA 109(19):1192–1200

Zhang QX, Sun Y, Hu HK, Chen B, Hong CT, Guo HP, Pan YH, Zheng BS (2012) Micropropagation and plant regeneration from embryogenic callus of *Miscanthus sinensis*. In Vitro Cell Dev Biol–Plant 48(1):50–57

Zhu XG, Portis AR, Long SP (2004) Would transformation of C-3 crop plants with foreign Rubisco increase productivity? A computational analysis extrapolating from kinetic properties to canopy photosynthesis. Plant Cell Environ 7:155–165

Zhu XG, de Sturler E, Long SP (2007) Optimizing the distribution of resources between enzymes of carbon metabolism can dramatically increase photosynthetic rate: a numerical simulation using an evolutionary algorithm. Plant Physiol 145:513–526

Zhu XG, Long SP, Ort DR (2010) Improving photosynthetic efficiency for greater yield. Annu Rev Plant Biol 61:235–261

# Chapter 13
# Greenhouse Gas Emissions

Michael T. Abberton

**Abstract** Plant-breeding approaches, allied to knowledge of animal nutrition, have considerable promise with respect to reducing emissions of both nitrous oxide and methane from livestock agriculture. Reduction in the requirement for nitrogenous fertilizer through enhanced nitrogen-use efficiency will also reduce nitrous oxide emissions from arable systems. Changes to the diet of a ruminant animal (e.g., in terms of sugar and lipid content, secondary metabolites, or protein levels) can affect the activity of rumen microbes and the efficiency of digestive functions, leading to a reduction in emissions of methane, a powerful greenhouse gas. Plant-breeding methods are also likely to be successful with respect to modifications of root structure and exudation in rice, again with the potential to reduce methane emissions. The application of modern genomic tools is beginning to unravel the genetic basis of key traits controlling nitrogen utilization from soil–plant–animal. Many agricultural systems have the potential to enhance carbon levels in soil, and this process is affected by plant traits amenable to the same process of genetic dissection. Thus, plant breeding can affect the greenhouse gas balance by enhancing carbon sequestration, with considerable potential in grassland systems.

## 13.1 Introduction

Many studies have looked at the potential to enhance adaptation by genetic improvement (e.g., drought tolerance), but much less work has been carried out on the ways in which plant breeding can contribute to climate change mitigation

M.T. Abberton (✉)
Genetic Resources Centre, International Institute of Tropical Agriculture (IITA), Ibadan, Nigeria
e-mail: M.Abberton@cgiar.org

C. Kole (ed.), *Genomics and Breeding for Climate-Resilient Crops*, Vol. 2,
DOI 10.1007/978-3-642-37048-9_13, © Springer-Verlag Berlin Heidelberg 2013

through reduced greenhouse gas (GHG) emissions and enhanced carbon (C) sequestration.

Human activities since the preindustrial era have resulted in an increase of methane from 714 to 1,770 parts per billion (ppb) and of nitrous oxide from 270 to 314 ppb. Agriculture is thought to be the main cause of these increased emissions. Povellato et al. (2007) found that agriculture (as a whole) and forestry account for 30 % of worldwide GHG emissions from human activities and 10 % in Europe.

A major focus of this chapter is grasslands and the livestock systems they support. These are major sources of GHG emissions, primarily methane and nitrous oxide, but grasslands also have considerable potential for carbon sequestration and hence climate change mitigation. Grasslands are one of the most important ecosystems and forms of land use in the world. They are crucial for food production and deliver ecosystem services. The changing considerations concerning the use and management of grasslands were reviewed by Kemp and Michalk (2007) and the impact of climate change on European grasslands and their role in climate change mitigation by Mannetje (2007a), Hopkins and Del Prado (2007), and Soussana and Luscher (2007). Morgan (2005) considered the global picture in terms of the response of grazing lands to increased atmospheric carbon dioxide, and Hopkins and Del Prado (2007) compiled the scenarios for climate change affecting European grasslands. In the latter, likely responses to predicted climate change include increased herbage growth, increased use of forage legumes particularly white and red clover and alfalfa (lucerne), reduced opportunities for grazing and harvesting on wetter soils, greater incidence of summer drought, and increased leaching from more winter rainfall. The key species of improved grasslands in substantial parts of Europe, Australasia, and to a lesser extent North and South America are forage grasses in the families *Lolium* (ryegrasses) and *Festuca* (fescues) and forage legumes in the genera *Trifolium* (clovers) and Medicago—particularly *Medicago sativa*, alfalfa (or lucerne). These species are the major source of grazed and conserved feed for dairy, beef, and sheep production. However, there is an increasing awareness of the important role they play in the delivery of ecosystem services and their value for leisure and amenity. Breeding programs in the major grassland species have an important role to play in reducing GHG emissions and enhancing carbon sequestration, and relevant aspects of these programs are outlined in this chapter.

However, livestock production is clearly not the only source of agricultural GHG. Other important areas are methane from rice production and fertilizer production and use, particularly nitrogenous fertilizers as major sources of nitrous oxide. These are also discussed briefly in this chapter with emphasis on what can be achieved from plant breeding.

## 13.2  Reducing Greenhouse Gas Emissions

### 13.2.1  Methane

The major sources of agricultural methane emissions are enteric fermentation in livestock, livestock manures, and rice production (Kebreab et al. 2006; Henry and Eckard 2009). It should be noted from the outset that estimates of methane emissions and IPCC emission factors are based on limited data and considerable work is ongoing to refine estimates for animal and herd emissions (e.g., Kennedy and Charmley 2012). Ruminant livestock can produce 250–500 L methane per day according to Johnson and Johnson (1995), and these authors described the major factors influencing methane emissions: level of feed intake, type of carbohydrate in the diet, addition of lipids or ionophores to the diet, and alteration of the ruminal microflora. These all, directly or indirectly, are amenable to plant breeding in combination with other approaches.

Bhatta et al. (2007) and Lassey (2007) reviewed methods of determining methane emissions, and the range of strategies to reduce methane emission from enteric fermentation was reviewed by Hopkins and Del Prado (2007). They categorized them as dietary changes, direct rumen manipulation, and systematic changes. The latter include considerations of breed, livestock numbers, and intensiveness of production. More intensive production may result in lower methane emission but may be less desirable in terms of other environmental impacts, highlighting the importance of a rigorous life cycle analysis in which trade-offs between different outcomes can be explicitly considered.

Direct rumen manipulation includes reducing protozoa numbers in the rumen (since protozoa parasitize methanogenic bacteria) and the addition of ionophores to enhance propionate levels. However, these approaches have drawbacks: reducing protozoa may lead to metabolic disease, and the main ionophores used are antibiotics such as monensin, where issues of resistance may limit utility.

Tamminga (1996) reviewed nutritional strategies for methane reduction. Dietary manipulations include the addition of organic acids (aspartate, malate, and fumarate) and yeast culture. These compounds encourage the production of propionate and butyrate in the rumen, which compete for hydrogen and reduce the ability of methanogenic microbes to produce methane. Research into the efficacy of these approaches and the optimum method of delivery of organic acids is ongoing. Animal selection approaches based on increased feed efficiency as a result of lower residual feed intake is an objective that may lead to "substantial and lasting methane abatement." Alford et al. (2006) and Hegarty et al. (2007) also demonstrated the potential for animal selection based on residual feed intake. This could reduce the methane costs of growth, i.e., increased efficiency.

Plant secondary metabolites such as tannins and saponins have been employed in attempts to reduce methane emissions from enteric fermentation. One such approach is the use of tannin-containing forages and breeding of forage species with enhanced tannin content. *Lotus corniculatus* (birdsfoot trefoil) and *Lotus uliginosus* (greater trefoil) possess secondary metabolites known as condensed tannins (CTs) in their leaves. CTs are flavonoid polymers, which complex with soluble proteins and render them insoluble in the rumen yet release them under the acidic conditions found in the small intestine, reducing bloat and increasing amino acid absorption. They are not present in the leaves of white or red clover but are present in the inflorescences. Methane production values were lower in housed fed sheep that fed on red clover and birdsfoot trefoil than on a ryegrass/white clover pasture (Ramirez-Restrepo and Barry 2005). A recent study has shown the extent of variation between and within varieties of *L. corniculatus* and *L. uliginosus* (Marley et al. 2006). Diverse germplasm is now available with CT content ranging from 20 mg/g DM to >100 mg/g DM for experiments to quantify the effect of CT content on methane in combination with other forage species. This will be more feasible using a high-throughput CT assay that enables rapid analysis of CT content in the numbers of genotypes required for a breeding program (Marshall et al. 2008). Rhizomatous lines of *L. corniculatus* with considerably improved persistence and contribution to mixed swards have been developed at IBERS. Waghorn and Shelton (1997) provided evidence that tannins in *L. corniculatus* grown in New Zealand improve protein utilization, with reductions in rumen ammonia concentrations by 27 % and increased absorption of essential amino acids from the small intestine of 50 %. Methane production values were lower in housed fed sheep that fed on red clover and birdsfoot trefoil than on a ryegrass/white clover pasture (Ramirez-Restrepo and Barry 2005).

Studies on perennial ryegrass (*Lolium perenne* L.) selected for high levels of water-soluble carbohydrate are ongoing at IBERS with respect to their potential to reduce methane as well as nitrous oxide emissions (see below).

It is clear that forage diet (and by extension plant breeding) can have a significant effect on methane emissions, although there is a need for careful interpretation and replication of results using respiration chambers. This is demonstrated by the work of Hammond et al. (2011) comparing white clover and perennial ryegrass in which results from using the sulfur hexafluoride technique suggesting that sheep fed fresh white clover had lower methane yields than those fed fresh perennial ryegrass were not borne out by detailed studies using respiration chambers. A similar outcome was recorded by Sun et al. (2011) comparing forage chicory (*Cichorium intybus*) and perennial ryegrass.

Wetland soils are a major biogenic source of methane release into atmosphere. This takes place through the microbial anaerobic degradation of complex organic molecules and methane production from methanogenic archaea. Methane produced can be oxidized to carbon dioxide by methanotrophic bacteria. Plants can promote methane emissions by supplying carbon—through root exudates and residues. Methane can be released from soils and is produced in layers under the topsoil under anaerobic conditions particularly during and after rainfall. Kammann et al. (2001) highlighted the importance of the topsoil aerobic layer in oxidizing methane and therefore reducing the amount released. This shows the value of reducing soil compaction, poaching, etc., and enhancing soil quality as measured by aeration and oxygen diffusivity. There is evidence that plant species differ in their visible effects on soil structure (Drury et al. 1991) and anecdotal reports have long supported a positive role for legumes in this respect. More detailed investigations of the process of soil structuring have been carried out on white clover (Mytton et al. 1993; Holtham et al. 2007) and red clover (Papadopoulos et al. 2006). It has been reported that the changes in soil structuring brought about by white clover resulted in improvements in water percolation rate (i.e., the soil became more freely drained) and in the extraction by plants of nutrients from the soil. Holtham et al. (2007) also reported evidence of local structuring of soil around white clover roots and greater drainage of water through soil cores under white clover than under perennial ryegrass monocultures. Similar benefits in terms of soil structure were noted for soil cores under red clover monocultures by Papadopoulos et al. (2006), although the effects were transient and were reversed when a cereal crop was sown the following year. Improved soil structure reduces the risk of soil compaction and water runoff, increases the soil's biological activity, and facilitates seedling establishment and root penetration. However, it appears likely that legume-driven improvements in soil structure and drainage also directly result in increased leaching of both fixed and applied nitrate in legume monocultures (Holtham et al. 2007).

Possible approaches to reducing GHG from rice production and the potential of biological nitrification inhibition have been recently summarized (Philippot and Hallin 2011) and are briefly discussed here. It is estimated that 20 % of global methane emissions are due to rice production. Rice production increased from 215 to 644 million tons between 1961 and 2006 and the global harvested area occupies around 14 % of the earth's arable land.

In rice 30–40 % of carbon fixed can contribute to methane production, and up to 90 % of methane emissions from rice during the growing period is mediated through the aerenchyma. Differences in rice variety can have a marked effect with up to 500 % difference in methane emissions reported (ButterbachBahl et al. 1997). Important variation in methanogenesis lies in the volume and nature of root exudates, the anatomy of aerenchyma, and gas diffusion capacity.

It seems clear that plant-breeding efforts focused on root exudates and on the relationships between root and soil exudates and on the relationships between roots and soil properties are likely to be of significant importance in reducing emissions as well as adaptation to climate change (e.g., responses to water stress).

## 13.2.2 Nitrous Oxide

Nitrous oxide has become the single most important ozone-depleting substance in the twenty-first century. It has a global warming potential (GWP) around 310 times that of carbon dioxide. Nitrous oxide emissions can arise directly from N inputs to soil, e.g., animal excreta, fertilizer, manure, crop residues, fixed nitrogen, and also indirectly from nitrates. Developments to reduce nitrate leaching or ammonia volatilization are also likely to reduce nitrous oxide emissions. Soil processes controlling nitrous oxide production, i.e., nitrification and denitrification, are affected not only by a range of abiotic factors (e.g., temperature, pH) but also by fertilizer addition and organic matter content (Mummey and Smith 2000; Hopkins and Del Prado 2007). Nitrous oxide emissions can be reduced by enhancing the competitiveness of plant uptake against that of competing soil microbial processes. Gregorich et al. (2005) found that emissions of nitrous oxide from soils increased linearly with the amount of mineral nitrogen fertilizer applied. The use of improved crop varieties was suggested by the Intergovernmental Panel on Climate Change (IPCC) as an alternative for mitigating GHG emissions, but no specifics were listed. The use of N fertilizer has increased by about 800 % since 1960, and around 120 million tons of nitrogen gas is now fixed from the atmosphere every year.

Wood and Cowie (2004) carried out a review of studies of GHG emissions from fertilizer production. Nitrogen fertilizer manufacture brings with it significant GHG emissions from the Haber-Bosch process of synthesizing ammonia and from nitric acid production. Synthesis of ammonia, the primary input for most nitrogen fertilizers, is very energy demanding with natural gas as the primary energy source. Nitric acid is used in the manufacturing of ammonium nitrate, calcium nitrate, and potassium nitrate. The oxidation of ammonia to give nitric oxide also produces a tail gas of nitrous oxide, nitric oxide, and nitrogen dioxide. Nitric acid production is the largest industrial source of nitrous oxide although clearly this is also used for purposes other than fertilizer manufacture. Estimates of nitrous oxide emissions from nitric acid manufacture are very variable: 550–5,890 $CO_2$ equiv./kg nitric acid.

Urea accounts for almost 50 % of world nitrogen fertilizer production and is synthesized from ammonia and carbon dioxide at high pressure to produce ammonium carbonate, which is then dehydrated by heating to give urea and water.

It is also worth noting at this point that the synthesis of phosphate fertilizers also results in GHG emissions (reviewed by Wood and Cowie 2004). Single superphosphate is produced from phosphate rock and sulfuric acid, triple superphosphate from phosphate rock and phosphoric acid. Wood and Cowie (2004) state that "more sulfuric acid is produced than any other chemical in the world and the largest single user is the fertilizer industry." Considerable variation in "net emissions" is seen according to method used and efficiency of plant—in some cases the heat generated in production of sulfuric acid is captured.

These figures must be interpreted cautiously as there is a large amount of variation in reports. However, they point to the GHG savings that result from an increased use of legume-fixed nitrogen. They also demonstrate the value of approaches to reduce fertilizer inputs by increasing the efficiency with which plants can utilize nitrogen (detailed above) and phosphorus.

It is estimated that 1.3–2.5 % of N applied as fertilizer is emitted as nitrous oxide. Direct evidence for the role of N fertilizer in nitrous oxide emissions has only recently been forthcoming (Park et al. 2012). Analysis of long-term trends in the isotopic composition of nitrous oxide in the atmosphere confirms for the first time that the rise in the atmospheric nitrous oxide is largely due to increased use of nitrogenous fertilizers.

A key area where genetic approaches can have an impact is in improving the nitrogen-use efficiency (NUE) of crops to allow lower fertilizer application and hence reduce nitrogenous emissions through the soil–plant–(animal)–soil cycle. NUE from soil to crop is typically lower for grass-based livestock than arable crop production, ranging from 10 % to 40 % for dairy compared with 40 % to 80 % for arable systems, on a whole-farm basis (Neeteson et al. 2004). More efficient use of N brings benefits to farmers both with respect to meeting regulatory requirements and in terms of cost savings from reduced fertilizer use. The case for increasing the efficiency of fertilizer nitrogen use on economic grounds is compelling, and global demand for fertilizers continues to rise. Eighty-five to ninety million tons (Mt) of N fertilizers is currently applied annually worldwide (Frink et al. 1999), and this is expected to rise to 240 Mt by 2050 (Tilman 1999). Breeding forage crops capable of using fertilizer inputs more efficiently offers a clean technology route to increased sustainability of livestock production, via lowering recommended fertilizer rates, reducing the agricultural footprint with respect to pollution, and reducing the wider consumption of nonrenewable resources. This is particularly so with respect to N, frequently the main determinant of both yield and environmental quality in agricultural systems.

Formal definition of NUE depends on the scope of the system, on the choice of mass balance or N flux approaches, and on whether N uptake, utilization, and retention within the plant are considered (Garnier and Aronson 1998; Good et al. 2004). For example, in agronomic terms, $NUE = NUpE \times NUtE$, where NUpE is the N uptake efficiency: the ratio between the amount of N absorbed by the plant and that supplied/available in the soil. NUtE is the utilization efficiency (the unit dry matter (DM) produced per unit N in the dry weight, or the DM flux per unit N flux in a whole stand in units of g biomass/mol of N). This approach has informed previous genetic improvement and mapping studies of N (e.g., Loudet et al. 2003) and can be employed in parallel with a flux-based approach to NUE, developed for ecosystem analysis (Berendse and Aerts 1987). This offers significant advantages for trait dissection of the NUtE component, given by the product of $aNP \times MRT$, where aNP is mean annual N productivity and MRT is mean residence time of N in the plant. Genetic variation for acquisition (Gorny and Sodkiewicz 2001) and utilization (Witt et al. 1999) has been demonstrated in a wide range of species. Quantitative trait loci (QTLs) for traits associated with NUE have been identified in *Arabidopsis* (Rauh et al. 2002; Loudet et al. 2003), maize (Gallais and Hirel 2004), barley (Mickelson et al. 2003), and ryegrass (Van Loo et al. 1997; Wilkins et al. 1997, 2000). Recent studies by Gaju et al. (2011) have detailed traits that may be important in improving NUE in wheat including delaying the onset of postanthesis senescence under low N supply.

Molecular genetic approaches are being applied to the dissection of NUE as a complex trait in perennial ryegrass. This trait can be broken down into the efficiency of uptake and efficiency of utilization. Analysis of a ryegrass mapping family under low and high N levels in flowing solution culture has allowed the characterization of QTLs for the efficiency of both uptake and utilization. A program of marker-assisted backcrossing is in place to introgress the major QTLs into elite germplasm. Similar experiments have been carried out to identify QTLs for phosphorus-use efficiency in both perennial ryegrass and white clover. This is important both with respect to climate change mitigation and environmental protection. The production and application of phosphorus fertilizers uses fossil fuel energy, and the diffuse phosphorus pollution of waterways is a major environmental challenge around the world. White clover, through the impact of nitrogen fixation on reduced N fertilizer application, can reduce GHG emissions by 20 % compared to a grass-only sward receiving 150 kg N/ha/year (Dawson et al. 2011). Similar results were reported by Ledgard et al. (2009).

Production of nitrous oxide in the soil is essentially a microbial process but it can be affected by plants. Nitrifying bacteria carry out the oxidation of ammonia to nitrous oxide and then to nitrate. Another pathway of nitrifier denitrification can produce both nitric and nitrous oxide under anaerobic conditions. The effects of plants can be mediated in different ways. Stimulation of nitrification can occur when increased organic matter turnover leads to increased N turnover. There is now a body of work on nitrification inhibitor chemicals derived from plants (biological nitrification inhibition, BNI) including a number of studies focused on chemicals derived from *Brachiaria humidicola* (Subbarao et al. 2009). A wild relative of wheat *Leymus racemosus* has a high BNI capacity, and an introgression program into wheat (*Triticum aestivum* Chinese Spring) has been developed (Subbarao et al. 2007), and studies of BNI have also been carried out on other crops (Philippot and Hallin 2011).

Ruminant animals (sheep, cattle, goats, etc.) convert plant matter into meat, milk, and other products. When sheep (MacRae and Ulyatt 1974) and cattle (Ulyatt et al. 1988) are given fresh forages, they can waste 25–40 % of forage protein. The efficiency of conversion of forage-N to microbial-N could be enhanced by (1) increasing the amount of readily available energy accessible during the early part of the fermentation and/or (2) providing a level of protection to the forage proteins, thereby reducing the rate at which their breakdown products are made available to the colonizing microbial population. One approach is to develop forage species with a better balance between water-soluble carbohydrate (WSC) and crude protein (CP), for example, by increasing the WSC content of the grass.

At IBERS, programs for breeding perennial ryegrass with elevated WSC and digestibility under grazing started some years ago. Varieties developed from this program have consistently higher levels of sugars than standard varieties throughout the growing season. Increased dry matter yield, digestibility, and intake are important corollaries of enhanced WSC in terms of impact on farm profitability as well as reduced emissions. In three zero-grazing experiments using early, mid-, and late lactation cows, N loss in urine was up to 24 % lower in animals fed the high WSC grasses than those fed controls (Miller et al. 2001), with concomitant increases in dietary protein reaching milk or body nitrogen.

The development of "high sugar grasses" was based on a detailed understanding of the biochemistry of fructan accumulation in ryegrass and a strong hypothesis based on knowledge of rumen function. The former underpinned the ability for rapid selection of higher sugar lines, while the latter pointed the way to experimental support for the potential for these grasses to simultaneously increase milk and meat production and reduce N in excreta. These grasses, or selections from them with even higher levels of fructans, are now also of interest with respect to the production of ethanol as a biofuel and a range of potential biorefinery applications.

Opportunities also exist within forages to select for other specific traits that can reduce protein loss. A good example of this approach is the emerging research on the enzyme polyphenol oxidase (PPO), which is at a particularly high level of activity in red clover in comparison with other species and has a role in protein protection (Owens et al. 2002). This enzyme converts phenols to quinines, which subsequently bind to protein and slow the rate of protein degradation. Thus, in silo, the protein made available for diffuse pollution of nitrogen, e.g., as ammonia, is reduced. Ensiling alfalfa (lucerne) leads to the degradation of 44–87 % of forage protein to non-forage protein (NPN). In comparison, red clover has up to 90 % less protein breakdown (Sullivan and Hatfield 2006). Increasing the level of PPO is a target for genetic improvement in red clover as a route to reduced nitrogenous pollution. Significant variation for PPO activity in red clover germplasm and differences in activity through the year has been shown in work at IBERS.

## 13.3 Enhancing Carbon Sequestration

It is widely accepted that there is more than twice as much carbon contained in the world's soils than in the atmosphere and that enhancing C sequestration into soils might offer a potentially useful contribution to climate change mitigation. Global stocks of carbon in aboveground vegetation are around 550 petagrams (PgC), whereas in the soil they are about 1,750 PgC (including peat) and in the atmosphere around 800 PgC (Royal Society 2001). There are considerable uncertainties in these land figures and they should be considered as broad approximations. Total C stocks in tropical savannas and temperate grasslands are estimated at 330 and 304 PgC, respectively; in comparison, tropical forests have stocks of 428 PgC, again with the same caveat. The Royal Society (2001) suggested that terrestrial vegetation and soils have been absorbing approximately 40 % of global $CO_2$ emissions from human activities. However, recent data indicate that the land has, on average over the last half century, been absorbing 28–29 % of the human emissions from fossil fuel burning, tropical deforestation, and cement manufacture. While there is a lot of year-to-year variation contributing to that mean, there is no indication yet of that mean fraction showing either a decreasing trend, as a result of saturation of natural terrestrial sinks (Canadell et al. 2007), or increasing as a result of purposeful biosequestration activities. The entire atmospheric $CO_2$ stock is exchanged every 10–15 years between land and atmosphere via net photosynthetic uptake by vegetation and organic matter oxidation via decomposition, herbivory, and fire. The size and location of land carbon sinks varies considerably annually.

According to the IPCC (2007), the global technical mitigation potential of agriculture (excluding fossil fuel offsets from biomass fuels) could be around 5.5–6.0 Gt $CO_2$ equiv. per year by 2030 of which approx. 1.5 Gt $CO_2$ equiv. is from grazing land management. Around 89 % of this proposed potential could be achieved by all forms of soil C sequestration. The economic potential in 2030 might be 1.5–1.6 Gt $CO_2$ equiv. per year with C price US\$ 20 per t $CO_2$ equiv. ranging up to 4.43 Gt $CO_2$ equiv. per year with a C price of US\$ 100 per t $CO_2$ equiv. About 70 % of this potential is expected to derive from developing countries (UNFCC 2008). Smith et al. (2008) divided the world into four "climatic types" based on the FAO Agro-Ecological Zones, climate zones within regions, and GIS: cool-dry, cool-moist, warm-dry, and warm-moist. Mean estimates of per area annual mitigation potentials for grassland management in each zone were given as follows ($tCO_2$/ ha/year): cool-dry 0.11, cool-moist 0.81, warm-dry 0.11, warm-dry 0.81.

Rangelands alone are the largest land-use type with around 40 % of land surface. In 28 countries, mainly in Africa, rangelands represent more than 60 % land area. There are 100–200 million pastoralist households in the world covering 5,000 Mha rangelands, in which are stored 30 % of the world carbon stocks (Tennigkeit and Wilkes 2008). These authors estimate improved rangeland management has the biophysical potential to sequester 1,300–2,000 Mt $CO_2$ equiv. worldwide to 2030.

Guo and Gifford (2002) described a global meta-analysis and concluded that although there was a decline in C stocks in moving from pasture to plantation (with an approximate mean change of 10 %), that was an expression only of a change when the plantation was coniferous, there being no change in soil C following transition from pasture to broadleaf tree species plantation or to naturally regenerated secondary forest. Although there was an increase in C stocks in moving from native forest to pasture (+8 %), and crop to pasture (+19 %), the forest result was expressed only in the 2,000- to 3,000-mm annual rainfall band, there being no significant change at lower or higher rainfall bands. The authors concluded with a working hypothesis that clearing of native forests for pastures would not in general lead to a decline in soil C. A recent UNFCC report (UNFCC 2008) concluded that "C sequestration in grasslands and agroforestry plantations has significant potential for C reductions from the agriculture sector at nonprohibitive costs."

Major stocks of C are seen in roots and soil, but organic C pools are not well characterized (Jones and Donnelly 2004). Techniques employed range from direct measurement, tracers, and use of modeling to estimate possible long-term changes. Byrne et al. (2007) combined the use of $CO_2$ flux measurements by eddy covariance with farm data for C imports and exports in a study of two dairy farms in SW Ireland. Both farms were apparently C sinks to the tune of around 1 t $CO_2$ equiv./ha/ year. McLauchlan et al. (2006) studied soil organic matter (SOM) accumulation in the Midwestern USA using a replicated chronosequence of former agricultural fields across 40 years after perennial grassland establishment on former cropland in the Great Plains. An increase in both labile and recalcitrant C pools was reported. Soil organic carbon (SOC) in the top 10 cm accumulated at 62 $g/m^2$/year and would return to levels in unplowed native prairies within 55–75 years if the rate were sustained. West and Six (2007) studied the rate and duration of soil C sequestration.

They found that equilibrium may be reached in around 33 years for grasslands, but the equilibrium level may increase further to maximum capacity or saturation with further management changes. C saturation of the soil mineral fraction is a phenomenon that is poorly understood with little empirical data Stewart et al. (2007). Skinner (2007) highlighted the importance of temporal variation in soil C fluxes in grazing pastures of Northeastern USA. During winter, when it is cool enough that plant growth ceases, pastures are net carbon sources. The major factors affecting C sequestration can also be considered as the routes through which management changes can make their impacts. Foremost among these, in the context of improved (sown) pasture systems, are changes to grazing and fertilization and the effects these have on primary production. This in turn impacts on litter inputs to the soil in terms of both structure and composition (decomposability). Edaphic factors in particular soil texture, structure, aggregation, cation exchange capacity, and organic matter content are important as is the microbial population of the soil. Clearly climate, particularly rainfall and temperature, also has an important role to play.

For extensive semiarid tropical non-sown rangelands in savanna and open grassy woodland ecosystems, where mineral fertilization is not an economic possibility, the levers for modifying soil carbon inventories are domesticated animal stocking rates; the management of wild and feral herbivory, wildfire, and plant species composition including the balance of annual and perennial herbaceous species; and woody species encroachment or thickening up in the ecosystem.

Skinner (2008) confirmed that conversion of plowed fields to pastures is one route to enhance C sequestration in temperate grasslands. However, for mature grasslands net C sinks may depend on the extent of (limitations on) biomass removal. Van Kessel et al. (2006) studied nitrogen (N)-fertilized grassland under free-air carbon dioxide enrichment (FACE) and compared fertilized perennial ryegrass (*Lolium perenne*), white clover (*Trifolium repens*), and a mixture of these species. Differences in $CO_2$ and N fertilizer rates and species had little impact on total soil C at 0–75-cm depth, and the results suggested limited use of such pastures as a net soil sink. Lynch et al. (2005) used modeling plus analysis to estimate the effects of improved grazing on SOC. A range of improvements were modeled to have significant potential impacts on increased SOC, but with some of them (e.g., use of fertilizer), gains were outweighed by $CO_2$ from energy use and nitrous oxide emissions such that only complementary grazing increased net returns to producers. Conant et al. (2001) carried out a review of 115 studies. The most frequent pasture improvements carried out in these studies were better grazing management and fertilization. Increases in soil C were seen and sequestration rates were highest in the first 40 years from improvement. The mean increase in rate of C sequestration with improvement was 0.54 MgC/ha/year. The exact nature and definition of "better grazing" remains problematic particularly in a rangelands context.

A number of studies have suggested that caution in the widespread use of nitrogenous fertilizers as an approach to increased productivity is appropriate given the potential for other emissions (Royal Society 2001). By contrast the role of legumes in supplying N through fixation is becoming increasingly to be seen as important and more beneficial in terms of overall GG balance than had once been thought. The introduction of legumes and their greater utilization as part of a pasture improvement process is therefore likely to be worthy of serious consideration in many circumstances. The Royal Society (2001) concluded that "raising fertility is possibly the most effective way of rapidly increasing carbon sink capacity."

In tropical America (Mexico, Central America, the Caribbean, and South America excluding Argentina, Chile, and Uruguay), there are 548 million ha of agricultural land. Pastoral areas (including silvopastoral) constitute 77 % of this including 250 million ha savanna (Mannetje 2007b; Amezquita et al. 2008a; Mannetje et al. 2008). Much of this is characterized by low productivity, degradation, and eroded soils. Inappropriate management is a major cause of these problems, e.g., cattle raising on slopes greater than 30 %. Many studies have shown productivity and economic benefits from pasture improvement, but until recently, little of this work has focused on carbon sequestration. Amezquita et al. (2008b) described studies of C sequestration using both long-established systems and short-term experiments. In established areas, soil C in silvopastoral systems represented 80 % of total C stock. Interactions between altitude, topography, and land-use change were studied with respect to changes in C sequestration. Improved grass and grass/legume mixtures showed sequestration of larger amounts of C in soil than did the use of "forage banks," which are belts of palatable shrubs and trees that can be browsed or lopped and eaten by stock in times of drought. Short-term experiments showed significant increases in C stocks over 3 years with quantities very variable, e.g., increments of 3–9 t/ha/year. The work also demonstrated the importance of deep-rooted grasses with legumes, e.g., *Brachiaria brizantha* with *Arachis pintoi*. The breeding of these crops has been reviewed by Jank et al. (2005).

Productive forage grasses such as *Festuca arundinacea* have been shown to increase the soil C pool by 17.2 % [equivalent to C sequestration of circa 3 mg C/ha/ year over a 6-year period (Lal et al. 1998)]. The amount of C retained by soils is influenced greatly by management practices with those that lead towards reduced soil disturbance and for increased crop persistency having the greatest benefits on C sequestration. It follows that change from arable to perennial grasslands leads to significant improvements in accumulation of soil C (Guo and Gifford 2002).

Elevated $CO_2$ levels have been demonstrated previously to increase photosynthesis of perennial ryegrass. Such an effect is likely to increase soil C inputs and microbial biomass through increased root exudates and turnover. However, the benefits of an increased production through elevated $CO_2$ may well be only short term and to be sustained will require complex interactions involving the continued availability of N in the soil, the NUE of the plants, and the uptake and loss of nutrients to maintain an effective balance of soil C and N necessary to deliver an effective decomposition process. To some extent, the negative impacts of increases in temperatures could be mitigated for by using multi-species communities that provide for complementary growth patterns and productivity.

The key plant traits likely to influence C sequestration (root depth, structure, and architecture; litter composition and amount) are reasonably well established and genetic variation is beginning to be characterized for many of them. Some early progress has been made at IBERS with regard to mapping of genes in perennial ryegrass for C sequestration, with effective C return in litter associated with loci on chromosomes 1 and 5.

## 13.4 System Level Studies and Trade-offs

Soussana et al. (2007) estimated a full GHG balance over 2 years for nine contrasted sites covering a major climatic gradient. The net ecosystem exchange of $CO_2$ (NEE) showed an average net sink of 240 ± 70 g $C/m^2$/year. However, C export (herbage) is greater than C import (manure), so average C storage (net biome productivity, NBM) was lower 104 ± 73, i.e., 435 of sink. Equivalent emissions of nitrous oxide, 14 ± 4.7, and methane, 32 ± 6.8, were seen, i.e., emissions of these two offset 19 % of NEE sink activity. Subtracting from NBP the plot emissions of nitrous oxide and methane and offplot emissions from digestion and fermentation gives an attributed GHG balance, which was on average not significantly different from zero. Levy et al. (2007) used a process-based model to estimate fluxes of major GHGs at 0.5° resolution across Europe, allowing spatially explicit modeling of total GWP. A 20-year simulation based on 10 biogeographical regions and typical current grassland management regimes showed that most grassland areas are net sources since nitrous oxide emissions from soil and methane emissions from livestock outweigh C sequestration.

It is important to place C sequestration in context with other changes occurring with high $CO_2$ concentrations. Kammann et al. (2008) showed that nitrous oxide ($N_2O$) emissions may rise from old, N-limited temperate grasslands. Over a 9-year FACE experiment additional $N_2O$ emissions corresponded to more than half a ton of $CO_2$/ha, which would need to be sequestered annually to result in no net effect on emissions. Improvements brought about to individual forage species should be seen within the context of the whole system at farm and catchment level and in terms of the balance between different outcomes, e.g., production, reduced pollution to water, and lower emissions to air. To this end the use of modeling approaches is likely to be extremely valuable. Modeling studies can consider the impact of dietary strategies and take into account the full range of economic and environmental attributes important for sustainability. Life cycle analysis is an emerging and increasingly important tool in the development of sustainable solutions to the delivery of multifunctional agriculture. It is likely that breeding approaches will prove both carbon and cost effective, but this needs to be rigorously established and compared with other potential approaches. However, many "alternative" strategies based on management change or animal selection may well prove to be complementary to plant genetic improvement.

In general, breeding approaches to increasing the efficiency of grassland agriculture can be characterized as:

(1) Accessible through seed without other inputs

(2) Lasting and cumulative impacts
(3) Bringing other benefits, e.g., varieties contributing to improved animal performance
(4) Easy to use and relatively inexpensive to the farmer
(5) Appropriate in the long term, representing sustainable "genetic"-based rather than "input"-based solutions

Plant breeding has been successful at increasing the yield, persistency, and stress tolerance of the major grasses and legumes of many grasslands in the world. These same approaches have considerable potential in altering plant traits to enhance the ecological efficiency of grassland agriculture. Plant-breeding approaches are cost effective, accessible to farmers through established routes, and show high rates of uptake in many parts of the world. Collaboration between plant breeders, animal scientists, and soil scientists is critical to future success. Progress is best evaluated within the framework of a comprehensive life cycle analysis approach that guards against emission swapping and develops solutions that consider economic as well as environmental sustainability.

# References

Alford AR, Hegarty RS, Parnell PF, Cacho OJ, Herd RM, Griffith GR (2006) The impact of breeding to reduce residual feed intake on enteric methane emissions from the Australian beef industry. Aust J Exp Agric 46:813–820

Amezquita MC, Amezquita E, Casasola F, Ramirez BL, Giraldo H, Gomez ME, Llanderal T, Velazquez J, Ibrahim MA (2008a) C stocks and sequestration. In: T'Mannetje L, Amezquita MC, Buurman P, Ibrahim MA (eds) Carbon sequestration in tropical grassland ecosystems. Academic, Wageningen, pp 49–67

Amezquita MC, Murgeitio E, Ramirez BL, Ibrahim MA (2008b) Introduction. In: Mannetje L, Amezquita MC, Buurman P, Ibrahim MA (eds) Carbon sequestration in tropical grassland ecosystems. Academic, Wageningen, pp 1–11

Berendse F, Aerts R (1987) Nitrogen-use-efficiency: a biologically meaningful definition? Funct Ecol 1:293–296

Bhatta R, Ehishi O, Kurihara M (2007) Measurement of methane production from ruminants. Asian-Aust J Anim Sci 20:1305–1318

ButterbachBahl K, Papen H, Rennenberg H (1997) Impact of gas transport through rice cultivars on methane emission from rice paddy fields. Plant Cell Environ 20:1175–1183

Byrne KA, Kiely G, Leahy P (2007) Carbon sequestration determined using farm scale carbon balance and eddy covariance. Agric Ecosyst Environ 121:357–364

Canadell JG, Pataki DE, Gifford R, Houghton RA, Luo Y, Raupach MR, Smith P, Steffen W (2007) Saturation of the terrestrial carbon sink, Chap 6. In: Canadell JG, Pataki D, Pitelka L (eds) Terrestrial ecosystems in a changing world, IGBP Series. Springer, Berlin, p 336

Conant RC, Paustian K, Elliott ET (2001) Grassland management and conversion into grassland: effects on soil carbon. Ecol Appl 11:343–355

Dawson LER, O'Kiely P, Moloney AP, Vipond JE, Wylie ARG, Carson A, Hyslop J (2011) Grassland systems of red meat production: integration between biodiversity, plant nutrient utilisation, green house gas emissions and meat nutritional quality. Animal 5(9):1432–1441

Drury CF, Stone JA, Findaly WI (1991) Microbial biomass and soil structure associated with corn, grasses and legumes. Soil Sci Soc Am J 55:805–811

Frink C, Waggoner PE, Ausubel JH (1999) Nitrogen fertilizer: retrospect and prospect. Proc Natl Acad Sci USA 96:1175–1180

Gaju O, Allard V, Martre P, Snape JW, Heumez E, LeGouis J, Moreau D, Bogard M, Griffiths S, Orford S, Hubbart S, Foulkes MJ (2011) Identification of traits to improve the nitrogen use efficiency of wheat genotypes. Field Crops Res 123:139–152

Gallais A, Hirel B (2004) An approach to the genetics of nitrogen use efficiency in maize. J Exp Bot 55:295–306

Garnier E, Aronson J (1998) Nitrogen-use efficiency from leaf to stand level. In: Lambers H, Poorter H, Van Vuuren MMI (eds) Inherent variation in plant growth. Physiological mechanisms and ecological consequences. Backhuys, Leiden, pp 515–538

Good AG, Shrawat AK, Muench DG (2004) Can less yield more? Is reducing nutrient input into the environment compatible with maintaining crop production? Trends Plant Sci 9:597–605

Gorny AG, Sodkiewicz T (2001) Genetic analysis of the nitrogen and phosphorus utilization efficiencies in mature spring barley plants. Plant Breed 120:129–132

Gregorich EG, Rochette P, VandenBygaart AJ, Angers D (2005) Greenhouse gas contributions of agricultural soils and potential mitigation practices in Eastern Canada. Soil Till Res 81:53–72

Guo LB, Gifford RM (2002) Soil carbon stocks and land use change: a meta analysis. Glob Change Biol 8:345–360

Hammond KJ, Hoskin SO, Burke JL, Waghorn GC, Koolaard JP, Muetzel S (2011) Effects of feeding fresh white clover (*Trifolium repens*) or perennial ryegrass (*Lolium perenne*) on enteric methane emissions from sheep. Anim Feed Sci Technol 166–167:398–404

Hegarty RS, Goopy JP, Herd RM, McCorkell B (2007) Cattle selected for lower residual feed intake have reduced daily methane production. J Anim Sci 85:1479–1486

Henry B, Eckard R (2009) Greenhouse gas emissions in livestock production systems. Trop Grassl 43:232–238

Holtham DAL, Matthews GP, Scholefield DS (2007) Measurement and simulation of void structure and hydraulic changes caused by root-induced soil structuring under white clover compared to ryegrass. Geoderma 142:142–151

Hopkins A, Del Prado A (2007) Implications of climate change for grasslands in Europe, impacts, adaptations and mitigation options: a review. Grass Forage Sci 62:118–126

IPCC (2007) Climate change 2007: mitigation. Contribution of working group II to the fourth assessment report of the intergovernmental panel on climate change. Cambridge University Press, Cambridge

Jank L, do Valle CB, Carvalho PF (2005) New grasses and legumes; advances and perspectives for the tropical zones of Latin America. In: Reynolds SG, Frame J (eds) Grasslands: developments, opportunities, perspectives. FAO 2005. Science, Plymouth, pp 55–79

Johnson KA, Johnson DE (1995) Methane emissions from cattle. J Anim Sci 73:2483–2492

Jones M, Donnelly A (2004) Carbon sequestration in temperate grassland ecosystems and the influence of management, climate and elevated $CO_2$. New Phytol 164:423–439

Kammann C, Gruenhage L, Jaeger H-J, Wachinger G (2001) Methane fluxes from differentially managed grassland study plots: the important role of $CH_4$ oxidation in grassland with a high potential for $CH_4$ production. Environ Pollut 115:261–267

Kammann C, Muller C, Grunhage L, Jager HJ (2008) Elevated $CO_2$ stimulates $N_2O$ emissions in permanent grassland. Soil Biol Biochem 40:2194–2205

Kebreab E, Clark K, Wagner-Riddke C, France J (2006) Methane and nitrous oxide emissions from Canadian animal agriculture: a review. Can J Anim Sci 86:135–158

Kemp DR, Michalk DL (2007) Towards sustainable grassland and livestock management. J Agric Sci 145:543–564

Kennedy PM, Charmley E (2012) Methane yields from Brahman cattle fed tropical grasses and legumes. Anim Prod Sci 52:225–239

Lal R, Henderlong P, Flowers M (1998) Forages and row cropping effects on soil organic carbon and nitrogen contents. In: Stewart BA (ed) Management of carbon sequestration in soil. CRC, Boca Raton, FL, pp 365–379

Lassey KR (2007) Livestock methane emission: from the individual grazing animal through national inventories to the global methane cycle. Agric For Meteorol 142:120–132

Ledgard S, Schils R, Eriksen J, Luo J (2009) Environmental impacts of grazed clover/grass pastures. Irish J Agric Food Res 48:209–226

Levy PE, Mobbs DC, Jones SK, Milne R, Campbell C, Sutton MA (2007) Simulation of fluxes of greenhouse gases from European grasslands using the DNDC model. Agric Ecosyst Environ 121:186–192

Loudet O, Chaillou S, Merigout P, Talbotec J, Daniel-vedele F (2003) Quantitative trait loci analysis of nitrogen use efficiency in Arabidopsis. Plant Physiol 131:345–358

Lynch DH, Cohen RDH, Fredeen A, Patterson G, Martin RC (2005) Management of Canadian prairie region grazed grasslands: soil C sequestration, livestock productivity and profitability. Can J Soil Sci 85:183–192

MacRae JC, Ulyatt MJ (1974) Quantitative digestion of fresh herbage by sheep. J Agric Sci 82:309–319

Mannetje L (2007a) Climate change and grasslands through the ages: an overview. Grass Forage Sci 62:113–117

Mannetje L (2007b) The role of grasslands and forests as carbon stores. Trop Grassl 41:50–54

Mannetje L, Amezquita MC, Buurman P, Ibrahim MA (eds) (2008) Carbon sequestration in tropical grassland ecosystems. Academic, Wageningen

Marley CL, Fychan R, Jones R (2006) Yield, persistency and chemical composition of Lotus species and varieties (birdsfoot trefoil and greater birdsfoot trefoil) when harvested for silage in the UK. Grass Forage Sci 61:134–145

Marshall AH, Bryant LG, Hauck B, Yott P, Morris P, Robbins M (2008) A high-throughput method for the quantification of proanthocyanidins in forage crops and its application in assessing variation in condensed tannin content in breeding programmes for Lotus corniculatus and Lotus uliginosus. J Agric Food Chem 56:974–981

McLauchlan KK, Hobbie SE, Post VM (2006) Conversion from agriculture to grassland builds soil organic matter on decadal timescales. Ecol Appl 16(1):143–153

Mickelson S, Deven S, Meyer FD, Garner JP, Foster CR, Blake TK, Fischer AM (2003) Mapping of QTL associated with nitrogen storage and remobilization in barley (Hordeum vulgare L.) leaves. J Exp Bot 54:801–812

Miller LA, Moorby JM, Davies DR, Humphreys MO, Scollan ND, MacRae JC, Theodorou MK (2001) Increased concentration of water-soluble carbohydrate in perennial ryegrass (Lolium perenne L.): milk production from late-lactation dairy cows. Grass Forage Sci 56:383–394

Morgan J (2005) Rising atmospheric $CO_2$ and global climate change: responses and management implications for grazing lands. In: Reynolds SG, Frame J (eds) Grasslands: developments, opportunities, perspectives. FAO 2005. Science, Plymouth, pp 235–260

Mummey DL, Smith JL (2000) Estimation of nitrous oxide emissions from US grasslands. Environ Manage 25:169–175

Mytton LR, Cresswell A, Colbourn P (1993) Improvement in soil structure associated with white clover. Grass Forage Sci 48:84–90

Neeteson JJ, Schroder JJ, Jakobsson C (2004) Drivers towards sustainability: why change? In: Hatch DJ, Chadwick DR, Jarvis SC, Roker JA (eds) Controlling nitrogen flows and losses. 12th international nitrogen workshop. Academic, Wageningen, pp 29–38

Owens VN, Albrecht K, Muck RE (2002) Protein degradation and fermentation characteristics of unwilted red clover and alfalfa silage harvested at various times during the day. Grass Forage Sci 57:329–341

Papadopoulos A, Mooney SJ, Bird NRA (2006) Quantification of the effects of contrasting crops in the development of soil structure: an organic conversion. Soil Use Manage 22:172–179

Park S, Croteau P, Boering KA, Etheridge DM, Ferretti D, Fraser KPJ, Kim K-R, Krummel PB, Langenfelds RL, van Ommen TD, Steele LP, Trudinger CM (2012) Trends and seasonal cycles in the isotopic composition of nitrous oxides since 1940. Nat Geosci 5:261–265

Philippot L, Hallin S (2011) Towards food, feed and energy crops mitigating climate change. Trends Plant Sci 16:476–480

Povellato A, Bosello B, Giupponi C (2007) Cost effectiveness of greenhouse gases mitigation measures in the European agro-forestry sector: a literature survey. Environ Sci Policy 10:474–490

Ramirez-Restrepo CA, Barry TN (2005) Alternative temperate forages containing secondary compounds for improving sustainable productivity in grazing ruminants. Anim Feed Sci Technol 120:179–201

Rauh BL, Busten C, Buckler E (2002) Quantitative trait loci analysis of growth response to varying nitrogen sources in *Arabidopsis thaliana*. Theor Appl Genet 104:743–750

Royal Society (2001) The role of land carbon sinks in mitigating global climate change. Policy document 10/01. http://www.royalsoc.ac.uk

Skinner H (2007) Winter carbon dioxide fluxes in humid-temperate pastures. Agric For Meteorol 144:32–43

Skinner RH (2008) High biomass removal limits carbon sequestration potential of mature temperate pastures. J Environ Qual 37:1319–1326

Smith P et al (2008) Greenhouse gas mitigation in agriculture. Philos Trans R Soc B 363:789–813

Soussana JF, Luscher A (2007) Temperate grasslands and global atmospheric change: a review. Grass Forage Sci 62:127–134

Soussana JF et al (2007) Full accounting of the greenhouse gas ($CO_2$, $N_2O$, $CH_4$) budget of nine European grassland sites. Agric Ecosyst Environ 121:121–134

Stewart CE, Paustian K, Conant RT, Plante F, Six J (2007) Soil carbon saturation: concept, evidence and evaluation. Biogeochemistry 8:619–631

Subbarao GV, Tomohiro B, Masahiro K, Osamu I, Samejima H, Wang HY, Pearse SJ, Gopalakrishnan S, Nakahara K, Hossain AK, Tsujimoto H, Berry WL (2007) Can biological nitrification inhibition (BNI) genes from perennial *Leymus racemosus* (Triticeae) combat nitrification in wheat farming? Plant Soil 299(1–2):55–64

Subbarao G, Nakahara K, Hurtado M, Ono H, Moreta DE, Salcedo A, Yoshihashi AT, Ishikawa T, Ishitani M, Ohnishi-Kameyama M, Yoshida M, Rondon M, Rao IM, Lascano E, Berry WL, Ito O (2009) Evidence for biological nitrification inhibition in Brachiaria pastures. Proc Natl Acad Sci USA 106:17302–17307

Sullivan ML, Hatfield RD (2006) Polyphenol oxidase and o-diphenols inhibit postharvest proteolysis in red clover and alfalfa. Crop Sci 46:662–670

Sun XZ, Hoskin SO, Muetzel S, Molano G, Clark H (2011) Effects of forage chicory (*Cichorium intybus*) on methane emissions *in vitro* and from sheep. Anim Feed Sci Technol 166–167:391–397

Tamminga S (1996) A review on environmental impacts of nutritional strategies in ruminants. J Anim Sci 74:3112–3124

Tennigkeit T, Wilkes A (2008) An assessment of the potential for carbon finance in rangelands. ICRAF China. ICRAF Working Paper No 68

Tilman D (1999) Global environmental impacts of agricultural expansion: the need for sustainable and efficient practices. Proc Natl Acad Sci USA 96:5995–6000

Ulyatt MJ, Thomson DJ, Beever DE (1988) The digestion of perennial ryegrass (*Lolium perenne* cv. Melle) and white clover (*Trifolium repens* cv. Blanca) by grazing cattle. Br J Nutr 60:137–149

UNFCC (2008) Challenges and opportunities for mitigation in the agricultural sector. Technical paper. UNFCCC Secretariat, Geneva

Van Kessel C, Boots B, de Graaf MA, Harris D, Blum H, Six J (2006) Total soil C and N sequestration in a grassland following 10 years of free air $CO_2$ enrichment. Glob Change Biol 11:2187–2199

Van Loo EN, Van Wijk AJP, Dolstra O, Marvin HJP, Snijders CHA (1997) Selection for nitrogen use efficiency in perennial ryegrass using hydroponics. In: Proceedings of the XVIII international grassland congress, Winnipeg and Saskatoon, Canada, June 1997, pp 4129–4130

Waghorn GC, Shelton ID (1997) Effect of condensed tannins in Lotus corniculatus on the nutritive value of pasture for sheep. J Agric Sci 128:365–372

West TO, Six J (2007) Considering the influence of sequestration duration and carbon saturation on estimates of soil carbon capacity. Clim Change 80:25–41

Wilkins PW, Macduff JH, Raistrick N, Collison M (1997) Varietal differences in perennial ryegrass for nitrogen use efficiency in leaf growth following defoliation: performance in flowing solution culture and its relationship to yield under grazing in the field. Euphytica 98:109–119

Wilkins PW, Allen DK, Mytton LR (2000) Differences in the nitrogen use efficiency of perennial ryegrass varieties under simulated rotational grazing and their effects on nitrogen recovery and herbage nitrogen content. Grass Forage Sci 55:69–76

Witt C et al (1999) Internal nutrient efficiencies of irrigated lowland rice in tropical and subtropical Asia. Field Crops Res 63:113–138

Wood S, Cowie A (2004) A review of greenhouse gas emission factors for fertiliser production. IEA Bioenergy. Task 38

# Index

Printed by Printforce, the Netherlands